GINKGO BILOBA

GINKGO BILOBA

Edited by

Teris A. van Beek
Laboratory of Organic Chemistry
Wageningen Agricultural University
The Netherlands

CRC PRESS

Boca Raton   London   New York   Washington, D.C.

FIRST INDIAN REPRINT, 2012

This book contains information obtained from authentic and highly regarded sources. Reprinted material is quoted with permission, and sources are indicated. A wide variety of references are listed. Reasonable efforts have been made to publish reliable data and information, but the author and the publisher cannot assume responsibility for the validity of all materials or for the consequences of their use.

Neither this book nor any part may be reproduced or transmitted in any form or by any means, electronic or mechanical, including photocopying, microfilming, and recording, or by any information storage or retrieval system, without prior permission in writing from the publisher.

Direct all inquiries to CRC Press LLC, 2000 N.W. Corporate Blvd., Boca Raton, Florida 33431.

© 2000 CRC Press, LLC

Trademark Notice: Product or corporate names may be trademarks or registered trademarks, and are used only for identification and explanation, without intent to infringe.

Visit the CRC Press Web site at www.crcpress.com

Printed and bound in India by
Replika Press Pvt. Ltd.

ISBN 10 : 90-5702-488-8
ISBN 13 : 978-90-5702-488-7

FOR SALE IN SOUTH ASIA ONLY.

CONTENTS

FOREWORD

It is a very encouraging and healthy sign that folk medicinal knowledge and practice has started to gain recognition among the populace and the medical professionals, although there still exist hard core opponents who only think in terms of so-called Western drugs. The revival in interest is reflected in the recent establishment of chairs in alternative medicine or natural medicine at universities even in the USA where recognition lagged far behind other continents. There is no doubt that folk medicinal herbs, which have been used for thousands of years, are effective and complementary to modern medicine. The knowledge and practice in natural medicine is comfortably accepted in the households of most countries independent of the stage of development of that nation. The interest has been traditionally high in countries such as China, Germany and Japan, and certainly in undeveloped areas healing through herb treatment is the norm. The official record of herbs, which started in China around 2800 BC, continues to provide the Chinese medicine and pharmacognosy community with a wealth of precious knowledge. Folk medicinal treatment is also widespread in the South American and African rain forests and similar areas, but here the knowledge is transmitted to the next generation through the medicine person or sharman and not through writing.

The mode of action of plant extracts and other folk medicines are far more complicated than that of a single compound because they are mixtures of numerous constituents (a crude plant extract may contain hundreds of constituents). It is probably true that only a few of the constituents are responsible for the healing effects, while others are merely present without exerting any effects. Understanding the mode of action of a Western drug is already complicated enough: what is the receptor(s) and what chain of events ensue after the interaction leading to healing? Rationalization and understanding of the effects of folk medicine is no doubt a most challenging but worthwhile target for future interdisciplinary studies. A broad interdisciplinary approach to clarify how folk medicine works was almost impossible until quite recently because of the lack of sophisticated technology and general knowledge. However, we can now start to be engaged in such approaches and disentangle the role of respective constituents and their synergism.

Ginkgo biloba, one of the oldest of the documented Chinese herbs, has a proven remedial record through its hundreds of years of use. At least one group of constituents, ginkgolide B and related terpenoid lactones have been shown by modern assays to selectively inhibit platelet activation although this has not led to a single-component drug. Nevertheless, the sale of the crude extract is increasing at a phenomenal rate throughout the world. Not many people would resist taking *G. biloba* extract when statistics have shown that it helps enhance memory and has a positive effect in slowing Alzheimer's disease, with no known adverse effects. It is most opportune that this plant has been selected to be included in the present series and that the editor has taken the time and effort to select the topics and contributors for an updated picture of this fantastic plant from nature.

Koji Nakanishi

PREFACE TO THE SERIES

There is increasing interest in industry, academia and the health sciences in medicinal and aromatic plants. In passing from plant production to the eventual product used by the public, many sciences are involved. This series brings together information which is currently scattered through an ever increasing number of journals. Each volume gives an in-depth look at one plant genus, about which an area specialist has assembled information ranging from the production of the plant to market trends and quality control.

Many industries are involved such as forestry, agriculture, chemical, food, flavour, beverage, pharmaceutical, cosmetic and fragrance. The plant raw materials are roots, rhizomes, bulbs, leaves, stems, barks, wood, flowers, fruits and seeds. These yield gums, resins, essential (volatile) oils, fixed oils, waxes, juices, extracts and spices for medicinal and aromatic purposes. All these commodities are traded world-wide. A dealer's market report for an item may say "Drought in the country of origin has forced up prices".

Natural products do not mean safe products and account of this has to be taken by the above industries, which are subject to regulation. For example, a number of plants which are approved for use in medicine must not be used in cosmetic products.

The assessment of safe to use starts with the harvested plant material which has to comply with an official monograph. This may require absence of, or prescribed limits of, radioactive material, heavy metals, aflatoxins, pesticide residue, as well as the required level of active principle. This analytical control is costly and tends to exclude small batches of plant material. Large scale contracted mechanised cultivation with designated seed or plantlets is now preferable.

Today, plant selection is not only for the yield of active principle, but for the plant's ability to overcome disease, climatic stress and the hazards caused by mankind. Such methods as *in vitro* fertilisation, meristem cultures and somatic embryogenesis are used. The transfer of sections of DNA is giving rise to controversy in the case of some end-uses of the plant material.

Some suppliers of plant raw material are now able to certify that they are supplying organically-farmed medicinal plants, herbs and spices. The Economic Union directive (CVO/EU No 2092/91) details the specifications for the **obligatory** quality controls to be carried out at all stages of production and processing of organic products.

Fascinating plant folklore and ethnopharmacology leads to medicinal potential. Examples are the muscle relaxants based on the arrow poison, curare, from species of *Chondrodendron*, and the antimalarials derived from species of *Cinchona* and *Artemisia*. The methods of detection of pharmacological activity have become increasingly reliable and specific, frequently involving enzymes in bioassays and avoiding the use of laboratory animals. By using bioassay linked fractionation of crude plant juices or extracts, compounds can be specifically targeted which, for example, inhibit blood platelet aggregation, or have antitumour, or antiviral, or any other required activity. With the assistance of robotic devices, all the members of a genus may be readily screened. However, the plant material must be **fully** authenticated by a specialist.

The medicinal traditions of ancient civilisations such as those of China and India have a large armamentarium of plants in their pharmacopoeias which are used throughout South East Asia. A similar situation exists in Africa and South America. Thus, a very high percentage of the world's population relies on medicinal and aromatic plants for their medicine. Western medicine is also responding. Already in Germany all medical practitioners have to pass an examination in phytotherapy before being allowed to practise. It is noticeable that throughout Europe and the USA, medical, pharmacy and health related schools are increasingly offering training in phytotherapy.

Multinational pharmaceutical companies have become less enamoured of the single compound magic bullet cure. The high costs of such ventures and the endless competition from me too compounds from rival companies often discourage the attempt. Independent phytomedicine companies have been very strong in Germany. However, by the end of 1995, eleven (almost all) had been acquired by the multi-national pharmaceutical firms, acknowledging the lay public's growing demand for phytomedicines in the Western World.

· The business of dietary supplements in the Western World has expanded from the Health Store to the pharmacy. Alternative medicine includes plant based products. Appropriate measures to ensure the quality, safety and efficacy of these either already exists or are being answered by greater legislative control by such bodies as the Food and Drug Administration of the USA and the recently created European Agency for the Evaluation of Medicinal Products, based in London.

In the USA, the Dietary Supplement and Health Education Act of 1994 recognised the class of phytotherapeutic agents derived from medicinal and aromatic plants. Furthermore, under public pressure, the US Congress set up an Office of Alternative Medicine and this office in 1994 assisted the filing of several Investigational New Drug (IND) applications, required for clinical trials of some Chinese herbal preparations. The significance of these applications was that each Chinese preparation involved several plants and yet was handled as a single IND. A demonstration of the contribution to efficacy, of each ingredient of each plant, was not required. This was a major step forward towards more sensible regulations in regard to phytomedicines.

My thanks are due to the staff of Harwood Academic Publishers who have made this series possible and especially to the volume editors and their chapter contributors for the authoritative information.

Roland Hardman

CONTRIBUTORS

Teris A. van Beek
Laboratory of Organic Chemistry
Phytochemical Section
Wageningen Agricultural University
Dreijenplein 8
6703 HB Wageningen
The Netherlands

Ezio Bombardelli
Indena SpA
Scientific Department
Viale Ortles 12
20139 Milano
Italy

Bernard Bulling
Aachener Strasse 312
50933 Köln
Germany

Fabrizio F. Camponovo
Research and Development
Pharmaton SA
6903 Lugano
Switzerland

Danielle Julie Carrier
Department of Agricultural and
    Bioresource Engineering
University of Saskatchewan
57 Campus Drive
Saskatoon, Saskatchewan
Canada S7N 5A9

Yves Christen
Institut IPSEN
24 rue Erlanger
75781 Paris Cedex 16
France

Norbert Clemens
Intersan GmbH
Einsteinstrasse 30
76275 Ettlingen
Germany

Amy Corzine
McAlpine Thorpe and Warrier Limited
50 Penywern Road
London SW5 9SX
UK

Aldo Cristoni
Indena SpA
Scientific Department
Viale Ortles 12
20139 Milano
Italy

Volker Dankers
Intersan GmbH
Einsteinstrasse 30
76275 Ettlingen
Germany

Cynthia L. Darlington
Department of Psychology and the
    Neuroscience Research Centre
University of Otago
Dunedin
New Zealand

Francis V. DeFeudis
Institute for Bioscience
153 West Main Street
Westboro, MA 01581
USA

Peter Del Tredici
Arnold Arboretum of Harvard University
125 Arborway
Jamaica Plain, MA 02130-3519
USA

Peter A.G.M. De Smet
Scientific Institute Dutch Pharmacists
Alexanderstraat 11
2514 JL Den Haag
The Netherlands

Martien C.J.M. van Dongen
Department of Epidemiology
Maastricht University
P.O. Box 616
6200 MD Maastricht
The Netherlands

Katy Drieu
Institut Henri Beaufour-IPSEN
24 rue Erlanger
75116 Paris
France

Frans M. van den Dungen
Biohorma B.V.
Medical Research
P.O. Box 33
8081 HH Elburg
The Netherlands

Kazuhiko Fukushima
School of Agriculture
Nagoya University
Nagoya 464-01
Japan

Marian A.J.T. van Gessel
Hamsestraat 79
4043 LG Opheusden
The Netherlands

Peter Halama
Berner Heerweg 175
22159 Hamburg
Germany

Andreas Hasler
Zeller AG Herbal Remedies
8590 Romanshorn
Switzerland

Robert Hoerr
Dr. Willmar Schwabe Pharmaceuticals
Clinical Research Department
Willmar Schwabe Strasse 4
76227 Karlsruhe
Germany

Hoon Huh
Laboratory of Pharmacognosy
College of Pharmacy
Seoul National University
56-1 Shinlim-Dong
Kwanak-Gu
Seoul 151-742
Korea

Hermann Jaggy
Dr. Willmar Schwabe Arzneimittel
Postfach 410925
7500 Karlsruhe 41
Germany

Paul Knipschild
Department of General Practice
Maastricht University
P.O. Box 616
6200 MD Maastricht
The Netherlands

Dominique Laurain
Laboratoire de Pharmacognosie
Faculté de Pharmacie
Université de Picardie Jules Verne
1 rue des Louvels
80037 Amiens Cedex 1
France

Johannes J. Lichius
Institut für Pharmazeutische Biologie
Philipps-Universität Marburg
Deutschhausstrasse 17A
35032 Marburg
Germany

Karyn Maclennan
Department of Pharmacology
School of Medical Sciences
University of Otago Medical School
Dunedin
New Zealand

Beat Meier
Zeller AG, Herbal Remedies
8590 Romanshorn
Switzerland

Volker Melzheimer
Botanischer Garten
Philipps-Universität Marburg
Karl von Frisch Strasse
35032 Marburg
Germany

Paolo Morazzoni
Indena SpA
Scientific Department
Viale Ortles 12
20139 Milano
Italy

Koji Nakanishi
Department of Chemistry
Columbia University
3000 Broadway
Mail Code 3114
New York, NY 10027
USA

Joe O'Reilly
Cara Partners
Little Island, Co. Cork
Ireland

Federico Peterlongo
Indena SpA
Laboratori Ricerca e Sviluppo
Via Don Minzoni 6
20090 Settala (Milano)
Italy

Erik van Rossum
Department of Epidemiology
Maastricht University
P.O. Box 616
6200 MD Maastricht
The Netherlands

Joerg Schulz
Klinikum Buch
Geriatrische Klinik
Zepernicker Strasse 1
13125 Berlin
Germany

Paul F. Smith
Department of Pharmacology
School of Medical Sciences
University of Otago Medical School
Dunedin
New Zealand

Fabio Soldati
Research and Development
Pharmaton SA
6903 Lugano
Switzerland

V. Srini Srinivasan
United States Pharmacopoeia
12601 Twinbrook Parkway
Rockville, MD 20852
USA

Otto Sticher
Department of Pharmacy
Federal Institute of Technology (ETH)
Winterthurerstrasse 190
8057 Zürich
Switzerland

Noritsugu Terashima
2-610 Uedayama, Tenpaku
Nagoya 468
Japan

Luisella Verotta
Dipartimento di Chimica Organica e
    Industriale
Università degli Studi di Milano
Via Venezian 21
20133 Milano
Italy

Keiji Wada
Faculty of Pharmaceutical Sciences
Health Sciences University of Hokkaido
Ishikari–Tobetsu
Hokkaido 061-02
Japan

Gopi Warrier
McAlpine Thorpe and Warrier Limited
50 Penywern Road
London SW5 9SX
UK

Herman J. Woerdenbag
University Centre for Pharmacy
Antonius Deusinglaan 2
9713 AW Groningen
The Netherlands

INTRODUCTION

TERIS A. VAN BEEK

*Laboratory of Organic Chemistry, Phytochemical Section,
Wageningen Agricultural University, Dreijenplein 8, 6703 HB
Wageningen, The Netherlands*

The Ginkgo tree has been of interest to mankind for more than 2000 years. Until 350 years ago interest was restricted to China, its natural habitat, Japan and Korea. There female trees were cultivated on a large scale to furnish the nuts which remain a popular delicacy in Eastern Asia until today. Most of the 100 Ginkgo specimens over 1000 years old that still exist in China are found near temples (Shan-An *et al.*, 1997). They were probably planted at such locations for their supply of nuts but maybe also for worship. Since more than 500 years Ginkgo fruits and to a lesser extent leaves are also medicinally used in China. After transportation of the first trees to Europe, on this continent the tree was at first solely admired for its beauty. Its unique and isolated botanical position ("living fossil") was recognized in the last century. Hundred years ago Hirase discovered the swimming sperm (spermatozoids) of *Ginkgo biloba* (Hirase, 1896). However a quantum jump in interest in this tree occurred in the sixties and seventies of this century when Schwabe Industries started to investigate and produce a special leaf extract for blood circulation problems (Kloss and Jaggy, 1971). From then onwards more and more research has been carried out on many different aspects of the Ginkgo tree and the end is not yet in sight. According to Christen (last chapter in this book) nowadays five scientific articles per week appear on Ginkgo. In 1997 a book on *Ginkgo biloba* was published by The Botanical Society of Japan which focussed primarily on botanical and phylogenetic aspects (Hori *et al.*, 1997). In that same year the first international Ginkgo symposium was held (Organizing Committee, 1997). Last year a second book was published with the emphasis on pharmacology and clinical studies (DeFeudis, 1998). The book which lies in front of you is a third book on *Ginkgo biloba* and the first book to cover both botany, processing of leaves, constituents, analytical chemistry, pharmacology, toxicology, clinical applications and other uses. In the remaining part of this introduction the different chapters will be briefly discussed to put them in an overall perspective.

The first chapter by Del Tredici provides a quick introduction into Ginkgo. It starts off with the evolutionary history. There is already fossil proof of Ginkgo species flourishing 180 million years ago. Maximum diversity was reached from 140 to 80 million years ago. The author is an expert on the question whether or not "wild" Ginkgo trees still exist in China. However a definite answer cannot yet be given. He further ponders on the type of animals involved in the dispersal of the large Ginkgo seeds both during the time of the dinosaurs and in our own time. Finally some attention is given to the cultivation of the trees for three different purposes: for nuts, for leaves and as an ornamental. The next chapter by Melzheimer and Lichius gives a thorough treatise of the many interesting botanical aspects: taxonomic classification, morphology, the peculiar embryology and karyology. In chapter 3 by Terashima and Fukushima the lignification of Ginkgo xylem cell walls is described in detail. Intricate chemical labelling experiments followed by micro-autoradiography and solid state NMR analysis shed light on the chemical composition and the 3-dimensional assembly of Ginkgo wood.

The cultivation of Ginkgo trees on a large scale for leaf production is an economically important issue nowadays. It is dealt with in chapter 4 by Laurain. She herself is involved in managing a Ginkgo plantation and could thus provide interesting inside information. Different types of propagation, pruning, fertilisation, weed and disease control and harvesting are all discussed. The same author together with Carrier is also responsible for the immediately following chapter on plant cell biotechnology of Ginkgo. Different culturing conditions, the various types of embryogenesis and the terpene trilactone content (ginkgolides and bilobalide) of cultures are discussed. Special attention is given to the latest developments in biosynthetic studies of these terpene trilactones. Recently it was discovered that these unique secondary metabolites are not produced from mevalonate but via an alternative "triose/pyruvate" pathway (Schwarz, 1994). The first part of the book ends with a chapter by O'Reilly who is responsible for the large scale extraction and processing of Ginkgo leaves in a commercial plant. He clearly points out the many technical, safety and environmental problems involved in running such a process. Current good laboratory practice (cGLP) plays a paramount role.

The chemical part of the book starts off with a thorough review of all constituents ever isolated from any part of the Ginkgo tree: among others volatile terpenes, terpene trilactones, sterols, carotenoids, polyprenols, flavonols, flavones, proantho-cyanidins, long chain hydrocarbons, organic acids and 6-hydroxykynurenic acid. The chapter is written by Hasler. Although it is obvious that Ginkgo is an amazingly versatile chemical factory, there are bound to be many more minor compounds – most likely pharmacologically relevant – which escaped attention so far. From all parts of the tree logically the leaves received most attention and the roots least. Most characteristic of all secondary metabolites found in Ginkgo are the ginkgolides. The structure of these diterpene trilactones was independently elucidated by two Japanese groups in 1967. In chapter 8 the leader of one of these groups, Nakanishi, describes in his own words the story behind the elucidation. According to him "it was the last classical and romantic structural study conducted in which numerous chemical reactions were carried out and structural clarification was accompanied by surprises and excitement, as if unfolding a mystery". The following chapter critically surveys all analytical assays ever published for the ginkgolides and their smaller 15-carbon sister bilobalide. The bottom line is that there is not yet a good, simple, validated quantitative assay available capable of analysing terpene trilactones in leaves, extracts and phytopharmaceuticals. The next chapter by Sticher, Meier and Hasler is similar to the ginkgolide chapter except that now the analysis of the important flavonoids is in the spotlight. For this group good methods are available. Normally the flavonol glycosides are quantified after hydrolysis through the resulting agly-cones. However it is shown by the authors that analysis of individual flavonol glyco-sides by means of RP-HPLC with a ternary gradient is also feasible. Lack of reference compounds and knowledge about which individual glycosides are impor-tant make such analysis complicated and the result difficult to interpret. During the last three years more and more attention is given to the concentration of the suppos-edly allergenic alkyl phenols (e.g. ginkgolic acid) in Ginkgo extracts and phytophar-maceuticals. These compounds occur in high concentrations in both Ginkgo leaves and the fleshy outer layer of Ginkgo fruits. In standardised extracts their concentra-tion is normally lower than 10 ppm due to various purification steps but in full extracts, like homeopathic mother tinctures, their concentration should be much higher and may pose a health problem. However proper analytical determinations and pharmacological studies on these potential health problems are so far lacking and this remains a matter of some controversy. Verotta, Morazzoni and Peterlongo summarize in chapter 11 the limited number of studies on these compounds. The sit-uation is similar to the ginkgolides and good validated quantitative assays have not been published. Qualitatively the situation is more clear. The last specific group of secondary metabolites to be highlighted are the polyprenols and this is done by Huh. These polyisoprenoid alcohols can be converted to another type of polyprenols occurring in cell membranes of animals and a significant amount of interest has been shown in them by a Japanese industry. RP-HPLC and SFC are the analytical methods of choice for separating the various isoprenoid homologues.

A pharmacopeia monograph for Ginkgo leaf is already being discussed for many years in Europe without much concrete results however. Europe has now been over-taken by the USA and a draft monograph for the US Pharmacopeia has been pub-lished (Anonymous, 1997). This is a very important and good step as it should lead to stricter quality control and thus a guaranteed minimum quality for the end-users, i.e. the patients taking the drug. The development of the monograph is discussed by Srinivasan, one of the scientists directly involved. Somewhat of a missed opportunity in this monograph is the lack of a quantitative assay for terpene trilactones but this must be due to the absence of a simple, reliable method. A somewhat similar chapter but then seen through industrial eyes has been prepared by Camponovo and Soldati who are responsible for the quality of Ginkgo drugs within a Swiss company. Chapter 15 serves both as a kind of intermediate summary of many of the preceding chapters and an introduction to the pharmacological part of the book. It discusses the history, development and constituents of the first, best known and most sold standardised Ginkgo extract namely EGb 761 developed by Schwabe and Ipsen. A lot of interesting background and inside information on this extract is provided by the authors Drieu and Jaggy from the Ipsen and Schwabe companies respectively.

Since the sixties and continuing until the present day an enormous amount of pharmacological research has been carried out to determine exactly how standard-ised Ginkgo extracts work and which compounds are responsible for the different activities. This is discussed in chapters 16 and 17 by DeFeudis and Drieu and in chapter 18 by Darlington, Smith and Maclennan. Some conclusions from these chapters are that standardised Ginkgo extracts show a polyvalent action and that the overall action cannot be explained by one or even two groups of compounds. The PAF-antagonistic activity of ginkgolides does play a role but it may not be of such paramount importance as once thought and these diterpenes certainly possess other activities than just anti-PAF. Further bilobalide which lacks anti-PAF activity, is not devoid of pharmacological activity and may be far more important than thought a few years ago. The flavonoids and proanthocyanidins are important too, but their *in vivo* radical scavenging activity may not run parallel to their *in vitro* activity because of limited bioavailability. Relatively new is that Ginkgo extracts possess anti-stress (anxiolytic) activity similar to that of benzodiazepines however without the usual side effects associated with benzodiazepines. It is certain that the pharmacological research will continue and perhaps in the far future we will be able to explain the clinical effects by the activity at the molecular and cellular level.

The section of the book dealing with clinical effects was considered on the one hand so important and on the other hand still so controversial that originally four chapters were foreseen. Two chapters by physicians directly involved with clinical trials on peripheral and central blood circulation problems respectively, one by an independent epidemiologist and one chapter by a prominent critic (Prof. dr. P.S. Schönhöfer, Bremen, Germany) of Ginkgo extracts (Schönhöfer, 1989). However this last author declined the invitation because according to him "... there is no scientific basis for including Ginkgo extracts into a serious book series on plant medicines ... and there are more urgent problems in clinical pharmacology than writing articles on useless drugs.". This should be considered somewhat of a missed opportunity as the reader could then have compared the various chapters and drawn his own conclusions. Nevertheless three very worthwhile chapters remain. The first by Schulz, Halama and Hoerr discusses eigthteen clinical trials for the treatment of early dementia and cerebral insufficiency. Their clear conclusion based on all the positive results is that standardised Ginkgo extracts are first-choice nootropics and an excellent choice for these diseases. The next chapter by Bulling, Clemens and Dankers focusses on the clinical trials against peripheral arterial occlusive disease (PAOD). Eight clinical trials are discussed. Their conclusion is evident: although physical exercise remains the basic therapy, Ginkgo is an effective drug for PAOD. The pain-free and maximum walking distance both increase significantly. To prove additional benefits more high quality clinical trials are indicated. The last chapter in this section, chapter 21 by van Dongen, van Rossum and Knipschild reviews the same studies as discussed in the two preceding chapters plus the recent positive study by Le Bars *et al.* (1997) for the treatment of Alzheimer's disease. However this time all the literature is seen through the eyes of outside epidemiologists and – perhaps not surprisingly – their conclusion is less positive. They conclude that although the evidence is rather convincing it is still debatable. If one reads between the lines their overall conclusion is that more and better controlled clinical trials are necessary to furnish unambiguous proof for the clinical effectiveness of Ginkgo special extracts. These authors also discuss clinical studies on less well known indications for Ginkgo drugs such as in the prevention of mountain sickness (highly significant effect), for the treatment of arterial erectile impotence (positive effects for the majority of patients), acute ischaemic stroke (no significant difference with a placebo) and congestive symptoms of the premenstrual syndrome (slightly better than a placebo). A very recent clinical trial with EGb761 for the indication mild dementia and age associated memory impairment did not give a statistically significant outcome in sharp contrast with 55 earlier trials (van Dongen, 1999).

In contrast to the clinical efficacy there is no debate on the (absence of any) toxicity and the extremely low occurrence of adverse effects of standardised Ginkgo extracts. This is the subject of the chapter by Woerdenbag and de Smet. Ginkgo seeds are toxic when too many are eaten especially for children and for instance in Japan people have died in the past from over-consumption. This is due to 4-O-methylpyridoxine as is shown in the chapter by Wada. This compound which is an antagonist of vitamin B6 also occurs in Ginkgo leaves but in such low concentrations that it poses no toxicological threat (Arenz *et al.*, 1996).

In the last section of the book a number of miscellaneous chapters have been brought together. Van den Dungen discusses the limited amount of material available on the homeopathic uses of Ginkgo leaf extracts. The homeopathic indications are quite different from the allopathic indications. Ginkgo mother tinctures are mainly prescribed for the treatment of migraine or sub-orbital headache on the left side and for writer's cramp. Bombardelli, Cristoni and Morazzoni deal in chapter 25 with the recently developed cosmetic application of certain Ginkgo extracts. The extracts stimulate the skin microcirculation and can be useful for the treatment of skin atrophism. The biflavonoids which are not present in normal standardised extracts, appear to play an important role. Essential is the inclusion of phosphadityl-choline for the formation of liposomes to facilitate the epicutaneous absorption. In chapter 26 compiled by van Gessel, all almost 200 patents filed for Ginkgo until 1997 are tabulated. It provides interesting reading to see for what purposes Ginkgo extracts and constituents may be used. It ranges from the use of Ginkgo fruit liquid for producing impregnants for electrical insulators, via many different medical applications and as an ingredient in chewing gum against dental plaque and caries to the use of Ginkgo compounds as insect repellants and as an additive to chicken feed for the production of low cholesterol eggs. In the next chapter Warrier and Corzine give an overview of the economic aspects of Ginkgo phytopharmaceuticals: turnover, cost of production, profit margins and market trends. Additionally they tabulated the main producers and product names. In the last chapter by Christen the term "ginkgology" is introduced to describe the large amount of research centering around Ginkgo. He makes it clear that this term is indeed appropriate and he further predicts in which direction the ginkgology field will move in the coming years. His predictions range from an increase in pharmacological research to *Ginkgo biloba* becoming a model for the study of biological senescence.

The editor hopes that this all-round book on Ginkgo with its mix of authors from Industry and Academia will not only be useful as a reference work for those already active in the "ginkgology" field but also for novices. For the latter group it should serve as a good starting point for literature research and for the generation of fresh ideas for new exciting research. This book proves that many secrets of

Ginkgo biloba have been discovered but many more are still waiting to be discovered.

REFERENCES

Anonymous (1997) *Ginkgo biloba* leaf – draft monograph. *Pharmacopeial Forum*, **23**, 3644–3647.

Arenz, A., Klein, M., Fiehe, K., Groß, J., Drewke, C., Hernscheidt, T., Leistner, E. (1996) Occurrence of neurotoxic 4'-O-methylpyridoxine in *Ginkgo biloba* leaves, Ginkgo medications and Japanese Ginkgo food. *Planta Med.*, **62**, 548–551.

DeFeudis, F.V. (1998) *Ginkgo biloba extract (EGb 761). From chemistry to the clinic*, Ullstein Medical, Wiesbaden.

van Dongen, M.C.J.M. (1999) Efficacy of *Ginkgo biloba* in dementia and cognitive delcine, Ph.D. Thesis, Universiteit Maastricht, Datawyse I Universitaire Pers Maastricht.

Hirase, S. (1896) Spermatozoids of *Ginkgo biloba*. *Bot. Mag. Tokyo*, **10**, 367–368. Cited from: Nagata, T. (1997) Scientific contributions of Sakugoro Hirase. In T. Hori, R.W. Ridge, W. Tulecke, P.D. Tredici, J. Trémouillaux-Guiller, and H. Tobe, (eds.), *Ginkgo biloba – a global treasure. From biology to medicine*, Springer, Tokyo, pp. 413–416.

Hori, T., Ridge, R.W., Tulecke, W., Del Tredici, P., Trémouillaux-Guiller, J. and Tobe, H. (eds.), *Ginkgo biloba – a global treasure. From biology to medicine*, Springer, Tokyo.

Kloss, P., Jaggy, H. (1971) Verfahren zur Herstellung eines Injektions- bzw. Infusionspräparates mit einem Gehalt an einem Wirkstoffgemisch aus den Blättern von *Ginkgo biloba*. Patent DE 2117429, Dr. Willmar Schwabe Arzneimittel.

Le Bars, P.L., Katz, M.M., Berman, N., Itil, T.M., Freedman, A.M., Schatzberg, A.F. (1997) A placebo-controlled, double-blind, randomized trial of an extract of *Ginkgo biloba* for dementia. *JAMA*, **278**, 1327–1332.

Organizing Committee for '97 International Seminar of Ginkgo (1997) *Proceedings of '97 International Seminar on Ginkgo*, Nov. 10–12, 1997, Beijing.

Schönhöfer, P.S., Schulte-Sasse, H., Manhold, C., Werner, B. (1989) Sind Extrakte aus den Blättern des Ginkgobaumes bei peripheren Durchblutungs- und Hirnleistungsstörungen im Alter wirksam? *Internist. Prax.*, **29**, 585–601.

Schwarz, M.K. (1994) Terpen-Biosynthese in *Ginkgo biloba*: eine überraschende Geschichte, Ph.D. Thesis, ETH, Zürich.

Shan-An, H., Gu, Y., Zi-Jie, P. (1997) Resources and prospects of *Ginkgo biloba* in China. In T. Hori, R.W. Ridge, W. Tulecke, P. Del Tredici, J. Trémouillaux-Guiller, and H. Tobe, (eds.), *Ginkgo biloba – a global treasure. From biology to medicine*, Springer, Tokyo, pp. 373–383.

1. THE EVOLUTION, ECOLOGY, AND CULTIVATION OF *GINKGO BILOBA*

PETER DEL TREDICI

Arnold Arboretum of Harvard University
125 Arborway
Jamaica Plain, Massachusetts 02130-3519, USA

EVOLUTIONARY HISTORY

Ginkgo biloba, which is not closely related to any other living plant, is generally classified in its own division, the Ginkgophyta. This taxon is distinguished from the Coniferophyta (conifers) on the basis of its reproductive structures, most notably its multiflagellated sperm cells, and from the Cycadophyta (cycads) on the basis of its vegetative anatomy (Wang and Chen, 1983; Gifford and Foster, 1987). Recent molecular analysis of the Ginkgo genome, while far from complete, suggests a much closer relationship to the cycads than to the conifers (Hasebe, 1997).

The fossil record of the genus Ginkgo is extensive, with numerous reports of "Ginkgophyte" foliage and wood from many stratigraphic regions in both the northern and southern hemispheres. Because most of this material is sterile, however, its precise relationship to the extant species has always been conjectural. This uncertainty has recently been resolved by the discovery in Henan Province, China of fossils from the Middle Jurassic (180 million years ago) that possessed Ginkgo-like ovule-bearing organs (Zhou and Zhang, 1989). These are the earliest, unequivocal representatives of the genus Ginkgo and have been described as a new species, *G. yimaensis*, that differs from *G. biloba* in having more highly dissected leaves and much smaller ovules clustered on branched peduncles (Figure 1). Another extinct "species," *G. adiantoides*, had an extensive distribution in the northern hemisphere from the lower Cretaceous through the Pliocene (141–1.8 mya), and many authors consider this taxon to be the likely ancestor of *G. biloba* because it had a similar leaf morphology and ovule structure (Tralau, 1968; Zhou, 1994).

The genus Ginkgo appears to have reached the peak of its diversity during the lower Cretaceous (141–98 mya), with several distinct species occupying a more or less circumpolar distribution in the northern hemisphere which also extended into several parts of the southern hemisphere. During the upper Cretaceous (98–65 mya), the fossil record for Ginkgo shows a decline in diversity and distribution, particularly toward the end of the period when worldwide temperatures decreased dramatically.

The diminution of Ginkgo's range continued into the Tertiary, and was particularly striking from the Oligocene (38–26 mya), when the genus disappeared from

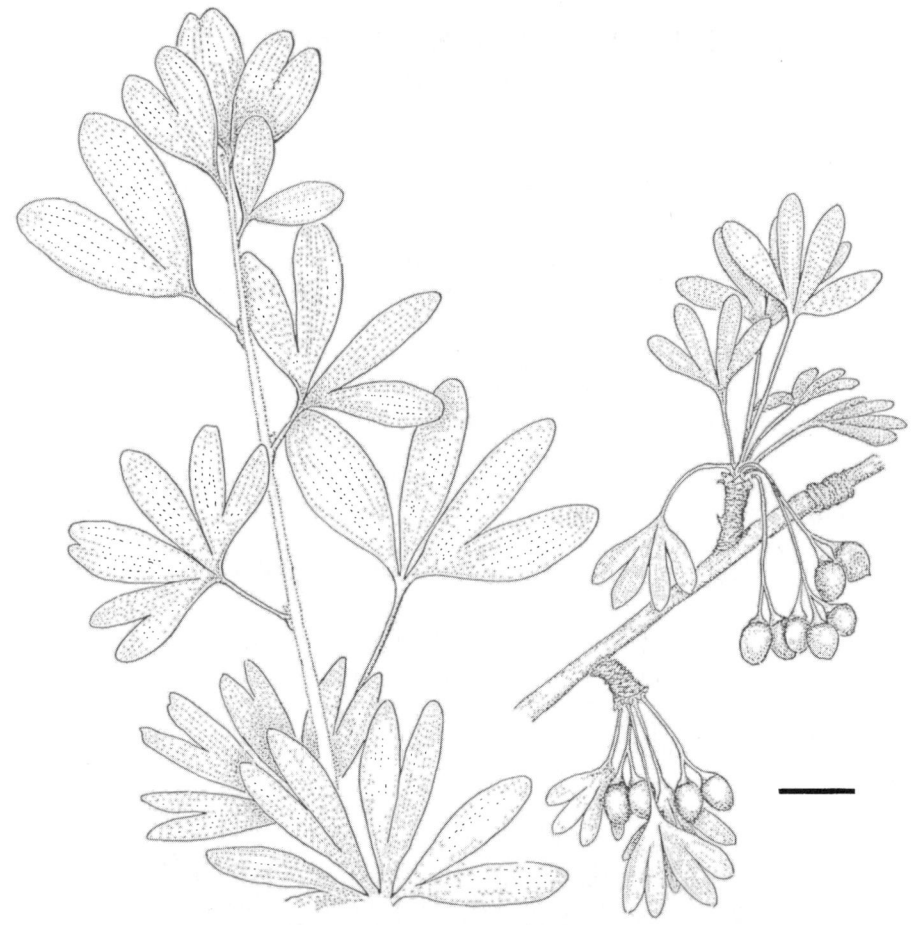

Figure 1 A reconstruction of *Ginkgo yimaensis*: left, leaves of various shapes and sizes believed to have been produced by a single long shoot; right, a portion of a long shoot with two dwarf shoots bearing both leaves and ovule-bearing organs. Bar at lower right equals three centimeters. Reprinted with permission from Zhou and Zhang, 1989.

polar areas, through the end of the Miocene (24–7 mya), when it disappeared from western North America. These dramatic changes were most likely the result of the extensive cooling that occurred throughout the Northern hemisphere during these time periods. The genus Ginkgo was gone from Europe by the end of the Pliocene (1.8 mya) as temperatures dropped and the rainfall regimen gradually shifted from one of summer-wet to one of summer-dry. The only known Pleistocene (1.8 mya to present) occurrences of the genus Ginkgo are from southwestern Japan. Based on their leaf characteristics, these fossils have been classified under the name of the extant species, *G. biloba* (Uemura, 1997). In general, the distribution of Ginkgo fossils indicates that the genus has been relatively consistent in its ecological toler-ances since the Cretaceous, preferring to grow in warm temperate climates characterized by moist summers and cool winters (Tralau, 1968; Uemura, 1997). As regards the morphology of its reproductive organs, the trend within the genus Ginkgo seems to be one of reduction, with the ovules decreasing in number while increasing in size, and the pedicels disappearing to leave sessile ovules connected directly to the peduncle (Zhou and Zhang, 1989; Zhou, 1991; 1994).

ECOLOGY

As a wild species, *Ginkgo biloba* is native to China and was probably a member of the mixed-mesophytic forest community that once covered the hill country bordering the Yangtze River valley. Most of this warm-temperate forest has now been cut down with the exception of a few remnants located in isolated valleys and on steep mountain slopes (Wang, 1961; Zheng, 1992a). One of Ginkgo's last wild refugia is thought to be in Zhejiang Province, China, on the west peak of Tianmu Mountain (Xitianmu Shan: 1506 meters elevation; 119° 25′ E; 30° 20′ N; mean annual rainfall 1767 millimeters; mean January temperature –3.2° C; mean July temperature 20.5° C). There are also reports of "wild" Ginkgo populations in other parts of China, including Guangxi, Guizhou and Sichuan Provinces, but these claims have yet to be substantiated by careful field research (Liang, 1993).

Botanists have long debated the "wildness" of the Ginkgo population growing on Tianmu Mountain. While the long history of human habitation in the area makes it difficult to determine whether or not the trees are truly wild, the exceptional species-richness of the surrounding forest and the large size of many of its trees suggests that they may well be (Del Tredici *et al.*, 1992; Zheng, 1992b). Recent isozyme studies on the population, indicate a relatively low degree of genetic diversity among the plants, an observation that led the authors to speculate that the population may be descended from cultivated trees (Wu *et al.*, 1992).

In 1984, the Tianmu Ginkgo population consisted of approximately 244 individuals with a mean diameter at breast height (DBH) of 45 centimeters and a mean height of 18.4 meters. Most of the trees were growing on disturbed sites such as along stream beds, on rocky slopes, and on the edges of exposed cliffs (Figure 2). Ginkgo seedlings were quite rare on Tianmu Mountain and are typically found in areas of the forest that have been opened up by disturbance, an observation that provides support for the idea that Ginkgo acts as a "pioneer" species in its native habitat (Del Tredici *et al.*, 1992).

Approximately 40% of the Tianmu Ginkgos possessed more than one trunk greater than 10 centimeters diameter at breast height (DBH). Most of these secondary trunks originated from lignotubers located at, or just below, ground level (Figure 3). Secondary trunk formation was most apparent in specimens that were damaged or under severe stress (Del Tredici *et al.*, 1992). The sprouting ability of Ginkgo is an important factor that has enabled the species to persist on the mountain's badly eroded slopes, and may well have played a role in the survival and morphological stability of the genus since the Tertiary (Figure 4).

Figure 2   The famous "living fossil" Ginkgo tree growing on Tianmu Mountain, Zhejiang Province, China, photographed in 1989. This ancient, ovulate specimen occupies an area of approximately 20 square meters and consists of 15 stems greater than 10 centimeters diameter. The largest trunk has a diameter of 110 centimeters. The Chinese describe this tree, perched on the edge of a steep cliff at 950 meters elevation, an "an old dragon trying to fly." The fence protecting the tree was built in 1980. Reprinted from Del Tredici *et al.*, 1992.

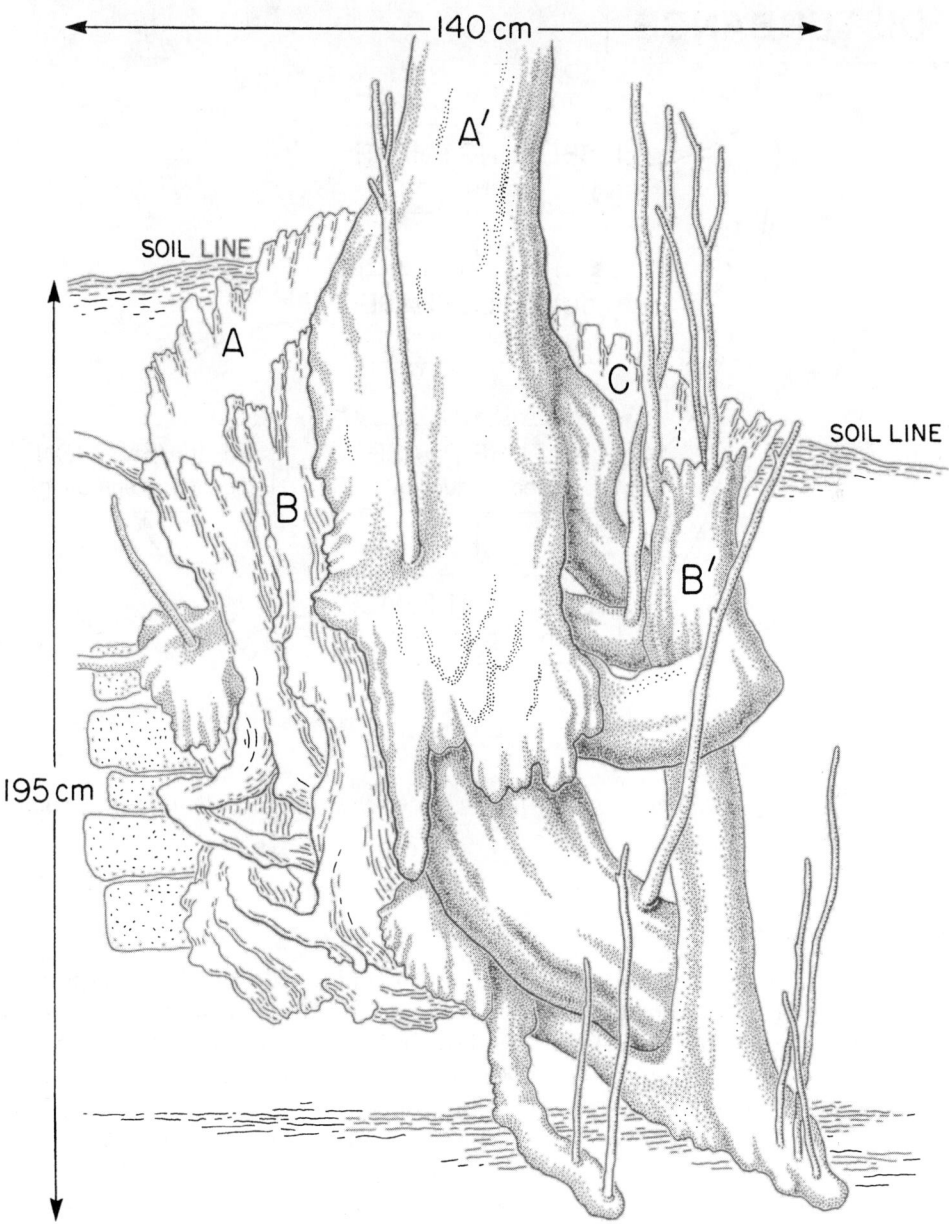

Figure 3   The lignotuber-developed shoot system of an old Ginkgo growing on top of a stone wall on Tianmu Mountain in Zhejiang Province, China. At least three generations of stems can be seen: the oldest represented by the cut trunks A, B, and C (with diameters of 55, 40 and 37 centimeters respectively); the second by the living trunks A′ and B′ (with diameters of 26 and 20 cm); and the third by suckers arising from the distal portions of the lignotuber (stippled). Reprinted from Del Tredici *et al.*, 1992.

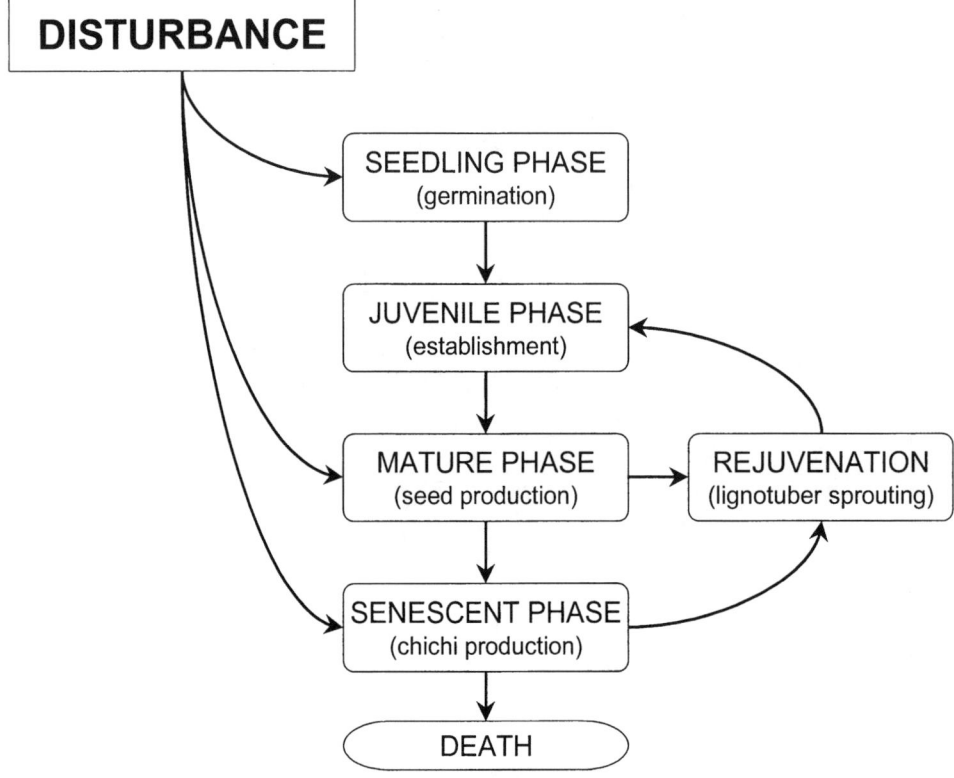

Figure 4 A schematic representation of the life cycle *of Ginkgo biloba* on Tianmu Mountain.

Sexual Reproduction

Ginkgo is a dioecious species, with separate male and female individuals occurring at a roughly 1:1 ratio, although occasional monoecious individuals are reported to occur (Santamour *et al.*, 1983b). Ginkgo shows a long juvenile period, typically not reaching sexual maturity until 20 to 30 years of age. Male and female sex organs are produced on short shoots, in the axils of bud scales and leaves. The male catkins emerge before the leaves and fall off immediately after shedding their pollen. Wind pollination occurs anywhere from early April in areas with mild winters to late May in areas with severe winters. As to pollination distance, it is difficult to say what the maximum is, but in the Boston area 400 meters between male and female trees does not inhibit seed set (Del Tredici, 1989).

Ginkgo ovules are 2 to 3 mm long and produced in pairs at the ends of stalks 1 to 1.5 centimeters long. When the ovule is receptive, it secretes a small droplet of mucilaginous fluid from its micropyle which functions to capture airborne pollen. Retraction of this droplet at the end of the day brings the pollen into the pollen chamber. Once inside the ovule, the male gametophyte commences a four-month long development period that culminates with the production of a pair of multiflagellated spermatozoids, one of which fertilizes a waiting egg cell (Friedman, 1987) while the ovules are still on the tree (Holt and Rothwell, 1997). Depending on the date of pollination, this union can occur anytime between late August to late September.

The mature seed of Ginkgo is relatively large (20–30 mm × 16–24 mm) and consists of an embryo embedded in the tissue of the female gametophyte surrounded by a thick seed coat. This seed coat consists of a soft, fleshy outer layer (the sarcotesta), a hard, stony middle layer, and a thin, membranous inner layer. The seed, devoid of the fleshy sarcotesta, is generally referred to as the Ginkgo "nut," with dimensions of 19–30 mm × 11–14 mm. The developing ovules are green until they mature in the autumn when, in response to cold temperatures, they turn the same yellow color that the leaves do. Typically they fall from the tree about a month after fertilization. The foul odor associated with Ginkgo seeds develops only when they are fully mature, and is the result of the presence of two volatile compounds, butanoic and hexanoic acids, localized in the sarcotesta (Parliment, 1995), also contains phenolic compounds known to cause contact dermatitis in humans (Kochibe, 1997).

Seed Dispersal and Establishment

The paucity of fossilized Ginkgo seeds has not deterred speculation as to what animals might have dispersed Ginkgo seeds over the course of its long evolution. Several authors have proposed that dinosaurs might have dispersed Ginkgo seeds, but none of them have provided any anatomical evidence that would support the claim or specified what type of dinosaur it might have been (Janzen and Martin, 1982; van der Pijl, 1982; Tiffney, 1984; Rothwell and Holt, 1997). A second proposal is that early mammals in the extinct family Mutituberculata could have been effective dispersal agents (Del Tredici, 1989). These marsupial-like creatures, often referred to as the "rodents of the Mesozoic," were widespread in the temperate parts of the northern hemisphere from the Late Jurassic through the Oligocene. Many multituberculates possessed teeth that were adapted to cracking hard food objects, such as seeds, and a skeletal structure that was adapted to living in trees (Krause and Jenkins, 1983).

In modern times, a number of different mammals have been observed feeding on, and presumably dispersing, the odoriferous, nutrient-rich seeds of *G. biloba*. In the order Rodentia, these include the red-bellied squirrel (*Callosciurus flavimanus* var. *ningpoensis*, family Sciuridae) on Tianmu Mountain (Del Tredici *et al.*, 1992), and the gray squirrel (*Sciurus carolinensis*, family Sciuridae) in eastern North America (Del Tredici, 1989). In the order Carnivora, potential dispersal agents include the masked palm civet (*Paguma larvata*, family Viveridae) on Tianmu Mountain (Del Tredici *et al.*, 1992), the leopard cat (*Felis bengalensis*, family Felidae) in Hubei Province, China (Jiang *et al.*, 1990), and the raccoon dog (*Nyctereutes procyonoides*, family Canidae) in Japan (Hori, 1996). The existence of three independent reports of carnivores consuming whole Ginkgo seeds and defecating intact nuts, raises the possibility that the foul smelling sarcotesta may be attracting these animals by mimicking the smell of rotting flesh (Del Tredici *et al.*, 1992). Projecting this line of speculation back into evolutionary timé, it seems likely that if dinosaurs were involved in the dispersal of Ginkgo seeds, then it was probably done by carrion feeding scavengers, with teeth adapted to tearing flesh, rather than by herbivores with dentition adapted to grinding vegetation (Del Tredici, 1989).

Ginkgo seeds are dormant when they fall from the tree because the embryo is not fully developed, being only about 4 to 5 millimeters in length. If seeds are collected shortly after dispersal, cleaned, and placed in a warm greenhouse, the embryo will grow to its full size – 10 to 12 millimeters in length – and germinate within eight to ten weeks (Li and Chen, 1934; Holt and Rothwell, 1997). This type of germination behaviour has important implications in terms of seedling establishment in the field. In warm-temperate climates Ginkgo seeds are shed in late summer or early fall, and the embryo is able to make considerable growth during the mild weather that follows. In cold-temperate climates, on the other hand, seeds are shed later in the season and the colder autumn temperatures delay full embryo development until the following spring. This differential timing of embryo maturation means that seeds in warm climates will be ready to germinate during the favourable conditions of mid- to late spring (April through early June), while those in cold climates will not germi- nate until later in the summer (late June through early August), when conditions for establishment are much less favorable (Del Tredici, 1991a). These phenological dif- ferences suggest that the innate germination requirements of Ginkgo seed may account for the species' warm-temperate distribution in recent times, as well as the warm-temperate distribution of its fossil ancestors.

Tree Architecture

Ginkgo is an extremely long-lived deciduous tree that, in cultivation in China, is capable or reaching ages in excess of a thousand years, with stem diameters between one and four meters (Lin, 1995). Depending on the growing conditions, mature trees typically reach heights of between 20 to 40 meters, although one exceptional speci- men in Korea has been reported with a height of 64 meters and a girth of 14 meters (Del Tredici, 1991b). The form of vigorous young Ginkgos is distinctly pyramidal, with a dominant central leader and widely spaced whorls of lateral branches that grow out at a diagonal orientation to the trunk. With the onset of sexual maturity, at around 25 years of age, height growth generally slows down and the tree fills in its sparsely branched juvenile structure, forming a broad, spreading crown (Gunkle *et al.*, 1949; Del Tredici *et al.*, 1991c).

Ginkgo trees produce two types of shoots: long shoots with widely spaced leaves that subtend axillary buds; and short, or spur, shoots with clustered leaves that lack both internodes and axillary buds. Long shoots are responsible for building up the basic framework of the tree and generating new growing points, while the short shoots produce the majority of leaves and all of the reproductive structures. In response to a variety of environmental and physiological factors, the growth pattern of these shoots can be reversed, such that a short shoot can proliferate into a long shoot, and the terminal growth of a long shoot can fail to elongate and become a short shoot (Gunkle *et al.*, 1949). This flexibility provides Ginkgo with a simple mechanism for modulating carbohydrate allocation between sexual reproduction and vegetative growth (Del Tredici, 1991c).

Under stressful growing conditions, Ginkgo is capable of producing secondary trunks at or just below ground level. Typically, these secondary stems originate from root-like, positively geotropic shoots known as lignotubers or "basal chichi." Anatomically lignotubers develop in all Ginkgo seedlings as part of their normal ontogeny from buds located in the axils of the two cotyledons (Figure 5). When stimulated by some traumatic event, one of these cotyledonary buds often grows out from the trunk to form a woody, positively-geotropic lignotuber which has the capability of producing both aerial shoots and adventitious roots (Del Tredici, 1992; 1997). Similar structures, known as "aerial chichi," often develop along the trunk and branches of very old Ginkgo trees in response to traumatic injury or environmental stress. These stalactite-like growths originate from embedded axillary buds and can produce both roots and shoots when they come in contact the soil (Del Tredici, 1992).

CULTIVATION AND UTILIZATION

While Ginkgo has been cultivated in China for several thousand years, the earliest written references to the plant that have been specifically cited in the literature date back only to Song dynasty documents of the early eleventh century (Li, 1956). According to these ancient texts, the tree came from an area south of the Yangtze River, in what is now the Ningguo District of southern Anhui Province. Most of these early writers praise the Ginkgo for the unique beauty of its leaves and for its edible, and medicinally active, nuts. Contrary to numerous reports in the literature, it is these attributes, rather than any special religious significance, that probably account for the tree's rapid spread in cultivation throughout China and into Korea (Li, 1956). It should be noted, however, that the oldest Ginkgos in China are generally found growing near Daoist and Buddhist temples, and that these ancient specimens have played an important role in the preservation and dissemination of the species (Figure 6).

Ginkgo was introduced into western Japan from eastern China about eight hundred years ago. As is the case in China, the largest trees are growing in the vicinity of Buddhist or Daoist temples and shrines (Tsumura *et al.*, 1992). From Japan, Ginkgo was introduced into Europe at the Botanic Garden in Utrecht, Netherlands, about 1730, and into Kew Gardens, near London, England, around 1754. From England the tree was imported into North America in 1784 at Philadelphia, Pennsylvania (Del Tredici, 1991b).

Ornamental Uses

Ginkgo is cultivated throughout the temperate zones of the world for ornamental purposes. This includes areas with a Mediterranean-type climate as well as those

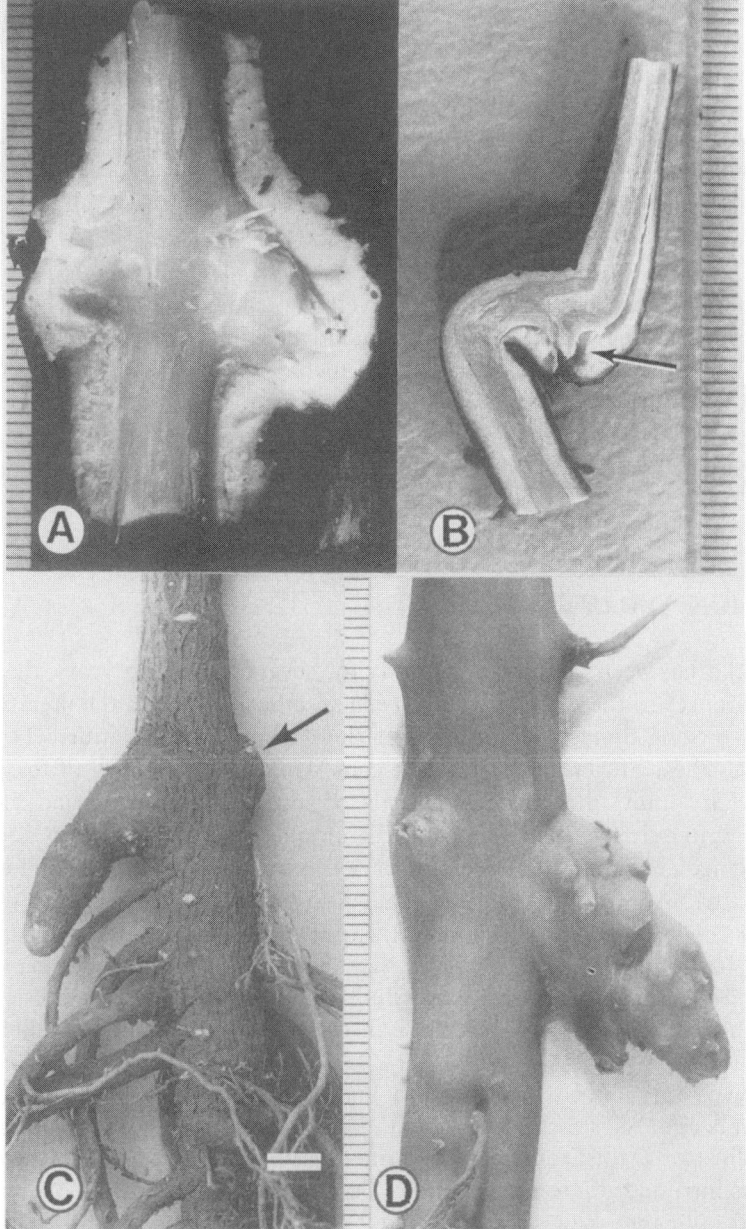

Figure 5 The early stages of lignotuber formation in two- to three-year-old Ginkgo seedlings. **A.** A partially debarked two year old seedling show the unequal development of the cotyledonary buds. Scale in millimeters. **B.** Longitudinal section of a partially debarked two-year-old seedling showing the strongly kinked stem that is often associated with lignotuber formation. Scale in millimeters. **C.** A three-year-old seedling in which one of the cotyledonary buds has formed a prominent lignotuber while the other has not (arrow). Bar equals one centimeter. **D.** A debarked three-year-old seedling showing the xylem traces of the numerous dormant shoot buds on a well-developed lignotuber. Scale in millimeters. Reprinted from Del Tredici, 1992.

Figure 6 A two thousand year-old Ginkgo growing on the grounds of an ancient temple in Xin Cun Village, Tan Chen County, Shandong Province, China. In 1985, the tree was 37 meters tall, with a trunk DBH of 2.3 meters. It is a staminate tree, with a large ovulate branch grafted onto it.

with a cold-temperate climate where minimum winter temperatures can reach
−30° C. While Ginkgo grows best when planted in full sun, it shows the ability to
persist indefinitely under low light and low nutrient conditions, such as when
planted along the streets of densely populated cities (Handa *et al.*, 1997). In the
eastern United States, Ginkgo grows rapidly within USDA hardiness zones 5 to 8, a
region characterized by abundant, year-round moisture, high temperatures during
the growing season, cold winters, and acid to neutral soils. Ginkgo does not perform
particularly well in subtropical climates or on soils that are overly wet or dry during
the growing season (Santamour *et al.*, 1983b; Del Tredici, 1991b; Liang, 1993).
Relative to other commonly cultivated trees, Ginkgo possesses a high degree of
resistance to insect damage and to fungal, viral, and bacterial diseases, as well as to
ozone and sulfur dioxide pollution, making it an excellent choice for planting in
urban areas (Sinclair *et al.*, 1987; Honda, 1997).

Numerous selections of Ginkgo have been made for ornamental purposes during
its long history. Older horticultural forms that are still cultivated include:
"Epiphylla" an ovulate tree which produces seeds attached to the leaves;
"Fastigiata" with a narrow, upright growth habit; "Pendula" with a spreading
growth habit composed entirely of horizontal branches; and "Variegata", a generally
unstable form with leaves striped with yellow or white (Santamour *et al.*, 1983a).
These, as well as other more recent cultivars can be propagated from rooted soft-
wood cuttings in summer, hardwood cuttings in winter, twig grafting in late winter,
or bud grafting in late summer (Del Tredici, 1991b; Liang, 1993). Because of
complex epigenetic effects, however, both rooted cuttings and grafted plants generally fail to reproduce the upright form of the seed-grown parent they were propagated from. Instead, they tend to produce low-branched, vase-shaped trees that
lack a dominant central leader (Figure 7) (Del Tredici, 1991b, 1991c). At the present
time, fastigiate male clones are the only vegetatively propagated selections that are
widely available from commercial nurseries in Europe and North America.

Seed Production

Each Ginkgo seed contains a single, thin-shelled "nut," that is traditionally consumed as both a food and a medicine throughout Asia. Dry nuts, which constitute
59% of the fresh nut weight, have a composition of roughly 6% sucrose, 68%
starch, 13% protein, and 3% fat (Duke, 1989). The commercial production of
Ginkgo nuts has been established in China for over 600 years, and at least 44 cultivars have been selected in that country based on the size and shape of their nuts,
and on their productivity (Santamour *et al.*, 1983a; Liang, 1993; He *et al.*, 1997).
These cultivars are usually propagated through grafting on seedling rootstocks, and
they begin producing nuts at about 5 years of age (Figure 7). For the widely grown
cultivar "Dafushon," annual yields in Jiangsu Province, China vary between 5 and
10 kilograms of nuts for 15-year-old trees and between 50 and 100 kilograms for
50-year-old trees. In general, Ginkgo produces a heavy crop of seeds every other
year, with relatively light crops in alternate years (Del Tredici, 1991b; Liang, 1993).
While precise figures are difficult to come by, one recent estimate of the Chinese

Figure 7 A grove of grafted Ginkgo trees cultivated for their edible nuts on Dongting Shan, Jiangsu Province, China. Note the spreading, vase-shaped form of the tree. Reprinted from Del Tredici, 1991b.

crop suggests that 700,000 to 800,000 trees produce an average of between six to seven thousand tons of dried nuts per year (He *et al.*, 1997).

Raw Ginkgo nuts, cleaned of their fleshy pulp, have long been used in traditional Chinese medicine to treat a variety of lung-related ailments such as asthma and bronchitis, as well as for the treatment of kidney and bladder disorders. They also display antibiotic effects against a variety of bacterial pathogens (Perry, 1980; Bensky and Gamble, 1986). Raw Ginkgo nuts contain a toxin that has been identified as 4-O-methylpyridoxine (MPN), whose primary mode of action is to antagonize the activity of vitamin B6 (Wada and Haga, 1997). Typically, children are much more susceptible to poisoning from Ginkgo nuts than adults.

For culinary purposes, Ginkgo seeds must be cooked to eliminate the potential for poisoning. They are usually steamed until the hard shell cracks open, allowing the kernel to be removed. In China, these kernels are either boiled in sugar water to make a sweet soup or roasted and eaten plain. Because of their potential toxicity, however, people are generally advised not to eat too many Ginkgo nuts at one sitting.

Leaf Production

Ginkgo leaves are known to contain a wide variety of medicinally active chemicals, most notably terpenoids (ginkgolides and bilobalide) and flavonoids (glycosides of kaempferol, quercetin, isorhamnetin, etc.) (Boralle *et al.*, 1988; DeFeudis, 1998). The use of Ginkgo leaf extract for pharmaceutical purposes was originally developed in Germany in 1965, and the first commercially available Ginkgo leaf extract was registered for human use in 1974 in France, under the code-name "EGb 761" (DeFeudis, 1998). This extract is made from dried Ginkgo leaves and has a standardized content of 22–27% flavonol glycosides and 5–7% terpene trilactones. This extract is taken internally for the treatment of cerebral and peripheral vascular diseases, as well as to alleviate some of the ailments associated with ageing, including dizziness, ringing in the ears, and short-term memory deterioration. So far as is known, the extract has only minimal side effects, even after prolonged use (DeFeudis, 1998; Juretzek, 1997). Contrary to numerous reports in the literature, Ginkgo leaf preparations have played a very minor role in the practice of traditional Chinese medicine, and have only recently been listed in official *Materia Medica*.

Since 1982, Ginkgo leaves have been cultivated on a large scale in both France (480 hectares) and the United States (460 hectares) specifically for the production of the EGb 761 extract. When grown for this purpose, seedlings are spaced 40 centimeters apart in rows 1 meters apart, producing a density of approximately 25,000 plants per hectare (Figure 8). Special pruning techniques are used to keep the trees below 3 meters tall, thereby allowing the use of mechanical harvesting equipment. Green leaves are generally harvested in mid- to late summer, after which the plants are cutback to near ground level once every four or five years. The new growth

Figure 8   The Ginkgo plantation in Sumter, South Carolina, in early spring. For scale, the individual segments of the irrigation system are about 45 meters long. Reprinted from Del Tredici, 1991b.

following such low cutbacks originates from buds located at the base of the stem or on basal lignotubers. When provided with an adequate supply of moisture and fertilizer during the growing season, the annual height growth from the point of cutting is typically a meter or more (Del Tredici, 1991b).

In China, yellow leaves are typically harvested at the same time that the nuts are collected, in early autumn, and sold for the manufacture of leaf extracts. These older leaves are not as rich in ginkgolides and other medicinally active compounds as green leaves, and consequently command a lower price in the international market. Since 1992, over 2000 hectares of Ginkgo seedlings have been planted in eastern China specifically for the production of green leaves. The first of these plantations began producing marketable quantities of leaves in 1996, and projections are for yields to increase dramatically over the next five to ten years (Balz, 1997; He *et al.*, 1997).

The use of Ginkgo leaf extract for therapeutic as well as other purposes has grown dramatically since its introduction in 1974. While accurate figures are difficult to come by, recent estimates of the world wide sales of Ginkgo leaf products suggest that it is around half a billion US dollars (see Warrier and Corzine, this volume), a truly remarkable figure given that the product has only been on the market for 25 years.

REFERENCES

Balz, J.-P. (1997) Agronomic aspects of *Ginkgo biloba* leaves production. In Proceedings of '97 International Seminar on Ginkgo, Nov. 10–12, 1997, Beijing, China, pp. 101–104.

Bensky, D. and Gamble, A. (1986) *Chinese Herbal Medicine: Materia Medica*. Eastland Press, Seattle, Washington (translated from the Chinese).

Boralle, N., Braquet, P., Gottlieb, O.R. (1988) *Ginkgo biloba*: a review of its chemical composition. In P. Braquet (ed.) *Ginkgolides–Chemistry, Biology, Pharmacology and Clinical Perspectives*, Vol. 1, J.R. Prous, Barcelona, pp. 9–25.

DeFeudis, F.V. (1998) *Ginkgo biloba Extract (EGb 761): From Chemistry to Clinic*, Ullstein Medical, Weisbaden.

Del Tredici, P. (1989) Ginkgos and multituberculates: evolutionary interactions in the Tertiary. *Biosystems*, **22**, 327–339.

Del Tredici, P. (1991a) Evolution and Natural History of *Ginkgo biloba*. PhD thesis, Boston Univ.

Del Tredici, P. (1991b) Ginkgos and people: a thousand years of interaction. *Arnoldia*, 51, 2–15.

Del Tredici, P. (1991c) The architecture of *Ginkgo biloba* L. In C. Edelin (ed.), L'Arbre, Biologie et Developpement. *Naturalia Monspeliansia* nº h.s., pp. 155–168.

Del Tredici, P. (1992) Natural regeneration of *Ginkgo biloba* from downward growing cotyledonary buds (basal chichi). *Am. J. Bot.*, **79**, 522–530.

Del Tredici, P. (1997) Lignotuber formation in *Ginkgo biloba*. In T. Hori, R.W. Ridge, W. Tulecke, P. Del Tredici, J. Tremouillaux-Guiller, and H. Tobe (eds.), *Ginkgo biloba* – A Global Treasure. Springer-Verlag, Tokyo, pp. 119–126.

Del Tredici, P., Ling, H., Yang, G. (1992) The Ginkgos of Tian Mu Shan. *Conserv. Biol.*, **6**, 202–209.

Duke, J.A. (1989) *CRC Handbook of Nuts*, CRC Press, Boca Raton, Florida.

Friedman, W.E. (1987) Growth and development of the male gametophyte of *Ginkgo biloba* within the ovule (in vitro). *Am. J. Bot.*, **74**, 1797–1815.

Gifford, E.M. and Foster, A.S. (1987) *Morphology and Evolution of Vascular Plants*, W.H. Freeman and Company, New York.

Gunkle, J.E., Thimann, K.V., and Wetmore, R.H. (1949) Studies of development in long shoots and short shoots of *Ginkgo biloba* L., part IV. Growth habit, shoot expression and the mechanism of its control. *Am. J. Bot.*, **36**, 309–316.

Handa, M., Iizuka, Y., and Fujiwara, N. (1997) Ginkgo landscapes. In T. Hori, R.W. Ridge, W. Tulecke, P. Del Tredici, J. Tremouillaux-Guiller, and H. Tobe (eds.), *Ginkgo biloba* – A Global Treasure. Springer-Verlag, Tokyo, pp. 259–283.

Hasebe, M. 1997. Molecular phylogeny of *Ginkgo biloba*: close relationship between *Ginkgo biloba* and cycads. In T. Hori, R.W. Ridge, W. Tulecke, P. Del Tredici, J. Tremouillaux-Guiller, and H. Tobe (eds.), *Ginkgo biloba* – A Global Treasure. Springer-Verlag, Tokyo, pp. 173–182.

He, S.-A., Yin, G., and Pang, Z.-J. (1997) Resources and prospects of *Ginkgo biloba* in China. In T. Hori, R.W. Ridge, W. Tulecke, P. Del Tredici, J. Tremouillaux-Guiller, and H. Tobe (eds.), *Ginkgo biloba* – A Global Treasure. Springer-Verlag, Tokyo, pp. 373–383.

Holt, B.F. and Rothwell, G.W. (1997) Is *Ginkgo biloba* (Ginkgoaceae) really an oviparous plant? *Am. J. Bot.*, **84**, 870–872.

Honda, H. (1997) Ginkgos and insects. In T. Hori, R.W. Ridge, W. Tulecke, P. Del Tredici, J. Tremouillaux-Guiller, and H. Tobe (eds.), *Ginkgo biloba* – A Global Treasure. Springer-Verlag, Tokyo, pp. 243–250.

Hori, T. (1996) Ginkgo to the Japanese people. *Microscopia*, 13, 184–185 (in Japanese).

Janzen, D.H. and Martin, P.S. (1982) Neotropical anachronisms: the fruits the gomphotheres ate. *Science*, **215**, 19–27.

Jiang, M., Jin, Y. and Zhang, Q. (1990) Preliminary study on *Ginkgo biloba* in Dahongshan region, Hubei. *J. Wuhan Bot. Res.*, 8, 191–193 (in Chinese).

Juretzek, W. (1997) Recent advances in *Ginkgo biloba* extract (Egb 761). In T. Hori, R.W. Ridge, W. Tulecke, P. Del Tredici, J. Tremouillaux-Guiller, and H. Tobe (eds.), *Ginkgo biloba* – A Global Treasure. Springer-Verlag, Tokyo, pp. 341–358.

Kochibe, N. (1997) Allergic substances of *Ginkgo biloba*. In T. Hori, R.W. Ridge, W. Tulecke, P. Del Tredici, J. Tremouillaux-Guiller, and H. Tobe (eds.), *Ginkgo biloba* – A Global Treasure. Springer-Verlag, Tokyo, pp. 301–307.

Krause, D.W. and Jenkins, F.A. (1983) The postcranial skeleton of North American multituberculates. *Bull. Mus. Comp. Zool.*, **150**, 199–246.

Li, H.L. (1956) A horticultural and botanical history of Ginkgo. *Bull. Morris Arb.*, **7**, 3–12.

Li, T.T. and Chen, S.M. (1934) Temperature and the development of the Ginkgo embryo. *Sci. Rep. Nat. Tsing Hua Univ.*, ser. B, **2**, 37–39.

Liang, L. (1993) *The Contemporary Ginkgo Encyclopedia of China*, Beijing Agric. Univ. Press (in Chinese).

Lin, J.-X. (1995) Old Ginkgo trees in China. International Dendrological Society Yearbook, 1995, pp. 32–37.

Parliment, T. (1995) Characterization of the putrid aroma compounds of *Ginkgo biloba* fruits. In R. Rouseff and M. Leahy (eds.) *Fruit Flavors: Biogenesis, Characterization, and Authentication*, Am. Chem. Soc. Symp. Ser., **596**, pp. 276–279.

Perry, L.M. (1980) *Medicinal Plant of East and Southeast Asia: Attributed Properties and Uses*, MIT Press, Cambridge, Massachusetts.

van der Pijl, L. (1982) *Principles of Dispersal in Higher Plants*, 3rd ed., Springer-Verlag, Berlin.

Rothwell, G.W. and Holt, B. (1997) Fossils and phenology in the evolution of *Ginkgo biloba*. In T. Hori, R.W. Ridge, W. Tulecke, P. Del Tredici, J. Tremouillaux-Guiller, and H. Tobe (eds.), *Ginkgo biloba* – A Global Treasure. Springer-Verlag, Tokyo, pp. 223–230.

Santamour, F.S., He, S.A., McArdle, A.J. (1983a) Checklist of cultivated Ginkgo. *J. Arboriculture* 9: 88–92.

Santamour, F.S., He, S.A., Ewert, T.E. (1983b) Growth, survival and sex expression in Ginkgo. *J. Arboriculture* 9: 170–171.

Sinclair, W.A., Lyon, H.H., Johnson, W.T. (1987) *Diseases of Trees and Shrubs*, Comstock Publishing Associates, Ithaca.

Tiffney, B.H. (1984) Seed size, dispersal syndrome and the rise of the angiosperms: evidence and hypotheses. *Ann. Missouri Bot. Gard.* 71: 551–576.

Tralau, H. (1968) Evolutionary trends in the genus Ginkgo. *Lethaia* 1: 63–101.

Tsumura, Y., Motoike, H., Ohba, K. (1992) Allozyme variation of old *Ginkgo biloba* memorial trees in western Japan. *Can. J. For. Res.* 22: 939–944.

Uemura, K. (1997) Cenozoic history of East Asia. In T. Hori, R.W. Ridge, W. Tulecke, P. Del Tredici, J. Tremouillaux-Guiller, and H. Tobe (eds.), *Ginkgo biloba* – A Global Treasure. Springer-Verlag, Tokyo, pp. 207–221.

Wada, K. and Haga, M. (1997) Food poisoning by *Ginkgo biloba* seeds. In T. Hori, R.W. Ridge, W. Tulecke, P. Del Tredici, J. Tremouillaux-Guiller, and H. Tobe (eds.), *Ginkgo biloba* – A Global Treasure. Springer-Verlag, Tokyo, pp. 309–321.

Wang, C.W. (1961) *The Forests of China*, Maria Moors Cabot Found., publ. 5. Harvard Univ., Cambridge, Mass.

Wang, F.H. and Chen, Z.K. (1983) A contribution to the embryology of Ginkgo with a discussion of the affinity of the Ginkgoales. *Acta Botanica Sinica*, 25, 199–211(in Chinese).

Wu, J., Cheng, P., and Tang, S. (1992) Isozyme analysis of the genetic variation of *Ginkgo biloba* L. population in Tian Mu Mountain. *J. Plant Resources Environment*, 1, 20–23 (in Chinese).

Zheng, C.Z. (1992a) A preliminary analysis of flora in Tianmu Mountain Reserve. In F. Yang (ed.) *Comprehensive Investigation Report on Natural Resource of Tianmu Mountain Nature Reserve*, Science and Technology Press, Hangzhou, pp. 89–93 (in Chinese).

Zheng, C.Z. (1992b) A catalogue of seed-plants in Tianmu Nature Reserve. In F. Yang (ed.) *Comprehensive Investigation Report on Natural Resource of Tianmu Mountain Nature Reserve*, Science and Technology Press, Hangzhou, pp. 94–128 (in Chinese).

Zhou, Z. (1991) Phylogeny and evolutionary trends of Mesozoic ginkgoaleans – a preliminary assessment. *Rev. Palaeobot. Palynol.*, 68, 203–216.

Zhou, Z. (1994) Heterochronic origin of *Ginkgo biloba*-type ovule organs. *Acta Palaeontologica Sinica*, 33, 131–139 (in Chinese).

Zhou, Z. and Zhang, B. (1989) A middle-Jurassic Ginkgo with ovule-bearing organs from Henan, China. *Palaeontographica*, B 211, 113–133.

2. *GINKGO BILOBA* L.: ASPECTS OF THE SYSTEMATICAL AND APPLIED BOTANY

VOLKER MELZHEIMER[1] and JOHANNES J. LICHIUS[2]

[1]*Botanischer Garten, Philipps-Universität Marburg,*
Karl-von-Frisch-Str., D-35032 Marburg, Germany
[2]*Institut für Pharmazeutische Biologie, Philipps-Universität Marburg,*
Deutschhausstr. 17 A, D-35032 Marburg, Germany

NOMENCLATURE AND SYSTEMATICS

The first definitive description of the genus and species of *Ginkgo biloba* L. was given by Linnaeus (1771). For this reason, other names like *Salisburia adiantifolia* Smith and *Pterophyllus salisburiensis* Nelson are either invalid or are regarded as synonyms according to the rules of nomenclature with particular reference to the rule of priority (Smith, 1797; Nelson, 1866). The spelling of proper names is also determined by these rules. Consequently, Mayr's correction of the spelling into Ginkgyo in 1906, which was justified in some respects, could not be accepted (Mayr, 1906). A German name is actually not in use. The English term "Maidenhair tree" is based on the similarity to the foliage of the "Maidenhair fern" (Adiantum). In Japan it is called "ginkyo", whereas the French names are "l'arbre aux quarante écus" and "noyer du Japon".

In older systems the genus Ginkgo was included in the family of Taxaceae. Originally this family included Podocarpaceae and Cephalotaxaceae and has always been considered very artificial. The inclusion of the genus Ginkgo in this family mainly resulted from ignorance. Hirase's discovery in 1895 that Ginkgo possesses multiciliated spermatozoids was the basis of Engler's classification of a particular family and class of Ginkgoopsida (Hirase, 1895; Pilger, 1926). This class, Ginkgo and Ginkgo-like precursors (Vozenin-Serra *et al.*, 1991), can be traced back to the Lower Permian (see Fig. 1) and is likely to extend into the Upper Devonian. According to Tralau (1967), the oldest reliable proof of the genus Ginkgo reaches back as far as the Lower Jurassic (see Fig. 2). With regard to the number of Ginkgo species, it is much smaller than has so far been believed because many fossil finds, each one of which has been regarded as a different species, can now be included in 2–3 different Ginkgo species (Tralau, 1967). Thus it is likely that the more highly developed representatives [progressive in that the ovules become sessile and reduced in number but increased in size (Zhou, 1991)] from the lower Cretaceous can be considered as the direct precursors of the recent *Ginkgo biloba* (also see Sporne, 1965). As far back as the Upper Palaeozoic there are a few primitive plants whose characteristics seem to align them with this family. During the Mesozoic, especially

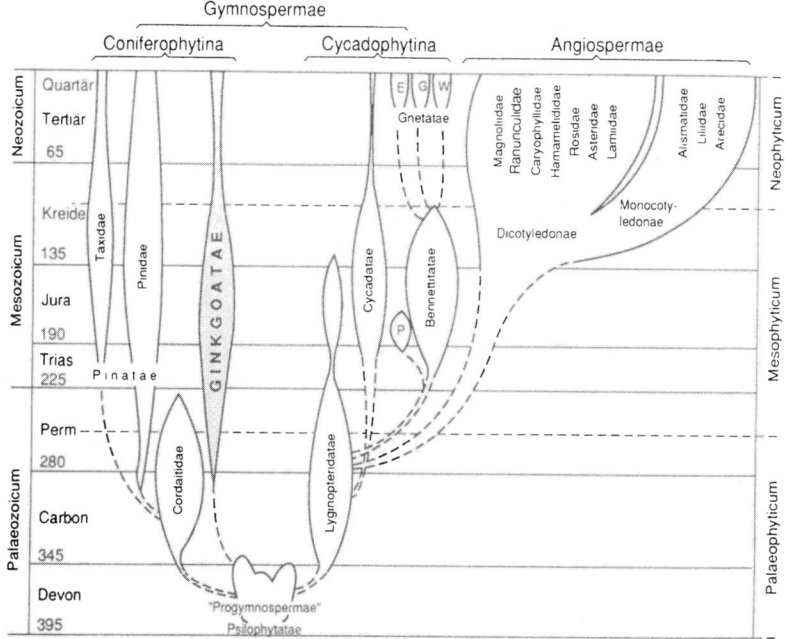

Figure 1 Possible phylogenetic connections between related groups of seed plants and their development during different geological eras (numbers at the beginning of geological formations stand for millions of years). Unreliable relations not documented by fossil finds are shown by broken lines or left white. (Strasburger, 1998)

P = Pentoxylidae, E = Ephedridae, G = Gnetidae, W = Welwitschiidae

during the Jurassic and the early part of the Cretaceous, Ginkgo attained its greatest prominence (see Fig. 2). This was not only true of the number of different species and genera occurring, but also of their nearly world-wide distribution and high number of habitats. Reaching its climax in the Upper Jurassic and Lower Cretaceous the group declined rather rapidly before the end of the Cretaceous. By the Oligocene all but two of the 19 genera with nearly 60 species of the family became extinct. From then on only a few species have persisted albeit in gradually lessening numbers and in gradually narrowing geographic range: Today they have disappeared from all but one continent, in which only one single species has survived (Dorf, 1958) (see Fig. 3). Whether or not this decline is due to natural or culture-related causes remains difficult to decide today (Del Tredici *et al.*, 1992). Thus *Ginkgo biloba* is the prime botanical example of a "living fossil", an expression first used by Darwin for the King crab (Limulus sp.), an animal alive today which is very similar to the fossil species *Mesolimulus sulcatus*. The currently used systematic classification into division and subdivision is as follows (Strasburger, 1998):

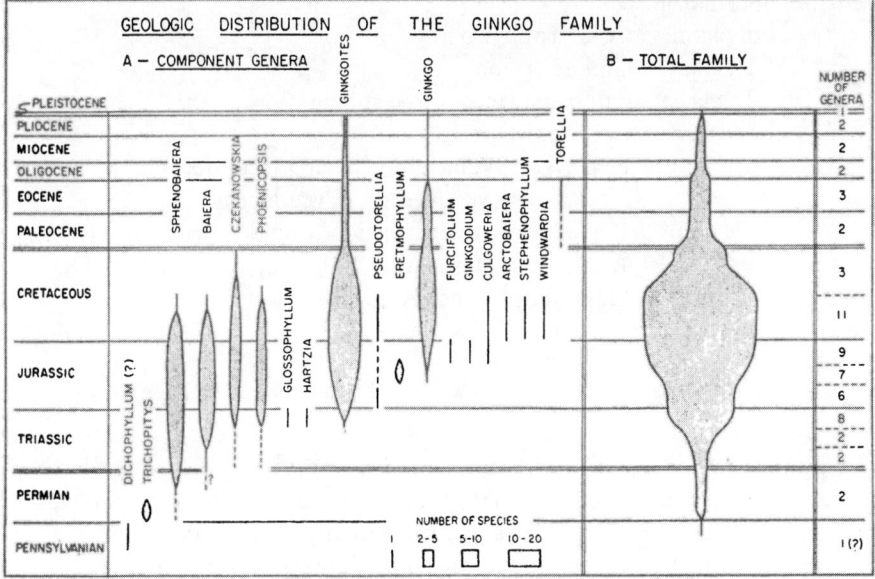

Figure 2   Chart of the geologic ranges of valid genera of the Ginkgo family, also showing the relative abundance of species within each genus, and the composite range and relative importance of the entire family in each successive geologic epoch (Dorf, 1958).

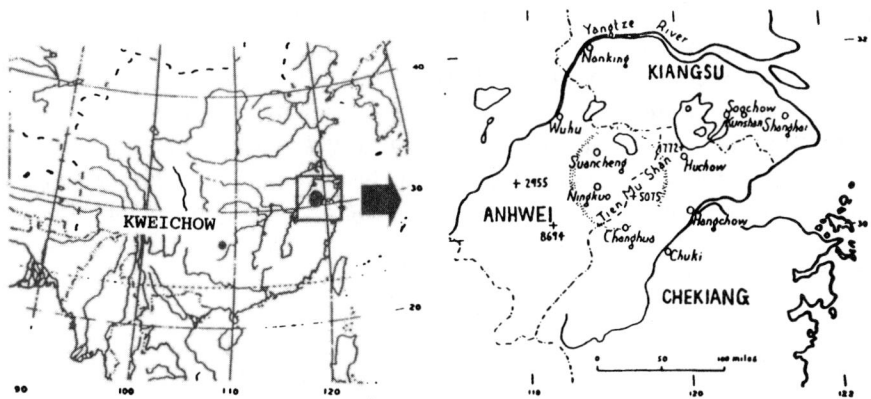

Figure 3   "Natural" occurrence of *Ginkgo biloba*. There are relatively dense growths of Ginkgo of different ages. The more important occurrence is in the provinces of Anhwei and Chekiang, here mainly in the Tian-Mu-Shan Mountains (according to Li, (1982), slightly modified).

17. division: Spermatophyta
      1. subdivision: Coniferophytina
            1. class: Ginkgoopsida
            2. class: Pinopsida (with three subclass)
      2. subdivision: Cycadophytina
            1. class   Lyginopteridopsida (seed-ferns)
            2. class   Cycadopsida (Cycadaceae a.o.)
            3. class   Bennettitopsida (only known in fossil form)
            4. class   Gnetopsida (Ephedra, Welwitschia, Gnetum)
      3. subdivision: Magnoliophytina (Angiospermae)

SHORT DESCRIPTION

Tree, height up to 40 m, multiform habit, slim and conical or spread out, bark grey, deeply furrowed on old trunks, alternate leaves on long shoots, fan-shaped, tough and leathery, 5–8 cm wide with long petiole, on the wider side often incised or lobed, parallel and forked nervation, colour: a fresh green, golden yellow in autumn before being shed, dioecious, seeds resembling small yellow plums, yellowish-green, about 2.5 cm long, fleshy outside, disagreeable smell, 2–3-edged stones, indigenous to China, frequently grown in Japan as temple tree, today also as roadside tree in Japan, Europe and America. Compared with other groups of plants, however, the Ginkgo has some peculiarities requiring a more detailed explanation.

VEGETATIVE ORGANS

The Ginkgo grows into a stately tree with a height of 30–40 m and a girth of 3–4 m; the bark is scabby, dark grey (for details of bark morphology see Kim *et al.*, 1992); at first of pyramidal growth, it later forms an expanding crown. The trees can reach an age of several hundred years. In Europe, the oldest are about 245 years old, whereas in East Asia there are said to be some specimens of 1000 years with a girth of up to 20 m measured 1 m above the ground (see Fig. 4). The famous Ginkgo of Sendai in Japan is allegedly 1200 years old. Its branches cover the enormous area of 250 m^2; it owes its fame to the voluminous "chichi" growing on the underside of the big branches (see Fig. 5). This word means "mother's breast", which is why Japanese women consider trees with such phenomena to be symbols of numerous offspring and good breast-feeding ability (Schmid, 1994). Anatomically these are strictly positive geotropic sprouting growths (origin and possible early development see Del Tredici, 1992), developing roots upon reaching the ground. The growth rings of such stalactite-like branches are usually narrower than normal ones and there are sometimes zones of irregularly oriented tracheids (Li and Lin, 1991). Subsequently, new lateral branches can develop on this "trunk". Similar phenomena are also known in different types of the genus Ficus (gum tree) where they help the trunk to expand, thus increasing the stability of the plant.

Figure 4   Giant Ginkgo in the Zensho-ji Temple District, Fuchizawa (Japan) (Kato, 1994).

Branching is achieved by short and long shoots (see Fig. 6). On the long shoots leaves are placed relatively far from each other in 2/5 or 3/8 spirals (also see Gunckel and Wetmore, 1946). The end bud, more often the origin of another long shoot rather than a short shoot, consists of about 15 scales, the exterior ones are small, broad, and almost entirely suberized and more or less densely pilose on the edge. There is a decreasing suberization of the interior scales towards their tips, so that the interior scales are green and not suberized; the innermost ones can have rudimentary blades as appendages. On the axial buds, normally the starting points of short shoots and rarely of long shoots, several pairs of decussated scales can be found. When the long shoots of old trees stop growing, there can be a spontaneous development of short shoots into long shoots. These short shoots can have leaves for many years up to several decades; their annual increase is low, however, and at an advanced age they appear warty on account of the densely placed bases of the shed

Figure 5   Branch with "Chichi" in the initial stage on the Ginkgo in Jizo-an near Itano (Kato, 1994).

leaves. According to earlier investigations the Ginkgo has no separation zone (which already exists in the family of Pinaceae), and the separation of the leaf stalk takes place in a layer of round cells between the stalk and the bark and is facilitated by segmentation at the base of the leaf.

The leaves of the Ginkgo are summer green, i.e. they are active only during one vegetation period. The form of the leaves varies considerably (see Fig. 7). The unlobed or bilobed early leaves of short shoots are preformed in the autumn bud, and their nearly synchronous-expansion in the spring is not accompanied by stem elongation. The others are multilobed late leaves which develop at intervals of several days and their production can go on possibly throughout the summer. According to Critchfield (1970), the developmental events strongly correlated with the pattern of auxin production or distinct concentration in short and long shoots respectively. The stalk is thin, about 4–9 cm long, more or less pilose, furrowed on the upper side and only a little broader at the base; near the base of the blade there is a short inconspicuous widening of the stalk, or else it becomes distinctly wedge-shaped. The blade itself is fan-shaped and approximately semi-circular, the edge has several irregular lobes and a narrow more or less deep incision opposite the stalk; or the blade is completely undivided and only the edge is retuse. The young leaves, on the other hand, have several deep incisions. After sprouting, the leaves are sparsely

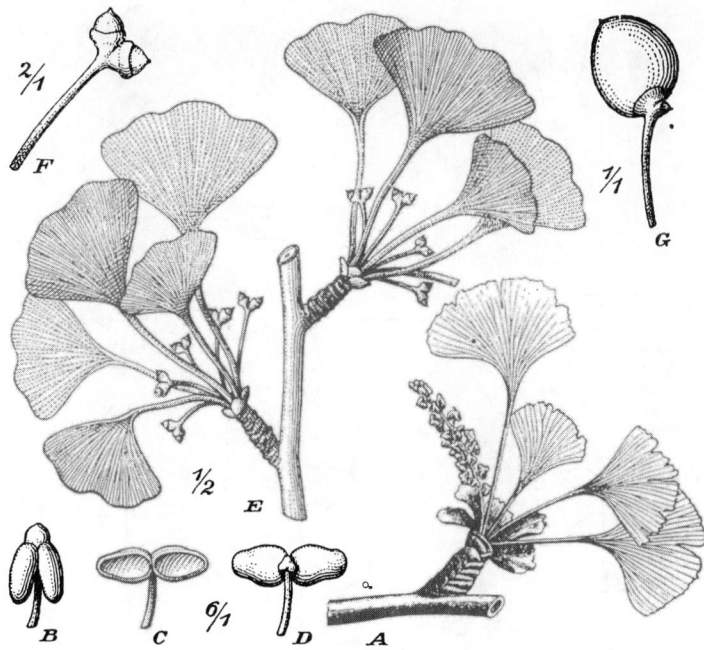

Figure 6 *Ginkgo biloba* L.: A branch with male blossoms on a short shoot; B-D staminodium; C with open sporangia; D rear view; E branch with female blossom on short shoot; F female blossom; G peduncle with developed seed-bud – fractions indicate magnification or reduction (Pilger, 1926).

hairy. The width of the blade is 7–10 (max. 12) cm for leaves of long shoots and 4–7 cm for those of short shoots.

Anatomy of the Vegetative Organs

As with all cormophytes, the primordial leaves of gymnosperms develop at the vegetative cone of equivalent initial cells (= initial field). Up until Strasburger's discoveries around 1870, the opinion was held that in gymnosperms the apical cell of a shoot was comparable to ferns (pteridophytes), i.e. the original meristem develops from one single apical cell. Ginkgo, however, has developed no tunica layer (= later epidermis). Some gymnosperms, as well as the monocotyledons, have a single-layered tunica whereas a very large number of dicotyledons have a double layered one. This so-called pinus type of primordial leaf is limited almost entirely to gymnosperms and is characterised by periclinal (in addition to anticlinal) divisions in the outer layer of cells of the vegetative cone (in an apical position as well as in the development of primordial leaves). As with all seed plants, the longitudinal growth of primordial leaves has two components: apical growth, which is typical of most ferns, and intercalary growth, which prevails in angiosperms. It can be generally

Figure 7 Outlines of Ginkgo leaves, the upper ones of long shoots, the lower ones of short shoots (1/3 of natural size) (Chamberlain, 1935).

said that intercalary growth begins before apical growth comes to an end. This is especially true of Ginkgo which – together with various Cycadaceae – is an exception among the gymnosperms because the development of the leaf stalk precedes the development of the blade. The stalk finishes its apical growth only when the growth of the leaves comes to an end altogether.

In the upper epidermal layer, the cells are more or less regularly arranged, especially above the longitudinal vascular bundles. In contrast, the cells of the lower layer are very irregularly arranged (Li *et al.*, 1989b). In a cross section, the epidermal cells on the upper and underside are approximately isodiametric; a top view shows slightly wavy walls. The epidermal cells of Ginkgo and a few other gymnosperms are bigger on the upper side than those on the underside. The exterior walls of the epidermal cells are covered by a more or less thin cuticle covered with wax structures (chemical composition see Guelz *et al.*, 1992). A palisade parenchyma is generally found – only the smaller leaves of flowering short shoots have no palisade parenchyma.

Gymnosperms are characterised by a wide variety of excretory organs, both with regard to the substances secreted in the metabolic process and the manner of excretion. Excreted substances are mainly resins, essential oils, gums, slime, tannins and calcium oxalate. Three kinds of excretory organs can be distinguished in the leaves of Ginkgo: 1. excretory organs of lysogenic origin alternating in rows of 1–7 mm in length with the vascular bundles. 2. secretory cells in curved and annular groups close to the vascular cylinder. 3. tannic idioblasts which can become thick-walled and very long by stretching or fusing. Excretory organs also occur in the leaf stalk, the primary bark and occasionally in the pith.

In most cases the stomata are arranged irregularly only on the underside of the leaf. Ginkgo thus differs from the longitudinal orientation of the stomata in the group of gymnosperms. The first stomata are still oriented parallel to the nerves which is no longer true for the later ones. Intact stoma mother cells on the upper side of the leaf are rare, more frequent are those which are not fully developed.

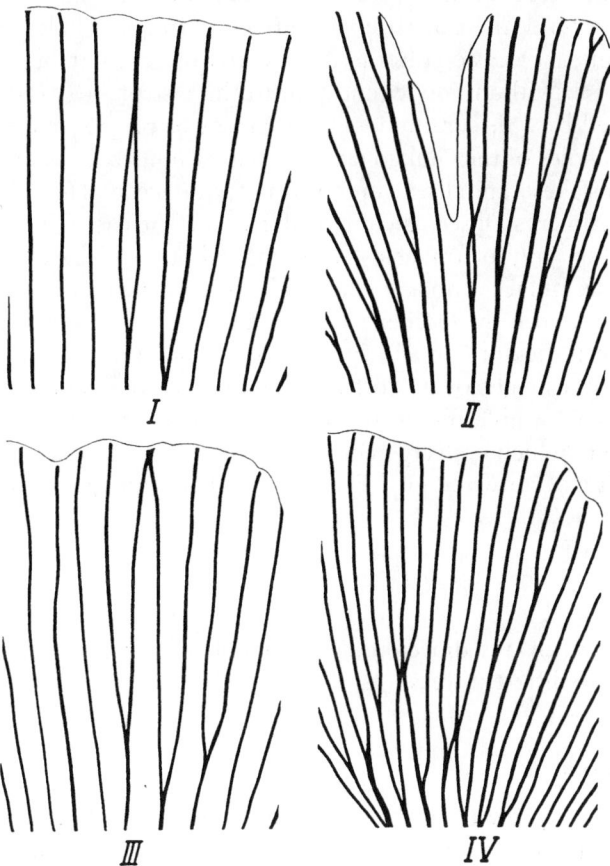

Figure 8   Anomalies in the vascular bundles of leaves.

I. After separating the two branches of a vascular bundle meet again and remain one until they reach the edge of the leaf.
II. After separating the two branches of a vascular bundle meet again and then separate again.
III. Two vascular bundles meet and remain one until they reach the edge of the leaf.
IV. Two vascular bundles only touch each other at one point.
(Arnott 1959; quoted from: Napp-Zinn, 1966)

These non-functioning stomata are interesting because fossil leaves of Ginkgoaceae from earlier geological periods are amphistomatic (= stomata on both sides) instead of hypostomatic (= stomata only on the underside of the leaves) as in *Ginkgo biloba* (Kanis and Karstens, 1963). On the other hand the stomata of the cotyledons are mainly found on the upper side of the leaf. The leaf stalk has stomata on all sides. The stomata are only a little retracted, and the exterior walls of the contiguous cells are on the same level as those of the other epidermal cells.

The leaf is fed by two vascular cylinders which, interestingly enough, stem from two different (!) cords of vascular bundles. These two vascular bundles converge with the xylem parts in the stalk; both ducts have bifurcations which develop dichotomously in the transitional zone between stalk and blade and feed one of the two halves of the blade. Each bifurcation normally develops only a few anastomoses (see Fig. 8). In general, these anastomoses are more frequent in leaves of long shoots than in short ones. Occasionally there are cords of vascular bundles which begin spontaneously, i.e. without any connection to the rest of the vascular system. This can be explained by a discontinuous development of protoxylem within the blade. The vascular bundles of the Ginkgo leaf are split by numerous parachymal pith rays; in gymnosperms these are absent only in the Cycadaceae. The cells of the phloem sometimes contain crystallised calcium oxalate. As is the case with all gymnosperms, the wood consists only of tracheids which have both a conductive and stabilising function. The secondary wood of the trunk forms growth rings. There are intercellular spaces between the tracheid. Bordered pits are also found on the tangential cell walls between the autumn and spring wood. The pith rays have a width of one row of cells and a height of up to five rows of cells; they contain starch in every section and remain viable for up to 30 years. Toward the lumen of the tracheids there are big pits which converge conically from a broad base so that a top view shows the picture of bordered pits. The pith rays continue in the bark.

BLOSSOMS

The blossoms are always dioecious, male and female blossoms always occurring separately on two different trees.

The Male Blossom

The male blossoms grow from the axil of scale leaves on short shoots (see Fig. 6); they are catkin-like, short-stalked, scattered, about 2 cm long, with irregularly spiral stamina placed at some distance from each other. The stalk has no leaves, only rarely with 1–2 prophylls. The stamina consist of a thin filament which has a short rotund apical extension or "knob", from which hang two (in some cases up to four) microsporangia (= pollen sac) (see Fig. 13). In connection with surplus microsporangia, this extension was formerly interpreted as a rudimentary microsporangium. Among other things, this interpretation is based on the fact that the extinct genus Baiera, which was closely related to Ginkgo, normally had stamina with several microsporangia. The microsporangia develop on the dorsal side of the staminodium; they hang down without contact next to each other in parallel rows and split open longitudinally on the sides turned towards each other; finally they open wide, almost horizontally, so that the pollen grains can be easily carried away by the wind. Contrary to the Cycadaceae and conifers which form an exothecium, the microsporangium also has an endothecium which is typical of angiosperms. Thus there is a thick-walled endothecium of three layers of cells under the epidermis of microsporangia (see Fig. 9). Towards the interior these are followed by parietal cells containing chlorophyll. The boat-shaped pollen grains measure 30 by 10 μm with a typical median groove along their monopodium (see Fig. 10). The outermost layer of exine, also called ektexine or sexine, is three-partite: a thick tectum, a narrow infratectum with small, irregular spaces, and a distinct footlayer. The inmost

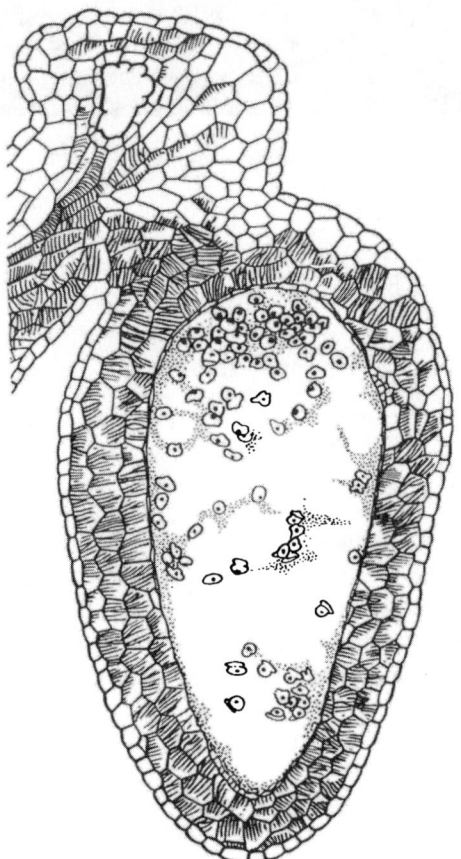

Figure 9 Microsporangium and part of the microsporophyll of *Ginkgo biloba*, longitudinal section with multilayered endothecium. (Jeffrey and Torrey, 1916; quoted from: Singh, 1978)

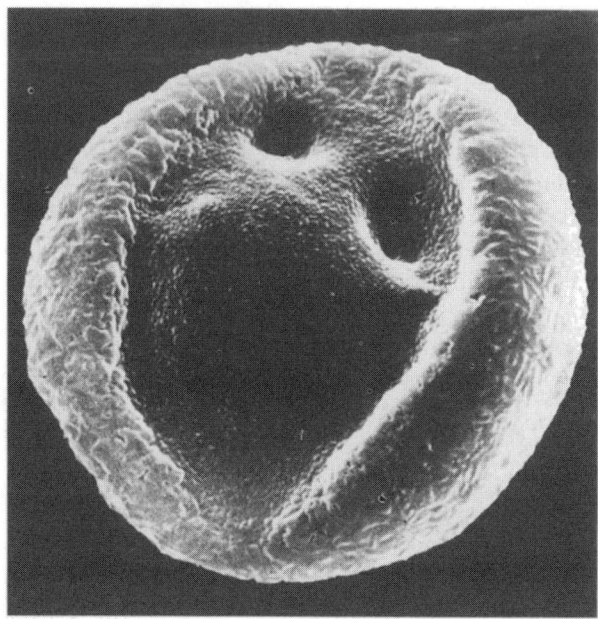

Figure 10   Pollen grain of *Ginkgo biloba* (magnification 4000 times) (Michel, 1986).

layer of exine, also known as endexine or nexine, is relatively thick and the lamellae are closely appressed to each other (Kurman, 1992).

The development of the male gametophyte

The development of the male gametophyte (see Fig. 11) begins with the four-cell-stage of the mature pollen grain, i.e. it is composed of 2 prothallial cells, one antheridial cell, and one vegetative pollen tube cell. When a pollen grain reaches the pollination droplet (in Central Europe approximately in May), a further develop-ment of the gametophyte begins (see below). The pollen grain absorbs water and other substances from the ovular liquid; in this way it becomes heavier and pen-etrates through the micropyle. On reaching the pollen chamber, the vegetative pollen tube cell germinates and expands. Rhizoid-like protuberances develop on the side turned away from the archegonium chamber, thus fixing the gametophyte in the nuclear tissue. During this morphological development of the male gametophyte of *Ginkgo biloba* Friedman (1987) distinguishes three distinct steps:

1) After germination a brief period of diffuse growth of the tube cell resulting in a bulbous protusion of the tube cell through the sulcus, so far only known in *Ginkgo biloba*.
2) In *Ginkgo biloba* the initiation of the tip growth and the formation of a tubula-gametophytic body (= normal in seed plants) is accompanied by an extensive degree of branching, giving rise to an extensive intercellular haustorial system.

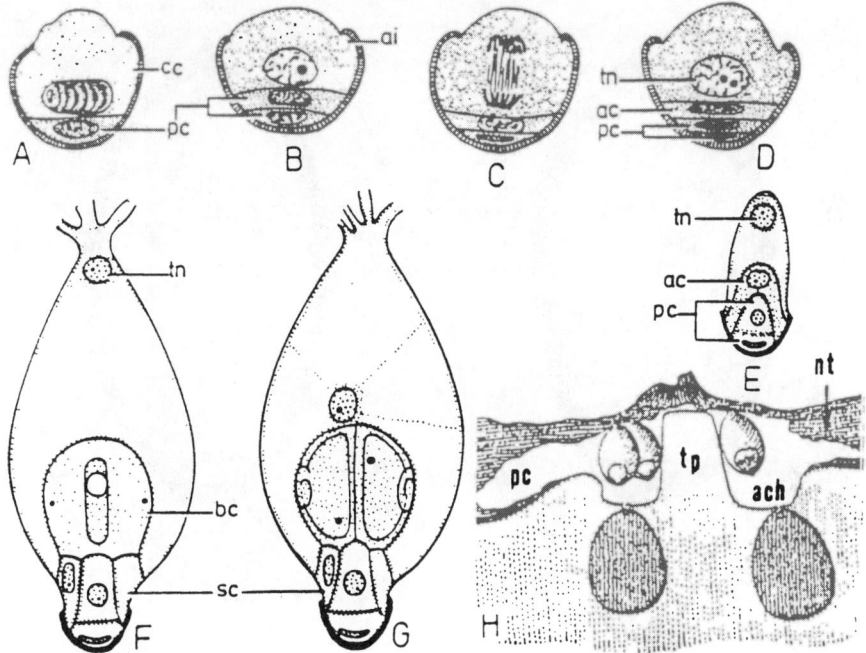

Figure 11   *Ginkgo biloba*, development of the male gametophyte. A-E First divisions up to the four-cell-stage (pc prothallial cells; cc central cell; ai antheridial initial; ac antheridial cell; tn tube nucleus; E-H development of the prothallium in the pollen chamber, in G the two spermatozoids have already developed (sc stalk cell, bc body cell, nt nuclear tissue, tp "tent pole", pc pollen chamber, ach archegoneal chamber). (A-G Chamberlain, 1935, Favre-Ducharte, 1956; quoted from: Singh, 1978; H according to Hirase; quoted from: Pilger, 1926).

3) The late radical swelling of the proximal end of the gametophyte (resulting in the metamorphosis of an originally tubular structure into a spherical or subspherical structure) appears to be unique in Ginkgo and cycads.

The antheridial cell divides into a stalk cell and a spermatogenic cell. Two spermatozoids of 70–90 μm finally develop from the spermagenic cell. Among the seed plants only Cycadaceae and Ginkgo have flagellated gametes. The flagellate apparatus shows the typical 9_2+2 substructure (Li *et al.*, 1989a). The discovery of the spermatozoids of Ginkgo by Hirase in 1895 and 1898 was a milestone in the history of gymnosperm embryology (Hirase, 1895; Hirase, 1898).

The Female Blossom

Although female blossoms grow individually in the axils of the leaves or in those of the topmost scale leaves of the spurs, there are always several blossoms placed together on one spur (see Fig. 6). A relatively thin stalk of several cm and somewhat thickened at its end has here two transversally placed sessile seed-buds. These are

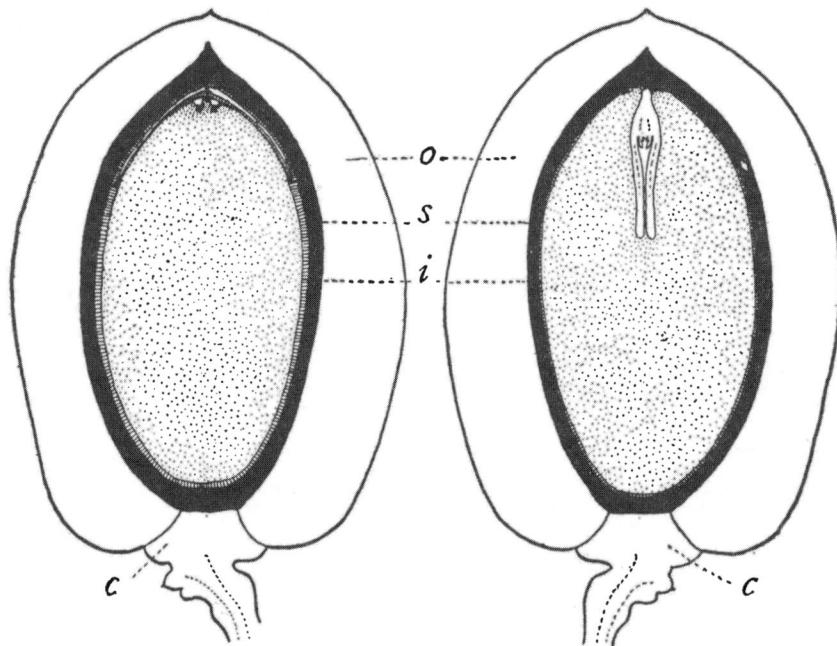

Figure 12   Seed-bud before and after fertilisation, longitudinal, o resin-like fleshy exterior layer (sacrotesta, white), s woody central layer (sclerotesta, black), i interior layer (endotesta, broken lines) still juicy on the left, on the right preserved only as a paper like skin; c collar (double size) (Chamberlain, 1935).

surrounded at their base by a swelling called "collar" (see Fig. 12). As a rule the stalk has two seed-buds; there are occasional exceptions as, for example, with only one or several additional seed-buds, either stemless or individually petiolate. It is interesting to see that the number of vascular bundles in the stalk is always twice the number of seed-buds on the stalk. In the past, this fact led to a great variety of interpretations – also in connection with the "collar" – regarding the question if the stalk and the seed-buds should be assigned to the leaf or the shoot.

The development of the female gametophyte

The seed-bud of gymnosperms consists of an integument and the nucellus included in it (bitegmental seed-buds occur in Ephedra, Welwitschia, and angiosperms; Ephedra has even three integuments). In young seed-buds the nucellus is not yet close-fitting (see Fig. 13); it is only by further cell divisions in the chalaza region that the nucellus is increased so that it closely fits the interior wall of the integument. A narrow interspace is only retained in the micropylar region. The meiosis of the megaspore mother cell results in a tetrad of four haploid daughter cells. The further development of the megaprothallial starts at the lowest cell. The remaining three cells degenerate. Approximately in July, i.e. two months after pollination, this

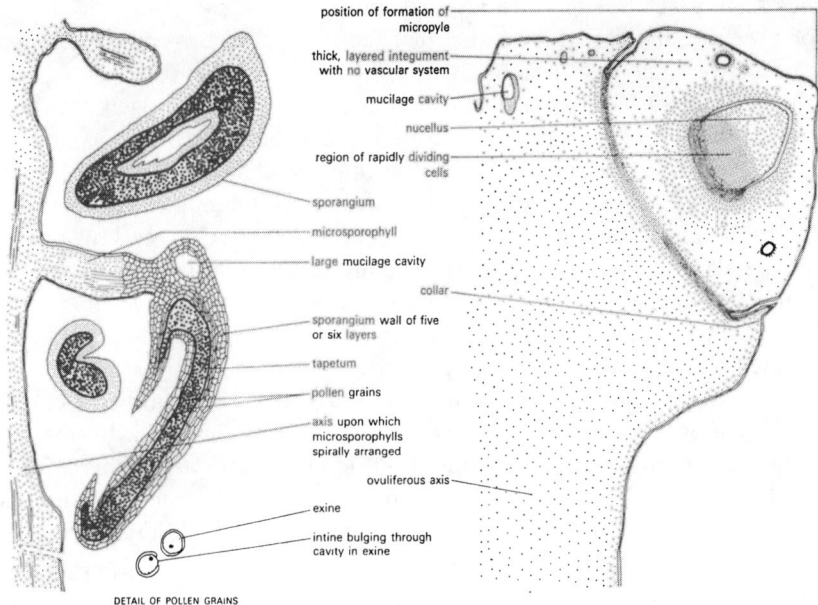

position of formation of micropyle

thick, layered integument with no vascular system

mucilage cavity

nucellus

region of rapidly dividing cells

sporangium

microsporophyll

large mucilage cavity

collar

sporangium wall of five or six layers

tapetum

pollen grains

axis upon which microsporophylls spirally arranged

ovuliferous axis

exine

intine bulging through cavity in exine

DETAIL OF POLLEN GRAINS

Figure 13 Left: longitudinal section through a male inflorescence (magnified 20 times). Right: longitudinal section through a young seed-bud (magnified 17 times) (Bracegirdle and Miles, 1973).

megaspore cell (= embryo sac cell) divides in a process of free nuclear division. The result is a syncytium, a multinuclear object. It is only at a level of 200 free nuclei that the development of the cell wall membranes begins. A generically typical pollen chamber develops in the nucellus at the apical end. The pollen chamber results from the degeneration of nucellus cells in the micropylar region. The interior cells dissolve first, the epidermis cells do so last (i.e. the pollen chamber does not result from the fragmentation of the nucellus tissue during the growth of the apical region!). During the ripening of the seed-buds the nucellar tissue is reduced to a thin membrane, now wrapping the big embryo sac. At the time of ripening (1-) 2 (-3), archegonia develop from a number of marginal or central cells. The archegonia consist of a central ovum, 2–4 neck cells and a ventral canalicular cell. It is worth noting that the cells of the embryo sac contain chlorophyll. So the female gametophyte of *Ginkgo biloba* is the only seed plant gametophyte known to contain chlorophyll (Friedman and Goliber, 1986). As a rule there are two potential seed-buds per pedice, only one of which fully develops in most cases.

EMBRYOLOGY, SEED, GERMINATION

During the ripening of the ovum, the pollen chamber is extended by the development of the archegonial chamber. What happens is that a column-like structure of cells called "tent pole" (see Fig. 11) develops in the embryo sac among the archegonia. It supports the nucellar cap above it, because the channel-like opening existing during pollination between the two fusing halves of the micropyle has disappeared. The two tips of the integument have formed the nucellar cap by growing together. After being set free, the two spermatozoids move through the pollen and archegonial chambers penetrating the ova where nuclear fusion takes place. There is another free nuclear division. Then parietal development begins, and an undifferentiated tissue, the proembryo, fills the ovum. Now there are special cell divisions at the end away from the micropyle (chalza region) and a massive accumulation of tissue penetrates into the endosperm without the development of a suspensory zone otherwise quite typical of gymnosperms. According to Schneckenburger (1989), however, no definite proof has been provided so far that the suspensor is not developed.

From this tissue a root, a small trunk-like structure of cells and two cotyledons, gradually develops. Now and then polyembryony can be observed. In this case two possible forms can be distinguished: polyzygotic polyembryony, if there are several fertilised archegonia. The developing germs occasionally reach maturity simultaneously. The second possibility is split polyembryony which has been repeatedly observed; in this case a second embryo forms on a fully developed embryo by budding. The accessory germs develop above all at the base of the original germ; together with the latter they can reach maturity and develop into intact plants. A second embryo also develops occasionally. With regard to the chronological sequence, it is sometimes claimed that fertilisation does not take place before the seed-buds are shed (see Fig. 14). This is admittedly possible, but it is not the rule; on the other hand, there is reliable proof of the fertilisation of seed-buds still on the tree. The seed-buds are shed in autumn, they have an integument whose exterior layer, the sarcotesta, has become a fleshy resin-like substance which, when ripe, exudes an unpleasant smell of butyric acid. Further inside, there is a ligneous layer called sclerotesta. It surrounds the stone which is about 2 cm long, ellipsoid, slightly compressed, and clearly 2–3-edged [in rare cases there can be one-edged or four-edged seeds (Kartsens, 1945)]. Adherent to the stony layer is an interior thin layer of thin-walled cells, the endotesta. In the seed-buds it is soft and juicy, dries up later on, and can be identified only as a paper-like skin in the ripe seed. The abundant nutritional tissue contains starch. The relatively big embryo has two cotyledons whose upper sides are placed close to each other. In this context *Ginkgo biloba* shows another feature. Ginkgo – and the cycads – do not show fixed dormancy, a characteristic of the seeds of higher gymnosperms (Dogra, 1992). During germination the cotyledons remain largely enclosed by the seed. The first two leaves following on the shoots of young plants are small and scaly; contrary to the later leaves, the first leaves have several deep incisions.

KARYOLOGY

In the case of Ginkgo the number of chromosomes is 2n = 24. This number is the same as that of other gymnosperms. Surprisingly, minimal differences have been

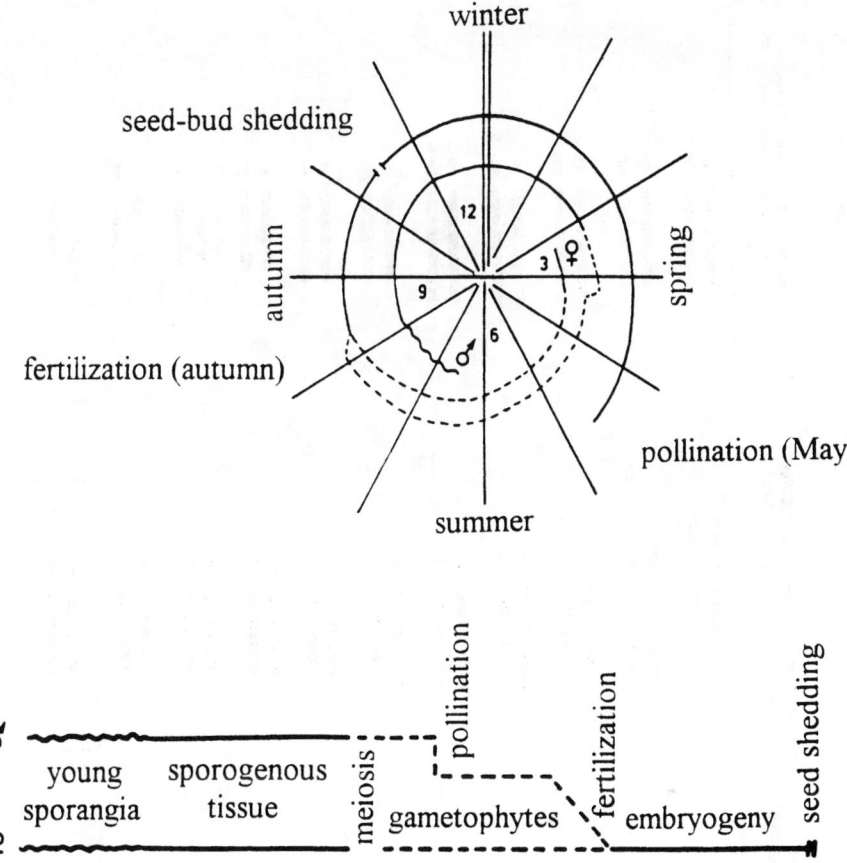

Figure 14   Diagram to illustrate the reproductive cycle: haploid phases in broken lines, diploid phases in solid lines. The total length for a 2-year-reproductive cycle is not exceptional in gymnosperms. The only unusual fact – this is not only true of Ginkgo – is that the embryonal development is completed only several months after the shedding of the seed-buds. (Chamberlain 1935; Favre-Ducharte 1962, quoted from: Singh, 1978, with additions and minor modifications).

found in the karyotype of one and the same plant. This also applies to the precise localisation of the centromeres and the satellites. Of course, Newcomer (1954) and Lee (1954), like other cytologists before them, tried to answer the question if perhaps a genotypical sex identification is available for the dioecious Ginkgo, i.e. if the karyogram reveals sex chromosomes of the type XY as e.g. in Melandrium (= Silene). According to Newcomer (1954), the two longest chromosomes, and according to Lee (1954), the two short satellite chromosomes (see Fig. 15), correspond to the assumed sex chromosomes of the XX and XY type. This did not yet provide a definite answer to the question of sex chromosomes. On the other hand, these different findings do not come as a surprise because the chromosomes of

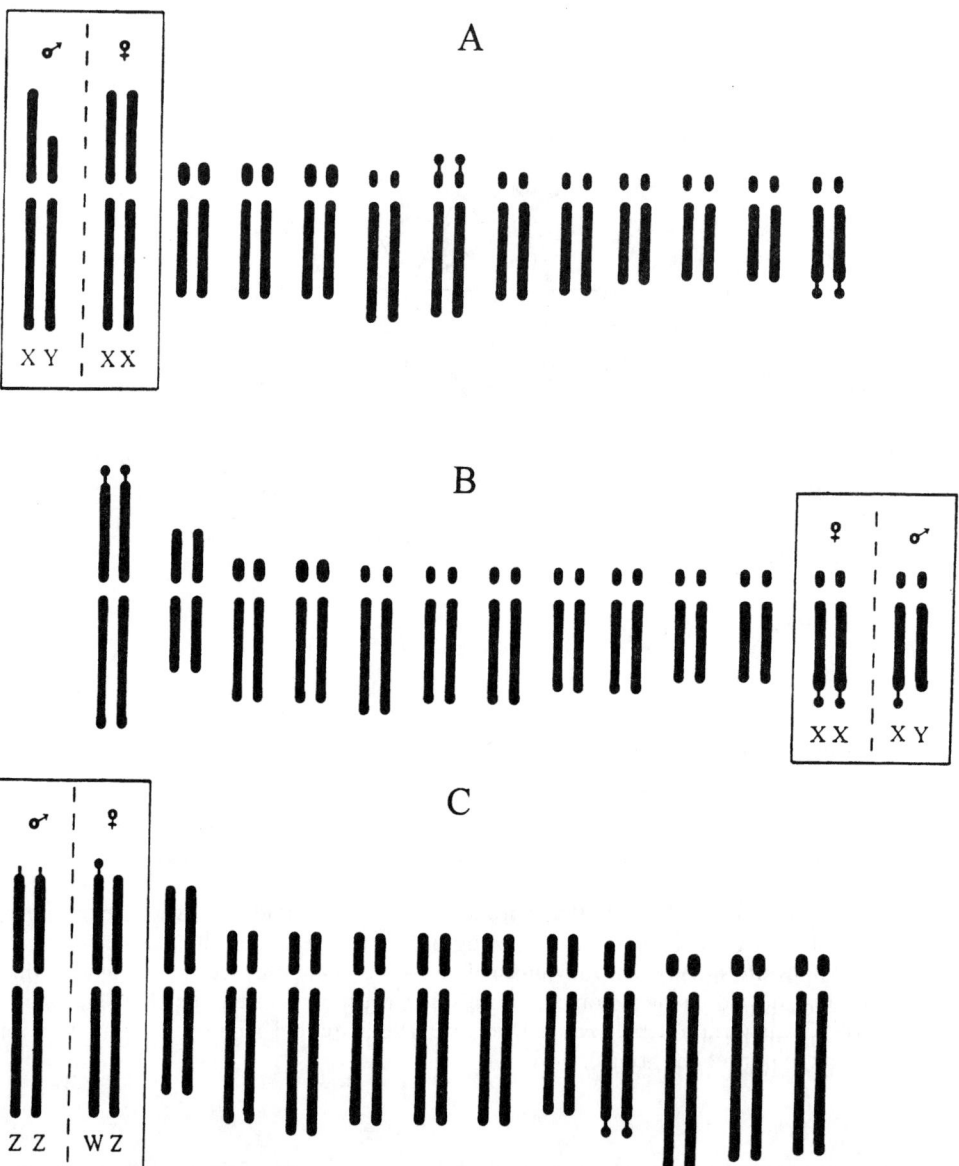

Figure 15 Different female karyograms of *Ginkgo biloba* with 2n = 24 chromosomes and their possible sex chromosomes. The possible sex chromosomes from the male karyogram are placed in a frame next to the possible female sex chromosomes.

A Karyogram according to Newcomer (1954). For better visual quality Newcomer's results were transferred into Lee's (1954) karyogram.

B Karyogram according to Lee (1954).

C Karyogram according to Chen *et al.* (1993).

Ginkgo are not easily made visible (Chen *et al.*, 1993; Kim, 1995). Quite recently, Chen and his collaborators have examined the chromosomes of Ginkgo with special methods (Giemsa C-banding and Ag-staining). They were able to confirm the geno-typical sex identification, but not the supposed XX/XY type. On the basis of their results, Chen *et al.* conclude that the sex determination mechanism of Ginkgo is explained by the WZ/ZZ type, i.e. the male set of chromosomes is 2n = 24 = 22A + ZZ-NOR, and the female set is 2n = 24 = 22A + WZ-NOR (NOR = Nucleolus Organisator Region). The short arm of one of the large pairs of chromosomes possesses only one obvious large satellite and NOR (named W-chromosome), while the other possesses a very small satellite and NOR (named Z-chromosome). This Z-chromosome is completely identical with the two large chromosomes (ZZ chromosomes) in the cell of the male plant. The update (Hizume, 1997) about the problem of sex-chromosomes shows that the satellites varied in occurrence and size in different trees (!). Using methods of molecular genetics Hizume pointed out that the hypothetical occurrence of sex chromosomes cannot be confirmed. It is clear that more investigations are necessary to confirm Hizume's opinion. Otherwise another explanation of sex determination remains to be found.

Several references (Ikeno, 1901; Miyoshi, 1931; Dietrich pers. communication, 1997), describing phenomena which cannot be definitely explained so far, can now be discussed in this context. The spontaneous growth of female branches on male trees without any outside influence has been observed (!). One such case, which has not yet been reported, was found in the Botanical Garden of Jena (Germany). There the development of a female branch on a 170-year-old male tree is under observation. The number of seed buds on this branch increases year by year (1993 = 34, 1994 = 132, 1995 = 144, 1996 = 538). Kato (pers. communication 1998) named three trees with female branches on male trees found in the province Iwate-ken, Japan.

This phenomenon might be explained by point-mutation or atavism (the Ginkgo and Ginkgo-like precursors were monoecious) (personal communication received in 1997 from Dietrich, curator of the Jena Botanical Garden, Germany). This unusual feature – seeds under a male tree have so far only been reported as occurring in male trees – is probably due to the fact that it is extremely conspicuous in contrast to what can be found in female trees.

PRACTICAL ASPECTS OF SYSTEMATICAL AND APPLIED BOTANY

Michel (1986) pointed out that the crushed seed-buds were used as washing powder. After removing or washing off the fleshy outer layer (which smells unpleasantly rancid because of its content of butyric acid) the seed-buds can easily be removed from the stone. Eaten raw, their taste resembles that of raw potatoes with a resin-like aftertaste. Roasted Ginkgo seed-buds are regarded as a delicacy all over the Far East and are used like pistachio nuts. However, as they have stronger traditional associations, it is customary to offer them at wedding ceremonies – in that case dyed red. The wood is bright-coloured, rather hard and brittle. It is only of negligible importance, although in China miscellaneous objects are made of it. Experiments have shown that with regard to the pollution of settled areas (cities), the Ginkgo can be reckoned among the relatively fume resistant plants (Kiermeier, 1984). That is why it is planted as a roadside tree more and more frequently in the USA, Europe and Japan. In the meantime, research carried out with a scanning electron microscope (Kim and Lee, 1990) has shown that the waxy cuticle of Ginkgo leaves is far less sensitive to pollution, e.g. to sulphur dioxide, than the needles of other gymnosperms (Kim and Lee, 1992; Kim and Lee, 1990). According to them, however, it is wrong to maintain that the Ginkgo is one of the most-planted roadside trees in the USA. The truth is that after several types of maple the Ginkgo holds a share of only 0.2–2% among the total number of roadside trees. Compared with European conditions this share is still quite considerable. Ginkgo is also often cultivated for its bright golden autumnal foliage. This is caused by changing optical properties associated with senescence which are due to the breakdown of chlorophyll in conjunction with a remarkably high retention of carotenoids (Matile et al., 1992). Depending on different possible uses, the cultivation and selection of certain varieties (especially with regard to growth) was begun as early as the middle of the last century. According to Kiermeier (1984), there are at present 18–20 European-American varieties and 25 Chinese fruit varieties which are of no importance to Europeans because there is no demand for roasted Ginkgo seeds. As the available seed-buds offer no clue whether or not the pollination, fertilisation, and embryonic development have taken place, a considerable loss has to be reckoned with. Consequently seed-buds of female branches grafted on male trees have a greater rate of success than when the two sexes stand at a great distance from one another. At the beginning the growth rate is quite high. Under favourable conditions young plants quickly develop a vertical axis with less robust horizontal branches. After 5 or 6 years the plants can reach a height of 2–3 m. Then the growth slows down, and it takes several decades before the tree achieves a certain size and beauty. The tree likes light and sunshine, is considered to be undemanding and takes a lot of heat and drought. Flesch et al. (1991) reported that higher light intensity and higher temperatures activate the growth of long shoots and the sprouting of lateral buds. The Ginkgo thrives in a great variety of soils such as sandy humus soils or heavy loamy loess. According to Kiermeier (1984), Ginkgo tolerates a soil of pH = 5.5–7.7. Young trees are more sensitive than older ones to late frost. In the literature on the subject it is underlined again and again that the species is largely resistant against diverse fungal diseases and that pest-infestation is hardly of any importance. Only in connection with propagation by cuttings can the roots be damaged by larvas (e.g. Bradysia paupera), but the damage can be reduced to a minimum by suitable countermeasures.

SUMMARY

Ginkgo biloba, a "living fossil" is the only recent representative of its genus, order, and class. The family developed its greatest morphological variety between the

Triassic and Cretaceous. Contrary to a widespread opinion, Ginkgo itself was probably only represented with 2–3 species which consequently can be regarded as the
direct precursors of the recent *Ginkgo biloba*. Definitely the dioecious Ginkgo can
morphologically and anatomously be included in the Coniferophytina. The spermatozoids of Ginkgo (within the Spermatophyta only to be found in Cycadatae) give it
an exceptional position. The reproductive cycle is characterised by chronological
delays: pollination in May, beginning of the ripening process of seed-buds and the
ova in July, fertilisation and beginning of the embryonic development in
September/October, October/November shedding of the ripe seed-buds with or
without embryo, continuing embryonic development on the ground, in late spring
germination of the seedlings. The problem of sex identification of non-blossoming
young plants has not been solved at present because reliable phenotypical sex characteristics are not known. Nor has genotypical sex identification according to the
XY mechanism been proved beyond all doubt and is thus not yet applicable to sex
identification. Consequently the safest method of obtaining young plants of the
desired sex is the unproblematical propagation by cuttings.

ACKNOWLEDGEMENTS

We wish to thank H. Reinstadler for his support in the formulation of this manuscript. The assistance in preparing the figures from M. Thumberger is gratefully
acknowledged. We appreciate the permission to use photos from the University of
Chicago Press, Intersan GmbH, Gebrüder Borntraeger Verlag, A. Kato (Fürth,
Germany), and Wagner Free Institute of Science (Philadelphia).

REFERENCES

Bracegirdle, B. and Miles, P.H. (1973) *An Atlas of Plant Structure*, Vol. 2., Heinemann
     Educational Books, London.
Chamberlain, J.C. (1935) *Gymnosperms, Structure and Evolution*, University of Chicago
     Press, Chicago – Illinois.
Chen, R.-Y., Song, W.-Q., Li, X.-L., An, Z.-P. (1993) A study on the sex chromosomes of
     Ginkgo biloba. *Cathaya*, 5, 41–48.
Critchfield, W.B. (1970) Shoot growth and Heterophylly in *Ginkgo biloba*. *Bot. Gaz.*, **131**,
     150–162.
Del Tredici, P. (1992) Natural regeneration of *Ginkgo biloba* from downward growing
     cotyledonary buds (basal chichi). *Am. J. Bot.*, **79**, 522–530.
Del Tredici, P., Ling, H., Yang, G. (1992) The Ginkgos of Tian Mu Shan. *Conserv. Biol.*, **6**,
     202–209.
Dogra, P.D. (1992) Embryogeny of primitive gymnosperms Ginkgo and Cycads – pro-embryo
     – basal plan and evolutionary trends. *Phytomorphology*, **42**, 157–184.
Dorf, E. (1958) The geological distribution of the Ginkgo family. *Bull. Wagner Free Inst. Sci.*,
     33, 1–10.
Ehrendorfer, F. (1998) Gymnospermae. In Strasburger, E., Ehrendorfer, F., *Lehrbuch der
     Botanik*, 34. edition, Fischer Verlag, Stuttgart, pp. 712–731.

Flesch, V., Jaques, M., Cosson, L., Petiard, V., Balz, J.P. (1991) Effects of light and temperature on growth of *Ginkgo biloba* cultivated under controlled long day conditions. *Ann. Sci. Forr. (Paris)*, **48**, 133–148. Cited from: *Biol. Abstr.*, **92**, AB–407, Ref. No. 51988.

Friedman, W.E. (1987) Growth and development of the male gametophyte of *Ginkgo biloba* within the ovule (in vivo). *Amer. J. Bot.*, **74**, 1797–1815.

Friedman, W.E., Goliber T.E. (1986) Photosynthesis in the female gametophyte of *Ginkgo biloba. Amer. J. Bot.*, **73**, 1261–1266.

Guelz, P.G., Mueller, E., Schmitz, K., Gueth, S. (1992) Chemical composition and surface structures of epicuticular leaf waxes of *Ginkgo biloba*, *Magnolia grandiflora* and *Liriodendron tulipifera. Z. Naturforsch.*, **47 C**, 516–526.

Gunckel, J.E., Wetmore, R.H. (1946) Studies of development in long shoots and short shoots of *Ginkgo biloba* L. II. Phyllotaxis and the organization of the primary vascular system, primary phloem and primary xylem. *Am. J. Bot.*, **33**, 532–543.

Hirase, S. (1895) Etude sur la fécondation et l'embryogenie du *Ginkgo biloba. Jour. Coll. Sci. Univ. Tokyo*, **8**, 307–822.

Hirase, S. (1898) Etude sur la fécondation et l'embryogenie du *Ginkgo biloba*, second memoire. *Jour. Coll. Sci. Univ. Tokyo*, **12**, 103–149.

Hizume, M. (1997) Chromosomes of *Ginkgo biloba*. In T. Hori, R.W. Ridge, W. Tulecke, P. Del Tredici, J. Trémouillaux-Guiller, H. Tobe, (eds.), *Ginkgo biloba – A global Treasure. From Biology to Medicine*, Springer, Tokyo, pp. 109–118.

Ikeno, S. (1901) Contribution à l'étude de la fécondation chez le *Ginkgo biloba. Ann. Sci. Nat. Bot.*, **8**, 305–318.

Kanis, A., Karstens, W.K.H. (1963) On the occurrence of amphistomatic leaves in *Ginkgo biloba* L. *Acta Bot. Neerl.*, **12**, 281–286.

Karstens, W.K.H. (1945) Variability of the female reproductive organs in *Ginkgo biloba* L. *Blumea*, **5**, 532–553.

Kiermeier, P. (1984) Zur Problematik stadtfester Gehölze. *Gartenamt*, **33**, 239–244.

Kim, S.I. (1995) Studies on the chromosome types of Ginkgo species. *J. Korean For. Soc.*, **84**, 131–144.

Kim, K., Whang, S.S., Sun B.Y. (1992) Bark morphology of some Korean gymnosperms. *Korean J. Bot.*, **35**, 339–358. Cited from: *Biol. Abstr.*, **96**, AB–694, Ref. No. 6327.

Kim, M.H., Lee S.W. (1992) Resistance function of woody landscape plants to air pollutants. *J. Korean For. Soc.*, **81**, 234–246. Cited from: *Biol. Abstr.*, **95**, AB–1061, Ref. No. 113187.

Kim, Y.S., Lee, J.K. (1990) Chemical and structural characteristics of conifer needles exposed to ambient air pollution. *Eur. J. Pathol.*, **20**, 193–200.

Kurmann H. (1992) Exine stratification in extant gymnosperms: a review of published transmission electron micrographs. *Kew Bulletin*, **47**, 25–39.

Lee, C.L. (1954) Sex Chromosomes in *Ginkgo biloba. Am. J. Bot.*, **41**, 545–549.

Li, H.L. (1982) A Horticultural and Botanical History of Ginkgo. In H.L. Li, (ed.), *Contributions to Botany*, Epoch Publishing, Taiwan, pp. 448–457.

Li, Y., Wang, F.H., Knox, R.B. (1989a) Ultrastructural analysis of the flagellar apparatus in sperm cells of Ginkgo biloba. *Protoplasma*, **149**, 57–63.

Li, Z.L., Lin, J. (1991) Wood anatomy of the stalactite-like branches of Ginkgo. *Int. Assoc. Wood Anat. Bull.*, **12**, 251–255. Cited from: *Biol. Abstr.*, **93**, AB–417, Ref. No. 27813.

Li, Z.L., He, X, Xu, B.W. (1989b) The epidermal structure of Ginkgo leaf. *Acta Bot. Sin.*, **31** 427–431. Cited from: *Biol. Abstr.*, **89**, AB–573, Ref. No. 49934.

Linnaeus, C. (1771) *Mantissa Plantarum altera*, vol. 2, p. 313.

Matile, P., Flach, B.M.-P., Eller, B.M. (1992) Autumn leaves of *Ginkgo biloba* L.: Optical properties, pigments and optical brighteners. *Bot. Acta*, **105**, 13–17.

Mayr, H. (1906) *Fremdländische Wald- und Parkbäume*, Parey, Berlin, p. 288.

Michel, P.F. (1986) *Ginkgo biloba, ein Baum besiegt die Zeit*, Intersan, Ettlingen.

Miyoshi, M. (1931) Merkwürdige *Ginkgo biloba* in Japan. *Mitt. Deut. Dendr. Ges.*, **Jahrbuch 1931**, 21–22.

Napp-Zinn, K. (1966) *Anatomie des Blattes: I. Blattanatomie der Gymnospermen*, Gebr. Borntraeger, Berlin.

Nelson, J. (1866) *Pinaceae*, Hatchard and Co., London, p. 163.

Newcomer, H. (1954) The karyotype and possible sex chromosomes of *Ginkgo biloba*. *Am. J. Bot.*, **41**,542–545.

Pilger, R. (1926) Ginkgoaceae. In A. Engler and K. Prantl, (eds.), *Die natürlichen Pflanzenfamilien*, Vol. 13., W. Engelmann Verlag, Leipzig, pp. 98–109.

Schneckenburger, S. (1989) Studien zur Embryogenese und Keimung verschiedener Gymnospermen unter besonderer Berücksichtigung der Suspensorbildung und Keimwurzelgenese. *Palmarum Hortus Francofortensis*, **1**, 56–68.

Singh, H. (1978) *Embryology of Gymnosperms*, Gebr. Borntraeger, Berlin.

Smith, J.E. (1797) *Trans. Linn. Soc.*, **3**, 330.

Sporne, K.R. (1965) *The Morphology of Gymnosperms*, Hutchinson and Co. Publ., London, pp. 164–171.

Strasburger, E., Noll, F., Schenck, H., Schimper, A.F.W. (founded) Sitte, P., Ziegler, H., Ehrendorfer, F., Bresinsky, A. (1998) *Lehrbuch der Botanik*, 34. edition, Fischer Verlag, Stuttgart, p. 697, 658–716.

Tralau, H. (1967) The phytogeographic evolution of the genus Ginkgo L. *Bot. Not.*, **120**, 409–422.

Vozenin-Serra, C., Broutin, J., Toutin-Morin, N. (1991) Permian woods of southwest Spain and southeast France: Implications for Paleozoic gymnospermous taxonomy and Ginkgophyta phylogeny, *Palaeontogr. Abt. B Palaeophytol.*, **221**, 1–26. Cited from: *Biol. Abstr.*, **91**, AB–809, Ref. No. 133453.

Zhou, Z. (1991) Phylogeny and evolutionary trends of Mesozoic Ginkgoaleans: A preliminary assessment. *Rev. Palaeobot. Palynol.*, **68**, 203–216. Cited from: *Biol. Abstr.*, **92**, AB-846, Ref. No. 90978.

3. LIGNIFICATION OF XYLEM CELL WALLS OF *GINKGO BILOBA*

NORITSUGU TERASHIMA[1] and KAZUHIKO FUKUSHIMA[2]

[1]2–610 Uedayama, Tenpaku, Nagoya 468-0001 Japan
[2]School of Agriculture, Nagoya University, Nagoya, 464-8601 Japan

INTRODUCTION

More knowledge about the chemical structure of lignin in plant cell walls and its formation mechanism can lead to a better understanding and developments in a wide area of science and technology related to woody plants as follows:

(1) From the viewpoint of plant science, lignin is one of the major cell wall polymers, which makes difference between trees (woody plants) and non-woody plants. The word "lignin" is derived from the Latin "Lignum" meaning wood. In the course of the evolution of plants, some species acquired a metabolic pathway to produce lignin, and their cell walls were endowed by lignification with characteristic physical, chemical and biological properties as follows: (a) lignification makes cell walls hydrophobic so that aqueous nutrients can be conducted through tissues, (b) lignification makes cell walls mechanically strong so that plants can grow higher and can extend branches to receive more sunlight, (c) lignification makes cell walls resistant to attacks by microorganisms and animals, and (d) lignification protects living cells from physico-chemical effects by sunlight. These unique properties of lignified cell walls enabled trees to grow and survive for hundreds of years, and the largest portion of organic substances (fixed carbon dioxide in stable form) on the earth including humic substances in the soil originate from lignified cell walls of woody plants.

(2) For the development in technology related to forestry and forest industry such as the wood, pulp and paper industry, the formation mechanism and structures of lignin and polysaccharides in the cell wall are the key fundamental knowledge, because physical, chemical and biological properties of wood and wood derived products depend primarily on the chemical structure of these polymers and their three dimensional (3D) assembly in the cell wall.

Ginkgo biloba is one of the oldest still living trees on earth. Because it retains primitive characteristic features of trees appearing in the early stage of evolution of trees, it is a suitable tree species for investigation of the lignification mechanism of plant cell walls and the chemical structure of lignin. Except for the irregularity in size and shape of tracheids, the anatomical features of vascular tissue and xylem tissue of this tree are similar to those of conifers which is one of the major classes of trees on earth (Timell, 1986). Except for some of the extraneous substances, the

49

major chemical components of Ginkgo wood are similar to those of coniferous wood (Timell, 1986) and lignification of Ginkgo cell walls and chemical structure of Ginkgo lignin are also similar to those of conifers.

In this chapter, the general feature of lignification of tree cell walls and its investigation methods will be introduced briefly, then lignification of Ginkgo xylem cell walls will be described in detail.

General Feature of Lignification in Tree Xylem Cell Walls

Pectin, hemicellulose, cellulose and lignin are the major components of cell walls of tree xylem. Lignified xylem cell wall is formed by successive deposition of these cell wall polymers to generate a composite in which component polymers are physically and chemically bound to each other in a biochemically regulated manner.

Evolution of plants involved an evolution of biosynthetic pathways of cell wall components. This evolution of biosynthetic pathways is especially extensive in secondary metabolites including lignin. Two aromatic amino acids in the primary metabolic pathway, phenylalanine (1) and tyrosine (2) are the precursors of various aromatic secondary metabolites (Figure 1). Lignin monomers (monolignols, 5,8 and 11) are derived from phenylalanine (1) and tyrosine (2), and polymerize to form a macromolecular lignin in the cell wall (Neish, 1968; Sarkanen, 1971; Higuchi, 1985; Haslam, 1993). Most gymnosperms, which appeared in an earlier stage of evolution than angiosperms, form p-hydroxyphenyl-guaiacyl lignin, a polymer of p-coumaryl alcohol (5) (minor monomer) and coniferyl alcohol (8) (major monomer). On the other hand, more evolved trees belonging to angiosperms form guaiacyl-syringyl lignin from coniferyl alcohol (8) and sinapyl alcohols (11).

Even in the course of differentiation of a cell wall, the type of monomers supplied to the lignifying cell walls changes with the stage of differentiation. At an early stage, p-hydroxyphenyl-guaiacyl lignin is formed in the compound middle lamella (combined layers of middle lamella between cells and primary wall), while at a later stage guaiacyl-syringyl lignin is formed in the secondary walls (Terashima and Fukushima, 1989).

The polymerization of these monomers is effected partly by laccase-type monolignol oxidase at an early stage and mainly by peroxidase/hydrogen peroxide in later stages of cell wall differentiation (Freudenberg, 1968; Sarkanen, 1971; Higuchi, 1985; Savidge and Udagama-Randeniya, 1992; Sterjiades et al., 1992; Bao et al., 1993; O'Malley et al., 1993; McDougall and Morrison, 1996). These monolignols are dehydrogenated to their phenoxy radicals, and coupling of their mesomeric radicals give polylignols in which monomeric units are bound by various types of linkages as shown in Fig. 2. (Freudenberg, 1968; Sarkanen, 1971). The distribution of these linkages is assumed not to be uniform in the lignin macromolecule.

METHODS FOR INVESTIGATING LIGNIFICATION

Non-destructive Approaches

Lignin exists as a 3D macromolecule in the cell wall, and its chemical structure is not homogeneous. The major part of lignin is intimately associated with poly-

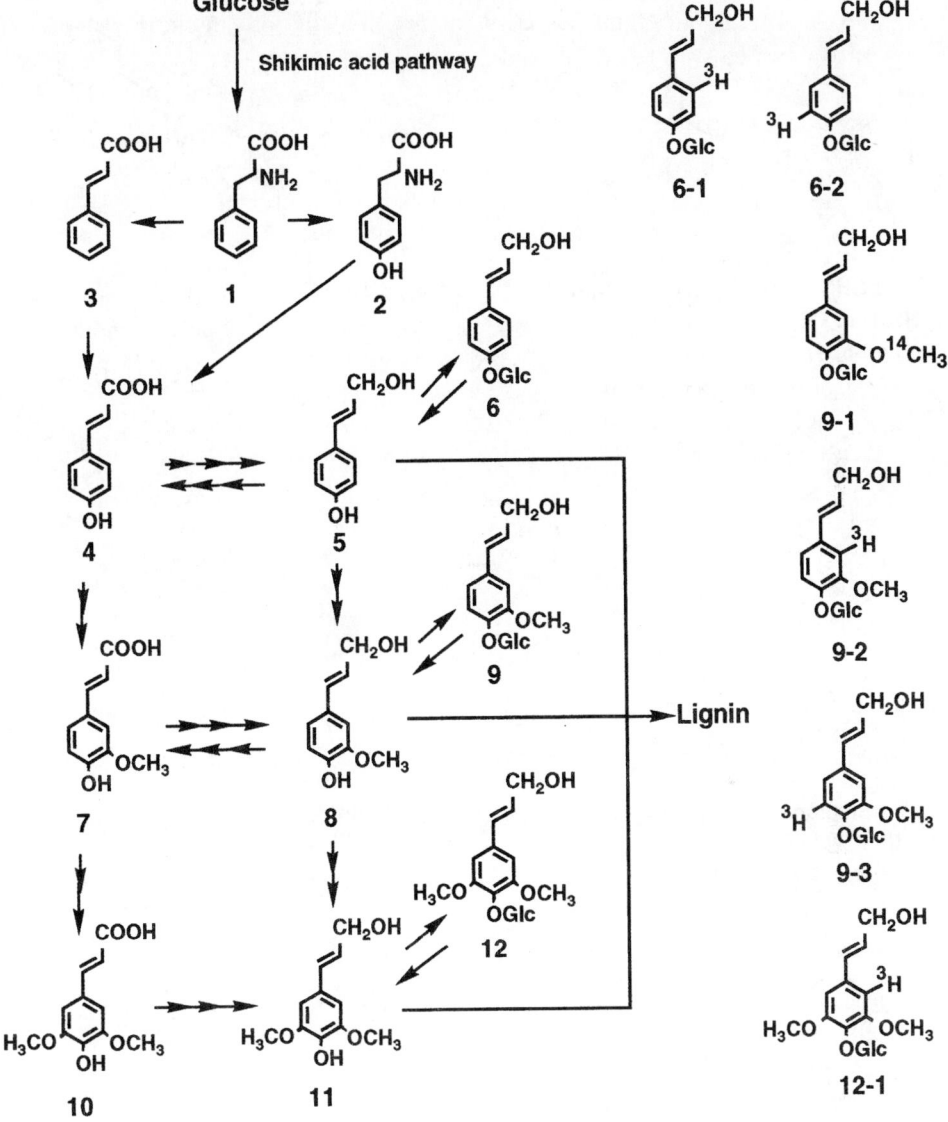

Figure 1 Biosynthetic pathway of lignin, and radio-labeled precursors used for specific labeling of monomeric units in Ginkgo lignin. 5: *p*-coumaryl alcohol, 8: coniferyl alcohol, 11: sinapyl alcohol.

saccharides by chemical and physical bonds. Therefore, it is impossible to isolate lignin from the cell wall retaining its 3D heterogeneous structure, and lignin as it is in the cell wall is often called protolignin to discriminate it from isolated lignin preparations. Because destructive analysis of protolignin causes losses in information on its 3D heterogeneous structure, investigation of the structure must be carried out by non-destructive methods. One of the effective approaches is to observe the

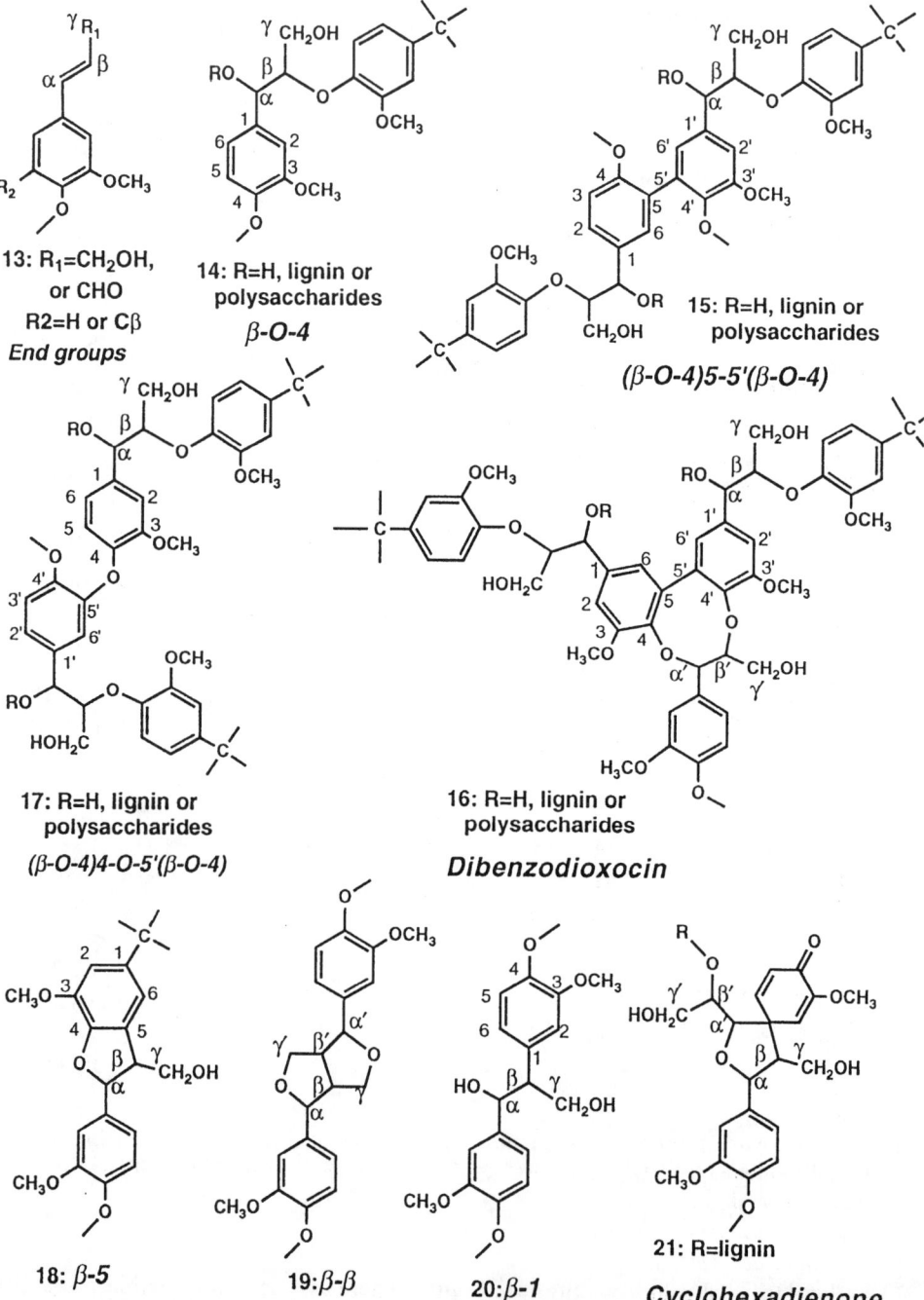

Figure 2   Lignin substructures showing various type of linkages between monomeric units.

lignification process of differentiating cell walls from the early stage to the final stage. Available non-destructive methods are in two categories: (1) microscopy-spectroscopic methods and (2) tracer methods. The former category includes various techniques such as electron microscopy, UV and visible light microspectroscopy, infrared and Raman microspectroscopy, and solid state NMR. The latter includes radio- and stable isotope tracer methods. Combinations of techniques in two categories provide useful information on the formation and structure of protolignin in the cell wall.

Tracer Methods Employing Radio- and Stable Isotopes

Tracer methods employing radioisotopes, ^{3}H or ^{14}C, and the stable isotope ^{13}C can provide useful information which cannot be obtained by any other methods. A technique of selective radio-labeling of protolignin or polysaccharides combined with a detection technique of the radioactivity by photomicrography is called micro-autoradiography. This enables one to visualize the deposition process of the labeled component during the formation of cell walls (Saleh *et al.*, 1967; Fujita and Harada, 1979; Takabe *et al.*, 1981; Terashima *et al.*, 1988; Terashima and Fukushima, 1989). Solid state ^{13}C-NMR analysis combined with a technique of selective ^{13}C-enrichment of a specific carbon in protolignin is also a useful non-destructive approach for elucidation of the chemical structure of protolignin in the cell wall (Lewis *et al.*, 1989; Eberhardt *et al.*, 1993; Terashima *et al.*, 1997a; 1997b).

Selective labeling of lignin

Among various precursors of lignin biosynthesis, the monolignol glucosides, *p*-glucocoumaryl alcohol (**6**), coniferin (**9**) and syringin (**12**) are found to be suitable precursors for selective labeling of protolignin in the cell wall. A specific hydrogen or carbon in the precursor is replaced by an isotope, and the labeled precursor is administered to a growing stem of a tree. The precursor is incorporated into the newly formed lignin in the differentiating cell wall, and the polymeric lignin is specifically labeled at a certain hydrogen or carbon corresponding to the administered monolignol.

Selective labeling of polysaccharides

Pectin and hemicelluloses can also be labeled selectively by administration of suitable precursors, and their incorporation into differentiating cell wall can be visualized by microautoradiography (Terashima *et al.*, 1988; Imai and Terashima, 1990; Imai *et al.*, 1996a; 1996b).

The deposition of cellulose microfibrils can be visualized by observation of a cross section of differentiating xylem with a polarized light microscope (Takabe *et al.*, 1981). Thus the order of deposition of major cell wall polymers, pectin, hemicellulose, cellulose microfibrils and lignin during the cell wall formation can be visualized by microautoradiography and polarized light microscopy.

LIGNIFICATION OF XYLEM CELL WALL OF GINKGO

Monolignol glucosides specifically labeled with ^{3}H or ^{14}C (Fig. 1, 6-1, 6-2, 9-1, 9-2, 9-3, 12-1) were administered to a growing stem of a Ginkgo tree to selectively label the corresponding position in the lignin. The process of incorporation of different monomer units and formation of different types of linkages were estimated by determination of the radioactivity in the differentiating cell walls using the technique of microautoradiography (Fukushima and Terashima, 1991).

Fig. 3 shows the microautoradiogram of differentiating xylem of Ginkgo after administration of coniferin-[OCH$_3$-^{14}C] (9-1). The silver grains indicate deposition of polylignol formed from coniferyl alcohol-[OCH$_3$-^{14}C] derived from the administered precursor. Because the major part of Ginkgo lignin is composed of a polymer of coniferyl alcohol (guaiacyl lignin) (Timell, 1986; Fukushima and Terashima, 1991), this microautoradiogram shows the process of formation of Ginkgo lignin during the cell wall differentiation.

The deposition of lignin occurs in three distinct stages (Fig. 3). The first deposition of lignin occurs at the cell corner and compound middle lamella regions just after the formation of the primary wall has finished and formation of the outer layer of secondary wall (S$_1$) has started (Fig. 3). It has been shown by microautoradiography that pectic substances are deposited as the major polysaccharide in the newly formed cell walls at the cambium region of pine (Terashima et al., 1988) and Ginkgo (Imai et al., 1996b) prior to the deposition of lignin. Therefore, the polymerization of monolignol to polylignol takes place in a gel of pectic substances. The lignification period in this region is rather short.

In the next stage, deposition of lignin is very slow. During this time, the major part of cellulose microfibrils and hemicellulose deposits in the middle layer of the secondary wall (S$_2$), and the cell wall becomes thick. It has been shown by microautoradiography that galactoglucomannan deposits at the same time as the cellulose microfibrils (Imai et al., 1996a), and hemicelluloses may associate with cellulose microfibrils to keep them separately in a hemicellulose gel (Terashima et al., 1993).

The third stage of lignin deposition occurs after the start of the formation of the inner layer of secondary wall, S$_3$ (Fig. 3). The major part of lignin deposits at this stage in the secondary wall. Polymerization of monolignols at this stage takes place in the mannan gel.

Fig. 4 shows the microautoradiograms of differentiating xylem of Ginkgo after administration of p-glucocoumaryl alcohol-[arom. ring-2-^{3}H] (6-1), coniferin-[arom. ring-2-^{3}H] (9-2), or coniferin-[arom. ring-5-^{3}H] (9-3). Because the tissue section labeled with the low-energy beta emitter ^{3}H can provide a high resolution microautoradiogram, it is possible to estimate semi-quantitatively the incorporation of labeled units in different morphological regions by counting the number of silver grains (Takabe et al., 1981). The distribution is shown in Fig. 5 (Fukushima and Terashima, 1991).

Incorporation of p-coumaryl alcohol into polylignol (p-hydroxyphenyl lignin moiety) occurs at an early stage of cell wall differentiation mainly in the middle

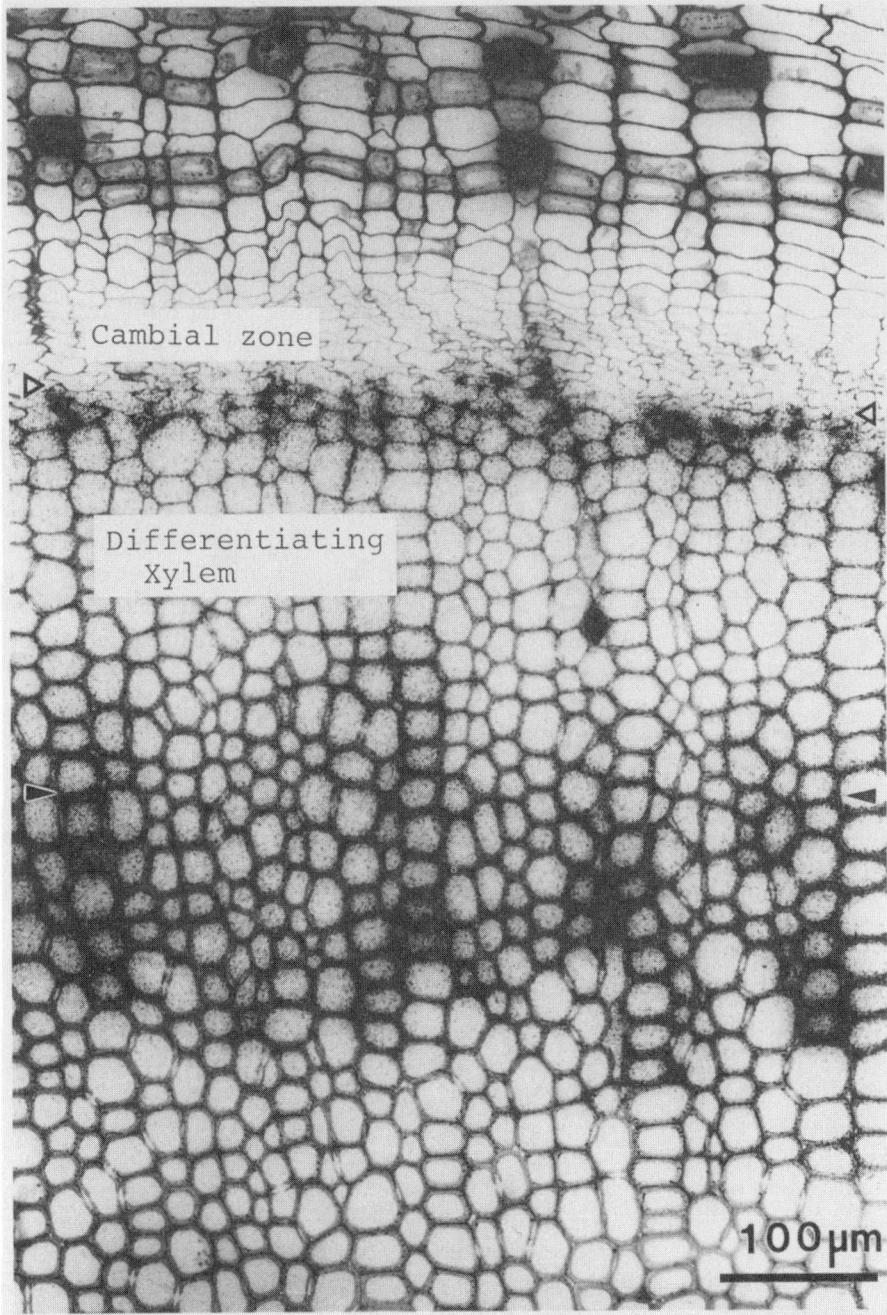

Figure 3 Microautoradiogram of differentiating xylem of *Ginkgo biloba* after administration of coniferin-[OCH_3-^{14}C]. Different stages of cell wall formation can be observed in a cross section. Deposition of lignin is visualized by silver grains on the autoradiogram. The first lignification starts at the middle lamella and the cell corners after the start of S_1 formation (◁). Intensive lignification occurs after the start of S_3 formation (◀).

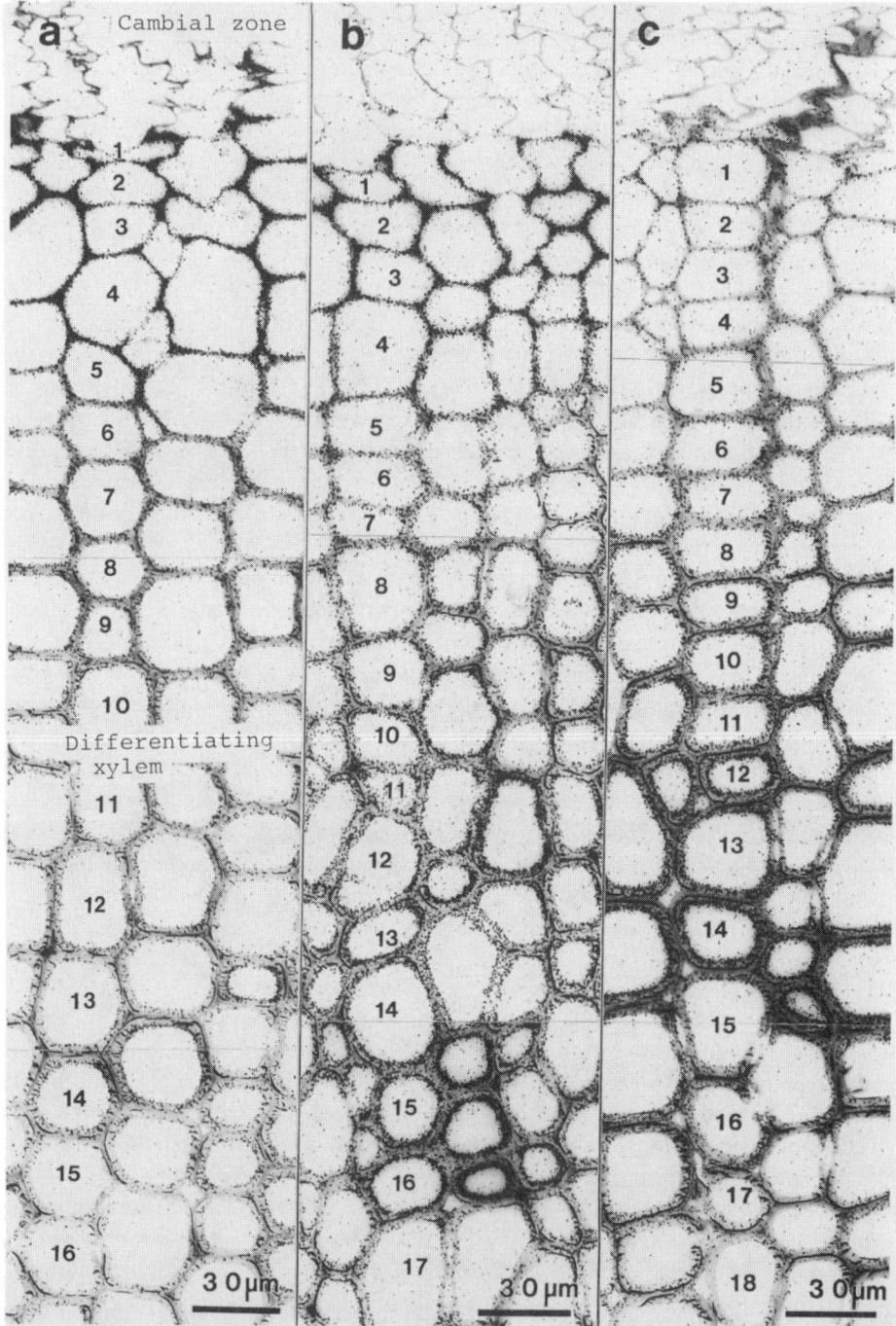

Figure 4 Microautoradiogram of differentiating xylem of *Ginkgo biloba* after administration of (**a**) *p*-glucocoumaryl alcohol-[arom. ring-2-^{3}H], (**b**) coniferin-[arom. ring-2-^{3}H], (**c**) coniferin-[arom. ring-5-^{3}H]. The number in the autoradiograms shows the number of cells counted from the cell in which formation of the S_1 layer started.

lamella region (Fig. 4a, and Fig. 5a). These observations are consistent with the results of determination of the methoxyl content (Whiting and Goring, 1982) and UV spectra (Fergus and Goring, 1970) of lignin in the middle lamella tissue fraction. It should be noticed that ^{3}H at position 2 of the aromatic ring of monolignol glucosides (6-1, 9-2, 12-1) is retained after polylignol is formed, because the new bond is scarcely formed at this position during the polymerization. On the other hand, ^{3}H at position 3 or 5 of the aromatic ring (6-2, 9-3) is removed when a new bond is introduced at this position to form condensed structural units (15, 16, 17, 18 in Fig. 2) or when a methoxyl is introduced to form guaiacyl units or syringyl units via the routes, 6-5-8 or 6-5-4-7-8, or 9-8-11 or 9-8-7-10-11 (Matsui et al., 1994). However approximately 75 and 97% of 6 and 9, respectively, were not methoxylated (Fukushima and Terashima, 1991). Therefore, the activity remaining after incorporation into lignin may be ascribed to the presence of a non-condensed structural moiety (13, 14, 19, 20 in Fig. 2). Differences in the distribution of silver grains in Fig. 5a and 5b indicate that a considerable part of p-hydroxyphenyl lignin in the middle lamella region may be of the condensed type. A high content of condensed p-hydroxyphenyl lignin in the middle lamella region is also consistent with the results that oxidative analysis of this tissue fraction gave only trace amounts of p-hydroxybenzaldehyde and p-hydroxybenzoic acid which are not produced from condensed units (Westermark, 1985). Fig. 4b and 4c show microautoradiograms of ginkgo xylem administered with coniferin-[aromatic ring-2-^{3}H] (9-2) and coniferin-[aromatic ring-5-^{3}H] (9-3). The distribution of silver grains is shown in Fig. 5c and 5d. It is obvious that deposition of total guaiacyl lignin (Fig. 4b and 5c) occurs in three distinct stages with two peaks of deposition as observed on the micro-autoradiogram of Ginkgo xylem administered with coniferin-[OCH$_3$-^{14}C] (9-1) (Fig. 3). Comparison of Fig. 5c and 5d indicates that removal of ^{3}H at position 5 of the guaiacyl ring due to the formation of condensed units (15, 16, 17, 18) is more frequent in the middle lamella region. When syringin-[aromatic ring-2-^{3}H] (12-1) was administered, the incorporation was low and mainly in the secondary wall (Fig. 5e).

Thus different types of monomer units are incorporated into lignin at the different stages of cell wall formation. As a result, the structure of Ginkgo lignin is heterogeneous with respect to the morphological regions in the cell wall.

When Ginkgo wood was subjected to alkaline nitrobenzene oxidation, combined yields of aldehydes and acids were as follows; p-hydroxybenzaldehyde and p-hydroxybenzoic acid derived from p-hydroxyphenyl lignin: 3.5%, vanillin and vanillic acid derived from guaiacyl lignin: 95.3%, syringaldehyde and syringic acid derived from syringyl lignin: 1.3% based on the total aromatic aldehydes and acids (Fukushima and Terashima, 1991). Pyrolysis-gas chromatography of Ginkgo wood also showed that the content of syringyl units in Ginkgo lignin is very low (Obst and Landucci, 1986). These results indicate that the major part of protolignin in Ginkgo wood consists of guaiacyl lignin containing a small amount of p-hydroxyphenyl lignin in the middle lamella region and a small amount of syringyl lignin in the secondary wall.

Solid state NMR analysis of Ginkgo xylem tissue in which the side-chain β-carbon of protolignin is specifically ^{13}C-enriched provides information on the type

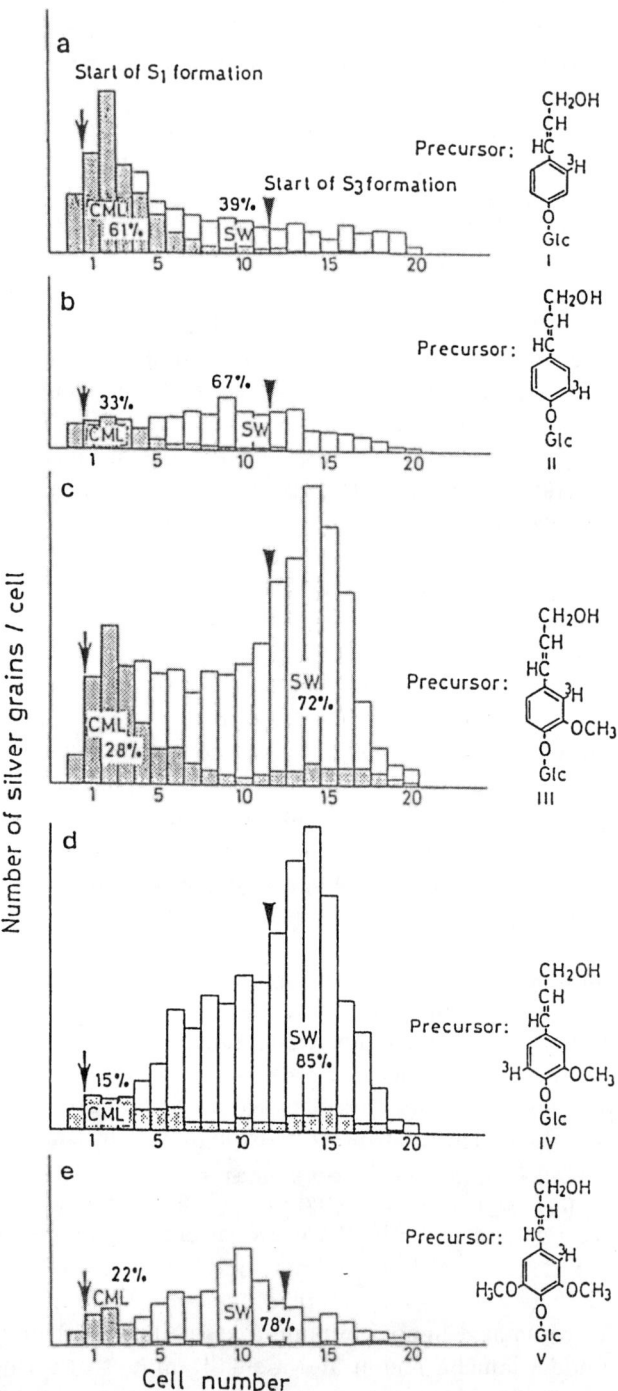

Figure 5   Distribution of silver grains in microautoradiograms of Ginkgo xylem after administration of labeled precursors. The cell number shows the number of cells counted from the cell in which the formation of the S_1 layer started.
CML: compound middle lamella, SW: secondary wall.

and content of linkages between monomeric units (Terashima *et al.*, 1997b). The content of β-O-4 substructure (**14**) including substructures (**15**), (**16**) and (**17**) in guaiacyl lignin was estimated to be $59 \pm 1.5\%$ and combined β-5 (**18**), β-β (**19**) and β-1 (**20**) substructures, $36 \pm 1.5\%$ and coniferyl alcohol/aldehyde end group (**13**), $4 \pm 1.5\%$. A part of β-1 (**20**) substructure may be present in wood as its precursor, cyclohexadienone structure (**21**) (Brunow and Lundquist, 1991). Because the β' carbon of **21** gives a signal in the same region as the β carbon of β-O-4 structures in solid state NMR, the above estimated percentage for β-O-4 includes one half of the content of structure **21** (Terashima *et al.*, 1997b).

These features on the heterogeneous formation and structure of Ginkgo lignin are very close to those observed for pine lignin (Terashima and Fukushima, 1988).

The reason why different kinds of labeled precursor were incorporated into different specific morphological regions may be explained partly by the substrate specificity of β-glucosidase which catalyzes the hydrolysis of monolignol glucosides to corresponding alcohols. It has been shown that the cell-wall bound β-glucosidase in spruce seedlings has different substrate specificity towards three kinds of monolignol glucosides (Marcinowski and Griesebach, 1978), and β-glucosidase specific for coniferin has been found in lignifying xylem of lodge-pole pine (Dharmawardhana *et al.*, 1995). Heterogeneity in lignification may be related to the development of biosynthetic pathways not only of lignin but also of polysaccharides during the differentiation of cell wall (Terashima *et al.*, 1993).

As shown above, studies on lignification of Ginkgo xylem provided important information on the lignification mechanism and the lignin structure in tree cell walls. Based on the results including those obtained with Ginkgo, a hypothetical general struc-ture of lignin has been proposed (Terashima *et al.*, 1998). In the future, determination of the frequency and distribution of bonds between lignin and polysaccharides will be important for better understanding the 3D assembly of major components in the cell wall. Evolution of plants also involved formation of compression wood at the lower side of the inclined stem or branch of trees belonging to conifers. Morphology and chemical composition of compression wood differs from normal wood, and the quality of compression wood is generally inferior for industrial use. Ginkgo also forms compression wood, however the difference between compression wood and normal wood is not as significant as the differences observed in pine (Timell, 1983; Tomimura *et al.*, 1980; Fukushima and Terashima, 1991). Detailed studies on the differences in lignification mechanism of compression wood between Ginkgo and other conifers may provide useful information for the improvement of forest trees in the future.

CONCLUSION

The process of lignification of Ginkgo xylem cell wall is similar to that of conifer xylem cell wall. The characteristic features of formation and structure of Ginkgo lignin are summarized as follows:

1. Lignified cell walls are formed by successive deposition of pectin, hemicellulose, cellulose and lignin in differentiating xylem cell walls. The kind of poly-saccharide and type of monolignol supplied to the cell wall changes with the age of the cell. Then the lignin formation takes place in different ways in the compound middle lamella and the secondary wall.
2. The major part of Ginkgo lignin is guaiacyl lignin. At an early stage of cell wall formation, guaiacyl lignin containing a small amount of p-hydroxyphenyl-propane units deposit rather quickly in a gel of pectic substances in the com-pound middle lamella region. At a later stage, the major part of guaiacyl lignin containing a small amount of syringylpropane units deposits rather slowly in mannan gel in the secondary wall.
3. The lignin formed in the compound middle lamella contains more condensed type substructures than the lignin formed in the secondary wall. About 60% of monomer units are bonded by a β-O-4 type linkage.
4. Thus the lignification of Ginkgo xylem proceeds heterogeneously with respect to the morphological region of the cell wall and with respect to the 3D macromole-cular lignin structure.

Further extensive studies on the lignification of Ginkgo cell walls in the future will promote better understanding in a wide area of science and technology related to forestry and forest industry.

REFERENCES

Bao, W., O'Malley, D.M., Whetten, R. and Sederoff, R.R. (1993) A laccase associated with lignification in loblolly pine xylem. *Science*, **260**, 672–674.

Brunow, G. and Lundquist, K. (1991) On the acid catalysed alkylation of lignins. *Holzforschung*, **45**, 37–40.

Dharmawardhana, D.P., Ellis, B.E. and Carlson, J.E. (1995) A β-glucosidase from lodge-pole pine xylem specific for the lignin precursor coniferin. *Plant Physiol.*, **107**, 331–339.

Eberhardt, T.L., Bernards, M.A., He, L-F., Davin, L.B, Wooten, J.B. and Lewis N.G. (1993) Lignification in cell suspension culture of *Pinus taeda*. In situ characterization of a gymnosperm lignin. *J. Biol. Chem.*, **268**, 21088–21096.

Fergus, B.J. and Goring, D.A.I. (1970). The distribution of lignin in birch wood as determined by ultraviolet microscopy. *Holzforschung*, **24**, 118–124.

Freudenberg, K. (1968) The constitution and biosynthesis of lignin. In K. Freudenberg and A.C. Neish, *Constitution and Biosynthesis of Lignin*, Springer-Verlag, Berlin, pp. 85–97.

Fujita, M. and Harada, H. (1979) Autoradiographic investigation of cell wall development II. Tritiated phenylalanine and ferulic acid assimilation in relation to lignification. *Mokuzai Gakkaishi*, **25**, 89–94.

Fukushima, K. and Terashima, N. (1991) Heterogeneity in formation of lignin. XIV. Formation and structure of lignin in differentiating xylem of *Ginkgo biloba*. *Holzforschung*, **45**, 87–94.

Haslam, E. (1993) *Shikimic acid-metabolism and metabolites*, John Wiley & Sons, Chichester, pp. 227–238.

Higuchi, T. (1985) Biosynthesis of lignin. In T. Higuchi (ed.), *Biosynthesis and Biodegradation of Wood Components*, Academic Press, Orlando, pp. 141–160.

Imai, T. and Terashima, N. (1990) Determination of the distribution and reaction of polysaccharides in wood cell walls by the isotope tracer technique. I. Selective radio-labeling of cell wall polysaccharides in *Magnolia kobus*. *Mokuzai Gakkaishi*, 36, 917–922.

Imai, T., Goto, N., Yasuda, S., and Terashima, N. (1996a) Visualization of deposition process of mannan in differentiating cell wall of Ginkgo and pine by microautoradiography. *Proc. 46th Conf. Wood Res. Soc. Japan*, Kumamoto, April 1996, p. 328.

Imai, T., Goto, N., Yasuda, S., and Terashima, N. (1996b) Selective labeling of non-cellulose polysaccharides of Ginkgo and pine. *Proc. 46th Conf. Wood Res. Soc. Japan*, Kumamoto, April 1996, p. 329.

Lewis, N.G., Razal, R.A., Yamamoto, E., Bokelman, G.H. and Wooten, J.B. (1989) Carbon-13 specific labeling of lignin in intact plants. In N.G. Lewis, and M.G. Paice, (eds.), *Plant cell wall polymers, biogenesis and biodegradation*; ACS Symposium Series 399, American Chemical Society, Washington DC, pp. 169–181.

Matsui, N., Fukushima, K., Yasuda, S. and Terashima, N. (1994) On the behavior of mono-lignol glucosides in lignin biosynthesis. II. Synthesis of monolignol glucosides labeled with ^{3}H at the hydroxymethyl group of side chain, and incorporation of label into Magnolia and Ginkgo lignin. *Holzforschung*, 48, 375–380.

McDougall, G.J. and Morrison, I.M. (1996) Extraction and partial purification of cell-wall-associated coniferyl alcohol oxidase from developing xylem of sitka spruce. *Holzforschung*, 50, 549–553.

Marcinowski, S. and Griesebach, H. (1978) Enzymology of lignification. Cell-wall-bound β-glucosidase for coniferin from spruce (*Picea abies*) seedling. *Eur. J. Biochem.*, 87, 37–44.

Neish, A.C. (1968) Monomeric intermediates in the biosynthesis of lignin. In K. Freudenberg and A.C. Neish (eds.), *Constitution and Biosynthesis of Lignin*. Springer-Verlag, Berlin, pp. 3–43.

O'Malley, D.M., Whetten, R., Bao, W., Chen, C-L. and Sederoff, R.R. (1993) The role of laccase in lignification. *Plant J.*, 4, 751–757.

Obst, J.R. and Landucci, L.L. (1986) The syringyl content of soft wood lignin. *J. Wood Chem. Technol.*, 6, 311–327.

Saleh, T.M., Leny, L. and Sarkanen, K.V. (1967) Radioautographic study of cotton wood, Douglas fir and wheat plants. *Holzforschung*, 21, 116–120.

Sarkanen, K.V. (1971). Precursors and their polymerization. In K.V. Sarkanen and C.H. Ludwig (eds.), *Lignins, Occurrence, Formation, Structure and Reactions*. Wiley-Interscience, New York, pp. 95–163.

Savidge, R. and Udagama-Randeniya, P. (1992). Cell wall bound coniferyl alcohol oxidase associated with lignification in conifers. *Phytochemistry*, 31, 2959–2966.

Sterjiades, R., Dean, J.F.D., and Ericksson, K.-E.L. (1992). Laccase from sycamore maple (*Acer pseudoplatanus*) oxidizes monolignols. *Plant Physiol.*, 99, 1162–1168.

Takabe, K., Fujita, M., Harada, H. and Saiki, H. (1981) Lignification process of Japanese black pine (*Pinus thunbergii* Parl) tracheids. *Mokuzai Gakkaishi*, 27, 813–820.

Terashima, N., Fukushima, K., Sano, Y. and Takabe, K. (1988) Heterogeneity in formation of lignin. X. Visualization of lignification process in differentiating xylem of pine by micro-autoradiography. *Holzforschung*, 42, 347–350.

Terashima, N. and Fukushima, K. (1988) Heterogeneity in formation of lignin. XI. An autoradiographic study of the heterogeneous formation and structure of pine lignin. *Wood Sci. Technol.* 22, 259–270.

Terashima, N. and Fukushima, K. (1989) Biogenesis and structure of macromolecular lignin in the cell wall of tree xylem as studied by microautoradiography. In N.G. Lewis and M.G. Paice, (eds.), *Plant Cell Wall Polymers, Biogenesis and Biodegradation*, ACS Symp. Ser. 399, Am. Chem. Soc., Washington, DC. pp. 160–168.

Terashima, N., Fukushima, K., He, L-F., and Takabe, K. (1993) Comprehensive model of the lignified plant cell wall. In H.G. Jung, D.R. Buxton, R.D. Hatfield and J. Ralph (eds.), *Forage Cell Wall Structure and Digestibility*, Am. Soc. Agronomy, Madison, Wisconsin, pp. 247–270.

Terashima, N., Atalla, R.H. and Van der Hart, D.L. (1997a) Solid state NMR spectroscopy of specifically ^{13}C-enriched lignin in wheat straw from coniferin. *Phytochemistry*, **46**, 863–870.

Terashima, N., Hafrén, J., Westermark, U. and Van der Hart, D.L. (1997b) Structure of lignin in Ginkgo wood determined by a combination of specific ^{13}C-enrichment technique and solid state NMR spectroscopy. In *Proc. 9th Intern. Symp. Wood and Pulping Chem.*, CPPA, Montreal, pp. H1-1-H1-5.

Terashima, N., Nakashima, J. and Takabe, K. (1998) Hypothetical structure of protolignin in plant cell walls. In N.G. Lewis and S. Sarkanen, (eds.), *Lignin and Lignan Biosynthesis*, ACS Symp. Ser. 697, Am. Chem. Soc., Washington, DC, pp. 180–193.

Timell, T.E. (1983) Origin and evolution of compression wood. *Holzforschung*, **37**, 1–10.

Timell, T.E. (1986) Origin and evolution of compression wood. In *Compression Wood in Gymnosperms*, Springer-Verlag, Berlin, Vol. 1, pp. 597–621.

Tomimura, Y., Yokoi, T. and Terashima, N. (1980) Heterogeneity in formation of lignin. V. Degree of condensation in guaiacyl nucleus. *Mokuzai Gakkaishi*, **22**, 259–270.

Westermark, U. (1985) The occurrence of *p*-hydroxyphenylpropane units in the middle lamella lignin of spruce (*Picea abies*). *Wood Sci. Technol.*, **19**, 223–232.

Whiting, P. and Goring, D.A.I. (1982) Chemical characterization of tissue fractions from middle lamella and secondary wall of black spruce tracheids. *Wood Sci. Technol.*, **16**, 261–267.

4. CULTIVATION OF *GINKGO BILOBA* ON A LARGE SCALE

DOMINIQUE LAURAIN

Laboratory of Pharmacognosy, Faculty of Pharmacy, 1 rue des Louvels, 80 037 Amiens Cédex 1, France

INTRODUCTION

Ginkgo biloba, the unique living representative of the order Ginkgoales which appeared in the Permian era, is an extraordinary tree because of its numerous botanical and physiological characteristics. Some of its properties, known since time immemorial in China, are nowadays exploited in two very different fields, i. e. in horticulture and in therapeutics.

Its interest for horticulture is linked to its wonderful foliage which takes on a bright yellow colour in autumn and to its resistance to all serious pests, bacteria, fungi, viruses and equally to its tolerance to air pollution (Rensselaer, 1969). Another example of its strength is shown by the regeneration of a shoot of Ginkgo the year following its destruction by the Hiroshima bomb (Michel, 1985). That is why this tree has become a favourite ornamental plant in parks and streets in Europe and the U.S.A.

Ginkgo is also of great interest for therapeutics because of its leaves which possess pharmaceutical properties such as radical scavenging, improved blood flow, vaso-protection and anti-PAF activity (DeFeudis, 1991; O'Reilly, 1993). Therefore the pharmaceutical industry needs huge quantities of leaves, two thousand tonnes of leaves are used each year in the world (Masood, 1997). China is the major producer in the world of Ginkgo leaves (Fusheng *et al.*, 1997). In this country leaves are harvested by hand from young trees growing for an ornamental use or from trees 5–20 m high (O'Reilly, 1993). Also fallen leaves can be collected during the harvesting of Ginkgo ovules (Fusheng *et al.*, 1997). These methods of harvesting pose however problems in terms of regular production and leaf quality. The chemical synthesis of some active molecules, like ginkgolide B (Corey *et al.*, 1988), could permit a production without trees but it is a long process and is therefore unprofitable.

So to meet the growing demand for *Ginkgo biloba* trees and leaves it is necessary to set up plantations (Fig. 1). Only few plantations of *Ginkgo biloba* with an indus-trial vocation are established in the world. In Europe, one sole plantation is estab-lished in the Bordeaux region (France) since 1982. Depending on the aim of the plantation, growing methods may be different, especially propagation, pruning and mineral nutrition. This chapter first deals with some aspects of the biology and the architecture of the tree that are instrumental in managing those plantations

Figure 1 *Ginkgo biloba* plantation in Champaubert (France).

efficiently. Then methods of propagating Ginkgo, those of culture in plantations and harvesting of leaves for pharmaceutical use are described.

BIOLOGY OF GINKGO

Ginkgo biloba is dioecious, possesses deciduous leaves and can exceed 30 m in height. The male reproductive organs consist of catkins situated at the level of short shoots which carry leaf clusters. The stamina carry two pollen bags and the anthesis takes place around mid-April in France. The female reproductive organs consist of spherical ovules in two's at the level of one petiole. Usually one of the two aborts before reaching maturity. Meiosis takes place in mid-April and gives rise to four haploid cells. Only the lowest of the four cells will evolve progressively toward a female gametophyte. Early in September, three layers of the ovule tegument differentiate inwards to form the sarcotesta, sclerotesta and endotesta. The reproductive organs appear in spring on trees that are at least 20 years old. However sexes can be distinguished before the trees are mature because leaves of male trees develop before the female ones and in autumn the males lose their leaves two or three weeks earlier than the female ones (Michel, 1985). Ginkgo trees have characteristic leaves with one or several lobes, dichotomic nervation and a long petiole. Trees present archaic characteristics with a zoidogamous fertilization and with a chlorophyll female gametophyte which accumulates reserves and parts with the sporophyte before fertilization (Favre-Duchartre, 1956). In France fertilization happens in September, 4.5 months after anthesis. The embryo develops immediately without a latent phase.

During the Quaternary era *Ginkgo biloba* was found only in China but nowadays it is planted in any temperate geographic region. This tree was introduced in Europe during the first half of the 18th century and it reached America in 1784 (Rohr, 1989).

ARCHITECTURE OF THE TREE

Ginkgo biloba trees raised from ovules are composed of an orthotropic monopodial trunk with plagiotropic branches per tier. Its architecture conforms to Massart's Model according to the classification system of Halle and Oldeman (1970). The trunk is formed by a single meristem that is active throughout the life of the tree. The form of vigorous young trees is quite conifer-like, with a dominant central

Table 1 Different cultivars of *Ginkgo biloba* used in horticulture and their properties (Nurseryman Adeline, La Chapelle Montlinard, France).

Cultivars	Properties
Albovariegata	Tree with a broad silhouette, slow growth, some leaves are either lined or with off-white spots.
Autumn Gold	Very regular conical silhouette, leaves particularly golden before they fall.
Barabit's Nana	Erected silhouette, small development, leaves fan-shaped and a little closed.
Fairmont	Erected and narrowed silhouette, fast growth.
Fastigiata	Pyramid-shaped, semi-erected branches.
Horizontalis	Very broad silhouette and low plant with big and jagged leaves, principal branches are in a horizontal position.
King of Donting	Female plant with a silhouette broader than that of a male tree.
Lakeview	Erected silhouette, big leaves deeply cracked.
Ohazuki	Dense silhouette forming a big ball, round fig leaves.
Pendula	Grafted on stem, spreaded and falling silhouette.
Saratoga	Silhouette of a very big bush, very long, whole or cracked leaves.
Tit	Dwarf form, very densely bush forming knots on branches where jagged leaves emerge.
Tremonia	Slender form, branches are in a horizontal position and are short. Leaves forming a fan more larger than the type-Ginkgo.
Tubifolia	All along the trunk branches are sloping. Half-closed leaves are little and funnel-shaped.
Umbrella	Erected silhouette with a ruffled aspect, very short space between each branch, identical leaves to the Ginkgo type or leaves with the form of a little tube.

leader and whorled lateral branches. As the tree ages the pyramidal shape is lost and a broad, spreading form develops (Del Tredici, 1991). Depending on cultivars the shape of the tree will be modified (Table 1).

The shoot system of *Ginkgo biloba* is composed of long and short shoots. Long shoots make up the woody framework of the tree, they increase in length from 2 cm to more than 1 m per year (Del Tredici, 1991). Leaves are arranged spirally in a 5/13 phyllotaxis (Gunckel and Wetmore, 1946). On the long shoots, leaves are alternate, they carry an axillary bud and a terminal bud is present at the top of the shoot. A short shoot is formed when an axillary bud of an auxiblast develops bringing about the formation of a rosette of leaves with a terminal bud (Camefort and Boue, 1969). The growth of these short shoots is limited to about 1 to 2 mm per year and the internodes are short. These produce the majority of leaves on the tree and when the Ginkgo is 25–30 years old, they bear the reproductive structures developed within the rosettes of leaves. These short shoot leaves lack axillary buds (Critchfield, 1970).

Under certain conditions, some short shoots are transformed into long shoots. This ability is frequent in the conifers when a reduction or a deletion of the carrier axis dominance appears after a traumatism (Nozeran, 1986). In *Ginkgo biloba* this event, which is in relation with a late partial reiteration (defined as the repetition of the seedling model), happens spontaneously on trees physiologically in senescence phase (Edelin, 1986). This phenomenon has been observed also on young trees of *G. biloba* in growing phase (Laurain, unpublished). The alteration of the auxin content of the shoot tip under extended light, high nutrition or other conditions could explain this phenomenon (Collins, 1903; Gunkel and Thimann, 1949).

Another mode of reiteration in *Ginkgo biloba* is the development of secondary stems directly from the base of old Ginkgos. According to Del Tredici (1991) this behaviour is common in China but not often seen in plants cultivated in Europe and in North America. However some cases have been observed in France, in particular on trees in the Plant Garden in Nantes and in private people's gardens in Mouleydier and in Creysse.

Another distinctive feature of Ginkgo trees growing in their natural habitat is the production of distinct organs called aerial "chichi" and basal "chichi" according to where they are located on the tree. Aerial "chichi" look like "air roots" or "burls" produced along the underside of large lateral branches of old trees (Del Tredici, 1991). Basal "chichi" can be defined as positively geotropic aggregates of suppressed shoot buds, they are located in the cotyledonary axils of all Ginkgo seedlings. "Chichi" constitute a particular system of vegetative rejuvenation of *Ginkgo biloba*. In fact they are capable of generating both aerial shoots and adventitious roots under appropriate conditions (Del Tredici, 1991) or when cut off from the parent trunk and planted upside down in soil (Hu, 1987; Li and Lin, 1990). Thus basal "chichi" could be a unique mechanism of clonal regeneration of this tree and interesting for commercial purposes (Del Tredici, 1992).

PROPAGATION

The establishment of large scale Ginkgo plantations for ornamental or pharmaceutical use requires a propagation system making it possible to get many new trees quickly. The *Ginkgo biloba* can be multiplied by vegetative propagation or sowing. Micropropagation by plant tissue culture systems has not yet been published for Ginkgo trees. *Ginkgo biloba* somatic embryos have been obtained from immature zygotic embryos (Laurain *et al.*, 1996) but they did not lead to plantlets.

Vegetative Propagation

In the horticultural field, male individuals are chosen for asexual reproduction because of the foul odor of ovules that contain butyric acid along with other compounds (Santamour *et al.*, 1983) making the female plant considerably less desirable. No relationship between plant sex and metabolite content in leaves has been shown. Concerning leaf production for the pharmaceutical industry, male or female plants are indifferently used. On the other hand vegetative propagation is effective for multiplying a hyperproductive tree which bears leaves rich in active components. Three vegetative propagation techniques have been described for Ginkgo: cutting, running and grafting.

Cutting is one technique of propagation but is not recommended for Ginkgo because it generally brings about weak vigour in trees which causes limited vegetative development. Several parameters have been shown to influence the rooting of cuttings. Thus the use of plant growth regulators has increased the rooting percentage and decreased the time required to achieve rooting (Huh and Staba, 1992). Treatments of basal stem cuttings with indolebutyric acid (50 mg/l for 23 hours) led them to root within 30 days whereas non-treated cuttings required 60 days to root (Doran, 1954). Skirvin and Chu (1979) have obtained similar results. The nature of shoots used as cutting material and the period of the year when cutting takes place influence the success of the vegetative multiplication. Short lateral shoots root well (Teuscher, 1951) but their growth is not satisfactory (Vermeulen, 1960). Apical shoots are better but less numerous. Woody cuttings gathered in March–April and treated with growth regulators give interesting results (Michel, 1985). The young age of the shoots is also a factor of success in cuttings. In this way cuttings, taken from a Ginkgo which had many young shoots after being drastically trimmed, have given the best results in rooting cuttings (Laurain, 1990). Rejuvenating mother plants will therefore be an effective means for obtaining cuttings.

Layering is unacceptable because layers do not grow in a straight line and trunks are twisted (Krussmann, 1968).

On the other hand, grafting is the technique used to ensure the propagation of Ginkgo cultivars known in horticulture (Table 1). This technique is also used in nurseries for propagating only male trees and the cuttings are taken exclusively from male trees. Grafting also makes it possible to obtain trees bearing male and female characteristics on a single tree. Putting together male and female sexes on a single tree improves the fertility when male trees are missing. In France, female Ginkgos bearing a grafted male branch have been observed in the "Jardin des Plantes" in Paris and in the "Jardin Botanique" in Tours. Grafting is variable: the age of the receiving tree and the type of transplant as well as the period of the year are all of influence. Grafting can be carried out in winter, after leaf fall and on sown potted plants during their first year of vegetation (Krussmann, 1968) or on older trees (2 to 3 years old). In mid-August shield-grafting or most often inlaid-grafting can be carried out at the level of the neck of a receiving tree in its second or third year of vegetation. The used graft should have two eyes. A scion one meter high is obtained within the year and the tree remains three years in a nursery before being transplanted.

Studies on vegetative multiplication of Ginkgo are few and date back several decades. However one may conclude that the most adapted technique for Ginkgo is grafting.

Sexual Reproduction

Through sexual reproduction of Ginkgo a large number of plants can be obtained quickly. Moreover, they present a faster vegetative development than trees from cutting or grafting. The main drawback with this propagation technique is the genetic variability concerning both the vegetative traits and the secondary metabolite content of the leaves.

Ginkgo biloba is dioecious and in order to propagate it through sowing, one must harvest ovules which come from female trees situated at a favourable distance from male trees for ovule fertilisation. In France, ovules are usually collected in November, after they have spontaneously fallen down from trees. It is equally possible to buy ovules imported from China, Korea and Japan. The price of 1 kg of ovules is variable according to the tradesman, it can reach 40 dollars. The size of the ovules is variable (0.76 to 4.17 g per ovule) depending on the trees which were bearing them and is not related to their germination potential. After the harvest, ovules are cleared of sarcotesta (yellow fleshy layer which emits a foul odour due to butyric acid) and washed with water. Usually the embryo cannot be seen with the naked eye, the fertilisation having only taken place in September. Then there is no latent phase, the embryo carries on developing according to external temperature: the higher the temperature, the faster the development. When the ovules have to wait for adequate temperature they must be cleaned and treated with a fungicide and kept in a mixture of moist peat and sand (80/20). Stocking takes place in the dark at a temperature over 4° C during winter until the time comes when they can be sown in the open-air after the last frosts in spring. In the north-east of France, sowing takes place at the end of April-beginning of May and the shoots appear in June. South of the Loire, in spring young shoots spontaneously burst out under the tree. Before sowing, soaking the ovules in hot water should provide injuries in the sclerotesta which should allow speedy germination (Krussmann, 1968).

Two sowing methods can be used for large scale cultivation of *Ginkgo biloba*. The first possibility consists of letting the plants remain for a year in the nursery before transferring them outside (Krussmann, 1968), but it is better to replant them after the second year of vegetation in countries where the winter is particularly rigorous. Ovules are sown in nurseries in soil which has been previously desinfected with methyl bromide (O'Reilly, 1993) every 10 cm in furrows 50 cm from one another. The principal drawbacks related to this method are the loss of plants after transplantation and slowing of vegetative development for several years compared to a non-transferred plant. It has been noted that rooting of transferred plants is different to that of plants which have not been transferred, following a cut in the radical pivot when it is uprooted (Laurain, 1990). One may note an emission of numerous roots producing a network of rootlets which is bigger and more ramified than those of plants with an uncut pivot. The health of transplanted plants is influenced by several external factors which are difficult to control: humidity, temperature and nature of soil. Replanting the trees is interesting because they can be spaced regularly at 50 cm on rows one meter from one another. However the space between the rows and the number of plants per linear meter can be modified, thus plantations contain 20,000 to 30,000 trees per hectare. One year old plantlets can be bought for 3 to 5 francs the piece according to the number bought and be planted directly in the plantation.

The second possibility is directly sowing on the plantation using a machine such as a garlic sower. The main drawback of this method is an irregular germination due to the variable origin of the ovules. A fertility rate of 70% has been observed for ovules harvested in France (Laurain, unpublished). Concerning ovules native from China, Korea and Japan, a 50 per cent germination was considered good (O'Reilly, 1993). This method has the advantage of limiting plant loss when replanting. Besides plant development is not disturbed. In Champaubert (France), plants reaching about 20 cm high in the first year of vegetation, 50 cm in the second year and a meter in the third year have been observed. On the other hand, after the second year of vegetation plant density must be regular, that is to say two plants per linear meter by uprooting superfluous plants.

PRACTICES OF CULTIVATION

Pruning

According to the aim of the plantation, the practice of prunning is different. Ornamental Ginkgos require no particular pruning. However, large urban Ginkgos may require drasting pruning according to size. This happened to 12 Ginkgos planted in 1864 in Saint Sulpice Laurière (France). In 1920, when the trees were 56 years old, they were topped at 7 meters from the ground and cut all around the trunk. After that operation, they developed several new shoots at the top of the trunk as well as numerous lateral shoots. One should note that this traumatism did not provoke new shoots at the base of the trunk. This is a good example of Ginkgo's resistance to pruning and its big ability to regenerate new stems. These properties are exploited in Ginkgo plantations whose aim is leaf production. Ginkgos are pruned every year in winter in order to limit vegetative development as it is realized for the upkeep of hornbeam hedges for example. Trees should not exceed a height of 1.20 meter to allow mechanized leaf harvesting. Trees can bear these annual prunings at least during 15 years, the older plantations, with an industrial vocation, being set up in 1982 (O'Reilly, 1993). Every five years a major pruning is carried out at the base of the tree in order to rejuvenate the plants and make the shoots vigorous. Between 2 to 5 new shoots can reach a height of 1.20 meter the first year after the major pruning. Likewise it has been noted that after the cutting of the hypocotyle or the stem of a young plant, one or two new shoots may develop again. In the years following the pruning, many new shoots grow but they do not reach a height of 1.20 meter. This large regenerative capacity might be explained by the presence of basal chichi reported by Del Tredici (1992) which are located in the cotyledonary axils of all Ginkgo seedlings and which are able to produce aerial stems in response to a traumatic injury.

Culture Conditions

Ginkgo biloba grows in a large variety of soils. However for best growth it should have a deep, fairly rich soil, limestone, silicious-argilo calcareous with neutral pH. It may also develop in acidic soils and tolerates alkaline soils although its growth is then reduced. It does not grow in soaked soils, however occasional deep watering gives best results and it needs sufficient rainfall, approximately 90 cm/year (Franklin, 1959).

Ginkgo also shows an extraordinary ability to adjust to a wide variety of climates. It grows in southern Canada and in other regions where winter temperature may drop to –25° C (Rensselaer, 1969). The tree is indigenous to China where natural populations supposedly exist in the Chekiang province of eastern China (Chen, 1933; Li 1956). Currently Ginkgo trees can be found in Asia, Europe, North America and in temperate regions of New Zealand and Argentina (Huh and Staba, 1992). It grows up to an altitude of 1000 meters (Michel, 1985).

Ginkgo is known to be a slow grower. For example the trees at St Sulpice Laurière (France) are 135 years old and have a circumference of between 1.65 and 3.10 meters which corresponds to an annual growth of about 1–2 cm. This result is comparable to oak-growth. However Ginkgo growth may be quicker. For example the Ginkgo in Mouleydier (France) is about 150 years old and has a circumference of 4.45 meters which corresponds to an average of 3 cm a year. Depending on the situation, climate and soil, Ginkgo's growth is more or less speedy. Fertilisers can improve the vegetative development of the tree. On the other hand, like fruit trees, a female Ginkgo which has an intense vegetative development has been noted to produce no ovules.

About the growth of Ginkgo, some criteria as the elongation speed of the main axis, the final size of the main axis, the plastochrone and the number of leaves formed, can be influenced by light and temperature under controlled long day conditions. On the other hand the alternation between long and short internodes and the insertion level of ramifications are independent (Flesch *et al.*, 1991).

Mineral Nutrition

Few studies have been published on the mineral nutrition of *Ginkgo biloba*. Observations on a field Ginkgo plantation and assays (Laurain, 1990) realized in the I.N.R.A. Agronomy Station in Angers (France) on plants in 2.5 l containers (containing a mixture of grinded pine-bark and yellow peat in equal quantities with the

Table 2   Composition (macronutrients) of nutritious solutions (Coïc Lesaint type) used for the mineral nutrition of *Ginkgo biloba* trees in containers.

Macronutrients (meq/l)	Nitrogen solutions				
	A	*B*	*C*	*D*	*E*
NO_3^-	1.80	3.60	7.10	10.70	14.30
PO_4^-	3.30	3.30	3.30	3.30	3.30
SO_4^{--}	4.90	4.90	2.00	1.30	1.50
Cl^-	2.50	1.50	2.00	0.20	0.50
NH_4^+	0.30	0.60	1.20	1.80	2.40
K^+	5.00	5.00	5.00	5.00	5.00
Na^+	0.00	0.00	0.00	0.00	2.50
Ca^{++}	4.00	4.50	5.00	5.50	6.00
Mg^{++}	1.00	1.00	1.00	1.00	1.50
N total in mg/l	29.4	58.8	116.2	175	233.8

pH adjusted to 6–6.3) have shown that *Ginkgo biloba* reacts well to additions of fertilizers. "Two year old Ginkgo plants in containers have received nutrient solutions of the type Coïc Lesaint (Table 2) containing increasing doses of nitrogen (29.4, 58.8, 116.2, 175, 233.8 mg/l) but with a constant ratio NO_3/NH_4 of 12/2 (approximately).". This study has shown an increase of plant growth as a function of increasing doses of nitrogen and particularly an increase of the number of lateral ramifications and a decrease of the apical dominance at the concentration of 116.2 mg/l of nitrogen. Following this study, the N, P, K, Ca and Mg contents in leaves and in one year old shoots have been analysed as a function of nitrogen doses. The results are presented in Table 3. In leaves the Mg content is stable at the different nitrogen doses while the N, P, K and Ca contents increase with the increasing doses of nitrogen. In shoots, the P and Ca contents do not increase, the Mg content decreases and the N and K contents increase with the increasing doses of nitrogen.

Nitrogen fertilization effects on foliar N and P dynamics and autumnal resorption in *Ginkgo biloba* have been studied (Brinkman and Boerner, 1994). Foliar N and P concentrations were more similar to those of angiosperms than of gymnosperms. On the other hand the levels of N and P resorption were similar to those reported for a wide range of gymnosperms and greater than those reported for most angiosperms.

Weed Control

The plantations are kept weed-free with a minimum use of herbicides in view to harvest leaves without pesticide residues. The soil of the nurseries and of the plantations are treated with a pre-emergent herbicide (diflufenicanil + isoproturon) before sowing. The same herbicides are sprayed in March before the leaves appear. During the growing season, the nurseries and the plantations are kept clean by burning and

Table 3 Contents (%) of N, P, K, Ca, Mg in leaves and in annual shoots of *Ginkgo biloba* trees two years old, cultivated in container, as a function of nitrogen doses.

	Nitrogen doses				
	29.4 mg/l	*58.8 mg/l*	*116.2 mg/l*	*175 mg/l*	*233.8 mg/l*
N					
Leaves	0.74	0.97	1.56	2.00	2.01
Annual Shoots	0.49	0.54	0.74	0.96	0.94
P					
Leaves	0.18	0.22	0.20	0.23	0.31
Annual Shoots	0.14	0.15	0.18	0.17	0.17
K					
Leaves	0.77	0.98	0.89	1.17	1.28
Annual Shoots	1.08	1.32	1.25	1.23	1.27
Ca					
Leaves	0.72	0.81	1.00	0.91	1.15
Annual Shoots	0.58	0.62	0.75	0.60	0.63
Mg					
Leaves	0.23	0.24	0.23	0.19	0.24
Annual Shoots	0.21	0.21	0.19	0.14	0.15

also weeding by hand. Local and pointed applications of glyphosate can be carried out during the growing period. Some herbicides have been tested on *Ginkgo biloba* (C.N.I.H., 1985), and the following active substances can be used in the plantations: alloxydim sodium, chloroxuron, lenacil + neburon, oxadiazon granules, oxadiazon + carbetamid, propyzamid, trifluralin. It is also possible to cultivate herbaceous plants between the rows occupied by Ginkgo trees in winter to keep down the weeds (O'Reilly, 1993).

Pests

Ginkgo is reputed to be resistant to all serious pests (Franklin, 1959). This tree has not acquired predator fauna in North America and it is unusual that the plant has no predator in its native habitat of East Asia (Wheeler, 1975).

However some predators have been identified in French Ginkgo plantations. They are different according to the age of the plant and the situation of the plantation. As soon as a shoot appears on the soil surface it runs the risk of being destroyed by birds, such as pigeons, doves and partridges. During the first years of vegetation, snails, slugs, mouses, rabbits and hares are principal predators. These pests eat the bark at the bottom of the tree which can cause death if the bark is removed all around the trunk. When the plantation is situated next to a forest, stags break branches and destroy the bark. To prevent attack by rodents it is advisable to put a plastic netting around each tree trunk up to a height of 50 cm or to fence off the plantation. It is recommended to cover nurseries with netting to protect young plants from birds and to spread metaldehyde, thiodicarb, mercaptodimethur or bensultap granules on the ground against slugs and snails.

Ginkgo is one of the least vulnerable trees as far as insects are concerned. They cause only minimal damage and none is specifically seen as a threat. The leaves are used in China and Japan as bookmarks to protect books from silverfish and larvae of other insects (Major, 1967). However a weasy insect, *Cacoecimorpha pronubana*, proved obnoxious to leaves (Martinez and Chambon, 1987). In addition some insects, *Brachytrupes portentosus*, *Agrotis ypsilon* and *Gulcula panterinaria*, were found to be the main causes of death of seedlings of *Ginkgo biloba* (Zhiquan *et al.*, 1991). The larvae of the corn pyral (*Pyrausta mubilalis*) can eat Ginkgo leaves but if an aqueous extract of the same leaves is given to the larvae, their growth is considerably inhibited (Major, 1967). In addition, Major and Tietz (1962) observed that Japanese beetles died of starvation rather than eat fresh leaves of *Ginkgo biloba*. Investigations have been carried out on the insecticidal properties of the tree. It has been shown that the leaves are very acidic and that 2-hexenal is produced when Ginkgo leaves are damaged in the presence of oxygen (Major *et al.*, 1963). Bevan *et al.* (1961) have reported that this aldehyde is an insect repellent. Ginkgolide A has been found the most active of several compounds which included bilobalide and ginkgolic acids, possessing antifeedant activity against larvae of the cabbage butterfly (*Pieris rapae crucivora*) (Matsumoto and Sei, 1987). Roots and stems of *G. biloba* are also toxic for insects (Major, 1967).

Diseases

Ginkgo is highly resistant to diseases. Only one phytopathological problem has been encountered in the plantation located near Bordeaux (France), a mould that grows at the bottom of the young plants (O'Reilly, 1993). This disease might be explained by soil fungi which infect ovules left on the ground for too long a time. Embryos grown from ovules taken immediately when they fall from the trees have been cultivated *in vitro* with no contamination, whereas embryos grown from ovules collected a long time after they have fallen on the ground were contaminated with mould despite ovule decontamination (Laurain, 1994). These results show how important it is to harvest healthy ovules. Ovule soil contamination would affect the germination rate and the development of healthy young plants. Death of *Ginkgo biloba* seedlings can be caused by some fungi (*Fusarium* sp., *Macrophomina phaseoli*) which generate root and stem rot (Zhiquan *et al.*, 1991). On the other hand Ginkgo, like the tree Eleagnus, appears to be resistant even after inoculation with the fungus *Verticillium dahliae* (Smith and Neely, 1979). The resistance of *G. biloba* to fungi could be explained by the 2-hexenal isolated from the steam distillate of its leaves that had an antifungal activity in low concentrations (Major *et al.*, 1960). But the unusual resistance of Ginkgo leaves to fungi cannot be explained completely on the basis of its 2-hexenal content (Major, 1967). Other investigations have shown that a waxy material in the cuticle of leaves reduced spore germination, inhibited germ-tube growth of certain fungi so that fungi did not penetrate the cuticle (Christensen, 1972a). However while the leaves did not appear to be damaged, the chemical nature of the epidermal wall of the leaves was changed (Adams *et al.*, 1962). Christensen (1972b) reported a phytoalexin production in Ginkgo leaves induced by *Botrytis allii* which selectively inhibited the process of cuticle penetration while allowing germination and growth of the fungi.

In addition the Ginkgo tree is remarkably free of bacteria and virus attack. The acidity of the leaves could explain the resistance to bacteria like *Escherichia coli*, *Pseudomonas phaseolicola*, *Xanthomonas phaseoli* and *Bacillus pumilus* (Major, 1967). A study had shown that extracts of the root of Ginkgo trees inhibited the growth of bean mosaic virus and tobacco mosaic virus (Major, 1967).

HARVESTING OF LEAVES AND QUALITY

Leaves are manually picked from trees. Concerning the few plantations specially established for the industrial culture of *Ginkgo biloba*, leaves are harvested with a modified cotton picker (O'Reilly, 1993). In France leaves are harvested in September/October before the first frost, which is responsible for the leaf fall, and before they turn yellow. Then they contain a maximum of active secondary metabolites. After the harvesting leaves have to be air dried within 24 hours in order to assure a good preservation. Leaves can be dried by mixing during a few minutes in propane heated rotary drum dryers. The moisture content of the leaves prior to drying is 75% and less than 10% after drying. 3.6 kg of green leaves correspond to 1 kg of dried leaves. Dried leaves are baled and can be stocked during months. The production of leaves from 4 to 5 years old plantations after transplanting is either 12–16 tons of fresh leaves per hectare or about 4 tons of dried leaves per hectare (Balz, 1997). The price of one kg of *Ginkgo biloba* dried leaves can reach 14 dollars in the medicinal plant trade.

Leaf quality varies according to harvest time. Van Beek *et al.* (1991) showed that the time of harvest plays an important role in the yield of terpenoids. The percentages of bilobalide and ginkgolides in Dutch Ginkgo leaves were lowest in spring and then gradually increased until a maximum was reached in late summer or early autumn (Fig. 2A, 2B). The concentration declined until leaf fall (van Beek and Lelyveld, 1992). The flavonol content is also seasonal with a highest flavonol glycoside content in the leaves in spring, decreasing throughout the season to a minimum before the leaves turn yellow (Fig. 2C) and then increasing after yellowing (Lobstein *et al.*, 1991; Sticher, 1993). The soil type and the environmental conditions in which the tree is grown influence the flavonol content too (O'Reilly, 1993). The age of the tree (van Beek and Lelyveld, 1992) and genetic diversity, also influence leaf quality. The content of flavonoids in leaves decreases with an increase of the age of the tree (Fusheng, 1997). Leaves harvested in September 1989 from a Dutch and a German Ginkgo showed a percentage of 0.196% and 0.032% terpene trilactones respectively (van Beek *et al.*, 1991). A ginkgolide content of 0.154, 0.244 and 0.136% has been obtained following different cultivation methods experimented in a plantation of Ginkgo in Champaubert (France) (Laurain, unpublished results). Moreover terpenoids have also been found in non-lignified shoots and in woody shoots two to

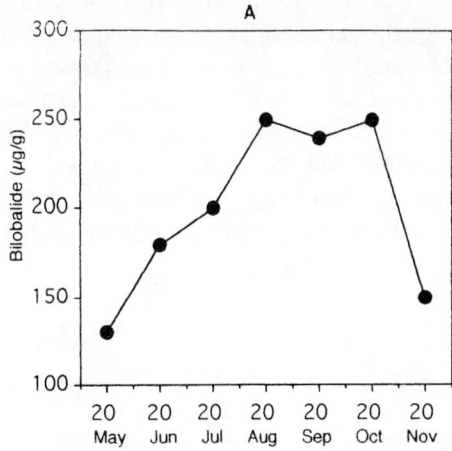

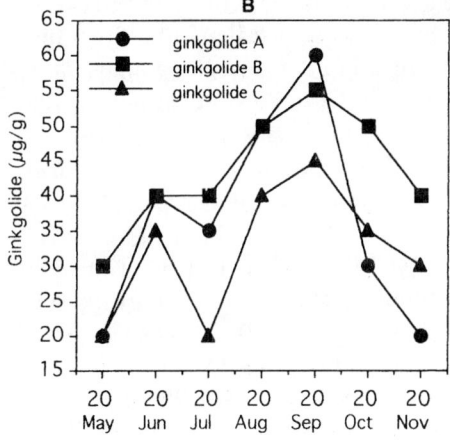

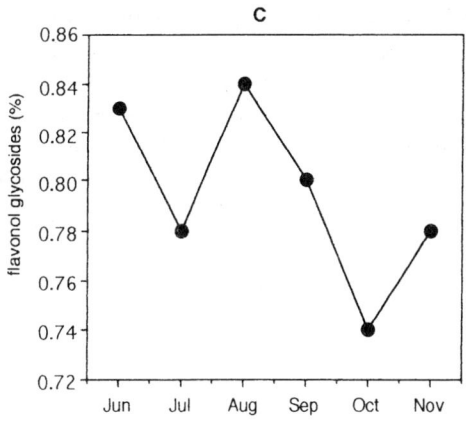

Figure 2   Relation between harvest time of leaves and bilobalide (A), ginkgolides (B) and flavonol glycoside (C) contents of female Ginkgo trees (van Beek and Lelyveld, 1992; Sticher, 1993).

four years old in variable amounts: 0.058% or 0.148% terpenoids in fresh shoots and 0.044% or 0.109% terpenoids in woody shoots according to the method of cultivation used. The nutritional quality of Ginkgo leaves has been analysed also. This study has been realised by the Laboratory of Nutrition and Feeding – National Veterinary School of Alfort (France) on dried leaves of *Ginkgo biloba*, harvested in June from trees growing in Champaubert (France). The dry matter, cellulose, ash, calcium and phosphorus content were analysed and were found to be 93.2, 11, 6.5, 1.8 and 0.5% respectively. These results showed a good nutritional value near that of wheat bran with a better Ca/P ratio.

CONCLUSION

Nowadays, *Ginkgo biloba* is an ornamental tree which can be found in numerous cities throughout the world and in recent decades the therapeutic use of its leaves in the West has given rise to the establishment of a few large-scale Ginkgo plantations (France, South Carolina and China). A report by a London-based management consultancy reports that some medicinal plants like *Ginkgo biloba* are species in urgent need of conservation because of the growing popularity of herbal medicines (Masood, 1997). In fact, demand of Ginkgo leaves in the world is increasing at 26 per cent a year. So urgent measures are needed to cultivate *Ginkgo biloba* at a large scale.

Trees of plantations established with the aim to harvest leaves are grown from ovules. On the other hand in horticulture, vegetative propagation by grafting is principally employed for obtaining on the one hand cultivars possessing desired ornamental characteristics and on the other hand for propagating only male trees, thus eliminating annoyances caused by ovules of female Ginkgos. The major problem in relation with Ginkgo multiplication by sowing is the diversity observed at the vegetative development level of young plants as well as at the level of their secondary metabolite content. Grafting could lead to a plantation made of homogeneous plants but it is a time-consuming technique and more expensive to achieve for millions of trees than multiplication by sowing. The first studies made on cultivation methods on a Ginkgo plantation (Laurain, unpublished) could solve the problem of secondary metabolite variability. They have shown an influence on ginkgolide content. Such studies have to be continued because they could lead to plantations with *Ginkgo biloba* trees, obtained by sowing, rich in active compounds.

ACKNOWLEDGEMENT

I would like to thank Mr. R. Laurain, Dr. J. Mattar, Prof. A. Jacquin and Dr. F. Beaujard for information and advice given during the preparation of this paper and Mrs. M. Pelabon for the translation into English.

REFERENCES

Adams, P.B., T. Sproston, H. Tietz and R.T. Major (1962). *Phytopathol.*, **52**, 233. Cited from: Major, R.T. (1967). The Ginkgo, the most ancient living tree. *Science*, **157**, 1270–1273.

Balz, J.P. (1997). Agronomic aspects of *Ginkgo biloba* leaves production. *Proceedings of '97 International Seminar on Ginkgo*, November 10–12, Beijing, China, 101–104.

Beek, T.A. van and Lelyveld, G.P. (1992). Concentration of ginkgolides and bilobalide in *Ginkgo biloba* leaves in relation to the time of year. *Planta Med.*, **58**, 413–416.

Beek, T.A. van, Scheeren, H.A., Rantio, T., Melger, W. Ch. and Lelyveld, G.P. (1991). Determination of ginkgolides and bilobalide in *Ginkgo biloba* leaves and phyto-pharmaceuticals. *J. Chromatogr.*, **543**, 375–387.

Bevan, C.W.L., Birch, A.J. and Caswell, H. (1961). An insect repellant from black cocktail ants. *J. Chem. Soc.*, **96**, 488.

Brinkman, J.A. and Boerner, R.E.J. (1994). Nitrogen fertilization effects on foliar nutrient dynamics and autumnal resorption in maidenhair tree (*Ginkgo biloba* L.). *J. Plant Nutr.*, **17**, 433–443.

Camefort, H. and Boue, H. (1969). Reproduction et biologie des principaux groupes végétaux. *Les Cormophytes ou Archégoniates*. Edition Doin, 45–50.

Chen 1933 cited from: Gaussen, H. (1946). Les Ginkgoales. In: Gaussen H. (ed.). Les gymnospermes actuelles et fossiles. Fac. Sci., Toulouse, 1–54.

Christensen, T.G. (1972a). Phytoalexin production in *Ginkgo biloba* in relation to inhibition of fungal penetration. *Phytopathol.*, **62**, 493–494.

Christensen, T.G. (1972b). Resistance of Ginkgo to fungi. Phytoalexin production induced by *Botrytis allii*. *Ciss. Abstr. Int. B.*, **32**, 4340.

Collins, G.N. (1903). Dimorphism in the shoots of the Ginkgo. *The Plant World*, **6**, 9–11.

C.N.I.H. (1985). Conseils pratiques pour le désherbage des pépinières. Comité National Interprofessionnel de l'Horticulture (ed.), Paris, 159 p.

Corey, E.J., M.C. Kang, N.C. Desai, A.K. Ghosh and I.N. Houpis. (1988). Total synthesis of (+)-ginkgolide B. *J. Am. Chem. Soc.*, **110**, 649–651.

Critchfield, W.B. (1970). Shoot growth and heterophylly in *Ginkgo biloba*. *Bot. Gaz.*, **131**, 2, 150–162.

De Feudis, F.V. (1991). *Ginkgo biloba extract EGb 761: Pharmacological activities and clinical applications*. Elsevier, Paris.

Del Tredici, P. (1991). The architecture of *Ginkgo biloba* L. *L'Arbre. Biologie et développement*. C. Edelin ed., Naturalia Monspeliensia, 155–167.

Del Tredici, P. (1992). Natural regeneration of *Ginkgo biloba* from downward growing cotyledonary buds (basal chichi). *Am. J. Botany*, **79**, 5, 522–530.

Doran, W.L. (1954). The vegetative propagation of Ginkgo. *J. For.*, **52**, 176–177.

Edelin, C. (1986). Stratégie de réitération et édification de la cime chez les conifères. Naturalia Monspeliensia. *Colloque International sur l'Arbre*. 139–158.

Favre-Duchartre, M. (1956). Contribution à l'étude de la reproduction chez le *Ginkgo biloba*. *Rev. Cytol. Biol. Veg.*, XVII, **1–2**, 214 p.

Flesch, V., Jacques, M., Cosson, L., Pétiard, V. and Balz, J.P. (1991). Influences de l'éclairement et de la température sur divers paramètres de croissance de jeunes *Ginkgo biloba* L. cultivés en conditions contrôlées de jours longs. *Ann. Sci. For.*, **48**, 133–147.

Franklin, A.H. (1959). *Ginkgo biloba* L.: Historical summary and bibliography. *Virginia J. Sci.* **10**, 131–176.

Fusheng, P., Wangou, L., Fugui, H., Shanan, H. and Zijie, P. (1997). Study on the optimizing of the resources of the *Ginkgo biloba* leaves. *Proceedings of '97 International Seminar on Ginkgo*, November 10–12, Beijing, China, pp. 105–112.

Gunckel, J.E. and Thimann, K.V. (1949). Studies of development in long shoots and short shoots of *Ginkgo biloba*. III. Auxin production in shoot growth. *Am. J. Bot.*, **36**, 145–151.

Gunckel, J.E. and Wetmore, R.H. (1946). Studies of development in long shoots and short shoots of *Ginkgo biloba* L. II. Phyllotaxis and the organization of the primary vascular system, primary phloem and primary xylem. *Am. J. Bot.*, **33**, 532–543.

Halle, F. and Oldeman, R. (1970). *Essai sur l'architecture et la dynamique de croissance des arbres tropicaux*. Masson, Paris.

Hu, Y.H. (1987). Chinese Penjing. Timber Press, Portland, Oregon.

Huh, H. and Staba, E.J. (1992). The botany and chemistry of *Ginkgo biloba* L. *J. Herbs Spices Med. Plants*, **1**, 91–124.

Krussmann, G. (1968). *La pépinière*. Tome 2: Organisation des exploitations. La Maison Rustique, Paris.

Laurain, D. (1990). Etude de la croissance, de la multiplication végétative et de la nutrition minérale chez le *Ginkgo biloba*. Mémoire de D.E.A., Station d'Agronomie, INRA Angers.

Laurain, D. (1994). Etablissement de cultures cellulaires haploïdes, diploïdes et transformées de *Ginkgo biloba* par diverses méthodes variabilisantes. Thèse de Doctorat, Université François Rabelais, Tours, France.

Laurain, D., Chénieux, J.C. and Trémouillaux-Guiller, J. (1996). Somatic embryogenesis from immature zygotic embryos of *Ginkgo biloba*. *Plant Cell Tissue Organ Culture*, **44**, 19–24.

Li, H.L. (1956). A horticultural and botanical history of Ginkgo. *Bull. Morris Arb.*, **7**, 3–12.

Li, Z. and Lin, J. (1990). Anatomical studies of the "burls" of Ginkgo (abstract). *I.A.W.A. Bull.*, **2**, 129.

Lobstein, A., Rietsch-Jako, L., Haag-Berrurier, M. and Anton, R. (1991). Seasonal variations of the flavonoid content from *Ginkgo biloba* leaves. *Planta Med.*, **57**, 430–433.

Major, R.T. (1967). The Ginkgo, the most ancient living tree. *Science*, **157**, 1270–1273.

Major, R.T., Marchini, P. and Boulton, A.J. (1963). Observations on the production of α-hexenal by leaves of certain plants. *J. Biol. Chem.*, **238**, 1813–1816.

Major, R.T., Marchini, P. and Sproston, T. (1960). Isolation from *Ginkgo biloba* of an inhibitor of fungus growth. *J. Biol. Chem.*, **235**, 3298–3299.

Major, R.T. and Tietz, H.J. (1962). Modification of the resistance of *Ginkgo biloba* leaves to attack by Japanese beetles. *J. Econ. Entomol.*, **55**, 272.

Martinez, M. and Chambon, J.P. (1987). Le point sur quelques ravageurs nouveaux, autochtones ou récemment introduits en France. In *Conférence internationale sur les ravageurs en agriculture*, T. **1**, Paris, ANPP, 9 pp.

Masood, E. (1997). Medicinal plants threatened by over-use. *Nature*, **66**, 570.

Matsumoto, T. and Sei, T. (1987). Antifeedant activities of *Ginkgo biloba* L. components against the larva of *Pieris rapae crucivora*. *Agric. Biol. Chem.*, **51**, 249–250.

Michel, P.F. (1985). *Ginkgo biloba-L'arbre qui a vaincu le temps*. Editions du Félin, Paris.

Nozeran, R. (1986). Le mouvement morphogénétique spécialement chez les végétaux supérieurs pérennes. Naturalia Monspeliensia. *Colloque International sur l'Arbre*. 415–428.

O'Reilly, J. (1993). *Ginkgo biloba* – Cultivation, extraction, and therapeutic use of the extract. In T.A. van Beek and H. Breteler, (eds.) Phytochemistry and Agriculture. Clarendon Press Oxford, pp. 253–270.

Rensselaer, M. (1969). The remarkable Ginkgo. *Plants Garden*, **25**, 50–53.

Rohr, R. (1989). Maidenhair Tree (*Ginkgo biloba* L.). In Y.P.S. Bajaj (ed.), *Biotechnology in Agriculture and Forestry*, Trees II, Springer Verlag, Berlin, 5, 574–590.

Santamour, F.S., He, S. and McArdle, A.J. (1983). Checklist of cultivated Ginkgo. *J. Arboric.*, 9, 88–92.

Skirvin, R.M. and Chu, M.C. (1979). Ginkgo: a beautiful tree with edible seeds. Res., Univ.: III *Agric. Exp. Stn.* 21, 10–11.

Smith, L.D. and Neely, D. (1979). Relative susceptibility of trees species to *Verticillium dahliae*. *Plant Disease Reporter*, 63, 328–332.

Sticher, O. (1993). Quality of Ginkgo preparations. *Planta Med.*, 59, 2–11.

Teuscher, H. (1951). *Ginkgo biloba* from cuttings. *Am. Nurs.*, 15, 7.

Vermeulen, J. (1960). Propagation of *Ginkgo biloba* by cuttings. *Proc. Plant Propag. Soc.*, 10, 127–130.

Wheeler, A.G. (1975). Insect associates of *Ginkgo biloba*. *Entomol. News*, 86, 37–44.

Zhiquan, Z., Dongrong, J., Guangquan, Z. and Yongmei L. (1991). A study on the causes of death of seedling of *Ginkgo biloba*. *Guihaia*, 11, 334–338.

5. PLANT CELL BIOTECHNOLOGY OF GINKGO

DANIELLE JULIE CARRIER and DOMINIQUE LAURAIN[1]

Department of Agricultural and Bioresource Engineering, University of Saskatchewan, 57 Campus Drive, Saskatoon, Saskatchewan, Canada, S7N 5A9.
[1]*Faculté de Pharmacie, Laboratoire de Pharmacognosie, 1 rue des Louvels, 80037 Amiens Cédex 1, France.*

INTRODUCTION

The *Ginkgo biloba* tree is the sole representative of the Ginkgoales order (Engler, 1954) and its origins trace back to the Jurassic period. Parts of this unique tree have been used as a phytotherapeutic agent since at least 1300 AD (Braquet, 1988; Michel and Hosford, 1988). This tree produces a wide range of secondary metabolites (Boralle *et al.*, 1988), among which, the terpene trilactones ginkgolide A, B, C, J and M further abbreviated as G-A, G-B, G-C, G-J and G-M, respectively.

Bilobalide is the major terpene of leaves (Nakanishi *et al.*, 1971); it has also been detected in roots (van Beek, unpublished). Ginkgolides have been detected in roots and leaves (*see* Nakanishi, 1988); their ecological significance as well as their site of synthesis within the *G. biloba* tree has yet to be resolved. Seasonal variation of their concentration has been reported (van Beek and Lelyveld, 1992; Flesch *et al.*, 1992). Ginkgolides have significant pharmacological properties (Braquet *et al.*, 1987). Although they have been chemically synthesized (Corey *et al.*, 1988), their commercial supply can only arise from biomass sources.

Ginkgolides have been detected in undifferentiated cell cultures (Carrier *et al.*, 1991; Huh and Staba, 1993; Jeon *et al.*, 1995). However, the low amounts of secondary metabolite obtained from tissue culture, from ng gdw^{-1} (Carrier *et al.*, 1991) to µg gdw^{-1} (Jeon *et al.*, 1995), preclude the commercial use of such methods when compared to the amounts of ginkgolides measured in the leaves of the tree: mg gdw^{-1} (Flesch *et al.*, 1992). Cellular differentiation is a factor which influences secondary metabolite synthesis. Galewsky and Nessler (1986) have reported that thebaine was not detected in *Papaver somniferum* undifferentiated cell cultures, but was detected in their somatic embryos. It is possible that *G. biloba* embryogenic cell cultures could lead to an increase in secondary metabolite production from such *in vitro* grown biomass.

In additon to knowledge on *G. biloba* differentiated cell cultures, an understanding of the ginkgolide biosynthetic pathway would, most likely, facilitate their production in an *in vitro* environment. Through their astute pioneering work, Nakanishi and Habaguchi (1971) have established that ginkgolides were of

81

terpenoid nature. Gaining further knowledge about the major biosynthetical steps, from carbohydrates to ginkgolides, would enable the understanding of the parameters affecting and controlling their biosynthesis. Once these controlling steps are understood, productive *in vitro* cultivation strategies could be contemplated.

In this work, a compilation of the most recent results pertaining to differentiated and undifferentiated *G. biloba* cell cultures is presented.

BIOTECHNOLOGICAL APPROACH

Cultivation of Undifferentiated Cells

Modifications of ammonium / nitrate medium composition ratio

Ginkgolides were detected in embryo-derived *G. biloba* cells cultivated in shake flasks and 6 L bioreactors using Murashige and Skoog (1962) medium (MS) (Carrier *et al.*, 1994). The detected quantities were extremely low, ranging in the ng gdw^{-1}. In an attempt to increase ginkgolide production, modifications were made to MS with respect to the ammonium / nitrate ratio. As a control, cells were first grown for 34 days in MS, where the ammonium / nitrate ratio was 1:2. Nutrient uptake was as previously described (Carrier *et al.*, 1991). The lyophilised cells were analysed, and yielded 14.7 and 12.3 ng gdw^{-1} of G-A and G-B, respectively. Cells were also cultured in a MS based medium modified with an ammonium / nitrate ratio of 1:1 (omission of KNO_3). Cells were cultured for 34 days, and were pale green and sometimes yellowish. Under these conditions, the cells did not entirely consume their extracellular glucose, where 5 g L^{-1} remained at the end of the cultivation cycle. Of the 20 mM of nitrate initially present, 3 mM were detected after 34 days. No traces of either G-A or G-B were detected in the lyophilised extracted biomass. Cells cultivated in an MS medium with an ammonium / nitrate ratio of 0:1 (omission of NH_4NO_3) were emerald green. Carbohydrates were entirely consumed by day 32, following the pattern of the control cultures. At the end of the culture period, 3 mM of extracellular nitrate was detected. The ginkgolide content of the lyophilised biomass was 2.6 and 7.2 ng gdw^{-1} of G-A and G-B, respectively. The highest concentration of ginkgolide was found in cells cultivated in MS. These results agree with those presented by Jeon *et al.* (1995) where the maximum G-B production was obtained in media with an ammonium / nitrogen ratio of 1:2 and 1:3.

Kinetics of cells cultured in Murashige and Skoog medium

The kinetics of ginkgolide production in MS in cells cultured in shake flasks and in a 2 L immobilisation bioreactor (Archambault *et al.*, 1990) cultures was evaluated. Cultivation vessels, either shake flasks or bioreactors, were simultaneously inoculated, but dismantled at different times throughout the cultivation period. The lyophilised biomass, either from the shake flask or from the bioreactor cultures, was taken for ginkgolide analysis. Fig. 1 presents biomass and ginkgolide concentrations as well as phosphate consumption of shake flask cultivated cells. Ginkgolides were detected concomitantly with dry weight increases. Extracellular phosphate depletion and the onset of ginkgolide biosynthesis occurred simultaneously. The onset of their production began with the depletion of extracellular phosphate, and peaked on day 21. The increase in extracellular phosphate concentrations can most likely be accounted for by release of phosphate during cell lysis.

Fig. 2 presents biomass and ginkgolide concentrations as well as phosphate consumption of cells cultivated in 2 L immobilization bioreactors. Bioreactors dismantled on days 20, 22, 27 and 29 yielded biomass concentrations of about 13 gdw L^{-1} and the maximum biomass production, of about 16 gdw L^{-1}, was obtained on day 31 and on day 34. Ginkgolides were first detected when phosphate was depleted, corresponding to the bioreactor dismantled on day 22. This possibly indicates that production of ginkgolides may commence when cells cease their division and initiate their increases in mass. Results presented by Jeon et al. (1995) also show that the maximum amount of ginkgolides produced corresponded to the middle of the exponential growth phase. Pépin et al. (1995) demonstrated that cultured plant cells undergo two distinct types of growth: a period of high cell division followed by a period of cell mass increase. The combined results of Carrier (1992), Pépin et al. (1995) and Jeon et al. (1995) show that there is a possible link between the offset of the high cell division period and ginkgolide production. When phosphate is depleted from the cultivation medium, cells may halt division and initiate secondary metabolite production.

Large scale undifferentiated Ginkgo cell cultures are not the method of choice to industrially produce ginkgolides. However, differentiated cultures may possibly be more amenable to such practices. Nakanishi and Habaguchi (1971) have detected ginkgolides in zygotic G. biloba embryos; hence, embryogenic Ginkgo cell cultures may yield higher ginkgolide concentrations than undifferentiated cell cultures.

Ginkgo biloba Embryogenesis

Ginkgo biloba haploid tissue culture was first reported by Tulecke in the early 1950's (Tulecke, 1997, and references therein). Yates (1986) investigated Ginkgo embryogenesis from mature zygotic embryos, and was successful in obtaining callus formation. Laurain et al. (1993a, 1993b, and 1996) obtained embryogenesis in Ginkgo from male (1993a) and female (1993b) tissues on one hand and from immature zygotic embryos on the other hand (1996). Variabilising methodologies, such as cellular clonings and methodologies capable of leading to embryogenesis, have been used and are presented below.

Type of explants leading to an embryogenesis

As research on somatic embryogenesis progressed, it became apparent that immature zygotic embryos seemed to be most suitable for initiating embryogenic cultures in both angiosperms (Vasil, 1987) and gymnosperms (Attree and Fowke, 1993). Recently, immature Ginkgo zygotic embryos were shown to be suitable for initiating somatic embryogenesis (Laurain et al., 1996). It was determined that the stage of

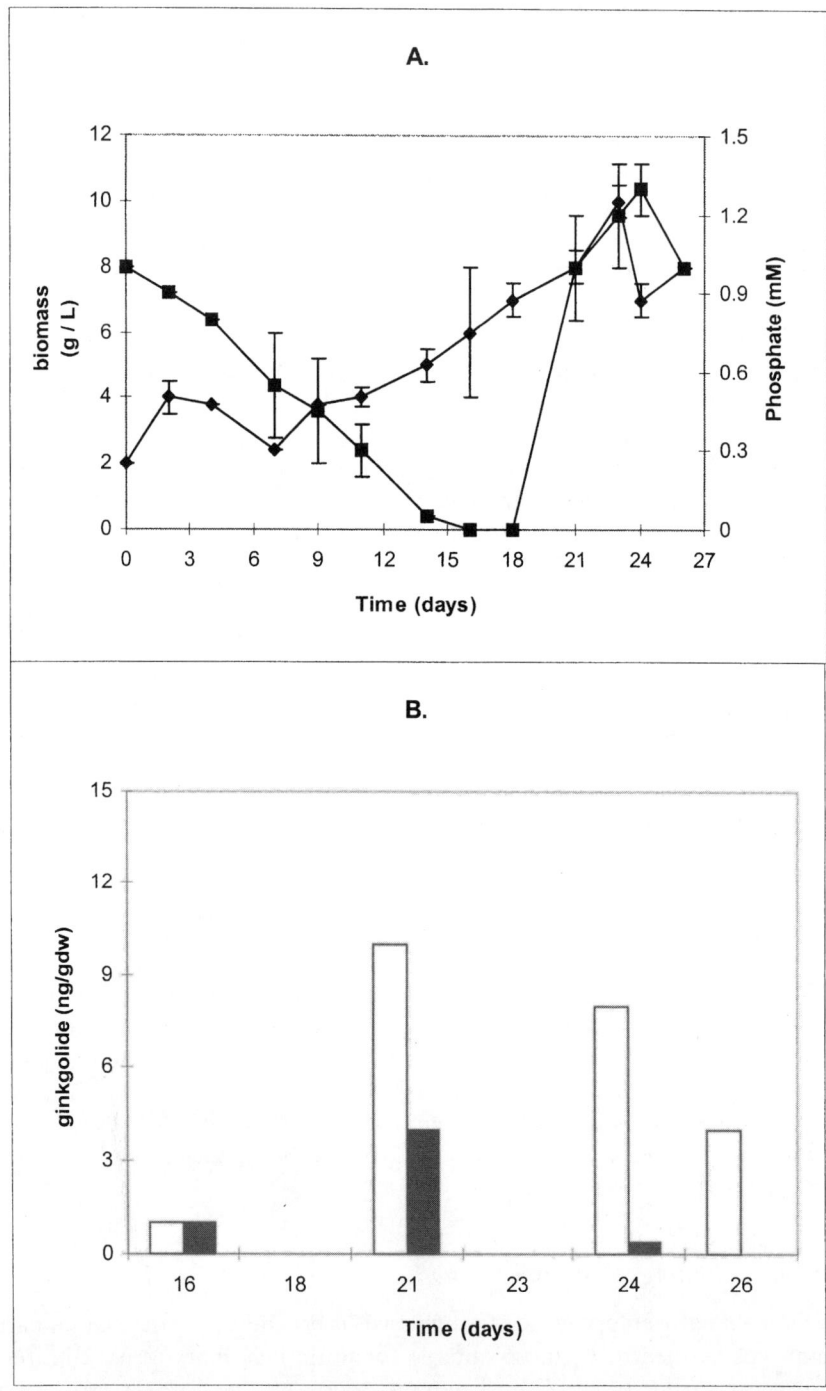

Figure 1 Embryo-Derived *G. biloba* Cells Cultivated in Shake Flasks. **A.** Biomass (◆) and phosphate (■). **B.** G-A (□) and G-B (■).

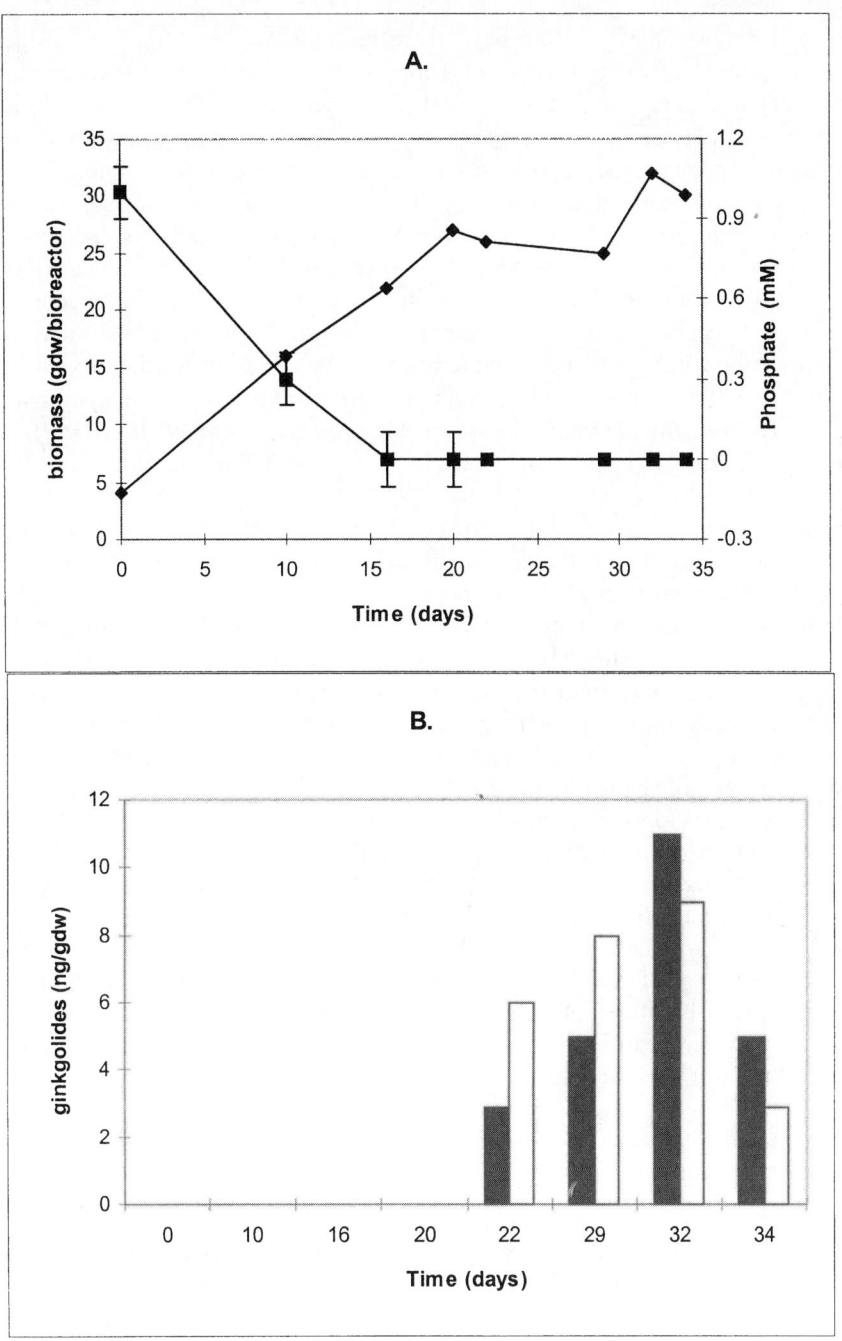

Figure 2   Embryo-Derived *G. biloba* Cells Cultivated in a 2 L Immobilization Bioreactor.
A. Biomass (◆) and phosphate (■). **B.** G-A (□) and G-B (■).

development of the immature zygotic embryo influenced the rate of induction of embryogenic lines.

The four stages of zygotic embryo development are: globular, torpedo, pre-cotyledonary and early cotyledonary. Of these four developmental periods, the pre-cotyledonary and early cotyledonary stages, characterized by the initiation and the development of cotyledons, respectively, were suitable for the initiation of embryo-genic tissue. As expected, cotyledonary stage embryos were longer than pre-cotyledonary stage embryos with lengths of 3.6 mm and 2.2 mm, respectively. Globular, torpedo, precotyledonary and cotyledonary stage embryos led to cellular proliferations of 5%, 69%, 98% and 99%, respectively. These frequencies included both embryogenic tissue and callus, of which the proportions varied according to the hormonal composition of the induction media. These results indicated that young cotyledonary stage embryos were found to be most appropriate for induction of *G. biloba* embryogenic tissues. Similar results were obtained for gymnosperms: *Picea glauca* (Lu and Thorpe, 1987), *Pinus elliottii* (Jain *et al.*, 1989), *Larix occidentalis* (Thompson and von Aderkas, 1992), and *Pinus patula* (Jones *et al.*, 1993).

Interest in embryogenic callus or direct embryogenesis, obtained from haploid tissue, is justified by the production of homozygous plants and by the isolation of mutants producing large quantities of medicinal substances (Sangwan-Norreel *et al.*, 1986). In Ginkgo, direct embryogenesis has been obtained from both male and female gametophytes. The rate of induction of haploid embryogenic tissue was limited by the degree of maturation of the explant material. For male Ginkgo gametophytic material, only microspores isolated at the uninucleate stage lead to embryogenesis (Laurain *et al.*, 1993a), forming embryogenic clusters that enabled the establishment of a pure male cell line (Trémouillaux-Guiller *et al.*, 1996). In Graminae, the developmental stage of the explant was also related to the success rate of the establishment of embryogenic clusters (Coumans *et al.*, 1989).

For the embryogenesis derived from female Ginkgo (Laurain *et al.*, 1993b), the prothallus used was characterized by a large cavity, which occupied its centre. At this stage of development, the prothallus was either at the coenocytic-stage or at the early septation-stage (Favre-Duchartre, 1956). Only such Ginkgo prothallus could produce haploid protoplasts, characterized by an ovoid hollow bag reaching a length of 0.5 to 3 mm. In France, ovules presenting such prothallus can be harvested from mid May to the end of June.

Somatic embryogenesis

Immature zygotic embryos were cultured on solid Murashige and Tücker (1969) medium prepared either at half strength (MT/2) or at full strength (MT) (Laurain *et al.*, 1996). The induction of embryogenic tissue was obtained by supplementing MT/2 or MT with either solely 10 μM Benzylaminopurine (BA) or with 5 μM (BA) combined with 5, 10 or 20 μM Naphthaleneacetic acid (NAA) (Table 1). After two weeks of culture, direct embryogenesis was observed on hypertrophic cotyledons cultivated on MT or MT/2 induction medium supplemented with solely 10 μM of

Table 1 Induction media used for obtaining somatic embryogenesis from *Ginkgo biloba* immature zygotic embryos (Laurain *et al.*, 1996).

Induction media (µM)	Development media (µM)	Mean number of somatic embryos per embryogenic explant (after 4 months)
MT/2 NAA(5) BA(5)	MT/2 BA(10)	2.6
MT/2 NAA(5)BA(5)	MT/2	0.8
MT/2 NAA(10) BA(5)	MT/2 BA(10)	1.4
MT/2 NAA(10) BA(5)	MT/2	2.2
MT/2 NAA(20) BA(5)	MT/2 BA(10)	0
MT/2 NAA(20) BA(5)	MT/2	1.7
MT/2 BA(10)	MT/2 BA(10)	9.6
MT/2 BA(10)	MT/2	5.3
MT NAA(5) BA(5)	MT BA(10)	1.7
MT NAA(5) BA(5)	MT	0.4
MT NAA(10) BA(5)	MT BA(10)	1.2
MT NAA(10) BA(5)	MT	0.9
MT NAA(20) BA(5)	MT BA(10)	0
MT NAA(20) BA(5)	MT	3.4
MT BA(10)	MT BA(10)	3.6
MT BA(10)	MT	6.5

MT: Murashige and Tücker (1969); MT/2: half strength mineral MT medium; BA: Benzylaminopurine; NAA: Naphthaleneacetic acid. Numbers in parenthesis indicate the µM concentration of the growth regulator.

BA. The latter combination yielded, on average, 9.6 somatic embryos per explant. This was the highest quantity of embryos generated in this study (Table 1). Unfortunately, direct embryogenesis was not observed for the majority of treatments; and, transfers on fresh MT/2 and MT media, devoid of or supplemented with 10 µM of BA, were necessary to obtain somatic embryos *via* indirect embryogenesis. Such embryos developed to the cotyledonary stage; however, they were unable to complete their development when isolated from their tissue of origin, and plated on MT medium devoid of growth regulators. As for the developing embryos, cultures could only be maintained for six months; after this time, they showed browning and necrosed cells were observed.

These results showed that immature zygotic *G. biloba* embryos could be induced to produce embryogenic material, using MT medium supplemented with solely BA. Cytokinins were also instrumental in inducing embryogenesis in the Abies genera (Schuller *et al.*, 1989; Norgaard and Krogstrup, 1991). In contrast to Abies, some varieties of woody species required exogenous auxins for the induction of somatic embryogenesis (Lu and Thorpe, 1987; von Arnold, 1987; Woods *et al.*, 1992). For Ginkgo, direct or indirect embryogenesis was observed depending on the nature of the exogenous growth regulators used in the induction medium. On hypertrophic cotyledons, direct embryogenesis was induced only in the presence of BA, while indirect embryogenesis seemed to be correlated with the presence of exogenous auxins. Embryogenic induction frequencies in the order of 90–95% were obtained, showing that Ginkgo is amendable to this technique. These results were high in comparison to those reported for *Pinus patula* (2.6%) (Jones *et al.*, 1993), *Larix occidentalis* (3%) (Thompson and von Aderkas, 1992), and *Pinus purgeus* (16%) (Afele *et al.*, 1992).

Gametophytic embryogenesis

Cell cultures stemming from both isolated uninucleate microspores and from protoplasts, isolated either from female prothallus or from microspore-derived cell lines, were amendable to the formation of embryos *via* direct embryogenesis.

G. *biloba* uninucleate microspores and protoplasts, isolated from female gametophytes, were cultured in Bourgin and Nitsch (1967) (BN) liquid medium devoid of growth regulators, but supplemented with coconut milk. The microspores were also cultured in BN medium supplemented with coconut milk and with Indoleacetic acid (IAA) and Kinetin (KIN). Protoplasts, isolated from female gametophytes, were also cultured in MT liquid medium devoid of ammonium ions; and supplemented with glutamine, BA and NAA. Protoplasts isolated from a microspore-derived cell line were cultured in either BN, MT/2, or Gamborg *et al.* (1976) (B5) supplemented with combinations of NAA, IAA, KIN, BA or Dichlorophenoxyacetic acid (2,4-D). The reader is referred to Trémouillaux-Guiller *et al.* (1996) and Laurain *et al.* (1993b) for exact media compositions.

Microspores as well as protoplasts isolated from female prothallus and from microspore-derived cell suspensions exhibited various modes of development before showing embryogenesis. In these three types of cultures, the gametophytic embryos were produced either from a unicellular microclone by endomitosis or from a microclone by typical divisions. A particularity, observed during embryogenesis from isolated microspores, was the formation of fourfold asymmetrically celled microspores. These asymmetrical divisions were comparable to those observed during the *in vivo* development of the pollen grain with the II prothallial cell, reproductive cell and germinal cell formation, while the I prothallial cell rapidly degenerated. The germinal cell increased in volume and in length, developing a pollen tube of variable length (up to 480 micron long), and occasionally provided rhizoid structures. Similarly, microspore-derived protoplasts produced long cytoplasmic extensions comparable to pollen tubes at the beginning the of culture cycle (Trémouillaux-Guiller *et al.*, 1996). The celled microspores could evolve to embryogenesis by ejecting enveloped nuclei into the extracellular medium through a phenomenon that is called inverted endocytosis (Laurain *et al.*, 1993b). Nuclei expulsions, producing small cells leading to microclones, have also been observed during cultivation of female haploid protoplast and microspore-derived cell suspension protoplasts.

In the presence or in the absence of exogenous growth regulators, microspores, female protoplasts and microspore-derived protoplasts isolated from G. *biloba* divided and formed microclones, which directly evolved into embryos.

The formation of pro-embryos, from microspore cells cultivated for four weeks, was observed in the presence or the absence of growth regulators. After ten weeks of culture, embryogenesis was observed with an efficiency factor (defined as the number of embryos divided by the number of microspores originally plated) of 1.8%. After cultivation for a period of four to five months in liquid media, slow growing embryo-clusters and embryos, isolated at different stages of development, were transferred onto BN solid media devoid of growth regulators or containing various combinations of KIN and IAA. Embryo-clusters showed signs of growth particularly when cultured on BN solid medium supplemented with 11.4 μM KIN; after one month, some of these embryo-clusters developed, exhibiting a club shape. Cell suspensions, established from these embryogenic clusters, produced embryogenic cells in MT liquid medium devoid of growth regulators or supplemented with 18.8 μM NAA (Trémouillaux-Guiller et al., 1996). Protoplasts were isolated from a six-day-old subculture of these cell suspensions. After two weeks of culture, microcolonies derived pro-embryos were obtained in the media preparations shown in Table 2 (MT, B5, and BN liquid media devoid of growth regulators or supplemented with various combinations of growth regulators). After four weeks of cultivation, the efficiency of pro-embryo and embryo formation ranged from 0.6% to 4%, depending on the media tested (Table 2).

Table 2 Efficiency of embryo formation obtained from female protoplast culture, protoplasts isolated from a microspore-derived cell suspension and microspore culture of *Ginkgo biloba* (Laurain et al. (1993b); Trémouillaux-Guiller et al. (1996)).

Female protoplasts		Protoplasts derived from a male cell line		Microspore culture	
Medium	Frequency of embryo formation after 3 months (%)	Medium	Frequency of pro-embryo and embryo formation after 4 weeks (%)	Medium	Frequency of embryo formation after 10 weeks (%)
MT' (1)	1	MT	1	–	–
MT' (2)	1.9	MT (1)	1.2	–	–
MT' (3)	0.7	MT (2)	1.2	–	–
MT' (4)	1.5	MT (3)	1.2	–	–
–	–	MT (4)	0.8	–	–
–	–	MT (5)	0.6	–	–
BN	0.2	BN	4	BN	1.8
–	–	BN (2)	2	BN (1)	1.8
–	–	BN (3)	3	–	–
–	–	B5	1.4	–	–
–	–	B5 (1)	1.5	–	–

MT: Murashige and Tücker (1969); MT': Murashige and Tücker (1969) basal medium modified by omitting NH_4NO_3; BN: Bourgin and Nitsch (1967); B5: Gamborg et al. (1976). The concentration of growth regulator is given in μM. MT (1): NAA (10.74); MT (2): NAA (10.74), KIN (0.93); MT (3): NAA (2.69), BA (8.87); MT (4): BA (8.87); MT (5): IAA (11.42), KIN (0.93); BN (2): IAA (11.42), KIN (0.46); BN (3): IAA (17.13), KIN (0.46); B5 (1) 2,4-D (0.45); MT' (1): NAA (26.85), BA (4.44); MT' (2): NAA (16.11), BA (4.44); MT'(3): NAA (10.74), BA (4.44); MT' (4): NAA (5.37), BA (4.44).

Between day 45 and day 60 of female protoplast culture, pro-embryos and embryos appeared in every Petri dish. Embryogenesis was observed with frequencies between 0.7% to 1.9% when using modified MT (without ammonium) liquid medium and of 0.2% when using BN liquid medium (Table 2). Most embryos had reached the oblong stage while some heart-shaped embryos were also present in the cultures. After three months, the number of embryos ranged from 165 to 1900 embryos mL^{-1} depending on the culture medium used and decreased to 50 to 470 embryos mL^{-1} one month later (Laurain et al., 1993b). The decline in embryo number could be related to browning which appeared after the addition of fresh medium. However, after five months of culture, a supply of fresh medium was no longer detrimental to the embryos. As for the embryos derived from the microspore cultures, embryos obtained from protoplasts of male and female gametophytes exhibited slow growth, rendering the transfer to solid media impossible.

These results show that isolated microspores and protoplasts isolated from both microspore-derived cell suspensions and from female prothallus did not require exogenously supplemented growth regulators for their culture. G. biloba female protoplasts produced new cell walls and divided in BN medium supplemented solely with coconut milk. G. biloba cell suspension-derived protoplasts divided in BN, MT and B5 with or without growth regulators. This result contrasts with the work of Zrÿd (1988) and references therein, but is similar to that observed by Bekkaoui et al. (1987) and Trémouillaux-Guiller et al. (1987).

Growth in media devoid of growth regulators may be due to the high endogenous levels of growth regulators. Similar results were observed during Zea mays androgenesis (Mitchell and Petolino, 1991; Coumans et al., 1989), and during Brassica campestris microspore-derived embryogenesis (Sato et al., 1989). It is possible, however, that the coconut milk, supplemented to the BN medium, contained cytokinins which acted as a supply, and contributed to the initiation of embryogenesis in G. biloba microspore and female prothallus protoplasts cell lines. Sangwan-Norreel et al. (1986) have reported that low levels of growth regulators, such as in coconut milk, stimulated increases in the number of gametophytic plants. This contrasts with Keller and Stringam (1978) who have shown that cytokinins were essential for maximum microspore response in Datura and Solanum.

It is well known that the application of electrical currents stimulates cellular division and differentiation, in particular for protoplasts of woody species regarded as recalcitrant to culture (Chand et al., 1988). Combinations of electric field (50 to 1000 V cm^{-1}), pulse-duration (20, 30 and 100 μs) and pulse-number (1 or 3 pulses) were applied to G. biloba microspores (Laurain et al., 1993a). The cells survived these high voltage treatments. After 24 hours of culture, the viability of the control and of the electrostimulated microspores was similar with values of 41%. Electrostimulated microspore cells showed embryogenesis earlier than that of the control. After four months of culture, embryos appeared in the electrostimulated microspore cultures, while only pro-embryos were present in the control cultures. These results showed that electrical pulses applied to G. biloba microspores cultures stimulated differentiation similar to the application of electrical currents on protoplasts (Davey and Power, 1987).

Ginkgolide content of various cell suspensions

Recently, Laurain *et al.* (1997) have established cell cultures from various *G. biloba* explants, of which, two were found to contain ginkgolides. The first one was established from roots, which developed from zygotic embryos inoculated with *Agrobacterium rhizogenes* A_4. The second cell suspension was derived from female prothallus. The combined amounts of bilobalide, G-A, G-B, G-C and G-J were determined to be 650 μg gdw^{-1} in the putatively transformed suspension culture and 870 μg gdw^{-1} in the female prothallus derived suspension. These results indicated that synthesis of bilobalide and ginkgolide in cell cultures is dependent on the nature of the initial explant, and that transformation with *A. rhizogenes* A_4 is favourable for the production of *G. biloba* terpene trilactones. Shunan *et al.* (1997) reported the appearance of hairy roots on Ginkgo leaves infected with *A. rhizogenes* A_4. Terpene trilactones were not quantified; however, these cultures were successfully grown in 2.5 L to 5 L flasks, indicating scale-up possibilities.

Ginkgolides and Bilobalide

Ginkgolide and bilobalide biosynthesis

Through their astute pioneering work, Nakanishi and Habaguchi (1971) established that ginkgolides were of terpenoid nature, assuming that they were formed following the classical acetate-pathway, namely through the biosynthesis of mevalonic acid and the primary terpenes (geranyl (GPP), farnesyl (FPP) and geranylgeranyl pyrophosphate (GGPP)). With the work of Schwarz (1994), important advances were made in the elucidation of the biosynthetical pathway of ginkgolide and bilobalide. This work has shown that, in *G. biloba*, parallel and mechanistically distinct biosynthetic pathways govern the formation of IPP. The IPP units required for the formation of sitosterol would be formed from the acetate-mevalonic acid steps, occurring in the cytosol. Surprisingly, ginkgolides and bilobalide would be synthesized from IPP units arising from a new pathway, namely the "triose/pyruvate"-pathway, which would possibly be located in the plastid. Schwarz (1994) suggested that bilobalide would not be a genuine sesquiterpene, but would be derived from the ginkgolide biosynthetical pathway. Carrier *et al.* (1996) reported that undifferentiated *G. biloba* cell cultures showed mainly FPP synthase activity. These results combined with those of Schwarz (1994) possibly suggest that the plastid "triose/pyruvate" pathway is not very active in undifferentiated cultured cells, justifying the minute ginkgolide concentrations detected in such cultures (Carrier *et al.*, 1991).

Recently, Neau *et al.* (1997), pursuing the work of Schwarz (1994), have elegantly shown, with the use of inhibitors of cytochrome P-450 dependent oxygenases, that ginkgolide and bilobalide biosynthesis proceeds via dehydroabietane. A precursor product relationship would exist between dehydroabietane and ginkgolides, including bilobalide. For the most recent schematics of the biosynthetic pathway of ginkgolides and bilobalide, the reader is referred to Neau *et al.* (1997).

Note added in proof: very recently an updated article on ginkgolide biosynthesis appeared (Schwarz and Arigoni, 1999).

Site of biosynthesis of ginkgolides and bilobalide

Analysis of aerial portions of G. *biloba* plants revealed that ginkgolides and bilobalide were present in equal concentration ratios. Of the ginkgolides, G-A occurred at the highest concentration, followed by relatively equal amounts of G-B and G-C (Flesch *et al.*, 1992; van Beek and Lelyveld, 1992; Huh and Staba, 1993; Cartayrade *et al.*, 1997; Carrier *et al.*, 1998). Interestingly, the underground portions of the plant, were practically free of bilobalide, and yielded in decreasing order of concentration, G-A, G-C, G-B, and G-J (Flesch *et al.*, 1992; Carrier *et al.*, 1998) (Table 3). This contrasted with G-C being detected as the most abundant ginkgolide in root bark of Gingko trees with trunks exceeding 30 cm in diameter (Nakanishi, 1988). The age of the plants extracted could possibly account for the differences in the ginkgolide concentration distribution patterns.

In an attempt to shed light on the possible site of ginkgolide biosynthesis, cell free extracts prepared from stem and root bark, root and root meristem, as well as terminal buds, rosettes and side branches of three year old G. *biloba* plants were prepared (Carrier *et al.*, 1998) (Table 4). The formation of farnesyl and geranylgeranyl pyrophosphate, were monitored by the incorporation of $[1-^{14}C]$-isopentenyl pyrophosphate ($[1-^{14}C]$-IPP) by these cell free extracts. FPP and GGPP were detected as the corresponding alcohols: farnesol and geranylgeraniol, respectively. Stem bark, terminal leaf as well as root and root meristem cell free extract preparations did incorporate the labeled substrate into terpenoid product(s). However, rosette leaf cell free extracts incubated with $[1-^{14}C]$-IPP did not yield terpenoid products (Carrier *et al.*, 1998).

Extracts from the terminal bud region contained high terpene concentrations and showed GGPP and FPP synthase activity. The production of GGPP in these growing tissues may be necessary for biosynthesis, of among others, gibberellic acid, carotenoids, the phytol chain of chlorophyll and diterpenes (Gray, 1987; Kleinig, 1989). Similarly, enzymatic activity of FPP synthase may be necessary for the synthesis of sterol derived constituents.

Interestingly, the concentration of Ginkgo terpene trilactones in leaves has been documented to increase throughout the summer season (Flesch *et al.*, 1992; van Beek and Lelyveld, 1992; Huh and Staba, 1993). In the work reported by Carrier *et al.* (1998), rosette leaves contained high concentrations of terpene trilactones, but yielded crude enzyme preparations incapable of incorporation of the labeled substrate. Results obtained with Capsicum (Kutz *et al.*, 1992) have shown that the activity of GGPP synthase and its mRNA expression varied as a function of plant development. Carrier *et al.* (1998) suggested that the high terpene concentration in rosette leaves could possibly be explained by the following or a combination of the following: (a) synthesized elsewhere in the plant and transported, (b) synthesized when the tissues were actively growing and then sequestered or (c) obtained from the catabolism of other moieties also derived from GGPP. Recently, Cartayrade *et al.* (1997), through intricate labeling experiments, shed light on the site of ginkgolide synthesis within the G. *biloba* plant. In distinct $^{14}CO_2$ and $(U-^{14}C)$ glucose labeling experiments, labeled ginkgolides were detected in roots before being traced in the stems and the leaves. Examination of the distribution pattern of the

Table 3 Bilobalide, G-A, G-B, G-C and G-J and total terpene content with standard deviation of various parts obtained from 3 year old G. biloba plants. Contents are expressed as mg terpene 100 mg^{-1} dry wt (Carrier et al., 1998). Numbers in italic represent standard deviations.

Plant part	Bilobalide	G-A	G-B	G-C	G-J	Total
Terminal	0.22 ± 0.04	0.18 ± 0.01	0.09 ± 0.02	0.09 ± 0.02	0.03 ± 0.00	0.61 ± 0.09
Side branch	0.16 ± 0.06	0.13 ± 0.01	0.06 ± 0.00	0.07 ± 0.01	0.01 ± 0.01	0.43 ± 0.09
Rosette	0.27 ± 0.12	0.19 ± 0.04	0.08 ± 0.01	0.08 ± 0.00	0.02 ± 0.01	0.64 ± 0.18
Root Meristem	0.00 ± 0.00	0.13 ± 0.01	0.05 ± 0.01	0.06 ± 0.01	0.03 ± 0.01	0.27 ± 0.04
Root	0.01 ± 0.02	0.13 ± 0.07	0.05 ± 0.02	0.05 ± 0.02	0.02 ± 0.01	0.26 ± 0.14
Root bark	0.00 ± 0.00	0.02 ± 0.02	0.01 ± 0.01	0.02 ± 0.00	0.01 ± 0.01	0.06 ± 0.04
Stem bark	0.01 ± 0.00	0.01 ± 0.01	0.00 ± 0.00	0.02 ± 0.00	0.01 ± 0.01	0.05 ± 0.02

Table 4 Formation of terpenoid products, monitored as the corresponding alcohols (Farnesol and Geranylgeraniol) in cell free extracts prepared from various parts obtained from 3 year old *Ginkgo biloba* plants (Carrier *et al.*, 1998).

	IPOH	GGOH	FOH
Terminal bud	416.5 ± 95.6	164.0 ± 56.6	32.0 ± 11.3
Side branch	717.3 ± 210.1	57.3 ± 11.5	33.3 ± 15.0
Rosette	748.5 ± 283.6	0 ± 0	0 ± 0
Root meristem	717.0 ± 35.4	12.0 ± 17.0	0 ± 0
Root	788.0 ± 152.7	24.0 ± 33.9	0 ± 0
Root bark	526.0 ± 195.2	22.5 ± 12.0	186.5 ± 78.5
Stem bark	531.5 ± 50.2	0 ± 0	73.0 ± 38.2

Numbers represent the peak area (counts per minute (CPM)/mg of protein). Numbers in italic represent standard deviations. IPOH = isopentenol, GGOH = geranylgeraniol, FOH = farnesol.

label, within a 27 day time frame, indicated that the terpenes were translocated from the roots to the stems and the leaves. Cartayrade *et al.* (1997) concluded that the aerial portion of the plant acted as a ginkgolide sink for the diterpene producing underground portions.

CONCLUSIONS AND PERSPECTIVES

In conclusion, undifferentiated *G. biloba* cell cultures can easily be scaled-up from shake flask to bioreactors; however, they produce trace amounts of ginkgolides, and are uninteresting from an industrial perspective. Higher ginkgolide yields can be obtained from differentiated cultures. The knowledge of the ginkgolide site of synthesis, being that of the root, combined to the expertise in differentiated cell culture techniques leads to interesting and exciting industrial prospects. However, many barriers remain to be crossed before this becomes reality. In the short term it is now imperative to further investigate cultivation protocols that pertain with root development.

REFERENCES

Afele, J.C., Senaratna, T., McKersie, B.D., Saxena, P.K. (1992) Somatic embryogenesis and plant regeneration from zygotic embryo culture in blue spruce (*Picea pungens* Engelman). *Plant Cell Rep.*, **11**, 299–303.

Archambault, J., Volesky, B., Kurz, W. (1990) Development of bioreactors for the culture of surface immobilized plant cells. *Biotechnol. Bioeng.*, **35**, 702–711.

von Arnold, S. (1987) Improved efficiency of somatic embryogenesis in mature embryos of *Picea abies* L. Karst. *J. Plant Physiol.*, **128**, 233–244.

Attree, S.M., Fowke, L.C. (1993) Embryogeny of gymnosperms: advances in synthetic seed technology of conifers. *Plant Cell Tissue Organ Cult.*, **35**, 1–35.

van Beek, T.A., Lelyveld, G.P. (1992) Concentration of ginkgolides and bilobalide in *Ginkgo biloba* leaves in relation to the time of year. *Planta Med.*, **58**, 413–416.

Bekkaoui, F., Saxena, P.K., Attree, S.M., Fowke, L.C., Dunstan, D.I. (1987) The isolation and culture of the protoplasts from an embryogenic cell suspension culture of *Picea glauca* (Moench) Voss. *Plant Cell Rep.*, 6, 476–479.

Boralle, N., Braquet and P., Gottlieb, O. (1988) *Ginkgo biloba*; a review of its chemical composition. In P. Braquet (ed.), *Ginkgolides – Chemistry, Biology, Pharmacology and Clinical Perspectives, Volume 1*, J.R. Prous Science Publishers, Barcelona, pp. 9–26.

Bourgin, J.P., Nitsch, J.P. (1967) Obtention de Nicotiana haploïdes à partir d'étamines cultivées *in vitro*. *Ann. Physiol. Vég.*, 9, 377–382.

Braquet, P. (1988) The ginkgolides. From Chinese pharmacopeia to a new class of pharmacological agents: The antagonists of platelet-activating factor. In P. Braquet (ed.), *Ginkgolides – Chemistry, Biology, Pharmacology and Clinical Perspectives, Volume 1*, J.R. Prous Science Publishers, Barcelona, pp. XV-XXXIV.

Braquet, P., Touqui, L., Shen, T., Vargaftig, B. (1987) Perspectives in platelet-activating factor research. *Pharmacol. Rev.*, 39, 97–145.

Carrier J., Chauret N., Coulombe P., Mancini M., Neufeld R., Weber M., Archambault J. (1991) Detection of ginkgolide A in *Ginkgo biloba* cell cultures. *Plant Cell Rep.*, 10, 256–259.

Carrier, D.J. (1992) Immobilization and ginkgolide production of *Ginkgo biloba* cells. PhD thesis, McGill University, Canada.

Carrier J., Chauret, N., Neufeld, R. and Archambault, J. (1994) *Ginkgo biloba* L. (maiden hair Tree): *In Vitro* culture and the formation of ginkgolides. In Y. Bajaj (ed.), *Biotechnology in Agriculture and Forestry, Volume 26*, Springer-Verlag, Berlin, pp. 136–145.

Carrier, D.J., Archambault, J., van der Heijden, R., Verpoorte, R. (1996) Formation of terpenoid products in *Ginkgo biloba* L. cultivated cells. *Plant Cell Rep.*, 15, 888–891.

Carrier, D.J., van Beek, T.A., van der Heijden, R., Verpoorte, R. (1998) Formation of terpenoid products in *Ginkgo biloba* L. cultivated cells. *Phytochemistry*, 48, 89–92.

Cartayrade, A., Neau, E., Sohier, C., Balz, J.P., Carde, J.P., Walter, J. (1997) Ginkgolide and bilobalide biosynthesis in *Ginkgo biloba*. I: Sites of synthesis, translocation and accumulation of ginkgolides and bilobalide. *Plant Physiol. Biochem.*, 35, 859–868.

Chand, P.K., Occhatt, S.J., Rech, E.L., Power, J.B., Davey, M.R. (1988) Electroporation stimulates plant regeneration from protoplasts of the woody medicinal species *Solanum dulcamara* L. *J. Exp. Bot.*, 39, 1267–1274.

Corey, E., Kang, M., Desai, C., Ghosh, A., Houpis, I. (1988) Total synthesis of (±) ginkgolide B. *J. Am. Chem. Soc.*, 110, 649–651.

Coumans, M.P., Sohato, S., Swanson, E.B. (1989) Plant development from isolated microspores of *Zea mays* L. *Plant Cell Rep.*, 7, 618–621.

Davey, M.R. and Power, J.B. (1987) Aspects of protoplasts and plant regeneration. In K.J. Puite, J.J. Dons, H.J. Huizing, A.J. Kool, M. Koornneef and F.A. Krens (eds.), *Progress in Plant Protoplast Research*, Kluwer Academic Publishers, Dordrecht. pp. 15–25.

Engler, A. (1954) Syllabus der Pflanzenfamilien, Ed. 12, Band 1. Gebrüder Borntraeger, Berlin.

Favre-Duchartre, M. (1956) Contribution à l'étude de la reproduction chez le *Ginkgo biloba*. *Revue Cytol. Biol Vég.*, 17, 1–212.

Flesch, V., Jacques, M., Cosson, L., Teng, B., Pétiard, V., Balz, J. (1992) Relative importance of growth and light level on terpene content of *Ginkgo biloba*. *Phytochemistry*, 31, 1941–1945.

Galewsky, S., Nessler, C.L. (1986) Synthesis of morphine alkaloids during opium poppy somatic embryogenesis. *Plant Science*, 45, 215–222.

Gamborg, O.L., Murashige, T., Thorpe, T.A., Vasil, I.K. (1976) Plant tissue culture media. *In vitro*, 12, 473–478.

Gray J. (1987) Control of isoprenoid biosynthesis in higher plants. *Adv. Bot. Res.*, 14, 25–91.

Huh, H., Staba, J. (1993) Ontogenic aspects of ginkgolide production in *Ginkgo biloba*. *Planta Med.*, **59**, 232–239.

Jain, S.M., Dong, N., Newton, R.J. (1989) Somatic embryogenesis in slash pine (*Pinus elliottii*) from immature embryos cultured *in vitro*. *Plant Science*, **65**, 233–241.

Jeon, M., Sung, S., Huh, H., Kim, Y. (1995) Ginkgolide B production in cultured cells derived from *Gingko biloba* L. leaves. *Plant Cell Rep.*, **14**, 501–504.

Jones, N.B., Vanstaden, J., Bayley, A.D. (1993) Somatic embryogenesis in *Pinus patula*. *Plant Physiol.*, **142**, 366–372.

Keller, W.A. and Stringam, G.R. (1978) Production and utilization of microspore-derived haploïd plants. In T.A. Thorpe (ed.), *Frontiers of Plant Cell culture*,. IAPTC, Canada. pp. 113–122.

Kleinig H. (1989) The role of plastids in isoprenoid biosynthesis. *Annual Rev. Plant Physiol. Plant Molec. Biol.*, **40**, 39–59.

Kutz, M., Romer, S., Suire, C., Hugueney, P., Weil, J., Schantz, R., Camara, B. (1992) Identification of a cDNA for the plastid-located geranylgeranyl pyrophosphate synthase from *Capsicum annuum*: correlative increase in enzyme activity and transcript level during fruit ripening. *The Plant J.*, **2**, 25–34.

Laurain, D., Trémouillaux-Guiller, J. Chénieux, J.C. (1993a) Embryogenesis from microspores of *Ginkgo biloba* L., a medicinal woody species. *Plant Cell Rep.*, **12**, 501–504.

Laurain, D., Chénieux, J.C., Trémouillaux-Guiller, J. (1993b) Direct embryogenesis from female haploid protoplasts of *Ginkgo biloba* L., a medicinal woody species. *Plant Cell Rep.*, **12**, 656–660.

Laurain, D., Chénieux, J.C., Trémouillaux-Guiller, J. (1996) Somatic embryogenesis from immature zygotic embryos of *Ginkgo biloba*. *Plant Cell Tissue Organ Cult.*, **44**, 19–24.

Laurain, D., Trémouillaux-Guiller, J. Chénieux, J.C, van Beek T.A. (1997) Production of ginkgolide and bilobalide in transformed and gametophyte derived cell cultures of *Ginkgo biloba*, *Phytochemistry*, **46**, 127–130.

Lu, C.Y., Thorpe, T.A. (1987) Somatic embryogenesis and plantlet regeneration in cultured immature embryos of *Picea glauca*. *J. Plant Physiol.*, **128**, 297–302.

Michel, P. and Hosford, D. (1988) *Ginkgo biloba*: from "living fossil" to modern therapeutic agent. In P. Braquet (ed.), *Ginkgolides – Chemistry, Biology, Pharmacology and Clinical Perspectives, Volume 1*, J.R. Prous Science Publishers, Barcelona, pp. 1–8 .

Mitchell, J.C., Petolino, J.F. (1991) Plant regeneration from haploid suspension and protoplast cultures from isolated microspores of maize. *J. Plant Physiol.*, **137**, 530–536.

Murashige, T., Skoog, F. (1962) A revised medium for rapid growth and bioassay with tobacco tissue cultures. *Physiologia Plantarum*, **28**, 473–497.

Murashige, T. and Tücker, D.P. (1969) Growth factor requirements of Citrus tissues cultures. In J.V. Chapman (ed.), *Proceedings of the First International Citrus Symposium*, University of California, Riverside, pp. 1155–1161.

Nakanishi, K. (1988) Ginkgolides-Isolation and structural studies carried out in the mid 1960's. In P. Braquet (ed.), *Ginkgolides – Chemistry, Biology, Pharmacology and Clinical Perspectives, Volume 1*, J.R. Prous Science Publishers, Barcelona, pp. 27–36.

Nakanishi, K, Habaguchi, K. (1971) Biosynthesis of ginkgolide B, its diterpenoid nature and origin of the *tert*-butyl group. *J. Am. Chem. Soc.*, **93**, 3544–3546.

Nakanishi, K., Habaguchi, K., Nakadaira, Y., Woods, M., Maruyama, M., Major, R., Alauddin, M., Patel, A., Weinges, K., Bähr, W. (1971) Structure of bilobalide, a rare *tert*-butyl containing sesquiterpenoid related to the C_{20}-ginkgolides. *J. Am. Chem. Soc.*, **93**, 3546–3547.

Neau, E., Cartayrade, A., Balz, J.P., Carde, J.P., Walter, J. (1997) Ginkgolide and bilobalide biosyntheisis in *Ginkgo biloba*. II: Identification of a possible intermediate compound by using inhibitors of cytochrome P-450-dependent oxygenases. *Plant Physiol. Biochem.* **35**, 869–879.

Norgaard, J.V., Krogstrup, P. (1991) Cytokinin induced somatic embryogenesis from immature embryos of *Abies nordmanniana* LK. *Plant Cell Rep.*, **9**, 509–513.

Pépin M.F., Archambault J., Chavarie C., Cormier C. (1995) Growth kinetics of *Vitis vinifera* cell suspension cultures: 1. Shake flask cultures. *Biotechnol. Bioeng.*, **47**, 131–138.

Sangwan-Norreel, B.S., Sangwan, R.S., Pare, J. (1986) Haploïdie et embryogénèse provoquée *in vitro. Bull. Soc. Bot. France*, **4**, 7–39.

Sato, T., Nishio, T., Harai, M. (1989) Plant regeneration from isolated microspore cultures of Chinese cabbage (*Brassica campestris* spp. *pekinensis*). *Plant Cell Rep.*, **8**, 486–488.

Schuller, A., Reuther, G., Geiert, T. (1989) Somatic embryogenesis from seed explants of *Abies alba. Plant Cell Tissue Organ Cult.*, **17**, 53–58.

Schwarz, M. (1994) Terpen-Biosynthese in *Ginkgo biloba*. Eine überraschende Geschichte. PhD thesis, Eidgenössischen Technischen Hochschule Zürich, Switzerland (ETH Nr. 10951).

Schwarz, M. and Arigoni, D. (1999) Ginkgolide biosynthesis. In D.E. Cane (ed.) *Isoprenoids including carotenoids and steroids*, Vol. 2 in D. Barton, K, Nakanishi and O. Meth-Cohn (eds.) *Comprehensive Natural Products Chemistry*. Elsevier, Amsterdam, pp. 367–400.

Shunan, L., Tianen, S., Genbao, L. (1997) Transformation of Ginkgo hairy root and establishment of its suspension culture clone. *Wuhan Univ. J. Nat. Sci.*, **2**, 493–495.

Thompson, R.G., von Aderkas, P. (1992) Somatic embryogenesis and plant regeneration from immature embryos of western larch. *Plant Cell Rep.*, **11**, 379–385.

Trémouillaux-Guiller, J., Andreu, F., Crèche, J., Chénieux, J.C., Rideau, M. (1987) Variability in tissues cultures of *Choisya ternata*. Alkaloid accumulation in protoclones and aggregate clones obtained from established strains. *Plant Cell Rep.*, **6**, 375–378.

Trémouillaux-Guiller, J., Laurain, D. and Chénieux J.C. (1996) Microspore and protoplast culture in *Ginkgo biloba*. In S.M. Jain, S.K. Sopory and R.E. Veilleux (eds.), *In vitro haploid Production in Higher Plants, Volume 3*, Kluwer Academic Publishers, The Netherlands, pp. 277–295.

Tulecke, W. (1997) Tissue culture studies on *Ginkgo biloba*. In T. Hori, R.W. Ridge, W. Tulecke, P. Del Tredici, J. Trémouillaux-Guiller and H. Tobe (eds), *Ginkgo biloba a global treasure. From biology to medicine*. Springer, Heidelberg, pp. 141–156.

Vasil, I.K. (1987) Developing cell and tissue culture systems for the improvement of cereal and grass crops. *J. Plant Physiol.*, **128**, 193–218.

Woods, S.H., Phillips, G.C., Woods, J.E., Collins, G.B. (1992) Somatic embryogenesis and plant regeneration from zygotic embryo explants in mexican weeping bamboo, *Otatea acuminata aztecorum. Plant Cell Rep.*, **11**, 257–261.

Yates, W.F. (1986) Induction of embryogenesis in embryo-derived callus of *Ginkgo biloba* L. *Proceedings of the VIth International Congress of Plant Tissue and Cell Culture*, University of Minnesota, pp. 43.

Zrÿd, J.P. (1988) *Cultures de cellules, tissus et organes végétaux*. Presses Polytech Romandes Lausanne, pp. 1–308.

6. *GINKGO BILOBA* – LARGE SCALE EXTRACTION AND PROCESSING

JOE O'REILLY

Cara Partners, Little Island, Co. Cork, Ireland

DEVELOPMENT OF EXTRACT OF *GINKGO BILOBA* – EGB 761

In 1965 Dr. Willmar Schwabe Arzneimittel discovered that extracts from the leaves of *Ginkgo biloba* were effective for the treatment of peripheral and cerebral arterial disturbances of the blood supply especially in elderly patients. As a result they proceeded to develop a special multistage process producing a standardized extract. This product was designated the code name EGb 761. A first patent was granted for the extract in 1971 in Germany and 1972 in France (Dr Willmar Schwabe, 1971 and 1972). The process was specifically designed to enrich the main active ingredients of the extract known at this time i.e. the flavonol glycosides to 24% (see Chapter 9 "Chemical Constituents of *Ginkgo biloba*" by A. Hasler). A further patent was granted in 1989 (Dr Willmar Schwabe, 1989) that describes the enrichment of terpene trilactones to 6% and also the removal of undesirable components from the extract such as the ginkgolic acids which are known allergens.

LABORATORY SCALE TO INDUSTRIAL SCALE

The process for the preparation of EGb 761 was first developed on laboratory scale. However when a process moves from laboratory scale to a commercial manufacturing operation, unexpected problems of a physical and chemical nature are often encountered. These problems multiply when dealing with a complex mixture such as EGb 761, where it is necessary to ensure that the uniqueness of the product is not compromised in the scale up.

In order to be successful, large scale production must process the leaves to produce an extract at planned production rates, at the projected manufacturing cost and to clearly defined quality standards. Manufacturing cost is not only based on obvious factors such as costs of raw materials, product yield and return on capital but also the overall safety of the operation with respect to personnel at the facility, the public and the environment.

Technology, Plant and Equipment

Some extraction and purification techniques for the preparation of medicines from natural sources go back more than 1000 years. However, in the pharmaceutical industry, innovation and new technology are essential for the continuous development of the process and result in optimized yield, lower operating costs, reduction in solvent usage and reduction of wastes and energy consumption. Similarly, in laboratory new and improved analytical techniques are developed and adopted to provide a better service to production and as a consequence higher specification of the product.

In order to ensure the integrity of EGb 761 during large scale extraction and processing, extensive studies were carried out on pilot plant scale using test equipment supplied by vendors. Data from these studies was used to specify full scale process equipment and materials of construction i.e. high quality stainless steel and borosilicate glass. Likewise mechanical seals on pumps and agitators and all gasket materials were specified to be compatible with the extract and all solvents used in the process.

Safety

Safety has a high priority. Personnel are trained in all aspects of chemical handling and supplied with the necessary protective clothing. They also undergo regular medical check-ups to ensure the adequacy of personnel protective equipment and procedures.

Large volumes of flammable solvents which are used in the process pose high risks and extensive controls are in place to guard against fire and explosion. These control measures include:

1. Elimination of ignition sources by use of explosion proof electrical equipment, earthing and bonding of equipment, "no smoking" regulations and wearing of anti-static shoes and gloves.
2. Elimination of flammable atmospheres by ventilation of process areas, sealed solvent handling systems, pressurisation of control rooms and explosion-proof electrical switch rooms.
3. Fire detection/protection systems which include heat and smoke detectors, Sprinkler systems, CO_2 suppression systems and automatic shut-off valves on solvent lines.
4. Solvent tanks are fitted with flame arrestors. Over-pressure relief devices are provided on vessels, vent headers and dust collection systems to safely vent any overpressure. These devices include bursting discs, relief valves, explosion panels and isolation valves.
5. Trained emergency response teams and first aiders are available on site in case of emergencies.

Environmental Protection

As with all manufacturing facilities, wastes are generated during the production of EGb 761 and it is essential that waste streams from the production facility be minimized. Waste leaves generated after extraction are used as mulch or compost. Waste water which may contain minute quantities of Ginkgo extract and solvents from the production facility is treated in an aerobic treatment plant for removal of organic substances. Also abatement systems are in place to reduce emissions to the atmosphere. Optimization of yield results in energy conservation and reduced solvent usage.

There is a fire water retention pond on site which is designed to retain all water from the Sprinkler Protection system in the event of a fire. The water collected in the pond is analysed and disposed of in an environmentally safe manner.

An environmental management programme is in place. The objective of this programme is to assess operations on site and to review all practical options for the use of cleaner technology and cleaner production to minimise material consumption and waste generation. The programme identifies areas for investigation and a proposed timetable for completion of specific projects and studies.

MANUFACTURE OF EGB 761

With the increasing demand for EGb 761 in the 1970's, Dr. Willmar Schwabe set up a joint venture with Laboratoires Beaufour of France for the production of EGb 761. The location chosen was a green-field site in Cork, Ireland. The process involves nine main steps (Diagram 1) each of which incorporates various industrial operations so that in all there are over 50 operations from start to finish.

The manufacturing process for EGb 761 has been developed to produce a standardized extract and the quality of the extract is defined by:

1. Quality of leaves
2. Process for extraction and purification of the extract using specific techniques and solvents.
    This gives a standardized extract containing a dosage of known active substances while also eliminating undesirable substances contained in the leaves and materials used during manufacturing. The main specifications for EGb 761 are set out in Table 1.

In order to meet these specifications the facility is operated to current Good Manufacturing Practices (cGMP) with emphasis on:

1. Leaf Quality
2. Quality Control of Raw Materials and Consumables
3. Quality Control of Process Operations
4. Quality Control of Finished Product

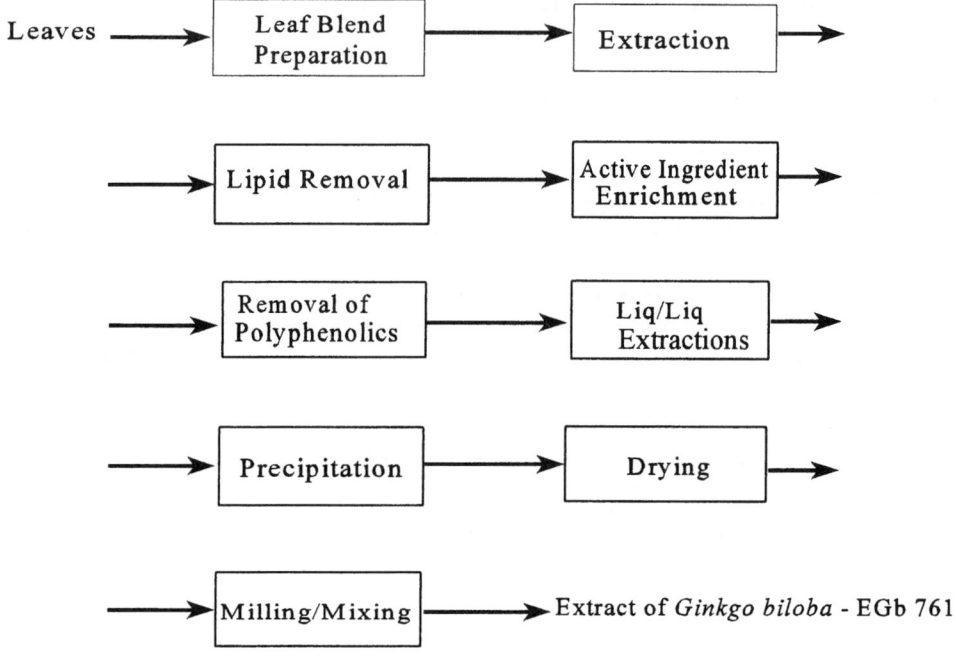

Diagram 1   Process Flow Diagram for Extract of *Ginkgo biloba* – EGb 761

Leaf Quality

The percentage active ingredients in the leaf varies significantly and can be influenced by natural and human factors. The natural influences which affect the quality of the leaves are:

(a) *The age of the plant*   The percentage active ingredients in young plants (1–3 years) is generally very high. However as the plant matures, there is a decrease in the active ingredients, levelling off after 6–8 years and thereafter remaining relatively constant.

(b) *Seasonal variation*   The percentage flavonol glycosides, ginkgolide A and bilobalide in the leaves decreases during the growing season whereas ginkgolide B remains relatively constant and ginkgolide C increases.

(c) *Type of Soil* The type of soil affects each of the active ingredients differently. As a result the relative ratios of active ingredients vary from area to area.

(d) *Climate* As with the type of soil, the climate affects the ratio of the different active ingredients in the leaves. Also during the growing season climatic conditions, e.g. drought or excessive rainfall affects the quality of the leaves.

The human influences which affect the quality of the leaves are:

(a) *Pruning* Large scale cultivation of *Ginkgo biloba* requires that the plant be pruned on a regular basis. The percentage active ingredients in the leaves is affected by the severity of the pruning.

(b) *Use of fertilizers* As well as increasing the yield of leaves, the use of fertilizers increases the percentage active ingredients in the leaves.

(c) *Drying of leaves* The leaves on harvesting contain 70–80% moisture. It is essential that leaves are dried within hours of harvesting thus preventing mould growth and fermentation which results in degradation of the active ingredients,

Due to the influences outlined above, the percentage active ingredients in the leaves varies considerably. In order to produce a high quality product and meet the specifications set out in Table 1 it is necessary to have the correct blend of leaves for production. From experience gained over the last 20 years the correct blend is achieved by irrigation, varying the time of harvest, use of fertilizer and pruning. As part of our quality control, the quantity and quality of our leaf stock is continuously monitored and the specifications for active ingredients in the leaves are set for the following harvest.

Table 1   Main specifications for extract of *Ginkgo biloba*–EGb 761

Flavonol Glycosides	21.6–26.4%
Terpene trilactones	5.4–6.6%
Ginkgolides A+B+C	2.8–3.4 %
Bilobalide	2.6–3.2%
HPLC Profile	Same as Ref. Std.
TLC	Same as Ref. Std.
Proanthocyanidins	< 9.5%
Ginkgolic Acids	< 5 ppm
Aspect	Conforms
Solubility	Soluble
Odour	Same as Ref. Std.
Water	< 3.0%
Solvents	< 250 ppm
Sulphate Ash	< 1.5%

Quality Control of Raw Materials and Consumables

To produce EGb 761 to a consistent quality it is necessary to have a quality system in place at all stages of manufacture from raw materials to finished goods. This system is based on FDA requirements for the preparation of Bulk Pharmaceuticals.

Quality control of leaves starts prior to commencement of the growing season with pruning and weeding carried out in the pre-growing season. The Beaufour/ Schwabe Group has 20 years experience in the planting, cultivation, harvesting, drying and baling of leaves. Agronomic engineers supervise operations on the plantations to ensure:

(a) Correct application of fertilizers
(b) There is no chemical treatment during the growing season
(c) Correct harvesting of leaves with particular attention to drying, baling and storage.

Figure 1 *Ginkgo biloba* Plantation

Figure 2   Leaf Preparation Area

The plant is highly resistant to pests and diseases and is unusually free from insect injury minimizing the requirement for pesticide control of the crop. However, as part of our quality system the leaves from the different areas are tested for pesticides.

On harvesting, the leaves are dried, baled and shipped to the production facility in Cork. On receipt of the leaves at the production facility the bales are offloaded, visually inspected and sampled, and placed in quarantine. A composite sample is analysed for moisture content, dry extract content and percentage active ingredients. On completion and acceptance of the analysis the leaves are passed and are ready for production. Similarly, quality control of solvents, inorganic salts, filter aids and filters used during processing is carried out and all consumables pass through quarantine/pass status prior to use in the process.

Quality Control of the Process

The production of EGb 761 has nine main steps (Diagram 1) which includes over fifty operations.

Based on the active ingredients in the leaf stock, a leaf blend is prepared. The relevant bales are taken from storage, the wrapping is removed and the bales crushed and broken. Twigs are separated from the leaves and they are now ready for extraction. The leaves are then exhaustively extracted with acetone/water ensuring complete recovery of the active ingredients. The extracted leaves are separated from the acetone/water solution and the solution is concentrated. The lipids are removed from the extract followed by active ingredient enrichment. The polyphenolic compounds are precipitated and removed by filtration. Further liquid/liquid extractions and final precipitation facilitate the removal of any remaining undesirable substituents.

On completion of the final precipitation, the extract is concentrated to remove solvents and dried in a belt dryer to give Extract of *Ginkgo biloba* – EGb 761. The extract is then sampled and analysed. The specifications on the active ingredients are 24.0% ± 10% i.e. 21.6%–26.4% for flavonol glycosides and 6.0% ± 10% i.e. 5.4%–6.6% for terpene trilactones. In order to produce a highly consistent product with active ingredients as close as possible to the mean value of 24% for flavonol

Figure 3   Leaf Extraction Area

glycosides and 6% for terpene trilactones, we have set internal specifications of 24 ± 5% for flavonol glycosides and 6.0% ± 5% for the terpene trilactones. To meet these specifications, the product is milled and mixed into its final bulk form producing a homogenous light brown powder.

It is important to note that all steps in the process are designed to:

(a) optimize the recovery of the active ingredients
(b) produce a consistent product
(c) remove undesirable substances.

This is achieved by a quality system which controls all operations during production. Specifications which include dry extract, percentage solvents, solvent ratios and temperature are set at each step of the process. Prior to progressing to the next step of manufacture, the product must meet the required specifications. As part of the cGMP each step in the process has been validated and is controlled by standard operating procedures. Each lot of finished product is traceable back to the leaves and all parameters during processing are recorded. Process operators are highly skilled and fully trained in all aspects of cGMP.

Quality Control of Finished Product

On completion of the final milling, the bulk pharmaceutical is sampled and a complete analysis carried out to ensure that the extract meets the required specifications. The laboratory operates to current Good Laboratory Practices (cGLP) and all standard operating procedures have been validated. The product is then shipped to the country of sale for formulation into the final form.

CONCLUSION

Since 1975 there has been a continuous increase in sales of EGb 761. The success of the product can be attributed in part to:

(a) Successful scale-up which maintained the uniqueness of the product, with emphasis on cost effectiveness, safety and environmental requirements.
(b) Continuous process optimization and the introduction of new technology which has improved yields and reduced wastes.
(c) Operation of current good manufacturing practices from leaf cultivation through to the finished form which ensures production of a high specification standardized extract.

REFERENCES

Dr Willmar Schwabe Arzneimittel, Karlsruhe (1971) Verfahren zur Herstellung eines Injektions- bzw. Infusionspräparates mit einem Gehalt an einem Wirkstoffgemisch aus den Blättern von *Ginkgo biloba*. German patent no DE 2117429 C3 (Inventors: Kloss P, Jaggy H).

Dr Willmar Schwabe Arzneimittel, Karlsruhe (1972) Procédé d'obtention d'un mélange de substances vasoactives des feuilles de *Ginkgo biloba* et médicaments renfermant ce mélange. French Patent no 72 12288.

Dr Willmar Schwabe GmbH & Co Karlsruhe (1989) Extrakt aus Blättern von *Ginkgo biloba*, Verfahren zu seiner Herstellung und den Extrakt enthaltende Arzneimittel. German Patent no. 3940095 A1 (Inventors: Schwabe, K.P.).

7. CHEMICAL CONSTITUENTS OF *GINKGO BILOBA*

ANDREAS HASLER

Zeller Ltd., Herbal Remedies, CH-8590 Romanshorn, Switzerland

INTRODUCTION

Ginkgo biloba is a unique plant, not only fascinating and interesting for artists but also for scientists. The Ginkgo tree was able to survive millions of years as the only member of a whole class of plants and its extracts of leaves are commonly used as phytomedicines in the treatment of peripheral and cerebral ischaemia (Hasler *et al.*, 1990c; Sticher *et al.*, 1991; Sticher, 1993; Joyeux *et al.*, 1995).

The first chemical investigations of *G. biloba* go back to Peschier (1818) and Schwarzenbach (1857) and have been continued in the second decade of the 20th century in Japan by Kawamura (1928) and Furukawa (1932). A series of phyto-chemical studies have started around 1980, using chromatography and spectroscopy. The leaves being used for phytomedicines in the Western world and the seeds, being roasted as food in Japan were of special interest to analyze. Further studies have been carried out on root bark, wood and on the sarcotesta.

Terpenes, flavonoids, organic acids, polyacetate derived compounds, carbohy-drates, miscellaneous organic compounds and inorganic compounds were found in *G. biloba* (Boralle *et al.*, 1988; Hölzl, 1992; Juretzek and Spiess, 1993). Most of the isolated compounds are found ubiquitously in the leaves of higher plants with the exception of certain flavonoids and the unique terpene trilactones. This chapter gives a complete overview of all constituents described so far from *G. biloba*.

TERPENES

Different groups of terpenes, namely terpene trilactones (ginkgolides, bilobalide), triterpenes (steroids, phytosterols), carotenoids, polyprenols and volatile terpenes (mono- and sesquiterpenes) can be found in Ginkgo. Up to date, the terpene trilactones were exclusively found in *G. biloba*.

Terpene Trilactones

Terpenes were first isolated by Furukawa (1932) from *G. biloba* root bark. Later, their structures were determined by Maruyama *et al.* (1967 a-d), Maruyama and Terahara (1967), Nakanishi (1967) and Woods *et al.* (1967) and called ginkgolides A, B, C and M. The ginkgolides A (Okabe *et al.*, 1967; Sakabe *et al.*, 1967),

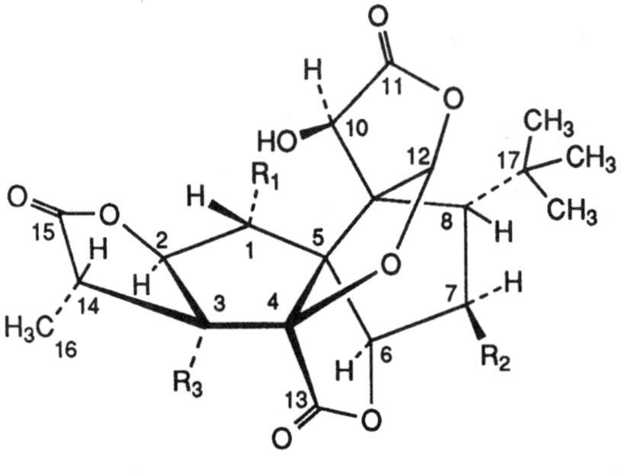

	R$_1$:	R$_2$:	R$_3$:	Ginkgolide:
	H	H	OH	A
	OH	H	OH	B
	OH	OH	OH	C
	H	OH	OH	J
	OH	OH	H	M

Figure 1   Ginkgolides

B (Okabe *et al.*, 1967), C (Okabe *et al.*, 1967) and J (Weinges *et al.*, 1987) were also found in the leaves of *G. biloba* (Figure 1).

The only differences between these compounds are the number and the position of hydroxyl groups which may be present on C1, C3 or C7 of the spirononane framework. The ginkgolide structures, which were mainly determined by chemical methods by Maruyama *et al.* (1967a-d), have been confirmed by high field proton and 1D and 2D NMR studies as well as x-ray crystallography (Dupont *et al.*, 1986a, b; Llabres *et al.*, 1989; van Beek and Lankhorst, 1996).

In addition to the ginkgolides (diterpenes), a compound named bilobalide (Major, 1967; Weinges and Bähr, 1969) was described. Bilobalide (Figure 2) is closely related to the ginkgolides and the structure was elucidated by Nakanishi and Habaguchi (1971) and Nakanishi *et al.* (1971).

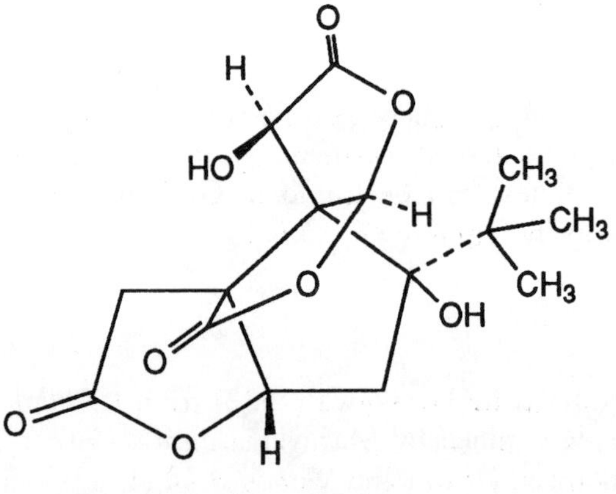

Figure 2   Bilobalide

Van Beek *et al.* (1991) found the concentration of bilobalide and the ginkgolides to be lowest in spring, gradually increasing to reach its maximum in mid-summer. When the leaves turned yellow in late autumn the concentration started to decline. Hasler and Meier (1992) found a significant fluctuation of the total terpene tri-lactone content during one vegetative season. This was mainly due to seasonal varia-tions of all single terpenes, except ginkgolide A. Ginkgolide A showed a decrease from July to November. In contrast to van Beek, Hasler and Meier (1993) have found a high content of the terpenes in the yellow and fallen leaves, comparable to that of green leaves. Furthermore, the content of ginkgolides in Ginkgo leaves showed differences – up to a factor of 300 – from tree to tree, compared to a factor of up to 160 for terpene lactones (van Beek *et al.*, 1991).

With the first trials to isolate terpene trilactones 10 g of ginkgolide A, 10 g of ginkgolide B, 20 g of ginkgolide C and 200 mg of ginkgolide M were obtained from 100 kg of root bark (equaling five trees 30 cm in diameter). Teng (1988) and van Beek and Lelyveld (1997) have described much more efficient extraction procedures to isolate pure ginkgolides A, B, C, J and bilobalide using column chromatography.

Flesch *et al.* (1992) have examined the relative importance of growth and light level on the terpene content. Huh and Staba (1993) have studied ontogenetic aspects of ginkgolide production by using *G. biloba* seedlings, greenhouse plants, young trees, mature trees cuttings, and plant tissue cultures and have found, that ginkgolides appeared to be independently biosynthesized in leaves and roots.

Triterpenes (steroids, phytosterols)

Furukawa (1932) detected sitosterol and a sitosterol glycoside. In 1970, Kircher also isolated β-sitosterol and traces of stigmasterol and Nes *et al.* (1977) identified campesterol and dihydrobrassicasterol (Figure 3). Dried leaves yielded approxi-mately 0.2% of phytosterols.

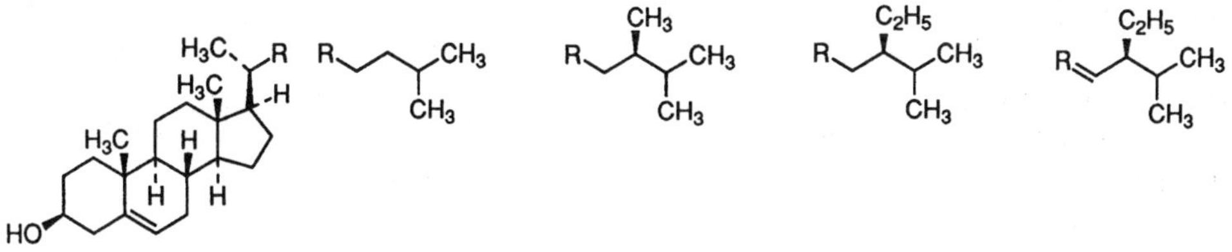

Cholesterol   Campesterol   β-Sitosterol   Stigmasterol

Figure 3   Triterpenes

Carotenoids

Karrer and Walker (1934) have described carotenoids (Figure 4) in the leaves of
G. *biloba* for the first time. Yadav *et al.* (1985a, b, 1987) have detected carotenoids
in the leaves and discussed their seasonal variation as well as their qualitative distribution pattern. Matile *et al.* (1992) have reported that senescent leaves of
G. *biloba* accumulate fluorescent compounds which may contribute to the golden
luminous appearance of the autumnal foliage and that the carotenoid esters
accumulate concomitant with chlorophyll breakdown.

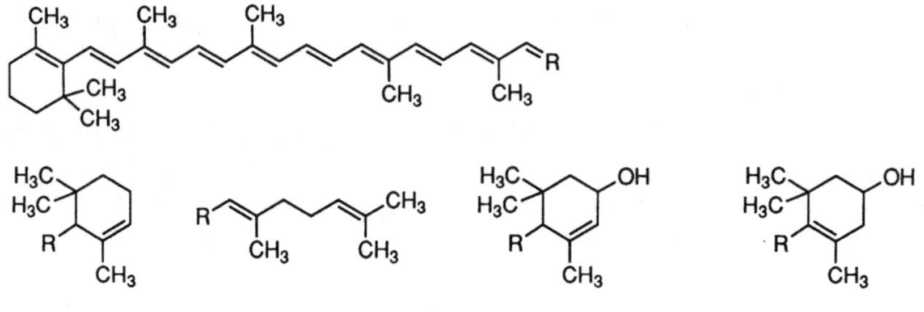

α-Carotene   γ-Carotene   Lutein   Zeaxanthin

Figure 4   Carotenoids

Polyprenols

Several types of polyprenols have been found in plants. Most of the polyprenols
isolated from angiosperms have the ω-t_3-C_n-OH structure. However, polyprenols
from many gymnosperms are known to have the ω-t_2-C_n-OH structure and are
therefore similar to the structure of the dolichols from animals. The precise physiological role of polyprenols in plants is not fully understood. They may have physiological roles similar to that of dolichols in animals. Dolichols are involved in the
transfer of sugar moieties during protein glycosylation.

Polyprenols from Ginkgo were first mentioned in a Japanese patent (Kuraray Co.
Ltd., 1982) as a possible pharmaceutical and cosmetic base and later described by
Ibata *et al.* (1983 and 1984a, b). The Ginkgo polyprenol mixture consisted of compounds with 14 to 22 isoprene units (Figure 5). 17, 18 and 19 isoprene-containing
compounds are predominant with a terminal dimethylallyl unit (ω-terminal), two
trans-isoprene residues, a sequence of 11–18 *cis*-isoprene residues and a terminal
hydroxylated or acylated isoprene unit (α-terminal). Takigawa *et al.* (1989) have
described the structures of G. *biloba* polyprenols as ω-t_2-C_n-OH.

The concentration of these polyprenols in leaves during the growing season
increased from 0.04% to 2.0%, according to HPLC. Huh *et al.* (1992) found the
content of polyprenols in Ginkgo leaves as determined by SFC assay to be higher
than the previously published values. In addition, the chromatogram of the highly

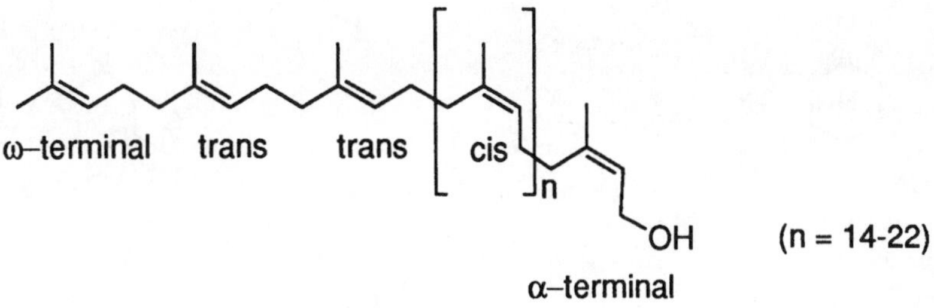

Figure 5 Polyprenols

concentrated leaf extract revealed the presence of an isoprene analog (C_{120}) not previously detected by HPLC methods.

In a recent study by Huh *et al.* (1993) ontogenetic aspects of the polyprenols in the leaves of Ginkgo seedlings, greenhouse plants and trees were investigated.

Volatile terpenes (constituents of essential oil)

The wood of *G. biloba* contains small amounts of essential oil, consisting of mono- and sesquiterpenes (Hölzl, 1992). Bilobanone (Kimura, 1962 and Kimura *et al.*, 1968), a sesquiterpene, is substituted on the ring framework by an isobutyl group. Irie *et al.* (1975) have isolated further sesquiterpenes called dihydroatlantones (Figure 6).

Hirao and Shogaki (1981) have analyzed the essential oil of Ginkgo leaves and found polycyclic aromatic hydrocarbons such as acenaphthene and 2,5,8-trimethyl-dihydronaphthalene. Furthermore *p*-cymene, *p*-tolyl-propylene and 1,4-dimethyl-2,5-diisopropylbenzene were found. Oxygenated compounds were *cis*-3-hexenol, *trans*-linalool oxide, α- and β-ionone, *cis*- and *trans*-4-hexenol, 2-isopropyl-phenol, thymol and heptadec-3,6,9-trien-1-ol (Figure 6).

FLAVONOIDS

Flavonoids are polyphenolic, low molecular weight compounds, probably found in all green plants. Currently, more than 30 genuine flavonoids are known in *G. biloba*. The great number of different flavonoids is not a result of the variability of the 2-phenylchromane framework but of the different glycosides found in Ginkgo. Nevertheless, only glucose and rhamnose can be found as sugar molecules and the variety of mono-, di- and triglycosides is a result of different binding patterns.

Furukawa (1932 and 1933) started investigations on flavonoids in *G. biloba*, Guth (1977) examined the biosynthesis of Ginkgo flavonoids, Schennen studied preparations of ^{14}C-marked flavonoids and Hasler (1990) isolated and characterized 21 flavonoids from Ginkgo leaves.

Bilobanone   (E)-10,11-Dihydroatlantone   (E)-10,11-Dihydro-6-oxo-
                                                                atlantone

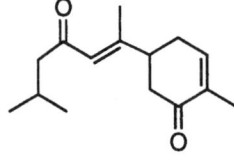

Acenaphthene   2,5,8-Trimethyl-dihydronaphthalene

β-Eudesmol   α-Ionone   trans-Linalool oxide

R₁:	R₂:	
CH_3	H	p-Cymene
H	OH	2-Isopropyl-phenol
CH_3	OH	Thymol

Figure 6   Constituents of essential oil

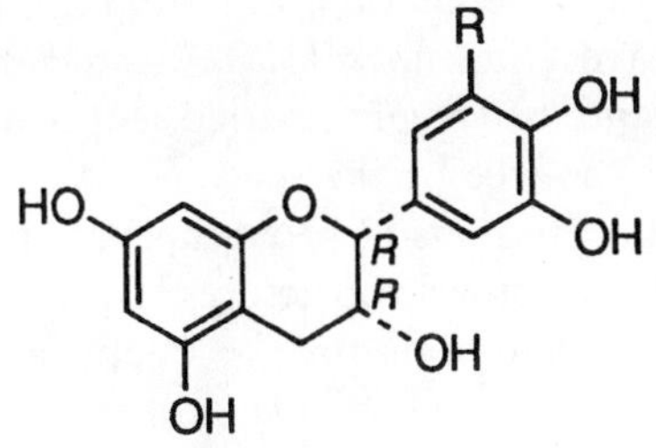

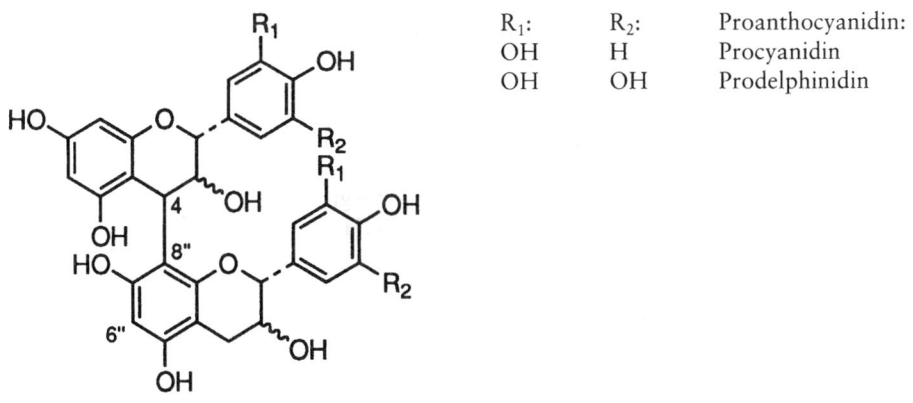

R: Flavanol: R: Flavanol:
H   (–)-Epicatechin   H   (+)-Catechin
OH   (–)-Epigallocatechin   OH   (+)-Gallocatechin

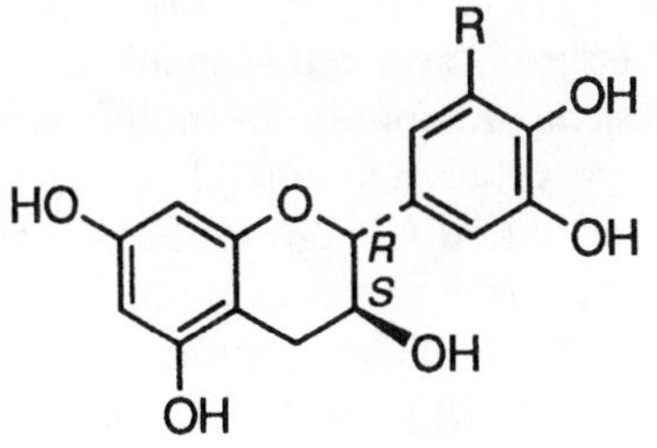

Figure 7   Flavanols

Flavanols

All flavan-3-ols have two chiral centers (C-atom 2 and 3) and therefore four optically active forms (Figure 7) and two racemates can be found. Only (+)-catechin and (–)-epicatechin occur in nature. The corresponding enantiomers are formed by epimerization during processing. Proanthocyanidins are oligomer or polymer flavan-3-oles, derivatives of the 2-phenylchroman. The monomer units are mostly linked by C4/C8″, rarely by C4/C6″. Whereas prodelphinidins are derivatives of gallocatechin or epigallocatechin, respectively, the procyanidins are derivatives of catechin or epicatechin, respectively. These findings were confirmed by Weinges *et al.* (1968a, b) who found the diastereomers (+)-catechin (C) and (–)-epicatechin (EC), as well as (+)-gallocatechin (G) and (–)-epigallocatechin (EG) in freshly cut leaves of *G. biloba*. Weinges *et al.* (1969) found in the ethanolic extract of Ginkgo leaves two proanthocyanidins (Figure 8). In 1986, Stafford *et al.* described four monomers (C, EC, G, EG), as well as four dimers (EC-C, C-C, GC-C, EG-C) in Ginkgo leaves.

R_1:	R_2:	Proanthocyanidin:
OH	H	Procyanidin
OH	OH	Prodelphinidin

Figure 8   Proanthocyanidins

Stafford *et al.* (1986) reported mainly dimeric and oligomeric proanthocyanidins in cell cultures and cell suspensions of Ginkgo leaves. Schennen (1988) stated that the content of proanthocyanidins is between 40 and 120 mg/g of dried leaves and that these values are not influenced by the sex of the tree or the season. Schennen summarized that Ginkgo leaves contain considerable amounts of proanthocyanidins, polymeric structures prevailing over dimeric and monomeric structures. Lang and Wilhelm (1996) have compared different analytical methods for the quantitative determination of proanthocyanidins in a purified extract of *G. biloba*.

Flavones

Weinges *et al.* (1968a) described a tricetin (= delphidenone) and a luteolin glycoside. Hasler *et al.* (1990a) characterized luteolin-3'-O-glucoside and apigenin-7-O-glucoside (Figure 9) and while screening extracts of *G. biloba* by HPLC apigenin and luteolin were detected as free aglycones (Hasler *et al.*, 1989, 1990b).

In addition, seven dimeric flavones, so called biflavones (Figure 10) of the amentoflavone-type were characterized. *G. biloba* is the first plant from which a biflavone (ginkgetin) was isolated (Furukawa, 1932), later described by Baker and Simmonds (1940) and ultimately identified by Nakazawa (1941). Bilobetin, isoginkgetin and sciadopitysin were isolated by Baker *et al.* (1959, 1963) and Seshadri (1969) gave an overview of biflavones. 5'-Methoxybilobetin was identified by Joly *et al.* (1980) and amentoflavone by Lobstein-Guth *et al.* (1988). 7-Methoxyamentoflavone was described by Gobbato *et al.* (1996) using thermospray liquid chromatography-mass spectrometry. All these biflavones bear the same 3',8″-linked nuclei with one to three methoxy substituents.

Hasler *et al.* (1990a) found apart from a relatively low content in May little variation in the total biflavone content during the growing season. Chang *et al.* (1993) have shown in their ontogenetic studies, that the total amount of biflavones increased during the growing season to reach its maximum in yellow autumn leaves. The concentration of biflavones in Ginkgo leaves varied from 0.4 to 1.9% (Juretzek and Spiess, 1993).

Pan *et al.* (1995) have isolated the known biflavones (5'-methoxybilobetin, bilobetin, ginkgetin, isoginkgetin, sciadopitysin) from the sarcotesta of Ginkgo.

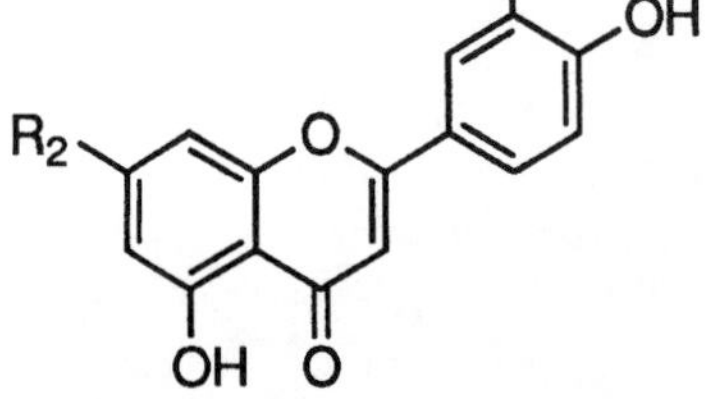

R₁:	R₂:	Flavone glucoside:
H	O–glc	Apigenin glucoside
O–glc	OH	Luteolin glucoside

Figure 9   Flavone glucosides

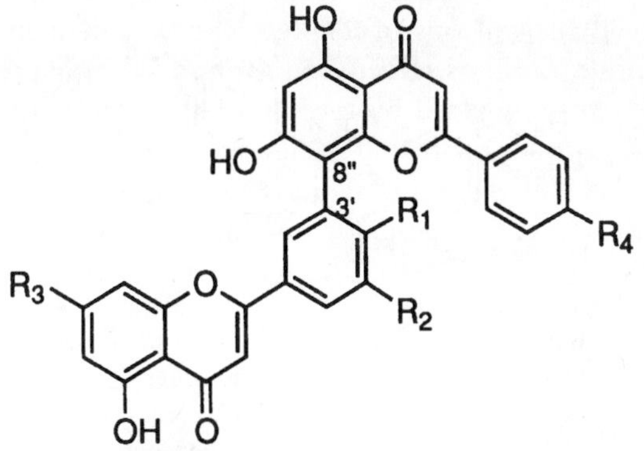

R_1:	R_2:	R_3:	R_4:	Biflavone
OH	OH	OH	OH	Amentoflavone
OCH_3	OH	OH	OH	Bilobetin
OCH_3	OH	OCH_3	OH	Ginkgetin
OCH_3	OH	OH	OCH_3	Isoginkgetin
OH	OH	OCH_3	OH	7-Methoxyamentoflavone
OCH_3	OCH_3	OH	OH	5'-Methoxybilobetin
OCH_3	OH	OCH_3	OCH_3	Sciadopitysin

Figure 10   Biflavones

Flavonols

The aglycones (Figure 11) quercetin, kaempferol and isorhamnetin were described by Fiesel (1965) and Hasler *et al.* (1990, 1991, 1992a) detected as well myricetin in *G. biloba* leaves using HPLC. Mono-, di- and triglycosides with glucose and rhamnose as sugar molecules can be found (Figure 11). Glycosides were described by Geiger and Beckmann (1965), Weinges *et al.* (1968a, b), Geiger (1979), Nasr *et al.* (1985), Vanhaelen and Vanhaelen (1988, 1989), Victoire *et al.* (1988), Schennen (1988) and Hasler (1990).

There are two different groups of diglycosides: the well-known rutinoside (3-O-(6″-O-(α-L-rhamnosyl)-β-D-glucosyl = 3-O-rutinoside) and the newly identified biloside (3-O-(2″-O-(β-D-glucosyl)-α-L-rhamnosyl = 3-O-biloside), a unique glycoside which is very characteristic for Ginkgo (Figure 12). The 3-O-triglycosides, determined by Schennen (1988), have a branched glycoside chain.

The acylated flavonols are an interesting group of compounds consisting of di- and triglycosides (Hasler *et al.*, 1992b, 1992c). The hydroxyl group of the C6‴ of the glucose unit is esterified with *p*-coumaric acid. Nasr *et al.* (1986, 1987) described for the first time acylated flavonol-diglycosides from Ginkgo leaves. Rhamnose and glucose are not linked by C1/C4, as reported by Nasr *et al.* (1986, 1987), but by C1/C2. The structures were determined by Hasler (1990) and confirmed by Kang *et al.* (1990). The sugar part consists of 3-O-biloside. The acylated triglycosides have a diglycoside unit (biloside) at C3 and a glucose molecule at C7 or C7‴″, respectively

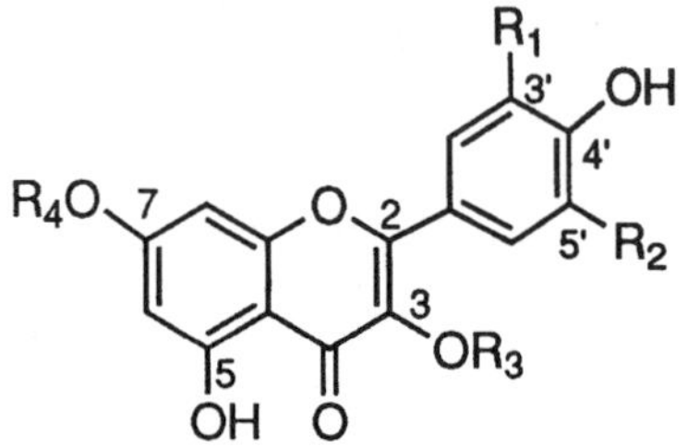

R_1 $\overset{3'}{\underset{}{}}$ OH
R_4O 7   O 2   4'
5' R_2
3 OR_3
5   OH O
OH

R_1:	R_2:	R_3:	R_4:	Flavonol aglycone:
H	H	H	H	Kaempferol
OH	H	H	H	Quercetin
OCH_3	H	H	H	Isorhamnetin
OH	OH	H	H	Myricetin
OCH_3	OH	H	H	3'-Methylmyricetin

R_1:	R_2:	R_3:	R_4:	Flavonol monoglycoside:
OCH_3	H	glc	H	3-O-(β-D-glucosyl)isorhamnetin
H	OH	glc	H	3-O-(β-D-glucosyl)kaempferol
H	OH	rha	H	3-O-(α-L-rhamnosyl)kaempferol
H	OH	H	glc	7-O-(β-D-glucosyl)kaempferol
OCH_3	OH	glc	H	3-O-(β-D-glucosyl)-3'-methylmyricetin
OH	H	glc	H	3-O-(β-D-glucosyl)quercetin
OH	H	rha	H	3-O-(β-D-rhamnosyl)quercetin

R_1:	R_2:	R_3:	R_4:	Flavonol diglycoside:
OCH_3	H	glc–rha	H	3-O-(6″-O-(α-L-rhamnosyl)-β-D-glucosyl)isorhamnetin
H	H	glc–rha	H	3-O-(6″-O-(α-L-rhamnosyl)-β-D-glucosyl)kaempferol
H	H	rha–glc	H	3-O-(2″-O-(β-D-glucosyl)-α-L-rhamnosyl)kaempferol
OH	OH	glc–rha	H	3-O-(6″-O-(α-L-rhamnosyl)-β-D-glucosyl)myricetin
OCH_3	OH	glc–rha	H	3-O-(6″-O-(α-L-rhamnosyl)-β-D-glucosyl)-3'-methylmyricetin
OH	H	glc–rha	H	3-O-(6″-O-(α-L-rhamnosyl)-β-D-glucosyl)quercetin
OH	H	rha–glc	H	3-O-(2″-O-(β-D-glucosyl)-α-L-rhamnosyl)quercetin
OCH_3	OCH_3	glc–rha	H	3-O-(6″O-(α-L-rhamnosyl)-β-D-glucosyl)syringetin

R_1:	R_2:	R_3:	R_4:	Flavonol triglycoside:
OCH_3	H	glc-(rha)$_2$	H	3-O-(2″-O,6″-O-bis(α-L-rhamnosyl)-β-D-glucosyl)isorhamnetin
H	H	glc-(rha)$_2$	H	3-O-(2″-O,6″-O-bis(α-L-rhamnosyl)-β-D-glucosyl)kaempferol
OH	H	glc-(rha)$_2$	H	3-O-(2″-O,6″-O-bis(α-L-rhamnosyl)-β-D-glucosyl)quercetin

Figure 11   Flavonol aglycones, mono-, di- and triglycosides

(Figure 13). Kaempferol- and quercetin-3-O-biloside, acylated flavonol derivatives and the biflavones are lead substances of *G. biloba* (Hasler, 1993).

Hasler *et al.* (1990c) investigated the influence of the season on the flavonoid concentration. In ontogenetic studies carried out between June and November, no significant differences in the flavonoid content were found. It was shown that dried Ginkgo leaves have an aglycone content of 0.2 to 1.4% w/w, corresponding to a calculated Ginkgo flavonol glycoside content of 0.5 to 3.5% w/w. Kang *et al.* (1993)

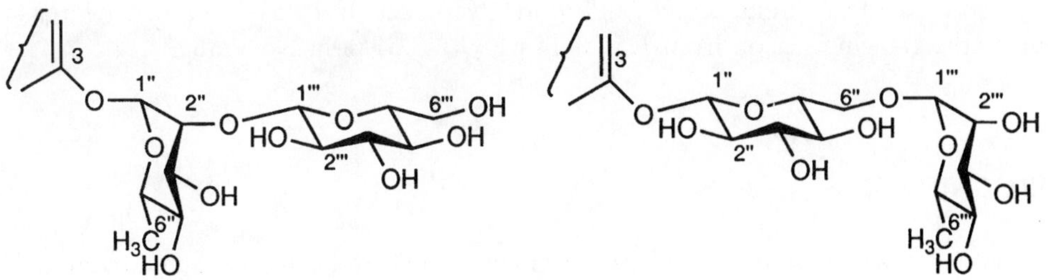

3-O-(2″-O-(β-D-glucosyl)-α-
L-rhamnosyl) = 3-O-biloside

3-O-(6″-O-(α-L-rhamnosyl)-β-
D-glucosyl) = 3-O-rutinoside

Figure 12   Rutinoside and biloside

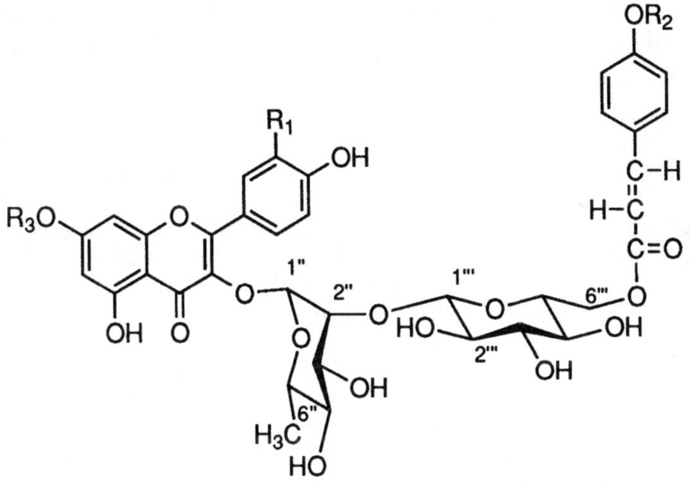

R₁: R₂: R₃: Acylated flavonol di- and triglycoside:

R_1	R_2	R_3	
H	H	H	3-O-(2″-O-(6‴-O-(*p*-hydroxy-*trans*-cinnamoyl)-β-D-glucosyl)-α-L-rhamnosyl)kaempferol
OH	glc	H	3-O-(2″-O-(6‴-O-(*p*-(β-D-glucosyl)oxy-*trans*-cinnamoyl)-β-D-glucosyl)-α-L-rhamnosyl)kaempferol
OH	H	H	3-O-(2″-O-(6‴-O-(*p*-hydroxy-*trans*-cinnamoyl)-β-D-glucosyl)-α-L-rhamnosyl)quercetin
OH	glc	H	3-O-(2″-O-(6‴-O-(*p*-(β-D-glucosyl)oxy-*trans*-cinnamoyl)-β-D-glucosyl)-α-L-rhamnosyl)quercetin
OH	H	glc	3-O-(2″-O-(6‴-O-(*p*-hydroxy-*trans*-cinnamoyl)-β-D-glucosyl)-α-L-rhamnosyl)-7-O-(β-D-glucosyl)quercetin

Figure 13   Acylated flavonol di- and triglycosides

have studied the content of six flavonoid glycosides and have found a decrease during the growing season from 1.57% in May to 0.39% in November.

ORGANIC ACIDS

Butyric acid was first described by Peschier (1818) and later by Schwarzenbach (1857), who studied the volatile compounds of ripe seeds. Griebel (1939) found acetic acid, butyric acid, formic acid, hexanoic acid and valeric acid after steam distillation of ripe seeds. Parliment (1995) has identified the odorous principles of mature gynoceum as butanoic (butyric) and hexanoic (caproic) acid.

Shikimic acid, a compound characteristic of gymnosperms was first discovered in Ginkgo leaves by Yamashita and Sato (1930). Tateoka (1963) has studied the biosynthetic pathway of shikimic acid in Ginkgo leaves and found, that the content shows a remarkable daily change and that the glycolysis might be of some importance in the biosynthesis of shikimic acid. Shikimic acid is important for the biosynthesis of tyrosine and phenylalanine, quercetin, cyanidine and lignin. In addition to shikimic acid, quinic acid (Weinstein *et al.*, 1962), ascorbic acid (Franke, 1965a, b), D-glucaric acid (Dittrich and Kandler, 1970), *p*-hydroxybenzoic acid, protocatechuic acid and vanillic acid, (Juretzek and Spiess, 1993) were also described (Figure 14).

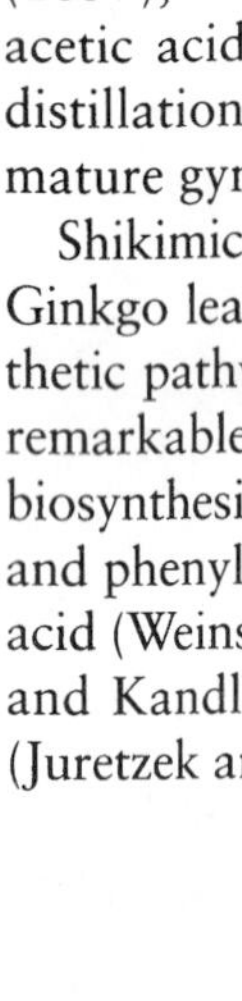

R:	
H	*p*-Hydroxybenzoic acid
OH	Protocatechuic acid
OCH_3	Vanillic acid

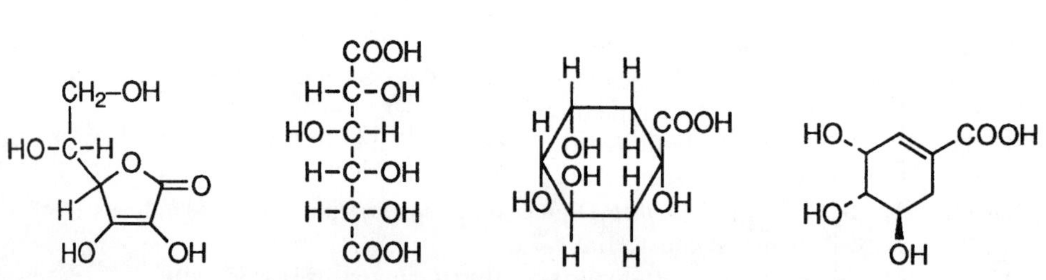

Ascorbic acid D-Glucaric acid Quinic acid Shikimic acid

Figure 14 Organic acids

POLYACETATE DERIVED COMPOUNDS

Alkyl Phenolic Acids and Alkyl Phenols

Being found in G. *biloba*, the anacardic acids (alkyl phenolic acids) are also called ginkgolic acids. These compounds belong to the group of 6-alkylsalicylic acids with alkyl residues of the size of 13–19 C-atoms and with up to three double bonds (Figure 15).

Ginkgolic acid was first described by Peschier (1818) and later by Schwarzenbach (1857) analyzing ripe seeds. Kawamura (1928) isolated from the sarcotesta ginkgolic acid and bilobol and in 1934, Furukawa described the same structures. Their final configurations were determined by Morimoto *et al.* (1968). Hydro-ginkgolinic acid was first isolated by Fu *et al.* (1962). Further alkyl derivatives (Figure 16) such as ginkgol and hydroginkgolic acid were detected in Ginkgo seeds. Novel phenolic lipids from Ginkgo seeds were identified by Gellerman *et al.* (1976a, b) as 4-hydroxyanacardic acids and represent the postulated precursors of cardols (5-alkylresorcinols). A phytochemical investigation by Adawadkar and El Sohly (1981) of the fresh sarcotesta extract of Ginkgo resulted in the isolation of an anacardic acid mixture (3.1% w/w), a bilobol mixture (0.53% w/w) and small amounts of cardanol. GC/MS analysis of the anacardic acids revealed the presence of $C_{13}H_{27}$, $C_{15}H_{29}$ and $C_{17}H_{33}$ side chains. Bilobols were found to have $C_{13}H_{27}$ and $C_{15}H_{29}$ side chains. Itokawa *et al.* (1987) and Itokawa and Takeya (1993) have isolated long-

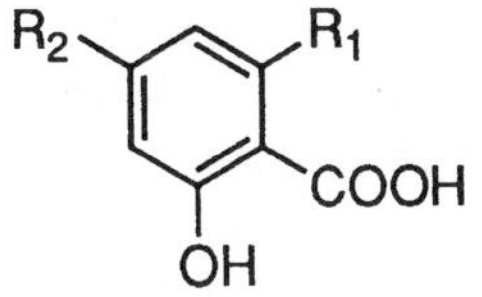

R_1:	R_2:	Anacardic acid:
$(CH_2)_{12}CH_3$	H	6-Tridecylsalicylic acid
$(CH_2)_{13}CH_3$	H	6-Tetradecylsalicylic acid (Hydroginkgolinic acid)
$(CH_2)_{14}CH_3$	H	6-Pentadecylsalicylic acid (Hydroginkgolic acid)
$(CH_2)_7(CH)_2(CH_2)_5CH_3$	H	6-[8-Pentadecenyl]salicylic acid (Ginkgolic acid)
$(CH_2)_{16}CH_3$	H	6-Hexadecylsalicylic acid
$(CH_2)_8(CH)_2CH_2(CH)_2(CH_2)_3CH_3$	H	6-[9,12-Heptadecadienyl]salicylic acid
$(CH_2)_7(CH)_2(CH_2)_7CH_3$	H	6-[8-Heptadecenyl]salicylic acid
		Resorcylic acid:
$(CH_2)_7(CH)_2(CH_2)_5CH_3$	OH	6-[8-Pentadecenyl]resorcylic acid
$(CH_2)_{12}CH_3$	OH	6-Tridecylresorcylic acid

Figure 15   Anacardic and resorcylic acids

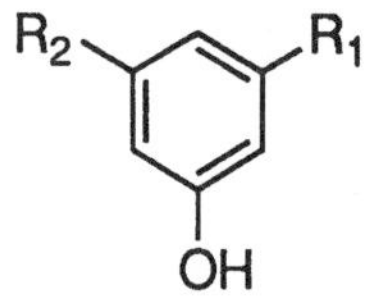

R₁: R₂: Cardanols:
$(CH_2)_{12}CH_3$ H   3-Tridecylphenol
$(CH_2)_{13}CH_3$ H   3-Tetradecylphenol
$(CH_2)_{14}CH_3$ H   3-Pentadecylphenol
$(CH_2)_{16}CH_3$ H   3-Heptadecylphenol
$(CH_2)_7(CH)_2(CH_2)_5CH_3$ H   Ginkgol

                                               Cardols:
$(CH_2)_{12}CH_3$ OH   5-Tridecylresorcinol
$(CH_2)_{13}CH_3$ OH   5-Tetradecylresorcinol
$(CH_2)_{14}CH_3$ OH   5-Pentadecylresorcinol
$(CH_2)_{16}CH_3$ OH   5-Heptadecylresorcinol
$(CH_2)_7(CH)_2(CH_2)_5CH_3$ OH   Bilobolol

Figure 16   Cardanols and cardols

chain phenols from the sarcotesta of *G. biloba*. Verotta and Peterlongo (1993) have developed a method for the selective extraction and analysis of alkyl phenolic acids and alkyl phenols by supercritical fluid extraction with carbon dioxide combined with off line GC/MS. Irie *et al.* (1996) have determined the amount of anacardic acids and found 35.4 mg/g dry sarcotesta and 0.02 mg/g dry nuts.

Furthermore, ginkgolic acids with alkyl residues of 13, 15 and 17 C-atoms respectively, are thought to be the main compounds of the leaves. Their derivatives, the ginkgols, are formed by decarboxylation of the ginkgolic acids and are characterized as phenols substituted with an alkyl residue (Figure 16). Ginkgolic acids were first described in the leaves of *G. biloba* by Gellerman and Schlenk (1968) and later, Gellerman *et al.* (1974) have studied the biosynthesis of anacardic acids. Chung *et al.* (1982) have isolated 3-(pentadec-8-enyl)phenol, which seems to be identical with ginkgol, a known constituent from the seeds of Ginkgo. Feistel (1994) found 0.88–12.43 ppm of ginkgolic acids in commercially available extracts of Ginkgo leaves. Irie *et al.* (1996) have isolated, identified and determined the content of four anacardic acids in the leaves, the sarcotesta and the nuts of *G. biloba*. The content of these four anacardic acids was 17.3 mg/g dry leaves with ginkgolic acid showing the highest amount up to 12 mg/g.

Alkyl coumarins

Searching for allergens in *G. biloba* pulp, Choukchou-Braham *et al.* (1994) have investigated non-polar compounds, present in trace amounts and identified optically active 8-hydroxy-3-alkyl-3,4-dihydroisocoumarins (Figure 17).

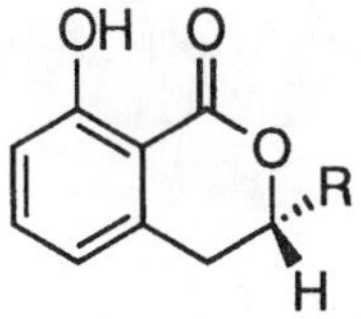

R: $C_{13}H_{27}$
      $C_{15}H_{29}$
      $C_{17}H_{31}$

Figure 17   Alkyl coumarins

Lipids

Tsujimoto (1940) detected linolenic acid in green and yellow colored Ginkgo leaves.

In 1969, Hata *et al.* have analyzed the suberin and found in the outer bark of *G. biloba* 11.4%. The fatty acids, phellonic acid and eicosanedicarboxylic acid, were isolated in crystalline form from sàponification products of Ginkgo suberin.

The main fatty acids of Ginkgo seeds were oleic and linoleic acid in the tri-glyceride fraction, and palmitic acid in the phospholipid fraction, respectively (Gellerman and Schlenk, 1969). The purified lipid fraction (1.26% on a wet weight basis) from nuts of *G. biloba* was found to consist of 90.6% neutral lipids, 7.5% polar lipids and very small amounts of glycolipids (Urakami *et al.* 1976). The composition of the steroid fraction is similar to that of soybeans: cholesterol, campesterol, stigmasterol and β-sitosterol (Figure 3). Nes *et al.* (1977) found, that the major sterols were, in decreasing quantities, β-sitosterol, campesterol, 22-dihydrobrassicasterol and cholesterol. These results were confirmed by Chung and Shin (1978). The total amount of glycolipids in *G. biloba* nuts (Kameyama and Urakami, 1979) was 3.12% of crude lipids, and the largest fraction of glycolipids was digalactosyldiglyceride (64.1%), followed by monogalactosyldiglyceride (31.2%) and cerebroside (4.7%). Tsuyuki *et al.* (1979) have studied the composition of lipids in Ginkgo seeds by means of TLC and GC analysis and found triglycerides (62.11%), different lipid compounds (12.14%), sterol esters (9.19%), sterols (6.05%), diglycerides (3.79%), monoglycerides (2.32%), free fatty acids (2.14%) and pigments (1.75%). GC-analysis by Takagi and Itabashi (1982) of fatty acids in the lipids from Ginkgo seeds showed that they contained nonmethylene-interrupted polyenoic acids as minor components and palmitic, oleic, linoleic and α-linolenic acids as major components. Hierro *et al.* (1996) have examined the fatty acid composition of seeds of *G. biloba* by a combination of capillary GC, silver ion HPLC and GC/MS. Some of the fatty acids identified are unusual in plants, e.g. an anteiso-methyl branched fatty acid, 14-methylhexadecanoic acid, 5,9-octadecadienoic acid, and 5,9,12-octadecatrienoic acid. Fourier-transformed infrared spectroscopy confirmed that all of the double bonds were cis.

Long Chain Hydrocarbons

Long chain hydrocarbons can be frequently and ubiquitously found in waxes covering leaves and fruits.

$CH_3(CH_2)_{25}OH$ 1-Hexacosanol

$CH_3(CH_2)_2(CH)_2CHO$ 2-Hexenal

$CH_3(CH_2)_8CHR(CH_2)_{18}CH_3$ R = H: n-Nonacosane
                                             R = O: n-Nonacosan-10-one (Ginnon)
                                             R = OH: n-Nonacosan-10-ol (Ginnol)

Figure 18   Long chain hydrocarbons and derivatives

Kawamura (1928) isolated ginnol from the pulp of Ginkgo seeds. Three years later, the same substance was found in apple peel and identified as (+)-10-nonacosanol (Figure 18) and the structure was confirmed by Seoane et al. (1969).

G. biloba leaf wax was analyzed by Ageta (1959), Hunneman and Eglinton (1972), Holloway et al. (1976), Casal and Moyna (1979), Nip et al. (1985) and Gülz et al. (1992). In 1932, ginnol was isolated by Furukawa and the structure was confirmed as (+)-10-nonacosanol. 1-Hexacosanol (cerylalcohol) was isolated in 1947 by Sosa. Ageta (1959) found n-nonacosane, n-nonacosan-10-one (ginnon) and n-octacosan-1-ol. While looking for fungistatic constituents in Ginkgo leaves, Major et al. (1960, 1963) found 2-hexenal, but only in a concentration of 75 ppm. Therefore, Major (1967) concluded, that other compounds are involved in the resistance against phytopathogenic organisms. 2-Hexenal is formed from linolenic acid (Major and Thomas, 1972), when freshly cut leaves are stored in the presence of air.

The cuticles of the Jurassic G. huttonii and those of a recent G. biloba were analyzed by means of Curie point pyrolysis-gas chromatography and Curie point pyrolysis-gas chromatography-mass spectrometry (Nip et al., 1985). Homologous series of straight-chain alkanes, alk-1-enes and α,ω-alkadienes formed from the highly aliphatic biopolymers in the pyrolysate are described for G. biloba cuticles as well as for Jurassic G. huttonii cuticles. The presence of cutin in G. biloba cuticles and its absence in the fossil G. huttonii indicates that cutin is sensitive to diagenesis.

CARBOHYDRATES

Ginkgo seeds consist of approximately 68% starch and 1.6% pentosans (Juretzek and Spiess, 1993).

Sucrose was found in Ginkgo leaves by Plouvier (1952).

Cyclites

The cyclic polyalcohols pinitol and sequoyitol (Figure 19) were found in Ginkgo leaves by Plouvier (1952, 1957). Biosynthetically, sequoyitol is a precursor of pinitol, and is only found in gymnosperms.

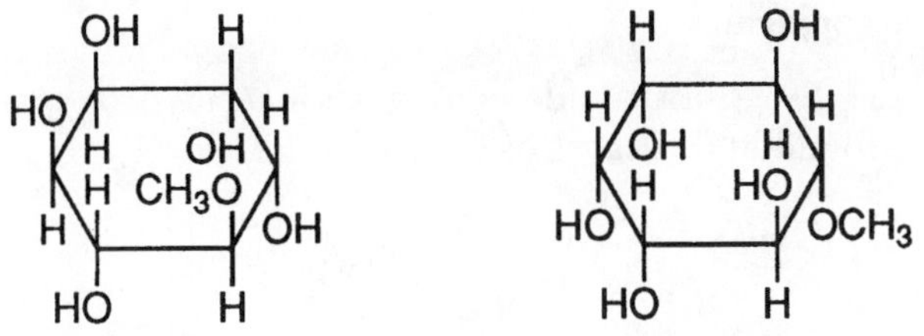

Pinitol   Sequoyitol

Figure 19   Cyclites

Polysaccharides

Mian and Timell (1960) have isolated an arabino-4-O-methylglucurono-xylan from Ginkgo wood and its structure was established. The hemicellulose contains a slightly branched framework of (1-4)-linked β-D-xylose residues attached with single side chains of (1-2)-linked 4-O-methyl-α-D-glucuronic acid and (1-3)-linked L-arabino-furanose residues.

Water- and alkali-soluble polysaccharides were isolated (Höllriegl *et al.*, 1986) from the leaves of *G. biloba* and separated into a low and a high molecular fraction. The water soluble fraction showed 30 to 50%, the alkali soluble fraction 92 to 97% of high molecular weight polysaccharides. After hydrolysis the high molecular poly-saccharide fraction consisted of galactose, arabinose, rhamnose and minor amounts of glucose, mannose and xylose as well as of galacturonic acid and glucuronic acid. In 1991, Kraus has isolated water-soluble polysaccharides from dried *G. biloba* leaves. The crude polysaccharide mixture was separated into a neutral and three acidic polysaccharide fractions. Besides low amounts of galactose, mannose and glucose, the neutral fraction (M_r 23,000) mainly consisted of arabinose indicating the presence of an arabinan. Two acidic fractions (M_r 500,000 and 24,000) were characterized by large amounts of mannose, rhamnose and glucuronic acid, as well as by low amounts of arabinose, glucose and galactose. The third acid polysaccha-ride fraction consisted of a large amount of galacturonic acid and equal amounts of arabinose, galactose, rhamnose and glucuronic acid.

MISCELLANEOUS ORGANIC COMPOUNDS

Glycerol derivatives

Kim *et al.* (1994) have characterized two glycerol derivatives (Figure 20) in the water-soluble and butanol-soluble fractions of *G. biloba* leaves.

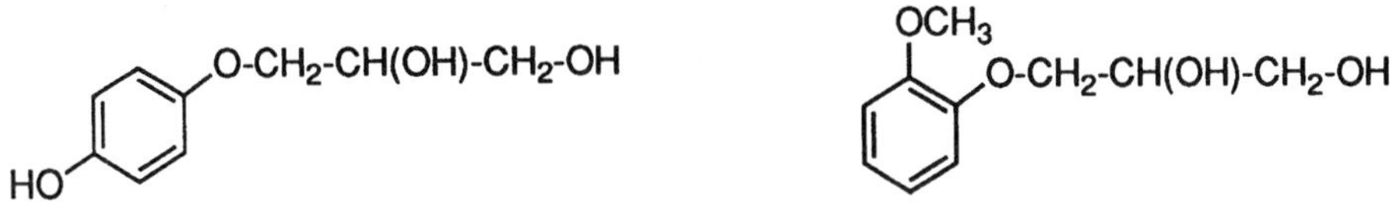

p-Hydroxyphenylglycerol   Guaiacylglycerol

Figure 20   Glycerol derivatives

1-Hydroxypyrene conjugates

β-O-Glucoside and β-O-glucuronide conjugates of 1-hydroxypyrene (Nakajima *et al.*, 1996) were found in the leaves of *G. biloba* (Figure 21).

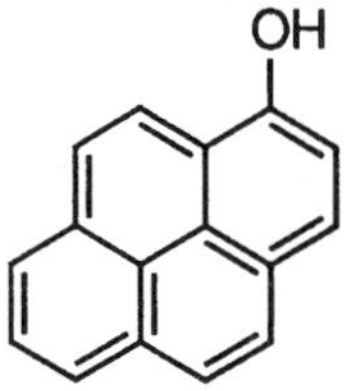

Figure 21   1-Hydroxypyrene

N-containing compounds

Amino acids and derivatives

Kawamura (1928) isolated from the pulp juice of Ginkgo gynoceum asparagine and Hatano (1956) described valine and ψ-amino butyric acid by paper chromatography extracted from Ginkgo seeds. Wada *et al.* (1988) have isolated from the albumen of the seeds 4-O-methylpyridoxine (Figure 22), which is mentioned together with "gin-nan food poisoning". Oh *et al.* (1995) have analyzed the non-protein and protein amino acids of the edible Ginkgo nut and have found glycine, pyroglutamic acid, asparagine, serine, glutamine, aspartic acid and traces of threonine and proline.

Ginkgo seeds consist for approximately 13% of proteins (Juretzek and Spiess, 1993).

Tsumura *et al.* (1992) have studied the allozyme variation of Ginkgo trees in western Japan. Yoshioka (1994) has investigated the biochemical changes in carbohydrates and sugar-metabolizing enzyme activities of Ginkgo nuts during growth and development.

Studying the amino acid metabolism during winter, Sagisaka (1993) has found that the buds and the living bark of *G. biloba* contain mainly proline, whereas proline and arginine coexisted in xylem.

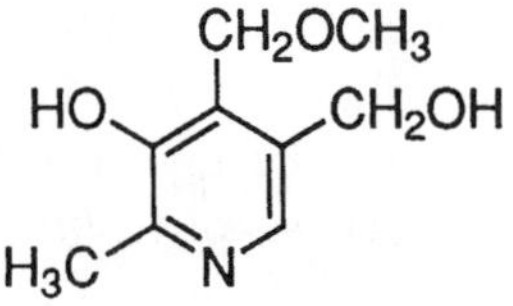

Figure 22   4-O-methylpyridoxine

Cytokinins

Cytokinin conjugates of zeatin (Figure 23), zeatin glucoside, ribosylzeatin, isopentenyladenine and their dihydro derivatives have been identified as the major

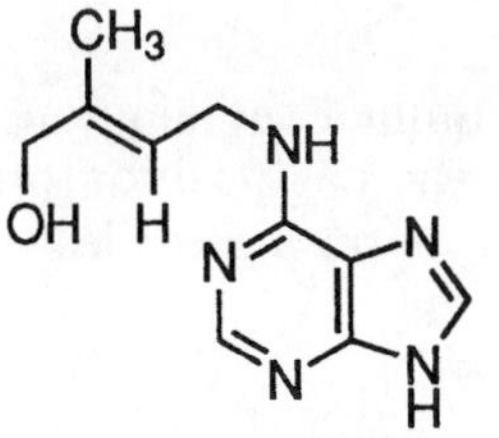

Figure 23   Zeatin

cytokinins present in mature *G. biloba* leaves. Ribosylzeatin was found in higher amounts than zeatin and dihydrozeatin. Most of the detected cytokinins showed polar properties and therefore could be easily hydrolyzed by β-glucosidase. It was concluded, that the glucose molecule is attached to the side chain of various derivatives (van Staden, 1978 and van Staden *et al.*, 1983). The qualitatively extremely complex cytokinin profile in mature leaves and the high amounts of cytokinins within the leaf tissue of *G. biloba* compared to other gymnosperms may be related to the development of the deciduous habit in a relatively primitive species (van Staden, 1996a). The metabolism of dihydro zeatin, a compound playing an important role in leaf processes, was studied by van Staden (1996b).

Hexosamines

Racusen and Foote (1974) have analyzed the content of hexosamines in Ginkgo leaves, which contained both polymer-bounded and 80% acetone-soluble forms of hexosamines. The total content was 3.5 µg hexosamines/mg protein. The hexosamines are suggested to be essential elements of membranes.

6-Hydroxykynurenic acid (6-HKA)

6-HKA (Figure 24) was the first nitrogen-containing secondary metabolite isolated from the leaves of *G. biloba* (Schennen and Hölzl, 1986). The 6-HKA content increased continuously throughout leaf development and reached a maximum value of 0.24% at leaf abscission in autumn. 6-HKA seems to be a main metabolite of tryptophane in *G. biloba*. Kang *et al.* (1995) have isolated and analyzed 6-HKA from yellow leaves and confirmed the structure by HMQC and HMBC techniques.

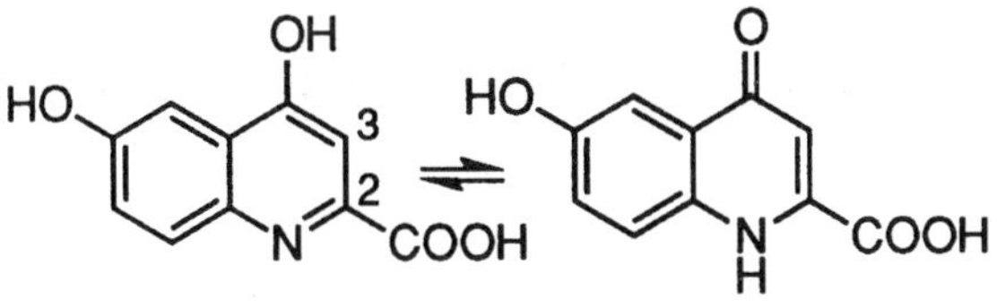

Figure 24   6-Hydroxykynurenic acid

Lectins

Clarke *et al.* (1978) have determined the amount of β-lectin (arabinogalactan proteins) in Ginkgo leaves and the composition of its protein part (amino acid analysis). The amount of β-lectins in fresh green leaves was 0.01%.

Pentadien-1,5-diyl-diphenol

Plieninger *et al.* (1986) have isolated this unusual natural product (Figure 25) from the leaves of *G. biloba*.

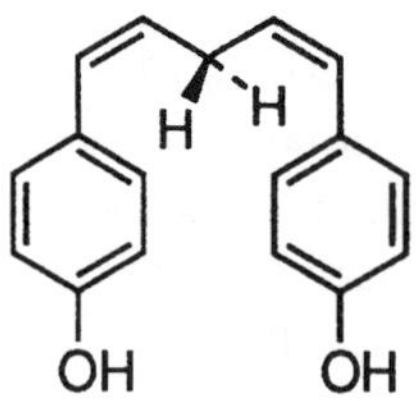

Figure 25   (Z,Z)-4,4′-(1,4-pentadien-1,5-diyl)diphenol

Tannins

In the bark of *G. biloba* Zhang *et al.* (1990) found condensed tannins and Kariyone *et al.* (1958) described (+)-sesamin (Figure 26) an ubiquitous lignan from the wood.

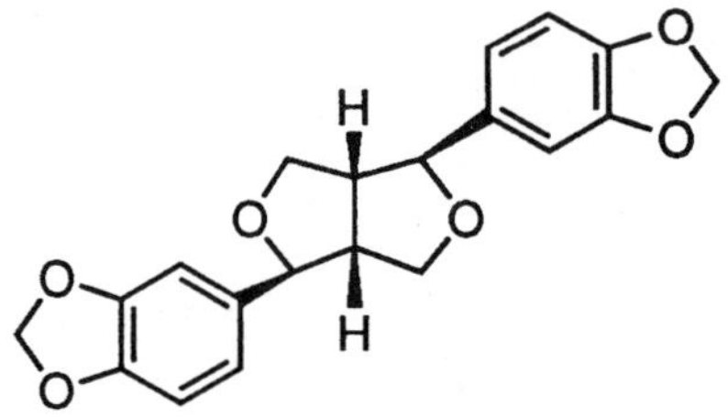

Figure 26   (+)-Sesamin

INORGANIC COMPOUNDS

Acids

Kawamura (1928) isolated from the pulp juice phosphoric acid.

Calcium oxalate

The object of the work of Umemoto and Hozumi (1972) was to look for a possible correlation between the degree of exposure to air pollution and any structural

Table 1   Constituents of root bark from *G. biloba*

TERPENES
Terpene trilactones
Diterpenes
Ginkgolide A (Furukawa, 1932; Maruyama *et al.*, 1967a-d; Maruyama and
Terahara, 1967; Sakabe *et. al.*, 1967; Woods *et al.*, 1967)
Ginkgolide B (Furukawa, 1932; Maruyama *et al.*, 1967a-d; Maruyama and
Terahara, 1967; Woods *et al.*, 1967)
Ginkgolide C (Furukawa, 1932; Maruyama *et al.*, 1967a-d; Maruyama and
Terahara, 1967; Woods *et al.*, 1967)
Ginkgolide M (Maruyama *et al.*, 1967a-d; Maruyama and Terahara, 1967;
Woods *et al.*, 1967)

POLYACETATE DERIVED COMPOUNDS
Lipids
Eicosanedicarboxylic acid (Hata *et al.*, 1969)
Phellonic acid (Hata *et al.*, 1969)

MISCELLANEOUS ORGANIC COMPOUNDS
Condensed tannins (Zhang *et al.*, 1990)

Table 2   Constituents of wood from *G. biloba*

TERPENES
Volatile terpenes (constituents of essential oil)
Monoterpenes
Cymene (Hölzl, 1992)
Ionone (Hölzl, 1992)
Isopropyl-phenol (Hölzl, 1992)
Linalool oxide (Hölzl, 1992)
Thymol (Hölzl, 1992)
Sesquiterpenes
Bilobanone (Kimura, 1962; Kimura *et al.*, 1968)
E- and Z- forms of 10,11-dihydroatlantone (Irie *et al.*, 1975)
E-form of 10,11-dihydro-6-oxoatlantone (Irie *et al.*, 1975)
Elemol (Hölzl, 1992)
Eudesmol (Hölzl, 1992)

CARBOHYDRATES
Polysaccharides
Arabino-4-O-methylglucurono-xylan (Mian and Timell, 1960)

MISCELLANEOUS ORGANIC COMPOUNDS
N-containing compounds
Amino acids and derivatives
Arginine, Proline (Sagisaka, 1993)
Tannins
(+)-Sesamin (Kariyone *et al.*, 1958)

Table 3   Constituents of leaves from *G. biloba*

TERPENES
  Terpene trilactones
    Diterpenes
      Ginkgolide A (Furukawa, 1932; Okabe *et al.*, 1967; Sakabe *et al.*, 1967)
      Ginkgolide B (Furukawa, 1932; Okabe *et al.*, 1967)
      Ginkgolide C (Furukawa, 1932; Okabe *et al.*, 1967)
      Ginkgolide J (Weinges *et al.*, 1987)
    Sesquiterpene
      Bilobalide (Major, 1967; Weinges and Bähr, 1969; Nakanishi *et al.*, 1971)
  Triterpenes (steroids, phytosterols)
    Campesterol (Nes *et al.*, 1977)
    22-Dihydrobrassicasterol (Nes *et al.*, 1977)
    Stigmasterol (Kircher, 1970)
    β-Sitosterol (Furukawa, 1932; Kircher, 1970)
    Sitosterol glucoside (Furukawa, 1932)
  Carotenoids (Matile *et al.*, 1992)
    α-Carotene (Karrer *et al.*, 1934; Yadav *et al.*, 1985)
    γ-Carotene (Karrer *et al.*, 1934; Yadav *et al.*, 1985)
    Lutein (Karrer *et al.*, 1934; Yadav *et al.*, 1985)
    Zeaxanthin (Karrer *et al.*, 1934; Yadav *et al.*, 1985)
  Polyprenols (Kuraray, 1982; Ibata *et al.*, 1983; 1984a, b; Takigawa *et al.*, 1989)
  Volatile terpenes (constituents of essential oil)
    Acenaphthene (Hirao and Shogaki, 1981)
    p-Cymene (Hirao and Shogaki, 1981)
    1,4-Dimethyl-2,5-diisopropylbenzene (Hirao and Shogaki, 1981)
    Heptadec-3,6,9-trien-1-ol (Hirao and Shogaki, 1981)
    cis-3-Hexenol (Hirao and Shogaki, 1981)
    cis- and trans-4-Hexenol (Hirao and Shogaki, 1981)
    α- and β-Ionone (Hirao and Shogaki, 1981)
    2-Isopropyl-phenol (Hirao and Shogaki, 1981)
    trans-Linalool oxide (Hirao and Shogaki, 1981)
    Thymol (Hirao and Shogaki, 1981)
    p-Tolyl-propylene (Hirao and Shogaki, 1981)
    2,5,8,-Trimethyl-dihydronaphthalene (Hirao and Shogaki, 1981)

FLAVONOIDS
  Flavanols
    Monomers
      Catechin (Weinges *et al.*, 1968a, b)
      Epicatechin (Weinges *et al.*, 1968a, b)
      Epigallocatechin (Weinges *et al.*, 1968a, b)
      Gallocatechin (Weinges *et al.*, 1968a, b)
    Dimers
      Catechin-Catechin (Stafford *et al.*, 1986)
      Epicatechin-Catechin (Stafford *et al.*, 1986)
      Epigallocatechin-Catechin (Stafford *et al.*, 1986)
      Gallocatechin-Catechin (Stafford *et al.*, 1986)
      Procyanidin (Weinges *et al.*, 1969)
      Prodelphinidin (Weinges *et al.*, 1969)
  Flavones
    Aglycones
      Apigenin (Hasler, 1990)
      Luteolin (Hasler, 1990)

Table 3 continued

Monoglycosides
　　7-O-[β-D-glucosyl]apigenin (Hasler, 1990)
　　3'-O-[β-D-glucosyl]luteolin (Hasler, 1990)
　　A delphidenone glycoside (Weinges *et al.*, 1968)
Biflavones
　　Amentoflavone (Lobstein-Guth *et al.*, 1988)
　　7-Methoxyamentoflavone (Gobbato *et al.*, 1996)
　　Bilobetin (Baker *et al.*, 1959, 1963)
　　5'-Methoxybilobetin (Joly *et al.*, 1980)
　　Ginkgetin (Furukawa, 1929; Baker and Simmonds, 1940; Nakazawa, 1941)
　　Isoginkgetin (Baker *et al.*, 1959, 1963)
　　Sciadopitysin (Baker *et al.*, 1959, 1963)
Flavonols
Aglycones
　　Isorhamnetin (Fiesel, 1965)
　　Kaempferol (Fiesel, 1965)
　　Myricetin (Hasler, 1990)
　　Quercetin (Fiesel, 1965)
Monoglycosides
　　3-O-[β-D-glucosyl]isorhamnetin (Geiger, 1965; Hasler, 1990)
　　3-O-[β-D-glucosyl]kaempferol (Geiger, 1979; Hasler, 1990)
　　7-O-[β-D-glucosyl]kaempferol (Nasr *et al.*, 1986)
　　3-O-[α-L-rhamnosyl]kaempferol (Hasler, 1990)
　　A myricetin glycoside (Geiger, 1979)
　　3-O-[β-D-glucosyl]quercetin (Weinges *et al.*, 1968)
　　3-O-[α-L-rhamnosyl]quercetin (Nasr *et al.*, 1987)
Diglycosides
　　3-O-[6-O-(α-L-rhamnosyl)-β-D-glucosyl]isorhamnetin (Geiger, 1979)
　　3-O-[2-O-(β-D-glucosyl)-α-L-rhamnosyl]kaempferol (Hasler, 1990)
　　3-O-[6-O-(α-L-rhamnosyl)-β-D-glucosyl]kaempferol (Geiger, 1965)
　　3-O-[6-O-(α-L-rhamnosyl)-β-D-glucosyl]myricetin (Hasler, 1990)
　　3-O-[6-O-(α-L-rhamnosyl)-β-D-glucosyl]-3'-methylmyricetin (Geiger 1979; Hasler, 1990)
　　3-O-[2-O-(β-D-glucosyl)-α-L-rhamnosyl]quercetin (Hasler, 1990)
　　3-O-[6-O-(α-L-rhamnosyl)-β-D-glucosyl]quercetin (Geiger, 1965)
　　3-O-[6-O-(α-L-rhamnosyl)-β-D-glucosyl]syringetin (Victoire *et al.*, 1988)
Acylated diglycosides
　　3-O-[2-O-(6-O-[p-hydroxy-trans-cinnamoyl]-β-D-glucosyl)-α-L-rhamnosyl]-kaempferol (Nasr *et al.*, 1986; Hasler, 1990; Kang *et al.*, 1990)
　　3-O-[2-O-(6-O-[p-hydroxy-trans-cinnamoyl]-β-D-glucosyl)-α-L-rhamnosyl]-quercetin (Nasr *et al.*, 1986; Hasler, 1990; Kang *et al.*, 1990)
Triglycosides
　　3-O-[2-O, 6-O-bis(α-L-rhamnosyl)-β-D-glucosyl]isorhamnetin (Schennen, 1988)
　　3-O-[2-O, 6-O-bis(α-L-rhamnosyl)-β-D-glucosyl]kaempferol (Schennen, 1988)
　　3-O-[2-O, 6-O-bis(α-L-rhamnosyl)-β-D-glucosyl]quercetin (Schennen, 1988)
Acylated triglycosides
　　3-O-[2-O-[6-O-(p-[β-D-glucosyl]oxy-trans-cinnamoyl)-β-D-glucosyl]-α-L-rhamnosyl]kaempferol (Hasler, 1990)
　　3-O-[2-O-[6-O-(p-[β-D-glucosyl]oxy-trans-cinnamoyl)-β-D-glucosyl]-α-L-rhamnosyl]quercetin (Hasler, 1990)
　　3-O-[2-O-[6-O-(p-hydroxy-trans-cinnamoyl)-β-D-glucosyl]-α-L-rhamnosyl]-7-O-[β-D-glucosyl]quercetin (Hasler, 1990)

Table 3   continued

ORGANIC ACIDS
>   *Ascorbic acid* (Franke, 1965a, b)
>   *D-glucaric acid* (Dittrich and Kandler, 1970)
>   *Hydroxybenzoic acid* (Juretzek and Spiess, 1993)
>   *Protocatechouic acid* (Juretzek and Spiess, 1993)
>   *Quinic acid* (Weinstein *et al.*, 1962)
>   *Shikimic acid* (Yamashita and Sato, 1930)
>   *Vanillic acid* (Juretzek and Spiess, 1993)

POLYACETATE DERIVED COMPOUNDS
Alkyl phenolic acids and alkyl phenols
>   *Anacardic and resorcylic acids* (Gellerman and Schlenk, 1968; Irie *et al.*, 1996)
>   *Cardanols and cardols* (Chung *et al.*, 1982)

Long chain hydrocarbons
>   *1-Hexacosanol* (Sosa, 1947)
>   *2-Hexenal* (Major *et al.*, 1960, 1963)
>   *n-Nonacosane* (Ageta, 1959)
>   *n-Nonacosan-10-ol* (= *ginnol*) (Furukawa, 1932)
>   *n-Nonacosan-10-one* (= *ginnon*) (Ageta, 1959)
>   *n- Octacosan-1-ol* (Ageta, 1959)

Lipids
>   *Linolenic acid* (Tsujimoto, 1940)

CARBOHYDRATES
>   *Sucrose* (Plouvier, 1952)
>   **Cyclites**
>   *Pinitol* (Plouvier, 1952)
>   *Sequoyitol* (Plouvier, 1957)
>   **Polysaccharides** (Höllriegl *et al.*, 1986; Kraus, 1991)

MISCELLANEOUS ORGANIC COMPOUNDS
>   **Glycerol derivatives** (Kim *et al.*, 1994)
>   **1-Hydroxypyrene conjugates** (Nakajima *et al.*, 1996)
>   **N-containing compounds**
>   Cytokinins
>   *Isopentenyladenine* (van Staden, 1978; van Staden *et al.*, 1983)
>   *Ribosylzeatin* (van Staden, 1978; van Staden *et al.*, 1983)
>   *Zeatin and dihydrozeatin* (van Staden, 1978; van Staden *et al.*, 1983)
>   *Zeatin glucoside* (van Staden, 1978; van Staden *et al.*, 1983)
>   *Hexosamines* (Racusen and Foote, 1974)
>   *6-Hydroxykynurenic acid* (Schennen and Hölzl, 1986)
>   *Lectins* (Clarke *et al.*, 1978)
>   *Pentadien-1,5-diyl-diphenol* (Plieninger *et al.*, 1986)

INORGANIC COMPOUNDS
>   *Calcium oxalate* (Umemoto *et al.*, 1972)
>   **Trace elements**
>   *Se, Cu, Zn, Cr, Fe, Mn* (Yu *et al.*, 1992)

Table 4   Constituents of gynoceum from *G. biloba*

FLAVONOIDS
  Flavones
    Biflavones
      Bilobetin (Pan et al., 1995)
      Ginkgetin (Pan et al., 1995)
      Isoginkgetin (Pan et al., 1995)
      5'-Methoxybilobetin (Pan et al., 1995)
      Sciadopitysin (Pan et al., 1995)

ORGANIC ACIDS
  Acetic acid (Griebel, 1939)
  Butyric acid (Peschier, 1818; Schwarzenbach, 1857; Griebel, 1939)
  Formic acid (Griebel, 1939)
  Hexanoic acid (Griebel, 1939)
  Valeric acid (Griebel, 1939)

POLYACETATE DERIVED COMPOUNDS
  Alkyl phenolic acids and alkyl phenols
      Anacardic and resorcylic acids (Peschier, 1818; Schwarzenbach, 1857; Kawamura, 1928; Fu *et al.*,
         1962; Morimoto *et al.*, 1968; Gellerman *et al.*, 1976a, b; Adawadkar and El Sohly, 1981;
         Irie *et al.*, 1996)
      Cardanols and cardols (Kawamura, 1928; Morimoto *et al.*, 1968; Adawadkar and El Sohly, 1981;
         Itokawa, 1987; Itokawa and Takeya, 1993)
      Alkyl coumarins (Choukchou-Braham *et al.*, 1994)
  Lipids
      Campesterol (Urakami *et al.*, 1976)
      Cholesterol (Urakami *et al.*, 1976)
      22-Dihydrobrassicasterol (Chung and Shin, 1978)
      Linoleic acids (Gellerman and Schlenk, 1969)
      14-Methylhexadecanoic acid (Hierro *et al.*, 1996)
      5,9-Octadecadienoic acid (Hierro *et al.*, 1996)
      5,9,12-Octadecatrienoic acid (Hierro *et al.*, 1996)
      Oleic acid (Gellerman and Schlenk, 1969)
      Palmitic acid (Gellerman and Schlenk, 1969)
      β-Sitosterol (Urakami *et al.*, 1976)
      Stigmasterol (Urakami *et al.*, 1976)
  Long chain hydrocarbons
    Ginnol (Kawamura (1928)

MISCELLANEOUS ORGANIC COMPOUNDS
  N-containing compounds
    Amino acids and derivatives
      ψ-Amino butyric acid (Hatano, 1956)
      Asparagine (Kawamura, 1928; Oh *et al.*, 1995)
      Aspartic acid (Oh *et al.*, 1995)
      Glutamine (Oh *et al.*, 1995)
      Glycine (Oh *et al.*, 1995)
      4-O-Methylpyridoxine (Wada *et al.*, 1988)
      Proline (Oh *et al.*, 1995)
      Pyroglutamic acid (Oh *et al.*, 1995)
      Serine (Oh *et al.*, 1995)
      Threonine (Oh *et al.*, 1995)
      Valine (Hatano, 1956)

INORGANIC COMPOUNDS
  Acids
    Phosphoric acid (Kawamura, 1928)

changes which might appear in plant tissue. A case in point proved to be Ginkgo leaves where significant changes in the distribution of inorganic crystal deposits can be observed. The crystal material is calcium oxalate, a fact which suggests a biochemical synthesis in the leaf of oxalic acid from inspired carbon monoxide, carbon dioxide, and water, with subsequent neutralization by calcium ions.

Trace elements

Yu *et al.* (1992) have studied the content of trace elements in Ginkgo leaves and found 0.1–2.17 ppm of Se, 2.8-6.9 ppm of Cu, 6.1–17.1 ppm of Zn, 1.1–6.6 ppm of Cr, 74–399 ppm of Fe and 15–73 ppm of Mn.

Acknowledgment

I would like to thank Annemarie Suter for performing literature searches and for editorial assistance.

Literature

Adawadkar, P.D., El Sohly, M.A. (1981) Isolation, purification and antimicrobial activity of anacardic acids from *Ginkgo biloba* fruits. *Fitoterapia*, **52**, 129–135.

Ageta, H. (1959) Chemical constituents of the plants of coniferae and allied orders. XXVI. Studies on plant waxes. 13. Waxes from leaves of *Ginkgo biloba* and from *Ephedra gerardiana. Yakugaku Zasshi*, **79**, 58–60.

Baker, W., Simmonds, W.H.C. (1940) Derivatives of 5:6:4′- and 5:8:4′-trihydroxyflavones, and a note on the structure of ginkgetin. *J. Chem. Soc.*, 1370–1374.

Baker, W., Finch, A.C.M., Ollis, W.D., Robinson, K.W. (1959) Biflavonyls, a new class of natural product. The structures of ginkgetin, isoginkgetin and sciadopitysin. *Proc. Chem. Soc.*, 91–92.

Baker, W., Finch, A.C.M., Ollis, W.D., Robinson, K.W. (1963) The structures of the naturally occurring biflavonyls. *J. Chem. Soc.*, 1477–1490.

van Beek, T.A., Scheeren, H.A., Rantio, T., Melger, W.Ch., Lelyveld, G.P. (1991) Determination of ginkgolides and bilobalide in *Ginkgo biloba* leaves and phytopharmaceuticals. *J. Chromatogr.*, **543**, 375–387.

van Beek, T.A., Lankhorst, P.P. (1996) Confirmation of the C–1 stereochemistry of ginkgolides by NMR. *Tetrahedron*, **52**, 4505–4514.

van Beek, T.A., Lelyveld, G.P. (1997) Preparative isolation and separation procedure for ginkgolides A, B, C and J and bilobalide. *J. Nat. Prod.*, **60**, 735–738.

Boralle, N., Braquet, P., Gottlieb, O.R. (1988) *Ginkgo biloba: a review of its chemical composition.* In Braquet, P. (ed.) *Ginkgolides – Chemistry, Biology, Pharmacology and Clinical Perspectives*, J.R. Prous Science Publishers, Barcelona, pp. 9–25.

Casal, H.L., Moyna, P. (1979) Components of *Ginkgo biloba* leaf wax. *Phytochemistry*, **18**, 1738–1739.

Chang, S.K., Youm, J.R., Kang, S.S. (1993) Seasonal variations of biflavone content from *Ginkgo biloba* leaves. *Saengyak Hakhoechi*, **24**, 54–57. Cited from: *Chem. Abstr.* (1993) **119**, 210366, 472.

Choukchou-Braham, N., Asakawa, Y., Lepoittevin, J.P. (1994) Isolation, structure determination and synthesis of new dihydroisocoumarins from *Ginkgo biloba* L. *Tetrahedron Lett.*, **35**, 3949–3952.

Chung, A.S., Shin, H.S. (1978) Studies on the lipid components of Ginkgo nut. *Korean J. Food Sci. Technol.*, **10**, 119–123.

Chung, B.Y., Won, L.S., Lee, B.R., Lee, C.H. (1982) A new chemical constituent of green leaves of *Ginkgo biloba* L. *Daehan Hwahak Hwoejee* (*J. Korean Chem. Soc.*), **26**, 95–98.

Clarke, A.E., Gleeson, P.A., Jermyn, M.A., Knox, R.B. (1978) Characterization and localization of β-lectins in lower and higher plants. *Aust. J. Plant Physiol.*, **5**, 707–722.

Dittrich, P., Kandler, O. (1970) D-Glucarsäure in Nadeln von *Larix decidua* Mill. und anderen Gymnospermen. *Z. Pflanzenphysiol.*, **62**, 116–123.

Dupont, L., Dideberg, O., Germain, G., Braquet, P. (1986a) Structure of ginkgolide B (BN 52021) monohydrate, a highly specific PAF/acether receptor antagonist isolated from *Ginkgo biloba* L. *Acta Cryst.*, **C42**, 1759–1762.

Dupont, L., Germain, G., Dideberg, O. (1986b) Crystal and molecular structure of BN52021, a PAF-acether antagonist. Comparison with the conformation of kadsurenone and related compounds. *Pharmacol. Res. Commun.*, **18**, 25–32

Feistel, B. (1994) *Untersuchungen zur analytischen Standardisierung von Ginkgo biloba Extrakten und deren galenischer Zubereitungen*, Dissertation, Martin-Luther-Universität, Halle, Deutschland.

Fiesel J. (1965) Kämpferol, Quercetin und Isorhamnetin aus grünen Blättern von *Ginkgo biloba* L. *Naturwissenschaften*, **52**, 592.

Flesch, V., Jacques, M., Cosson, L., Teng, B.P., Petiard, V., Balz, J.P. (1992) Relative importance of growth and light level on terpene content of *Ginkgo biloba*. *Phytochemistry*, **31**, 1941–1945.

Franke, W. (1965a) Über den Vitamin-C-Gehalt der Herbstblätter. 1. Mitteilung: Der Vitamin-C-Gehalt bis zum Laubabwurf. *Z. Pflanzenphysiol.*, **53**, 289–319.

Franke, W. (1965b) Über den Vitamin-C-Gehalt der Herbstblätter. 2. Mitteilung: Der Vitamin-C-Gehalt nach dem Laubabwurf. *Z. Pflanzenphysiol.*, **53**, 289–319.

Fu, F.Y., Yu, D.Q., Sung, W.L., Jai, Y.F., Sun, N.C. (1962) The chemical study of hydroginkgolinic acid. A new constituent of *Ginkgo biloba* L. *Hua Hsueh Hsueh Pao*, **28**, 52–56.

Furukawa, S. (1932) Studies on the constituents of the "*Ginkgo biloba*" L., leaves. Part I and II. *Scientific Papers Institute Physical Chemical Research*, **19**, 27–42.

Furukawa, S. (1933) Studies on the constituents of the "*Ginkgo biloba*" L., leaves. Part III and IV. *Scientific Papers Institute Physical Chemical Research*, **21**, 273–285.

Furukawa, S. (1934) Studies on the constituents of the *Ginkgo biloba* L., fruits, parts I, II and III. *Scientific Papers Institute Physical Chemical Research*, **24**, 304–324.

Geiger, H., Beckmann, S. (1965) Über das Vorkommen von Rutin und Kämpferol–3-rhamnoglucosid in *Ginkgo biloba* L. *Z. Naturforsch.*, **20b**, 1139–1140.

Geiger, H. (1979) 3'-O-Methylmyricetin–3-rhamnoglucosid, ein neues Flavonoid aus dem Herbstlaub von *Ginkgo biloba* L. *Z. Naturforsch.*, **34c**, 878–879.

Gellerman, J.L., Schlenk, H. (1968) Methods for isolation and determination of anacardic acids. *Anal. Chem.*, **40**, 739–743.

Gellerman, J.L., Schlenk, H. (1969) Preparation of ^{14}C-labeled fatty and anacardic acids from *Ginkgo biloba*. *Lipids*, **4**, 484–487.

Gellerman, J.L., Anderson, W.H., Schlenk, H. (1974) Biosynthesis of anacardic acids from acetate in *Ginkgo biloba*. *Lipids*, **9**, 722–725.

Gellerman, J.L., Anderson, W.H., Schlenk, H. (1976a) 6-(Pentadec-8-enyl)–2,4-dihydroxy-benzoic acid from seeds of *Ginkgo biloba*. *Phytochemistry*, **15**, 1959–1961.

Gellerman, J.L., Anderson, W.H., Schlenk, H. (1976b) Synthesis of anacardic acids in seeds of *Ginkgo biloba*. *Biochim. Biophys. Acta*, **431**, 16–21.

Gobbato, S., Griffini, A., Lolla, E., Peterlongo, F. (1996) HPLC quantitative analysis of biflavones in *Ginkgo biloba* leaf extracts and their identification by thermospray liquid chromatography-mass spectrometry. *Fitoterapia*, **67**, 152–158.

Griebel, C. (1939) Beiträge zur Kenntnis der Ginkgosamen. *Dtsch. Apoth. Ztg.*, **54**, 603–608.

Gülz, P.G., Müller, E., Schmitz, K., Marner, F.J., Güth, S. (1992) Chemical composition and surface structures of epicuticular leaf waxes of *Ginkgo biloba*, *Magnolia grandiflora* and *Liriodendron tulipifera*. *Z. Naturforsch.*, **47c**, 516–527.

Guth, H. (1977) *Untersuchungen zur Biosynthese der Flavonoide in Ginkgo biloba und deren Steuerung durch exogene Faktoren*, Dissertation, Universität Karlsruhe, Deutschland.

Hasler, A., Meier, B., Sticher, O. (1989) HPLC – Analysis of 5 widespread flavonoid aglycones. *Planta Med.*, **55**, 538.

Hasler, A. (1990) *Flavonoide aus Ginkgo biloba L. und HPLC-Analytik von Flavonoiden in verschiedenen Arzneipflanzen*, Dissertation 9353, Eidgenössische Technische Hochschule (ETH), Zürich, Schweiz.

Hasler, A., Meier, B., Sticher, O. (1990a) HPLC-analysis of 5 widespread flavonoid aglycones. *J. Chromatogr.*, **508**, 236.

Hasler, A., Meier, B., Sticher, O. (1990b) Quantitative HPLC-analysis of flavonoid aglycones in different medicinal plants, *Planta Med.*, **56**, 575.

Hasler, A., Meier, B., Sticher, O. (1990c) *Ginkgo biloba* – Botanische, analytische und pharmakologische Aspekte. *Schweiz. Apoth. Ztg.*, **128**, 341–347.

Hasler, A., Meier, B., Sticher, O. (1991) Qualitative HPLC-analysis of flavonoids of *Ginkgo biloba* L. *Planta Med.*, **57**, Suppl. 2A, 120.

Hasler, A., Meier, B., Sticher, O. (1992a) Identification and determination of the flavonoids from *Ginkgo biloba* by high-performance liquid chromatography. *J. Chromatogr.* **605**, 41–48.

Hasler, A., Gross, G.A., Meier, B., Sticher, O. (1992b) Complex flavonol glycosides from the leaves of *Ginkgo biloba*. *Phytochemistry*, **31**, 1391–1394.

Hasler, A., Meier, B., Sticher, O. (1992c) HPLC-analysis of the main Ginkgo acylflavonoids. *Planta Med.*, **58**, A688.

Hasler, A., Meier, B. (1992) Determination of terpenes from *Ginkgo biloba* L. by capillary gas chromatography. *Pharm. Pharmacol. Lett.*, **2**, 187–190.

Hasler, A. (1993) Ginkgo, ein modernes Phytopharmakon? *Schweiz. Apoth. Ztg.*, **131**, 315–316.

Hasler, A., Meier, B. (1993) *Ginkgo biloba* – Content of flavonoids and terpenes from leaves during the harvest time and from full extracts determined by chromatographic and biologic methods. *Planta Med.*, **59**, A632.

Hata, K., Sogo, M., Fukuhara, T., Hochi, M. (1969) On the suberin in the outer bark of some Japanese tree species. *Kagawa Daigaku Nogakubu Gakuzyutu Hokoku*, **20**, 112–119.

Hatano, K.I. (1956) A trial to paper chromatography. *Nippon Ringaku Kaishi*, **38**, 425–427.

Hierro, M.T.G., Robertson, G., Christie, W.W., Joh, Y.G. (1996) The fatty acid composition of the seeds of *Ginkgo biloba*. *J. Am. Oil Chem. Soc.*, **73**, 575–579.

Hirao, N., Shogaki, T. (1981) On the essential oil of *Ginkgo biloba* L. (Icho). *Kinki Daigaku Rikogakubu Ken Kyu Hokoku*, **47**, 47–50.

Höllriegl, H., Koehler, H., Franz, G. (1986) Über höhermolekulare Polysaccharide aus den Blättern von *Ginkgo biloba. Sci. Pharm.*, **54**, 321–330.

Hölzl, J. (1992) Inhaltsstoffe von *Ginkgo biloba. Pharmazie u. Zeit*, **21**, 215–223.

Holloway, P.J., Jeffree, C.E., Baker, E.A. (1976) Structural determination of secondary alcohols from plant epicuticular waxes. *Phytochemistry*, **15**, 1768–1770.

Huh, H., Staba, E.J., Singh, J. (1992) Supercritical fluid chromatographic analysis of polyprenols in *Ginkgo biloba* L. *J. Chromatogr.*, **600**, 364–369.

Huh, H., Staba, E.J. (1993) Ontogenetic aspects of ginkgolide production in *Ginkgo biloba. Planta Med.*, **59**, 232–239.

Huh, H., Singh, J., Staba, E.J. (1993) Ontogenetic aspects of Ginkgo polyprenols. *Planta Med.*, **59**, 379–380.

Hunneman, D.H., Eglinton, G. (1972) The constituent acids of gymnosperm cutins. *Phytochemistry*, **11**, 1989–2001.

Ibata, K., Mizuno, M., Takigawa, T., Tanaka, Y. (1983) Long-chain betulaprenol-type polyprenols from the leaves of *Ginkgo biloba. Biochem. J.*, **213**, 305–311.

Ibata, K., Mizuno, M., Tanaka, Y., Kageyu, A. (1984a) Long-chain polyprenols in the family Pinaceae. *Phytochemistry*, **23**, 783–786.

Ibata, K., Kageyu, A., Takigawa, T., Okada, M., Nishida, T., Mizuno, M., Tanaka, Y. (1984b) Polyprenols from conifers: multiplicity in chain length distribution. *Phytochemistry*, **23**, 2517–2521.

Irie, H., Ohno, K., Ito, Y., Uyeo, S. (1975) Isolation and characterisation of 10,11-dihydro-atlantone and related compounds of *Ginkgo biloba* L. *Chem. Pharm. Bull.*, **23**, 1892–1894.

Irie, J., Murata, M., Homma, S. (1996) Glycerol–3-phosphate dehydrogenase inhibitors, anacardic acids from *Ginkgo biloba. Biosci. Biotech. Biochem.*, **60**, 240–243.

Itokawa, H., Totsuka, N., Nakahara, K., Takeya, K., Lepoittevin, J.P., Asakawa, Y. (1987) Antitumor principles from *Ginkgo biloba* L. *Chem. Pharm. Bull.*, **35**, 3016–3020.

Itokawa, H., Takeya, K., (1993) Antitumor substances from higher plants. *Heterocycles*, **35**, 1467–1501.

Joly, M., Haag-Berrurier, M., Anton, R. (1980) La 5′-méthoxybilobétine, une biflavone extraite du *Ginkgo biloba. Phytochemistry*, **19**, 1999–2002.

Joyeux, M., Lobstein, A., Anton, R., Mortier, F. (1995) Comparative antilipoperoxidant, anti-necrotic and scavenging properties of terpenes and biflavones from Ginkgo and some flavonoids. *Planta med.*, **61**, 126–129.

Juretzek, W., Spiess, E. (1993) Ginkgo. In Hänsel, R., Keller, K., Rimper, H., Schneider G., (Hrsg.) *Hagers Handbuch der Pharmazeutischen Praxis*, 5. vollständig neubearbeitete Auflage, Springer Verlag, Berlin, pp. 269–292.

Kameyama, H., Urakami, C. (1979) Glycolipids isolated from Ginkgo nut (*Ginkgo biloba*) and their fatty acid composition. *J. Am. Oil Chem. Soc.*, **56**, 549–551.

Kang, S.S., Kim, J.S., Kwak, W.J., Kim, K.H. (1990) Structures of two acylated flavonol glucorhamnosides from *Ginkgo biloba* leaves. *Arch. Pharm Res.*, **2**, 207–210.

Kang, G.S., Youn, J.R., Kang, S.S. (1993) Seasonal variations of the flavonol glycoside content from *Ginkgo biloba* leaves. *Saengyak Hakhoechi*, **24**, 47–53. Cited from: *Chem. Abstr. (1993)* **119**, 233790.

Kang, S.S., Koh, Y.M., Kim, J.S., Lee, M.W., Lee, D.S. (1995) Phytochemical analysis of *Ginkgo biloba* yellow leaves. *Kor. J. Pharmacogn.*, **26**, 23–26.

Kariyone, T., Kimura, H., Nakamura, I. (1958) Studies on the components of *Ginkgo biloba* L. I. Isolation of bilobanone. *J. Pharmaceut. Soc. Japan*, 78, 1152–1155.

Karrer, P., Walker, O. (1934) Pflanzenfarbstoffe LVI. Untersuchungen über die herbstlichen Färbungen der Blätter. *Helv. Chim. Acta*, 17, 43–54.

Kawamura, J. (1928) Ueber die chemischen Bestandteile der Frucht von *Ginkgo biloba*. *Japanese J. Chem.*, 3, 89–108.

Kim, B.E., Ban, S.H., Woo, S.H., Chung, S.K. (1994) A study on the composition of *Ginkgo biloba* leaves. *Yongu Nunmun*, 8, 105–114. Cited from: *Chem. Abstr. (1994)* 121, 17799.

Kimura, H. (1962) Studies on the component of *Ginkgo biloba* L. IV. Structure and reduction of bilobanone by lithium in ethylenediamine. *Yakugaku Zasshi*, 82, 214–218.

Kimura, H., Irie, H., Ueda, K., Uyeo S. (1968) The constituents of the heartwood of *Ginkgo biloba* L. V. The structure and absolute configuration of bilobanone. *Yakugaku Zasshi*, 88, 562–572.

Kircher, H.W. (1970) β-Sitosterol in *Ginkgo biloba* leaves. *Phytochemistry*, 9, 1879.

Kraus, J. (1991) Water-soluble polysaccharides from *Ginkgo biloba* leaves. *Phytochemistry*, 30, 3017–3020.

Kuraray Co. Ltd. (1982) Patent. Polyprenyl compounds from Ginkgo for pharmaceutical and cosmetic bases. *Jpn. Kokai Tokyo Koho JP*, 91, 931. Cited from: *Chem. Abstr. (1982)* 97, 222939.

Lang, F., Wilhelm, E. (1996) Quantitative determination of proanthocyanidins in *Ginkgo biloba* special extracts. *Pharmazie*, 51, 734–737.

Llabres, G., Baiwir, M., Sbit, M., Dupont, L. (1989) Ginkgolides A, B and C: a 1D and 2D NMR study. *Spectrochim. Acta A. Molecular Spectros.* 45, 1037–1045.

Lobstein-Guth, A., Briançon-Scheid, F., Victoire, C., Haag-Berrurier, M. (1988) Isolation of amentoflavone from *Ginkgo biloba*. *Planta Med.*, 54, 555–556.

Major, R.T., Marchini, P., Sproston, T. (1960) Isolation from *Ginkgo biloba* L. of an inhibitor of fungus growth. *J. Biol. Chem.*, 235, 3298–3299.

Major, R.T., Marchini, P., Boulton, A.J. (1963) Observations on the production of α-hexenal by leaves of certain plants. *J. Biol. Chem.*, 238, 1813–1816.

Major, R.T. (1967) The Ginkgo, the most ancient living tree. *Science*, 157, 1270–1273.

Major, R.T., Thomas, M. (1972) Formation of 2-hexenal from linolenic acid by macerated Ginkgo leaves. *Phytochemistry*, 11, 611–617.

Maruyama, M., Terahara, A., Itagaki, Y., Nakanishi, K. (1967a) The ginkgolides. I. Isolation and characterization of the various groups. *Tetrahedron Lett.*, 4, 299–302.

Maruyama, M., Terahara, A., Itagaki, Y., Nakanishi, K. (1967b) The ginkgolides. II. Derivation of partial structures. *Tetrahedron Lett.*, 4, 303–308.

Maruyama, M., Terahara, A., Nakadaira Y., Woods, M.C., Nakanishi, K. (1967c) The ginkgolides. III. The structure of the ginkgolides. *Tetrahedron Lett.*, 4, 309–313.

Maruyama, M., Terahara, A., Nakadaira, Y., Woods, M.C., Takagi, Y., Nakanishi, K. (1967d) The ginkgolides. IV. Stereochemistry of the ginkgolides. *Tetrahedron Lett.*, 4, 315–319.

Maruyama, M., Terahara A. (1967) The NMR data of the ginkgolides and derivatives. *Sci. Rep. Tôhoku Univ. Series I*, 50, 92–99.

Matile, Ph., Flach, B.M.P., Eller, B.M. (1992) Autumn leaves of *Ginkgo biloba* L.: optical properties, pigments and optical brighteners. *Bot. Acta*, 105, 13–17.

Mian, A.J., Timell, T.E. (1960) Studies on *Ginkgo biloba* L., 2. The constitution of an arabino-4-O-methylglucorono-xylan from the wood. *Svensk Papperstidning Arg.*, 63, 769–774.

Morimoto, H., Kawamatsu, Y., Sugihara, H. (1968) Sterische Struktur der Giftstoffe aus dem Fruchtfleisch von *Ginkgo biloba* L. *Chem. Pharm. Bull.*, 16, 2282–2286.

Nakajima, D., Kojima, E., Iwaya, S., Suzuki, J., Suzuki, S. (1996) Presence of 1-hydroxy-pyrene in woody plant leaves and seasonal changes in their concentrations. *Environ. Sci. Technol.*, 30, 1675–1679.

Nakanishi, K. (1967) The ginkgolides. *Pure Appl. Chem.*, 14, 89–113.

Nakanishi, K., Habaguchi, K. (1971) Biosynthesis of ginkgolide B, its diterpenoid nature, and origin of the *tert*-butyl group. *J. Am. Chem. Soc.*, 93, 3546–3547.

Nakanishi, K., Habaguchi, K., Nakadaira, Y., Woods, M.C., Maruyama, M., Major, R.T., Alauddin, M., Patel, A.R., Weinges, K., Bähr, W. (1971) Structure of bilobalide, a rare *tert*-butyl containing sesquiterpenoid related to the C–20 ginkgolides. *J. Am. Chem. Soc.*, 93, 3544–3546.

Nakazawa, J. (1941) *J. Pharm. Soc. Japan*, 61, 174.

Nasr, C., Haag-Berrurier, M., Lobstein-Guth, A., Anton, R., (1985) Flavonol glycosides of *Ginkgo biloba* L. *Acta Agron. Hung.*, 34 (Suppl.), 73.

Nasr, C., Haag-Berrurier, M., Lobstein-Guth, A, Anton, R. (1986) Kaempferol coumaroyl glucorhamnoside from *Ginkgo biloba*. *Phytochemistry*, 25, 770–771.

Nasr, C., Lobstein-Guth, A, Haag-Berrurier, M., Anton, R. (1987) Quercetin coumaroyl glucorhamnoside from *Ginkgo biloba*. *Phytochemistry*, 26, 2869–2870.

Nes, W.R., Krevitz, K., Joseph, J., Nes W.D., Harris, B., Gibbons, G.F., Patterson, G.W. (1977) The phylogenetic distribution of sterols in tracheophytes. *Lipids*, 12, 511–527.

Nip, M., Tegelaar, E.W., Brinkhuis, H., de Leeuw, J.W., Schenck, P.A., Holloway, P.J., (1985) Analysis of modern and fossil plant cuticles by Curie point Py-GC and Curie point Py-GC-MS: Recognition of a new, highly aliphatic and resistant biopolymer. *Org. Geochem.*, 10, 769–778.

Oh, C.H., Kim, J.H., Kim, K.R., Mabry, T.J. (1995) Rapid gas chromatographic screening of edible seeds, nuts and beans for non-protein and protein amino acids. *J. Chromatogr.*, 708A, 131–141.

Okabe, K., Yamada, K., Yamamura, S., Takada, S. (1967) Ginkgolides. *J. Chem. Soc. (C)*, 2201–2206.

Pan, J., Zhang, H., Tang, W., Hong, M. (1995) Biflavones from the testa of *Ginkgo biloba* L. *Zhiwu Ziyuan Yu Huanjing*, 4, 17–21. Cited from: *Chem. Abstr.* (1995) 123, 165078.

Parliment, T.H. (1995) Characterization of the putrid aroma compounds of *Ginkgo biloba* fruits. ACS *Symp. Ser.*, 59b, 276–279.

Peschier (1818) Recherches analytiques sur le fruit du Ginkgo. *Bibl. uni. Genève, Sc. Arts.*, 7, 3me année, 29–34.

Plieninger, H., Schwarz, B., Jaggy, H., Huber-Patz, U., Rodewald, H., Irngartinger, H., Weinges, K. (1986) Isolierung, Strukturaufklärung und Synthese von (Z,Z)-4-4'-(1,4-Pentadien-1,5-diyl)diphenol, einem ungewöhnlichen Naturstoff aus den Blättern des Ginkgo-Baumes (*Ginkgo biloba* L.). *Liebigs Ann. Chem.*, 1772–1778.

Plouvier, M.V. (1952) Sur la recherche du pinitol chez quelques conifères et plantes voisines. *Comptes Rendues*, 234, 362–364.

Plouvier, V.M. (1957) Sur la recherche du séquoyitol et du pinitol chez quelques gymnospermes. *Comptes Rendues*, 245, 2377–2379.

Racusen, D., Foote, M. (1974) The hexosamine content of leaves. *Can. J. Bot.*, **52**, 2111–2113.

Sagisaka, S. (1993) Amino acid metabolism in nongrowing environments in higher plants. *Amino Acids*, **4**, 141–155.

Sakabe, N., Takada, S., Okabe, K. (1967) The structure of ginkgolide A, a novel diterpenoid trilactone. *Chem. Commun.*, 259–261.

Schennen, A., Hölzl, J. (1986) 6-Hydroxykynurensäure, die erste *N*-haltige Verbindung aus den Blättern von *Ginkgo biloba*. *Planta Med.*, **52**, 235–236.

Schennen, A. (1988) *Neue Inhaltsstoffe aus den Blättern von Ginkgo biloba L. sowie Präparation* [14]*C-markierter Ginkgo-Flavonoide*, Dissertation, Universität Marburg, Deutschland.

Schwarzenbach (1857) Untersuchungen des Fruchtfleisches von *Salisburia adiantifolia* (*Ginkgo biloba*). *Wittsteins Vierteljahresschr. prakt. Pharm.* **6**, 424.

Seshadri T.R. (1969) A useful development in the chemistry of flavonoids as natural products. *Proc. Natn. Inst. Sci. India*, **35B**, 438–445.

Seoane, E., Arino, P. Ahsan, A.M. (1969) Identity of celidoniol and ginnol. *Anales Quimica*, **65**, 303–304.

Sosa, A. (1947) Sur quelques constituants du *Ginkgo biloba* L. *Bull. Soc. Chim. Biol.*, **29**, 833–836.

van Staden, J. (1978) A comparison of the endogenous cytokinins in the leaves of four gymnosperms. *Bot. Gaz.*, **139**, 32–35.

van Staden, J., Hutton, M.J., Drewes, S.E. (1983) Cytokinins in the leaves of *Ginkgo biloba*. *Plant Physiol.*, **73**, 223–227.

van Staden, J. (1996a) Changes in foliar cytokinins of *Salix babylonica* and *Ginkgo biloba* prior to and during leaf senescence. *S. Afr. J. Bot.*, **62**, 1–10.

van Staden, J., (1996b) Metabolism of [3H]-DHZ in mature leaves of *Ginkgo biloba* and *Salix babylonica*. *S. Afr. J. Bot.*, **62**, 111–116.

Stafford, H.A., Kreitlow, K.S., Lester, H.H. (1986) Comparison of proanthocyanidins and related compounds in leaves and leaf-derived cell cultures of *Ginkgo biloba* L., *Pseudotsuga menziesii* Franco, and *Ribes sanguineum* Pursh. *Plant Physiol.*, **82**, 1132–1138.

Sticher, O., Hasler, A., Meier, B. (1991) *Ginkgo biloba* – eine Standortbestimmung. *Dtsch. Apoth. Ztg.*, **131**, 1827–1835.

Sticher, O. (1993) Quality of Ginkgo preparations. *Planta Med.*, **59**, 2–11.

Takagi, T., Itabashi, Y. (1982) *cis*–5-Olefinic unusual fatty acids in seed lipids of gymnospermae and their distribution in triacylglycerols. *Lipids*, **17**, 716–723.

Takigawa, T., Ibata, K., Mizuno, M. (1989) Synthesis of mammalian dolichols from plant polyprenols. *Chem. Phys. Lipids*, **51**, 171–182.

Tateoka, T.N. (1963) On the synthetic pathway of shikimic acid in *Ginkgo biloba*. *Bot. Mag.*, **76**, 391–394.

Teng B.P. (1988) *Chemistry of ginkgolides*. In Braquet, P., (ed.) *Ginkgolides – Chemistry, Biology, Pharmacology and Clinical Perspectives*, J.R. Prous Science Publishers, Barcelona, pp. 37–41.

Tsujimoto, M. (1940) On the occurrence of linolenic acid in the leaves of land plants. *J. Soc. Chem. Ind. Japan*, **43**, 208–209.

Tsumura, Y., Motoike, H., Ohba, K. (1992) Allozyme variation of old *Ginkgo biloba* memorial trees in western Japan. *Can. J. For. Res.*, **22**, 939–944.

Tsuyuki, H., Itoh, S., Nakatsukasa, Y. (1979) Studies on the Lipids in Ginkgo Seeds. *Nihon Daigaku Nojuigakubu Gakujutsu Kenkyu Hokoku*, **36**, 156–162.

Umemoto, K., Hozumi, K. (1972) Correlation between the degree of air pollution and the distribution of calcium oxalate crystals in the Ginkgo leaf. *Microchem. J.*, **17**, 689–702.

Urakami, C., Oka, S., Han, J.S. (1976) Composition of the neutral and phospholipid fractions from Ginkgo nuts (*Ginkgo biloba*) and fatty acid composition of individual lipid classes. *J. Am. Oil Chem. Soc.*, **53**, 525–529.

Vanhaelen, M., Vanhaelen-Fastre, R. (1988) Countercurrent chromatography for isolation of flavonol glycosides from *Ginkgo biloba* leaves. *J. Liquid Chromatogr.*, **11**, 2969–2975.

Vanhaelen, M., Vanhaelen-Fastré, R. (1989) Flavonol triglycosides from *Ginkgo biloba*. *Planta med.*, **55**, 202.

Verotta, L., Peterlongo, F. (1993) Selective extraction of phenolic components from *Ginkgo biloba* extracts using supercritical carbon dioxide and off-line capillary gas chromatography/mass spectrometry. *Phytochem. Anal.*, **4**, 178–182.

Victoire, C., Haag-Berrurier, M., Lobstein-Guth, A., Balz, J.P., Anton, R. (1988) Isolation of flavonol glycosides from *Ginkgo biloba* leaves. *Planta Med.* **54**, 245–247.

Wada, K., Ishigaki, S., Ueda, K., Take, Y., Sasaki, K., Sakata, M., Haga, M. (1988) Studies on the constitution of edible and medicinal plants. I. Isolation and identification of 4-O-methylpyridoxine, toxic principle from the seed of *Ginkgo biloba* L.. *Chem. Pharm. Bull.*, **36**, 1779–1782.

Weinges, K., Bähr, W., Kloss, P. (1968a) Übersicht über die Inhaltsstoffe aus den Blättern des Ginkgo-Baumes (*Ginkgo biloba* L.). *Arzneim.-Forsch./ Drug Res.*, **18**, 537–539.

Weinges, K., Bähr, W., Kloss, P. (1968b) Die phenolischen Inhaltsstoffe aus den Blättern des Ginkgo-Baumes (*Ginkgo biloba* L.). *Arzneim.-Forsch./ Drug Res.*, **18**, 539–543.

Weinges, K., Bähr, W., Theobald, H., Wiesenhütter, A., Wild, R., Kloss, P. (1969) Über das Vorkommen von Proanthocyanidinen in Pflanzenextrakten. *Arzneim.-Forsch./Drug Res.*, **19**, 328–330.

Weinges, K., Bähr, W. (1969) Bilobalid A, ein neues Sesquiterpen mit tert.-Butyl-Gruppe aus den Blättern von *Ginkgo biloba* L. *Liebigs Ann. Chem.*, **724**, 214–216.

Weinges, K., Hepp, M., Jaggy, H. (1987) Isolierung und Strukturaufklärung eines neuen Ginkgolids. *Liebigs Ann. Chem.*, **742**, 521–526.

Weinstein, L.H., Porter, C.A., Laurencot, H.J. (1962) Biosynthesis of uniformly-labeled shikimic and quinic acids in leaves of *Ginkgo biloba* L. *Contribution Boyce Thompson Inst.*, **21**, 439–445.

Woods, M.C., Miura, I., Nakadaira, Y., Terahara, A., Maruyama, M., Nakanishi, K. (1967) The ginkgolides V. Some aspects of their NMR spectra. *Tetrahedron Lett.*, **4**, 321–326.

Yadav, S., Dwivedi, P.D., Tewari, M., Rana, R. (1985a) Photosynthetic pigments in maidenhair tree *Ginkgo biloba*. *Environ. Ecol.*, **3**, 442–443.

Yadav, S., Dwivedi, P.D., Tewari, M., Rana, R. (1985b) Seasonal variation of carotenoid contents in maidenhair tree *Ginkgo biloba*. *Environ. Ecol.*, **3**, 446–447.

Yadav, S., Ralhan, P.K., Singh, S.P. (1987) Qualitative distribution pattern of carotenoids in three selected gymnosperms. *Current Sci.*, **56**, 354–359.

Yamashita, T., Sato, F. (1930) Shikimisäure aus den Blättern von *Ginkgo biloba* L. *J. Pharm. Soc. Jap.*, **50**, 113–117.

Yoshioka, K. (1994) Changes in saccharides and enzyme activities related to metabolism of saccharides during growth and development of ginkgo nut (*Ginkgo biloba* L.) on the tree. *Nippon Shokuhin Kogyo Gakkaishi*, **41**, 259–264. Cited from: *Chem. Abstr. (1994)*, **121**, 78500.

Yu, X., Gu, L., Zhuang, X., Shou, H., Fang, Y. (1992) Study on trace elements in *Ginkgo biloba* leaves in Zhejiang Province. *Fenxi Ceshi Tongbao*, **11**, 69–71. Cited from: *Chem. Abstr. (1994)* **120**, 50174.

Zhang, J., Takahashi, K., Kono, Y., Suzuki, Y., Takeuchi, S., Shimizu, T., Yamaguchi, I., Chijimatsu, M., Sakurai, A. (1990) Bioactive condensed tannins from bark: chemical properties, enzyme inhibition and anti-plant-viral activities. *Nippon Noyaku Gakkaishi*, **15**, 585–591. Cited from: *Chem. Abstr. (1991)* **114**, 118544.

8. A PERSONAL ACCOUNT OF THE EARLY GINKGOLIDE STRUCTURAL STUDIES

KOJI NAKANISHI

Department of Chemistry, Columbia University, New York, N.Y. 10027, USA

INTRODUCTION

I moved to Tohoku University, Sendai, from Tokyo Kyoiku University (currently Tsukuba University) in spring 1963 to succeed Professor S. Fujise and started working on the isolation of the ginkgolides which he had started in 1960. This became an exciting and rewarding study which came to a conclusion in September 1966 after the finding of many extraordinary reactions as well as the encounter with the NMR intramolecular nuclear Overhauser effect (noe), unknown when we first observed it. Although no special biological activity was found for these molecules at that time, it is quite extraordinary that later they were found to be potent and selective antagonists of platelet activating factor. Even more striking is that *Ginkgo biloba* is already mentioned in the Chinese Materia Medica 5,000 years ago (Deng, 1988). The exponentially growing annual sales of the crude extract as a phytopharmaceutical and dietary supplement, which is reputed to improve memory and sharpen mental focus, was ca. 0.9 billion dollars in 1997 (Germany 280 M$, other Europe 200 M$, US 205 M$, Asia 200 M$) and projected to exceed 1 billion dollars in 1998 (Pharmanex, 1997; see also Grunwald, 1994).

The Ginkgo tree (*Ginkgo biloba L.*) is the last surviving member of a family of trees which appeared more than 250 million years ago (Paleozoic) and which reached their climax during the Jurassic period. During the last few million years all species except *G. biloba*, which is believed to have remained unchanged for the last 150 million years, have become extinct.

The studies were headed by M. Maruyama and performed by A. Terahara, Y. Nakadaira, the late M.C. Woods, Y. Itagaki, I. Miura, Y. Takagi, the late K. Habaguchi, and assisted by undergraduates Y. Hirota, Y. Sugawara, T. Dei and M. Miyashita (Maruyama *et al.*, 1967a-d; Woods *et al.*, 1967; Nakanishi 1967, 1974, 1988). In 1964 we received permission to fell five typhoon-damaged Ginkgo trees; 100 kg of the root bark gave after extraction, chromatography, and a tedious 10–15 step fractional recrystallization, 10 g of ginkgolide A (GA), 10 g of GB, 20 g of GC and 200 mg of GM. Purification of ginkgolides was seriously hampered by their remarkable tendencies to exhibit polymorphism and form mixed crystals; recently, van Beek and Lelyveld (1997) have devised a convenient preparative scale isolation procedure using medium-pressure liquid chromatography. These were pre-HPLC days, when 100 MHz (or 100 Mc) NMR was the most powerful and high resolution MS (HRMS) was not common; carbon NMR of course was unheard of. About 50 derivatives were prepared and submitted to detailed NMR studies by Maruyama and Terahara (1967). We all derived great fun through the intense day and night research, and as far as I am concerned it is the last classical and romantic structural study conducted in which numerous chemical reactions were carried out and structural clarification was accompanied by surprises and excitement, as if unfolding a mystery. Such an experience is seldom encountered today because of the development in isolation and spectroscopic techniques. A brief description of the highlights of structural studies follows.

UNIQUENESS OF THE GINKGOLIDE SKELETON

It is a diterpene with an aesthetically beautiful cage skeleton consisting of six 5-membered rings, i.e. a spiro[4.4]nonane carbocyclic ring, three lactones and a tetrahydrofuran (Figure 1). It is probably still the only terpenoid to carry a t-butyl group. Although the NMR (all spectra measured in the mild solvent TFA) always showed a 9-proton singlet at 1.2–1.3 ppm, presence of the t-Bu group was established only after detection of a strong t-Bu peak in the HRMS at 57.074 (calcd. 57.070) and isolation of pivalic acid, characterized as its crystalline p-bromophenacyl ester, upon chromic acid oxidation of ginkgolide C (GC). The bilobalide structure was established later in 1971 in collaboration with the groups of Professors Major and Bähr at Virginia and Heidelberg, respectively (Nakanishi et al., 1971).

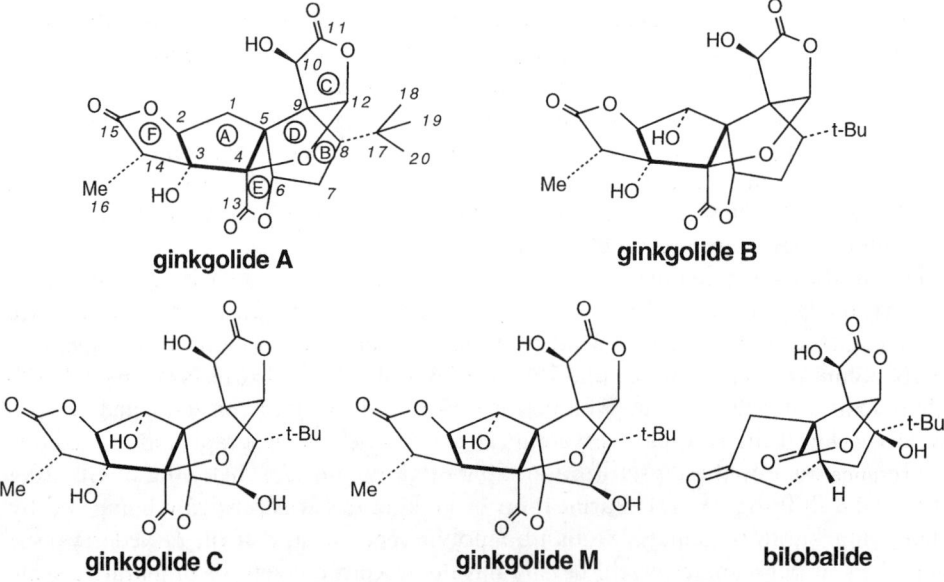

Figure 1   The ginkgolides.

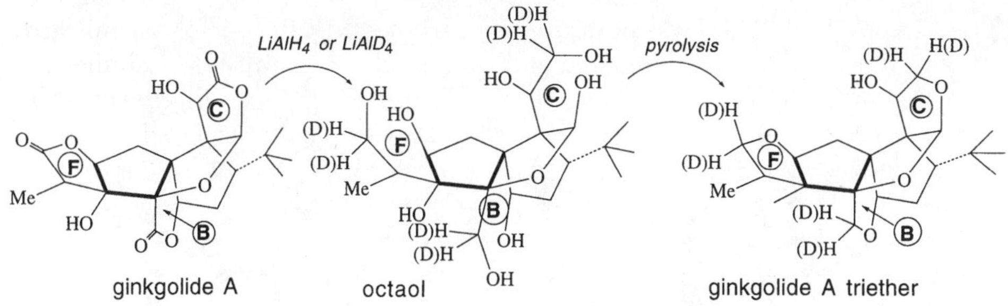

Figure 2   Formation of ginkgolide "triether".

THE TRILACTONES

Because the connectivity of proton groups in the ginkgolide structures are blocked by the intervening quaternary carbons (Figure 1), the ginkgolide NMR are overly simple so that they do not impart much information (see Figure 3 for ginkgolide A). Without too much thought regarding the outcome, GA was reduced with $LiAlH_4$ to see what would happen; not surprisingly this resulted in a useless syrup, an octaol. However, drying the syrup, which accidentally contained a trace of acid, gave rise to a solid which became *the critical compound* for structural studies, i.e. "GA triether" (because of ring D, it is in fact a tetraether). Later the triether was purified by vacuum pyrolysis. A thorough study of the NMR of GA triether by Vyn Woods and Iwao Miura led to the observation of what is now understood as NOE, which at that time was unknown (see Fig. 3) (Woods *et al.*, 1967). A comparison of the complex triether and simple GA spectra show that except for peaks designated by small case letters cc', ee' and ff', the two are quite similar; namely, the three doublet of doublet methylene peaks were introduced by $LiAlH_4$ reduction. This was confirmed by the "GA triether-d_6" spectrum resulting from $LiAlD_4$ reduction in which all peaks denoted by small case letters have disappeared. Moreover, similarity in the GA and triether-d_6 NMR show that the original ring structure is retained despite the difference between three lactones and three ether rings ; this was the first indication of the presence of three lactone rings.

The IR of GB showed one broad band at 1790 cm^{-1}; this band was due to lactone rings since it was replaced by carboxylate bands upon evaporation of an aqueous NaOH solution of GB and making a KBr disk out of the residual mixture of GB sodium salt and NaOH; dissolution of the disk and acidification gave unchanged GB. However, titration of aqueous solutions of ginkgolides in 0.1 N NaOH showed the presence of only two lactone rings while other results, e.g. incorporation of six additional hydrogens in GA triether (Fig. 3), suggested the number to be three. The correct number three was finally achieved by evaporating the alkali solutions to dryness, then adding water and carrying out the titration. Apparently, the high base concentration achieved during the evaporation process opens all three lactone rings.

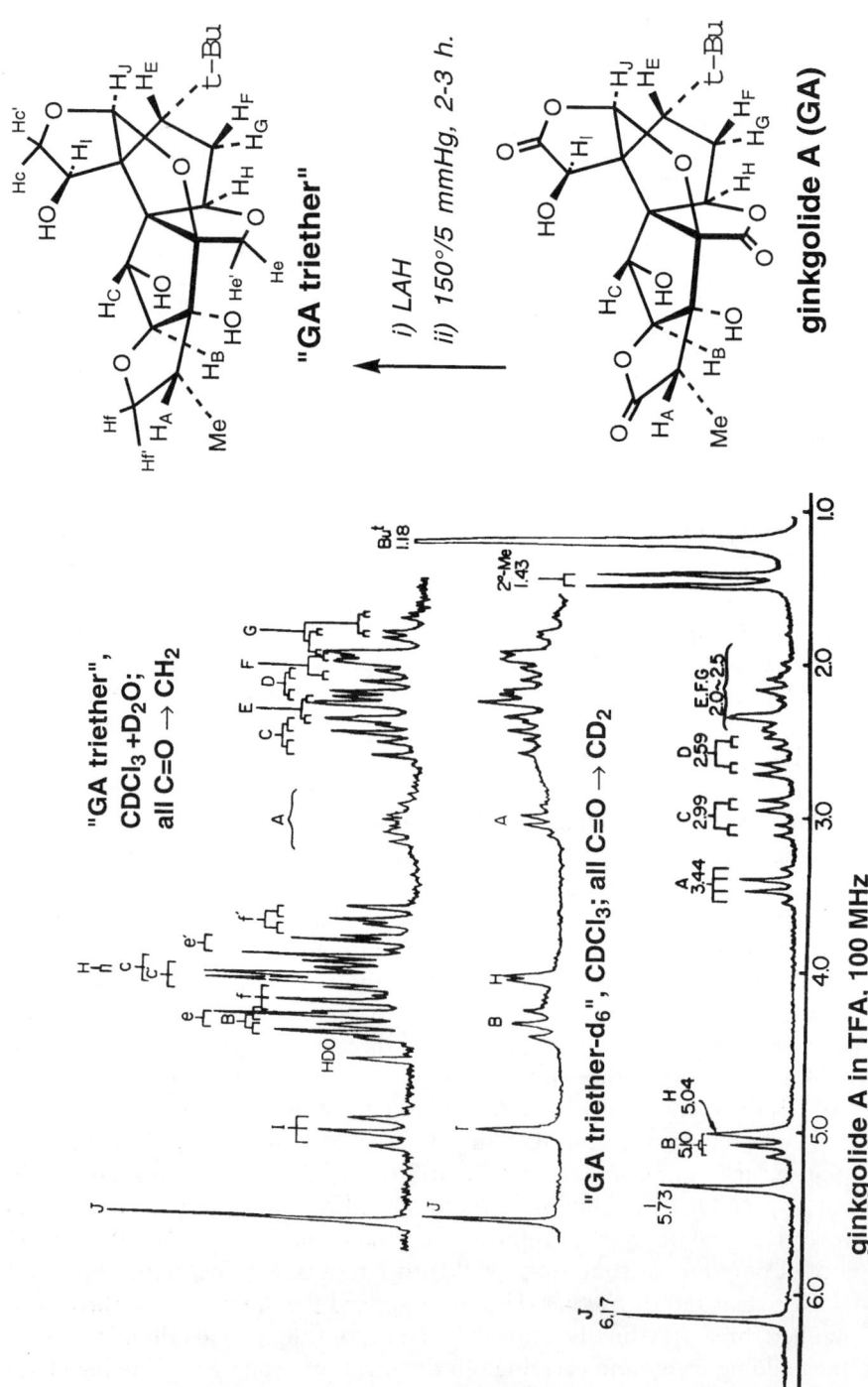

Figure 3   NMR of ginkgolide A, GA triether, and triether-d₆.

FIRST ENCOUNTER WITH NOE

During detailed NMR measurements of the triether, Vyn and Iwao noticed that irradiation of the t-Bu peak led to 10–33% enhancements in the area of certain proton signals (Figure 4). Instead of dismissing this as an integrator error, they repeated the measurements until they were convinced of the increments and thought that it may be related to the Overhauser effect which was known in electron spin resonance spectroscopy. Shortly after he had interpreted the results correctly, the first NOE paper by Anet and Bourn (1965) dealing with dimethylacrylic acid and another cage molecule appeared. Vyn's uncompromising attitude had led to the finding of NOE, unknown at that time, and which naturally concluded the final stages of structure determination. However, we were not fully aware of the power of NOE and the title of our first paper dealing with NOE was "The ginkgolides. V. Some aspects of their NMR spectra." (Woods *et al.*, 1967). As depicted in Figure 4, irradiation of the t-Bu group in GC tetraacetate enhanced the integrated area of protons I and J by 20%; however, for proton H there was no increase.

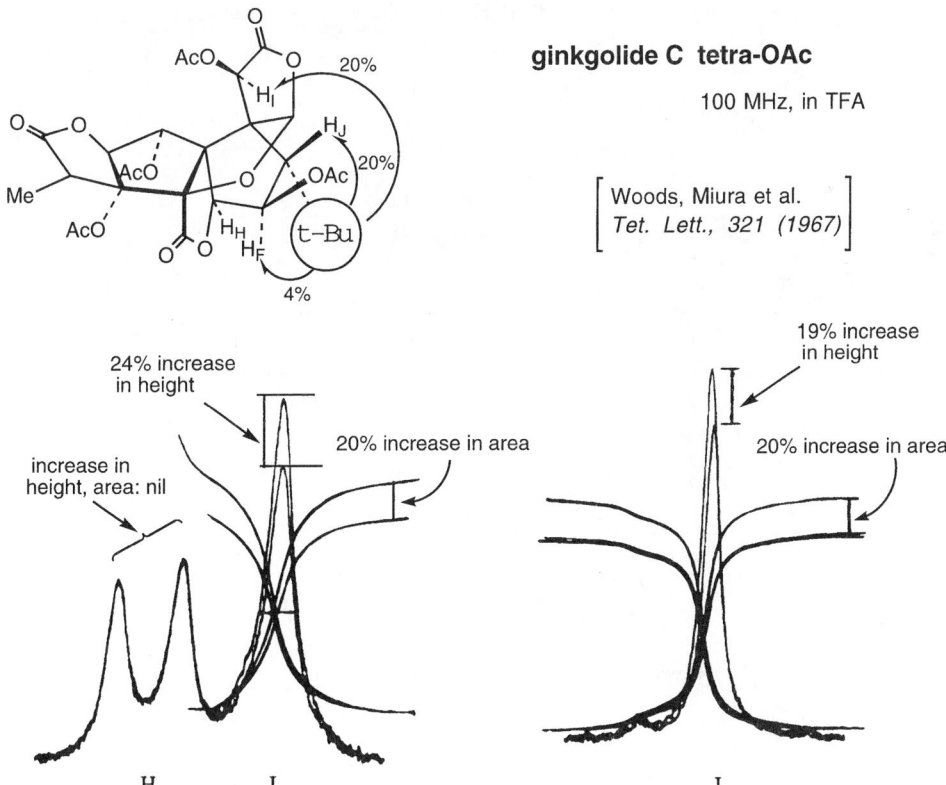

Figure 4   Intramolecular nuclear Overhauser effects.

BISNOR GA AND PHOTODEHYDRO GA

The fact that LiAlH$_4$ reduction of GA followed by pyrolysis of the octaol resulted in recyclization to the original cage skeleton demonstrates the unique stability of the beautiful cage structure. The following also shows the remarkable stability of the ginkgolide structure. Thus alkali fusion of 300 mg GA merely resulted in expulsion of two carbons, captured as oxalic acid, and generation of the hemiacetal, bisnor GA.

Treatment of an ice-cooled solution of GA in conc. sulfuric acid with a solution of sodium dichromate in 50% sulfuric acid gave a green solution which upon pouring into ice-water gave crystals. GA (2 g) gave 1.7 g of crude crystals of dehydro GA, which were recrystalized from dilute acetone to give fine needles, 670 mg, sublimed at 180° /10 mm Hg. However, dehydro GA, the oxolactone, was a source of confusion which led us to consider a wrong structure. Namely, as we had expected the presence of a hydroxy lactone, the oxidation was to yield an oxolactone. The NMR of dehydro-GA was measured immediately and therefore showed the familiar 9-proton singlet which was present in all ginkgolide spectra. However, if this were an alpha-keto lactone it should demonstrate weak but distinct UV bands and accompanying ORD (optical rotatory dispersion) Cotton effects around 400 nm, which was not the case. It was thus concluded that the oxidation product must be a

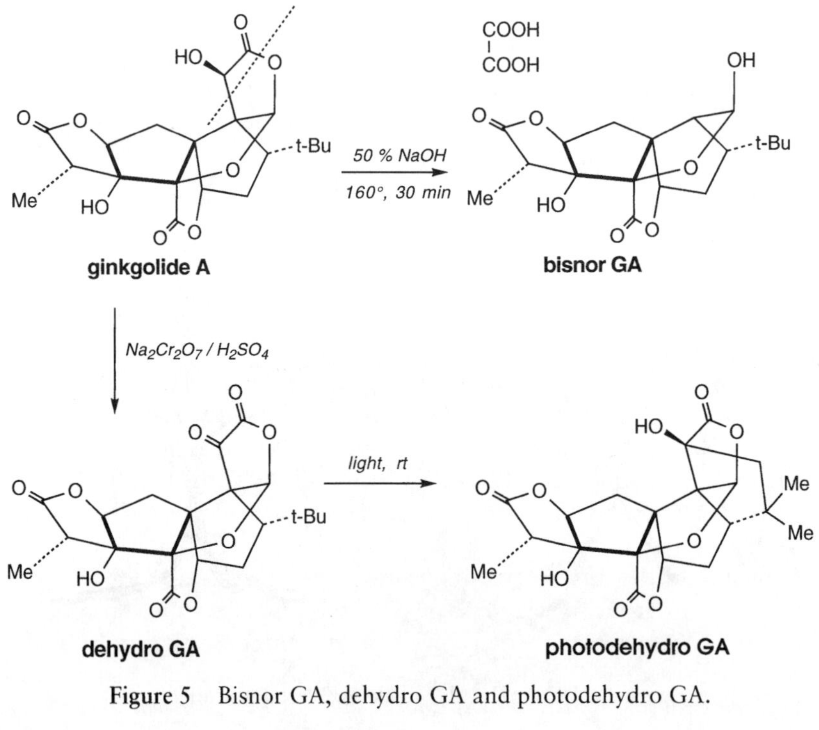

Figure 5   Bisnor GA, dehydro GA and photodehydro GA.

tetralactone. What happened was that ORD and UV, because of the psychological barrier in measuring them (the samples had to be weighed!) relative to NMR, were measured with a sample that had been left for some time, during which period the photocyclization to photodehydro-GA had taken place; the photocyclization occurs even during TLC. Only after the NMR of a dehydro GA sample left around had been remeasured in a different solvent to gain further information did it become clear that the t-Bu group had disappeared and had been replaced by a gem-dimethyl group! The unusual spectroscopic and photochemical properties of dehydro-GA were subsequently clarified (Nakadaira et al., 1969 ab).

Because the ginkgolides can be obtained in several large polymorphic crystalline forms and because of their extreme stability, we finally thought they might be nice as pendants. However, they are not to be licked because they will be very bitter tasting.

BIOSYNTHESIS

We succeeded in growing undifferentiated cells of Ginkgo biloba by adding 2,4-dichlorophenoxyacetic acid to an agar culture of its embryo containing glucose, glutamine and inorganic salts. Preliminary biosynthetic studies with labeled precursors demonstrated the gross biosynthetic pathway as well as the surprising fact that the t-Bu group is formed by cleavage of a diterpenoid ring A containing the 4-gem-dimethyl group, loss of one C and reintroduction of one C from methionine (Nakanishi and Habaguchi, 1971). However, more recent biosynthetic studies performed in Professor Arigoni's group in Zürich have shown that the ancient ginkgolides are one of the first members of a newly emerging class of terpenoids which do not follow the mevalonate pathway (Schwarz, 1994; Schwarz and Arigoni, 1999).

X-RAY CRYSTALLOGRAPHY

We did not spend too much effort in making crystals suitable for X-ray crystallography because I was more interested in developing methods for structural studies rather than determining the final structure. However, whatever the reason, we did not succeed in preparing suitable derivatives. At the very final stage of our structural studies, we heard that Sakabe et al. had determined the structure by X-ray (Sakabe et al., 1967). As we did not want our several years' effort to get lost simply because of the final structure, we asked Professor H. Kakisawa (then at Tokyo Kyoiku University) to act as mediator; both parties communicated the structure over the phone (pre-FAX days) to him, who built molecular models and found that both structures, including the absolute configuration, were identical. We presented the ginkgolide structures at the 10th Symposium on the Chemistry of Natural Products, late October, Tokyo, 1966, and submitted the five communications in mid-October (Maruyama et al., 1967a–d; Woods et al., 1967). It is rewarding to see that the ginkgolides are attracting world-wide interest.

REFERENCES

Anet, F.A.L., Bourn, A.L.R. (1965) Nuclear magnetic resonance spectral assignments from nuclear Overhauser effect. *J. Am. Chem. Soc.*, **87**, 5250–5251.

van Beek, T.A., Lelyveld, G.P. (1997) Preparative isolation and separation procedure for ginkgolides A, B, C, and J and bilobalide. *J. Nat. Prod.*, **60**, 735–738.

Deng, Q. (1988) Chinese Medicine, The Dawn, the founders, the first pharmacopoeia. *Drug Use Perspective*, **1**, 57–58.

Grunwald, J. (1994) The sales in Europe only for 1993 was $300 million: European phytomedicine market. *HerbalGram*, **34**, 60–65.

Maruyama, M., Terahara, A., Itagaki, Y., Nakanishi, K. (1967a) The Ginkgolides I: Isolation and characterization of the various groups. *Tetrahedron Lett.*, 299–302.

Maruyama, M., Terahara, A., Itagaki, Y., Nakanishi, K. (1967b) The Ginkgolides II: Derivation of partial structures. *Tetrahedron Lett.*, 303–308.

Maruyama, M., Terahara, A., Nakadaira, Y., Woods, M.C., Nakanishi, K. (1967c) The Ginkgolides III: The structure of the ginkgolides. *Tetrahedron Lett.*, 309–313.

Maruyama, M., Terahara, A., Nakadaira, Y., Woods, M.C., Takagi, Y., Nakanishi, K. (1967d) The Ginkgolides IV: Stereochemistry of the ginkgolides. *Tetrahedron Lett.*, 315–319.

Maruyama, M., Terahara, A. (1967) The NMR Data of the ginkgolides and derivatives, Science Reports of Tohoku University, Series I, **Vol L**, No. 2, pp. 92–99.

Nakadaira, Y., Hirota, Y., Nakanishi, K. (1969a) Spectroscopic properties of two non-enolizable alpha-keto-gamma-lactones, dehydroginkgolide A and 3,3-dimethyl-2-oxo-gamma-butyrolactone. *J. Chem. Soc. Chem. Commun.*, 1467–1469.

Nakadaira, Y., Hirota, Y., Nakanishi, K. (1969b) Photochemical reactions of non-enolizable alpha-keto-esters derived from dehydroginkgolide A. *J. Chem. Soc. Chem. Commun.*, 1469–1470.

Nakanishi, K. (1967) The Ginkgolides. *Pure Appl. Chem.*, **14**, 89–113.

Nakanishi, K., Goto, T., Ito, S., Natori, S., Nozoe, S. (eds.) (1974) *Natural Products Chemistry* Vol. I, Kodansha/Academic Press, p. 295–300.

Nakanishi, K., Habaguchi, K., Nakadaira, Y., Woods, M.C., Maruyama, M., Major, R.T., Alauddin, M., Patel, A.R., Weinges, K., Bähr, W. (1971) Structure of bilobalide, a rare tert-butyl containing sesquiterpenoid related to the C_{20}-ginkgolides. *J. Am. Chem. Soc.*, **93**, 3544–3546.

Nakanishi, K., Habaguchi, K. (1971) Biosynthesis of ginkgolide B, its diterpenoid nature and origin of the t-Bu Group. *J. Am. Chem. Soc.*, **93**, 3546–3547.

Nakanishi, K. (1988) Ginkgolides – Isolation and structural studies carried out in the mid 1960's. In *Ginkgolides – Chemistry, Biology, Pharmacology and Clinical Perspectives*, P. Braquet, (ed.), J.R. Prous Science Publishers, Barcelona, pp. 27–36.

Pharmanex Inc., Simi Valley, Ca 93065, estimate: private communication.

Sakabe, N., Takada, S., Okabe, K. (1967) The structure of ginkgolide A, a novel diterpenoid trilactone. *Chem. Commun.* 259–261.

Schwarz, M.K. (1994) Ph. D. Thesis, Eidgenössische Technische Hochschule, Zürich, Switzerland.

Schwarz, M. and Arigoni, D. (1999) Ginkgolide biosynthesis. In D.E. Cane (ed.) *Isoprenoids including carotenoids and steroids*, Vol. 2 in D. Barton, K, Nakanishi and O. Meth-Cohn (eds.) *Comprehensive Natural Products Chemistry*. Elsevier, Amsterdam, pp. 367–400.

Woods, M.C., Miura, I., Nakadaira, Y., Terahara, A., Maruyama, M., Nakanishi, K. (1967) The ginkgolides V: Some aspects of their NMR Spectra. *Tetrahedron Lett.*, 321–326.

9. CHEMICAL ANALYSIS OF GINKGO TERPENE TRILACTONES

TERIS A. VAN BEEK

Laboratory of Organic Chemistry, Phytochemical Section,
Wageningen Agricultural University, Dreijenplein 8, 6703 HB
Wageningen, The Netherlands

INTRODUCTION

Among the most important and characteristic constituents of the tree *Ginkgo biloba* is a small group of five compounds collectively indicated with the name "ginkgolides". The structures of four of these diterpenoids (ginkgolides A, B, C and M) were independently elucidated in the sixties by two Japanese groups (Maruyama *et al.*, 1967; Okabe *et al.*, 1967). In 1987 one more representative (ginkgolide J) was identified. Their structures and that of a closely related sesquiterpene trilactone, named bilobalide, are given below. In the remainder of this chapter their names will abbreviated to G-A, G-B, G-C, G-M, G-J and BB. For more than 15 years these compounds were considered to be devoid of any clear pharmacological activity or toxicity. However research carried out in the mid eighties showed that they were potent and selective antagonists of platelet activating factor (PAF) which fact greatly stimulated the interest in these remarkable compounds. Now they are considered to be at least partially responsible for the beneficial activity of Ginkgo phytopharmaceuticals.

Bilobalide (BB)

G-A: R_1 = OH, R_2 = R_3 = H
G-B: R_1 = R_2 = OH, R_3 = H
G-C: R_1 = R_2 = R_3 = OH
G-J: R_1 = OH, R_2 = H, R_3 = OH
G-M: R_1 = H, R_2 = R_3 = OH

Their chemical analysis has been difficult from the very start until the present day. This is caused by their lack of a good chromophore, the absence of any reactive groups, their relatively low concentrations in leaves ($\approx$ 0.05%), the difficult separation of some ginkgolides and the occurrence of many other UV-active substances in high concentrations in Ginkgo leaves. In this chapter the various analytical methods used for the separation, detection and quantitation of these terpene trilactones are reviewed and discussed. The first technique to be discussed is TLC, which was already used back in the sixties. Until recently it has been solely used for qualitative purposes. The most widely employed technique for quantitative purposes is HPLC which is the subject of the second section. After derivatisation also GC can be employed for a quantitative determination. This is discussed in the next section. In the last section several miscellaneous techniques for the analysis of Ginkgo terpene trilactones are summarised.

THIN LAYER CHROMATOGRAPHY (TLC)

The various groups initially involved with ginkgolides had only TLC at their disposal for the qualitative analysis of fractions or supposedly pure substances. They used without exception silica gel as stationary phase in combination with relatively apolar solvents like $CHCl_3$–MeOH (8:2) (Okabe et al., 1967) and benzene–Me_2CO (8:2) (Weinges and Bähr, 1969). Two problems were encountered. The first problem was that the pairs G-A & G-B and G-J & G-C have almost identical R_f values on silica gel. This is caused by a strong internal hydrogen bonding between the 1- and 10-hydroxyl groups in G-B and G-C. This hydrogen bonding more or less neutralises the presence of the additional hydroxyl in comparison with G-A and G-J respectively. Thus G-B and G-C are considerably less polar than a priori expected. The second problem is that the ginkgolides are very stable, unreactive compounds without any olefinic double bonds. This makes them difficult to detect on TLC plates. For this reason initially charring by concentrated sulphuric acid was used.

Other solvents mentioned more recently in the literature which also do not resolve G-A & G-B and G-J & G-C well, are toluene–Me_2CO (7:3) (Weinges et al., 1987; Steinke et al., 1993), cyclohexane–EtOAc (1:1) (Lobstein-Guth et al., 1983), EtOAc–toluene–Me_2CO–hexane (4:3:2:1) (Tallevi and Kurz, 1991) and $CHCl_3$–Me_2CO–HCO_2H (75:16.5:8.5) (Steinke et al., 1993). In 1993 silica gel impregnated with NaOAc was introduced as a new stationary phase for a better resolution of G-A & G-B and G-J & G-C (van Beek and Lelyveld, 1993). With MeOAc as solvent the R_f values were reported as 0.58, 0.41, 0.25, 0.12 and 0.06 for BB, G-A, G-B, G-J and G-C respectively. As explanation for the much improved separation a disruption of the internal hydrogen bonding by NaOAc was given. Also tailing diminished which contributed to a better resolution (van Beek and Lelyveld, 1993). At the same time the acetate ion catalysed the effect of the best TLC spray reagent for terpene trilactones so far: Ac_2O followed by heating. The Ac_2O reagent was originally introduced by Lobstein-Guth et al. (1983) and produces rather faint orange-brown fluorescent spots after irradiation with light of 366 nm. In combination with NaOAc impregnated plates all spots were more intense and detection limits as low as 0.1 μg were claimed with this reagent (van Beek and Lelyveld, 1993). An example chromatogram is shown in Fig. 1. This reagent is considered to be superior over other reagents used such as NH_2OH–$FeCl_3$ (Weinges et al., 1987), spraying with H_2O followed by heating (Teng 1988) or just simply heating at 150° (Tallevi and Kurz, 1991). This impregnation method has been incorporated in the draft monograph on Ginkgo folium of the United States Pharmacopeia (see chapter by S. Srinivasan in this book).

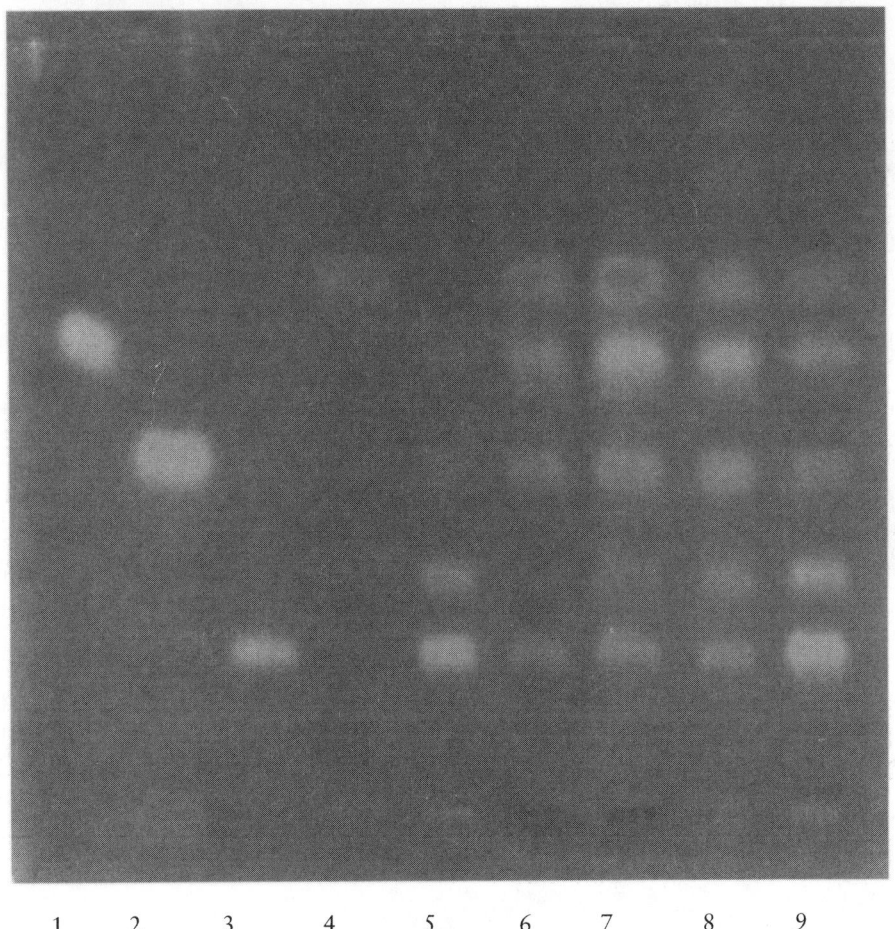

1   2   3   4   5   6   7   8   9

Figure 1 Silica gel TLC plate impregnated with NaOAc showing spots of ginkgolides and bilobalide viewed under UV 366 nm after spraying with acetic anhydride and heating. Saturated chamber; horizontal development 8 cm; solvent EtOAc–MeOAc (1:1). Applied from left lane to right lane: (1) G-A; (2) G-B; (3) G-C; (4) BB; (5) mixture of G-C and G-J; (6) crude mixture of Ginkgo terpene trilactones without clean-up; (7) Indena standardised G. biloba extract after clean-up; (8) Tanakan phytopharmaceutical after clean-up; (9) mixture of BB, G-A, G-B, G-J and G-C.

The impregnation technique stood also at the basis of a new preparative isolation method for terpene trilactones. These compounds are not commercially available in a purity >95% while quality control labs do need them for calibration curves. Thus there is a demand for simple isolation procedures. By using an MPLC column filled with silica gel impregnated with NaOAc, the use of expensive HPLC columns for the separation of G-A & G-B and G-J & G-C is no longer necessary (van Beek and Lelyveld, 1997).

QUANTITATIVE ANALYSIS

High Pressure Liquid Chromatography (HPLC)

One disadvantage of TLC in its most simple form is, that it lends itself less well for quantitative purposes. However in the eighties a need for a Ginkgo terpene tri-lactone assay slowly developed. Therefore another method was sought and – because of the non-volatile nature of ginkgolides – HPLC was chosen. Lobstein-Guth *et al.* (1983) were the first to determine the optimal solvent composition for the separation of BB, G-A, G-B and G-C on a standard C18 RP-HPLC column. The best solvent was a mixture of MeOH–THF–H_2O (1:3:15). Retention times of 9.2, 11.9, 13.6 and 19.3 min were reported for G-C, G-A, BB and G-B respectively. Detection was accomplished by UV at 220 nm. In the same study they also tried to apply this technique for the analysis of leaves. However even after a labour-intensive sample clean-up involving 35 partitions, no analysable chromatogram resulted. In later studies by others it was demonstrated that the cause was probably the UV detection. Due to the absence of any maximum with a significant extinction coefficient above 200 nm, these terpene trilactones cannot be analysed by UV except when they are totally pure. Even minor impurities (<1%) in an enriched extract give rise to an unintelligable chromatogram (van Beek *et al.*, 1991; Camponovo *et al.*, 1995; Camponovo 1996).

For this reason other detection techniques were investigated. Two groups independently reported on the use of refractive index (RI) detection. Komoda *et al.* (1988) obtained an analysable extract after a relatively simple sample clean-up which is given in Table 1. As solvent they used H_2O–MeOH (67:33) which is the best THF-free solvent for the separation of Ginkgo terpene trilactones on C18 HPLC columns. Unfortunately this solvent in combination with their 10 cm column

Table 1   HPLC method used by Komoda *et al.* (1988).

(1) extraction of Ginkgo leaves with MeOH–H_2O (7:3)
(2) evaporation of MeOH and extraction of aqueous layer with cyclohexane (discard)
(3) extraction of remaining aqueous layer with MeEtCO
(4) evaporation of MeEtCO layer and redissolve in HPLC compatible solvent
(5) HPLC analysis on a 4 μm 100 × 8 mm C18 column, 2 ml/min MeOH–H_2O (33:67), RI detection

was not capable of separating BB and G-J. No validation experiments which are essential for a quantitative determination were published so the method is presently unsuitable for quantitative work. Also Teng (1988) used RI detection. He used an improved version of the optimal HPLC solvent published by Lobstein-Guth et al. (1983): H_2O–MeOH–THF (7:2:1). In combination with a 12.5 cm C18 column a nice baseline separation of G-J, G-C, BB, G-A and G-B within 10 min was achieved. His clean-up method must have been much better than that of Komoda et al. (1988) because no peaks other than those of the Ginkgo terpene trilactones are visible in his chromatogram. Unfortunately his efficient clean-up method was never published in detail making it impossible for others to reproduce his results. It was only mentioned that the extraction could be carried out with warm aqueous acetone and that interfering lipids and polyphenols could be removed by extraction with apolar organic solvents and normal-phase chromatography respectively (Teng, 1988).

One year later a more complex method was published (Wagner et al., 1989). It included among others an initial extraction with EtOAc (5x), a filtration step, an intermediate extraction step with Et_2O, a column chromatography step (silica gel in combination with cyclohexane-EtOAc) and a final investigation of several CC fractions by means of RP-HPLC with UV detection at 220 nm. Because of losses due to the lengthy clean-up procedure and the poor method of detection (vide supra), the terpene trilactones still present in the final fractions were probably obscured by more intense peaks in the chromatogram. Possibly some compounds got lost as (1) Et_2O is a rather poor solvent for dissolving terpene trilactones and (2) low recoveries of the column chromatographic step. Chauret et al. (1991) have reported losses for G-B as high 70% after column chromatography on silica gel with cyclohexane-EtOAc as solvent. Unfortunately Wagner et al. (1989) failed to carry out the necessary recovery experiments and then made some inappropriate conclusions about the absence of terpene trilactones in certain phytopharmaceuticals and the lack of pharmacological importance of these compounds. These findings and their methods were subsequently heavily criticised (Stumpf, 1989) and later it was shown by others that terpene trilactones are present in these phytopharmaceuticals (van Beek et al., 1991).

Pietta et al. (1990) have published a two-column sample clean-up method for Gingko extracts. Details of this method are presented in Table 2. The first step is a partitioning between water and diethyl ether. Neither water nor ether are very good solvents for terpene trilactones so probably some amounts are lost in this step. Especially the polar G-C is at risk. The second step uses alumina on which according to some authors bilobalide is degraded (van Beek and Taylor, 1996). Another group

Table 2 HPLC method used by Pietta et al. (1990).

---

(1) 600 mg Ginkgo extract is dissolved in 10 ml Me_2CO-H_2O (5:5)
(2) after dilution with water to 50 ml, 20 ml is brought on an Extrelut column
(3) elute with 100 ml Et_2O, evaporate and dissolve in 10 ml MeOH
(4) 2 ml of this methanolic solution is brought on an alumina column, elute with 8 ml MeOH
(5) evaporate and dissolve in 1 ml MeOH and filter through a 0.45 μm membrane filter
(6) HPLC on a 3 μm 100 × 4.6 mm C18 column, 1 ml/min iPrOH–H_2O (1:9), UV at 220 nm

---

did not report such a degradation (Yao *et al.*, 1997). The purity of terpene trilactone peaks was checked with diode-array detection. Unfortunately no recovery experiments were carried out. Using external standardisation they found a good RSD of 2.8% for extracts in spite of the fact that the published chromatogram does not look particularly clean (baseline does not return to zero). The authors preferred UV detection over RI detection because of a better sensitivity. They gave as lowest detectable amount 100 ng. This value is indeed better than for RI detection, unfortunately the selectivity of UV is much worse.

The first fully validated assay for terpene trilactones was published in 1991 (van Beek *et al.*, 1991). The method – involving two different SPE columns – was not very simple but gave good results. As final method of detection RI was used. Details are presented in Table 3. Several different Ginkgo leaf samples and standardised extracts were analysed. The reproducibility was dependent on the concentration and varied from 2–15%. The minimum concentration that could be analysed in leaves was 10 μg/g dry leaves. An example chromatogram of a phytopharmaceutical showing the much better performance of the RI detector over the UV detector is shown in Fig. 2. This method was applied in the analysis of a large number of different leaf samples during an entire growing season (van Beek and Lelyveld, 1992). In this study large individual differences between trees were found. Maximum concentrations were detected in early autumn. This was later confirmed in several other studies (*vide infra*) but contradicted by O'Reilly's study (1993) where a maximum content was reported early in the season. From an analytical point of view, the authors concluded that their assay was good but time-consuming and that sometimes a certain batch of C18 SPE columns showed breakthrough. Later this was also noted by others who remedied the problem by increasing the amount of stationary phase from 500 to 1000 mg (Hasler and Meier, 1992).

The effect of age, organ (leaves, root, shoot) and light level on the terpene trilactone concentration of young Ginkgo trees was studied by Flesch *et al.* (1992). The assay used by this group is summarised in Table 4. The most surprising fact of this assay is, that no sample clean-up by means of either partitioning or SPE is described even though a relatively universal RI detector is used. The only sample clean-up consisted of a defatting step. Regrettably no method validation or a

Table 3 HPLC method used by van Beek *et al.* (1991).

---

(1) reflux 600 mg Ginkgo leaves twice with 5 ml MeOH–H_2O (1:9) during 15 min

(2) filter over Büchner, quantitatively collect aqueous extracts and apply to a 500 mg polyamide SC6 column connected in series with a 500 mg C18 SPE column

(3) wash columns with 15 ml and 5 ml of 2% and 5% MeOH in H_2O respectively, suck dry with air

(4) disconnect polyamide column (discard) and wash C18 column with 6 ml hexane (discard)

(5) elute C18 SPE column with 7 ml hexane–MeOAc (6:4) and evaporate solvent

(6) dissolve in MeOH, add internal standard (benzyl alcohol) and inject into HPLC

(7) HPLC on a 5 μm 250 × 4.6 mm C18 column, 1 ml/min MeOH–H_2O (33:67), RI detection

---

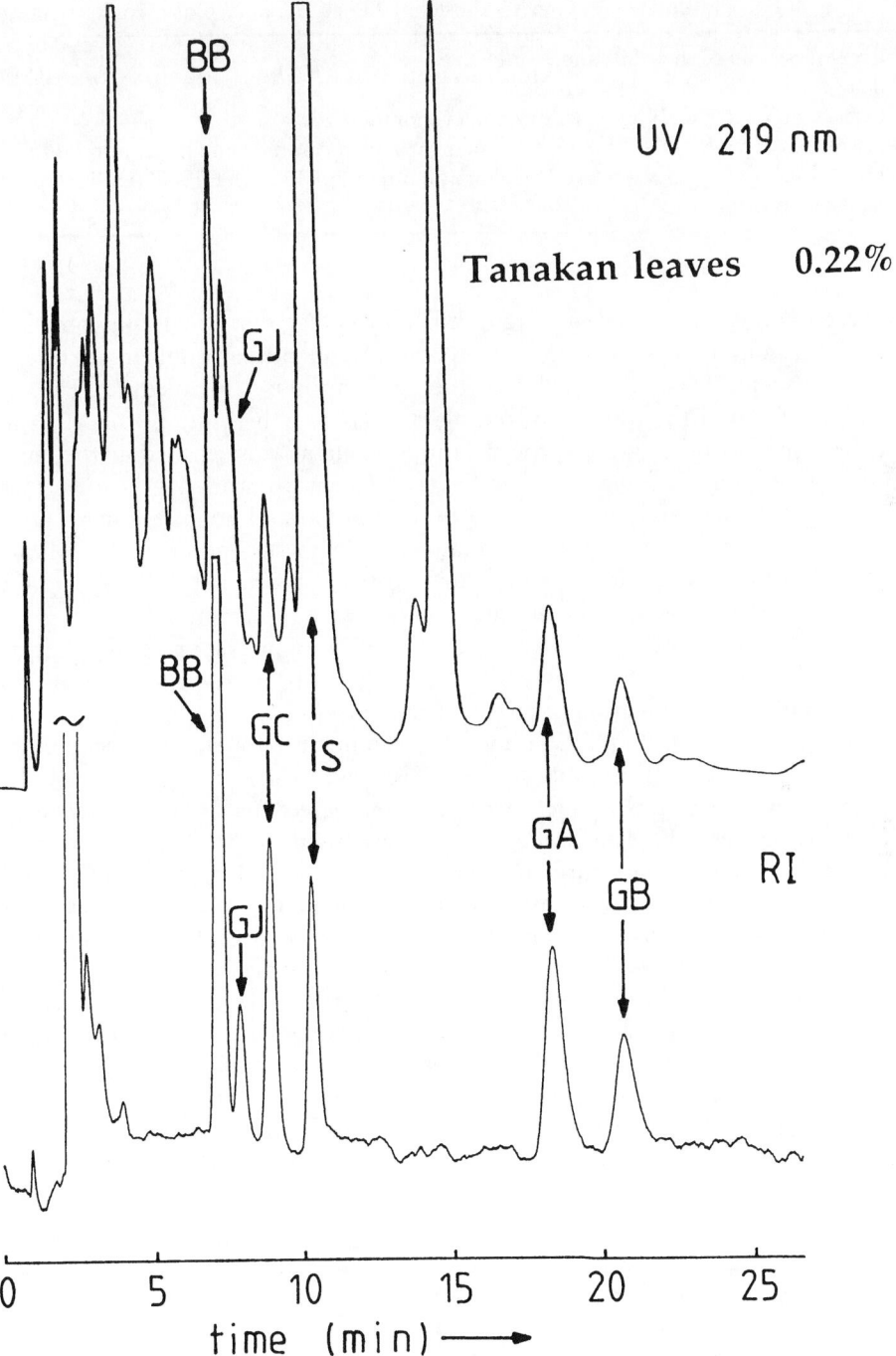

Figure 2 HPLC traces of 1.00 ml of purified Tanakan phytopharmaceutical with (upper trace) UV detection at 219 nm and (lower trace) RI detection. Internal standard (IS) is benzyl alcohol.

Table 4   HPLC method used by Flesch *et al.* (1992).

(1) dry leaves in an oven at 80° and grind
(2) defat the leaves with trichloroethylene
(3) extract with Me_2CO–H_2O (4:1), dry and evaporate *in vacuo*
(4) dissolve in H_2O–MeOH–THF (7:2:1), filter through membrane and inject into HPLC
(5) HPLC on a 5 μm 125 × 4 mm C18 column, 0.5 ml/min H_2O–MeOH–THF (7:2:1),
     RI detection

chromatogram was presented making it difficult to judge this potentially interesting
procedure. Additionally some details (temperatures, quantities of solvent, extraction
times) are lacking making it hard for others to reproduce this method. The concentrations found in these 1–3 year old plants were very high, up to 0.8%. In line
with the results of other groups, the maximum content was reached at the end of
summer or beginning of autumn. The lowest concentration (3–4 fold lower) was
found in spring. The level may be linked to the amount and quality of light. Of the
different organs investigated, leaves contained the highest concentrations followed
by roots. During the first three years of their existence, the age of the trees had little
effect on the total or individual concentration of terpene trilactones. Five years later
the same assay was used by Cartayrade *et al.* (1997) to analyse the ginkgolide and
bilobalide content of *G. biloba* seedlings in their study on the sites of synthesis,
translocation and accumulation of terpene trilactones.

A very simple one column SPE sample clean-up method for standardised extracts
and phytopharmaceuticals (Table 5) was described by Pietta *et al.* (1992). It is
obvious that alumina gives a much better clean-up effect than silica gel because in
spite of the fact that UV detection at 220 nm was used the chromatograms are as
clean as the chromatograms purified over silica and analysed by RP-HPLC with RI
detection. UV detection at 220 nm is essentially unsuitable for the detection of
terpene trilactones because of their low ϵ-value (*vide supra*). This also results in
poor detection limits. In this study G-B could hardly be analysed and G-J was below
the detection limit. No validation was published for this sample clean-up. Later it
was found that bilobalide rapidly degrades on all types of alumina (including the
one used in this study) and that the method is thus unsuitable for the analysis of
bilobalide (van Beek and Taylor, 1996).

Maillard *et al.* (1993) have published a semi-quantitative HPLC-MS method for
ginkgolides. The separation was carried out by means of RP-HPLC on a Novapak

Table 5   HPLC method used by Pietta *et al.* (1992)

(1) dissolve 80 mg standardised extract in 2 ml MeOH
(2) apply 0.5 ml to SPE column filled with alumina prewashed with MeOH
(3) elute with MeOH and collect first 2 ml, filter through 0.45 μm membrane
(4) HPLC on a 5 μm 250 × 4 mm C8 column, 1 ml/min iPrOH–THF–H_2O (1:13:86), UV at
     220 nm

C18 column. A 12 min gradient of 25–52% aqueous MeOH was used at a flow of 1 ml/min. After leaving the column, 0.2 ml/min 0.5 M NH$_4$OAc was added to the eluent. The final concentration of the volatile buffer which was needed for ion evaporation ionization was 0.08 M. In the thermospray type mass spectrometer (= TSP-MS) only strong quasi molecular ions M + NH$_4^+$ at m/z 344, 426, 442, 442 and 458 for BB, G-A, G-B, G-J and G-C respectively were observed and no other adducts or fragmentation were visible. Optimization of the TSP ionization showed that the vaporizer and source temperatures had great influence while the repeller voltage had none. Best results were obtained at 100° and 190° respectively (Camponovo et al., 1995). Under these conditions HPLC-TSP-MS was found to be very efficient although one problem was noted. The linearity range was limited because saturation occurred at high solute concentrations. The concentrations should be strictly between 35–1500 ng/10 μl of sample. Within these concentrations good linearity was observed (coefficient of correlation = 0.98). Detection limits (40–66 ng) and quantitative limits (120–200 ng) were equal to those obtainable with GC-FID. The relative standard deviation for analysis of the same solution during one day was 5–7%. The day to day variation was larger and thus a new calibration curve had to be prepared every day. The authors compared their method with the GC-FID method proposed by Hasler and Meier (1992). As samples they used Tanakene® tablets and Gincosan® capsules. The samples were stirred for 24 h in 50% MeOH and after filtration a 10 ml aliquot was evaporated. After solvation in MeOH this solution was either directly submitted to LC-TSP-MS or purified according to a procedure proposed by van Beek et al. (1991) and modified by Hasler and Meier (1992). After silylation, the derivatised ginkgolides were analysed by GC.

The same group introduced the evaporative light scattering detector (ELSD) as an alternative for the RI detector (Camponovo et al., 1995). In a detailed study they compared the results of this detector for a Ginkgo leaf sample and a number of different phytopharmaceuticals with those of GC-FID and HPLC-MS. They checked the necessity of sample clean-up, reproducibility (RSD), linearity and accuracy. Linearity was good between 0.8–2.8 mg/ml, the limit of detection was 1.1 μg, the RSD was 1.9–5.4% and accuracy varied from –0.8% to +2.2%. While precision and sensitivity are thus slightly lower for LC-ELSD as for GC-FID, no cumbersome derivatisation is necessary. Therefore HPLC-ELSD holds considerable promise for routine quality control of Ginkgo samples and appears to be superior over HPLC-RI. For instance ELSD allows the use of gradients or THF-containing solvents which is impossible respectively difficult with RI detection (Fig. 3). The baseline stability and selectivity are also better. The authors showed that the two column sample clean-up developed by van Beek et al. (1991) and modified by Hasler and Meier (1992) could be replaced by a partitioning step between EtOAc and H$_2$O without a significantly different outcome. Once more the chromatograms published in their paper show the total unsuitability of UV detection for Ginkgo terpene trilactones.

HPLC has been used for determining the concentration of terpene trilactones in transformed and gametophyte derived cell cultures of G. biloba (Laurain et al., 1997). The procedure is given in Table 6. A one column sample preparation step was used. Unusual about this clean-up is, that silica gel is used as a support for

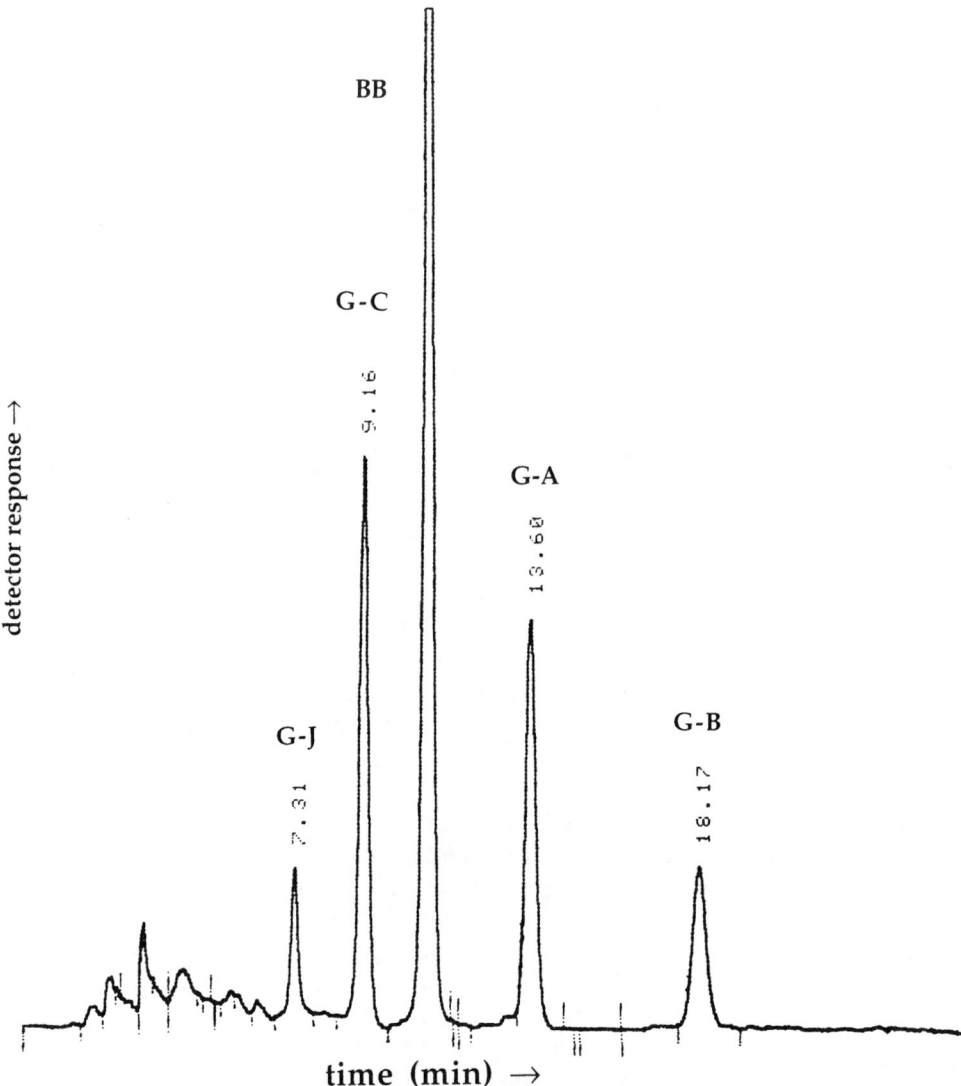

Figure 3 HPLC-ELSD profile of a methanolic solution of a supercritical fluid extract of a standardised Ginkgo extract. Phenomenex column 250 × 4.6 mm filled with Spherisorb 5 ODS(2), solvent H_2O–THF–MeOH (68.5:10.5:21) 1.0 ml/min, Varex ELSD, N_2 as nebulizer gas at 2.06 l/min, drift tube 107° (van Beek and Taylor, 1996).

liquid-liquid partition chromatography. No validation details have so far been published about this clean-up step. The extraction and quantitation procedure was very similar to a validated procedure published earlier van Beek *et al.* (1991). In two cultures relatively high concentrations of 0.065 and 0.087% terpene trilactones were detected.

Table 6   HPLC method used by Laurain *et al.* (1997).

(1) 1 g lyophilized powder is twice refluxed with 10 and 5 ml H$_2$O–MeOH (9:1) resp. for 15 min
(2) after Büchner filtration, the combined aqueous extracts are evaporated *in vacuo*
(3) after adding 5 ml H$_2$O, the solution is refluxed for 15 min
(4) the solution is quantitatively transferred to a silica gel column (10 g) which is then eluted with 40 ml MeOAc–hexane (3:1); after drying, the eluate is evaporated *in vacuo*
(5) after addition of internal standard (benzyl alcohol), the solution was analysed by HPLC
(6) HPLC on a 5 µm 25 cm × 4.6 mm C18 column, 0.95 ml/min MeOH–H$_2$O (31:69), RI detection

Yao *et al.* (1997) have published an assay for terpene trilactones in different plant parts and standardised Ginkgo extracts which uses a variation of the sample clean-up proposed by Pietta *et al.* (1992) (dissolving in MeOH, apply on an alumina column and elute with MeOH) and the quantitation method published by van Beek *et al.* (1991) (HPLC-RI with benzyl alcohol as internal standard). An important difference in comparison with Pietta *et al.* might be that in this study *acidic* alumina was used. Various validation experiments have been performed. The recoveries of G-A, G-B, G-C and BB varied from 97–103%, while the RSD was between 1.2–2.9%. The detection limit was < 1 µg for G-C and 4.2 µg for BB. As advantages of their method over the two column method proposed by van Beek *et al.* (1991), the authors mentioned time gain, better reproducibility, higher recovery and lower solvent consumption. A direct comparison of the two methods with the same sample gave similar results except for G-C.

Carrier *et al.* (1998) have performed a biosynthetic study on 3-year old Ginkgo plants in which they tried to correlate the occurrence of terpene trilactones in various parts of the plants with the capacity of those same parts to convert radio-active isopentenyl pyrophosphate to farnesyl and geranylgeranyl pyrophosphate. In the course of this study a substantial number of assays were performed. The assay consisted of a combination of the extraction and quantitation (HPLC-RI) method proposed by van Beek *et al.* (1991) (*vide supra*), a partitioning between EtOAc and H$_2$O and the one column SPE sample clean-up of Lolla *et al.* (1998) (*vide infra*). No validation of the entire procedure has been published. It appears rather long, the intermediate partitioning is probably not necessary. The highest terpene trilactone concentrations were found in leaves from the rosettes and terminal buds. The highest geranylgeranyl pyrophosphate formation was found in the terminal buds. No definitive conclusions could be drawn with regard to the site of terpene trilactone synthesis.

Very recently Chen *et al.* (1998) published a method for the industrial control of Ginkgo extracts. Their procedure is summarised in Table 7. The sample preparation takes around 4 hr which can be considered quite time-consuming. Various parameters were validated. Reproducibility varied between 1.5–5%. Detector response was linear between 4–60 µg of terpene trilactone. The detection limit varied between 0.5 and 0.2 µg. The recovery of the extraction procedure was more than

Table 7   HPLC method used by Chen *et al.* (1998).

(1) 300 mg standardised extract is dissolved in 3 ml MeOH in an ultrasonic bath, add 12 ml H_2O
(2) add 12 ml EtOAc, shake for 40 min and centrifuge, repeat 4 more times
(3) combine EtOAc layers, evaporate with N_2 at 50° and dissolve in 2 ml MeOH
(4) HPLC on a 4 μm 150 × 3.5 mm C18 column, 1.5 ml/min MeOH–H_2O (23:77), RI detection

96.5%. According to the authors the precision and accuracy was such that an internal standard was not necessary. The method was useful for following the effect of the different steps in the production of Ginkgo extracts.

In the chapter "Industrial quality control of Ginkgo products" in this book by Camponovo and Soldati, some information is presented about the method used by the Pharmaton company for the quantitation of terpene trilactones in their extracts. The sample clean-up takes place on a silica gel cartridge after which the terpene trilactones are separated on a 150 × 4.6 mm Supelcosil C18 column with as eluent MeOH-H_2O (28:72) at a flow rate of 1 ml/min and a temperature of 45°. Detection is carried out by RI. The whole separation takes less than 17 min and the chromatogram presented looks very nice in terms of cleanliness and baseline stability. For external calibration a well characterised reference extract is used. Although validation experiments have certainly been carried out, no information is provided on this aspect nor on the exact extraction and clean-up procedure.

Gas Chromatography (GC)

Although Ginkgo terpene trilactones are totally involatile and start to melt only at temperatures over 300°, they can be separated by GC once they have been silylated. The first report on the preparation of volatile trimethylsilyl ethers was made by Braquet (1988). Essentially this was only a chromatogram showing that the derivatives were stable enough for GC analysis. As detector the flame ionisation detector (FID) was used. Carrier and Chauret were the first to apply this method for the analysis of Ginkgo samples derived from cells cultured *in vitro* (Carrier *et al.*, 1990; Carrier *et al.*, 1991; Chauret *et al.*, 1991). Their method is presented in Table 8.

The purification step was necessary to avoid contamination of the injector, column and MS source by the crude extract. The recovery of the purification step

Table 8   Gas chromatographic method used by Carrier *et al.* (1990, 1991).

(1) extract lyophilized Ginkgo cells (20 g) with 1 l acetone during 16 hr at room temperature
(2) filter over 0.45 μm, evaporate *in vacuo* and redissolve in ≈5 ml cyclohexane–EtOAc (1:1)
(3) purification over silica gel with a cyclohexane–EtOAc gradient
(4) derivatisation of evaporated G-A & G-B containing fraction with BSA-DMF at 75° for 1 hr
(5) GC-MS analysis of fractions on a DB-1 column (30 m × 0.25 mm), oven temp. 250° to 300°

was approximately 90% and 35% for G-A and G-B respectively. Derivatised culture samples had to be analysed within one day because of decomposition. Reference samples were stable for at least four days. For a selective detection of the small amounts of G-A and G-B in these culture samples GC-MS was necessary. G-A and G-B were detected at m/z 537 and 625 (M–Me$^+$). Other terpene trilactones were not determined. For quantitative work, deutero-trimethylsilylated G-A and G-B were added as internal standards. The concentration in the cultured cell samples varied from 5 to 50 pg/g dry weight for both G-A and G-B (Chauret *et al.*, 1991).

More recently a good and validated quantitative gas chromatographic method was published for the analysis of G-A, G-B, G-C, G-J and BB in dried leaves and extracts (Hasler and Meier, 1992). For the isolation and purification a slightly modified version of the sample clean-up method used by van Beek (*vide supra*) was used. Especially when high amounts of terpene trilactones were present, Hasler *et al.* found a better recovery when the amount of stationary phase in the SPE column was doubled. After evaporation of the SPE elution solvent, the residue was dissolved in DMF containing internal standard (cholesterol) and subsequently silylated with N,O-bis-(trimethylsilyl)-trifluoroacetamide (BSTFA) containing 1% trimethyl-chlorosilane (TMCS) at 120°. Other silylation reagents gave worse results. Most likely the tertiary hydroxyl is silylated only with difficulty. No removal of the silylation reagent was attempted to avoid problems with desilylation. A standard dimethylpolysiloxane column with a temperature gradient up to 310° gave an excellent separation of all terpene trilactones. They eluted in the order: BB, G-A, G-J, G-B and G-C. The detection limit was 100 ng for BB and 50 ng for the ginkgolides. The reproducibility of the method was excellent with RSD's from 1–2%. Apart from the somewhat unpleasant and time-consuming derivatisation step, the gas chromatographic method is clearly superior over HPLC methods with either UV or RI detection.

Huh and Staba recently studied the ontogenic aspects of ginkgolide production (Huh and Staba, 1993). For the quantitation of ginkgolides in their seedlings, greenhouse plants, young trees, mature trees cuttings and plant tissue cultures they used a rather long (15 partitions) purification procedure modified after Carrier *et al.* (1990). The procedure is depicted in Table 9. As ginkgolides are not well soluble in both water and diethyl ether, some of the more polar ginkgolides like G-C and G-J

Table 9   Gas chromatographic method used by Huh and Staba (1993).

(1) dry Ginkgo plants at 60° for 48 hr and ground through 20 mesh sieve
(2) extract plant material (1 g) 4 × with 20 ml H$_2$O–Me$_2$CO (1:1) while stirring and filter over paper
(3) remove acetone *in vacuo*, wash 5 × with 20 ml hexane and bring aqueous phase at pH=2
(4) extract 5 × with 20 ml EtOAc and evaporate organic layer *in vacuo*, suspend in 20 ml water
(5) extract aqueous phase 5 × with 20 ml Et$_2$O, dry organic layer and evaporate *in vacuo*
(6) redissolve in MeOH, add Trisil BSA and DMF and heat at 80° for 1 hr
(7) GC analysis on a capillary DB-5 column (no column specifications given), oven temp. 295°, FID

Table 10   Gas chromatographic method used by Steinke *et al.* (1993).

---

(1) 100 mg extract is dissolved in 1 ml pyridine and iBuMeCO is added until 10 ml total volume
(2) after 5 min of shaking any precipitate is removed by filtration or centrifugation
(3) 200 μl clear solution is derivatised with 80 μl BSTFA and heated at 100° for 10 min
(4) GC analysis on an Ultra 2 column (25 m × 0.32 mm × 0.52 μm), oven temp. 50° to 300°, FID

---

may be partially lost in this partition step. The final purified residue was subsequently derivatised with Trisil BSA and then the trimethylsilyl derivatives of G-A, G-B and G-C were analysed by GC-FID. Reported retention times were 8.25, 8.90 and 9.20 min for G-A, G-B and G-C respectively. The same assay was also used by Jeon *et al.* (1995) and Sung *et al.* (1994) for the analysis of ginkgolides in embryo-derived Ginkgo plantlets and for the determination of seasonal and sexual variation in Ginkgo leaves respectively. Leaves of plantlets without roots contained less G-B than plantlets with roots suggesting that ginkgolides may be synthesized in the root and transported to the aerial part. The seasonal maximum occurred in August and declined thereafter confirming similar results of other groups.

Steinke *et al.* (1993) have published a GC assay for Ginkgo extracts. They used this chemical assay to compare the results of their biological assay (*vide infra*). The procedure is summarised in Table 10. The assay is relatively simple without any column chromatographic clean-up necessary. No internal standard was used. The reproducibility varied from 4–5%. The recovery was more than 95% for G-A, G-B, G-C and BB. Small amounts were detectable in the removed precipitate. Up to the maximum concentrations tested all calibration curves were linear. The limit of detection was 10 ppm for any terpene trilactone.

Fourtillan *et al.* (1995) have studied the pharmacokinetics of BB, G-A and G-B in humans. They studied blood and urine samples. Concentrations of the three terpene trilactones were quantitatively determined by means of GC-MS using negative chemical ionization. According to the authors the method is very sensitive allowing plasma concentrations as low as 0.2 ng/ml of each compound to be measured. Unfortunately no detail whatsoever about this method is presented in their paper nor a chromatogram. This makes it impossible for others to reproduce their results. A warning was issued by them about the instability of BB under neutral or basic conditions. Where ginkgolides are very stable, BB decomposes irreversibly to another product.

Recently van Beek and Taylor (1996) reported on the use of Supercritical Fluid Chromatography (SFE) as a sample clean-up method for standardised Ginkgo extracts containing ≈24% flavonoid glycosides and ≈6% Ginkgo terpene trilactones. It was found that ginkgolides and bilobalide could be selectively extracted at 335 atm and 45° with carbon dioxide containing 10% methanol. The procedure is given in Table 11. Flavonol glycosides were retained by the silica gel present in the extraction cell. Thus extraction and clean-up are combined in one step. Quantitation took place by GC-FID although it was shown that the final extracts could also be

Table 11   Supercritical extraction and clean-up method used by van Beek and Taylor (1996).

(1) apply ≈18 mg standardised Ginkgo extract in MeOH on 2 g silica gel and 0.5 g sand in a
     thimble
(2) extract supercritically at 335 atm and 45° with 10% MeOH in CO_2; first 5 min static,
     then 40 min dynamic at 1.5 ml/min, nozzle and trap both 80°; solid trap consists of
     400 mg silica gel
(3) stop flow, cool nozzle and trap to 50° and 30° respectively and wash trap with 2.5 ml
     MeOAc
(4) collect first 1.7 ml, evaporate, add i.s. (octacosane), evaporate, add 0.25 ml DMF and
     0.25 ml BSTFA containing 1% TMCS and heat 45 min at 120°
(5) GC analysis on an Ultra 1 column (25 m × 0.2 mm × 0.33 μm), oven temp. 230° to
     280°, FID

analysed by HPLC-ELSD. A representative gas chromatogram can be found in
Fig. 4. When the SFE results were compared with the results obtained by a classical
solvent extraction and SPE clean-up (according to the method described by Lolla
et al., *vide infra*), no differences were seen with regard to the cleanliness of the
extracts, recovery and amount of organic solvent needed. The SFE method was

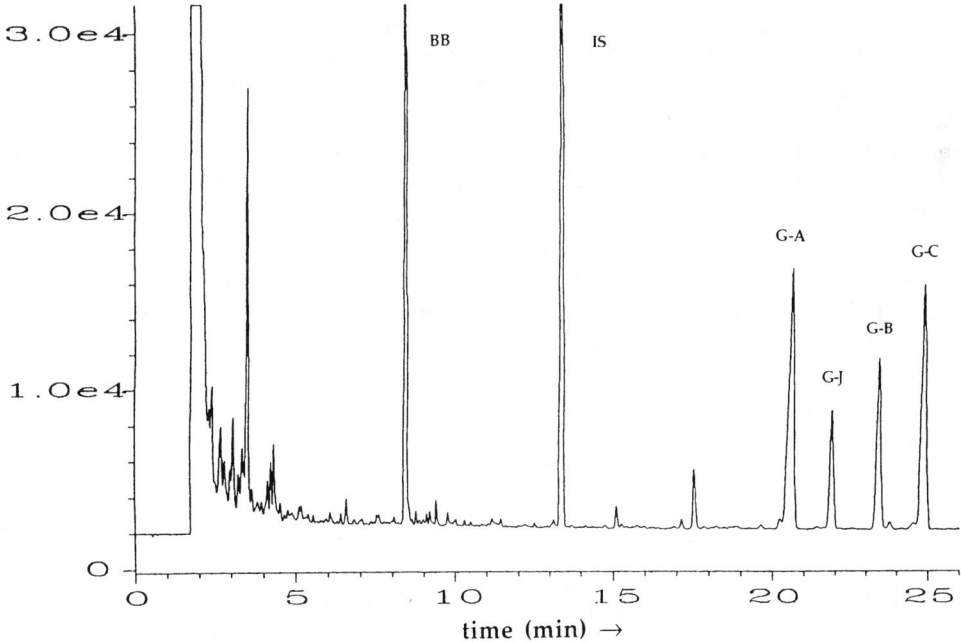

Figure 4   GC-FID profile of standardised Ginkgo leaf extract after supercritical fluid
extraction (SFE) and silylation (van Beek and Taylor, 1996). Internal standard = octacosane.
For comparison see Fig. 5 for a chromatogram of the same extract after SPE clean-up over
silica gel analysed under identical chromatographic conditions.

Table 12   Sample clean-up method used by Lolla *et al.* (1998).

---

(1) extract 5 g powdered leaves in a Soxhlet with hexane for 4 hr, discard hexane and dry leaves in an oven

(2) extract leaves with 120 ml $Me_2CO–H_2O$ (6:4) overnight at room temperature with stirring

(3) filter, evaporate *in vacuo* and dissolve in 5 ml MeOH under sonication

(4) transfer 0.5 ml to an SPE column containing 1.2 g silica gel; dry column at 60° *in vacuo* for 4 hr

(5) elute column with 10 ml toluene–Me_2CO (7:3), add internal standard (squalane) to eluate

(6) evaporate under N_2 at 60°, add pyridine and 1% TMCS in BSTFA and heat at 60° for 30 min

(7) GC analysis on a DB-1 column (15 m × 0.25 mm × 0.25 μm), oven temp. 120° to 300°, FID

---

considered superior in terms of labour investment and reproducibility. Especially for industrial quality control the method may prove interesting because the extraction can be fully automated. Disadvantages of this method are that it requires an expensive supercritical fluid extractor and is unsuitable for the analysis of leaves.

A very elegant one column sample clean-up method was very recently published by Lolla *et al.* (1998). Their procedure for Ginkgo leaves is presented in Table 12. Clean-up took place over an SPE silica gel column while the quantitation was performed by GC-FID after silylation. As internal standard squalane was chosen. The problem that the proper solvent for extracting or dissolving terpene trilactones from Ginkgo leaves or extracts is often too polar for a chromatographic separation of the same compounds on silica gel was nicely solved by an intermediate evaporation of the solvent used (MeOH) from the silica gel column. The method can be used both for leaves and standardised extracts. In the case of extracts, the defatting and extraction can be omitted and one can immediately start with the SPE clean-up. Validation experiments were carried out with respect to peak purity (by GC-MS), linearity (linear over a factor 20), reproducibility (RSD = 2.5–3.5%) and recovery (99–99.5% for BB, G-A and G-B). This assay appears very rugged and has already been used by others with similar results (van Beek and Taylor, 1996). An example chromatogram can be found in Fig. 5.

Electron capture detection (ECD) can also be used for Ginkgo terpene trilactones after GC separation. Two different types of derivatives have been analysed. The more common trimethylsilyl ethers, probably because of the large number of oxygens present in the molecule, can be detected more sensitively by ECD than by FID (van Beek *et al.*, 1998). An example chromatogram of a silylated evaporated aqueous extract of Ginkgo leaves without any sample clean-up is shown in Fig. 6. The other type of derivatives contain fluorine and can be detected much more sensitively. Examples are trifluoroacyl and heptafluorobutyryl. An example chromatogram of heptafluorobutyrylated terpene trilactones is given in Fig. 7 (van Beek, unpublished results). The detection limit for these ginkgolide derivatives is very low: < 1 pg. The peak for BB appears too small which indicates a poor stability of the derivative. Unusual is further that the elution order for G-A, G-B and G-C is reversed with respect to the silyl ethers, probably because of the unusual interaction

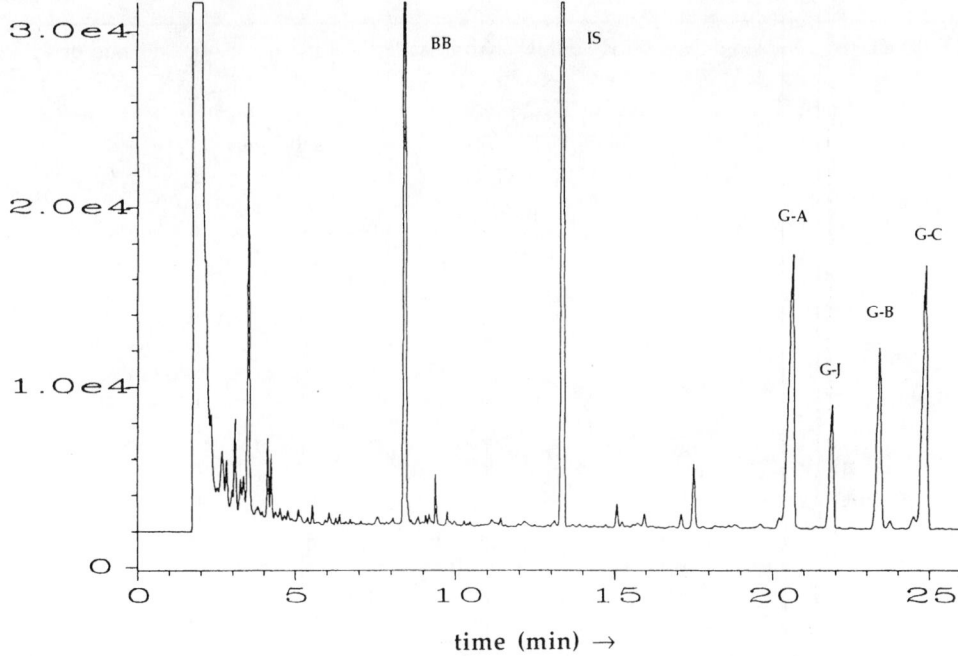

Figure 5   GC-FID profile of standardised Ginkgo leaf extract after methanol extraction, sample clean-up by SPE and silylation according to Lolla *et al.* (1998) and van Beek and Taylor (1996) respectively. Internal standard = octacosane.

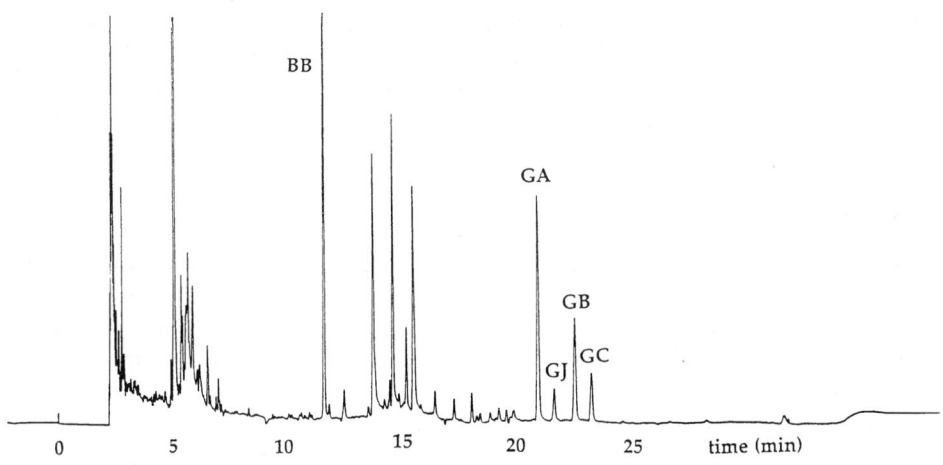

Figure 6   GC-ECD profile of a silylated evaporated aqueous extract of Ginkgo leaves without any sample clean-up (van Beek *et al.*, 1998).

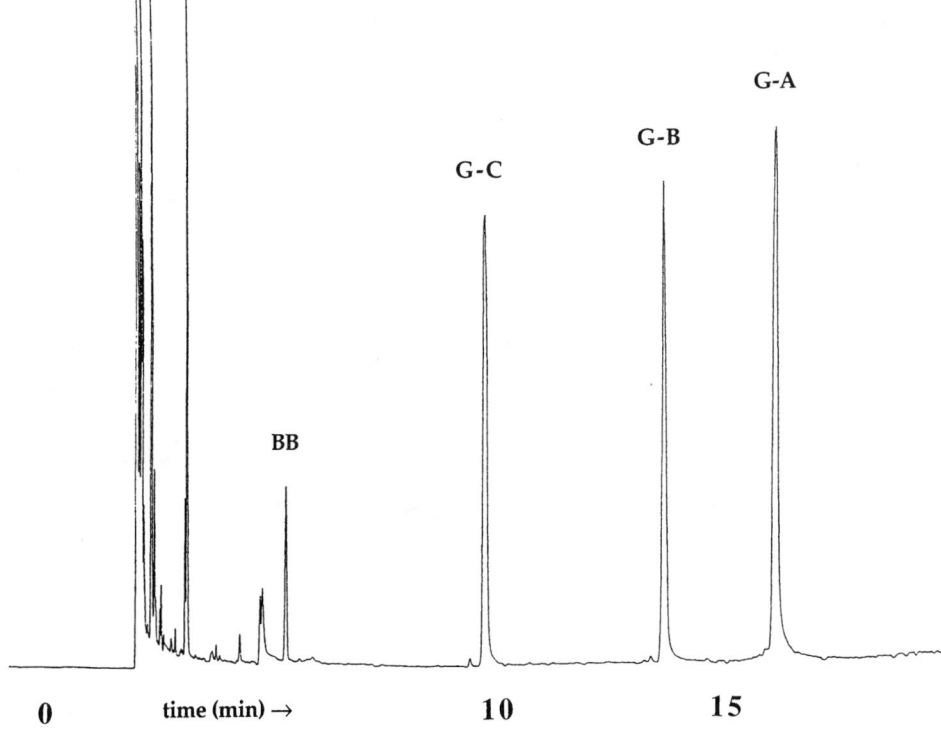

Figure 7 GC-ECD profile of a toluene extract of BB, G-C, G-B and G-A reference substances after reaction with heptafluorobutyric anhydride. Each peak corresponds with approximately 10 pg of underivatised terpene trilactone (van Beek, unpublished). After 14.5 min attenuator value is halved.

of the large number of fluor atoms (28 for G-C) with the apolar stationary phase. For quality control purposes this derivatisation is of no practical use, but it may prove useful for pharmacokinetic studies where it could possibly replace GC-MS.

Miscellaneous Methods

Thin Layer Chromatography (TLC)

Based on the good separation achievable with sodium acetate impregnated silica gel plates, recently a quantitative densitometric assay of terpene trilactones was published (Xie *et al.*, 1997). The total procedure is presented in Table 13 and appears rather time-consuming and labour-intensive. The validation showed that the linear ranges were 1–8 μg, 1–50 μg, 2–6 μg and 2–16 μg for G-A, G-B, G-C and BB respectively. As advantages of this assay, the authors claim cost efficiency, flexibility of post-chromatographic derivatization and reliability. The TLC scan profile in their paper shows a baseline separation of BB, G-A and G-B. Only G-J and G-C are not fully separated. There was good agreement between the results of this method for a

Table 13   Details of densitometric assay used by Xie *et al.* (1997).

(1) ultrasonically extract 160 mg extract in 10 ml MeOH during 15 min, centrifuge; repeat twice

(2) combine MeOH and evaporate on water bath to dryness; dissolve residue in H_2O

(3) transfer to polyamide column and wash with 150 ml H_2O, evaporate H_2O *in vacuo* to dryness

(4) dissolve residue in MeOH, add 1.5 g silica gel, mix and dry over P_2O_5 for 4 hr

(5) transfer the silica gel to silica gel column (5 g) and elute with 200 ml EtOAc, evaporate EtOAc

(6) dissolve in 2 ml MeOH and apply 3 μl to TLC plate made of silica gel impregnated with a solution of 4% NaOAc and 0.5% carboxymethyl cellulose in H_2O

(7) develop over 8 cm with toluene–EtOAc–Me_2CO–MeOH (2:1:1:0.12) (rel. humidity 47%)

(8) dry the plate and place in a sealed tank with Ac_2O vapour at 160° during 30 min

(9) scan the TLC plate with λ_{ex} 366 nm in fluorescent mode, calibrate using polynomial regression

standardised extract and the HPLC data supplied by the extract manufacturer. The method was applied to standardised extracts and Ginkgo phytopharmaceutical (tablets, oral solution, solutions for injection), but not leaves.

Supercritical Fluid Chromatography (SFC)

Recently it was shown that SFC is well suitable for the separation of Ginkgo terpene trilactones (Strode *et al.*, 1996). On a Deltabond deactivated aminopropyl HPLC column (15 × 4.6 mm, 5 μm) with 12% MeOH in CO_2 as fluid (280 atm, 3.5 ml/min, 40°) a baseline separation of bilobalide and all four ginkgolides could be achieved within 9 min. Detection occurred through evaporative light scattering detection (ELSD). A detection limit of around 10 ng was claimed. The selectivity of the system appears much higher than that of RP-HPLC. The explanation of this phenomenon given by the authors was that the separation is essentially a normal phase one. As most of the impurities present in Ginkgo extracts are more polar in nature than the ginkgolides, the latter elute first and the impurities remain on the column. Although a sample clean-up over silica did give cleaner chromatograms, some standardised extracts and phytopharmaceuticals could be analysed without any clean-up (Fig. 8). Thus SFC could have some merit for the analysis of Ginkgo extracts: quick, lowered consumption of organic solvent and less demanding in terms of sample clean-up. However first more experience and validation is necessary.

Capillary electrophoresis

Oerhle (1995) has demonstrated that ginkgolides A and B and bilobalide can be separated by micellar electrokinetic capillary electrophoresis (Oehrle, 1995). The separation was achieved in a capillary of 60 cm × 75 μm at 30° C with a buffer consisting of 25 mM phosphate and 90 mM SDS. The applied voltage was not

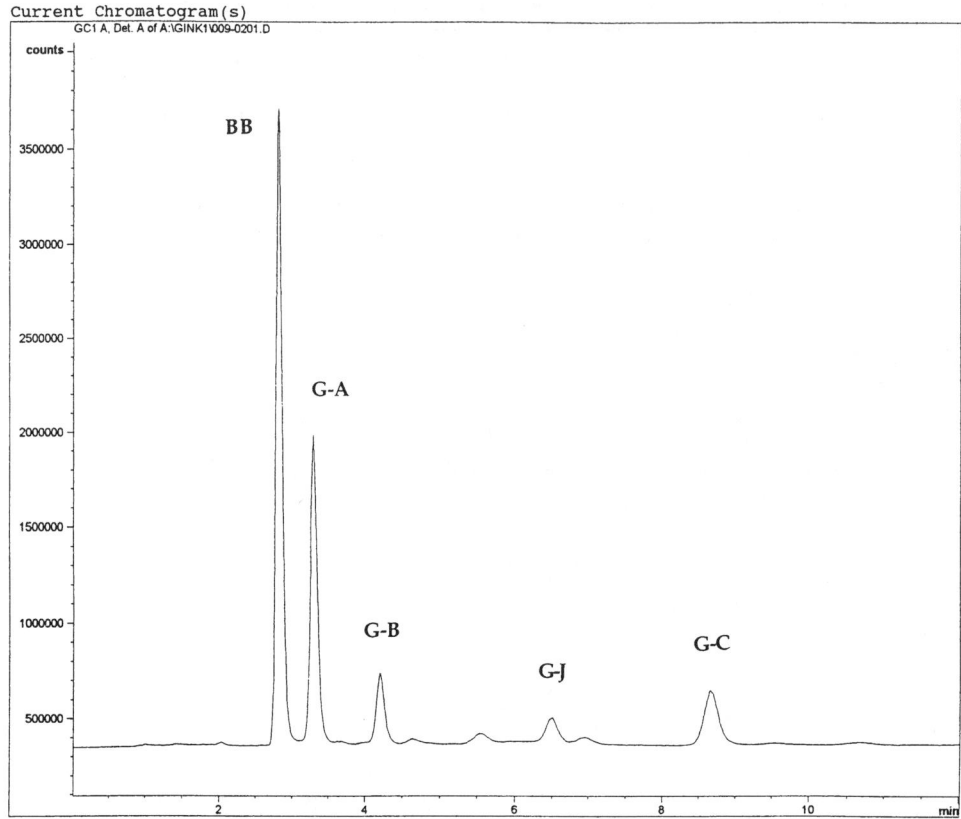

Current Chromatogram(s)

Figure 8 SFC-ELSD profile of a methanol solution of Indena standardised Ginkgo extract without any prior sample clean-up. Deltabond Amino 2 column, 5 μm 15 cm × 4.6 mm. Pressure 280 atm, flow 3.5 ml/min, fluid 12% MeOH in CO_2, 40° (Strode *et al.*, 1996).

given. Detection occurred by UV at 185 nm. There was a good separation between bilobalide (10 min) and the two ginkgolides. G-A and G-B were not baseline separated and eluted at 16.5 and 16.7 min respectively. No quantitative data or chromatograms of leaf extracts were included in this paper. Unless the separation considerably improves, it is clear that capillary electrophoresis cannot yet compete with HPLC, GC or SFC for the separation of Ginkgo terpene trilactones.

Biological control methods

An original approach was followed by Steinke *et al.* (1993). They focussed on the potent and selective Platelet-Activating Factor (PAF) antagonism of the various ginkgolides (Wagner and Steinke, 1991). In their *in vitro* biological assay they measured the inhibition of PAF-induced platelet aggregation by a Ginkgo extract. The assay showed an acceptable variation: for 70% of all measurements the absolute variation was smaller than 12% inhibition. The assay is also specific for ginkgolides, no other compounds present in the extract showed significant PAF antagonism. The assay is reasonably sensitive as 50% inhibition occurs at a G-B concentration of 3.5 μM. The authors were able to recalculate the biological outcome to a total % ginkgolides. When this value was compared with the results of a chemical assay, the results were in good agreement for some extracts but for others the biological assay gave more than 100% higher values. As explanation, errors in both methods and a non-additivity of individual ginkgolides were given. The method has two severe disadvantages which will most likely preclude any practical use in quality control labs. One is that bilobalide is not a PAF antagonist is thus not assayed. Especially in recent years more and more pharmacological evidence accumulates that bilobalide is an important compound in Ginkgo extracts. The second disadvantage is that fresh rabbit blood is needed for the assay. This is not regularly available in chemical quality control labs.

A principally similar method has been used by Li and Wong (1997) to determine the bioavailability of terpene trilactones of a "27 + 7" standardised Ginkgo extract (BioGinkgo) in rabbits. The bioavailability was compared with a more usual "24 + 6" standardised Ginkgo extract (Ginkgoba). The "24 + 6" stands for extracts containing 24% flavonol glycosides and 6% terpene trilactones. Ginkgolides in serum were assayed by their ability to inhibit the binding of PAF to its platelet membrane receptor *in vitro*. The inhibition was measured by incubating the platelets with rabbit serum and radioactively labelled PAF during 40 min after which free and bound PAF was separated by filtration. Subsequently the radioactivity was determined in the dried filters. There were significant differences between the "24 + 6" and "27 + 7" extract. The bioavailability of the latter was greater and while the "24 + 6" extract showed one peak in its plasma concentration after 3 hr the "27 + 7" extract showed two peaks after 1 and 5 hr respectively. The greater bioavailability is probably due to the slightly higher total ginkgolide content and the relatively higher content of G-B in the "27 + 7" extract. The authors speculated that the second peak might be caused by the formation of a ginkgolide metabolite that is unique to BioGinkgo 27/7. The relevance of this data for the clinical situation in humans remains to be established.

Nuclear Magnetic Resonance (NMR)

A rather unusual method for the analysis of Ginkgo terpene trilactones is based on quantitative NMR (van Beek *et al.*, 1993). After a relatively simple one column sample preparation step (refluxing in water, transfer to Extrelut column, elution with hexane-MeOAc (1:2), drying), maleic acid (internal standard) was added to the sample after which the solution was evaporated *in vacuo*. The sample was dissolved in 500 μl of Me$_2$CO-d6 – pyridine-d5 – MeOD-d4 (18:6:1) and an NMR spectrum was recorded. Because the H12 protons of G-A, G-B, G-C, G-J and BB are in a slightly different environment due to the different substitution patterns, each terpene trilactone has its own characteristic H12 signal in an NMR spectrum of a mixture. Because H12 has no neighbouring protons they appear as sharp

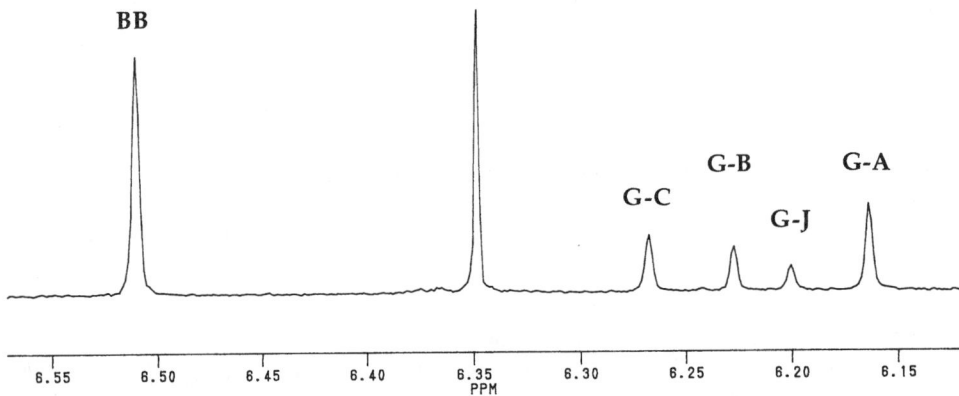

Figure 9 Relevant part of a 200 MHz NMR spectrum (128 scans) of a purified Tanakan sample (French phytopharmaceutical drug). Peak at 6.35 ppm = internal standard (maleic acid).

singlets as well as the protons of maleic acid. Because the six signals appear between 6.15–6.55 ppm – generally a non-crowded part of NMR spectra – there is relatively little interference from other compounds. An example spectrum is presented in Fig. 9. After careful integration, the amount of each terpene trilactone present can be easily calculated because the amount of maleic acid is precisely known. The area of each signal is proportional to the molar concentration of the molecule the proton belongs to. To obtain good quantitative data a relaxation delay of 10 sec was used after each NMR pulse. The minimum amount that could be assayed with a 200 MHz spectrometer within 30 min, was 0.1 mg. This translated to a detection limit of 0.005% in dry leaves or standardised extracts which is comparable with HPLC-RI. Also the reproducibility was similar to HPLC-RI. A significant disadvantage is that an expensive NMR spectrometer is necessary. The biggest advantage of the method is, that no reference substances or calibration curves are needed. It can also be used to determine the absolute purity of reference substances for HPLC or GC determinations which is difficult to accomplish by other means as 100% pure standards of terpene trilactones are currently not available.

CONCLUSIONS

Since the appearance of the first quantitative paper 16 years ago, significant development have taken place in the field of Ginkgo terpene trilactone analysis. For rapid qualitative analysis, TLC is still the way to go. By impregnating silica gel plates with sodium acetate, an excellent resolution of all five terpene trilactones occurring in leaves is nowadays possible. Additionally the impregnation increases the sensitivity of the acetic anhydride spray reagent which is the best reagent found so far. This method has been incorporated as an identification test in the first Pharmacopeia monograph on Ginkgo leaves (Anonymous, 1997).

If many quantitative routine analysis of Ginkgo leaves, extracts or phyto-pharmaceuticals have to be performed, the first choice is between high pressure liquid chromatography (HPLC) and gas chromatography (GC). The biggest advantage of HPLC over GC is that no time-consuming derivatisation step with an aggressive reagent is necessary as the compounds of interest can be analysed as such. Also the more simple the assay, the less chance there is for losses. However GC gives a better separation in less time than HPLC and HPLC lacks an equivalent of the universal flame ionisation detector (FID) of GC. Sensitive and selective detection of terpene trilactones in LC remains therefore somewhat of a problem. The originally used UV detector has been abandoned because many compounds present in G. biloba leaves have ϵ-values which are orders of magnitude larger than those of the terpene trilactones at their absorption maximum (219 nm). Thus even traces of impurities can overshadow the peaks of interest. Surprisingly in a very recent paper UV detection reappeared once again (Itoh et al., 1998). Refractive index (RI) detection, although less sensitive, is much better suited for the analysis of relatively pure Ginkgo samples and was introduced about 10 years ago. It has been successfully used and is still used by several industries for controlling the quality of their Ginkgo products. Main shortcomings are the rather poor sensitivity, poor baseline stability and the impossibility of some models to work with THF. Two more LC detectors have been introduced in the nineties: the mass spectrometer (LC-MS) and the evaporative light scattering detector (ELSD). LC-MS combines selectivity with sensitivity. Unfortunately the reproducibility for quantitative purposes could be better and the price tag is still too high for routine analysis. However if the price comes further down with a simultaneous increase in ruggedness, the mass spectrometer might become a routine detector for ginkgolides in ten years time. Currently much more interesting is ELSD which could be viewed as a sort of improved RI detector: more stable baseline, compatible with all volatile solvents including THF, compatible with gradient elution, robust and approximately one order of magnitude more sensitive. The price is only slightly higher than that of an RI detector. If no detector has yet been selected, ELSD would be an excellent choice if many analysis have to be performed.

For GC the detector choice is not an issue, FID is the way to go. It is reliable, sensitive, cheap and more reproducible than any LC detector available. The only other detector which has been experimentally used, is the electron capture detector (ECD). It is radioactive, has a limited linear range, is not as widespread as the FID and has only the advantages of a somewhat greater selectivity and a much higher sensitivity. To achieve such a high sensitivity, a prior derivatisation of the terpene trilactones with a halogen containing reagent is necessary. The only application of ECD might be in future pharmacokinetical studies where such a high sensitivity is indeed necessary.

Most of the other quantitative methods which have been proposed are rather specialised and unsuitable for routine analysis. A biological method based on the antagonism of PAF by ginkgolides requires experimental animals and does not assay bilobalide. A TLC method requires a densitometer and otherwise offers no real advantages over existing HPLC or GC methods. The only published capillary electrophoresis (CE) separation is worse than the best LC or GC separations and requires UV for detection. Quantitation by NMR is also unsuitable for routine analysis and requires the presence of an expensive NMR spectrometer. Its one big advantage over all other methods is, that it does not require pure terpene trilactones for calibration purposes. The only alternative separation/detection method which could compete with either LC or GC for routine analysis is supercritical fluid chromatography (SFC) in combination with ELSD. SFC is capable of achieving a baseline separation of all five terpene trilactones within 10 min and with a greater selectivity than RP-HPLC because of its normal phase-like character. The analysis of some standardised extracts without any prior sample clean-up looks promising although more validation is indicated.

The real problems in terpene trilactone analysis however do not lie in the separation and detection of the final sample solutions but in the initial extraction and even more so in the sample clean-up. As extraction solvent usually mixtures of water with methanol, alcohol or acetone are used. As long as the water content is ⩾50%, the solubility of terpene trilactones in such mixtures does not vary much and is sufficient to warrant a good extraction (Stumpf, 1997). This does not hold true for many other primary and secondary metabolites occurring in Ginkgo leaves and extracts, and thus the water/organic solvent ratio strongly influences the type and amount of co-extracted constituents (e.g. flavonol glycosides, biflavonoids, alkyl-phenols, chlorophyll). Generally speaking the amount of such compounds increases with an increasing percentage of organic solvent with the exception of the polar proanthocyanidins. An illustration of this phenomenon are the percentages of 10.4% and 0.023% of unwanted allergenic ginkgolic acids in initial leaf extracts prepared with pure acetone and water respectively (Aye and Müller, 1991). Some authors increase the selectivity of the initial extraction medium further by decreasing the organic content to 10% or even 0%. To obtain a good extraction of G-B (solubility in cold water only 0.011%) refluxing is then necessary. Camponovo (1996) compared the efficiency of cold methanol, cold 50% aqueous methanol and boiling water and found that although all three solvents extracted the terpene trilactones equally well, 50% aqueous methanol gave the most clean extract. No matter which solvent is chosen, experiments to determine the extraction efficiency and recovery are necessary. The addition of a small percentage of acid could be considered to prevent possible decomposition of the less stable bilobalide and to diminish co-extraction of chlorophyll.

As can be seen from the Tables 1–13, many different approaches have been followed to remove unwanted impurities from the primary extracts prior to the separation and analysis. Some groups use a defatting step, others one or more partitioning steps and still others use one or two solid phase extractions (SPE) columns. Some procedures are suitable only for extracts while others can be used for both leaves and extracts. Many of the procedures have not been validated and should be used with caution. Some publications of an industrial group lack many analytical details and are thus not up to current standards for scientific articles because they make it impossible for others to reproduce their work (Teng 1988; Flesch et al., 1992; Fourtillan et al., 1995; Cartayrade et al., 1997). The final conclusion is that one simple and rapid sample clean-up method which gives 100% recovery of all ginkgolides and bilobalide, and which is suitable for both leaves and extracts does not exist.

Simple and rapid means no time-consuming, error-prone and labour-intensive defatting and partitioning steps but preferably a one column SPE step. To maximize the selectivity of the clean-up procedure for the later GC or RP-HPLC separations, an adsorption chromatography step on silica gel is probably optimal. Terpene trilactones can be eluted from such columns with mixtures of hexane and ethyl acetate which do not elute more polar impurities like flavonol glycosides. An additional advantage of such 100% organic solvents is that they are easy to evaporate and that the final samples for GC or HPLC can thus be rapidly prepared. Alumina has been proposed as an improvement over silica gel. Although it is certain that alumina offers greater selectivity for terpene trilactones than silica gel, its use is disputable. At least one author has reported rapid decomposition of bilobalide by different brands of alumina. It should therefore only be used after careful validation experiments. A problem with normal phase materials like silica gel is, that the optimal solvent for elution is usually too apolar for transferring the initial extract in a small volume to the top of the column. This problem was elegantly solved by Lolla et al. (1998) who transferred the initial extract dissolved in methanol to the column after which the methanol was removed in vacuo at 60°. Their approach appears one of the best published so far. An alternative for the adsorption step could be a partitioning step on an SPE column filled with diatomaceous earth. This has already been proposed by several groups but a good, validated method has not yet appeared. An advantage of such columns over silica gel is that they require to be loaded with highly aqueous solutions. They are thus compatible with the initial extracts with their often high percentage of water. Elution can be achieved with mixtures of hexane and ethyl acetate.

The only automated extraction/purification procedure for Ginkgo extracts published so far uses supercritical fluid extraction (SFE). By the presence of silica gel in the extraction chamber and solid trap, relatively clean extracts are obtained comparable with samples purified by SPE on silica gel. Although the method requires an expensive SFE apparatus with autosampler, it could prove interesting for pharmaceutical industries. The manpower input is very low and the reproducibility and 24 hr sample throughput good due to the automation. The final extract is obtained in a 2 ml autosampler vial in organic solvent and is amenable both to GC and HPLC analysis. Regrettably the method is not applicable to Ginkgo leaves.

REFERENCES

Anonymous (1997) *Ginkgo biloba* leaf – draft monograph. *Pharmacopeial Forum*, **23**, 3644–3647.

Aye, R.D., Müller, B. (1991) Ginkgo-Extrakte – Chromatographische Standardisierung und Einflüsse der Herstellung auf die Zusammensetzung. *Münch. med. Wschr.*, **133**, S58–S60.

van Beek, T.A., Bombardelli, E., Morazzoni, P., Peterlongo, F. (1998) *Ginkgo biloba* L. *Fitoterapia*, **69**, 195–238.

van Beek, T.A., Lelyveld, G.P. (1992) Concentration of ginkgolides and bilobalide in *Ginkgo biloba* leaves in relation to the time of year. *Planta Med.*, **58**, 413–416.

van Beek, T.A., Lelyveld, G.P. (1993) Thin layer chromatography of bilobalide and ginkgolides A, B, C and J on sodium acetate impregnated silica gel. *Phytochem. Anal.*, **4**, 109–114.

van Beek, T.A., Lelyveld, G.P. (1997) Preparative isolation and separation procedure for ginkgolides A, B, C, and J and bilobalide. *J. Nat. Prod.*, **60**, 735–738 (1997).

van Beek, T.A., Scheeren, H.A., Rantio, T., Melger, W.C., Lelyveld, G.P. (1991) Determination of ginkgolides and bilobalide in *Ginkgo biloba* leaves and phytopharmaceuticals. *J. Chromatogr.*, **543**, 375–387.

van Beek, T.A., Taylor, L.T. (1996) Sample preparation of standardized extracts of *Ginkgo biloba* by supercritical fluid extraction. *Phytochem. Anal.*, **7**, 185–191.

van Beek, T.A., van Veldhuizen, A., Lelyveld, G.P., Piron, I., Lankhorst, P.P. (1993) Quantitation of bilobalide and ginkgolides A, B, C and J by means of nuclear magnetic resonance spectroscopy. *Phytochem. Anal.*, **4**, 261–268.

Braquet, P. (1988) The ginkgolides. From Chinese pharmacopeia to a new class of pharmacological agents: the antagonists of platelet-activating factor. In P. Braquet, (ed.), *Ginkgolides – Chemistry, Biology, Pharmacology and Clinical Perspectives*, **Vol. I**, J.R. Prous Science Publishers, Barcelona, pp. xv-xxxiv.

Camponovo, F. (1996) Thesis: *Mise au point de procédés pour l'analyse phytochimique et étude comparative de quelques médicaments à base de Ginkgo biloba, Panax ginseng et Harpagophytum procumbens*, Université de Lausanne.

Camponovo, F.F., Wolfender, J.-L., Maillard, M.P., Potterat, O., Hostettmann, K. (1995) Evaporative light scattering and thermospray mass spectrometry: two alternative methods for detection and quantitative liquid chromatographic determination of ginkgolides and bilobalide in *Ginkgo biloba* leaf extracts and phytopharmaceuticals. *Phytochem. Anal.*, **6**, 141–148.

Carrier, D.J., Chauret, N., Mancini, M., Coulombe, P., Neufeld, R., Weber, M., Archambault, J. (1991) Detection of ginkgolide A in *Ginkgo biloba* cell cultures. *Plant Cell Rep.*, **10**, 256–259.

Carrier, D.J., Coulombe, P., Mancini, M., Neufeld, R., Weber, M., Archambault, J. (1990) Immobilized *Ginkgo biloba* cell cultures. In H.J.J. Nijkamp, L.H.W. van de Plas, and J. Aartrijk, (eds.), *International congress on plant tissue and cell culture*, Kluwer, Amsterdam, pp. 614–618.

Carrier, D.J., van Beek, T.A., van der Heijden, R., Verpoorte, R. (1998) Distribution of ginkgolides and terpenoid biosynthetic activity in *Ginkgo biloba*. *Phytochemistry*, **48**, 89–92.

Cartayrade, A., Neau, E., Sohier, C., Balz, J.-P., Carde, J.-P., Walter, J. (1997) Ginkgolide and bilobalide biosynthesis in *Ginkgo biloba*. I: Sites of synthesis, translocation and accumulation of ginkgolides and bilobalide. *Plant Physiol. Biochem.*, **35**, 859–868.

Chauret, N., Carrier, J., Mancini, M., Neufeld, R., Weber, M., Archambault, J. (1991) Gas chromatographic-mass spectrometric analysis of ginkgolides produced by *Ginkgo biloba* cell culture. *J. Chromatogr.*, **588**, 281–287.

Chen, P., Su, N.-L., Nie, L.-H., Yao, S.-Z., Zeng, J.-G. (1998) Analysis of ginkgolides and bilobalides in *Ginkgo biloba* L. extract for its production process control by high-performance liquid chromatography. *J. Chromatogr. Sci.*, **36**, 197–200.

Flesch, V., Jacques, M., Cosson, L., Teng, B.P., Petiard, V., Balz, J.P. (1992) Relative importance of growth and light level on terpene content of *Ginkgo biloba*. *Phytochemistry*, **31**, 1941–1945.

Fourtillan, J.B., Brisson, A.M., Girault, J., Ingrand, I., Decourt, J.P., Drieu, K., Jouenne, P., Biber, A. (1995) Propriétés pharmacocinétiques du bilobalide et des ginkgolides A et B chez le sujet sain après adminstrations intraveineuses et orales d'extrait de *Ginkgo biloba* (EGb 761). *Thérapie*, 50, 137–144.

Hasler, A., Meier, B. (1992) Determination of terpenes from *Ginkgo biloba* L. by capillary gas chromatography. *Pharm. Pharmacol. Lett.*, 2, 187–190.

Huh, H., Staba, E.J. (1993) Ontogenic aspects of ginkgolide production in *Ginkgo biloba*. *Planta Med.*, 59, 232–239.

Itoh, K., Shiraishi, K., Takechi, Y., Miyata, M., Tachibana, S. (1998) Production of ginkgo-lides by callus cultures of Icho (*Ginkgo biloba*) leaves. *Mokuzai-Gakkaishi*, 44, 1–8.

Jeon, M.H., Sung, S.H., Huh, H., Kim, Y.C. (1995) Ginkgolide B production in cultured cells derived from *Ginkgo biloba* L. leaves. *Plant Cell Rep.*, 14, 501–504.

Komoda, Y., Nakamura, H., Uchida, M. (1988) HPLC analysis of ginkgolides by using a differential refractometer. *Iyo Kizai Kenkyusho Hokoku (Reports of the Institute for Medical & Dental Engineering)*, 22, 83–85.

Laurain, D., Trémouillaux-Guiller, J., Chénieux, J.-C., van Beek, T.A. (1997) Production of ginkgolide and bilobalide in transformed and gametophyte derived cell cultures of *Ginkgo biloba*. *Phytochemistry*, 46, 127–130.

Li, C.L., Wong, Y.Y. (1997) The bioavailability of ginkgolides in *Ginkgo biloba* extracts. *Planta Med.*, 63, 563–565.

Lobstein-Guth, A., Briançon-Scheid, F., Anton, R. (1983) Analysis of terpenes from *Ginkgo biloba* L. by high-performance liquid chromatography. *J. Chromatogr.*, 267, 431–438.

Lolla, E., Paletti, A., Peterlongo, F. (1998) Gas chromatographic determination of ginkgolides and bilobalide in *Ginkgo biloba* leaves and extracts. *Fitoterapia*, 69, 513–519.

Maillard, M.P., Camponovo, F., Wolfender, J.-L., Hostettmann, K. (1993) Semi-quantitative determination of terpenes from *Ginkgo biloba* by LC-TSP-MS. Poster 138, *International symposium of the Phytochemical Society of Europe "Phytochemistry of plants used in traditional medicine"*, 29 Sept.–1 Oct. 1993, Lausanne, Switzerland.

Maruyama, M., Terahara, A., Nakadaira, Y., Woods, M.C., Takagi, Y., Nakanishi, K. (1967) The ginkgolides. IV. Stereochemistry of the ginkgolides. *Tetrahedron Lett.*, 315–319.

O'Reilly, J. (1993) *Ginkgo biloba* – cultivation, extraction, and therapeutic use of the extract. In T.A. van Beek, and H. Breteler, (eds.), *Phytochemistry and Agriculture*, Vol. 34 in Proceedings of the Phytochemical Society of Europe, Clarendon Press, Oxford, pp. 253–270.

Oehrle, S.A. (1995) Analysis of ginkolides and biobalides by capillary electrophoresis. *J. Liq. Chromatogr.*, 18, 2855–2859.

Okabe, K., Yamada, K., Yamamura, S., Takada, S. (1967) Ginkgolides. *J. Chem. Soc.*, 2201–2206.

Pietta, P., Mauri, P., Rava, A. (1992) Rapid liquid chromatography of terpenes in *Ginkgo biloba* L. extracts and products. *J. Pharm. Biomed. Anal.*, 10, 1077–1079.

Pietta, P.G., Mauri, P.L., Rava, A. (1990) Analysis of terpenes from *Ginkgo biloba* L. extracts by reversed phase high-performance liquid chromatography. *Chromatographia*, 29, 251–253.

Steinke, B., Müller, B., Wagner, H. (1993) Biologische Standardisierungsmethode für *Ginkgo*-Extrakte. *Planta Med.*, 59, 155–160.

Strode III, J.B.T., Taylor, L.T., van Beek, T.A. (1996) Supercritical fluid chromatography of ginkgolides A, B, C and J and bilobalide. *J. Chromatogr.*, 738 A, 115–122.

Stumpf, H. (1989) Ginkgolide in Tebonin forte. *Deutsche Apoth. Ztg.*, 129, 2794.

Stumpf, K.-H. (1997) Pharmaceutical quality and extract specification of EGb 761®, *Proceedings of the International Seminar of Ginkgo '97*, Organizing committee for '97 International Seminar on Ginkgo, Beijing, pp. 39–56.

Sung, S.H., Jeon, S.H., Moon, Y.S., Lee, H.S., Huh, H., Kim, Y.C. (1994) Seasonal and sexual variation of ginkgolides contents in Ginkgo leaves. *Yakhak Hoeji*, **38**, 20–23.

Tallevi, S.G., Kurz, W.G.W. (1991) Detection of ginkgolides by thin-layer chromatography. *J. Nat. Prod.*, **54**, 634–625.

Teng, B.P. (1988) Chemistry of ginkgolides. In P. Braquet, (ed.), *Ginkgolides – Chemistry, Biology, Pharmacology and Clinical Perspectives*, Vol. 1, J.R. Prous Science Publishers, Barcelona, pp. 37–41.

Wagner, H., Bladt, S., Hartmann, U., Daily, A., Berkulin, W. (1989) *Ginkgo biloba*. DC- und HPLC-Analyse von Ginkgo-Extrakten und Ginkgo-Extrakte enthaltenden Phytopräparaten. *Dtsch. Apoth. Ztg.*, **129**, 2421–2429.

Wagner, H., Steinke, B. (1991) Biologische Wirkwertbestimmung von Ginkgo-Extrakten. *Münch. med. Wschr.*, **133**, S54-S57.

Weinges, K., Bähr, W. (1969) Bilobalid A, ein neues Sesquiterpen mit tert.-Butyl-Gruppe aus den Blättern von *Ginkgo biloba* L. *Liebigs Ann. Chem.*, **724**, 214–216.

Weinges, K., Hepp, M., Jaggy, H. (1987) Isolierung und Strukturaufklärung eines neuen Ginkgolids. *Liebigs Ann. Chem.*, 521–526.

Xie, P., Yan, Y., Qian, H. (1997) Determination of terpenoid lactones in *Ginkgo biloba* extract and its preparations by means of quantitative fluorophotometric TLC analysis, *Proceedings of the International Seminar of Ginkgo '97*, Organizing committee for '97 International Seminar on Ginkgo, Beijing, pp. 174–177.

Yao, W., Yang, C., Tian, Y., Liu, T., Xu, Y. (1997) Rapid determination of ginkgolides and bilobolide in *Ginkgo biloba* leaves and its extracts by HPLC, *Proceedings of the International Seminar of Ginkgo '97*, Organizing committee for '97 International Seminar on Ginkgo, Beijing, pp. 182–188.

10. THE ANALYSIS OF GINKGO FLAVONOIDS

OTTO STICHER,[1] BEAT MEIER[2] and ANDREAS HASLER[2]

[1]*Department of Pharmacy, Swiss Federal Institute of Technology (ETH) Zurich, CH-8057 Zürich, Switzerland*
[2]*Zeller AG, Herbal Remedies, CH-8590 Romanshorn, Switzerland*

INTRODUCTION

The use of Ginkgo preparations for the treatment of cough, bronchial asthma, irritable bladder and alcohol abuse can be traced back to the origins of Chinese herbal medicine. But the modern use of Ginkgo phytomedicines is not derived from traditional medicine. Research on *Ginkgo biloba*, done by the pharmaceutical company of Dr. W. Schwabe, Karlsruhe, Germany, led to the introduction of Ginkgo leaf extracts in 1965 for the treatment of circulatory diseases resulting from older age. Today, preparations containing Ginkgo leaf extracts are among the best selling phytopharmaceuticals in Europe, particularly in France and Germany. Compared to the importance of Ginkgo preparations, there is only a very small number of publications dealing with quality control of Ginkgo and Ginkgo preparations. To date, only a few general quality standards for Ginkgo leaves and Ginkgo phytomedicines have been published, although various laboratories are actively involved in the analysis of Ginkgo constituents. It appears that the marketing strategies of most companies which produce Ginkgo preparations do not permit the publication of the company-developed quality control standards. A monograph entitled "Ginkgo folium" is in preparation for the European pharmacopoeia while a similar monograph for the United States Pharmacopeia has been published in draft form (see chapter "Considerations in the development of the U.S. Pharmacopeia's monograph on *Ginkgo biloba* L." by V.S. Srinivasan in this volume).

First of all, standardisation of medicinal plant extracts is used to assure quality of phytomedicines. Standardisation is mostly done on the active compounds, which are in the case of Ginkgo leaves flavonoids and terpene trilactones. In this chapter suitable methods for the analysis of flavonoid glycosides in Ginkgo leaves and Ginkgo preparations will be presented and discussed. Earlier reviews on analytical aspects of Ginkgo flavonoids have been published by Sticher (1992, 1993a, 1993b) while reviews on the isolation and occurrence have been published by Yoshitama (1997) and Hasler (see chapter "Chemical constituents of *Ginkgo biloba*" in this volume).

FLAVONOIDS OF *GINKGO BILOBA* LEAVES

The flavonoids of Ginkgo leaves (Fig. 1) represent a great variety of flavonol glycosides based on kaempferol and quercetin and are found as mono-, di- and

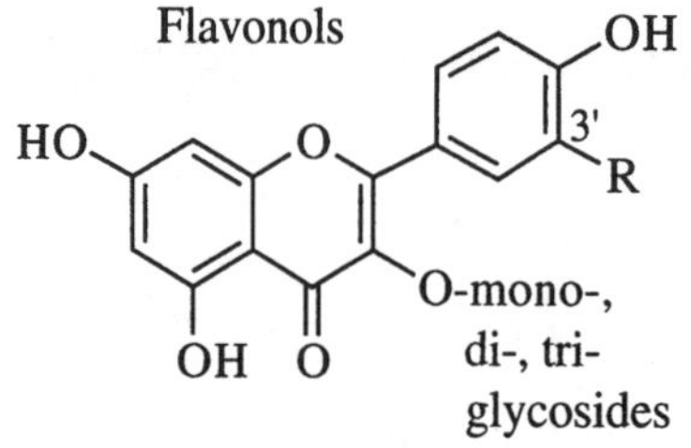

R	
H	Kaempferol derivative
OH	Quercetin derivative
OCH_3	Isorhamnetin derivative

Coumaric acid esters of flavonols

R	
H	Kaempferol derivative
OH	Quercetin derivative

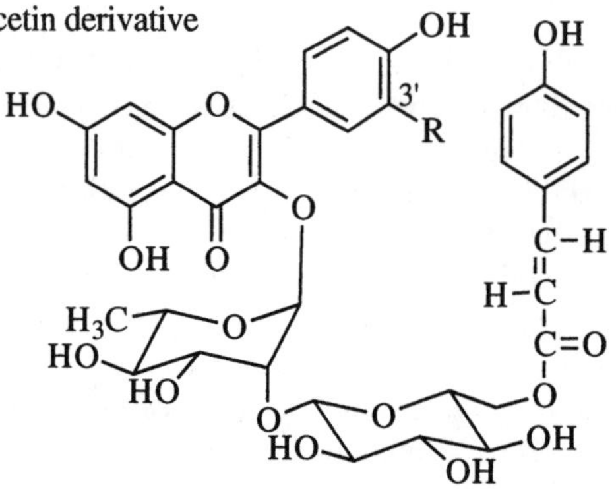

Figure 1 Structures of flavonol glycosides from leaves of *G. biloba.*

triglycosides. Minor flavonoid compounds are derived from isorhamnetin, myricetin, 3′-methylmyricetin (with flavonol structures) as well as from apigenin and luteolin (with flavone structures). Structurally interesting compounds include flavonol glycosides esterified with *p*-coumaric acid. These acylated flavonol glycosides are useful lead compounds in the quality control of Ginkgo preparations. In addition, non-glycosidic biflavonoids (Fig. 2) and proanthocyanidins have been isolated. The flavonol glycosides and the proanthocyanidins contribute to the free radical scavenging activity of Ginkgo extracts (DeFeudis, 1991 and references cited therein).

ANALYSIS OF FLAVONOIDS

Analysis of Flavonol Glycosides as their Aglycones after Hydrolysis

Spectrophotometric determination

Hydrolysis of the glycosides, followed by a spectrophotometric determination of the aglycones as an aluminium chloride chelate complex, is the method currently used to determine the flavonoid concentration in herbal drugs (Fig. 3). This method,

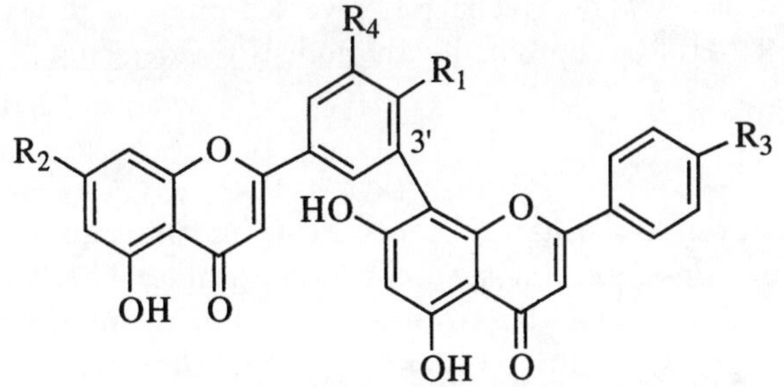

R_1	R_2	R_3	R_4	
OH	OH	OH	H	Amentoflavone
OCH_3	OH	OH	H	Bilobetin
OCH_3	OCH_3	OH	H	Ginkgetin
OCH_3	OH	OCH_3	H	Isoginkgetin
OCH_3	OCH_3	OCH_3	H	Sciadopitysin
OCH_3	OH	OH	OCH_3	5'-Methoxybilobetin

Figure 2   Structures of biflavones from leaves of *G. biloba*.

described in several pharmacopoeias, is not very specific and only permits an approximate estimation of the total flavonoids. A detailed determination of the qualitative and quantitative composition of the aglycones is not possible. For Ginkgo leaves, this method is not reproducible due to an incomplete extraction of the polar flavonol glycosides (acetone extraction), and probably due to a large amount of interfering proanthocyanidins (Hasler, 1990). Today, the method of choice for a selective analysis of

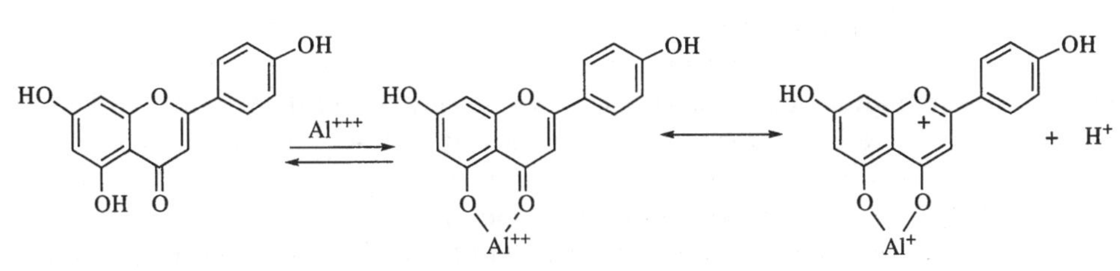

Figure 3   Structure of the aluminium chloride chelate complex of quercetin.

both flavonoid glycosides and aglycones is reversed-phase high-performance liquid chromatography (RP-HPLC) combined with diode-array detection.

Reversed phase HPLC

A direct quantification of the naturally occurring flavonol glycosides, part of the active principles of Ginkgo extracts, would be desirable. The flavonoid profile of Ginkgo is very complex, and therefore analysis during pharmaceutical quality control is rather tedious. Additionally, most of the reference compounds are not commercially available. In the pharmaceutical industry Ginkgo extracts have been standardised on flavonol glycosides since a long time. The great variety of genuine flavonol glycosides is reduced by hydrolysis to the corresponding aglycones. The UV spectra of the aglycones allow a selective detection at wavelengths above 300 nm. HPLC methods which, after hydrolysis with methanolic HCl of the flavonol glycosides, quantify the main aglycones kaempferol, quercetin, isorhamnetin and the minor aglycones myricetin and luteolin instead of more than 20 single flavonol glycosides have been presented by Schennen (1988) and Wagner et al. (1989). Schennen as well as Wagner et al. were able to separate the aglycone mixtures, after pre-purification over Sephadex LH-20, using a LiChroCart RP-18 column with RP-8 guard column and methanol-water-formic acid (55:41:4) (Schennen, 1988) or a Nucleosil C_{18} (5 μm) column and a methanol-water gradient (Wagner et al., 1989), respectively, as mobile phase. Wagner et al. determined the aglycone content using external standards. The calculation was based on an acyl glycoside with a mean molecular weight of 745 or on rutin (MW = 665). Schennen used rutin as an internal standard without giving detailed information about the calculation procedure and applied the proposed method for biological investigations, whereas Wagner et al. quantitatively determined flavonoids in G. biloba.

Kim et al. have determined the stability of the major flavonol glycosides in aqueous solutions of Ginkgo extract (Kim et al., 1989) at various pH values. The optimal pH for storage was found to be 6.5. The shelf life ($T_{90\%}$) at pH = 6.5 at 20° C was extrapolated to be four years. The formed aglycones were analysed by HPLC on LiChrosorb RP-18 (10 μm) with acetonitrile-isopropanol-water-citric acid (50:6:100:1) as solvent at 2.0 ml/min and UV detection at 365 nm.

More recently two selective HPLC methods which allow the analysis of flavonol glycosides in Ginkgo leaves, extracts, and phytomedicines were developed. The first method (Method A) allows the qualitative and quantitative determination of the aglycones after hydrolysis of the flavonol glycosides (Hasler, 1990; Hasler et al., 1990a, 1990b, 1992). This method for the standardisation of Ginkgo extracts is simple, rapid and reproducible. The work-up procedure consists of two basic steps: extraction and hydrolysis of the glycosides, and sample clean-up (Fig. 4). Extraction and hydrolysis are performed by refluxing the pulverised plant material (4 g) or a plant extract (2 g) with 10 ml hydrochloric acid (25%) in 70 ml methanol for 60 min. The use of methanol helps to efficiently extract the more polar di- and tri-glycosides. The addition of hydrochloric acid allows a direct hydrolysis of the glycosides during extraction. Sample clean-up is carried out using C_{18} solid-phase extraction cartridges (e.g. Bond

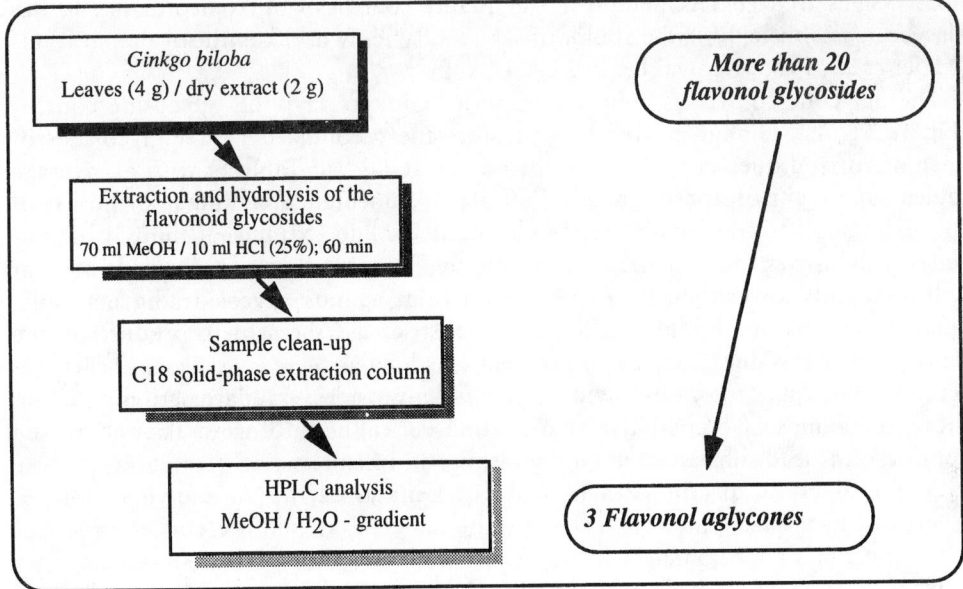

Figure 4 Work-up procedure and HPLC determination of flavonoid aglycones (Hasler *et al.*, 1992).

Elut). Finally, an aliquot (10 ml) of the purified extract is subjected to HPLC analysis. The flavonol aglycones can be easily analysed by RP-HPLC (e.g. Hypersil ODS; 5 μm) using a methanol-water gradient with 0.5% (v/v) *ortho*-phosphoric acid in water and UV detection at 370 nm (Table 1). The reduction of the genuine glycosides by

Table 1 Chromatographic parameters for the determination of flavonoid aglycones from *Ginkgo biloba* (Method A) (Hasler, 1990; Hasler *et al.*, 1990b)

| – Column: | Hypersil ODS, 5 μm (100 × 4 mm i.d.) |
| – Mobile phase: | A = methanol, B = 0.5 % *ortho*-phosphoric acid in water; Gradient elution: |

Time (min)	Solvent A (%)	Solvent B (%)
0.01	38.0	62.0
12.00	48.2	51.8
12.01	100.0	0.0
17.00	100.0	0.0
17.01	38.0	62.0

– Flow:	2.0 ml/min
– Temperature:	25° C
– Detection:	UV at 370 nm
– Injection volume:	10 μl

hydrolysis is an accepted method in the quality control of phytomedicines and has also been established for the standardisation of willow and hawthorn preparations (Meier *et al.*, 1985; Rehwald *et al.*, 1994).

The aglycone content is converted into a "Ginkgo flavonol" glycoside content (Figure 5). As "Ginkgo flavonol" glycosides the *p*-coumaroyl ester glycosides of kaempferol and quercetin (= flavonols; no. 21 and 22 in Table 5) with an average molecular weight of approximately 760 are documented. According to the list of specifications accompanying the standardised Ginkgo extracts of many pharmaceutical industries this is a standard procedure. The true flavonol glycoside content will be slightly lower than the experimental value as most glycosides have a molecular weight below 760. However as all industries use the same procedure, in the daily practice this does not lead to problems. As long as there is no exact knowledge on the biological activity of individual flavonol glycosides, standardisation based on these compounds is acceptable from a pharmaceutical point of view. They should be considered as lead substances in Ginkgo leaves and extracts.

A typical HPLC profile of a hydrolysed Ginkgo extract is shown in Fig. 6. Generally, kaempferol and quercetin are the main peaks and the concentration of isorhamnetin is approximately five times lower. The minor peaks represent further aglycones, such as apigenin and luteolin. At higher concentrations an analysis of these aglycones would also be possible. Investigations have shown that dried Ginkgo

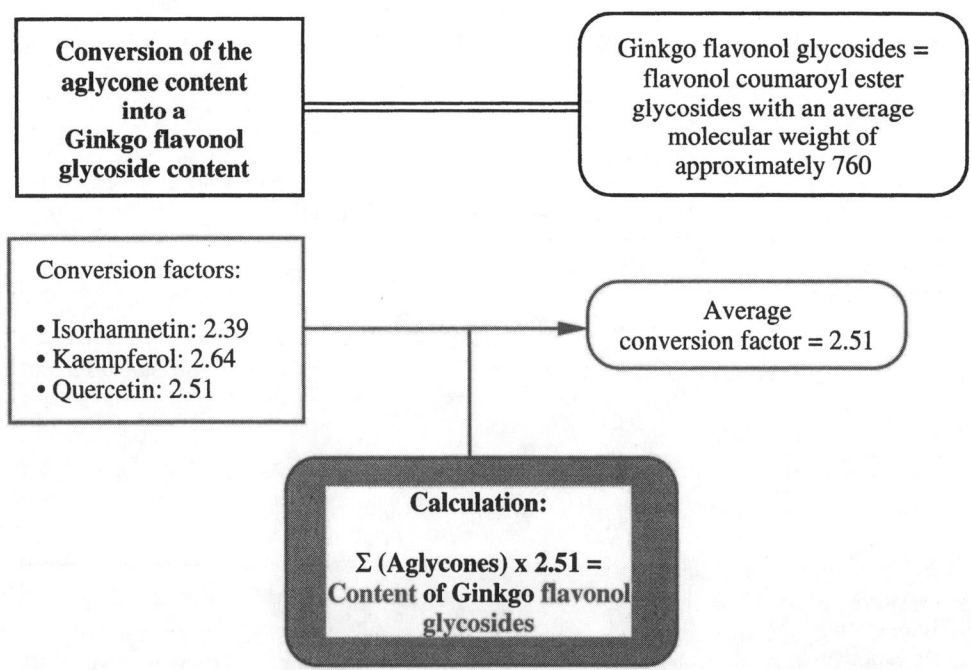

Figure 5   Calculation of the Ginkgo flavonol glycoside content.

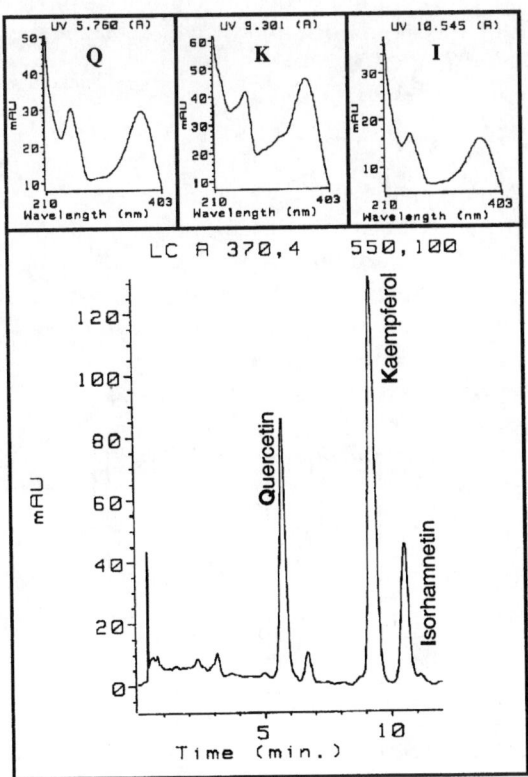

Figure 6   HPLC-UV profile of a Ginkgo leaf extract after hydrolysis of the flavonol glycosides (λ = 370.4 nm).

leaves from the commercial market have an aglycone content of 0.2 to 0.4% w/w, corresponding to a calculated Ginkgo flavonol glycoside content of 0.5 to 1% w/w (Hasler, 1990; Hasler *et al.*, 1992). Green leaves are considered to be of higher quality. However, no significant differences in the flavonoid concentration were found in ontogenetic studies except for the month of May (Fig. 7). The self-collected and self-dried leaves showed a higher concentration of flavonoids and a different ratio of kaempferol to quercetin compared to commercially available leaves (Table 2) (Hasler, 1990; Hasler *et al.*, 1992).

To confirm the accuracy of the quantitation of flavonols after hydrolysis by method A, a second method (Method B) was developed using a methanol-tetrahydrofuran gradient with 0.5% (v/v) *ortho*-phosphoric acid in water (Table 3). Using tetrahydrofuran as an organic solvent the elution order of isorhamnetin and kaempferol is reversed (Hasler, 1990; Hasler *et al.*, 1992). Identical values resulted for both methods (Hasler *et al.*, 1992). Yao *et al.* (1995a) and Chen *et al.* (1997) have used similar procedures to analyse Ginkgo leaves. The extraction procedure used by the former group is unusual. They used supercritical fluid extraction (SFE) to remove the flavonoids from the leaves. Optimum results were obtained at a

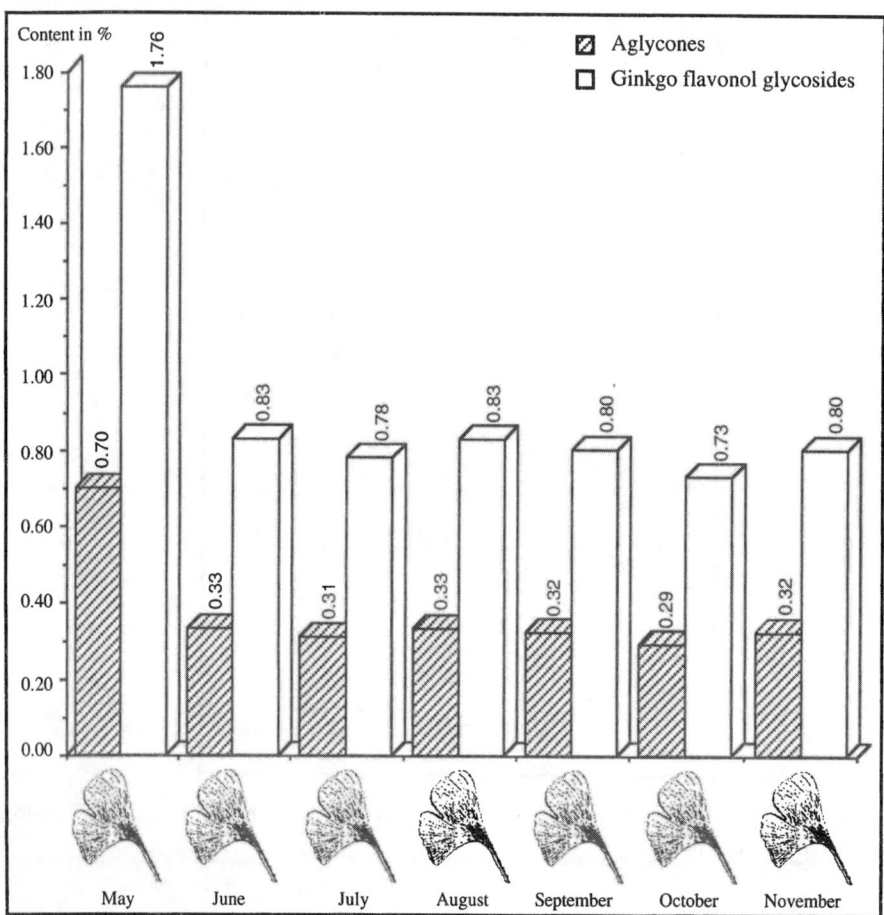

Figure 7 Concentration of flavonol aglycones and Ginkgo flavonol glycosides in dependence of the harvest time (leaves of a female tree, Zurich, Mythenquai; calculated on a dry weight basis).

pressure of 28 MPa, 40° C and 13% methanol in carbon dioxide. The average recovery was 87.3% with RSD = 5.4%. Flavonoid percentages of up to 1.4% were found in their leaves which is higher than usual. Claimed advantages are the lack of solvent residues, high extraction efficiency and mild conditions (no decomposition).

RP-HPLC with diode array detection (DAD) was used to study the metabolic fate of the flavonol glycosides present in standardised Ginkgo extracts after administration to rats (Pietta *et al.*, 1995) and humans (Pietta *et al.*, 1997). After sample clean-up by means of SPE, the blood, urine and faeces samples of rats were analysed and analysed for residual flavonol glycosides and aglycones. No glycosides or aglycones were detectable but several phenylalkyl acids could be detected. Human blood and urine samples were treated similarly. Also in this case, no intact flavonoids could be detected. Only in urine hydroxybenzoic acids could be detected as metabolites.

Table 2 Determination of the aglycones in Ginkgo leaves (Method A)[a] (Hasler, 1990; Hasler *et al.*, 1992)

Harvest month, 1988	Isorhamnetin %	Kaempferol %	Quercetin %	Σ (Aglycones) %
May	0.14 (2.5)	0.38 (1.8)	0.18 (1.2)	0.70
June	0.06 (3.2)	0.18 (2.5)	0.09 (2.0)	0.33
July	0.06 (2.1)	0.17 (2.6)	0.08 (1.7)	0.31
August	0.07 (2.1)	0.17 (2.0)	0.09 (0.9)	0.33
September	0.06 (2.5)	0.17 (2.8)	0.09 (1.2)	0.32
October	0.06 (3.0)	0.16 (3.0)	0.07 (2.2)	0.29
November[b]	0.07 (2.8)	0.17 (3.2)	0.08 (2.8)	0.32
Commercial sample				
Hänseler 912050	0.05 (0.8)	0.10 (3.1)	0.10 (3.2)	0.25
Dixa 38747	0.04 (3.1)	0.13 (2.7)	0.10 (2.7)	0.27
Siegfried 147974-02	0.05 (3.8)	0.11 (2.8)	0.09 (3.6)	0.25
Extract				
Flachsmann 6 L 035	0.10 (2.5)	0.39 (2.8)	0.51 (3.0)	1.00
Zeller 32/89	0.14 (2.9)	0.42 (1.9)	0.55 (2.8)	1.11

[a]Values in parentheses are relative standard deviations (%) ($n = 3$).
[b]Yellow and fallen leaves.

Table 3 Chromatographic parameters for the determination of flavonoid aglycones and biflavones from *Ginkgo biloba* (Method B) (Hasler, 1990; Hasler *et al.*, 1992)

- Column: Nucleosil 100-C_{18}, 3 μm (100 × 4 mm i.d.)
- Mobile phase: A = methanol, B = tetrahydrofuran, C = 0.5% *ortho*-phosphoric acid in water; Gradient elution:

Time (min)	Solvent A (%)	Solvent B (%)	Solvent C (%)
0.01	22.5	22.5	55.0
9.50	22.5	22.5	55.0
9.51	25.0	25.0	50.0
17.00	25.0	25.0	50.0
17.01	30.0	30.0	40.0
20.00	30.0	30.0	40.0
20.01	100.0	0.0	0.0
25.00	100.0	0.0	0.0
25.01	22.5	22.5	55.0

- Flow: 1.0 ml /min
- Temperature: 30° C
- Detection: UV at 370 nm
- Injection volume: 10 μl

Determination of Biflavonoids

During the past 15 years, several investigations on the separation and quantitative determination of Ginkgo biflavonoids were reported. Briançon-Scheid *et al.* (1982, 1983) were the first to present an HPLC separation of biflavonoids. Similar HPLC separations were reported by Song (1986) and Pietta *et al.* (1988). Briançon-Scheid *et al.* (1982) have evaluated various stationary and mobile phases and concluded that an optimal separation is achieved with normal phase chromatography. The separation of the four biflavones bilobetin, isoginkgetin, ginkgetin and sciadopytisin could be successfully performed using a LiChrosorb diol column (10 μm) in combination with a tetrahydrofuran-chloroform gradient. On the other hand, this system was not suitable for the quantitative determination of the biflavones in crude leaf extracts (Briançon-Scheid *et al.*, 1983). According to the same authors quantitation should be possible with a LiChrosorb diol column (5 μm) and a combination of an isocratic separation and a two step gradient with hexane-chloroform (25:75) and tetrahydrofuran after sample clean-up using a Lichroprep diol pre-column. However the HPLC chromatogram in the original publication clearly shows that the proposed separation system is neither suitable for the separation nor for the quantitation of the biflavones. Both the sample clean-up and the separation system are not selective enough. Pietta *et al.* (1988) criticized the long analysis time of 75 minutes. They developed an isocratic RP-HPLC separation system using a Novapak C_{18} column with a Guard-Pak μBondapak C_{18} pre-column and tetrahydrofuran-propanol-water (21:10:69) as mobile phase. This system allows a rapid (15 minutes) and reproducible separation of all four biflavonoids. Both the sample clean-up and the separation system are selective enough (Fig. 2 in ref. Pietta *et al.*, 1988) and quantitative determination is possible using external standardisation. Two further isocratic HPLC systems using LichroCart RP-18 250-4 cartridges with an RP-8 guard-column and methanol-water-formic acid (75:25:4) or methanol-water-formic acid (80:20:4) as respective mobile phase were described by Schennen (1988). With these systems peak identification or purity control of isolated biflavonoids is possible but they are less suitable for quantitative assays. The pair ginkgetin/isoginkgetin is not base-line separated.

In this decade, two further RP-HPLC methods for the separation and determination of biflavonoids were reported. Lobstein *et al.* (1991) used a LiChroCART 125-4, Superspher 100 RP-18 column with a Lichroprep RP-18 guard column and gradient elution with acetonitrile and water containing 10% phosphoric acid (0.1 N). Bilobetin, ginkgetin/isoginkgetin and sciadopitysin could be separated and determined. Gobbato *et al.* (1996) used a SupelcoSil LC 18 column (5 μm) equipped with an RP-18 Newguard guard column (7 μm) and gradient elution with acetonitrile-water containing 0.3% phosphoric acid. This system allows a satisfactory separation of six biflavones, with good reproducibility and without strong peak tailing. The resolution of ginkgetin from isoginkgetin, being the most difficult pair to separate, is good and allows their quantitative evaluation. Ginkgetin was used to calculate recoveries as part of the method validation. Identification of the main biflavone constituents of *G. biloba* was also performed by HPLC-MS using a thermospray interface. Thin layer chromatographic methods, which will not be discussed in this paper, have been described by Schennen (1988), Vanhaelen and Vanhaelen-Fastre (1988) and Wagner *et al.* (1989).

Biflavonoids in contrast to the ubiquitous flavonol aglycones are characteristic lead compounds for the identification of Ginkgo leaves and leaf extracts (Hasler, 1990; Hasler *et al.*, 1992). The method B discussed earlier can also be used for the qualitative identification of the crude drug. After hydrolysis, the three main aglycones quercetin, kaempferol and isorhamnetin and the biflavones bilobetin, ginkgetin, isoginkgetin, and sciadopitysin can be separated in the same run (Fig. 8). The obtained chromatogram ensures that authentic Ginkgo leaves were used as starting material for the extraction (Hasler, 1990; Hasler *et al.*, 1992). This HPLC method is advantageous compared to TLC as it also allows the separation of the pair ginkgetin/isoginkgetin. However, the qualitative method is not useful for special extracts such as EGb 761 (Dr. Willmar Schwabe, Karlsruhe) or LI 1370 (Lichtwer Pharma, Berlin) because biflavones have been removed from such extracts. Recently Zhong and Xu (1995) published a new method for the analysis of Ginkgo biflavonoids. For the final separation and quantitation they used HPLC with internal standardisation (i.s. anthracene). The column was LiChrosorb RP-18 (250 × 4 mm, 10 μm) with as solvent methanol-water-formic acid (85:15:0.8) at

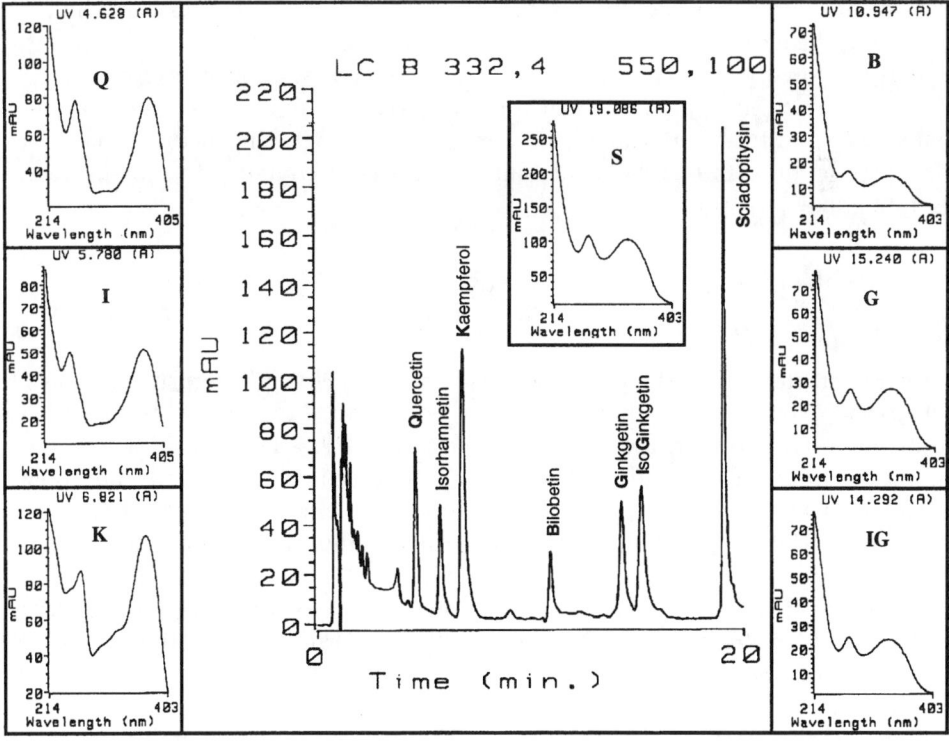

Figure 8   HPLC profile and UV spectra of the aglycones and biflavones of a crude Ginkgo leaf extract after acid hydrolysis.

1.0 ml/min. Detection occurred at 330 nm. Ginkgetin, isoginkgetin and bilobetin were well separated. The recovery of isoginkgetin was 98.8% and RSD's varied from 0.3–1.8%.

Good analytical methods for Ginkgo biflavonoids may also be useful for clearing up biological questions (Briançon-Scheid et al., 1983; Song, 1986; Hasler, 1990; Sticher et al., 1991; Lobstein et al., 1991). Biflavonoids were thought not to possess any clear pharmacological activities and therefore there was no need for standard-isation. But recently, free radical scavenging properties as well as protective effects against peroxidative reactions were reported (Joyeux et al., 1995). Additionally complexes of biflavonoids and phospholipids have shown to increase the skin microcirculation after topical application (Bombardelli et al., 1996).

Proanthocyanidins

Ginkgo leaves are rich in polyphenols of the catechin type (Stafford et al., 1986). In comparison with the true flavonoids, relatively little has been published on these polymeric flavan-3-ols. Recently four different methods for the analysis of this class were compared (Lang and Wilhelm, 1996). The non-specific Folin Ciocalteus and hide powder DAC assays gave much too high values due to the presence of inter-fering flavonoids. Acidic hydrolysis with the aid of iron salts, followed by meas-uring of the formed anthocyanidins gave a value of 2.3%. The fourth method, a reaction of the terminal flavanol groups under acidic conditions followed by meas-uring the colored products with catechin as a reference substance gave a value of 0.9%. The conclusion is that these methods give only an approximation of the true content and that further research is necessary. Chromatographic methods may give more precise answers but the availability of reference compounds and sufficient resolution may pose problems.

Analysis of the Genuine Ginkgo Flavonoids

Reversed phase HPLC

Further characterisation of Ginkgo extracts is possible with a fingerprint analysis of the flavonoids. A chromatographic system was developed to separate all flavonoids occurring in G. biloba (Hasler, 1990; Hasler et al., 1992). Because of the complex mixture of very polar (triglycosides), polar (mono- and diglycosides) and apolar (biflavones) flavonoids, the proposed separation within 30 minutes requires a sophisticated HPLC procedure including a three-pump system and a diode-array detector. The extraction is done twice with 80% ethanol. This guarantees a total extraction of the polar glycosides and the apolar biflavones. The complex mixture is analysed with RP-HPLC using Nucleosil 100-C_{18} (3 μm) and a ternary gradient with as eluents (A) water containing 0.5% ortho-phosphoric acid, (B) isopropanol/tetrahydrofuran (25:65) and (C) acetonitrile and UV detection at 350 nm (Table 4). In leaves and extracts of G. biloba it was possible to identify unambiguously 22 flavonoid glycosides, six flavonoid aglycones and five biflavones from the elution order and UV spectra. An elution profile and the chromatographic results of the

Table 4 Chromatographic parameters for the fingerprint analysis (Method C) (Hasler, 1990; Hasler *et al.*, 1992)

– Column:	Nucleosil 100-C_{18}, 3 μm (100 × 4 mm i.d.)
– Mobile phase:	A = isopropanol-tetrahydrofuran (25:65), B = acetonitrile, C = 0.5% *ortho*-phosphoric acid in water; Gradient elution:

Time (min)	Solvent A (%)	Solvent B (%)	Solvent C (%)
0.01	15.0	1.5	83.5
7.00	15.0	1.5	83.5
7.01	12.0	5.0	83.0
12.00	12.0	5.0	83.0
16.00	15.0	13.0	72.0
20.00	5.0	25.0	70.0
20.01	5.0	30.0	65.0
24.00	8.0	42.0	50.0
24.01	0.0	48.0	52.0
30.00	0.0	78.0	22.0
30.01	0.0	100.0	0.0
35.00	0.0	100.0	0.0
35.01	15.0	1.5	83.5

– Flow:	1.0 ml / min
– Temperature:	30° C
– Detection:	UV at 350 nm
– Injection volume:	10 μl

separation of reference compounds and of a Ginkgo leaf extract are shown in Table 5 and Fig. 9. The reference run was stored with UV spectra in the data system and could be used for peak assignment. Fingerprint analysis is especially useful for performing stability tests. It was shown that the flavonoid glycosides and the biflavones are relatively stable, and that the ratios of these compounds do not change when leaves or extracts are properly stored. Thus an increase in the aglycone content and a decrease in the glycoside content indicate undesired degradation processes, i.e. poor storage. In order to separate 9 and 10 a more selective system is required. The UV spectra of the tested extracts showed a dominance of 10.

The fingerprint analysis allowed the identification of the characteristic flavonol coumaroyl ester glycosides 21 and 22 (Fig. 9). Both are well separated from each other and from other compounds. Meier *et al.* (1992) presented a modification of the fingerprint method mentioned above for rapid detection and determination of the two flavonol coumaroyl ester glycosides (peaks 21 and 22 in Fig. 9). Using the suggested HPLC conditions the two compounds can be analysed within only 10 minutes (Table 6, Fig. 10).

Fingerprint chromatography was introduced several years ago and subsequently accepted by the WHO (World Health Organization, 1991) for quality control of herbal medicines. The possibility of on-line diode-array detection offered real

Table 5   The gradient elution profile of the naturally occurring flavonoids of *Ginkgo biloba* L. (Hasler, 1990; Hasler *et al.*, 1992)

No.	t_R	Flavonoid
1	3.21	3-O-[2-O-(6-O-{p-Coumaroyl}-β-D-glucosyl)-α-L-rhamnosyl]-7-O-[β-D-glucosyl]quercetin
2	3.47	3-O-[2-O, 6-O-Bis(α-L-rhamnosyl)-β-D-glucosyl]quercetin
3	4.09	3-O-[2-O, 6-O-Bis(α-L-rhamnosyl)-β-D-glucosyl]isorhamnetin
4	4.71	3-O-[2-O, 6-O-Bis(α-L-rhamnosyl)-β-D-glucosyl]kaempferol
5	4.97	3-O-[6-O-(α-L-Rhamnosyl)-β-D-glucosyl]myricetin
6	5.96	3-O-[6-O-(α-L-Rhamnosyl)-β-D-glucosyl]-3′-methylmyricetin
7	6.53	3-O-[2-O-(6-O-{p-[β-D-Glucosyl]coumaroyl}-β-D-glucosyl)-α-L-rhamnosyl]quercetin
8	7.19	3-O-[6-O-(α-L-Rhamnosyl)-β-D-glucosyl]quercetin
9	8.73	3-O-[2-O-(6-O-{p-[β-D-Glucosyl]coumaroyl}-β-D-glucosyl)-α-L-rhamnosyl]kaempferol
10	8.73	3-O-[6-O-(α-L-Rhamnosyl)-β-D-glucosyl]isorhamnetin
11	9.68	3-O-[β-D-Glucosyl]quercetin
12	10.37	3-O-[6-O-(α-L-Rhamnosyl)-β-D-glucosyl]kaempferol
13	10.74	3-O-[2-O-(β-D-Glucosyl)-α-L-rhamnosyl]quercetin
14	11.39	3-O-[β-D-Glucosyl]isorhamnetin
15	13.00	3-O-[β-D-Glucosyl]kaempferol
16	13.58	7-O-[β-D-Glucosyl]apigenin
17	14.58	3-O-[2-O-(β-D-Glucosyl)-α-L-rhamnosyl]kaempferol
18	15.22	3-O-[α-L-Rhamnosyl]quercetin
19	16.65	3′-O-[β-D-Glucosyl]luteolin
20	17.05	3-O-[α-L-Rhamnosyl]kaempferol
21	17.60	3-O-[2-O-(6-O-{p-Coumaroyl}-β-D-glucosyl)-α-L-rhamnosyl]quercetin
22	18.51	3-O-[2-O-(6-O-{p-Coumaroyl}-β-D-glucosyl)-α-L-rhamnosyl]kaempferol
23	18.76	Myricetin
24	20.60	Luteolin
25	21.50	Quercetin
26	22.22	Apigenin
27	22.76	Isorhamnetin
28	22.91	Kaempferol
29	24.02	Amentoflavone
30	25.06	Bilobetin
31	27.00	Ginkgetin
32	27.17	Isoginkgetin
33	29.35	Sciadopitysin

progress in fingerprint chromatography (Meier and Sticher, 1986). A fingerprint HPLC separation of flavonoids from Ginkgo leaves was also described by Wagner *et al.* (1989) who detected 20 peaks within 55 min using a LiChroCart 125-4 column with a Guard-Pak RP-18 pre-column and an acetonitrile-water gradient as mobile phase, but only two flavonoids (rutin and astragalin), and four biflavones (bilobetin, ginkgetin/isoginkgetin and sciadopitysin), ginkgol, shikimic acid, and

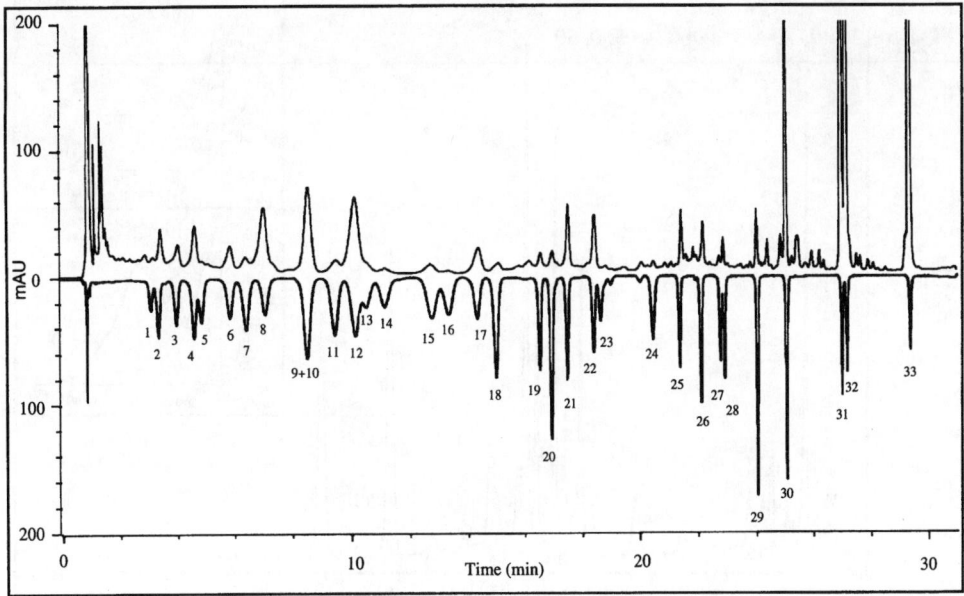

Figure 9 Fingerprint chromatogram of a crude ethanolic Ginkgo leaf extract (upper chromatogram) and of reference compounds (lower chromatogram).

Table 6 Chromatographic parameters for a rapid detection and determination of the two main flavonol coumaroyl ester glycosides 21 (quercetin derivative) and 22 (kaempferol derivative) from *Ginkgo biloba* (Meier *et al.*, 1992)

– Column:	Nucleosil 100-C_{18}, 5 μm (125 × 4 mm i.d.)
– Mobile phase:	A = Isopropanol-tetrahydrofuran (25:65), B = acetonitrile, C = 0.5% *ortho*-phosphoric acid in water; Gradient elution:

Time (min)	Solvent A (%)	Solvent B (%)	Solvent C (%)
0.01	12.0	5.0	83
2.00	12.0	5.0	83
2.01	12.5	15.5	72
5.00	12.5	15.5	72
5.01	12.5	15.5	72
10.00	5.0	25.0	70

– Flow:	1.0 ml/min
– Temperature:	30° C
– Detection:	UV at 314 nm
– Injection volume:	10 μl

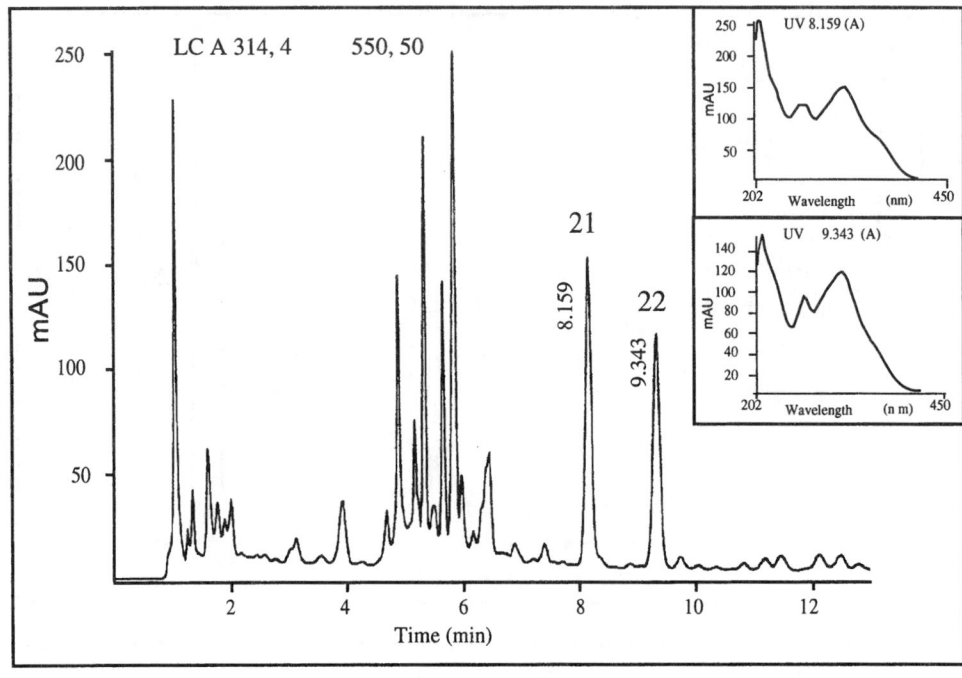

Figure 10 HPLC-UV profile of an ethanolic Ginkgo leaf extract for a rapid detection and determination of the two main flavonol coumaroyl ester glycosides 21 (8.159 min) and 22 (9.343 min) (Table 6).

6-hydroxykynurenic acid could be assigned. Pietta *et al.* (1991a) reported an HPLC method using a C_8 Aquapore RP-300 (7 μm) column and a linear gradient from (A) water-isopropanol (95:5) to (B) isopropanol-tetrahydrofuran-water (40:10:50) for the separation of 15 known Ginkgo flavonoids within 50 min. The assignment was done with six flavonoid reference compounds. The other flavonoid glycosides were assigned by their UV spectra using a diode-array detector, although their absorptions were less than 10 milliabsorption units (mAU). Such assignments, without any other investigations, are very speculative. In both papers (Wagner *et al.*, 1989; Pietta *et al.*, 1991a), the separation and identification of the flavonoids are not very developed and not complete. Lobstein *et al.* (1991) used a LiChroCART 125-4, Superspher 100 RP-18 column with a Lichroprep RP-18 guard column and gradient elution with acetonitrile and water containing 10% phosphoric acid (0.1 N) to separate flavonoids and biflavones within 50 min. No diode-array detector was coupled to the HPLC system. Additionally, the lack of reference compounds prevented complete peak assignment. Kaempferol- and quercetin 3-O-coumaroyl glucorhamnoside and biflavones in leaves were determined. Molnar *et al.* (1991) used part of a Ginkgo leaf extract to illustrate the capacity of current software to find optimal elution schemes with only a limited number of trial runs. For Ginkgo flavonol glycosides optimal resolution was found on a Nucleosil C18 column with a gradient from

15 to 40% B in 25 min. The (A) solvent was water adjusted to pH=3.5 with *ortho-*phosphoric acid while the (B) solvent was pure acetonitrile. The correlation between the simulated run and the final experiment was satisfactory. No peaks were assigned. As part of their studies to find a simple and selective purification method for the preparative isolation of flavonoids from Ginkgo leaf extracts, Yoon *et al.* (1997) used RP-HPLC to analyse their samples. A C18 μBondapack column (300 × 3.9 mm) was used in combination with a gradient from 70% water in methanol to 30% water in methanol in 70 min. The UV absorption was measured at 260 nm. Peak identity was confirmed by diode array detection. XAD-7 resin was found to be suitable for the purification of flavonol glycosides.

Capillary electrophoresis

Pietta *et al.* (1991b) have described the use of micellar electrokinetic capillary chromatography for the analysis of Ginkgo flavonol glycosides. With a 72 cm fused silica capillary (50 μm i.d.), a 20 mM sodium borate buffer (pH = 8.3) – 50 mM sodium dodecyl sulphate (SDS) analysis buffer and +20 kV voltage at 27° C, 4 nL of a 0.5% Ginkgo extract solution were well separated within 15 min. The detection took place at 260 nm. The resolution was similar to RP-HPLC but the separation took only half as much time. The elution order was slightly different. Amounts as low as 6 pg could be detected.

ONTOGENETIC STUDIES

As shown earlier, no significant differences in the flavonoid concentration of Ginkgo leaves were found during ontogenetic studies (Fig. 7). However, there are differences in the concentration of single flavonoids. Hasler (1990) has shown that the concentration of the two main coumaroyl derivatives (21 and 22, Table 5) gradually halved in the period June to November (Fig. 11). In May the concentration was much higher and in contrast to the rest of the year there was also more of the quercetin glycoside than of the kaempferol glycoside. The concentration of the three main rutinosides decreased from May to July and continuously increased from July to November (Fig. 12). The behaviour of the biflavones was different from that of the glycosides (Fig. 13). Significantly lower biflavone concentrations were generally found in May. In green leaves the concentration of biflavones increased until August and decreased again from August to October. It is interesting to note that in November (yellow, fallen leaves) there was again a considerably higher concentration of biflavones compared to the preceeding month (Hasler, 1990).

Similar results were obtained by Lobstein *et al.* (1991). Their investigations on leaves of a single tree revealed the highest concentration of acyl flavonol glycosides in the leafbuds in March and the highest concentration of flavonoid glycosides in April, gradually decreasing to a minimum in October and increasing again in November. The study by Yao *et al.* (1995b) shows the same trend except that no increase in November was found. The biflavonoid concentration slowly increased

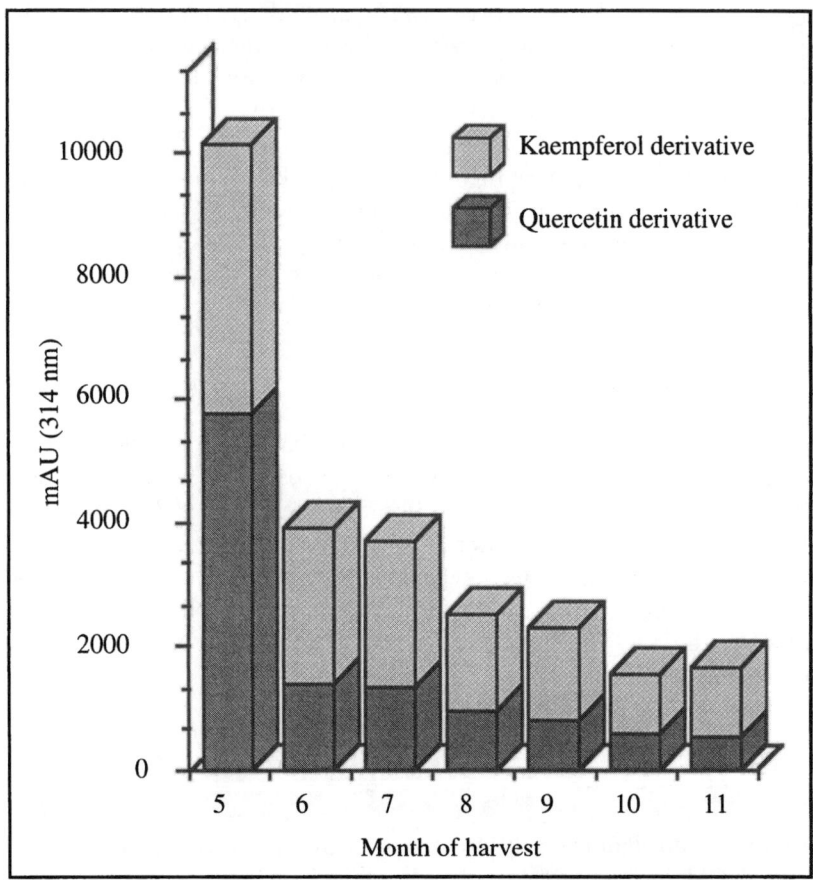

Figure 11   Flavonol coumaroyl ester glycosides in leaves of *G. biloba* (female tree, Zürich, Mythenquai).

over the course of the vegetation period without reaching high values (Lobstein *et al.*, 1991). To confirm these results a large-scale field study would be necessary, investigating leaves from various habitats, from trees of different age and sex, as well as from different vegetation periods.

STANDARDISATION OF GINKGO EXTRACTS AND COMMERCIAL PRODUCTS

Two types of Ginkgo extract are available: Ginkgo full or total extracts (incl. homeo-pathic mother tinctures) and enriched or special extracts (Fig. 14). Full extracts are mainly produced using ethanol-water for extraction. The result is a complex mixture of polar to fairly non-polar natural compounds consisting of active principles and inert plant constituents. In the case of Ginkgo, 40–60% ethanol is recommended to

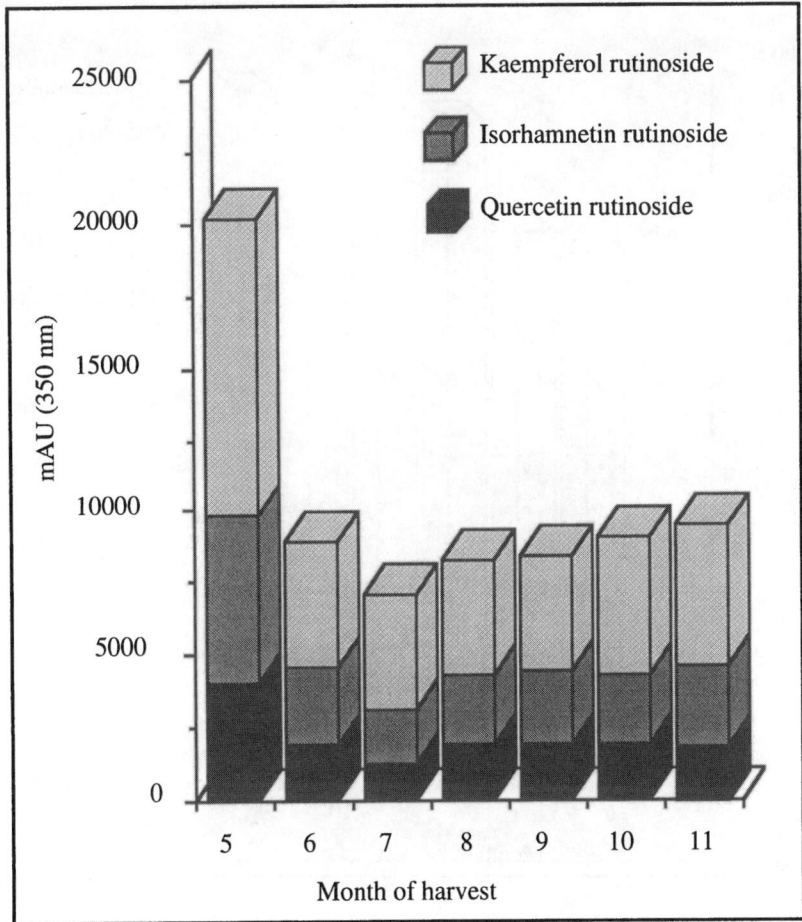

Figure 12   Rutinoside derivatives in leaves of *G. biloba* (female tree, Zürich, Mythenquai).

obtain a high flavonoid yield. An extract of this type generally contains 2 to 4% flavonol glycosides. Commercial products based on total extracts are sometimes standardised to varying concentration levels of flavonol glycosides, and exceptionally also to a minimum terpene trilactone content. They are licensed in a number of European countries for over-the-counter sale.

Enriched extracts require several steps of purification or separation after extraction and thus possess a higher content of flavonol glycosides and terpene trilactones, whereas several other constitutents are largely eliminated. Enriched extracts are obtained using a relatively tedious work-up and purification procedure consisting of various steps such as liquid/liquid extraction, precipitation and concentration (Drieu, 1991). Therefore, high-molecular-weight constituents such as tannins, proteins and polysaccharides, and in addition slightly soluble and lipophilic compounds like biflavones and ginkgolic acids are removed or reduced. These substances are

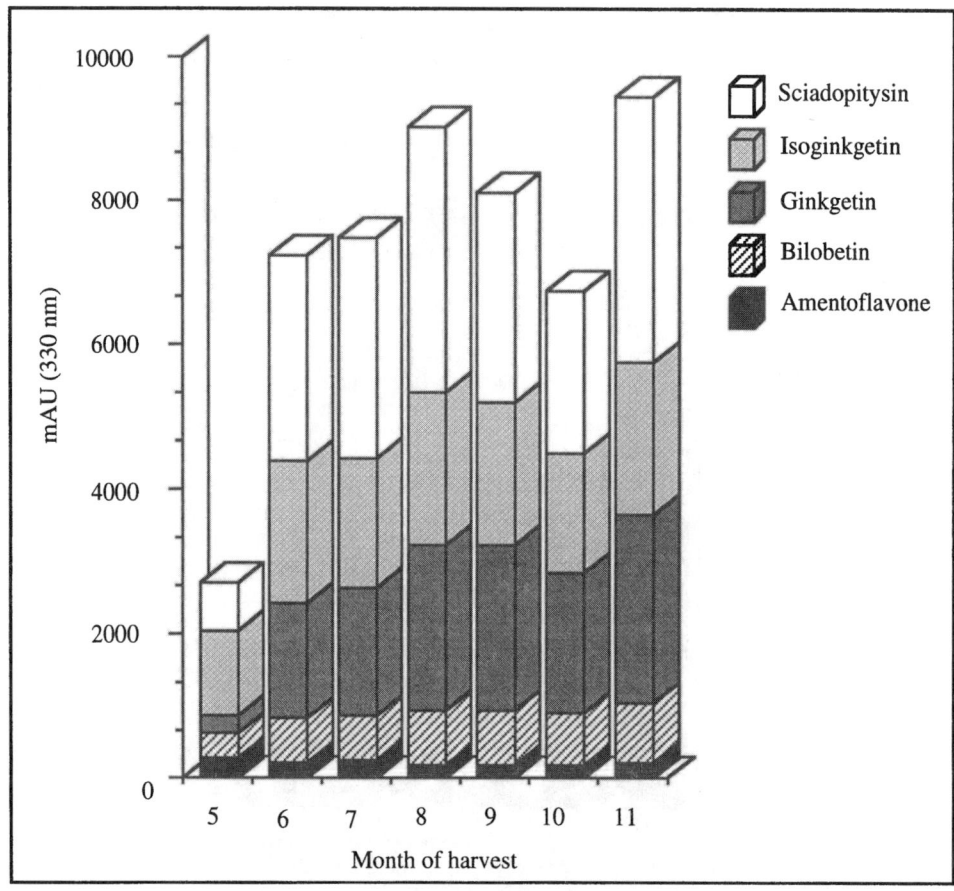

Figure 13   Biflavones in leaves of G. *biloba* (female tree, Zürich, Mythenquai).

undesirable in some liquid dosage forms because complex formation with other compounds may occur. Furthermore, an up to now not really documented allergenic potential of ginkgolic acid was postulated (Lepoittevin *et al.*, 1989; Koch *et al.*, 1992). The active principles, namely the flavonol glycosides and the terpene tri-lactones, are thus enriched. Products based on the special extract EGb 761 or products based on extracts manufactured in a similar way are standardised to a Ginkgo flavonol glycoside content of 16 to 26% (w/w) and a terpene trilactone (ginkgolides and bilobalide) content of 5 to 7%. Considering the available pharmaceutical and clinical evidence, as well as the indications claimed for commercial products based on such special extracts, usually a prescription is required. Generally, the view of therapeutic equivalence should be focused more on the daily dosage rather than on the percentage of active compounds in the extract. For preparations based on EGb 761 extracts, three dosages of 40 mg of extract per day are proposed, corresponding to 28.6 mg of Ginkgo flavonol glycosides and 7.2 mg of terpene trilactones. With

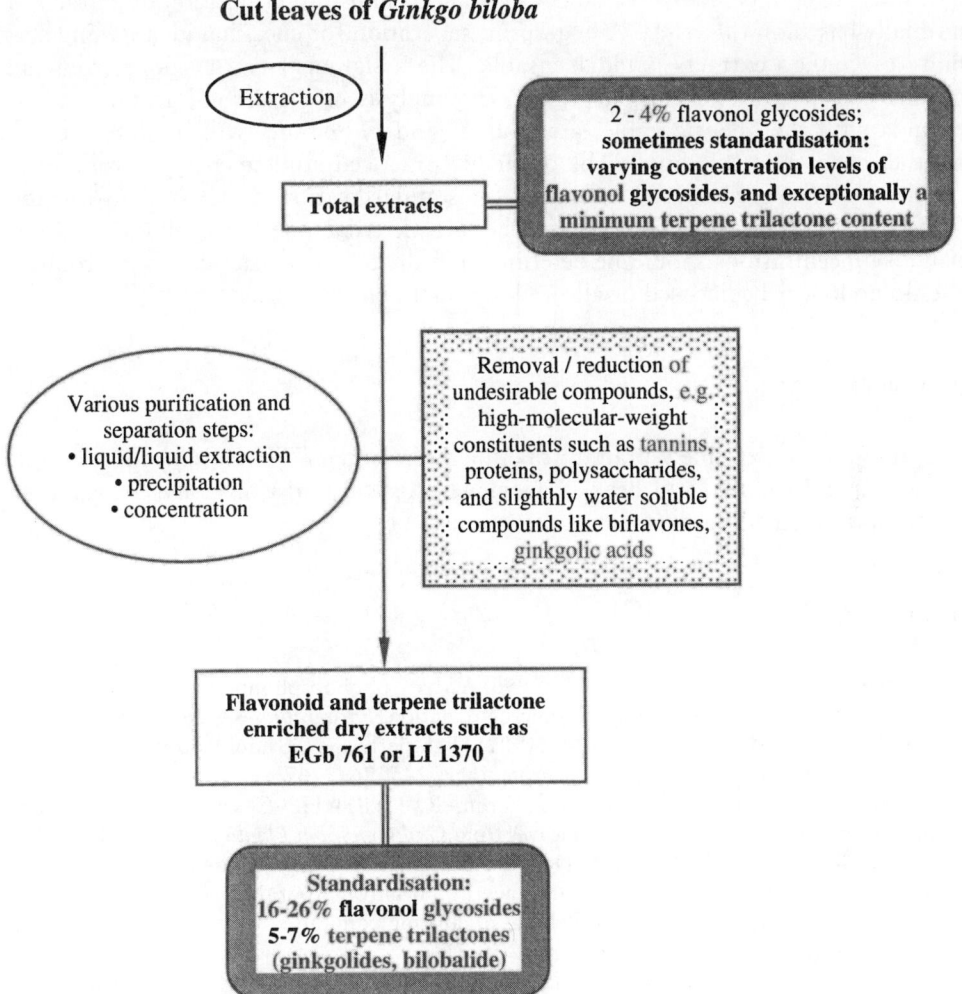

Figure 14 Extraction procedure for and standardisation of Ginkgo preparations.

regard to the use of active constituents for standardisation, the quantitative determination of flavonol glycosides and terpene trilactones is not only desirable, but absolutely essential.

CONCLUSIONS

New guidelines for phytopharmaceuticals demand better developed analytical methods for quality control. There is a large number of flavonoids in *Ginkgo biloba* leaves, mainly derivatives of isorhamnetin, kaempferol and quercetin. After hydrolysis of the flavonoid glycosides determination of the aglycones is performed. The flavonols are commercially available as standards and standardisation of extracts is normally based on this assay. A fingerprint separation for checking identity and stability of Ginkgo extracts is indispensable. The assignment can be done from the on-line UV spectra and elution profile. The analysis of the second group of compounds with therapeutic value, ginkgolides and bilobalide, will be described in another chapter of this book. The originally practiced adjustment of flavonol glycosides no longer satisfies the demand for standardisation of active constituents. Consequently, both the flavonoid and the terpene trilactone (ginkgolides and bilobalide) concentrations should be determined in the future. Unstandardised products should no longer be licensed or allowed to remain on the market.

ACKNOWLEDGEMENT

The authors are extremely grateful to Annemarie Suter, ETH Zurich, for editorial assistance and Dr. T.A. van Beek, Wageningen Agricultural University, for valuable suggestions for the improvement of the manuscript.

REFERENCES

Bombardelli, E., Cristoni, A., Curri, S.B. (1996) Activity of phopspholipid-complex of *Ginkgo biloba* dimeric flavonoids on the skin microcirculation. *Fitoterapia*, 67, 265–273.

Briançon-Scheid, F., Guth, A., Anton, R. (1982) High-performance liquid chromatography of biflavones from *Ginkgo biloba* L. *J. Chromatogr.*, 245, 261–267.

Briançon-Scheid, F., Lobstein-Guth, A., Anton, R. (1983) HPLC separation and quantitative determination of biflavones in leaves from *Ginkgo biloba*. *Planta Med.*, 49, 204–207.

Chen, Y., Hu, W., Xie, B., Yiang, Z. (1997) Determination of the flavonol glycosides from *Ginkgo biloba* leaves by high performance liquid chromatography. *Proceedings of '97 International Seminar on Ginkgo*, Beijing, organizing committee for the '97 international seminar on ginkgo, November 1997, pp. 206–211.

DeFeudis, F.V. (1991) *Ginkgo biloba Extract (EGb 761): Pharmacological Activities and Clinical Applications*. Editions Scientifiques Elsevier, Paris.

Drieu, K. (1991) Herstellung und Definition von Rökan. In F.H. Kemper, and H. Schmid-Schönbein, (eds.), Rökan – *Ginkgo biloba* EGb 761, Springer, Berlin, pp. 13–18.

Gobbato, S., Griffini, A., Lolla, E., Peterlongo, F. (1996) HPLC quantitative analysis of biflavones in *Ginkgo biloba* leaf extract and their identification by thermospray liquid chromatography-mass spectrometry. *Fitoterapia*, 67, 152–158.

Hasler, A. (1990) *Flavonoide aus Ginkgo biloba L. und HPLC-Analytik von Flavonoiden in verschiedenen Arzneipflanzen*. Thesis No. 9353, ETH Zürich.

Hasler, A., Meier, B., Sticher, O. (1990a) *Ginkgo biloba*. Botanische, analytische und pharmakologische Aspekte. *Schweiz. Apoth. Ztg.*, 28, 342–347.

Hasler, A., Sticher, O., Meier, B. (1990b) High-performance liquid chromatographic determination of five widespread flavonoid aglycones. *J. Chromatogr.*, 508, 236–240.

Hasler, A., Sticher, O., Meier, B. (1992) Identification and determination of the flavonoids from *Ginkgo biloba* by high-performance liquid chromatography. *J. Chromatogr.*, 605, 41–48.

Joyeux, M., Lobstein, A., Anton, R., Mortier, F. (1995) Comparative antilipoperoxidant, anti-necrotic and scavenging properties of terpenes and biflavones from Ginkgo and some flavonoids. *Planta Med.*, **61**, 126–129.

Kim, C.-K., Park, M.-K., Lee, E.-J., Hwang, S.-J. (1989) Stability of ginkgoflavonglycoside in Ginkgo extract aqueous solution. *J. Kor. Pharm. Sci.*, **19**, 213–217.

Koch, E., Chatterjee, S.S., Jaggy, H. (1992) Immunotoxicological effects of constituents of leaves from *Ginkgo biloba* L. on the popliteal lymph node reaction in mice. *Naunyn-Schmiedebergs Arch. Pharmacol.*, **345 (Suppl.)**, R95.

Lang, F., Wilhelm, E. (1996) Quantitative determination of proanthocyanidins in *Ginkgo biloba* special extracts. *Pharmazie*, **51**, 734–737.

Lepoittevin, J.-P., Benezra, C., Asakawa, Y. (1989) Allergic contact dermatitis to *Ginkgo biloba* L.: relationship with urushiol. *Arch. Dermatol. Res.*, **281**, 227–230.

Lobstein, A., Rietsch-Jako, L., Haag-Berrurier, M., Anton, R. (1991) Seasonal variations of the flavonoid content from *Ginkgo biloba* leaves. *Planta Med.*, **57**, 430–433.

Meier, B. and Sticher, O. (1986) The use of a high speed spectrophotometric detector (diode-array) in the HPLC-analysis of medicinal plants. *Pharm. Ind.*, **48**, 87–91.

Meier, B., Sticher, O., Bettschart, A. (1985) Weidenrinden-Qualität. Gesamtsalicinbestimmung in Weidenrinden und Weidenpräparaten mit HPLC. *Dtsch. Apoth. Ztg.*, **125**, 341–347.

Meier, B., Hasler, A., Sticher, O. (1992) HPLC-analyses of the main Ginkgo acylflavonol glycosides. *Planta Med.*, **58** (Suppl. 1), 688A–689A.

Molnar, I., Gober, K.H., Christ, B. (1991) Fast development of a robust high-performance liquid chromatographic method for *Ginkgo biloba* based on computer simulation. *J. Chromatogr.*, **550**, 39–49.

Pietta, P., Mauri, P., Rava, A. (1988) Reversed-phase high-performance liquid chromatographic method for the analysis of biflavones in *Ginkgo biloba* L. extracts. *J. Chromatogr.*, **437**, 453–456.

Pietta, P., Mauri, P., Bruno, A., Rava, A., Manera, E., Ceva, P. (1991a) Identification of flavonoids from *Ginkgo biloba* L., *Anthemis nobilis* L. and *Equisetum arvense* L. by high-performance liquid chromatography with diode-array UV detection. *Planta Med.*, **57** 430–433.

Pietta, P.G., Mauri, P.L., Rava, A., Sabbatini, G. (1991b) Application of micellar electro-kinetic capillary chromatography to the determination of flavonoid drugs. *J. Chromatogr.*, **549**, 367–374.

Pietta, P.G., Gardana, C., Mauri, P.L., Maffei-Facino, R., Carini, M. (1995) Identification of flavonoid metabolites after oral administration to rats of a *Ginkgo biloba* extract. *J. Chromatogr.*, **673B**, 75–80.

Pietta, P.G., Gardana, C., Mauri, P.L. (1997) Identification of *Ginkgo biloba* flavonol metabo-lites after oral administration to humans. *J. Chromatogr.*, **693B**, 249–255.

Rehwald, A., Meier, B., Sticher, O. (1994) Qualitative and quantitative reversed-phase high-performance liquid chromatography of flavonoids in Crataegus leaves and flowers. *J. Chromatogr.*, **677A**, 25–33.

Schennen, A. (1988) *Neue Inhaltsstoffe aus den Blättern von Ginkgo biloba L. sowie Präparation ¹⁴C-markierter Ginkgo-Flavonoide*. Thesis, Universität Marburg.

Song, Y. (1986) Chemical composition and utilization of *Ginkgo biloba* L. *Chem. and Ind. of Forest Prod.*, **6**,1–4.

Stafford, H.A., Kreitlow, K.S., Lester, H.H. (1986) Comparison of proanthocyanidins and related compounds in leaves and leaf-derived cell cultures of *Ginkgo biloba* L., *Pseudotsuga menziesii* Franco, and *Ribes sanguineum* Pursh. *Plant Physiol.*, **82**, 1132–1138.

Sticher, O. (1992) *Ginkgo biloba* – Analytik und Zubereitungsformen. *Pharmazie in unserer Zeit*, **21**, 253–265.

Sticher, O. (1993a) Quality of Ginkgo Preparations. *Planta Med.*, **59**, 2–11.

Sticher, O. (1993b) *Ginkgo biloba* – Ein modernes pflanzliches Arzneimittel. *Vierteljahrsschrift der Naturforschenden Gesellschaft in Zürich*, **138**, 125–168.

Sticher, O., Hasler, A., Meier, B. (1991) *Ginkgo biloba* – eine Standortbestimmung. *Dtsch. Apoth. Ztg.*, **131**, 1827–1835.

Vanhaelen, M., Vanhaelen-Fastre, R. (1988) Countercurrent chromatography for isolation of flavonol glycosides from *Ginkgo biloba* leaves. *J. Liq. Chromatogr.*, **11**, 2969–2975.

Wagner, H., Bladt, S., Hartmann, U., Daily, A., Berkulin, W. (1989) *Ginkgo biloba*. DC- und HPLC-Analyse von Ginkgo-Extrakten und Ginkgo-Extrakte enthaltenden Phytopräparaten. *Dtsch. Apoth. Ztg.*, **129**, 2421–2429.

World Health Organization (1991) *Guidelines for the Assessment of Herbal Medicines*, Munich, 28.6.1991, WHO, Geneva.

Yao, W.X., Cheng, X.M., Zhang, Y.Q. (1995a) Determination of flavonoid compounds in *Ginkgo biloba* leaves supercritical fluid extraction and high performance liquid chromatography (SFE/HPLC). *Chin. Chem. Lett.*, **6**, 589–592.

Yao, W., Zhang, Z., Song, Z. (1995b) Study of ecologic characteristics of the flavonoids in Ginkgo by supercritical fluid extraction and high-performance liquid chromatography (SFE/HPLC). *Proc. of Int. 6th Beijing Conf. and Exhib. of Instrum. Analysis*, D155-D156.

Yoon, S.Y., Choi, W.J., Park, J.M., Yang, J.-W. (1997) Selective adsorption of flavonoid compounds from leaf extract of *Ginkgo biloba* L. *Biotechnol. – Techniques*, **11**, 553–556.

Yoshitama, K. (1997) Flavonoids of *Ginkgo biloba*. In T. Hori, R.W. Ridge, W. Tulecke, P. Del Tredici, J. Trémouillaux-Guiller, and H. Tobe, (eds.), *Ginkgo biloba – A global treasure. From biology to medicine*, Springer, Tokyo, pp. 287–299.

Zhong, Y.Q., Xu, L.S. (1995) Extraction, isolation and HPLC determination of biflavones in *Ginkgo biloba* L. *Acta Pharm. Sinica*, **30**, 694–697.

11. OCCURRENCE AND ANALYSIS OF ALKYL PHENOLS IN *GINKGO BILOBA*

LUISELLA VEROTTA,[1] PAOLO MORAZZONI[2] and
FEDERICO PETERLONGO[3]

[1]*Dipartimento di Chimica Organica e Industriale, Università degli
Studi di Milano, via Venezian 21, 20133 Milano, Italy*
[2]*INDENA SpA, Direzione Scientifica, Viale Ortles, 12–20139
Milano, Italy*
[3]*INDENA SpA, Laboratori Ricerca e Sviluppo, Via Don Minzoni,
6–20090 Settala (Milano), Italy*

INTRODUCTION

The phenolic lipids are a comparatively little known group of compounds which may be considered as biogenetically derived from fatty acids and containing a benzene ring, one to two phenolic groups and zero to one carboxyl group on the benzene ring. Some of them have had an applied artistic use for centuries for the preparation of Japanese and Chinese lacs, and others, e.g. the Cashew Nut Shell Liquid (CNSL) play a vital role in certain modern technical uses for chemical treatments and industrial utilizations (Attanasi *et al.*, 1996).

Historically, most of the analytical work on alkyl phenols has been carried out on *Anacardium occidentale*, because of its commercial value, and the acquired experience was translated to other alkyl phenols containing plants. This paper reviews the literature concerning the characterization of these compounds in *Ginkgo biloba* plant materials and in pharmaceutical preparations mainly derived from the leaves of the plant. In fact, side effects concerning the allergenic properties of this class of compounds, have been described (Hill *et al.*, 1934; Sowers *et al.*, 1965; Lepoittevin *et al.*, 1989; Koch *et al.*, 1992), and a number of industrial processes have been set up in order to avoid their occurrence in phytopharmaceuticals. A small review on the chemistry and biology of ginkgo alkylphenols has appeared recently (Jaggy and Koch, 1997).

NATURAL DISTRIBUTION, NOMENCLATURE AND BIOSYNTHESIS

The non-isoprenoid phenolic lipids exist in plants from a number of different families, which occur in tropical and temperate climates, notably the Anacardiaceae (Anacardium and Rhus), the Proteaceae (Grevillaea and Hakea) and Ginkgoaceae (*Ginkgo biloba*). They have also been found in certain bacterial sources. It is convenient to classify the long chain phenols into (1) phenolic acids, (2) dihydroxy and (3) monohydroxy phenols. Certain species contain all three phenolic types, others only two and certain sources only a single group (Tyman, 1979). Thus this classification can be used for chemotaxonomical purposes.

As with most groups of natural products, both trivial and systematic names are used, mostly related to the source of origin. A useful concise system which takes account of the chain length and the number of double bonds and can denote the position and configuration of the double bonds is exemplified with C_{15} anacardic acid *cis* monoene, which becomes (15:1) 8'(Z)-anacardic acid. It has been reported as ginkgolic or ginkgoic acid, anacardic acid monoene, Δ^8 15:1 anacardic acid and 2-hydroxy-6-pentadecenyl benzoic acid. The systematic name 6-[8'(Z)-pentadecenyl] salicylic acid is best used for this compound; in this chapter trivial or simplified names for rapid comprehension are used, accompanied by systematic names.

Anacardic acids are salicylic acids substituted in position 6 with saturated or unsaturated long-chain alkyl groups; cardanols (also known as ginkgols) are 3-alkyl phenols; bilobols (also known as cardols) are 5-alkyl resorcinols.

anacardic acids   bilobols   cardanols

R= saturated or unsaturated *n*-alkyl chain from C_{13} to C_{17}

The biosynthesis of long chain phenols takes into account the condensation of an activated unsaturated (or saturated) fatty acid with a polyketide derived from three or four units of malonyl-CoA and subsequent cyclization (Figure 1). The similarity of alkyl chains both in anacardic acids and cardanols, allows one to postulate cardanols as decarboxylated derivatives of the precursor anacardic acids.

OCCURRENCE AND ANALYSIS

It is apparent from the literature that a systematic study of anacardic and similar phenolic lipids was hampered by lack of efficient preparative and analytical methods. This may be caused by the combination of phenolic and lipid character. Thus the compounds are hybrids in terms of techniques and interests.

Anacardic acids are obtained as mixtures with fatty acids and other lipids by extraction of the source materials or by saponification of the extracted lipids. They can be separated from most of the other lipid classes by adsorption chromatography.

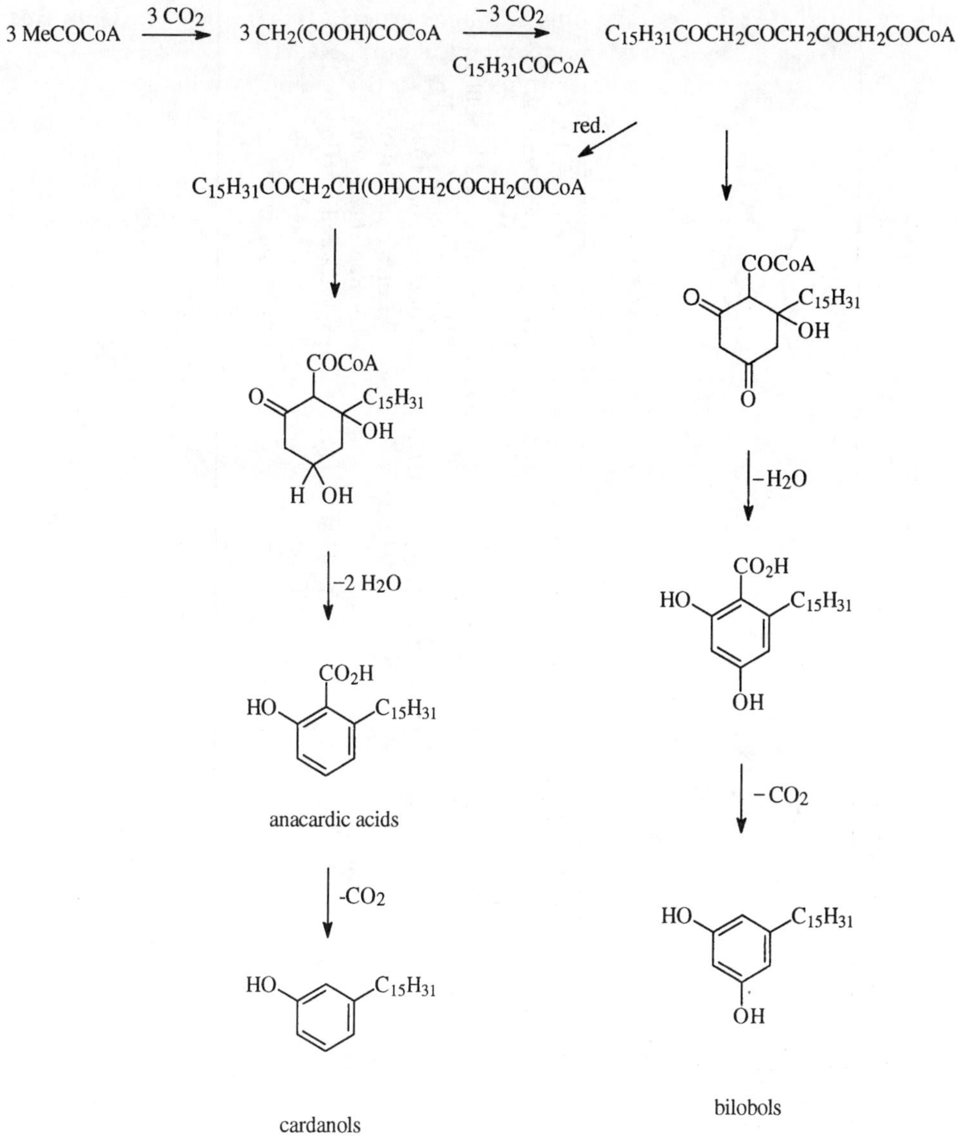

$$3\ MeCOCoA \xrightarrow{3\ CO_2} 3\ CH_2(COOH)COCoA \xrightarrow[C_{15}H_{31}COCoA]{-3\ CO_2} C_{15}H_{31}COCH_2COCH_2COCH_2COCoA$$

red.

$$C_{15}H_{31}COCH_2CH(OH)CH_2COCH_2COCoA$$

anacardic acids

cardanols

bilobols

Figure 1   Biogenetic pathway of *Ginkgo biloba* alkyl phenols (from Tyman, 1979)

However, for a long time, an efficient method for the separation of anacardic from fatty acids did not exist. Mixtures of anacardic acids have been fractionated by crystallization, and argentation chromatography has been applied for a separation according to their degree of unsaturation. Gas-liquid chromatography (GLC) for separation and identification of anacardic acids has not met with success.

Data so far published on the alkyl phenol composition of *Ginkgo biloba* are reported in Table 1. As a completion of the work performed by Kawamura (1928)

Table 1 Anacardic acids, bilobols and cardanols from *Ginkgo biloba* plant material

Plant material	R	%	R	%	R	%	References
Fruits	$C_{15}H_{29}$ Δ^8	0.33	$C_{15}H_{29}$ Δ^8	0.12	$C_{15}H_{29}$ Δ^8	0.06	Morimoto *et al.*, 1968
Leaves	$C_{13}H_{27}$, $C_{15}H_{29}$ Δ^8, $C_{15}H_{29}$ Δ^{10}, $C_{15}H_{31}$*, $C_{17}H_{31}$*, $C_{17}H_{33}$ Δ^{12}						Gellerman and Schlenk, 1968
Fruits	$C_{13}H_{27}$, $C_{15}H_{29}$, $C_{17}H_{33}$	3.1	$C_{13}H_{27}$, $C_{15}H_{29}$	0.53		traces	Adawadkar and El Sohly, 1981
Leaves	$C_{15}H_{29}$, $C_{17}H_{33}$						Matsumoto and Sei, 1987
Sarcotesta	$C_{13}H_{27}$, $C_{15}H_{29}$, $C_{17}H_{33}$		$C_{15}H_{29}$, $C_{17}H_{33}$		$C_{15}H_{29}$, $C_{17}H_{33}$		Itokawa *et al.*, 1987
Leaves and fruits					$C_{15}H_{29}$ Δ^8		Wagner *et al.*, 1989
Leaves	$C_{15}H_{29}$ Δ^8, $C_{17}H_{33}$ Δ^{10}						Fu-shun *et al.*, 1990
Leaves and fruits	$C_{13}H_{27}$, $C_{15}H_{29}$ Δ^8, $C_{15}H_{29}$ Δ^{10}, $C_{15}H_{31}$, $C_{17}H_{31}$, $C_{17}H_{33}$		$C_{15}H_{29}$†		$C_{13}H_{27}$, $C_{15}H_{29}$, $C_{15}H_{31}$, $C_{17}H_{33}$		Verotta and Peterlongo, 1993
Leaves, sarcotesta and fruits	$C_{13}H_{27}$, $C_{15}H_{29}$ Δ^8, $C_{17}H_{33}$ Δ^8, $C_{17}H_{31}$ Δ^9 Δ^{12}		‡				Irie *et al.*, 1996

*: traces

†: detected only in fruits

‡: content in Table 2

who isolated one ginkgolic acid and one bilobol from *G. biloba* fruits, Morimoto *et al.* (1968) extracted the fresh fruits of *G. biloba* isolating, beside the two reported products, a third compound which turned out to be a cardanol. The three compounds were identified and characterized using spectroscopic techniques. In particular, ^{1}H-NMR and IR experiments established that all these three constituents contained the same side chain (Δ^8 15:1) and that the double bond configuration was *cis*. The yields of the extraction from the fresh fruits were 0.33% for the ginkgolic acid, 0.12% for the bilobol and 0.06% for the cardanol. The mild extraction conditions (temperature never higher than 20° C) suggested that the cardanol was not an artefact coming from decarboxylation of the ginkgolic acid, but a natural substance contained in the plant material.

Gellerman and Schlenk (1968) described the isolation of anacardic acids from leaves of *G. biloba*. Extraction of the plant material was carried out with $CHCl_3/CH_3OH$, 2:1 (v/v). After saponification and removal of non-saponifiables, phenolic acids were recovered together with fatty acids. Separation of these two classes of compounds was obtained by means of chromatography on SiO_2 of the fraction previously treated with acidic methanol. In fact fatty acids are rapidly esterified to the corresponding methyl esters while phenolic hydroxyl groups remain unchanged. Anacardic acids and fatty esters were then separated by chromatography. The enriched phenolic fraction was subsequently treated with diazomethane and the anacardic acids were separated by reversed-phase liquid-liquid chromatography as ester derivatives on a larger scale. Anacardic acid structures were established by mass and NMR spectra and also by means of ozonolysis procedures. Terminal double bonds were determined by the analysis of the corresponding aldehydes by GLC. The position of other aliphatic double bounds was determined by ozonolysis-oxidation of dimethyl anacardic ether esters. The anacardic acids fraction obtained from *G. biloba* leaves turned out to contain mainly Δ^8 and Δ^{10} 15:1 anacardic acid {6-[8'(Z)- and 6-[10'(Z)-pentadecenyl]salicylic acid} (52%) and Δ^{12} 17:1 anacardic acid {6-[12'(Z)-heptadecenyl]salicylic acid} (40%), beside a small amount of the saturated 13:0 anacardic acid {6-tridecanyl salicylic acid} (8%).

Following the interesting results concerning the antibacterial activity of extracts obtained from the fruits of *G. biloba*, Adawadkar and El Sohly (1981) extracted and isolated the constituents responsible for the activity. As a result of this work a mixture of anacardic acids, a mixture of bilobols and small amounts of cardanols were obtained. The anacardic acids represented the major components of the extracts (3.1% w/w of the fresh deseeded fruits), while the bilobol fraction constituted the 0.53%. Only traces of cardanols were isolated. In addition, the occurrence of cardanols with an alkyl side chain different from those of the anacardic acids demonstrated that cardanols are natural products contained in the plant material and not only obtained as by-products from the corresponding anacardic acid by decarboxylation at high temperatures during the extraction procedure. Chromatographic analysis of the anacardic acid mixture showed the presence of $C_{13}H_{27}$, $C_{15}H_{29}$, $C_{17}H_{33}$ side chains, while bilobols turned out to contain mainly $C_{13}H_{27}$ and $C_{15}H_{29}$ side chains.

During the search for pest insect control agents based on natural products, Matsumoto and Sei (1987) examined *Ginkgo biloba* which is relatively free from insect attack. The crude methanol extract obtained from *G. biloba* leaves was found to possess antifeedant activity against the third-instar larvae of the cabbage butterfly (*Pieris rapae crucivora*) by using a leaf disk bioassay. Further separation of this extract was carried out in order to understand the nature of the compounds responsible for the feeding deterrent action. The separation was monitored using the bioassay method described above. The crude methanolic extract was dispersed into water and the aqueous suspension was extracted with chloroform and then with methyl ethyl ketone. The chloroform extract, showing 80% inhibition power, after further fractionation over silica gel and/or C_{18} reversed phase columns, gave, among other compounds, two anacardic acids, 15:1 anacardic acid (syn. 6-pentadecenylsalicylic acid) and 17:1 anacardic acid (syn. 6-heptadecenylsalicylic acid).

The discovery that the methanol extract from the sarcotesta of *Ginkgo biloba* showed remarkable activity in preliminary antitumour screening tests, led to the isolation and the identification of the compounds responsible for this activity (Itokawa *et al.*, 1987). Seven long-chain phenols were isolated and three of them showed antitumour activity. The products were isolated from a total methanol extract of the sarcotesta of *G. biloba* which was taken up with water and partitioned successively with chloroform, ethyl acetate and *n*-butanol. The antitumor activity was concentrated in the chloroform fraction. Pure compounds were obtained by submitting the chloroform fraction first to silica gel and/or alumina and then to reversed phase (C18) column chromatography. The structures were elucidated on the basis of spectroscopic data (MS and NMR), as mixtures of anacardic acids (60.8% of the chloroform extract), bilobols (9.8%) and cardanols (1.8%). MS fragmentations, supported by ozonolysis experiments, identified them as 6-[tridecanyl] salicylic acid (7.5% of the anacardic acids mixture), 6-[8'(Z)-pentadecenyl] salicylic acid (43.2%), 6-[10'(Z)-heptadecenyl] salicylic acid, 3-[pentadecenyl] phenol (9.7% of the bilobols mixture), 3-[heptadecenyl] phenol (0.15%), 5-[pentadecenyl] resorcinol (1.4% of the cardanols mixture) and 5-[heptadecenyl] resorcinol (0.4%).

More recently, Wagner *et al.* (1989), in the course of work devoted to the identification of the main constituents in *G. biloba* containing phytopharmaceuticals, developed an HPLC and a TLC method for the determination of both the flavonoid and the terpene trilactone fractions in purified extracts and commercial products. Using these analytical methods they were also able to identify a cardanol (named ginkgol) in extracts prepared from the leaves and the fruits of *G. biloba*, the side chain of which was Δ^8 15:1. Two other cardanols were detected by TLC but their structures were not determined. These phenolic components were not observed to be present in the phytopharmaceutical preparations analysed.

As already reported by Matsumoto and Sei (1987), Fu-shun *et al.* (1990) described the isolation and partial identification of some Ginkgo leaf constituents and their efficacy as feeding deterrent for the oligophagous larva of *Pieris brassicae* L. As a result of a detailed extraction procedure applied on *G. biloba* leaves, Fu-shun *et al.* were able to test 10 fractions for feeding responses. From the fraction showing the strongest effect, two anacardic acids were isolated and identified as Δ^8

15:1 anacardic acid and as Δ^{10} 17:1 anacardic acid. Because the antifeedant properties of *Ginkgo* leaves appeared to be distributed over different fractions, it was concluded that several substances are involved.

A detailed investigation of the phenolic components contained in *G. biloba* plant material was performed by Verotta and Peterlongo (1993), who reported on the lipid characterization of leaves and fruits. An *n*-hexane extract of *G. biloba* leaves was purified by absorption on an anion exchange resin. The retained material was analysed by capillary GC-MS and compared with the total methanol extract, in order to detect selective retention of certain classes of lipids on the resin. The analysis revealed the presence, beside the fatty acids palmitic and linolenic acid, of cardanols and anacardic acids while no bilobols could be detected. The total methanolic extract obtained from the fruits showed the same distribution pattern in cardanols and anacardic acids in both their relative intensities and isomeric composition, but bilobols were also present. The use of the mass spectrometer as a specific detector for the gas chromatograph provided a powerful analytical technique for the identification of the extract components. The phenolic lipids present in *G. biloba* plant material contain derivatizable functions so that their GC analysis as trimethylsilyl derivatives increases the accuracy of the analysis by improving resolution and peak symmetry. In this way, beside the most abundant components, also the minor ones were identified (see GC-MS profile of the fruit extract in Figure 2). The

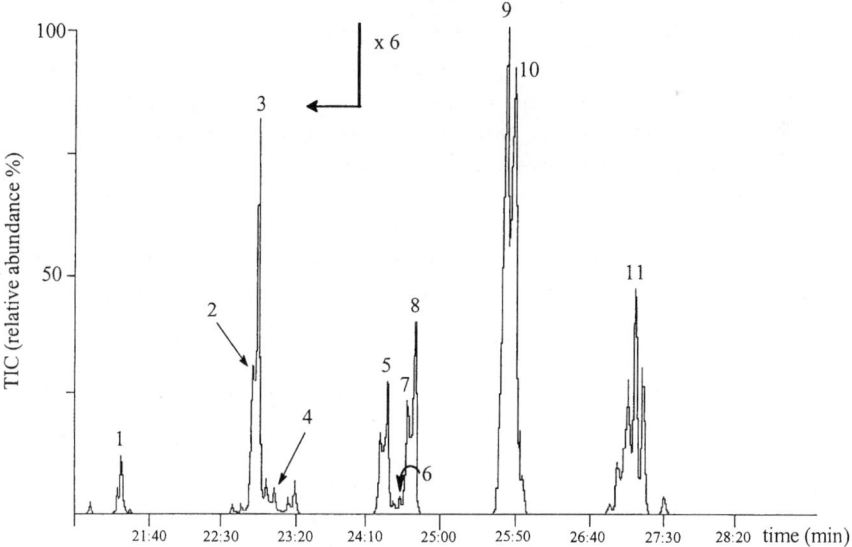

Figure 2 Expanded region of the capillary GC EI-MS profile of *G. biloba* fruit extract after trimethylsilylation obtained on an SPB-1 column (Supelco), 30 m × 0.32 mm I.D., 0.25 μm film thickness. 13:0 Cardanol (peak 1), 15:1 Cardanol (peak 2), 15:1 Cardanol (peak 3), 15:0 Cardanol (peak 4), 13:0 Anacardic Acid (peak 5), 17:1 Cardanol (peak 6), 15:1 Bilobol (peak 7), 15:1 Bilobol (peak 8), 15:1 Anacardic Acid (peak 9), 15:1 Anacardic Acid (peak 10), 17:1 Anacardic Acid (peak 11) (Verotta and Peterlongo, 1993).

GC-MS analysis of the TMS derivatives of the components of Ginkgo extracts showed that the structure of the different components can be determined from their EI mass spectra. In spite of the fact that these mass spectra showed poor fragmentations and low intensity molecular ions, phenolic lipids can be identified as anacardic acids, bilobols or cardanols because of the presence of strong ions at m/z 219, 268 and 180, respectively.

m/z 219   *m/z 268*   *m/z 180*

Since the purification of the phenolic fraction from the fatty acids has been demonstrated to be more difficult than the separation of the anacardic acids and cardanols themselves, Verotta and Peterlongo (1993) developed a method for the selective extraction of phenolic components from crude extracts of both the leaves and fruits of *G. biloba* by dynamic supercritical fluid extraction with carbon dioxide. The selective extraction was performed by varying both temperature and pressure and the analysis of the obtained extracts was performed through off-line capillary GC-MS. The anacardic fractions were then purified by reversed phase medium pressure chromatography allowing the isolation of pure 13:0, 15:1 and 17:1 anacardic acids. Ozonolysis experiments demonstrated that the 15:1 anacardic acid was composed of Δ^8 and Δ^{10} side chains, in conformity with the previously reported data (Gellerman and Schlenk, 1968). This method could be scaled-up and was applicable to isolation and identification of toxic phenolic components in commercial extracts of *G. biloba*.

During the search for new effective inhibitors of triacylglycerol synthesis, *G. biloba* acetone extracts were found to inhibit the glycerol-3-phosphate dehydrogenase, a key enzyme in the synthesis of triacylglycerol (Irie *et al.*, 1996). The components responsible for this activity were isolated from an *n*-hexane extract of *G. biloba* leaves through Sephadex LH-20 and preparative RP-HPLC and found to be four anacardic acids, Ana-a, -b, -c and -d (Table 2). Identification of the isolated compounds was performed by means of a combination of NMR and MS. ¹H-NMR and ¹³C-NMR gave information about the lengths of the alkyl chains and the number of double bonds. The positions of the last double bonds in the chain were located by oxidative degradation. Each anacardic acid was treated with a KMnO₄ and NaIO₄ mixture after which the produced fatty acid was methylated by methanol-BF₃ and analysed by GC. Ana-b was found to be the most abundant

Table 2   Distribution of anacardic acids in *Ginkgo biloba* (Irie *et al.*, 1996). HPLC analysis was obtained by using a YMC pak R-ODS-10 (Yamamura, i.d. 4.6 × 250 mm) column, eluent (CH$_3$CN/5% AcOH = 80/20 v/v), detector UV at 280nm, flow rate 1.0 mL/min.

Compounds	Contents (mg/g dry weight)		
	Leaf	*Sarcotesta*	*Nut*
Ana-a (13:0)	0.40 ± 0.08	1.90 ± 0.20	n.d.[*]
Ana-b (Δ^8 15:1)	12.0 ± 0.09	31.0 ± 2.5	n.d.[*]
Ana-c (Δ^{9}Δ^{12} 17:2)	0.50 ± 0.05	0.30 ± 0.05	0.02 ± 0.01
Ana-d (Δ^8 17:1)	4.40 ± 0.04	2.20 ± 0.60	n.d.[*]

[*]: not detected

anacardic acid isolated from the *n*-hexane extract of *G. biloba* leaves (1.2% of the dried material). The distribution of anacardic acids in *G. biloba* leaves, sarcotesta and nuts was determined by HPLC analysis (Table 2).

PHYTOPHARMACEUTICALS

Ginkgo biloba Leaves

According to the different extraction procedures adopted for the preparation of commercial phytopharmaceuticals, the content of active principles is dramatically variable. A clear example of the compositional variation of primary extracts is reported by Aye and Müller (1991). The authors investigated the influence of the industrial procedures on the extract composition. Regarding the alkylphenols content, extracts obtained through the use of different ratios of acetone/water as primary extraction solvent give a content from 230 ppm (pure water) to 104,000 ppm (pure acetone).

Because alkylphenols are reported to be responsible for toxic effects and especially for causing strong allergies and contact dermatitis (Hill *et al.*, 1934; Lepoittevin *et al.*, 1989), procedures were developed in order to obtain alkylphenol-free commercial extracts.

Dried, crushed leaves from cultivated (France and USA) or wild plants (China, Japan, Northern and Southern Korea) are industrially extracted with acetone-water, or ethanol-water. The extract is concentrated at low pressure, diluted with water and filtered. Under these conditions, the alkylphenols, the chlorophyll, the fatty acid derivatives and the biflavones precipitate and can be separated by filtration. The content of alkylphenols at this stage is less than 10 ppm.

O'Reilly and Jaggy (1990) and Schwabe (1989), describe the preparation of selective extracts where the alkylphenol content is further reduced to less than 10 ppm, by submitting an aqueous ethanol solution of the recombined extract fractions according to the desired final product (flavonol glycosides and ginkgolides or flavonol glycosides and bilobalide) to a multistep liquid-liquid extraction with *n*-heptane.

Bombardelli *et al.* (1989) developed an industrial process for the preparation of *Ginkgo biloba* extracts which avoids the use of polluting lead salts for facilitating the extraction. The content of alkylphenols (analysed as ginkgolic acids) in the final product was found to be lower than 5 ppm.

Ginkgo biloba Fruit

Baiguo, Semen Gingko, is the dry ripe seeds of *Ginkgo biloba*, collected during fall. This crude drug is officially listed in the Chinese Pharmacopoeia and used in Traditional Chinese medicine as an antiasthmatic and against polyuria. Before the extraction, the "samen" are washed and eliminated of the external pulp which is responsible for fruit toxicity. Cases of strong allergic reactions after contact with Ginkgo fruits are known (Sowers *et al.*, 1965; Nakamura, 1985). Serious mucosal disturbances after eating Ginkgo fruits have also been described (Becker and Skipworth, 1975). After drug consumption, cramps can arise. This is ascribed to the presence of 4-O-methylpyridoxine. It is proposed that this toxic principle causes food poisoning not only by antagonizing vitamin B6 in the body, but also by inhibiting the formation of 4-aminobutyric acid from glutamate in the brain (Wada *et al.*, 1988).

Ginkgo biloba hom. HAB1, *Ginkgo biloba* hom. PFX and *Ginkgo biloba* hom. HPUS88

Homeopathic preparations do not give any data on alkylphenol content.

CONCLUSIONS

Modern analytical approaches allow characterization of the family of phenolic lipids both to elucidate their profile in *G. biloba* plant material and to control their presence in phytopharmaceuticals obtained mainly from the leaves of the plant. A specific approach is the use of GC/MS analyses. Nevertheless, because this technique demands prior derivatisation and is not widely available in pharmaceutical quality control labs, a more convenient and relatively inexpensive approach is the application of HPLC. This is the method of choice for the quality control of the industrial production.

Phenolic lipids are reported to play a critical role in inducing allergic reactions while, apparently, they do not play a significant role in the clinical properties of *G. biloba* derivatives.

Nevertheless, it is worthwhile to underline that the class of phenolic lipids according to recent literature appears to be endowed with interesting biochemical and pharmacological properties which have to be considered for possible future exploitation in fields besides the classical application of *G. biloba* in the therapy of cerebral insufficiency. These include molluscicidal (Kubo *et al.*, 1986), antifeedant (Matsumoto and Sei, 1987, Fu-shun *et al.*, 1990), antitumour (Kubo *et al.*, 1993), antibiotic (Himejima and Kubo, 1991) and antiacne (Kubo *et al.*, 1994a) properties of anacardic acids, cardols and cardanols.

Anacardic acid monoene was found to be a good inhibitor of 5-lipoxygenase, relating this fact to the natural resistance of anacardic acid producing plants to insects (Shobha *et al.*, 1994). In addition, 16 phenolic compounds (anacardic acids, bilobols and cardanols, all possessing a C15 alkyl chain) have been shown inhibitory activity on mushroom tyrosinase, a key enzyme in the insect molting process (Kubo *et al.*, 1994b). The antibacterial activity of natural and synthesized anacardic acids has been extensively studied by Kubo *et al.* (1995), who found them exhibiting a narrow spectrum of activity against Gram-positive bacteria, dependent on the alkyl chain. More recently, the anacardic acids from G. *biloba* were found to inhibit glycerol-3-phosphate dehydrogenase (GPDH), a key enzyme in the synthesis of triacylglycerol and cholesterol (Irie *et al.*, 1996). The regulation of the amount of these lipids in biological tissues is important in the control of obesity.

REFERENCES

Adawadkar, P.D. and El Sohly, M. (1981) Isolation, purification and antimicrobial activity of anacardic acids from *Ginkgo biloba* fruits. *Fitoterapia*, **52**, 129–135.

Attanasi, O.A., Buratti, S. and Filippone, P. (1996) Cardanol: a versatile natural fine chemical largely available today. *La Chimica e l'Industria*, **78**, 693–696 and references cited herein.

Aye, R.D. and Müller, B. (1991) *Ginkgo*-Extrakte. Chromatographische Standardisierung und Einflüsse der Herstellung auf die Zusammensetzung. *Münch. med. Wschr.*, **133**, S58–S60.

Becker, L.E. and Skipworth, G.B. (1975) Ginkgo tree dermatitis, stomatitis and proctitis. *J. Am. Med. Assoc.*, **231**, 1162–1163.

Bombardelli, E., Mustich, G. and Bertani, M. (1989) New extracts of *Ginkgo biloba* and their methods of preparation. App. INDENA S.p.A. EP 0 360 556 B1.

Fu-shun, Y., Evans, K.A., Stevens, L.H., van Beek, T.A. and Schoonhoven, L.M. (1990) Deterrents extracted from the leaves of *Ginkgo biloba*: effects on feeding and contact chemoreceptors. *Entomol. exp. appl.*, **54**, 57–64.

Gellerman, J.L. and Schlenk, H. (1968) Methods for isolation and determination of anacardic acids. *Anal. Chem.*, **40**, 739–743.

Hill, G.A., Mattacotti, V. and Graham, W.D. (1934) The toxic principle of the poison ivy. *J. Am. Chem. Soc.*, **56**, 2736–2738.

Himejima, M. and Kubo, I. (1991) Antibacterial agents from the cashew *Anacardium occidentale* (Anacardiaceae) nut shell oil. *J. Agric. Food Chem.*, **39**, 418–421.

Irie, J., Murata, M. and Homma, S. (1996) Glycerol-3-phosphate dehydrogenase inhibitors, anacardic acids, from *Ginkgo biloba*. *Biosci. Biotech. Biochem.*, **60**, 240–243.

Itokawa, H., Totsuka, N., Nakahara, K., Takeya, K., Lepoittevin, J.P. and Asakawa, Y. (1987) Antitumor principles from *Ginkgo biloba* L. *Chem. Pharm. Bull.*, **35**, 3016–3020.

Jaggy, H. and Koch, E. (1997) Chemistry and biology of alkylphenols from *Ginkgo biloba* L. *Pharmazie*, **52**, 735–738.

Kawamura, J. (1928) Chemical constituents of the fruit of *Ginkgo biloba*. *Japan J. Chem.*, **3**, 89–108.

Koch, E., Chatterjee, S.S. and Jaggy, H. (1992) Immunotoxicological effects of constituents of leaves from *Ginkgo biloba* L. on the popliteal lymph node reaction in mice. *Naunyn-Schmiedebergs Archives of Pharmacology*, **345** (Suppl.), R95.

Kubo, I., Komatsu, S. and Ochi, M. (1986) Molluscicides from the cashew *Anacardium occidentale* and their large-scale isolation. *J. Agric. Food Chem.*, **34**, 970–973.

Kubo, I., Ochi, M:, Viera, P.C. and Komatsu, S. (1993) Antitumor agents from the cashew (*Anacardium occidentale*) apple juice. *J. Agric. Food Chem.*, **41**, 1012–1015.

Kubo, I., Muroi, H. and Kubo, A. (1994a) Naturally occurring antiacne agents. *J. Nat. Prod.*, **57**, 9–17.

Kubo, I., Kinst-Hori, I. and Yokokawa, Y. (1994b) Tyrosinase inhibitors from *Anacardium occidentale*. *J. Nat. Prod.*, **57**, 545–551.

Kubo, I., Muroi, H. and Kubo A. (1995) Structural functions of antimicrobial long-chain alcohols and phenols. *Bioorg. Med. Chem.*, **3**, 873–880.

Lepoittevin, J.P., Benezra, C. and Asakawa, Y. (1989) Allergic contact dermatitis to *Ginkgo biloba* L.: relationship with urushiol. *Arch. Dermatol. Res.*, **281**, 227–230.

Matsumoto, T. and Sei, T. (1987) Antifeedant activities of *Ginkgo biloba* L. Components against the Larva of *Pieris rapae crucivora*. *Agric. Biol. Chem.* **51**, 249–250.

Morimoto, H., Kawamatsu, Y. and Sugihara, H. (1968) Sterische Struktur der Giftstoffe aus dem Fruchtfleisch von *Ginkgo biloba*. *Chem. Pharm. Bull.*, **16**, 2282–2286.

Nakamura, T. (1985) Ginkgo tree dermatitis. *Contact Dermatitis*, **12**, 281–282.

O'Reilly, J. and Jaggy, H. (1990) Active components concentrates and new active component combinations from *Ginkgo biloba* leaves, their method of preparation and pharmaceuticals containing the active component concentrates or the active component combinations. (Patent) App. Montana Ltd. EP 0 436 129 A1.

Schwabe, K.P. (1989) Extract aus Blättern von *Ginkgo biloba*. (Patent) App. Dr. Willmar Schwabe GmbH & Co. DE 39 40 091 A1.

Shobha, S.V., Ramadoss, C.S. and Ravindranath, B. (1994) Inhibition of soybean lipoxygenase–1 by anacardic acids, cardols, and cardanols. *J. Nat. Prod.*, **57**, 1755–1757.

Sowers, W.F., Weary, P.E., Collins, O.D. and Cawley, E.P. (1965) *Ginkgo*-tree dermatitis. *Arch. Dermatol.*, **91**, 452–456.

Tyman, J.H.P. (1979) Non-isoprenoid long chain phenols. *Chem Soc. Rev.* **8**, 499–537.

Verotta, L. and Peterlongo, F. (1993) Selective extraction of phenolic components from *Ginkgo biloba* extracts using supercritical carbon dioxide and off-line capillary gas chromatography/mass spectrometry *Phytochem. Anal.*, **4**, 178–182.

Wada, K., Ishigaki, S., Ueda, K., Take, Y., Sasaki, K., Sakata, M. and Haga, M. (1988) Studies on the constitution of edible and medicinal plants. I. Isolation and identification of 4-O-methylpyridoxine, toxic principle from seed of *Ginkgo biloba* L. *Chem. Pharm. Bull.*, **36**, 3016–3020.

Wagner, H., Bladt, S., Hartmann, U., Daily, A. and Berkulin W. (1989) DC- und HPLC-Analyse von *Ginkgo*-Extrakten und *Ginkgo*-Extrakte enthaltenden Phytopraeparaten. *Dtsch. Apoth. Ztg.*, **129**, 2421–2429.

12. OCCURRENCE AND ANALYSIS OF GINKGO POLYPRENOLS

HOON HUH

Laboratory of Pharmacognosy, College of Pharmacy,
Seoul National University
56-1 Shinlim-Dong, Kwanak-Gu, Seoul 151-742, Korea

INTRODUCTION

Polyprenoid alcohols are linear polymers of isoprene, consisting of 6–22 units with a terminal primary alcohol group. In the past, polyprenoid alcohols have been referred to collectively as polyprenols, regardless of whether the terminal isoprene unit (α-isoprene unit) is saturated or unsaturated. Among the polyisoprenoid alcohols, compounds which contain a saturated α-isoprene unit (Figure 1) are usually referred to as dolichols. The term dolichol (from the Greek *dolichos* meaning long) was first applied to a mixture of polyisoprenoid alcohols isolated from pig liver. These compounds are normally found in animal tissue, but have also been found in plants. At present, the term polyprenols is restricted to α-unsaturated compounds in order to distinguish them from dolichols. The generalized structure of polyisoprenoid alcohols in terms of *cis*, *trans* conformation is shown in Table 1 (top). Certain polyprenoid alcohols differ from dolichols, belonging to Family I, since they have more than 2 internal *trans* isoprene units at the ω-end of the chain (Family II) or they have all the double bonds in the *trans* configuration (Family III).

DISTRIBUTION OF PLANT POLYPRENOLS

Since the initial report by Wellburn and Hemming (1966), over 30 years ago, of the presence of long-chain, mainly *cis* polyprenols in plants, these compounds are now

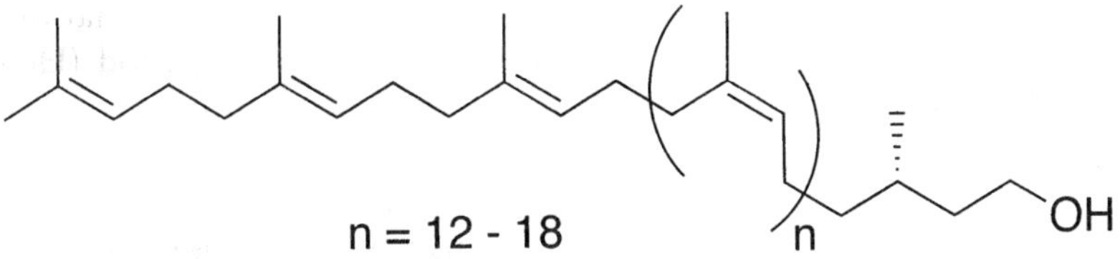

n = 12 - 18

Figure 1 Structure of dolichols

215

Table 1 Stereochemical classification of polyisoprenoid alcohols

| ω-terminal | trans | cis | α-terminal |

Family
  I. Di-*trans*, poly-*cis*
     (a) $\omega(t)_2(c)_{7\text{-}20}\text{-OH}$
     (b) $\omega(t)_2(c)_{11\text{-}20}S\text{-OH}$
 II. Tri-*trans*, poly-*cis*
     $\omega(t)_3(c)_{6\text{-}13}\text{-OH}$
III. All-*trans*
     $\omega(t)_{8\text{-}9}\text{-OH}$

Table from Rip *et al.* (1985)
The value for x ranges from 2 to 9, while that for y from 0 to over 20.
ω = omega unit; t = trans unit; c = cis unit; n = total number of isoprene units.
The α-isoprene unit may or may not be saturated (S). Family Ia, II and III are together known as "polyprenols". Family Ib is known as "dolichols".

known to be present in numerous plant species. Two groups of polyprenols are found in plants. The all-*trans*-polyprenols (Family III) are found mostly in the tissues of higher plants, while *cis*, *trans*–polyprenols (Families Ia and II) occur in a wide range of plants, and have even been reported for a green algae (Rip *et al.*, 1985). The majority of polyprenols isolated from the leaves of angiosperms have structures corresponding to Family II (Chojnacki and Vogtman, 1987). In contrast to angiosperms, the polyprenols from gymnosperms have structures corresponding to Family Ia (Ibata *et al.*, 1984a; Ibata *et al.*, 1984b; Swiezewska and Chojnacki, 1988) which contain long chains of poly-*cis* units. The content and the chain length distribution of polyprenols from gymnosperms are shown in Table 2. The polyprenols isolated from the leaves of *Ginkgo biloba* and other conifers have also been found to belong to Family Ia (Figure 2). The same arrangement of *trans* and *cis*-isoprene units is observed for dolichol having a saturated terminal isoprene unit (Family Ib).

BIOSYNTHESIS

The exact pathways and mechanisms for the biosynthesis and regulation of the levels of polyisoprenoids are at present not completely understood (Hemming, 1983). The biosynthesis of dolichols and polyprenols starts with acetyl-CoA. The initial steps in the biosynthesis of dolichols and polyprenols are identical with the initial steps in the biosynthesis of cholesterol and ubiquinone up to all-*trans*-farnesyl pyrophosphate (Chojnacki and Dallner, 1988). The terminal steps of polyisoprenoid synthesis begin with the condensation reactions in which the sequential addition of

Table 2 Chain length distribution of polyprenyl acetates

Origin	Composition (% wt)																	
	11#	12	13	14	15	16	17	18	19	20	21	22	23	24	25	26	27	28
Ginkgo biloba				0.9	1.6	6.5	24.9	36.5	17.7	6.7	2.9	1.6	0.8					
Pinus strobus				1.8	6.4	21.5	38.8	22.3	6.7	1.8	0.9							
Pinus sylvestris		2.0	3.0	7.8	24.2	35.1	18.8	6.4	2.8									
Pinus densiflora		1.1	2.0	4.4	10.2	28.3	30.0	14.3	5.2	2.3	1.3	0.8						
Pinus thunbergii	0.5	0.8	1.5	2.6	10.6	32.3	32.3	12.4	4.3	2.0	0.8							
Picea abies		0.8	2.8	18.4	39.6	26.9	8.6	2.2	0.8									
Cedrus deodora				0.9	1.9	8.0	25.7	38.4	17.7	4.8	1.4	0.7	0.4					
Crytomeria japonica				0.6	1.4	6.6	21.5	21.9	5.6	3.1	9.4	16.5	10.2	2.6	0.5			
Metasequoia glyptostroboides						1.7	10.1	15.7	8.3	9.0	25.2	22.6	6.4	1.0				
Scidopitys verticillata[a]			0.5	1.1	5.0	18.7	30.1	23.0	7.3	2.9	3.5	4.5	2.2	0.9	0.4			
Chamaecyparis obtusa				0.5	2.9	6.7	12.1	10.8	7.0	10.6	21.5	18.8	7.1	1.6	0.4			
Juniperus chinensis				0.2	1.0	5.4	16.3	22.8	18.4	16.5	12.0	5.7	1.7					
Juniperus rigida				0.6	3.2	13.1	26.5	19.8	12.0	10.2	9.1	3.9	1.6					
Araucaria brasiliana[a]			0.5	1.4	4.7	12.0	12.2	6.0	4.5	4.5	7.5	15.0	19.2	9.8	2.1	0.6		
Podocarpus macrophylla[b]			0.4	1.5	5.2	10.8	9.5	4.9	2.0	1.0	2.3	7.1	19.9	23.7	9.1	2.0	0.6	
Podocarpus nagi[b]			0.3	0.6	1.2	2.1	2.5	2.1	1.3	0.9	2.7	16.6	34.7	22.1	8.0	3.0	1.2	0.6
Cephalotaxus harringtonia subsp. nana					1.7	9.4	28.3	35.6	14.5	4.9	2.6	1.8	1.2					
Taxus cuspidata				0.5	1.0	4.6	16.5	20.8	9.3	9.8	14.0	12.7	6.9	2.9	1.0	0.3		
Torreya nucifera				0.6	1.4	9.3	33.4	41.5	11.7	1.7	0.5							

Table from Takigawa et al. (1989)

#Number of isoprene units

aPolyprenols obtained from an unfractionated mixture by saponification

bPolyprenols obtained from a polyprenyl acetate fraction by saponification

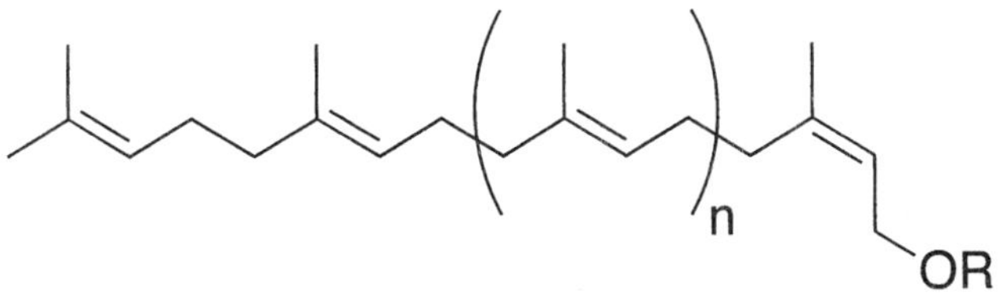

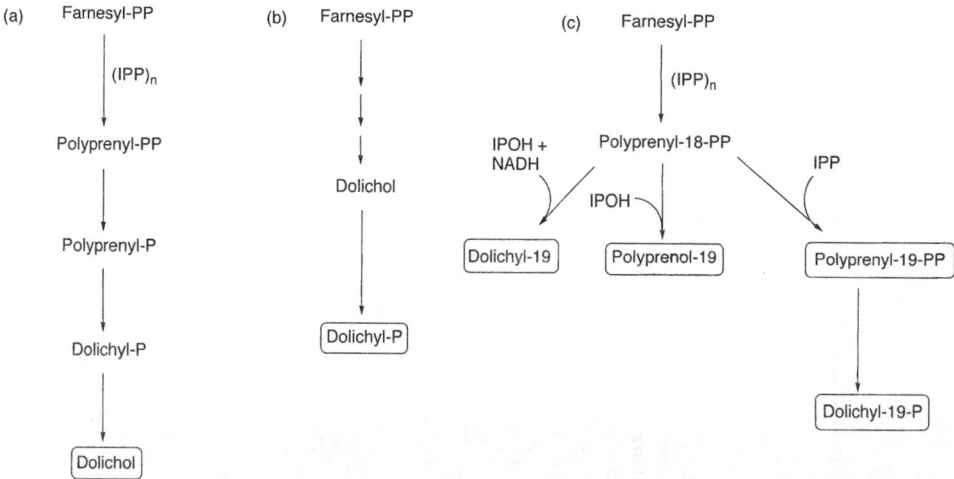

2a   R = H
2b   R = Ac

n = 11 - 19

Figure 2   Structure of Ginkgo polyprenols and their esters

2a: Polyprenols, 2b: Acyl esters

cis-isopentenyl pyrophosphate to *trans-trans*-farnesylpyrophosphate takes place (van Desssel *et al.*, 1990). After the appropriate isoprenoid chain length has been attained by the condensation reactions, the α-unsaturated pyrophosphate derivative is dephosphorylated to the monophosphate, followed by α-saturation to yield dolichyl phosphate (Figure 3a). When dolichol and dolichyl phosphate (dolichyl-P) were studied in sea urchin embryos using radioactively labeled precursors, the results indicated that the principal end product of *de novo* synthesis is the free alcohol and that this compound is subsequently phosphorylated to give dolichyl-P (Rossignol *et al.*, 1981; Rossignol *et al.*, 1983). It has been suggested that the level of dolichyl-P is regulated

Figure 3   Proposed biosynthetic pathways for dolichols and polyprenol.
Numbers represent the numbers of isoprene units.
Abbreviations : -P, monophosphate ; -PP, pyrophosphate ; IPOH, isopentenol

by dolichol kinase and dolichol phosphatase activities (Eggens *et al.*, 1988). However, Ekstrom *et al.* (1984) reported that they found no evidence that dolichol serves as a precursor for dolichyl-P in a study using rat liver tissue (Figure 3b). It has also been suggested that the immediate precursors of dolichol and dolichyl-P are the α-unsaturated polyprenol and the corresponding polyprenyl phosphate, respectively (Keller, 1986). Data of Dallner's group (Ekstrom *et al.*, 1987; Ericsson *et al.*, 1988; Chojnacki and Dallner, 1988) also suggested that the ultimate condensation step involves isopentenol and not isopentenyl pyrophosphate. In summary, it may be concluded that, when the appropriate chain length has been attained, an isopentenol is added to the previous isoprenologue pyrophosphate to yield a specific polyprenol (Figure 3c).

FUNCTION AND UTILIZATION OF POLYPRENOLS

Function of Dolichols and Polyprenols

Polyisoprenoid alcohols are involved as essential intermediates in the formation of cell wall peptidoglycans (polyprenols: Hemming, 1974; Rip *et al.*, 1985) and asparagine-linked glycoproteins (dolichol-P: Kornfeld and Kornfeld, 1985; Struck and Lennarz, 1980). The function involves the assembly of an oligosaccharide on phosphorylated forms of dolichol or polyprenol and their subsequent transfer to a polypeptide or protein.

These polyisoprenoid compounds are present in all tissues and intracellular membranes, either as a free alcohol, esterfied with a fatty acid and/or in phosphorylated form. Dolichols of different chain lengths have been isolated from animals (Dallner and Hemming, 1981), plants and microorganisms (Rip *et al.*, 1985). However, the level of dolichols in animal tissues is very low (less than 0.03% of the wet weight), except for human internal organs (0.01–0.7% of wet weight) (Rupar and Carrol, 1978; Tollbom and Dallner, 1986), thus making it difficult to obtain sufficient amounts of dolichol for comprehensive investigations.

The function of the free alcohol in biological membranes has not yet been elucidated. However, on the basis of research in model membranes, it has been proposed that dolichol influences membrane properties, such as fluidity and permeability (Valtersson *et al.*, 1985; Boscoboinik *et al.*, 1985; Schroeder *et al.*, 1987; Monti *et al.*, 1987; Wood *et al.*, 1989). Numerous studies also show that tissue dolichols may increase with age in various tissues of experimental animals and in human brain (Keller and Nellis, 1986; Crick and Rip, 1989). Furthermore, higher levels of tissue dolichol have been found in brain tissue of patients with Alzheimer's disease and cerebral neuronal lipofuscinosis (Ng Ying Kin *et al.*, 1983; Keller *et al.*, 1984). Because dolichol levels are elevated in the urine of alcoholics, urinary dolichols could serve as an important laboratory marker for alcoholism (Roine *et al.*, 1987).

The sugar carrier property of the polyisoprenoid alcohols has been largely studied in animal systems and, for these systems, the function of dolichol has been well established. A few studies have shown that the α-unsaturated polyprenol may also participate in glycosylation reactions (Stoll *et al.*, 1988). Since plants contain polyprenols rather than dolichol as their major polyisoprenoid alcohols, one may speculate that plants utilize polyprenols as sugar carriers in protein glycosylation, but no decisive studies have yet been reported.

Polyprenols as Precursors of Synthetic Dolichols

It has been shown that plant polyprenols are valuable natural sources of structural analogues of mammalian dolichols for biological studies and valuable precursors for the chemical synthesis of dolichol (Chojnacki *et al.*, 1987). Because of their important functions in protein glycosylation, reasonable quantities of individual dolichols are required for biochemical investigations. Since the content of individual dolichols in animal tissue is very low, the purification of dolichols from animal tissues is very elaborating. Therefore, even with extensive extraction and purification processes, substantial amounts of dolichols cannot be easily obtained and thus synthetic approaches have been made (Suzuki *et al.*, 1983; Imperiali and Zimmerman, 1988; Takigawa *et al.*, 1989; van Houte *et al.*, 1994). Higher plants are attractive sources, because of their usually high content of polyprenols (Sasak and Chojnacki, 1973; Chojnacki *et al.*, 1987). Most polyprenols isolated from the leaves of angiosperms have three *trans*-isoprene units (ω-t_3-c_n-OH), and, thus, are structurally different from the dolichols (Stone *et al.*, 1967; Wellburn *et al.*, 1967). In contrast to angiosperms, the majority of the polyprenols from gymnosperms examined thus far contain the ω-t_2-c_n-OH structure with a long chain of poly-*cis* units (Sasak *et al.*, 1976; Zinkel and Evans, 1972). Due to their high content and similar length of isoprene units, polyprenols from *Ginkgo biloba*, *Cedrus deodara* and *Torreya nucifera* are especially attractive starting materials for the synthesis of mammalian dolichols (Ibata *et al.*, 1983; Takigawa *et al.*, 1989; Table 2).

ANALYSIS OF GINKGO POLYPRENOLS

Analysis of Polyprenols and Dolichols

Most of the analytical studies have concentrated on the extraction and purification of mammalian dolichols because of their important physiological activities in animals. Plant polyprenol analysis was not a great concern of phytochemists until the plant polyprenols were found to represent a good alternative source of animal dolichols. The polyisoprenoid alcohols occur naturally in a variety of derivatized states, including acyl and monophosphoesters of the free alcohols, as well as in a number of glycosylated forms. Since the simultaneous analysis of all these derivatives would be a complex task, it is convenient to convert them into a smaller number of assayable species, namely free alcohols and the monophosphoesters.

For animal tissues, this is generally accomplished by direct saponification, a process which completely dissolves the tissues and converts all the dolichol derivatives to dolichol and dolichyl phosphate. In the case of plant tissues, which often contain large amounts of interfering substances such as wax or suberinized cellulose, pectin and/or tannin, it is more desirable to extract the polyisoprenoid alcohols with appropriate organic solvents prior to saponification. Polyprenols and dolichols are stable to treatment with strong alkali, allowing the use of saponification conditions during purification without serious losses. In contrast, mild acid degrades polyprenyl phosphate, while dolichyl phosphate is stable under these conditions (Wedgwood *et al.*, 1974).

Extraction is usually accomplished with n-hexane alone or with a combination of acetone and n-hexane. Plant tissue may be pulverized for a more efficient and faster extraction. Extraction time is dependent upon the end use of the product. If the extraction is on a preparative scale it may require up to a month (Ibata *et al.*, 1984a). However for analytical work, a one hour extraction time is usually sufficient, especially when an appropriate internal standard such as a polyprenol palmitate (Swiezewska and Chojnacki, 1989) is added to the plant tissues prior to extraction. The extracted sample is evaporated and saponified by treatment with alcoholic KOH or K_2CO_3. Although a few laboratories (Adair and Keller, 1985) added pyrogallol routinely as was reported by Burgos in 1963, it may not be necessary for plant material. Saponification of animal tissues, however, may require pyrogallol to precipitate interfering substances. However, no other laboratories have reported difficulties in the saponification of plant polyprenol derivatives in the absence of pyrogallol. Saponification may not require an extensive treatment time. If a high temperature is used, a treatment time from one to several hours appears to be sufficient for complete saponification. At room temperature, hydrolysis periods of overnight or longer have been used.

The hydrolysate is usually extracted several times with a non-polar organic solvent, such as n-hexane or petroleum ether. Various column chromatography techniques can be employed for clean-up, prior to HPLC. These include chromatography on adsorbents such as silica gel or hydrophobic resins as Lipidex (hydroxy-alkoxypropyl-Sephadex, Packard-Becker) and Sephadex LH-20. Esters and free alcohol forms are easily separated by eluting the column with solvents of different polarity. In the case of silica gel, the polyprenyl acetate fraction is eluted first with lipophilic solvents and the free polyprenol fraction is eluted with solvents having a higher polarity. The resulting polyprenyl acetate fraction can be further purified by gel permeation chromatography on a preparative column packed with high resolution styrene/divinyl benzene gel using chloroform as the eluent (Ibata *et al.*, 1983; Ibata *et al.*, 1984a). Elution is easily monitored by TLC on silica gel in EtOAc/benzene or in n-hexane/EtOAc. Reverse phase C-18 plates may also be used with acetone or acetone/hexane as the eluent. Multiple development may be required for better resolution (Swiezewska and Chojnacki, 1989). The analytical methods for plant polyprenols are summarized in Table 3.

HPLC

The semi-purified extract can be further purified by HPLC for analysis and quantification of isoprenologues. HPLC of polyprenols can be carried out on a silica

Table 3 Major plant polyprenol analytical techniques

	Ibata et al., 1983	Ibata et al., 1984a	Ibata et al., 1984b	Swiezewska and Chojnacki, 1988
Extraction from plant tissues	acetone/n-hexane (1:1, v/v)	acetone/n-hexane (1:1, v/v)	acetone/n-hexane (1:1, v/v) for 1 month	acetone/n-hexane (1:1, v/v)
Saponification				1 ml of 60% KOH 1 ml of 5% methanolic pyrogallol 0.5 ml benzene 1 h on a boiling water bath
Extraction after saponification				diethyl ether/hexane (1:1, v/v)
Column Chromatography	silica gel NW-6201 n-hexane/diethyl ether (19:1, v/v)–acetate n-hexane/diethyl ether (17:3, v/v)–free alcs	silica gel NW-6201 n-hexane/diethyl ether (19:1, v/v)–acetate n-hexane/diethyl ether (17:3, v/v)–free alcs	silica gel NW-6201 n-hexane/diethyl ether (19:1, v/v)–acetate n-hexane/diethyl ether (17:3, v/v)–free alcs	silica gel in hexane 3% ether in hexane–esters 10% ether in hexane–free alcs
TLC	silica gel 60 F$_{254}$ n-hexane/diethyl ether (19:1, v/v)–acetate n-hexane/diethyl ether (17:3, v/v)–free alcs	silica gel 60 F$_{254}$ n-hexane/diethyl ether (19:1, v/v)–acetate n-hexane/diethyl ether (17:3, v/v)–free alcs	silica gel 60 F$_{254}$ n-hexane/diethyl ether (19:1, v/v)–acetate n-hexane/diethyl ether (17:3, v/v)–free alcs	silica gel G EtOAc/benzene (5:95, v/v) or RP-18 acetone/hexane (95:5, v/v)
HPLC	RP Nucleosil 5C-18 acetone/methanol (9:1, v/v), 3 ml/min RID	RP-partition Scortex–OT acetone/methanol (9:1, v/v), 3 ml/min UV 210 nm		Resolve C-18 MeOH/isopropanol/water (60:40:5), 1.5 ml/min UV 210 nm
Gel Permeation CC	styrene/divinyl benzene in chloroform	styrene/divinyl benzene in chloroform	styrene/divinyl benzene in chloroform	
SFC				

Table 3 continued

	Swiezewska and Chojnacki, 1989	Takigawa et al., 1989	Hub et al., 1992	van Houte et al., 1994	Anonymous, 1995
Extraction from plant tissues	acetone/n-hexane (9:1, v/v)	n-hexane	n-hexane for 1h	acetone/n-hexane (5:1, v/v)	acetone/n-hexane (1:1, v/v)
Saponification	benzene/EtOAc/H_2O (3:15:2, 15% KOH) 1 h at 90° C	active carbon treatment K_2CO_3 in MeOH 12 h	K_2CO_3 in MeOH 12 h	15% KOH in MeOH 3 h at 55° C	3 kg KOH 2 l water 100 g pyrogallol 2 l EtOH under nitrogen
Extraction after saponification	n-hexane	n-hexane	n-hexane	pet. ether	
Column Chromatography	Lipidex–5000 in acetone silica gel 60 in hexane hexane + acetone	silica gel n-hexane/EtOAc (9:1, v/v)		silica gel n-hexane/diethyl ether (9:1, v/v)	Lipidex–5000 MeOH/acetone (1:1, v/v) acetone + MeOH; silica gel EtOAc/benzene (1:19, v/v) or RP-18 acetone
TLC	silica gel G EtOAc/benzene (5:95, v/v) or RP-18 acetone/hexane (95:5, v/v)	silica gel 60 F254 n-hexane/diethyl ether (19:1, v/v)–acetate n-hexane/diethyl ether (17:3, v/v)–free alcs			
HPLC	Resolve C-18 MeOH/isopropanol/water (60:40:5), 1.5 ml/min UV 210 nm	RP Nucleosil 5C-18 acetone/methanol (9:1, v/v), 3 ml/min, RID			
SFC			SB-Phenyl-50 column Eluent : SF CO_2 Pressure : 200 atm to 400 atm, (20 atm/min) Detector : FID, 325° C Oven : 100° C		

or octadecyl-silica column, depending on the information desired. For those instances in which direct quantification as a single peak is preferred, a silica gel column with 0.3% alcohol in hexane as solvent may be used. Detection of prenols is performed by on-line UV-monitoring at 210 nm. The lower limit of detection is 10 ng. If the isoprene distribution of the prenol sample is desired, reversed-phase HPLC on a 5 μm C-18 column is carried out. The solvent systems are mixtures of iso-propanol/methanol/water, or acetone/methanol (Swiezewska and Chojnacki, 1989; Takigawa et al., 1989). At a flow rate of 1.0–3.0 ml/min, the isoprenologues elute within approximately 20 min with baseline resolution. If HPLC shows contaminants interfering with polyprenol analysis, further purification is necessary. Prior purification on a C-18 Sep-Pak is preferred by many laboratories because of its simplicity and high efficiency. However, for plant material, the use of C-18 solid phase extraction may not be appropriate, since many plants contain prenols which contain less than 15 isoprene units. Adair and Keller (1985) claimed that these are not effectively retained by Sep-Pak columns, even when applied in various mixtures of methanol and water.

SFC

For polyisoprenoid alcohol analysis, HPLC has been the method of choice along with TLC as a supplemental qualitative method (Ibata et al., 1984a; Swiezewska and Chojnacki, 1989). However, the relatively new technique of supercritical fluid chromatography (SFC) may represent a good alternative to HPLC due to its wide detector availability and good resolution in multihomologue analysis.

A substance becomes supercritical when both pressure and temperature exceed its critical pressure and temperature. Supercritical fluid chromatography is a chromato-graphic method wherein supercritical fluids, rather than normal liquids or gases are used as mobile phase. Carbon dioxide, which becomes supercritical at 73 atm and 31° C is, by far, the most widely used supercritical fluid. The unique feature of supercritical fluids as solvents is that the solvating strength of the fluid is directly related to density, which can be easily varied as a function of pressure and tempera-ture. Some physical properties of the supercritical fluids, such as density and solute diffusion, fall between those of gases and liquids. In general, the solvating properties of supercritical fluids are more similar to those of liquids, while their viscosities are closer to those of gases. Because of the relatively high molecular weight of the Ginkgo polyprenols, a high density mobile phase is required for analysis.

Huh et al. (1992, 1993) demonstrated a rapid separation of polyprenol homo-logues present in Ginkgo leaves by using SFC. The extraction and saponification techniques employed were similar to those described by Takigawa et al. (1989) except for the purification steps involving adsorbent charcoal treatment and column chromatography. Briefly, pulverized Ginkgo leaves spiked with 500 μg of the inter-nal standard dodecaprenol from Rhus typhina were extracted for 1 h in n-hexane with occasional stirring. The resulting n-hexane extract was washed three times with 90% aqueous methanol and then vigorously stirred with potassium carbonate and methanol for 12 h at room temperature. After a further washing, once with water and twice with saturated sodium chloride solution, the resulting n-hexane solution was dried by passing through sodium sulfate. The eluted n-hexane solution was evaporated under nitrogen. The residue was reconstituted with 1 ml of n-hexane prior to separation of the polyprenols by SFC. Samples were subjected to capillary supercritical fluid chromatography on a chemically bonded SB-Phenyl-50 capillary column (10 m x 50 μm id, film thickness 0.25 μm). Separation of the polyprenols was achieved by means of a pressure gradient in which the initial pressure of 200 atm was increased at a rate of 20 atm/min to a final pressure of 400 atm. The oven temperature was maintained at 100° C.

The SFC analytical procedure described above showed a broad range of linearity and baseline separated all of the known polyprenol homologues. The SFC assay results showed that the content of polyprenols in Ginkgo leaves varied with the season in which they were collected (Huh *et al.*, 1993) and the total content was higher than the previously published values (Huh *et al.*, 1992). In addition, the chromatogram of the concentrated leaf extract revealed the presence of an isoprenologue (C_{120}) which has not previously been detected by HPLC methods.

CONCLUSION

Polyprenols are structural analogues of animal dolichols which are known carriers of sugars. Many gymnosperms contain large amounts of polyprenols which have the same *cis-* and *trans-* configuration of their isoprene units as in animal dolichols. Due to their high content and similar chain length and stereochemical configuration of isoprene units, polyprenols from *Ginkgo biloba* are especially attractive as starting materials for the synthesis of mammalian dolichols.

A variety of analytical techniques are employed for Ginkgo polyprenol analysis. For polyisoprenoid alcohol analysis, HPLC has been the method of choice along with TLC as a supplemental qualitative method. Saponified plant extracts need to be first subjected to column chromatography prior to HPLC for pre-purification. The relatively new supercritical fluid chromatography represents a promising alternative to HPLC, due to its wide detector availability, relatively easy sample pre-purification and good resolution in multihomologue analysis.

REFERENCES

Adair, W.L. and Keller, R.K. (1985) Isolation and assay of dolichol and dolichyl phosphate. In *Steroids and Isoprenoids Part B*, J.H. Law, and H.C. Rilling, (eds.) *Methods in Enzymology* 111, Academic Press, New York, pp. 201–215.

Anonymous (1995) Dolichols, polyprenols and derivatives, In *Collection of polyprenols*, 1995, Institute of Biochemistry and Biophysics, Polish Academy of Sciences (ed.), Warsaw.

Boscoboinik, D.O., Feliz, M., Disalvo, E.A. and Belocopitow, E. (1985) The effect of dolichol on the permeability properties of phosphatidylcholine bilayers. *Chem. Phys. Lipids*, **38**, 343–352.

Burgos, J., Hemming, F.W., Pennock, J.F. and Morton, R.A. (1963) Dolichol: a naturally-occurring C_{100}-isoprenoid alcohol. *Biochem. J.*, **88**, 470–474.

Chojnacki, T. and Dallner, G. (1988) The biological role of dolichol. *Biochem. J.*, **251**, 1–9.

Chojnacki, T. and Vogtman, T. (1984) The occurrence and seasonal distribution of C_{50}-C_{60} polyprenols and of C_{100}- and similar long chain polyprenols in leaves of plants. *Acta Biochim. Polon.*, **31**, 115–126.

Chojnacki, T., Swiezewska, E. and Vogtman, T. (1987) Polyprenols from plants – structural analogues of mammalian dolichols. *Chem. Scripta*, **27**, 209–214.

Crick, D.C. and Rip, J.W. (1989) Age-associated changes in dolichol and dolichyl phosphate metabolism in the kidney and liver of mice. *Biochim. Biophys. Acta*, **1004**, 180–186.

Dallner, G. and Hemming, F. (1981) Lipid carriers in microsomal membranes. In *Mitochondria and microsomes*, C. Lee, G. Schatz, and G. Dallner, (eds.), Addison-Wesley, Reading, pp. 655–680.

van Dessel, G., De Wolf, M., Hilderson, H.J., Lagrou, A. and Dierick, W. (1990) Intracellular and extracellular flow of dolichol. In *Intracellular transfer of lipid molecules*, H.J. Hilderson, (ed.), Plenum Press, New York, pp. 227–278.

Eggens, I., Ericsson, J. and Tolbom, O. (1988) Cytidine 5′-triphosphate dependent dolichol kinase and dolichol phosphate activities and levels of dolichyl phosphate in microsomal fractions from highly differentiated human hepatomas. *Cancer Res.*, **48**, 3418–3424.

Ekstrom, T.J., Chojnacki, T. and Dallner, G. (1984) Metabolic labeling of dolichol and dolichyl phosphate in isolated hepatocytes. *J. Biol. Chem.*, **259**, 10460–10468.

Ekstrom, T.J., Eggens, I. and Dallner, G. (1987) The α-saturation and terminal events in dolichol biosynthesis. *J. Biol. Chem.*, **262**, 4090–4097.

Ericsson, J., Ekstrom, T., Chojnacki, T. and Dallner, G. (1988) The terminal reactions in polyisoprenoid biosynthesis. *Acta Chem. Scand.*, **B42**, 202–205.

Hannus, K. and Pensar, G. (1974) Polyisoprenols in *Pinus sylvestris* needles. *Phytochemistry*, **13**, 2563–2566.

Hemming, F.W. (1974) Lipids in glycan biosynthesis. In *MTP Int. Rev. Sci. Biochem. Lipids, Biochem Ser. One*, T.W. Goodwin, (ed.), Butterworths, London, pp. 39–97.

Hemming, F.W. (1983) Biosynthesis of dolichols and related compounds. In *Biosynthesis of isoprenoid Compounds*, J. Porter, and S. Spurgeon, (eds.) John-Wiley & Sons, Chichester, England, pp. 305–354.

van Houte, H.A., Busson, R.H., Parmentier, G.G., Declercq, P.E., van Veldhoven, P.P, Mannaerts, G.P. and Eyssen, H.J. (1994) Synthesis of [1–^{14}C] dolichoic acid. *Chem. Phys. Lipids*, **72**, 103–107.

Huh, H., Singh, J. and Staba, E.J. (1992) Supercritical fluid chromatographic analysis of polyprenols in *Ginkgo biloba* L. *J. Chromatography*, **600**, 364–369.

Huh, H., Singh, J. and Staba, E.J. (1993) Ontogenic aspects of Ginkgo polyprenols. *Planta Med.*, **59**, 379–380.

Ibata, K., Mizuno, M., Takigawa, T. and Tanaka, Y. (1983) Long-chain betulaprenol-type polyprenols from the leaves of *Ginkgo biloba*. *Biochem. J.*, **213**, 305–311.

Ibata, K., Kageyu, A., Takigawa, T., Okada, M., Nishida, T., Mizuno, M. and Tanaka, Y. (1984a) Polyprenols from conifers: multiplicity in chain length distribution. *Phytochemistry*, **23**, 2517–2521.

Ibata, K., Mizuno, M., Tanaka, Y. and Kageyu, A. (1984b) Long-chain polyprenols in the family of Pinaceae. *Phytochemistry*, **23**, 783–786.

Imperiali, B. and Zimmerman, J.W. (1988) Synthesis of dolichols via asymmetric hydrogenation of plant polyprenols. *Tetrahedron Lett.*, **29**, 5343–5344

Keller, R.K. (1986) The mechanism and regulation of dolichyl phosphate biosynthesis in rat liver. *J. Biol. Chem.*, **261**, 12053–12059.

Keller, R.K., Armstrong, D., Crum, F. C. and Koppang, N. (1984) Dolichol and dolichyl phosphate levels in brain tissue from English setters with ceroid lipofuscinosis. *J. Neurochem.* **42**, 1040–1047.

Keller, R.K. and Nellis, S.W. (1986) Quantitation of dolichyl phosphate and dolichol in major organs of the rat as a function of age. *Lipids*, **21**, 353–355.

Kornfeld, R. and Kornfeld, S. (1985) Assembly of asparagine-linked oligosaccharides. *Annu. Rev. Biochem.*, **54**, 631–644.

Monti, J.A., Christian, S.T. and Schutzbach, J.S. (1987) Effects of dolichol on membrane permeability. *Biochim. Biophys. Acta*, **905**, 133–142.

Ng Ying Kin, N M.K., Palo, J., Haltia, M. and Wolfe, L.S. (1983) High levels of brain dolichols in neuronal ceroid-lipofuscinosis and senescence. *J. Neurochem.*, **40**, 1465–1473.

Rip, J.W., Rupar, C.A., Ravi, K. and Carrol, K.K. (1985) Distribution, metabolism and function of dolichol and polyprenols. *Prog. Lipid Res.*, **24**, 269–309.

Roine, R.P., Turpeinen, U., Ylikahri, R. and Salaspuro, M. (1987) Urinary dolichol-a new marker of alcoholism. *Alcohol. Clin. Exp. Res.* **11**, 525–527.

Rossignol, D.P., Lennarz, W.J. and Waechter, C.J. (1981) Induction of phosphorylation of dolichol during embryonic development of the sea urchin. *J. Biol. Chem.*, **256**, 10538–10542.

Rossignol, D.P., Scher, M.G., Waechter, C.J. and Lennarz, W.J. (1983) Metabolic interconverion of dolichol and dolichyl phosphate duing development of the sea urchin embryo. *J. Biol. Chem.*, **258**, 9122–9127.

Rupar, C.A. and Carrol, K.K. (1978) Occurrence of dolichols in human tissues. *Lipids*, **13**, 291–294.

Sasak, W. and Chojnacki, T. (1973) Long-chain polyprenols of tropical and subtropical plants. *Acta Biochim. Polon.*, **20**, 343–350.

Sasak, W., Mankowski, T., Chojnacki, T. and Daniewski, W.M. (1976) Polyprenols in *Juniperus communis* needles. *FEBS Lett.*, **64**, 55–58.

Schroeder, F., Gorka, C., Williamson, L.S. and Wood, W.G. (1987) The influence of dolichols on fluidity of mouse synaptic plasma membranes. *Biochim. Biophys. Acta.*, **902**, 385–393.

Stoll, J., Rosenwald, A.G. and Krag, S.S. (1988) A Chinese hamster ovary cell mutant F2A8 utilizes polyprenol rather than dolichol for its lipid-dependent asparagine-linked glycosylation reactions. *J. Biol. Chem.*, **263**, 10774–10782.

Stone, K.J., Wellburn, A.R., Hemming, F.W. and Pennock, J.F. (1967) The characterization of ficaprenol–10, –11, and –12 from leaves of *Ficus elastica* (decorative rubber plant). *Biochem. J.*, **102**, 325–330.

Struck, D.K. and Lennarz, W.J. (1980) In *The biochemistry of glycoproteins and proteoglycans*, W.J. Lennarz, (ed.) Plenum Press, New York, pp. 35–73.

Suzuki, S., Mori, F., Takigawa, T., Ibata., Ninagawa, Y., Nishida, T., Mizuno, M. and Tanaka, Y. Synthesis of mammalian dolichols from plant polyprenols. *Tetrahedron Lett.*, **24**, 5103–5106

Swiezewska, W. and Chojnacki, T. (1988) Long-chain polyprenols in gymnosperm plants. *Acta Biochim. Polon.*, **35**, 131–147.

Swiezewska, W. and Chojnacki, T. (1989) The occurrence of unique, long-chain polyprenols in the leaves of Potentilla species. *Acta Biochim. Polon.*, **36**, 143–159.

Takigawa, T., Ibata, K. and Mizuno, M. (1989) Synthesis of mammalian dolichols from plant polyprenols. *Chem. Phys. Lipids*, **51**, 171–182.

Tollbom, O. and Dallner, G. (1986) Dolichol and dolichylphosphate in human tissues. *Br. J. Exp. Path.*, **67**, 757–764.

Valtersson, C., van Duayn, G., Verkleij, A.J., Chojnacki, T., Kruijff, B. and Dallner, G. (1985) *J. Biol. Chem.*, **260**, 2742–2751.

Wedgwood, J.F., Warren, C.D. and Strominger, J.L. (1974) Transfer of sugars from nucleoside diphosphosugar compounds to endogenous and synthetic dolichyl phosphate in human lymphocytes. *J. Biol. Chem.*, **249**, 6316–6324

Wellburn, A.R. and Hemming, F.W. (1966) The occurrence and seasonal distribution of higher isoprenoid alcohols in the plant kingdom. *Phytochemistry*, **5**, 969–975.

Wellburn, A.R., Stevenson, J., Hemming, F.W. and Morton, R.A. (1967) The characterization of properties of castaprenol–11, –12 and –13 from the leaves of *Aesculus hippocastanum* (horse chestnut). *Biochem. J.*, **102**, 313–324.

Wood, W.G., Sun, G.Y. and Schroeder, F. (1989) Membrane properties of dolichol in different age groups of mice. *Chem. Phys. Lipids*, **51**, 219–226.

Zinkel, D.F. and Evans, B.B. (1972) Terpenoids of *Pinus strobus* cortex tissues. *Phytochemistry*, **11**, 3387–3389.

13. CONSIDERATIONS IN THE DEVELOPMENT OF THE U.S. PHARMACOPEIA'S MONOGRAPH ON *GINKGO BILOBA* L.

V. SRINI SRINIVASAN

United States Pharmacopeia, 12601 Twinbrook Parkway, Rockville, MD 20852, U.S.A.

INTRODUCTION

It is amazing that the Ginkgo tree, *Ginkgo biloba* L., the last remaining member of the Ginkgoaceae family has suddenly engaged the interest of Western medicine. Worldwide use of Ginkgo leaf extracts is reported to be on the rise in view of the reported beneficial effects of the Ginkgo as an antiasthmatic and on the circulatory system (Schilcher, 1988). Is it more appropriate to say that Western medicine has at last considered the medicinal properties of *Ginkgo biloba* whose fruits and seeds have been valued in China for their medicinal properties for the last several hundred years? Though the leaves of *Ginkgo biloba* were sparsely used, in comparison to the fruits and seeds, in ancient medicines in the East, the Western medicine as practiced extensively in Europe uses the leaves.

Senna leaves are used directly in the preparation of tea, but traditionally extracts of Ginkgo leaves are used directly in the Western medicine preparations. United States Pharmacopeia's (USP) interest and involvement in the development and establishment of public standards began in 1820 with the publication of its first edition. The first edition of the United States Pharmacopeia included monographs on botanicals. Thereafter for the next 120 years or so the USP Committee of Revision were busy adding more botanicals monographs to the USP. At the turn of the 20th century, USP had included monographs for over 600 botanicals and their preparations. However, with the arrival of isolated active ingredients, and later of synthetic organic medicinal substances, the focus of USP Committee of Revision turned to the more well-defined and easily characterized single organic chemicals. Also the use of tablets as a precise and convenient dosage form made botanical extracts less appealing. As a result gradually over the next 45 years or so the USP Committee of Revision replaced the older monographs on botanicals with those of newer synthetic organic medicinal substances and purified active principles of botanicals and their dosage forms. The result is that today at the time of writing this chapter the current edition of U. S. Pharmacopeia (USP 23–NF 18 1995) has only about 30 or so monographs on botanicals and their preparations out of more than 3500 monographs (e.g. *Rauwolfia serpentina*, Digitalis, Psyllium husk, Senna leaves etc.).

The recent resurgence of interest in the use of botanicals by the American public is contributing to the rapid change in the healthcare scene in the United States of America. Those who are disillusioned with modern healthcare which uses modern synthetic medicines are seeking alternative therapies. These alternative therapies include botanical preparations. The U.S. Pharmacopeial Convention, in response to this evolving change in the healthcare system, adopted a resolution at its March 1995 Convention which encouraged the U. S. Pharmacopeia to explore the feasibility and advisability of developing and establishing public standards for dietary supplements. Even though the resolution did not specifically mention botanicals, in the context of the definition provided in the Dietary Supplement Health Education Act of 1994, dietary supplements include botanicals. The United States Pharmacopeia thus is going back to its roots.

CONSIDERATIONS IN THE SELECTION OF BOTANICALS FOR DEVELOPMENT WORK.

In regard to the selection and prioritization of botanicals for compendial standards development work due consideration should be given to those botanicals widely used in the U.S.A. and which are also significant in international usage. This is immediately followed by other considerations such as harmonization of standards with other major pharmacopeias such as the European Pharmacopeia and Japanese Pharmacopeia. In following the above, the USP Committee of Revision prioritized a list of about 20 botanicals which includes *Ginkgo biloba*. Irrespective of whether all the active constituents of a botanical have been identified (and established), the strategy of the USP Committee of Revision is to develop a definition of the plant part. Since this book is exclusively devoted to a discussion of Ginkgo this chapter will deal with issues involved and considered in the development of pharmacopeial standards for *Ginkgo biloba* (leaves).

ISSUES IN THE STANDARDIZATION

Pharmacopeial monographs imply shelf-life standards. In other words the product purported to conform to the USP must be in conformity with the Pharmacopeial specifications when analyzed by the methods specified therein till the last date of expiry claimed on the product label. It should be kept in mind that pharmacopeial standards are not routine quality control standards but are referee standards. The General Notices of the U. S. Pharmacopeia (USP 23–NF 18 1995) allow use of alternate methods i.e. determination of compliance with the pharmacopeia may be done by the use of alternate methods which are chosen for advantages in accuracy, sensitivity, precision, selectivity or adaptability to automation provided these alternate procedures are validated. However since pharmacopeial standards and procedures are interrelated, where a difference appears or in the event of dispute only the result obtained by the procedure given in the Pharmacopeia is conclusive.

Therefore one can expect the standards setting organization to choose only those methods that would definitively determine the extent of decomposition of the botanical article under test. In developing a pharmacopeial monograph for articles of botanical origin considerations should be given to the way the botanical parts have been used in the past or moving in the international commerce. In reviewing the pharmacopeial monographs of yesteryear one can see clearly three categories namely (1) plant parts as a whole (e.g. subterranean parts, leaves, flowers etc.) or their powdered forms, (2) simple extracts of the plant parts derived from simple percolation procedures or (3) extracts of the plant parts that have undergone extensive processing. The standards for the above would obviously differ considerably from each other.

The purpose of this chapter is to highlight the issues and considerations in the development of pharmacopeial standards for botanical material as it relates to *Ginkgo biloba* leaf. Since a draft monograph on *Ginkgo biloba* leaf prepared by the USP Committee of Revision has just been published for public comments, references are made to the draft throughout this chapter (*Ginkgo biloba* leaf, 1997). It is to be understood that any or all aspects of the proposed draft monograph may undergo further changes as comments received from interested parties are reviewed by the members of the Committee of Revision [see Appendix for draft monograph].

Having said the above, the initial draft monograph on *Ginkgo biloba* developed by the USP Committee of Revision, whom the author has the privilege to work with, defines the Ginkgo leaves. Even though Ginkgo leaves as such are not used like Senna leaves, pharmacopeial practice of defining the raw material used in the preparation of their extracts or infusions or powders is a *sine qua non* for the quality assurance of the dosage forms derived therefrom. A future focus of the pharmacopeias is likely to be development of monographs on plant extracts and dosage forms where the issues involved are somewhat more complex.

Definition

A pharmacopeial monograph on a botanical has to be very specific in its definition by specifying the genus and species thereby indicating that the standards apply to that particular plant part of the stated genus and species. The draft monograph defines Ginkgo Leaf as consisting of the dried leaf of *Ginkgo biloba* belonging to the family Ginkgoaceae. In keeping with the customary practice of the U. S. Pharmacopeia to state, after the name of the species, the name of the botanist, Linnaeus, who first described the species (or the current commercially dominant description), is specified in the definition.

Packaging, Storage and Labeling

As stated above since pharmacopeial specifications are shelf-life specifications, the type of container in which the article is to be stored as well as the temperature conditions of storage need to be specified in the monograph. Under the stated conditions the pharmacopeial articles are expected to retain their stated potency till the expiry date claimed on the product label.

The name in Latin binomial form is the only recourse we have at present to save the public from the prevailing hopeless situation in the market place where manufacturers and distributors of botanicals use several different names on the product thus causing confusion in the minds of the consumers. The product label of a pharmacopeial article is expected to specify, in addition to the common name Ginkgo, the name in Latin binomial form as given in the definition. The insistence of the name in Latin binomial form is both a regulatory issue and a consumer issue. Thus when a product labeled as Ginkgo USP or NF (National Formulary) (assuming the monograph on Ginkgo would be adopted officially by the USP in the near future) is seized by regulatory authorities for determination of compliance, all that would be needed is the product label and the USP book. The label enables the consumers to compare the product purchased to other species of botanicals they have been purchasing.

Identification

A Pharmacopeial monograph specifies *Identification* tests to ensure confirmation that the identity of the article under examination is in agreement with what is stated in the definition of the article. Histology and morphology alone are not sufficient for positive identification of botanicals (Grady, 1994). This does not mean magnifying lenses and microscopes, which were the only tools available for the 19th and early 20th century scientists for positive identification of botanicals, have no place in modern pharmacopeias. In addition to the microscopic examination, pharmacopeial monographs would employ two other types of identification tests. For chemical methods of identifying the active ingredients or marker substances either as a class (e.g. alkaloids, terpenes) or by their functional group (e.g. ketones, phenolic hydroxyl), color tests or spectrophotometric techniques such as UV, IR are specified. Use of separation techniques such as thin-layer chromatography, high pressure liquid chromatography etc. permit not only identification of a marker or active ingredient but can detect contamination with other foreign substances. The use of the last two identification tests above have evolved during the last fifty years and are powerful, reliable and critical sources for positive identification.

Botanical characteristics

Macroscopical examination of fresh or dried whole Ginkgo leaf shows it to be ty-pically fan-shaped but occasionally sub-reniform and sides somewhat concave thus distinguishing it from any adulterant botanical. Histology of the Ginkgo leaf reveals numerous lignified elements derived from the lamina and petiole including xylem vessels with annular thickening, tracheids and vessels with bordered pits. The extent of lignification, particularly in the petiole, increases with age of the leaf and mature leaves may show the presence of very rare polygonal to circular starch granules approximately 20 μm in diameter with a central hilium and exhibiting a marked maltese cross under crossed-polaroids. While macroscopic and microscopic exam-ination results in rich information enabling positive identification, most modern analytical laboratories do not have analysts trained in histological examination. Thus the necessity for other tests for positive identification remains.

Chemical methods

Analytical methods involving chemical reactions contribute to a more unequivocal identification of botanicals (Schorn, 1993). However since Ginkgo leaves contain a variety of compounds such as isoprenoids (sterols, terpene trilactones), aliphatic alcohols and ketones, organic acids, polysaccharides, flavonol glycosides etc., a single chemical reaction yielding a bright color characteristic of flavonol glycosides or terpene trilactones does not appear to be promising. Hence the best course for a pharmacopeial monograph would be to specify methods involving separation techniques.

Separation techniques

Constituents responsible for the beneficial properties of Ginkgo leaves are reported to be flavonol glycosides and terpene trilactones: the sesquiterpene bilobalide and diterpenic ginkgolides. These compounds are reported to show effects on vascular and cerebral metabolic processes and ginkgolides inhibit platelet activating factor (PAF). Quantitative determinations of these by liquid chromatographic techniques have been reported in the literature (Sticher, 1993). It is therefore logical that pharmacopeial identification tests should focus on tests that would positively identify the presence of the above compounds in Ginkgo leaves. It could be argued that biflavones present in Ginkgo leaves are also characteristic *marker substances* but they do not lend themselves for standardization at least at this point of time though there are literature reports of attempted quantification of these compounds (Pietta *et al.*, 1988).

The critical aspect of identification of botanical materials by separation techniques is the use of reference standards since it provides an unambiguous calibration at the time of use. The reference standards used in pharmacopeial monographs are invariably those that are present in the material under test. When individual reference standards are difficult to develop and/or are prohibitively expensive, other substances which display suitable characteristics for the intended purpose can be occasionally employed. With regard to the separation technique for the positive identification of flavonol glycosides and terpene trilactones in Ginkgo leaves the natural choice is thin-layer chromatography (TLC) since it is the most simple and widely used technique available at sources of small scale manufacturers and ports of imports. There is a need to use two different solvent systems for the identification of flavonol glycosides and terpene trilactones because of the different nature of functional groups present in these compounds. Accordingly the TLC system chosen for the identification of flavonol glycosides in the draft monograph is practically identical to the one that is under consideration for the European Pharmacopeia. This will achieve the objective of harmonization as well. The test involves comparison of the TLC of the test solution, prepared by methanol extraction of the Ginkgo leaves, with the chromatogram of the standard solution containing rutin and chlorogenic acid run in a developing solvent consisting of a mixture of ethyl acetate, water, anhydrous formic acid and glacial acetic acid (67.5 : 17.5 : 7.5 : 7.5). The spray reagents used are a methanolic solution of diphenylboryloxyethylamine and a methanolic solution of polyethylene glycol 400. Visualization is under ultraviolet light (365 nm). Though neither rutin nor chlorogenic acid is present in Ginkgo leaves, their spots provide a reference point for easy location of the spots due to the major flavonol glycosides.

   With regard to the method for the identification of terpene trilactones, the lack of a suitable UV-absorbing functional group and their very low concentrations (0.01–0.1%) along with a host of other compounds in Ginkgo leaves make the analysis of these terpenes trilactones very difficult. Although all the five terpene trilactones in Ginkgo leaves have been separated by HPLC using a refractive index detector (van Beek et al., 1991) it is not a practical procedure for a quick identification check because of extensive clean-up steps involved in the procedure. Thus the best alternative to HPLC is TLC. Several papers have appeared in the literature on the thin-layer separation of terpene trilactones of Ginkgo biloba leaves but unfortunately a clear separation and subsequent detection of ginkgolides A and B and ginkgolides C and J on a thin-layer chromatographic plate can be seen as a major problem in these publications (Lobstein-Guth et al., 1983; Okabe et al., 1967; Tallevi and Kurz, 1991; Teng, 1988; Weinges et al., 1987). The best published TLC separation of all relevant terpene trilactones appears to be the one involving the use of silica gel plates impregnated with sodium acetate (van Beek and Lelyveld, 1993). Use of a silica gel impregnated with sodium acetate as reported by van Beek and Lelyveld gives an excellent separation of all major terpene trilactones in Ginkgo leaves using the solvent system ethyl acetate and hexane (9:1) contained in a non-saturated chamber or by using methyl acetate as the solvent in a saturated chamber. The detection of all the terpene trilactones is achieved by spraying the plate with acetic anhydride followed by heating at 105° and viewing under ultraviolet light at 254 nm and 366 nm. The method is reported to detect all the four ginkgolides and bilobalide in quantities less than 1 μg. The above procedure does not involve any known toxic solvent and uses readily available chemicals. Since the United States Pharmacopeia tends to use such simple procedures with advantages at all fronts, the above procedure was considered for adoption in the draft for public comments.

Microbial Limits

It is well known that medicinal plant materials generally are contaminated with a great number of bacteria and moulds arising from the soil. Additional contamination is normally caused by practices of harvesting, handling and processing. An index of good harvesting practices and good manufacturing practices can be expected to come from a determination of E. coli and yeast and moulds counts. One would expect the botanical materials to be decontaminated by post-harvest methods (e.g. chemical treatment, gamma irradiation) at the source of export (e.g. in the field, customs shed). However this does not assure that the material would arrive at the port of import free from any contamination. The General Notices of the USP 23–NF 18 under Foreign substances and impurities states "it is manifestly impossible to include in each monograph a test for every impurity, contaminant, or adulterant that might be present, including microbial contamination" (USP 23–NF 18, 1995).

In view of this, USP monographs would not be expected to specify tests for every known microorganism.

No one would disagree on the pharmacopeial requirements of the absence of pathogens such as *Salmonella* in the articles of commerce. However in regard to the total aerobic count, mould count and *E. coli* limits, the end use should dictate these rather than a single set of values for all botanicals. One can tolerate rather generous limits for the above organisms for botanicals that will undergo extensive processing such as boiling followed by extraction with solvents since the processing can be expected to reduce the bio-burden. Conversely where no processing other than boiling with water (e.g. senna leaves for tea preparation) is involved, relatively tighter limits may be called for. In regard to the determination of moulds count there are divergent views and there are arguments that microbiological methods alone are not satisfactory. Instrumental methods for the determination of aflatoxins (e.g. TLC, HPLC) are a *sine qua non* for testing botanicals since aflatoxins present in plant materials can cause health hazards if absorbed even in very small amounts. A more practical approach for the testing of aflatoxins would be to require it depending on the parts of the plant material used; for example subterranean organs such as rhizomes, roots, stolons are the ones more likely to be contaminated with aflatoxins than the aerial parts such as leaves, or flowers. Alternatively requirements of an aflatoxin test may be specified if the mould counts exceed certain limits irrespective of the plant parts used. In this regard Grady's comments are noteworthy "One possibility for testing for moulds would be a screening test based on a serological response to *Aspergillus* antigen, which could be done by contract laboratories and then, only if this were positive, one could look for aflatoxins" (Grady, 1994).

Given the fact that pharmacopeias have been giving due consideration to harmonization of standards for articles moving in the international commerce, the USP draft monograph on Ginkgo leaves proposed the following limits for microbial contaminants. In turn these limits have been proposed by the World Health Organization (WHO, 1995): total combined yeasts and moulds counts does not exceed 100/g, total bacterial count does not exceed 10,000/g, and meets the requirements for the absence of *Salmonella, E. coli, and Staphylococcus aureus*.

Quantitative Determination of "Active" Ingredients

Pharmacopeial monographs of synthetic organic medicinal substances or purified active principles of plant extracts, whose pharmacological activity is predictable, always include a quantitative determination for the active ingredient under *Assay*. The United States Pharmacopeia tends to use stability-indicating liquid chromatographic procedures involving comparison with a well-defined Reference Standard of known purity (usually 100%) which provide a stable accuracy system transferable to routine quality control work. Pharmacopeial *Assay* procedures, prior to adoption, are subject to extensive validation requirements-accuracy, specificity, precision, limit of detection, limit of quantitation, linearity, and range-outlined in the chapter *Validation of Compendial methods* (USP 23–NF 18, 1995). Such validation parameters have been and continue to be successfully used in the validation of pharmacopeial procedures containing multiple active ingredients (perhaps up to five) which are well-defined substances. In dealing with botanicals such as *Ginkgo biloba* leaves that contain multitudes of compounds – a mixture of flavonol glycosides, terpene trilactones, polysaccharides, etc. – a different approach is called for. This is necessitated by the fact that in many of the botanicals it is not known with certainty what the active ingredients responsible for the therapeutic properties are. Historical data on these have shown in most cases it is not a single compound that is responsible for the biological properties but a group of compounds – some of them may be known and others unknown – due perhaps to synergistic effects. Under these circumstances perhaps the best option open for a pharmacopeial monograph on *Ginkgo biloba* is to specify a procedure for the quantitative determination of the *Content of flavonol glycosides and terpene trilactones* since available evidence suggests the assignment of the pharmacological properties of Ginkgo to these groups of compounds. A more practical approach, albeit less desirable, here would be not to place too much emphasis on validation of the chosen procedure for all the parameters mentioned above that are a *sine qua non* for the Assay of synthetic organic medicinal substances. It is hoped that tests such as identification by either TLC, HPLC, or gas chromatography and colorimetric procedures would identify any decomposed and decomposition products thereby providing good quality control of the material under test.

In view of the different nature of the flavonol glycosides and terpene trilactones, two separate analytical procedures are called for. It is clear from studies reported in the literature that the Ginkgo leaves collected in the spring season contain the highest concentration of flavonol glycosides and a very low concentration of terpene trilactones. Leaves collected in the early fall season contain the lowest concentration of flavonol glycosides while the concentration of the terpene trilactones is the highest (Lobstein *et al.*, 1991; van Beek and Lelyveld, 1992). Pharmacopeial monographs would be expected to specify a minimum content for the ingredients which are serious candidates for the observed pharmacological properties and in some cases a minimum-maximum content where dose responses are known.

Content of flavonol glycosides

The general procedure, followed in the past, for the determination of flavonol glycosides in plant materials, involves hydrolysis of the glycosides to free aglycones which are subsequently complexed with aluminium chloride before spectrophotometric measurements. However this spectrophotometric procedure is not specific and quantitation of individual glycosides is not possible. It is reported that the spectrophotometric procedure when applied to *Ginkgo biloba* leaves was not reproducible, possibly due to the high concentration of proanthocyanidins (Sticher, 1993). This necessitated the search for a reproducible and specific procedure for the quantitative determination of at least the major glycosides – kaempferol, quercetin, and isorhamnetin in *Ginkgo biloba* leaves. In this connection the liquid chromatographic procedure validated and reported by Sticher *et al.* appeared to be the most promising one (Sticher *et al.*, 1990). It was thought this procedure with suitable modifications could serve the compendial requirement of an appropriate procedure and eventually be adopted in the draft. The procedure involves hydrolysis of glycosides to free aglycones which are injected into the HPLC equipped with a 370 nm UV detector and a column containing a packing of octadecylsilane chemically bonded to porous silica or ceramic micro-particles. Quantitation of kaempferol, quercetin and isorhamnetin in the sample under test is achieved by using quercetin as the reference standard. The choice of quercetin as the external reference standard is dictated by closeness of its molar absorptivity to those of the other two. Further quercetin can be expected to be made available for worldwide distribution by a USP Reference Standard program.

Content of terpene trilactones

Several HPLC procedures have been reported in the literature for the determination of ginkgolides and bilobalide using ultraviolet detection at 220 nm (Lobstein-Guth *et al.*, 1983; Pietta *et al.*, 1990). Some researchers have reported the use of refractive index detectors in LC procedures (Teng, 1988; Komoda *et al.*, 1988; van Beek *et al.*, 1991). A more attractive and perhaps a suitable method for routine quality control work appears to be the gas-chromatographic method that involves derivatization and use of a capillary column and flame-ionization detector (Hasler and Meier, 1992). In all these publications, ginkgolides and bilobalide isolated from the leaves are subjected to purification procedures prior to their injection into the chromatograph. Recently Camponovo *et al.* (1995) reported a quantitative determination of ginkgolides and bilobalide by using a liquid chromatograph coupled with an evaporative light scattering detector and a thermospray-mass spectrometer. All of the above methods use pure reference compounds of ginkgolides and bilobalide which are difficult to obtain commercially, are of limited purity and in addition very expensive and this therefore makes these procedures less attractive for compendial adoption. In this regard a more attractive procedure for the quantitative determination of terpene trilactones appears to be the one which uses nuclear magnetic resonance (NMR) (van Beek *et al.*, 1993). The NMR procedure has the advantage that no reference standards of terpene trilactones are required and is based on the comparison of the integral of each H-12 proton of the five Ginkgo terpene trilactones with that of the olefinic protons of maleic acid used as an internal standard. However a 200 MHz NMR spectrometer is very expensive and therefore is not likely to be a routine quality control instrument at the source of supply. At this point of time, it appears that there is no suitable analytical method, from the compendial perspective for the determination of terpene trilactones that could be considered for adoption. This may be an area, from the pharmacopeial perspective, for researchers in *Ginkgo biloba* to develop a suitable liquid chromatographic procedure using a refractive index detector and a readily available substance for comparison.

TOXIC IMPURITIES IN BOTANICAL MONOGRAPHS AND THEIR LIMITS

The pharmacopeial approach is to specify limits for any identified toxic impurity or impurities in individual monographs and botanical monographs will be no exception. However the botanicals present a totally different challenge to pharma-copeial committees. Toxic impurities may either be inherently present in the botanical under test or could conceivably come from contamination with other plant materials such as related species or due to treatment with chemicals such as ethylene oxide or pesticides. Risks due to the latter, one would hope, could be minimized to a great extent by following certain good harvesting practices, by minimizing the use of pesticides and herbicides and by appropriate identification – botanical and chemical – tests.

Pesticides Limits

In view of the enormous variety of pesticides used world-wide in the cultivation of botanicals and considering these pesticides are toxic, a limit for pesticides is con-sidered appropriate in all botanical monographs. Here a more practical approach would be to require testing for lists of pesticides depending on the country of origin in addition to those that are in use at the importing country. It is to be kept in mind that analysis of pesticides requires special skills and expensive instrumentation facil-ities. To minimize the analytical work load, efforts on the development of a list of world-wide banned pesticides and a list of only those pesticides that can be used should be speeded up.

Other Requirements

Pharmacopeial monographs on botanicals are specific in defining the article by stating the part of the plant (e.g. leaves). This therefore necessitates specifying a visual test and limits for excluding contamination with other parts of the plant such as twigs, stems etc. as well as other extraneous matter under *Foreign organic matter*. In addition other pharmacopeial tests such as *Residue on ignition, Heavy metals etc.* are essential to monitor contamination during harvesting and/or processing.

CONCLUSION

Development of modern pharmacopeial monographs for botanicals poses the great-est challenge to the USP Committee of Revision and certainly dwarfs the issues and efforts involved in the development of monographs on modern synthetic organic medicinal compounds. The fact that the USP Committee of Revision could develop a draft monograph on *Ginkgo biloba* leaf attests to the availability in the literature of research publications on the chemical identification and quantitative determination of certain key chemical constituents present in it. The aim of protecting the public health though development and establishment of a pharmacopeial monograph would be hard to achieve without the contributions of those scientists, who have contributed to our understanding of the profile of *Ginkgo biloba* leaf, and the work of the USP Committee of Revision in reviewing those publications and arriving at appropriate specifications and test procedures for regulatory control. The task of establishing standards for *Ginkgo biloba* leaf has just begun with the development of a draft monograph. Since USP has a continuous revision policy, as new research evidence surfaces and new analytical techniques are developed for better assessment of the quality of *Ginkgo biloba*, one can expect the USP Committee of Revision to deal with the situation appropriately.

ACKNOWLEDGMENTS

The author wishes to express his gratitude to Dr. Lee T. Grady and Dr. Claudia C. Okeke for their valuable comments and constructive criticism during the preparation of the manuscript.

REFERENCES

Beek, T.A. van, Scheeren, H.A., Rantio, T., Melger, W. Ch. and Lelyveld, G.P. (1991) Determination of ginkgolides and bilobalide in *Ginkgo biloba* leaves and phytopharmaceuticals. *J. Chromatogr.*, 543, 375–387.

Beek, T.A. van and Lelyveld, G.P. (1992) Concentration of ginkgolides and bilobalide in *Ginkgo biloba* leaves in relation to the time of year. *Planta Med.*, 58, 413–416.

Beek, T.A. van and Lelyveld, G.P. (1993) Thin-layer chromatography of bilobalide and ginkgolides A, B, C and J on sodium acetate impregnated silica gel. *Phytochem. Anal.*, 4, 109–114.

Camponovo, F.F., Wolfender, J.-L., Mailard, M.P., Potterat, O. and Hostettmann, K. (1995) Evaporative light scattering and thermospray mass spectrometry: two alternative methods for detection and quantitative liquid chromatographic determination of ginkgolides and bilobalide in *Ginkgo biloba* leaf extracts and phytopharmaceuticals. *Phytochem. Anal.*, 6, 141–148.

Ginkgo biloba leaf-draft monograph (1997) *Pharmacopeial Forum* 23(2) [Mar.–Apr.], 3644–3647.

Grady, L.T. (1994) Worldwide harmonization of botanical standards. A Pharmacopeial view. Lecture presented at the Conference on Botanicals in American Healthcare, December 15–17 (1994) Washington, D.C.

Hasler, A. and Meier, B. (1992) Determination of terpenes from *Ginkgo biloba* L. by capillary gas chromatography. *Pharm. Pharmacol. Lett.*, 2, 187–190.

Komoda, Y., Nakamura, H. and Uchida, M. (1988) HPLC analysis of ginkgolides by using a differential refractometer. *Iyo Kizai Kenkyusho Hokoku*, 22, 83–85.

Lobstein-Guth, A., Briançon-Scheid, F. and Anton, R. (1983) Analysis of terpenes from *Ginkgo biloba* L. by high-performance liquid chromatography. *J. Chromatogr.*, 267, 431–438.

Lobstein, A., Rietsch-Jako, L., Haag-Berrurier, M. and Anton, R. (1991) Seasonal variations of the flavonoid content from *Ginkgo biloba* leaves. *Planta Med.*, 57, 430–433.

Okabe, K., Yamada, K., Yamamura, S. and Takada, S. (1967) Ginkgolides. *J. Chem. Soc.*, 2201–2206.

Pietta, P.G., Mauri, P.L. and Rava, A. (1988) Reversed-phase high performance liquid chromatographic method for the analysis of biflavones in *Ginkgo biloba* L. extracts. *J. Chromatogr.*, 437, 453–456 and references cited therein.

Pietta, P.G., Mauri, P.L. and Rava, A. (1990) Analysis of terpenes from *Ginkgo biloba* L. extracts by reverse-phased high-performance liquid chromatography. *Chromatographia*, **29**, 251–253.

Schilcher H. (1988) *Ginkgo biloba* L. Untersuchungen zur Qualitaet, Wirkung, Wirksamkeit und Unbedenklichkeit. *Z. Phytother.*, **9**, 119–127.

Schorn, P.J. (1993) Identification of herbal form, derived drugs and formulations. *Pharm. Industrie*, **55**, 268–271.

Sticher, O., Hasler, A. and Meier, B. (1990) High-performance liquid chromatographic determination of five widespread flavonoid aglycones. *J. Chromatogr.*, **508**, 236–240.

Sticher, O. (1993) Quality of Ginkgo preparations. *Planta Med.*, **59**, 2–11.

Tallevi, S.G., and Kurtz W.G.W. (1991) Detection of ginkgolides by thin-layer chromatography. *J. Nat. Prod.*, **54**, 624–625.

Teng, B.P. (1988) Chemistry of ginkgolides. In P. Braquet (ed.), *Ginkgolides – Chemistry, Biology, Pharmacology and Clinical Perspectives*, **Vol. 1**, J.R. Prous Science publishers, Barcelona, pp. 37–41.

United States Pharmacopeia 23–National Formulary 18 (1995) Published by the United States Pharmacopeial Convention Inc., 12601 Twinbrook Parkway, Rockville, MD 20852.

Weinges, K., Hepp, M. and Jaggy, H. (1987) Isolierung und Strukturaufklärung eines neuen Ginkgolids. *Liebigs Ann. Chem.*, 521–526.

APPENDIX

Ginkgo Leaf.* This new monograph is based on information retrieved from a survey of the literature conducted on various online databases, such as *Chemical Abstracts* and *Medicine*. The liquid chromatographic procedure used in the test for *Content of flavone glycosides* was described by Otto Sticher *et al.* in the *Journal of Chromatography*, Vol. 508 (1990). The method is based on analyses performed with the Hypersil ODS brand of L1 column. Typical retention times reported for quercetin, kaempferol, and isorhamnetin are about 6. 9.5, and 10.5 minutes, respectively.

Ginkgo Leaf*

<<Ginkgo leaf consists of the dried leaf of *Ginkgo biloba* Linn (Fam. Ginkgoaceae).

Packaging and storage—Preserve in well-closed containers, protected from light and moisture.

Labeling—The label states that it is Ginkgo Leaf and states also the genus and species.

USP Reference standards—<11> *USP Quercetin RS, USP Rutin RS*.

Botanic characteristics—

Macroscopic: Dried whole, folded or fragmented leaves, with or without attached petiole, varying from khaki green to greenish-brown in color, often more brown at the apical edge, and darker on the adaxial surface. Lamina broadly obcuneate (fan shaped), 2 to 12 cm in width and 2 to 9.5 cm in length from petiole to apical margin; mostly 1.5 to 2 times wider than long. Base margins entire, concave; apical margin sinuate, usually truncate or centrally cleft, and rarely multiple cleft. Surface glabrous, with wrinkled appearance due to prominent dichotomous venation appearing parallel and extending from the lamina base to the apical margin. Petiole of a similar color to leaf, channeled on the adaxial surface, 2 to 8 cm in length.

Histology—

Transverse section of lamina: A thin but marked cuticle occurs over a single layer of epidermal cells on both surfaces. Stomata are present on the lower surface only, with guard cells sunken with respect to adjacent epidermal cells. Palisade elements, elongated at right angles to the surface and often irregular in appearance occur just below the upper epidermis. Vascular bundles occur at intervals along the width of the blade, with adjacent cluster crystals of calcium oxalate. Cells of the mesophyll are smaller than the palisade cells, elongated parallel to the leaf surface and separated by large intercellular spaces.

Powdered lamina and petiole: Under the microscope, transverse fragments of the leaf display a smooth cuticle, present on both leaf surfaces and staining pinkish-orange with Sudan III TS. In surface view, cells of the upper epidermis are elongated and wavy-walled, with abundant yellow droplets 2 to 12 μm in diameter visible in mature and old leaves but not in young leaves. Cells of the lower epidermis are similar in shape but have straighter walls and are interrupted by anisocytic stomata. Numerous lignified elements derived from the lamina and petiole are present, including xylem vessels with annular thickening, tracheids and vessels with bordered pits. The extent of lignification, particularly in the petiole, increases with age of leaf. Calcium oxalate crystals are numerous, present scattered or associated with vessels, ranging in size from 5 to 50 μm in young leaves to 15 to 100 μm in mature leaves. Under crossed polaroids, numerous smaller prism- or tear-shaped shiny features of indeterminate nature may be present. Very occasional, highly elongated, uniseriate, covering trichomes with no obvious cross walls and smooth or warty surfaces may be seen. Mature leaves may show the presence of very rare, polygonal to circular starch granules approximately 20 μm in diameter with a central hilum and exhibiting a marked maltese cross under crossed polaroids.

Identification—

A: Transfer 0.2 g of finely powdered Ginkgo Leaf to a test tube, add 10 mL of methanol, and heat on a water bath at 65° for 10 minutes. Shake the mixture frequently during the heating. Allow to cool to room temperature, filter, concentrate the filtrate on a hot water bath at 60° to half its volume, and cool. Apply separately, as bands, 20 μL each of the test solution and a Standard solution of USP Rutin TS and a chlorogenic acid in methanol containing about 0.6 mg per mL and 0.2 mg per mL, respectively, to a suitable thin-layer chromatographic plate (see *Chromato-*

graphy <621>) coated with a 0.25-mm layer of chromatographic silica gel, and allow the bands to dry. Develop the chromatograms in a mixture of ethyl acetate, water, anhydrous formic acid, and glacial acetic acid (67.5:17.5:7.5:7.5) until the solvent front has moved about 10 cm from the origin. Remove the plate from the chromatographic chamber, and dry it in a circulating air oven at 100° to 105°. Immediately spray the hot plate with a solution of diphenyl boryloxyethylamine in methanol containing 10 mg per mL, then spray with a solution of polyethylene glycol 400 in alcohol containing 50 mg per mL. Allow the plate to cool for 30 minutes, and examine it under long-wavelength ultraviolet light. The chromatogram of the Standard solution shows in its middle part, with increasing R_f values, the yellowish brown fluorescent zone due to rutin and a light blue fluorescent zone due to chlorogenic acid as well as two greenish brown-yellow fluorescent zones, located above. Other, less intense zones may be seen in the chromatogram of the test solution.

  B: Prepare a suitable thin-layer chromatographic plate (see *Chromatography* <621>) coated with a 0.50-mm layer of chromatographic silica gel as follows. Immerse the plate for 20 seconds in a solution of sodium acetate in methanol containing 1 g per 10 mL. Allow the excess coating liquid to drip from the plate, and dry in a forced-air oven at 70° for 30 minutes. Cool in a dessicator. Transfer 0.8 g of the dried test specimen retained from the test for *Loss on drying* to a suitable flask fitted with a reflux condenser, add 5 mL of a mixture of methanol and water (1 in 10), and heat under reflux for 15 minutes. While still hot, filter the contents of the flask with the aid of vacuum. Rinse the flask and the test specimen with 2 mL of a mixture of methanol and water (2 in 100), and transfer the rinsings to the filter with the aid of vacuum. Return the powdered Ginkgo Leaf to the flask, add 4 mL of a mixture of methanol and water (1 in 10), and repeat the extraction. After filtration, wash the residue of powdered Ginkgo Leaf twice with 1.5 mL of a mixture of methanol and water (2 in 100). Combine the filtrates, and transfer the combined filtrates (about 12 mL) to a solid-phase extraction column containing L1 packing with a sorbent mass-to-column volume ratio of 1000 mg per 3 mL or equivalent. [NOTE—Initially pass 10 mL of methanol and then 10 mL of a mixture of methanol and water (2 in 100) through the column to condition it, Do not allow the column to dry.] Collect the eluate at the rate of 1 drop per second. Evaporate the eluate to dryness, and dissolve the residue in 2 mL of methanol. Separately apply several 10-μL spots of the test solution to the impregnated plate, and allow the spots to air-dry. Develop the plate in a mixture of ethyl acetate and methyl acetate (1:1) in a chromatographic chamber without filter paper attached to the walls until the solvent front travels about three-fourths of the length of the plate. Remove the plate from the chamber, mark the solvent front, and dry in an oven at 105° for 15 minutes. Spray the plate with acetic anhydride, and heat in an oven at 140° for 25 minutes. Cool, and examine the plate under short- and long-wavelength ultraviolet light. [NOTE—The compounds present in high concentrations may be visible in daylight as light brown spots.] The presence of terpene lactones in the test solution is shown by the following spots detected in the chromatogram at both the short and long wavelengths: bilobalide (R_f about 0.75), ginkgolide A (R_f

about 0.68), ginkgolide B (R_f about 0.52), ginkgolide J (R_f about 0.39), and ginkgolide C (R_f about 0.27). Other spots of varying intensities also may be seen.

Stems and other foreign organic matter <561>: not more than 3.0% of stems and not more than 2.0% of other foreign organic matter.

Loss on drying <731>—Dry 1.0 g of finely powdered Ginkgo Leaf at 105° for 2 hours: it loses not more than 11.0% of its weight. Reserve the dried test specimen for use in *Identification* test B.

Total ash <561>: not more than 11.0% determined on 1.0 g of finely powdered Ginkgo Leaf.

Heavy metals <231>— [To come]

Pesticide residues— [To come]

Microbial limits <61>—The total bacterial count does not exceed 10,000 per g, the total combined molds and yeasts count does not exceed 100 per g, and it meets the requirements of the tests for absence of *Salmonella* species, *Escherichia coli*, and *Staphylococcus aureus*.

Content of flavonol glycosides—
   Extraction solvent—Prepare a 60% (w/w) solution of acetone in water.
   Hydrochloric acid solution—Transfer 10.0 mL of hydrochloric acid to a 50-mL volumetric flask, dilute with water to volume, and mix.
   Mobile phase—Prepare a mixture of citric acid solution (0.6 in 100), acetonitrile, and isopropyl alcohol (100:47:5). Make adjustments if necessary (see *System suitability* under *Chromatography* <621>).
   Standard solutions—Dissolve an accurately weighed quantity of USP Quercetin RS in methanol to obtain a solution having a known concentration of about 0.8 mg per mL (*Standard stock solution*). To three separate 100-mL volumetric flasks, transfer 3.0 mL, 5.0 mL, and 10.0 mL of *Standard stock solution*, and add 10 mL of water followed by 30 mL of *Hydrochloric acid solution* to each flask. Dilute the contents of each flask with methanol to volume, and mix to obtain *Standard solutions A, B,* and *C* having known concentrations of 0.024 mg, 0.04 mg, and 0.08 mg per mL, respectively.
   Test solution—Transfer about 2.5 g of Ginkgo Leaf, finely powdered and accurately weighed, to a suitable flask fitted with a reflux condenser. Add 50 mL of *Extraction solvent*, and heat on a hot water bath, under reflux, for 30 minutes. Allow to cool, filter, and collect the filtrate in a 100-mL volumetric flask. Extract the residue on the filter a second time in the same manner using 40 mL of *Extraction solvent*, and collect the filtrate in the same 100-mL volumetric flask. Dilute the contents of the flask with *Extraction solvent* to volume, and mix. Evaporate 50.0 mL of the solution to dryness, and transfer the residue to a 50-mL volumetric flask with

the aid of 30 mL of methanol. Add 4.4 mL of *Hydrochloric acid solution*, dilute with water to volume, mix and centrifuge. Transfer 10.0 mL of the supernatant liquid to a suitable amber-glass vial, close with a rubber seal and an aluminum cap, heat in a boiling water bath for 25 minutes and allow to cool to room temperature.

Chromatographic system (see *Chromatography* <621>)—The liquid chromatograph is equipped with a 370-nm detector and a 4.6-mm × 12.5-cm column that contains packing L1. The flow rate is about 1.5 mL per minute. Chromatograph *Standard solution A*, and record the peak responses as directed under *Procedure*: the relative standard deviation for replicate injections is not more than 2.0%.

Procedure—Separately inject equal volumes (about 10 μL) of each of the *Standard solutions* and the *Test solution* into the chromatograph, record the chromatograms, measure the areas of the responses for the major peaks, and add the peak areas of all the major peaks due to quercetin, kaempferol and isorhamnetin in the chromatogram of the *Test solution*. The relative retention times of the flavone glycosides of interest are about 1.0 for quercetin, 1.6 for kaempferol, and 1.7 for isorhamnetin. [Note—Isorhamnetin sometimes co-elutes with kaempferol]. Plot the peak areas of responses of the major peaks of the *Standard solutions* versus concentrations, in mg per mL, of USP Quercetin RS and draw the straight line best fitting the three plotted points. From the graph so obtained determine the concentration, in mg per mL, of flavone glycosides in the *Test solution*. Multiply the values obtained by a mean molecular mass factor of 2.514 and calculate the percentage of flavonol glycosides, as quercetin, with a mean molecular mass of 756.7: not less than 0.8% of flavone glycosides is found.

14. INDUSTRIAL QUALITY CONTROL OF GINKGO PRODUCTS

FABRIZIO F. CAMPONOVO and [1] F. SOLDATI[1]

[1]*Research & Development, Pharmaton SA, 6903-Lugano, Switzerland*

INTRODUCTION

The marketing of pharmaceutical products in Europe requires an authorisation, according to the directives 65/65 EEC and 75/318 EEC. Phytomedicines are included in this general legislation, and are subjected to the same treatment as medicinal products in general. Thus, evidence of the product's safety, efficacy and quality has to be submitted before a marketing authorisation will be granted.

In this chapter the focus will be on the parameters and the analytical methods to evaluate the quality of *Ginkgo biloba* leaves extracts and finished products. Earlier experience gained at the University of Lausanne (Camponovo, 1996) and at Pharmaton for the standardisation of GK 501 *Ginkgo biloba* extract Pharmaton and for the finished product Gincosan® and Memfit® Pharmaton, plays an important role in this.

CONTROL OF STARTING MATERIAL (*Ginkgo biloba* leaves)

A consistent quality for products of vegetable origin can only be guaranteed if the starting materials are rigorously defined. But what does a good quality of Ginkgo leaves mean? So far, there are no pharmacopoeia monographs on *Ginkgo biloba* leaf to which one can refer. Each producer supplies its own monograph with some specifications that are more or less detailed and dependent on the technical capacity of the supplier.

From a medical point of view, one can assume that the quality of *Ginkgo biloba* leaves, and generally of all vegetal raw materials, depend on the content of markers/active substances and on the absence of contaminants and toxic secondary metabolites. In *Ginkgo biloba*, the active and toxic substances are well known and as will be seen, the requirements for high quality leaves are well defined and accepted by the whole scientific community. The secondary metabolite content of Ginkgo leaves, depends on the site of collection as well as the time of harvesting and the stage of growth of the plant. Little changes on these parameters can drastically influence and modify the ratio of compounds present in the plant and consequently change the final quality of the extract. Thus, a main point to assure the consistency of the quality of the vegetal material, is the standardisation of these parameters.

According to the literature the terpene trilactones show the biggest variations between the two main classes of active compounds. Van Beek *et al.* (1991) have found that the content of single ginkgolides can vary by a factor of more than 100 between leaves having a different origin. The flavonoids, an other class of active compounds, have shown a smaller variation (Lobstein *et al.*, 1991; Hasler *et al.*, 1992). Hasler *et al.* (1990) investigated the flavonol glycoside content in different batches of leaves obtained from the market and they found a content varying from 0.5 to 1% w/w. Therefore, to avoid serious problems with the standardisation of the final extract, control of the amounts of these compounds in Ginkgo leaves is an important step before starting the extraction procedure. However, from experience it is known that if the growth parameters are well controlled and the leaves are harvested yearly from cultivated plants in a specific place, the variation in concentration of active substances is not a critical issue (Camponovo and Soldati, unpublished results). Usually, *Ginkgo biloba* leaves are harvested from July to September, when the leaves are completely developed and just before the green colour begins to turn yellow.

In 1991, Pharmaton investigated the variation of the content of terpene trilactones and flavonoids, during the harvesting period, in leaves of male and female trees. The leaves used for the study were collected from four male and female trees at Peace Memorial Park in Hiroshima by Professor K. Yamasaki. The results reported in Figures 1 to 4 show significant differences between male and female trees especially

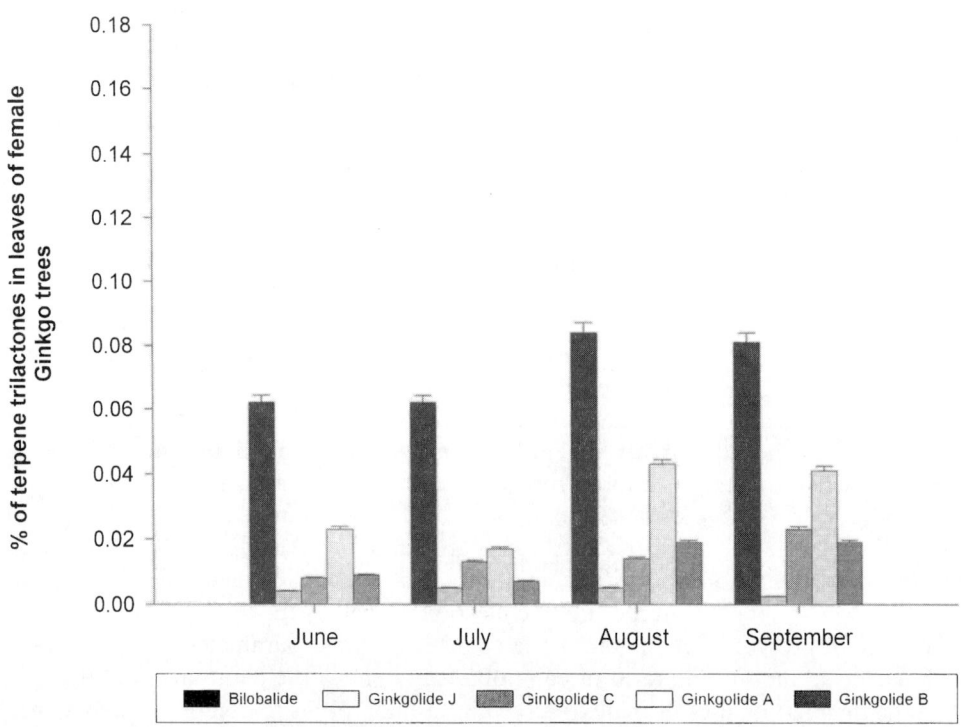

Figure 1   Concentration of terpene trilactones in leaves of female *Ginkgo biloba* trees as a function of the harvest time (Camponovo and Soldati, unpublished).

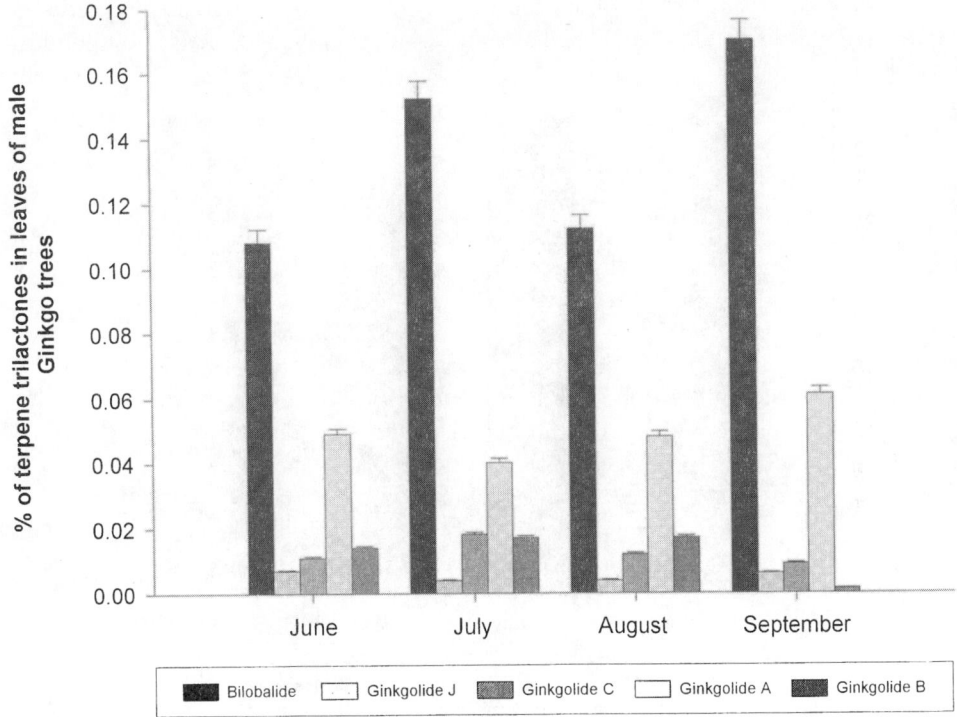

Figure 2 Concentration of terpene trilactones in leaves of male *Ginkgo biloba* trees as a function of the harvest time (Camponovo and Soldati, unpublished).

in the ratio between individual terpenes. This means that even the sex of the trees has to be taken into account during the collection of the leaves for the manufacturing of the extract. In the leaves collected from Ginkgo trees of the same gender, however, only little variation was observed during the harvesting period from July to September.

Identification

The authentication of the vegetal raw material is an important step, therefore, in addition to the macroscopic and microscopic examination, a chromatographic identification of characteristic substances has to be performed. The relevant macroscopic and histological characteristics are described in the USP pharmacopoeia monograph preview on Ginkgo Leaf (1997).

The terpene trilactones and the main Ginkgo flavonol glycosides can be used as characteristic substances for a chromatographic identification of the drug. The identification can be performed by means of published TLC or HPLC methods (Hasler and Sticher, 1990; Lobstein *et al.*, 1991; Tallevi and Kurz, 1991; Hasler *et al.*, 1992; van Beek and Lelyveld, 1993; Camponovo, 1996).

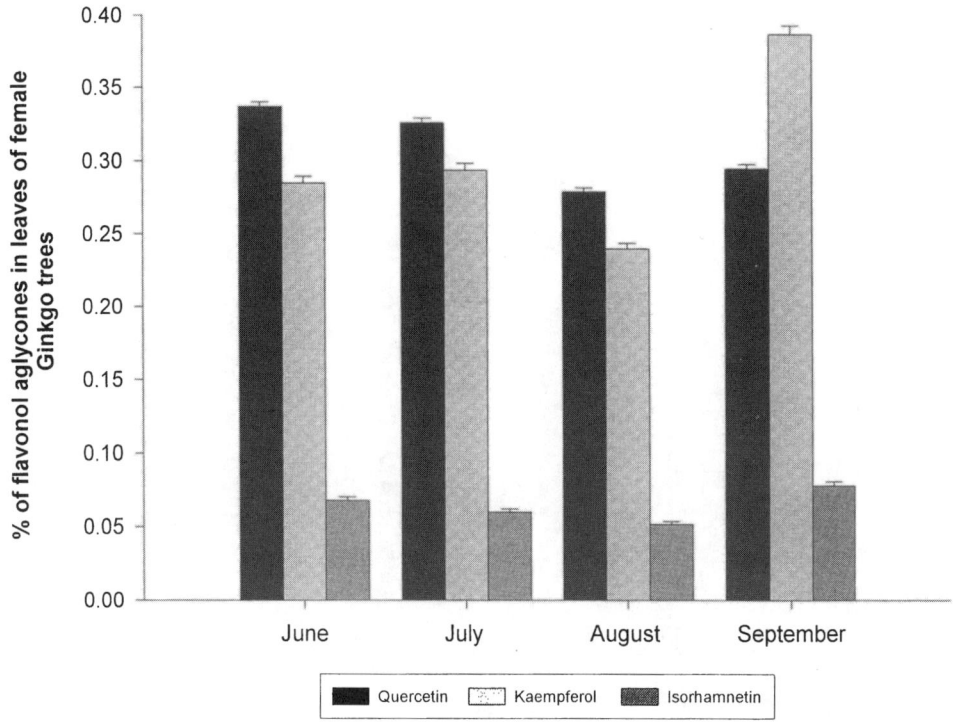

Figure 3   Concentration of flavonol aglycones in hydrolysed leaf extracts of female *Ginkgo biloba* trees as a function of the harvest time (Camponovo and Soldati, unpublished).

Quantitative Determination of Markers/Active Substances

The identification of the drug is followed by a quantitative determination of the known active constituents. In the case of *Ginkgo biloba* leaves, the total amount of flavonoids, ginkgolides and bilobalide has to be quantified. The analytical methods that can be used will be briefly discussed in the section dealing with the analysis of extracts.

Foreign Matter

The plant material should neither present any pathogenic organism, nor any possible extraneous materials, e.g. micro-organisms, undesirable parts of plants, insects, animal excreta and so on. The presence of extraneous materials can be checked according to Ph. Eur. III, 2.8.2.

Determination of Ash

The aim of this test is the determination of the amount of material present after ignition. This remaining material is partially derived from the plant tissue itself but also, possibly from extraneous matters (e.g. sand and soil).

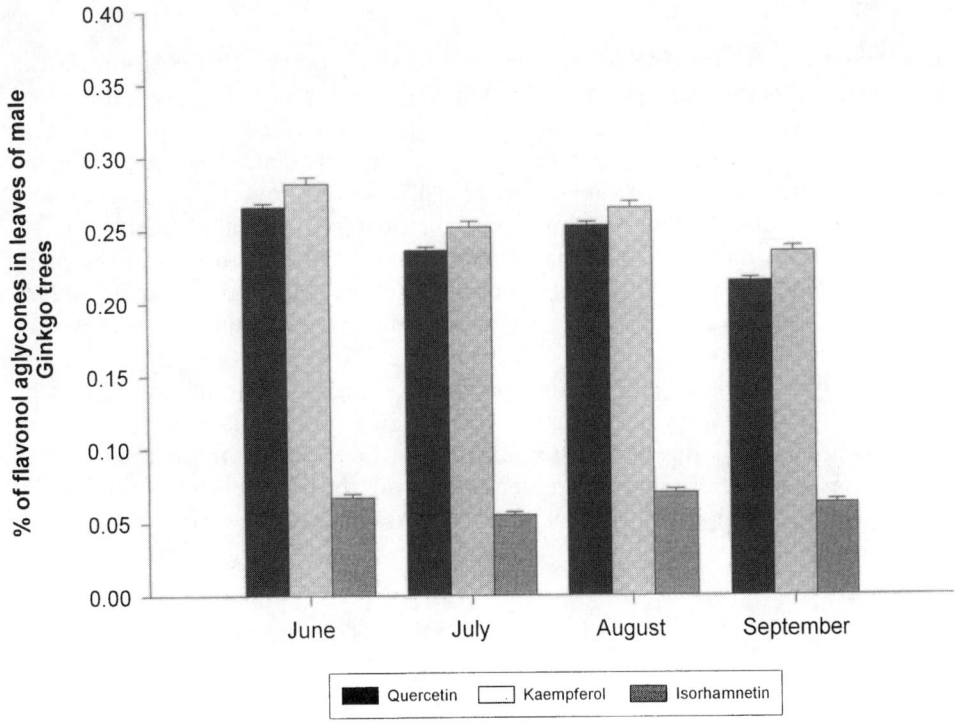

Figure 4 Concentration of flavonol aglycones in hydrolysed leaf extracts of male *Ginkgo biloba* trees as a function of the harvest time (Camponovo and Soldati, unpublished).

For Ginkgo leaves the sulphated ash can be determined according to the method of Ph. Eur. III, 2.4.14.

Loss on Drying

This test determines both water and volatile matters. According to the USP pharmacopoeia preview on Ginkgo leaf (1997) it can be determined by weight after 2 hours at 105° C.

Test for Aflatoxins

This test is designated to detect the possible presence of aflatoxins B_1, B_2, G_1 and G_2, which are highly toxic contaminants. The analytical method is described in the section reporting on the analysis of the intermediate product (the extract of leaves). The test on the plant material is very important, as any deteriorated raw material can be eliminated before starting the manufacturing procedure, thus avoiding a waste of time and money.

Pesticide Residues

The analysis of the pesticide residues is another important test to perform on the vegetal material before starting the manufacturing procedure. The analytical method used, and the classes of pesticides analysed, will be described later, in the section dealing with the analysis of the intermediate product. Experience has shown that the amount of pesticides found in Ginkgo leaves is always very low (Figini, unpublished results). This is not surprising because due to the extreme resistance of Ginkgo trees, the cultivators do not need to use large amounts of pesticides to protect the plant from insects, moulds or fungi. Furthermore, the leaves in contrast to other organs change every year and the risk to have an accumulation of contaminants is very low.

Other tests, such as heavy metals and microbiological contamination, can be performed but if the aim is the production of an extract it is more convenient to perform these tests directly on the extract. In fact, during the manufacturing procedure, some contaminants are removed and the possible microbiological contamination is eliminated by the organic solvents used for the extractions.

CONTROL TESTS ON INTERMEDIATE PRODUCT (extract of *Ginkgo biloba* leaves)

Only one official monograph on a standardised extract of *Ginkgo biloba* leaves is available so far. It was drawn up in 1994 by the German Commission E ("Bundesanzeiger Nr. 133", Vol. 46, July 1994 (pages 413–415). This monograph contains only a few analytical specifications and these are practically the same as those of EGb 761 and GK 501 extracts:

> 22–24% of Ginkgo flavonol glycosides
> 5–7% of terpene trilactones (of which 2.8–3.4% of ginkgolides A, B, C
> and 2.6–3.2% of bilobalide)
> < 5 ppm of ginkgolic acid

The routine methods for the quality assessment of the extract have to be chosen according to new developments in analytical chemistry but should also consider routine laboratory feasibility.

In this chapter the methods that according to experience (Camponovo and Soldati, unpublished) should be adopted for the quality control of extracts and the requirements for the release of products will be briefly described.

Identification of Markers/Active Substances

HPLC and TLC are two adequate techniques for the identification of the main active principles and some markers. HPLC can also be used for quantitative analyses and it will be discussed in the next section.

The identification by TLC is particularly interesting because it is a simple and inexpensive technique that gives good results. The same procedures used for the identification of leaves can be adopted for the extract.

Quantitative Analysis of Active Substances and Other Markers

Ginkgo flavonol glycosides

In principle, a direct quantification of naturally occurring active principles would be desirable. But this often is not possible because most reference standards are not commercially available. Additionally, in the case of Ginkgo, the flavonoid profile is very complex; thus the quantification of each flavonoid cannot be performed routinely. However, according to Hasler et al. (1990) after acidic hydrolysis of the extract, the 3 main aglycones are obtained that can be easily quantified and correlated with the total flavonol glycoside content.

In practice, the work-up procedure consists of two steps. First, the Ginkgo flavonol glycosides contained in the standardised GK 501 Ginkgo biloba extract are extracted and hydrolysed with a mixture of methanol / hydrochloric acid. Then the free aglycones quercetin, kaempferol and isorhamnetin are quantified by HPLC against an external standard, after separation with an isocratic system on a reverse phase column. The aglycones content is then converted into Ginkgo flavonol glycosides content using the ratio between the corresponding molecular weights. The detection is performed at 370 nm and in less than 20 min the peaks corresponding to the 3 main aglycones can be separated and integrated without interference of the matrix (Fig. 5).

This analytical procedure which is simple, rapid and reproducible, is especially advantageous for routine analyses in quality control and stability tests of phytomedicines. The absence of tedious sample preparation and the isocratic chromatographic conditions allow checking the quality of several different batches in just one day.

Terpene trilactones

Due to the poor UV-absorption and the low concentration, the analysis of terpene trilactones is much more difficult than of flavonoids. In the literature, only few quantitative methods have been described. Most of them are based on an HPLC separation with UV or RI detection (Lobstein-Guth et al., 1983; Teng, 1988; Komoda et al., 1988; Pietta et al., 1990; van Beek et al., 1991; Pietta et al., 1992). Ginkgolides have also been analysed after derivatisation using GC-FID and GC-MS (Chauret et al., 1991; Hasler and Meier, 1992; Huh and Staba, 1993). In addition, biological standardisation using the PAF-antagonist properties of ginkgolides (Steinke et al., 1993) and ^{1}H-NMR (van Beek et al., 1993) have been proposed.

In general all the methods cited above need a tedious sample preparation consisting of a derivatisation or a purification procedure that is not convenient for routine analysis. Recently, Camponovo et al. (1995) have proposed two new methods based on HPLC with MS and Evaporative Light Scattering Detection (ELSD) that permit

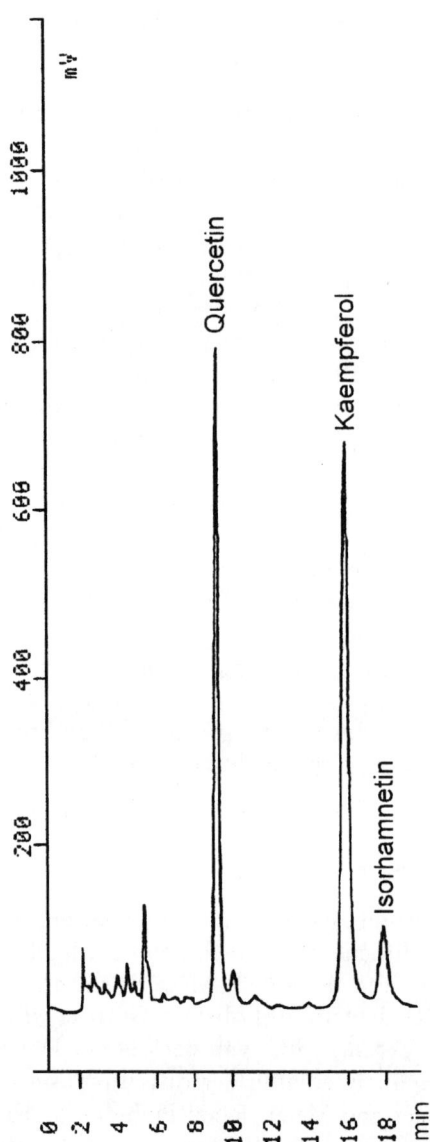

Figure 5   HPLC-UV determination of flavonol aglycones in the standardised GK 501 *Ginkgo biloba* extract. Column: Shim-Pack® CLC-ODS (RP C18, 150 × 6 mm I.D., Shimadzu), eluent: methanol–water–glacial acetic acid (50:50:3), flow rate: 1.5 ml/min, detection: UV 370 nm.

the analysis of *Ginkgo biloba* leaves and phytopharmaceuticals after a relatively simple and inexpensive sample preparation.

The LC-MS method appears to be particularly suitable for rapid and sensitive qualitative and semi-quantitative control of Ginkgo extracts. However, due to the price of this sophisticated apparatus it is not suitable for routine analyses. In contrast, the low cost, accuracy and reproducibility of the results and the drastic simplification of sample preparation makes the LC-ELSD procedure, highly promising for routine quality control quantitation of ginkgolides and bilobalide in Ginkgo leaves and phytopharmaceuticals.

Actually, following the Pharmaton procedure for the quantification of terpene trilactones in the standardised GK 501 *Gingko biloba* extract, a method based on an HPLC separation and an RI detection is used. The separation of the terpenes occurs on a reverse phase column kept at 45° C, after a simple purification of the sample on a silica gel cartridge. The quantification is performed against a standard GK 501 extract with a known content of bilobalide and ginkgolides A, B, C (Fig. 6). The same method, but based on the quantification of terpene trilactones in the extract against the single pure terpenes, is used to titrate the working standard GK 501 extract.

Analysis of Contaminants

The control of contaminants is an important factor to guarantee the safety of phytomedicines. The plant extract used in the manufacturing of the finished product has to be tested for all possible contaminants. As a general rule, a vegetable drug must not be subjected to ethylene oxide sterilisation or to treatment with ionising radiation. If all suppliers of the drug have been expressly informed of this requirement, a test for ethylene oxide and its reaction products is not necessary. Usually, the following possible contaminants are routinely checked:

Ginkgol and ginkgolic acid

These resorcinol derivatives are responsible for strong allergies and contact dermatitis (Courio, 1986; Lepoittevin *et al.*, 1989). As far as the safety is concerned, these alkylphenols must be practically absent in the commercial extract. For that purpose the manufacturers have developed different special procedures to provide an extract substantially free of alkylphenols.

In the literature some methods for the analysis of resorcinol derivatives (Cheng-Yu *et al.*, 1980; Kubo *et al.*, 1986; Tyman *et al.*, 1989; Nagabhushana and Ravindranath, 1995) can be found but not one is specific for extracts *of Ginkgo biloba* leaves. Following a Pharmaton internal procedure, the test for these potential impurities is applied as a limit test using an HPLC method especially developed and validated for the standardised GK 501 *Ginkgo biloba* extract. Ginkgol and ginkgolic acid, are extracted with n-hexane, separated by HPLC on a RP8 column and detected at 210 nm. The detection limit is approximately 1 ppm and the impurity content is expressed as total amount of ginkgol and ginkgolic acid. The efficacy of

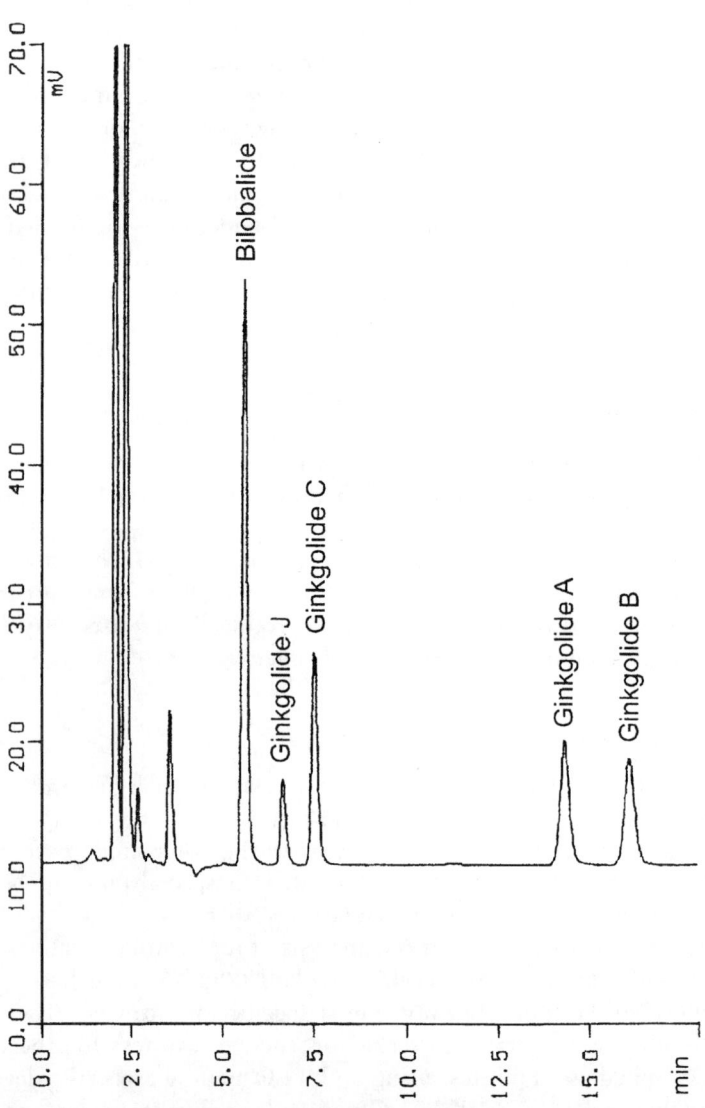

Figure 6   HPLC-RI determination of terpene trilactones in the standardised GK 501 *Ginkgo biloba* extract. Column: Supelcosil® (RP C18, 150 × 4.6 mm I.D., Supelco), eluent: methanol–water (28:72), flow rate: 1 ml/min, detection: RI.

the extraction procedure for the impurity to be analysed has been checked through a quantitative determination of a spiked blank sample. Pure reference substances of ginkgol and ginkgolic acid, needed for this analysis, are difficult to isolate and commercially not available. Nevertheless, they are quite stable and can be stored for several years under vacuum at 5° C.

Heavy metals

Heavy metals are in widespread industrial use and, when released into the air or into rivers, distort the naturally occurring distribution of metals. Plants extract elements from the soil in which they grow, and they can concentrate these undesirable trace elements up to toxic levels. Heavy metals in this context mean: mercury, cadmium and lead in the first place. In a broader sense, other toxic elements as arsenic (coming from certain pesticides) and barium are included. Determination of heavy metals is generally performed by atomic absorption spectrophotometry after acid digestion of the sample.

From experience it is known that, in a *Ginkgo biloba* extract less than 20 ppm of total heavy metals occurs (Camponovo and Soldati, unpublished).

Microorganisms

Vegetal materials normally carry a great number of bacteria and moulds coming from the soil and from the naturally occurring microflora of plants. During the manufacturing procedure with organic solvents, these organisms are practically totally destroyed, but during the storage of the final extract they can restart to grow. For this reason, the microbiological purity of the standardised extract must be checked just before the manufacturing of the finished product. The specifications fixed for the extract GK 501 are in conformity with the requirements of Ph. Eur. III, 5.1.4., category 3A. The method is described in Ph. Eur. III, 2.6.12. (plate count).

Aflatoxins

Among the various mycotoxins, the aflatoxins have been subject of the most intensive research because of the high toxicity of aflatoxin B1. They are produced by *Aspergillus flavus*, a mould that grows on improperly stored crops and food products. Precautions to help the control of aflatoxin involve the prevention of mould formation by proper drying and storage of crops, removal of damaged material before storage or processing, and providing adequate moisture and humidity control of stored material. Generally, aflatoxins occur in susceptible crops as a mixture of aflatoxins B_1, B_2, G_1, and G_2, with only aflatoxins B_1 and G_1 demonstrating carcinogenicity (Kotsonis *et al.*, 1996).

In accordance with the German requirements of the "Aflatoxin-Verordnung", dated November 30, 1976 and modified on May 6, 1991, a limit of 4 μg/kg for the total aflatoxins B_1, B_2, G_1, G_2 and 2 μg/kg for aflatoxin B_1 that is the most hepatocarcinogenic can be set.

A quantitative determination of these compounds based on an HPLC method has been developed and validated for the standardised GK 501 *Ginkgo biloba* extract. The sample is extracted with chloroform and purified over a Sep-Pak® Florisil cartridge, then diluted with buffer and put over a conditioned Aflatest® extraction column. The aflatoxins are eluted with methanol, separated by HPLC, derivatised with electrochemically generated bromine and determined quantitatively using a fluorescence detector. A similar method has been used by Kok *et al.* (1986) for the determination of aflatoxins in cattle feed.

Pesticides

The use of pesticides has greatly reduced the presence of insects, fungi and moulds in cultivated medicinal plants. Ideally the injurious action of these compounds would be highly specific for undesirable target organisms and non-injurious to desirable, non-target organisms. In fact, however, most of these compounds are not selective enough and they are generally toxic to many non-target species including humans. Medicinal plants are liable to be affected by pesticide residues, which accumulate during the cultivation, but also during the storage due to the fumigant used to prolong the conservation time.

The extent to which pesticides will remain in soils after application depends upon a number of factors: such as soil type, temperature, pH, microorganism content and the degradability of the pesticide itself. In general, the organochlorine insecticides and a few organophosphorus pesticides are the most persistent in soils. Therefore the medicinal plant material and the extracts, should be tested for the presence of these two classes of pesticides.

The analysis of pesticides in complex materials usually requires several purifications and trace enrichment steps, to obtain the quantitative separation of the analytes from coextracted materials in the samples under analysis. During the last years, the purification procedure has been simplified by the introduction of the solid phase extraction technique.

Concerning the analytical technique, gas chromatography remains the basic technique for the determination of pesticides in most standard multiresidue methods (Holland and Malcolm, 1992). GC detectors such as ECD, NPD and MS have the high sensitivity and selectivity required for residue work. Furthermore, the advent of narrow-bore capillary columns for High Resolution Gas Chromatography allows the separation and quantification of many pesticides in only one run.

For the analysis of pesticide residues in *Ginkgo biloba* extracts, two different methods are suitable: one for the organochlorine pesticides and the other for the organophosphorus pesticides. At Pharmaton, the sample preparation of both methods is performed with a Bond-Elut® C_{18} column and the quantification is done by means of capillary gas chromatography. The organochlorine pesticides can be detected with a ^{63}Ni Electron Capture Detector (ECD) and the organophosphorus with a Nitrogen Phosphorus Detector (NPD). These two methods allow the separation and quantification of 31 organochlorine (Fig. 7) and 16 organophosphorus pesticides (Fig. 8) that have been chosen according to the requirement of Ph. Eur. III,

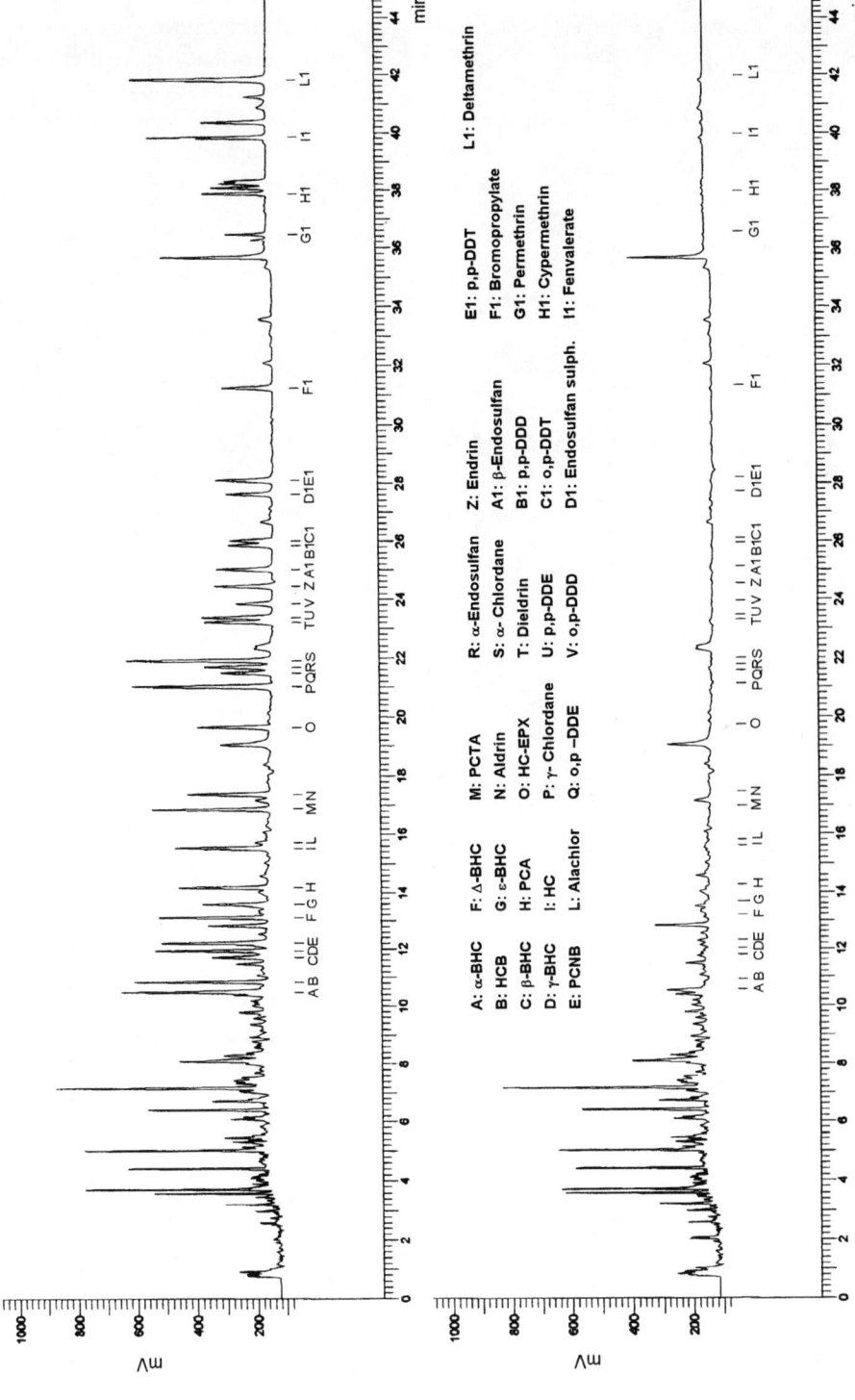

Figure 7 GC-ECD determination of organochlorine pesticides.
Below: chromatogram of the standardised GK 501 Ginkgo biloba extract; above chromatogram of the standardised GK 501 Ginkgo biloba extract spiked with 100 ppb of each pesticide.

A: α-BHC
B: HCB
C: β-BHC
D: γ-BHC
E: PCNB

F: Δ-BHC
G: ε-BHC
H: PCA
I: HC
L: Alachlor

M: PCTA
N: Aldrin
O: HC-EPX
P: γ-Chlordane
Q: o,p–DDE

R: α-Endosulfan
S: α-Chlordane
T: Dieldrin
U: p,p-DDE
V: o,p-DDD

Z: Endrin
A1: β-Endosulfan
B1: p,p-DDD
C1: o,p-DDT
D1: Endosulfan sulph.

E1: p,p-DDT
F1: Bromopropylate
G1: Permethrin
H1: Cypermethrin
I1: Fenvalerate

L1: Deltamethrin

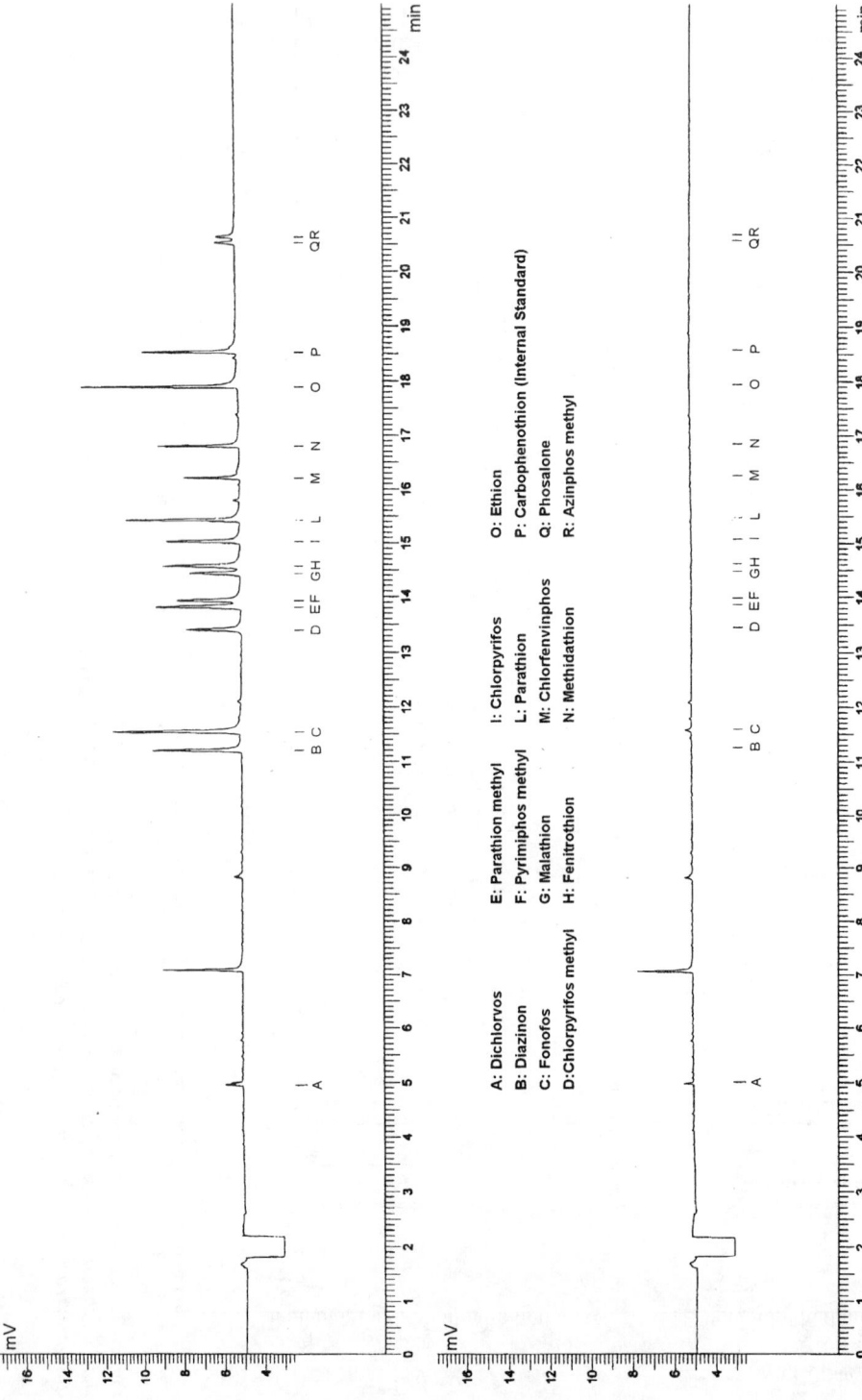

Figure 8   GC-NPD determination of organophosphorus pesticides.
Below: chromatogram of the standardised GK 501 Ginkgo biloba extract; above: chromatogram of the standardised GK 501 Ginkgo biloba extract spiked with 300 ppb of each pesticide

2.8.13 and as a function of the utilisation in the country where the plant was grown. The validation of these procedures has been performed using the standard addition method, starting with a *Ginkgo biloba* extract free of pesticides to which known amounts of the analysed pesticides were added.

Solvent residues

Residual solvents are typically determined using chromatographic techniques such as headspace gas chromatography. The residues of all the organic solvents used during the manufacturing procedures, that are not completely removed have to be quantified and kept below the limit levels established according to the ICH guidelines Q3C (1997). Usually the residues of the following solvents have to be determined: ethanol, methanol, toluene, n-butanol, n-hexane, acetone, cyclohexane and 2-butanone.

Stability Tests for the Standardised *Ginkgo biloba* Extract

The purpose of the tests is to study how the quality of the extract varies in time under the influence of a variety of environmental factors such as temperature, humidity and light. The result enables to establish recommended storage conditions and the shelf life of the raw material.

Stability tests are usually performed on different batches of extract stored in the original container at 25° C/60% r.h. The length of the study should be sufficient to cover storage, shipment and subsequent use of the plant extract.

According to the ICH guidelines Q1A (1993), the test frequency for long term stability is every three months, over the first year, every six months over the second year and then annually. At each test point the most important chemical, physical and microbiological parameters are checked and a stability characteristic of the extract is established. Since the entire plant is regarded as the active ingredient, the stability of the extract should also be determined by appropriate fingerprint chromatograms and not only by the stability of the known active principles.

To evaluate the impact of briefly exceeding the label storage conditions, three batches can be stored for accelerated testing at 40° C/75% r.h. and analysed after one, three and six months.

For the standardised GK 501 *Ginkgo biloba* extract, stability tests have been performed on different batches, stored protected from light, in double polyethylene bags contained in three-walled cardboard boxes, in presence of a drying agent (silica gel), at 25 °C/60% r.h. over a period of 48 months. During these tests, microbiological purity, physical parameters and the content of terpene trilactones and Ginkgo flavonol glycosides were periodically checked. From the obtained results and from the previous analytical and stability experience, the standardised extract GK 501 can be certified to be stable for at least 48 months, stored as described above.

CONTROL TESTS ON FINISHED PRODUCT

The finished phytopharmaceutical product (galenic form) should comply with general requirements for particular dosage forms. Product specifications and tests for release have to be established in conformity with the requirements of the pharmacopoeia monographs. A method of identification and quantification of the plant extract in the finished product should be developed and validated. Usually, the analytical methods can be similar to those for intermediate product but they have to be adapted to eliminate the interference of excipients. The quantification of the standardised GK 501 *Ginkgo biloba* extract in the finished product Gincosan®, for example, can be performed by the same HPLC method described above for the extract, but the sample preparation has to be modified (Fig. 9 and 10). The finished product, in this case a hard gelatine capsule, should also comply with the other tests, specific for this galenic form (disintegration, microbiological purity, water absorption). Furthermore, all the excipients and the permitted colouring agent used during the manufacturing procedure have to be identified. The preservatives have to be quantified and their antimicrobial activity should be demonstrated.

Recently a comparative study of the terpene trilactones content in some finished products containing *Ginkgo biloba* extract was performed (Camponovo, 1996). The results (Fig. 11) have shown that only the administration of Tanakene® and Gincosan® at the recommended dosages can provide a daily intake of terpene trilactones similar to that used in the published clinical studies (Kleijnen and Knipschild, 1992a,b; Herrschaft, 1992; Itil *et al.*, 1996; Kwiecinski *et al.*, 1997). Terpene trilactones were detected in all other phytopharmaceuticals investigated, albeit at a much lower, and therefore not therapeutic dosage (Camponovo, 1996).

Before any drug can enter the market, the stability characteristics of the finished product have to be studied according to the ICH guidelines Q1A (1993). For this purpose, accelerated and long-term stability tests on batches of the same dosage form and in the containers proposed for the marketing have to be performed. The result enables one to establish recommended storage conditions and the shelf life of the finished product. The range of testing should not only cover chemical, physical, biological and microbiological stability but also particular properties of the dosage form (e.g. disintegration time for oral solid dosage form).

CONCLUSION

Consumers, regulating agencies as well as the pharmaceutical industry are all interested in some level of phytopharmaceutical regulation. In some countries adulterated or low potency phytopharmaceuticals have been put on the market and there are also problems with contaminants like pesticides. The level of product standardisation should not be determined by its scientific feasibility, but rather by the need for safety and the therapeutic relevance of the chemical components of the natural products.

In the case of *Ginkgo biloba* extracts, recommended limits of active and toxic substances have been published by the German Authorities. The quality of an extract is not only a matter of control of finished or intermediate products. It already starts during the selection of (cultivated) plants. With chromatographic

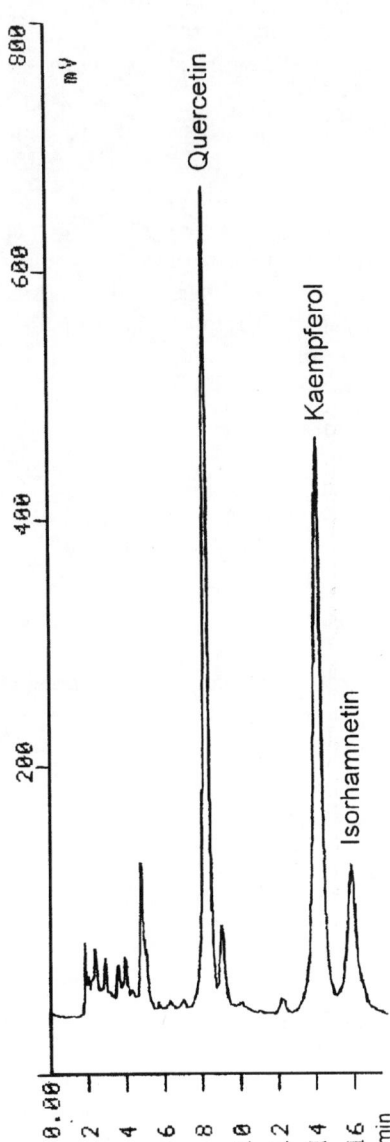

Figure 9 HPLC-UV determination of flavonol aglycones in the finished product Gincosan Pharmaton®. Column: Shim-Pack® CLC-ODS (RP C18, 150 × 6 mm I.D., Shimadzu), eluent: methanol–water–glacial acetic acid (50 : 50 : 3), flow rate: 1.5 ml/min, detection: UV 370 nm.

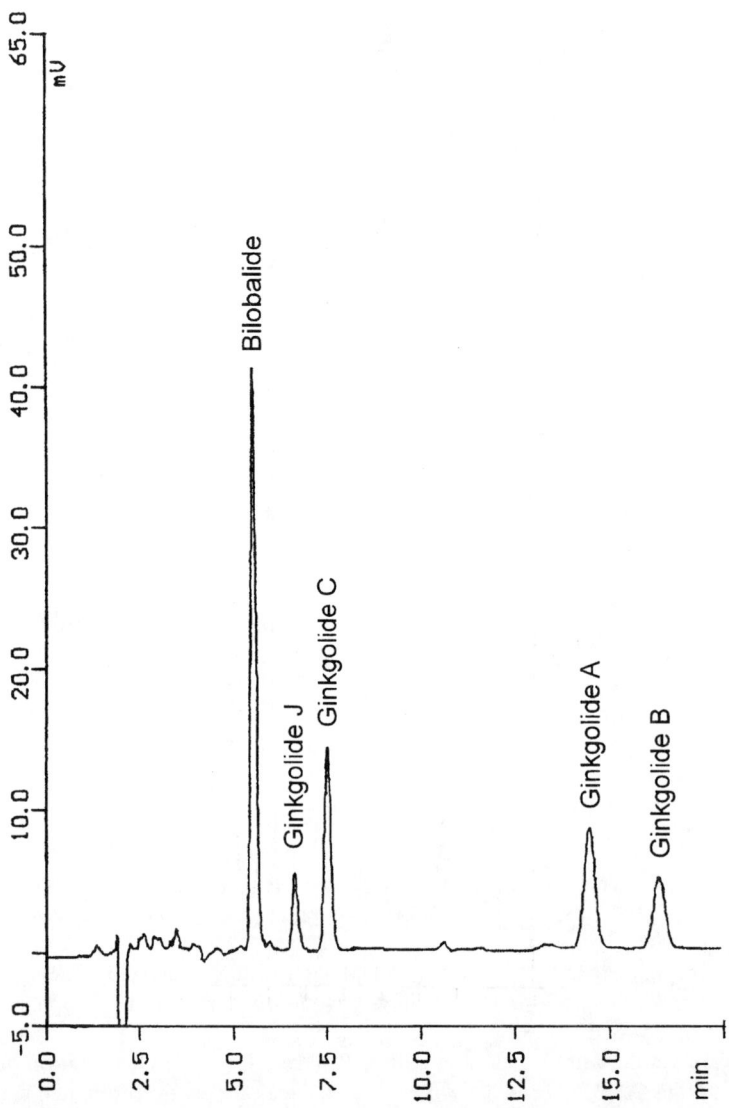

Figure 10   HPLC-RI determination of terpene trilactones in the finished product Gincosan Pharmaton®. Column: Supelcosil® (RP C18, 150 × 4.6 mm I.D., Supelco), eluent: methanol–water (28 : 72), flow rate: 1 ml/min, detection: RI.

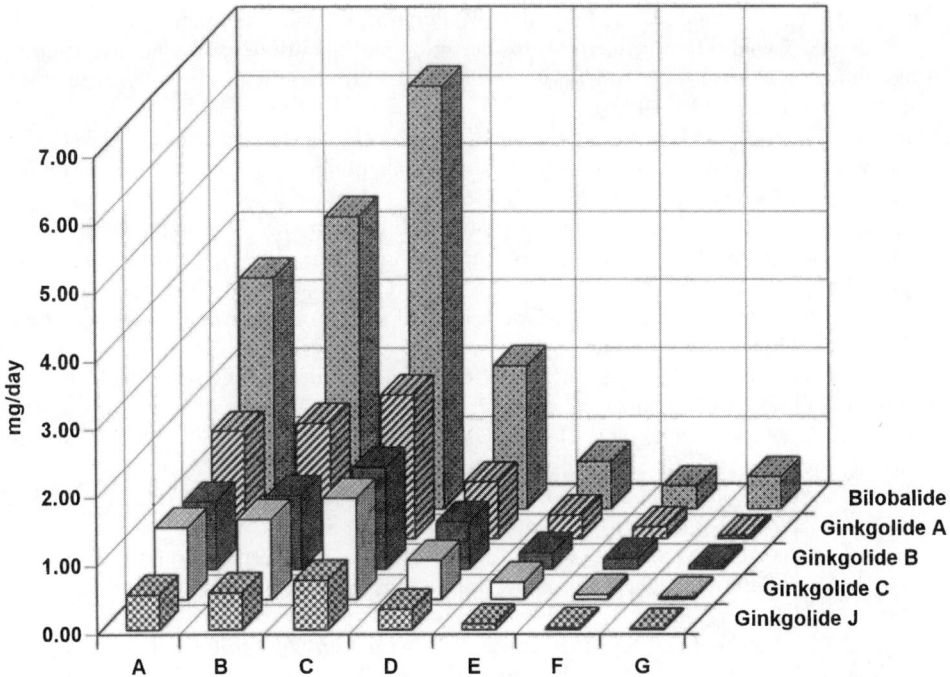

Figure 11 Terpene trilactone content in the maximal recommended daily posology of different marketed preparations (Camponovo, 1996). *A: Tanakène drag.; B: Tanakène drops; C: Gincosan®; D: Ginkgo Vital Tonic®; E: Arkogélules®; F: Allium Plus®; G: Gériaforce®.*

methods such as HPLC and GC one can determine active (Gingko flavonoids, terpene trilactones) and toxic (ginkgolic acid) constituents as well as contaminants (pesticides, aflatoxins). The validation of the analytical methods is essential, and international guidelines like the ICH Q2A (1994) and ICH Q2B (1996) should be followed.

REFERENCES

Beek, T.A. van, Lelyveld, G.P. (1993) Thin layer chromatography of bilobalide and ginkgolides A, B, C, and J on sodium acetate impregnated silica gel. *Phytochem. Anal.*, **4**, 109–114.

Beek, T.A. van, Scheeren, H.A., Rantio, T., Melger, W.Ch., Lelyveld, G.P. (1991) Determination of ginkgolides and bilobalide in *Ginkgo biloba* leaves and phytopharmaceuticals. *J. Chromatogr.*, **543**, 375–387.

Beek, T.A. van, Veldhuizen, A. van, Lelyveld, G.P., Piron, I. (1993) Quantitation of bilobalide and ginkgolides A, B, C, and J by means of nuclear magnetic resonance spectroscopy. *Phytochem. Anal.*, **4**, 261–268.

Camponovo, F.F. (1996) Mise au point de procédés pour l'analyse phytochimique et étude comparative de quelques médicaments à base de *Ginkgo biloba, Panax ginseng* et *Harpagophytum procumbens. Thesis*, Univ. Lausanne, Switzerland.

Camponovo, F.F., Wolfender, J.L., Maillard, M.P., Potterat, O., Hostettmann, K. (1995) ELSD and TSP-MS: Two alternative methods for detection and quantitative LC determination of ginkgolides and bilobalide in *Ginkgo biloba* leaf extract and phytopharmaceuticals. *Phytochem. Anal.*, 6, 141–148.

Chauret, N., Carrier, J., Mancini, M., Neufeld, R., Weber, M., Archambault, J. (1991) Gas chromatographic-mass spectrometric analysis of ginkgolides produced by *Ginkgo biloba* cell culture. *J. Chromatogr.*, 588, 281–287.

Cheng-Yu, M., Elsohly, M.A., Baker, J.K. (1980) High-performance liquid chromatographic separation of urushiol congeners in poison ivy and poison oak. *J. Chromatogr.*, 200, 163–169.

Courio, H. (1986) Phytodermites allergiques de contact dues aux anacardiacées et proteacées. *Thesis*, Univ. Montpellier I, France.

Hasler, A., Meier, B. (1992) Determination of terpenes from *Ginkgo biloba* L. by gas chromatography. *Pharm. Pharmacol. Lett.*, 2, 187–190.

Hasler, A., Meier, B., Sticher, O. (1990) *Ginkgo biloba*: Botanische, analytische und pharmakologische Aspekte. *Schweiz. Apoth. Ztg.*, 128, 342–347.

Hasler, A., Sticher, O. (1990) High-performance liquid chromatographic determination of five widespread flavonoid aglycones. *J. Chromatogr.*, 508, 236–240.

Hasler, A., Sticher, O., Meier, B. (1992) Identification and determination of flavonoids from *Ginkgo biloba* by high performance liquid chromatography. *J. Chromatogr.*, 605, 41–48.

Herrschaft, H. (1992) Zur klinische Anwendung von *Ginkgo biloba* bei dementiellen Syndromen. *Pharm. Unserer Zeit*, 21, 266–275.

Holland, T.P. and Malcolm, C.P. (1992) Multiresidue analysis of fruits and vegetables. In T. Cairns and J. Sherma, (eds.), *Emerging Strategies for Pesticides Analysis*, CRC Press, Boca Raton, USA, pp. 71–98.

Huh, H., Staba, E.J. (1993) Ontogenic aspects of ginkgolide production in *Ginkgo biloba*. *Planta Med.*, 59, 232–239.

ICH (July 1997) Q3C: (step 4, consensus guideline) Impurities: residual solvents. *International conference on harmonisation of technical requirements for registration of pharmaceuticals for human use.*

ICH (November 1996) Q2B: (step 4 consensus guideline) Validation of analytical procedures: methodology. *International conference on harmonisation of technical requirements for registration of pharmaceuticals for human use.*

ICH (December 1993) Q1A: Stability testing of new drugs substances and products. *International conference on harmonisation of technical requirements for registration of pharmaceuticals for human use.*

ICH (November 1994) Q2A: Validation of analytical methods: definitions and terminology. *International conference on harmonisation of technical requirements for registration of pharmaceuticals for human use.*

Itil, T.M., Eralp, E., Tsambis, E., Itil, K.Z., Stein, U. (1996) Central nervous system effects of *Ginkgo biloba*, a plant extract. *Am. J. Ther.*, 3, 63–73.

Kleijnen, J., Knipschild, P. (1992a) *Ginkgo biloba. The Lancet*, 340, 1136–1139.

Kleijnen, J., Knipschild, P. (1992b) *Ginkgo biloba* for cerebral insufficiency. *Br. J. Clin. Pharmacol.*, 34, 352–358.

Kok, Th.W., Neer, Th.C.H. van, Traag, W.A., Tuinstra, L.G.M.Th. (1986) Determination of aflatoxins in cattle feed by liquid chromatography and post-column derivatization with electrochemically generated bromine. *J. Chromatogr.*, 367, 231–236.

Komoda, Y., Nakamura, H., Uchida, M. (1988) HPLC analysis of ginkgolides by using a differential refractometer. *Iyo Kizai Kenkyusho Hokoku*, 22, 83–85.

Kotsonis, F.N., Burdock, G.A. and Flamm, W.G. (1996) Food Toxicology. In C.D. Klaassen (ed.), *Casarett and Doull's Toxicology: the basic science of poisons*, 5th ed., McGraw-Hill, New York, USA, pp. 909–949.

Kubo, I., Komatsu, S., Ochi, M. (1986) Molluscicides from the Cashew *Anacardium occidentale* and their large-scale isolation. *J. Agric. Food Chem.*, **34**, 970–973.

Kwiecinski, H., Lusakowska, A., Mieszkowski, J. (1997) Improvement in concentration following treatment with Ginseng/*Ginkgo biloba* combination in patients with chronic cerebrovascular disorders: a double-blind, placebo-controlled study. *Eur. J. Clin. Res.*, **9**, 59–67.

Lepoittevin, J.-P., Benezra, C., Asakawa, Y. (1989) Allergic contact dermatitis to *Ginkgo biloba* L. : relationship with urushiol. *Dermatol. Res.*, **281**, 227–230.

Lobstein, A., Rietsch-Jako, L., Haag-Berrurier, M., Anton, R. (1991) Seasonal variations of the flavonoid content from *Ginkgo biloba* leaves. *Planta Med.*, **57**, 430–433.

Lobstein-Guth, A., Briançon-Scheid, F., Anton R. (1983) Analysis of terpenes from *Ginkgo biloba* L. by high-performance liquid chromatography. *J. Chromatogr.* **267**, 431–438.

Nagabhushana, K.S., Ravindranath, B. (1995) Efficient medium-scale chromatographic group separation of anacardic acids from solvent extracted cashew nut (*Anacardium occidentale*) shell liquid. *J. Agric. Food Chem.*, **43**, 2381–2383.

Pietta, P.G., Mauri, P.L., Rava, A. (1990) Analysis of terpenes from *Ginkgo biloba* L. extracts by reversed phase high-performance liquid chromatography. *Chromatographia*, **29**, 251–253.

Pietta, P.G., Mauri, P.L., Rava, A. (1992) Rapid liquid chromatography of terpenes in *Ginkgo biloba* L. extracts and products. *J. Pharm. Biomed. Anal.*, **10**, 1077–1079.

Steinke, B., Müller, B., Wagner, H. (1993) Biologische Standardisierungsmethode für Ginkgo-Extrakte. *Planta Med.*, **59**, 155–160.

Tallevi, S.G., Kurz, W.G.W. (1991) Detection of ginkgolides by thin-layer chromatography. *J. Nat. Prod.*, **54**, 624–625.

Teng, B.P. (1988) Chemistry of ginkgolides. In P. Braquet, (ed.), *Ginkgolides-Chemistry, Biology, Pharmacology and Clinical Perspectives*, Vol. **1**, J.R. Prous Science Publ., Barcelona, Spain, pp. 37–42.

Tyman, J.H.P., Johnson, R.A., Muir, M., Rokhgar, R. (1989) The extraction of natural cashew-nut shell liquid from the cashew nut (*Anacardium occidentale*). *J. Am. Oil Chem. Soc.*, **68**, 553–557.

USP (1997) Ginkgo leaf. *Pharmacopeial Forum*, **23** (2), 3644–3647.

15. HISTORY, DEVELOPMENT AND CONSTITUENTS OF EGb 761

KATY DRIEU[1] and HERMANN JAGGY[2]

[1]Institut Henri Beaufour-IPSEN, 24 rue Erlanger, 75116 Paris, France

[2]Dr. Willmar Schwabe Arzneimittel, Postfach 41 09 25, 7500 Karlsruhe 41, Germany

INTRODUCTION

Ginkgo biloba was introduced in Europe in the 18th century, the one surviving in the botanical garden of Utrecht probably being among the first ones that arrived (around 1730). On September 27, 1815, Goethe wrote a poem in Heidelberg which he attributed to the glory of this tree "entrusted by the Orient to his garden" and of his beloved friend Marianne Von Willemer, and which was inspired by the duality that could be reunited in a single entity (Goethe, 1819).

In contrast to the material contained in certain publications (Michel 1988–Hobbs 1991), the prescribing of *Ginkgo biloba* leaves for internal medical use in China seems to have been first mentioned in the book of Liu Wen-Tai in 1505 A.D. (see Del Tredici, 1991). However, although the leaves were never introduced into the Pents'ao (systematic records of Traditional Chinese Materia Medica during the last two millenia), the fruits ("Baigo") have been used for a very long time (Administration of Traditional Chinese Medicine; communication to K. D., 1996). A leaf extract of *Ginkgo biloba* was introduced into the Medical Dictionary of the Republic of China (1977) after such an extract had been introduced in Europe, and words such as "quercetin", "kaempferol" and "rutin" appear clearly in the Chinese text (see Fig. 1).

BIRTH OF EGb 761

Ginkgo biloba leaf preparations were launched for the first time in modern medicine by Dr. Willmar Schwabe in 1965. The Dr. Willmar Schwabe Company has been working on plant-derived remedies since 1866 and initiated a systematic ethno-medicinal research program in the 1950s. Dr. Willmar Schwabe III, the grandson of the founder of the company, travelled all around the world to find medicinal plants used in traditional medicine. Although he visited repeatedly Middle and South America, Africa and South and East Asia, he had never been to China, where the Ginkgo tree has survived. By chance, or by being attracted by the exceptional shape of the Ginkgo leaf, he took some leaves from a local ornamental Ginkgo tree and

1436 白果叶 ^{bái guǒ yè}（《品汇精要》）

【基原】 为银杏科植物银杏 Ginkgo biloba L. 的叶。原植物详"白果"条。

【采集】 9～10 月间采收，晒干。

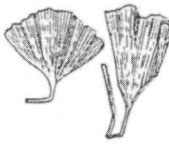

白果叶药材

【药材】 干燥叶片，大多折叠或已破碎，完整者呈扇形。上缘有不规则波状缺刻，有时中间凹入，基部楔形，叶脉为射出数回二分叉平行脉，细而密，光滑无毛，易纵向撕裂。气清香，味微涩。以叶色黄绿、整齐不破者为佳。

【成分】 叶含异鼠李素(Isorhamnetin)、山柰酚(Kaempferol)[1]、山柰酚-3-鼠李葡萄糖甙[2]、槲皮素(Quercetin)[1]、芸香甙(Rutin)[2]、槲皮甙(Quercetin)[1]、白果双黄酮(Ginkgetin)[4]、异白果双黄酮(Isoginkgetin)[5]、白果苦内酯 A、B、C (Ginkgolides A、B、C)[6]、儿茶素、表儿茶素、没食子儿茶精等鞣质类成分[2]。

白果叶含脂 0.7～1%，其中 10% 是酸性成分，15% 是蜡酯，75% 是非酯成分[7]；蜡的三要成分是白果醇(Ginnol)[4,7]、廿九烷、廿九烷-10-酮、廿八烷醇[7]。白果叶中尚含莽草酸(Shikimic acid)[8]、α-己烯醛[9]、亚麻酸[10]、β-谷甾醇和豆甾醇豆甾醇[11]等。

【药理】 ①对心血管的影响 从白果叶中提出的黄酮醇 Flavonol (为槲皮素、山柰酚、异鼠李素之混合物)[1~3]，对豚鼠离体心脏灌流可引起冠状血管扩张，注射于豚鼠后肢灌流之动脉内，可扩张后肢血管[2,3]，叶中之双黄酮对大鼠后肢血管亦有扩张作用[4]。叶的粗提物（每毫升含总黄酮约 4.263 毫克）可对抗肾上腺素所致离体兔耳的血管收缩作用，总甙之作用较弱；对麻醉兔肌肉注射临床用量的 20～40 倍，对血压、呼吸均无影响[1]。乙醇提取物与黄酮醇一样，对猫、兔血压、心脏、呼吸亦无影响；如应用大于血管扩张剂量 100～1000 倍的黄酮醇，注射于豚鼠，可引起中等度的降压，心率及呼吸的增加；但这些改变与麻醉、动物兴奋、溶剂的影响有关，无实际价值[2,3]。叶的提取制剂口服或静脉注射，可使巴金森氏病患者的脑血流增加，改善脑的营养[5]。

②对平滑肌的作用 黄酮醇对豚鼠离体肠管有解痉作用，并能对抗组织胺及氯化钡引起的痉挛，其作用强度与罂粟碱相似，但较持久[2,3]；叶的乙醇提取物对豚鼠离体气管和回肠，能拮抗组织胺和乙酰胆碱的致痉作用，在肠管实验中还可对抗氯化钡的作用，腹腔注射可制止组织胺所致之豚鼠哮喘[6]，叶中的双黄酮对心动徐缓素 Bradykinin 所致胶管痉挛有解痉作用[4,7]。

③其他作用 含黄酮、银杏素、白果双黄酮的片剂，对高胆甾醇血症患者似能降低血清胆甾醇水平、升高磷脂，改善C/P比值，对高血压病人似有一定的降压作用，但幅度不大[8,9]。初步试管试验对绿脓杆菌、金黄色葡萄球菌、痢疾杆菌有抑菌作用[10]。从银杏树皮中分离出之些生物碱具有收缩血管及扩张瞳孔的作用[11]。

毒性 叶提取物 0.5～1 毫升/公斤(1毫升相当 0.5 克生药)给家兔静脉注射连续 10 天，对血象、肝、肾功能及病理检查均无改变，用比人用量大 10～40 倍的药物，给狗连续静脉注射 1 周，出现流涎、恶心、呕吐、腹泻、食欲减退现象；组织切片检查，小肠粘膜分泌亢进，麻醉狗、兔肠蠕动增

加，注射局部血管变硬[1]。黄酮醇对血象、凝血系统无影响，但更高剂量可妨碍血液凝固[2,3]。

【性味】 《中药志》："甘苦涩，平。"

【功用主治】 益心敛肺，化湿止泻。治胸闷心痛，心悸征忡，痰喘咳嗽，泻痢，白带。

①《品汇精要》："为末和面作饼，煨熟食之，止泻痢。"

②《中药志》："敛肺气，平喘咳，止带浊。治痰喘咳嗽，白带白浊。"

③苏医《中草药手册》："治象皮腿。"

【用法与用量】 内服：煎汤，1.5～3 钱；或研末。

【宜忌】 《中药志》："有实邪者忌用。"

【临床报道】 用于治疗冠状动脉硬化性心脏病。初步观察对心绞痛有较好效果，对降低胆甾醇及血压也有一定作用[1~9]。㈠用银杏叶的提取物制成片剂内服，每次 2 片(每片含黄酮 2 毫克)，每日 3 次；或制成针剂肌内注射，每次 2～4 毫升(每 2 毫升含黄酮 1.3～1.7 毫克)，每日 2 次。据 139 例的观察，显效(心绞痛消失或明显减轻或发作次数明显减少，心电图恢复正常或好转)占 34.5%，有效(心绞痛程度减轻或发作次数减少，心电图好转，但仍有波动)占 43.2%。大部分病人在用药后 3～10 天内生效，少数在 20～40 天才出现疗效。停药后，部分病例症状复发，但程度较治疗前为轻，继续用药仍然有效。约近半数病例心电图恢复正常或有改善，但与心绞痛的改善并不完全一致。血压高的病例，多数用药后有不同程度的下降，但单独应用不及与降压药合并使用效果显著。49 例高胆甾醇的患者，治疗后下降 20～40 毫克%或 40 毫克% 以上的 25 例，占 51.02%，上升 20 毫克%的 18 例，无变化 6 例[4]。另据 100 例血清胆甾醇过高症的治疗结果，胆甾醇降低者 88 例，磷脂升高者 65 例，胆甾醇与磷脂之比例有改善者 71 例；3 项综合评定：显效者 49 例，有效者 50 例，无效 1 例[8]。也有通过 23 例观察，未见血压和胆甾醇有明显变化[3]。㈡用银杏甙元含片(每片含银杏甙元 0.5 毫克)舌下含服，每次 2 片，每日 3 次。治疗心绞痛 41 例，总有效率 70.1%，其中显效占 34.1%。治疗高血压的 32 例患者，治疗后 15 例有效，9 例停药。约半数病人治疗前心电图显示有冠状动脉供 血不足，9 周后复查改善情况与临床疗效相符[1]。有用银杏叶双黄酮片剂舌下含服，每次 2 片(含 1.2 毫克)，每日 3 次。24 例心绞痛患者经治疗 6 周后，1 例显效，9 例改善，14 例基本无效；原来用硝酸甘油的 21 例，治疗后停用者 1 例，减量者 9 例；心电图有异常的 18 例中，有 1 例好转。疗效发生时间，最早在用药后 2 周，多数在 4～5 周[1]。㈢银杏叶、川芎、红花各 5 钱，制成片剂，每日 3 次分服。60 例心绞痛患者，服药 8 周后总有效率为 75.5%，其中显效率为 11.7%。见效时间在 2～5 周。半数以上病例可以停用或减少硝酸甘油类药物的用量[5]。另有用银杏叶、首乌、钩藤制成片剂内服，每日量均为 1.5 钱，治疗心绞痛 129 例，服药后 4 周的近期有效率为 60.4%，显效率占 8.5%，疗效大多出现在服药 2 周内[6]。此外，还有用银杏叶配合杜仲、首乌[7]，或配合锦鸡儿[8]，制成片剂或针剂，对心绞痛、高血压，及降低胆甾醇方面均有一定效果。银杏叶的副作用较少，但也有少数患者有食欲减退、恶心、稀便、腹胀、口干、鼻塞及头晕、头痛、耳鸣等，个别病例出现过敏性皮疹[1~8]。如在治疗过程中出现血压明显下降，心绞痛加重，特别是心功能不全加重时，需减量或停药[1]。

Figure 1 A page of the text of the "Pharmacopeia of the People's Republic of China" (1977).

used them to produce an extract. Employing pharmacological screening tests, he detected its powerful effects in some *in vitro* preparations; e.g., guinea pig isolated limb and the so-called "Langendorff heart" (isolated heart of guinea pig) (see e.g., Peter *et al.*, 1966). By concentrating the flavonoid content of this extract he could enhance its pharmacological activity, and therefore Dr. Schwabe used this concentrated extract for clinical research. In open clinical trials, the extract provided effective treatment for disturbances in peripheral and cerebral blood flow which led to an early market entry of Tebonin®. Further purification and concentration of this extract led to the production of an injectable form.

It was around that time (1968) that Dr. A. Beaufour and his brother met Drs. Willmar and Wolfgang Schwabe and established a collaboration that endures to the present day. The Beaufour and Schwabe groups chose to develop the more purified and concentrated form of the extract prepared initially for injections. A pharmaceutical product, "Ginkor®", composed of troxerutin and heptaminol in association with *Ginkgo biloba* extract and destined for use in venous pathology, was first developed by the Beaufour Laboratories and was commercialized in 1973. Simultaneously, pharmacological and clinical studies on the concentrated *Ginkgo biloba* extract, termed "EGb 761", resulted in the registration of Tanakan® in 1974 which was introduced to the market in 1975 in France by IPSEN, a subsidiary of the Beaufour group that was created specifically for launching this product. In identical dosage, two other products bearing the trade names Rökan® and Tebonin forte®, were commercialized in Germany, the first one by INTERSAN and the second one by the Schwabe Laboratories.

CULTIVATION OF GINKGO TREES IN THE FAR-EAST

Due to the rapid commercial success achieved with Tanakan®, it became imperative to ensure a constant and controlled supply of large quantities of high quality Ginkgo leaves. The first supplies of leaves were obtained from South Korea and Japan, countries in which large amounts of *Ginkgo biloba* seeds were available. It was in these two countries that contacts were made which resulted in the installation in 1978 of a large number of small plantations supervised by European agricultural engineers. Studies of the conditions for cultivation of the trees, conducted with cooperative groups in Japan and with peasants in Korea, permitted rapid harvesting of the leaves of young trees under optimal conditions. Two units for drying were installed in Japan and three in Korea. The leaves were dried on site employing rotatory hot-air driers imported from France. The method for dehydration that was developed did not modify the leaf constituents. The dry leaves were then compressed in the form of bales (about 120 kg each) and shipped to Europe. Studies were then undertaken in plant nurseries to examine and optimize the cultivation conditions for the trees.

MANUFACTURING OF EGb 761

Patent applications for complete extraction and standardization procedures for EGb761 were initially filed and obtained in Germany in 1971 (DE 2117429 C3) and in France in 1972 (72.12288); and patents for an improved extract were obtained for all European countries in 1994 (EP 0 431 535 B1) and for the US in 1995 (Patent 5,399,348). These later patents describe the invention of the procedures used to obtain EGb761 from the leaves of *Ginkgo biloba*, and relate to an improved extract which guarantees that ginkgolic acids are reduced to a few ppm, while the content of flavonoids and terpene trilactones remain unaltered: the ginkgolic acids which belong to the family of anacardic acids, have allergenic properties. (Jaggy and Koch, 1997)

The extract of *Ginkgo biloba* had been initially manufactured in Karlsruhe, Germany by the Schwabe Laboratories. At the end of 1977, the Beaufour and Schwabe groups agreed to construct a manufacturing plant for the extract; this factory is located near Cork, Ireland and was inaugurated in January, 1980. From the beginning, all work on Ginkgo extracts has been conducted in accord with "Good Manufacturing Practices and Good Laboratory Practices", as required by the regulatory authorities, including those of the Food and Drug Administration (U.S.A.). It was only after the extract of *Ginkgo biloba* had been subjected to particularly rigorous studies of standardization and control that it was given the code name "EGb 761". This code name was officially given to the extract in the early eighties and was arbitrarily chosen by the Beaufour-Ipsen and Schwabe groups, in order to differentiate it from all other Ginkgo extracts.

EGb 761 is an water-acetone extract whose quality and activity depend upon a well defined, patented manufacturing process. In addition to its content of "active substances", consisting of controlled amounts of 24% Ginkgo-flavonol glycosides and 6% terpene trilactones (3.1% ginkgolides and 2.9% bilobalide), EGb 761 contains other substances that also depend upon its method of production and whose constancy is very important in ensuring its high quality (see O'Reilly, this volume). These other substances consist of a controlled amount of proanthocyanidins (less than 9%), carboxylic acids, non-flavon-glycosides, and some high molecular-weight components.

An extension of the factory in Cork was established in 1992 to increase production capacity. The manufacture of the same extract has always been continued in Karlsruhe in the Schwabe Laboratories. At these two sites (Cork and Karlsruhe), about 4000 tons of dry leaves were transformed into EGb 761 in 1996.

CULTIVATION OF GINKGO TREES IN WESTERN COUNTRIES

In 1980, upon considering the possible risks associated with being dependent on countries in the Far-East for supplying leaves, the Beaufour and Schwabe groups decided to establish their own plantations. After organizing nurseries to conduct studies for selecting ideal conditions of soil and climate, the first plantation was established in the southwest region of France at Saint-Jean d'Illac (near Bordeaux) in 1982 (see Fig. 2), and then a second one was established in South Carolina (U.S.A.) in 1984. A third plantation has been more recently established in France, not far from the first one. Together, these three plantations comprise slightly more than

Figure 2   The *Ginkgo biloba* plantation in Saint-Jean d'Illac (France). (Institut Henri Beaufour-IPSEN photograph)

1500 hectares (15 km^2) and about 25 million Ginkgo trees, all of which have been grown from fertilized seeds imported from the Far-East.

Although the leaves could be harvested manually from young trees of the nurseries and also from old trees in the Far-East, it became essential to develop another collection procedure in the Western plantations. Preliminary studies had shown that when Ginkgo trees are pruned every year they become "bush-like" in appearance and have branches that are not longer than two meters. Thus, it became possible to envisage an automated method for harvesting the leaves. This was accomplished by developing harvesting machines that were modified cotton-pickers. Calculations were made for optimal spacing of the trees, and they were aligned (by laser beam) to assure the lines of planting. Agricultural machinery was also adapted with respect to the dimension of the plants and maintenance of soil. An entirely new machine was developed for weeding with gas burners in order to eliminate the use of herbicides during the period of development of the leaves. Remaining amounts of potentially detrimental weeds were removed by hand-weeding during the few weeks preceding the harvest. Automatic irrigation and watering systems were installed to ensure that the amount of water required for good development was available to the plants.

The leaves are harvested during slightly different periods of the year, depending upon the climatic conditions and the geographic location of the plantation. The exact time that is chosen is based on punctual analyses of the leaves. Leaves that are collected during night and day are dried as rapidly as possible using drying units identical to those installed previously in South Korea and Japan. The dried leaves are then pressed and packed into bales of about 120 kg (60 × 60 × 80 cm) and sent in containers to the place of extraction.

Since 1996, joint ventures have been in operation between the Beaufour and Schwabe groups on one hand, and the Chinese administrations of two adjacent districts of the provinces of Shandong and Jiangsu, on the other hand. In this latter region of China, where Ginkgo trees had already existed traditionally for the production of their fruits, the Central Administration of China had ordered the peasants to plant a number of hectares of young Ginkgo trees for three or four years. They were discovered by the Beaufour/Schwabe group in 1995 and two Sino-European companies were founded. The leaves were collected by hand, and driers were installed so that those leaves that were of good quality could be conveniently dried. In 1996, about 1400 tons of dry leaves were harvested in this region.

MOLECULAR CONSTITUENTS OF EGB 761

The most specific and significant scientific research on EGb 761 was initiated in the late 1970s when suitable pathological models became available, since only slight effects of the extract can be shown under normal conditions. Initially, Ginkgo-flavonol glycosides were thought to be responsible for the effects of EGb 761. Fractions obtained by chromatographic fractionation of the extract by one of us (H. J.) and special extracts lacking one or several constituents were tested, leading to the accumulation of a large amount of unpublished data which was useful for understanding why it is important to use *the total EGb 761 preparation for therapeutic purposes.*

The flavonoid fraction of EGb 761 is known to play an important role in regulating capillary permeability. Also, numerous experimental studies have shown that the extract, likely due to its flavonoidic constituents, has *in vitro* free radical-scavenging activity, although at rather high concentrations (see e.g. Packer *et al.*, 1996).

The research efforts of Dr. S.S. Chatterjee (Schwabe Laboratories) have contributed significantly to the understanding of the actions of bilobalide. Comparison of the flavonoid and non-flavonoid fractions, using models in which EGb 761 had been shown to be active, led in 1982 to the discovery that bilobalide was responsible for the protective effect of EGb 761 on triethyltin-induced edema (Chatterjee *et al.*, 1985). These results were later confirmed by Sancesario and Kreutzberg (1986). More recent studies have indicated that bilobalide can counteract the uncoupling of oxidative phosphorylation, thereby optimizing ATP formation and glucose utilization (Spinnewyn *et al.*, 1995; Janssens *et al.*, 1995). These potent biochemical effects of bilobalide may explain its antagonism of the action of triethyltin since this toxic substance is an uncoupler of oxidative phosphorylation. Bilobalide is also capable of facilitating glucose entry into cells and cellular glycogen synthesis (Rapin *et al.*, 1994, 1997).

Dr. P. Braquet and colleagues, working in the research laboratories of the Beaufour-IPSEN group, discovered and published findings which indicated that the ginkgolide constituents of EGb 761 are antagonists of platelet-activating factor (PAF-acether) at the level of its receptor, ginkgolide B being the most potent (see Braquet et al., 1985). This finding initiated a significant amount of research and led to a large number of publications (see e.g. Braquet, 1988, 1989). Some of the published data may be questionable since, due to its insolubility in aqueous media, ginkgolide B was dissolved in dimethylsulfoxide in certain in vitro and in vivo experiments (F.V. DeFeudis, personal communication, 1989).

Later, an injectable form of ginkgolide B was developed for clinical use. The regulatory studies for a complete registration file were performed. A pharmaco-clinical study (Duchier et al., 1989) permitted determination of the dose of ginkgolide B that would inhibit skin wheals induced by PAF in humans. This dose was 120 mg via intravenous injection every 12 hours. Several clinical studies were conducted using this dose in patients afflicted with pathological conditions in which PAF was believed to be involved: patients suffering from burns, hematological disorders occurring during cardiopulmonary bypass surgery (Nathan et al., 1994), multiple sclerosis (Brochet et al., 1995), post-transplant renal failure (Grino, 1994) and septic shock (Dhainaut et al., 1994).

In all of these clinical studies, ginkgolide B failed to show significant activity, perhaps because PAF is not a key (initiating) factor in these pathological states. However, a secondary analysis of the double-blind study on septic shock (Dhainaut et al., 1994) did reveal that ginkgolide B treatment was effective in a sub-group of 120 of the 262 patients who were afflicted with gram-negative sepsis. Thus, a second double-blind study was performed by these same investigators in which the recruitment of gram-negative infested patients was favoured. However, this more recent study, although perfectly monitored, led to a negative result (Dhainaut et al., 1998).

After obtaining these negative results, all research efforts concerning the PAF-antagonistic effect of ginkgolide B were terminated at the Beaufour-IPSEN and Schwabe laboratories. However, more recent studies conducted along different lines have shown that ginkgolide A and B possess other extremely interesting properties that may underlie the pharmacological activity and therapeutic applications of EGb 761. These substances possess antioxidant activity, demonstrated both in vivo and in vitro at concentrations as low as 50 nanograms/ml, which involves inhibition of free radical formation rather than direct free radical-scavenging (Pietri et al., 1997a; see also Drieu and DeFeudis, this volume). This protection against free radicals has also been recently demonstrated in humans with EGb 761 after preventive oral administration in patients undergoing coronary bypass surgery (Pietri et al., 1997b). Furthermore, ginkgolides can down-regulate the peripheral benzodiazepine receptor (PBR) of adrenal mitochondria, and therefore may regulate circulating glucocorticoid levels under stressful conditions (Papadopoulos et al., 1997). This latter molecular mechanism is likely to play a role in mediating the "anti-stress" activity of EGb 761 (see Drieu and DeFeudis, this volume).

The internal IHB-IPSEN code names for the terpene trilactone constituents of EGb 761 are BN 52020 for ginkgolide A, BN 52021 for ginkgolide B, BN 52022 for ginkgolide C and BN 52023 (also Lab Schwabe code: HE 134) for bilobalide. Pharmacological experiments have not been performed with ginkgolide C because comparison of intravenous and oral routes in a bioavailability study in humans indicated that it is not absorbed, whereas ginkgolide A and B are absorbed to the extent of 85–95% (Fourtillan *et al.*, 1995).

Collectively, these studies have shown that many of the constituents of EGb 761 play a rôle in relation to its pharmacological and therapeutic activities, acting in a complementary and/or synergistic manner to confer to the extract its regulatory and protective actions, and therefore the extract must be considered as an equilibrated "totum".

BIOTECHNOLOGICAL STUDIES ON *GINKGO BILOBA*

Considering the potential therapeutic uses of ginkgolide B, a program has been initiated to devise methods, based on the selection of highly productive trees or on cell or organ culture, for ensuring the availability of large amounts of pure EGb 761 constituents. This program was led by V. Petiard (Centre de Recherche Nestlé, Tours, France).

With regard to the selection and the propagation of highly productive trees, it was first surprisingly observed that the variability of the terpene trilactones content in leaves is very low compared to what could have been expected for a dioecious species. A thousand trees sampled in many different areas showed a very small diversity in their ability to biosynthesize bilobalide and the ginkgolides. This could be due to a very narrow genetic origin. The controlled crosses (partial diallelic assay) of various trees did not lead to a significant increase in variability even though preliminary data indicated an inheritance of terpene trilactones composition.

Then came the problem of true-to-type propagation of selected trees. Vegetative propagation, whether conducted with rooted cuttings or bud-grafting, was not particularly efficient and therefore *in vitro* propagation appeared as the most likely usable technique. Micropropagation experiments were successful. However this method was not economical in comparison to seed multiplication. If a greater diversity of terpene trilactones content had been observed, it may have been practical to use this technique on very rich trees despite the fact that additional effort was required to improve the field behavior of micropropagated trees. An attempt to achieve somatic embryogenesis, which would have been a powerful and economically feasible way to propagate selected trees, also failed. Based on published scientific articles, true somatic embryos have never been produced. Some primordial leaf-like structures resembling somatic embryos have been obtained but they never developed into seedlings. Some gametic embryos were obtained but they are of no interest with respect to propagation although they could be useful in a long-term breeding program.

Concerning the use of cell cultures, many thriving cell strains were obtained and screened for their ability to produce ginkgolides. Only trace amounts were produced in spite of many attempts to diversify the strains and the culture conditions. Cultures were successfully scaled up to the 600 liters bioreactor scale. According to the finding that the terpenes trilactones are biosynthetized in the roots and are translocated in the leaves (Cartayrade, 1997), root cultures were also examined. These produce terpene-trilactones at about the same concentration as the whole plants, but they grew very poorly and therefore could not be used for the economic production of these substances. Attempts at genetic transformation with *Agrobacterium rhizogenes* was successful, but this did not lead to a rapid growth of hairy roots.

Metabolic studies conducted at Nestlé Research & Development (Tours), as well as in the laboratory of Prof. D. Arigoni at the Swiss Institute of Technology in Zürich, have shown that the ginkgolides are biosynthesized in the roots of the tree and then accumulated in the leaves (Schwarz, 1994). Isolated leaves or non-rooted micro-cuttings do not possess a higher content of ginkgolides than the primary explant. However, the ginkgolide content increases when regenerated roots have been induced. In addition, Prof. Arigoni's group has shown that the biosynthetic pathway of mevalonic acid in *Ginkgo biloba* is unusual for higher plants and possibly very "archaic" (Schwarz, 1994).

CONCLUSION

EGb 761 has been extensively studied for nearly three decades. Following the improvement of existing methods and the availability of more sophisticated technology, a supreme effort was made to increase the knowledge concerning all scientific aspects of EGb 761, from the biology of the tree and its cultivation to optimizing the quality of the extract and discovering the secrets of its mechanisms of action. Due to the overwhelming success of EGb 761 in Europe and the increasing interest in "natural drugs", a great number of *Ginkgo biloba* extracts (other than EGb 761) have been commercialized, both in the United States under the "Dietary Supplement Health and Education Act" and in Europe as "dietetic, homeopathic, and/or phytotherapeutic products". Although it is claimed that these other extracts are useful for the clinical indications of authentic EGb 761 and that their efficacy is supported by the studies performed with EGb 761, most of them are manufactured by different procedures which are not stringently controlled, leading to poor reproducibility of the extracts. Although a few manufacturers in Europe provide extracts that do not differ substantially from EGb 761, most of the other extracts are currently imported from China where there exist between 50 and 100 manufacturers that do not comply with any of the regulatory requirements of "Good Manufacturing Practice".

It is intended to continue to increase the knowledge on EGb 761, and convince others of the intriguing properties of this amazing and exceptional extract. Those who may consider that any kind of Ginkgo extract will have the same properties as the authentic preparation are cautioned.

REFERENCES

Braquet, P., Spinnewyn, B., Braquet, M., Bourgain, R.H., Taylor, J.E., Etienne, A. and Drieu, K. (1985) BN 52021 and related compounds: A new series of highly specific PAF-acether antagonists isolated from *Ginkgo biloba* L. *Blood Vessels*, **16**, 558–572.

Braquet, P. (Ed.) (1988) *Ginkgolides: Chemistry, Biology, Pharmacology and Clinical Perspectives.*, **Vol. 1.** J.R. Prous, Barcelona.

Braquet, P. (Ed.) (1989) *Ginkgolides: Chemistry, Biology, Pharmacology and Clinical Perspectives.*, **Vol. 2.** J.R. Prous, Barcelona.

Brochet, B., Guinot, P., Orgogozo, J.M., Confavreaux, C., Rumbach, L., Lavergne, V. and the Ginkgolide Study Group in Multiple Sclerosis (1995) Double blind placebo controlled multicentre study of ginkgolide B in treatment of acute exacerbations of multiple sclerosis. *J. Neurol. Neurosurg. Psychiatry*, **58**, 360–362.

Cartayrade, A., Neau, E., Sohier, C., Balz, J.P., Carde, J.P. and Walter, J. (1997) Ginkgolide and bilobalide biosynthesis in *Ginkgo biloba*. I: Sites of synthesis, translocation and accumulation of ginkgolides and bilobalide. *Plant Physiol. Biochem.*, **35**(11) 859–868

Chatterjee, S.S., Gabard, B.L. and Jaggy, H.E.W. (1985) *Beschreiben sind Bilobalid enthalthende Arzneimittel, die zur Bekämpfung von nervälen Erkränküngen eingesetzt werden können.* German Patent No. DE 3338995 A1 (May 9, 1985).

Del Tredici, P. (1991) Ginkgos and people : A thousand years of interaction. *Arnoldia*, **51**, 2–15

Dhainaut, J.F.A., Tenaillon, A. and the BN 52051 Sepsis Study Group (1994) Platelet-activating factor receptor antagonist BN 52021 in the treatment of severe sepsis: a randomized, double-blind, placebo-controlled, multicenter clinical trial. *Crit. Care Med.*, **22**, 1720–1728.

Dhainaut, J.F.A., Tenaillon A. and the BN 52021 Sepsis Study Group (1998) The confirmatory platelet-activating factor receptor antagonist trial in patients with severe gram-negative bacterial sepsis: a phase III randomized, double-blind, placebo-controlled, multicenter trial. *Crit. Care Med.*, **26**, 1963–1971.

Duchier, J., Cournot, A., Hosford, D., Bonvoisin, B., Berdah, L. and Guinot, Ph. (1989) Clinical studies on the tolerance and the effects of BN 52021 on PAF-induced platelet aggregation and skin reactivity in healthy volunteers. In P. Braquet (ed.) *Ginkgolides: Chemistry, Biology, Pharmacology and Clinical Perspectives.*, **Vol. 2**, J.R. Prous, Barcelona, pp. 897–904.

Fourtillan, J.B., Brisson, A.M., Girault, J., Ingrand, I., Decourt, J.P., Drieu, K., Jouenne, P. and Biber, A. (1995) Propriétés pharmacocinétiques du bilobalide et des ginkgolides A et B chez le sujet sain après administrations intraveineuses et orales d'extrait de *Ginkgo biloba* (EGb 761). *Thérapie*, **50**, 137–144.

Goethe, J.W. (1819) *West-Ostlicher Divan*, Ed. Cotta, Tübingen.

Grino, J.M. (1994) BN 52021, a platelet activating factor antagonist, to prevent post-transplant renal failure. *Ann. Intern. Med.*, **121**, 345–347.

Hobbs, C. (1991) Ginkgo, elixir of youth. *Botanica press* (California).

Jaggy, H. and Koch, E. (1997) Chemistry and biology of alkylphenols from *Ginkgo biloba* L. *Pharmazie*, **52**, 735–738.

Janssens, D., Michiels, C., Delaive, E., Eliaers, F., Drieu, K. and Remacle, J. (1995) Protection of hypoxia-induced ATP decrease in endothelial cells by *Ginkgo biloba* extract and bilobalide. *Biochem. Pharmacol.*, **50**, 991–999.

Medical Dictionary of the Republic of China. (1977) Medical Dictionary Commission of the Health Ministry of China.

Michel P.M., Hosford D. (1988) *Ginkgo biloba* : from "living fossil" to modern therapeutic agent. In P. Braquet (ed.) *Ginkgolides: Chemistry, Biology, Pharmacology and Clinical Perspectives*. vol. 1 J.R. Prous, Barcelona. pp. 1–8.

Nathan, N., Mercury, P., Denizot, Y., Cornu, E., Laskar, M., Arnoux, B. and Feiss, P. (1994) Effects of the platelet-activating factor antagonist BN 52021 on hematologic variables and blood loss during and after cardiopulmonary bypass. *Anesth. Analg.*, 79, 205–211.

Packer, L., Marcocci, L., Haramaki, N., Kobuchi, H., Christen, Y. and Droy-Lefaix, MT. (1996) Antioxidant properties of *Ginkgo biloba* extract EGb761 and clinical Implications In L. Packer, M.G. Traber, Weinjuan Xin (eds), Proceedings of the International Symposium on Natural Antioxidants Molecular Mechanisms and Health Effects AOCS Press, pp. 472–498.

Papadopoulos, V., Amri, H., Boujrad, N., Cascio, C., Culty, M., Garnier, M., Hardwick, M., Li, H., Vidic, B., Brown, A.S., Reversat, J.L., Bernassau, J.M. and Drieu, K. (1997) Peripheral benzodiazepine receptor in cholesterol transport and steroidogenesis. *Steroids*, 62, 21–28.

Peter, H., Fisel, J. and Weisser, W. (1966) Zur Pharmakologie der Wirkstoffe aus *Ginkgo biloba*. *Arzneim.-Forsch./Drug Res.*, 16, 719–725.

Pietri, S., Maurelli, E., Drieu, K. and Culcasi, M. (1997a) Cardioprotective and antioxidan-effects of the terpenoid constituents of *Ginkgo biloba* extract (EGb 761). *J. Mol. Cell. Cardiol.*, 29, 733–742.

Pietri, S., Séguin, J.R., d'Arbigny, P., Drieu, K. and Culcasi, M. (1997b) *Ginkgo biloba* extract (EGb 761) pretreatment limits free radical-induced oxidative stress in patients undergoing coronary bypass surgery. *Cardiovasc. Drugs Therapy*, 11, 121–131.

Rapin, J.R., Provost, P., DeFeudis, F.V. and Drieu, K. (1994) Effects of repeated treatments with an extract of *Ginkgo biloba* (EGb 761) and bilobalide on glucose uptake and glycogen synthesis in rat erythrocytes: An *ex vivo* study. *Drug. Dev. Res.*, 31, 164–169.

Rapin, J.R., Yoa, R.G., Bouvier, C. and Drieu, K. (1997) Effects of repeated treatments with an extract of *Ginkgo biloba* (EGb 761) and bilobalide on liver and muscle glycogen contents in the non-insulin-dependent diabetic rat. *Drug Dev. Res.*, 40, 68–74.

Sancesario, G. and Kreutzberg, G.W. (1986) Stimulation of astrocytes affects cytotoxic brain edema. *Acta Neuropath.*, 72, 3–14.

Schwarz, M.K. (1994) *Terpen-Biosynthese in Ginkgo biloba: Eine überraschende Geschichte*. Doctoral Thesis, Swiss Federal Institute of Technology.

Spinnewyn, B., Blavet, N. and Drieu, K. (1995) Effect of *Ginkgo biloba* extract (EGb 761) on oxygen consumption by isolated cerebral mitochondria. In Y. Christen, Y. Courtois and M.T. Droy-Lefaix (eds.) *Advances in Ginkgo biloba Extract Research, Vol. 4. Effects of Ginkgo biloba Extract (EGb 761) on Aging and Age-Related Disorders*, Elsevier, Paris, pp. 17–22.

16. *IN VITRO* STUDIES OF THE PHARMACOLOGICAL AND BIOCHEMICAL ACTIVITIES OF *GINKGO BILOBA* EXTRACT (EGb 761) AND ITS CONSTITUENTS

[1]FRANCIS V. DeFEUDIS and [2]KATY DRIEU

[1]*Institute for BioScience, 153 West Main Street, Westboro, MA 01581, U.S.A. and*
[2]*Institut Henri Beaufour-IPSEN, 24 rue Erlanger, 75116 Paris, France*

INTRODUCTION

It has been long known that the *Ginkgo biloba* tree possesses medicinal properties. The seeds of this tree were mentioned in Li Tung-wan's *Shi Wu Ben Cao* (Edible Herbal; 1280–1368 A.D.) and in Wu-rui's *Ri Yong Ben Cao* (Herbal for Daily Usage; 1350 A.D.) (see Del Tredici, 1991; Foster and Chongxi, 1992). The internal use of *Ginkgo biloba* leaves for medical purposes was first mentioned by Liu Wen-Tai in the *Ben Cao Pin Hue Jing Yaor* (1505 A.D.). Both the leaves and fruit of *Ginkgo biloba* are still recommended for treating dysfunctions of the heart and lungs in contemporary Chinese pharmacopoeias (see Chang and But, 1987; Foster and Chongxi, 1992). At the present time, preparations of the leaf extract, termed "EGb 761", are among the most widely prescribed medicinal plant products in the world, and are more commonly associated with medicine in Europe than in China. However, this situation may change since EGb 761-containing products are now being introduced into China. EGb 761 is composed of flavonol glycosides (24%), terpene trilactones (ginkgolides, bilobalide; total of 6%), proanthocyanidins, organic acids and other constituents, several of these substances being responsible for its therapeutic effects (see Drieu, 1986; DeFeudis, 1991, 1998). Clinical indications for treatment with EGb 761 include cerebral and peripheral vascular insufficiency states, age-related cognitive impairment, neurosensory problems (e.g., tinnitus, vertigo) and eye ailments (see e.g., DeFeudis, 1991).

This chapter is centered on *in vitro* studies that have been performed with the authentic EGb 761 preparation. In certain exceptional cases, *ex vivo* experiments will also be mentioned. *Ex vivo* results will be discussed in greater depth in the companion chapter (Drieu and DeFeudis, this volume). The limitations of *in vitro* experimentation are realized, but it is maintained that such studies are indispensable since in many cases they represent the only possible means for detecting "potential mechanisms of action" of active substances that might pertain to their *in vivo* and clinical actions.

It might be noted that although both 3,4-dihydroxybenzoic acid and 3-methoxy-4-hydroxybenzoic acid, which are present in EGb 761 and which may be flavonoid metabolites (see DeFeudis, 1991) are relatively potent inhibitors of hydrogen peroxide (H_2O_2) production by stimulated human neutrophils (Limasset et al., 1990). After oral administration of EGb 761 to rats, some flavonoid metabolites (likely formed via degradation of the extract) have been found in the blood (Pietta et al., 1995; see companion chapter). However, the intact flavonoid heteroside constituents of EGb 761 may not circulate after oral (in vivo) administration of the total extract to humans or experimental animals. If this latter contention is validated with further advances in technology, since methods for determining the bioavailability of intact flavonoid heterosides are not currently available, then it would seem probable that many of the effects of the flavonoid constituents of EGb 761 that have been shown in vitro may not occur in vivo. However, as mentioned in the companion chapter (Drieu and DeFeudis, this volume), methods for determining the pharmacokinetics of ginkgolides A and B and bilobalide (terpene constituents of EGb 761) have been devised and these substances do appear to be bioavailable after oral administration of EGb 761. Therefore, if they are found to be active in in vitro experiments, especially in low (ng/ml) concentrations, it seems probable that they could contribute to the pharmacological effects of EGb 761 that are observed in vivo and in the clinical setting.

EFFECTS OF EGb 761 ON CARDIOVASCULAR MECHANISMS

Cardiovascular mechanisms, both central and peripheral, are primary targets for the pharmacological actions of EGb 761. The extract acts on all of the major components of the cardiovascular system; i.e., arteries, veins, capillaries, and formed elements of the blood. This "vasoregulatory" action of EGb 761 involves a spasmolytic action on the arterial wall, increased tissue perfusion, reinforcement of capillary resistance, and enhanced venous return.

Experiments with Isolated Blood Vessels

Rabbit isolated aorta

Studies conducted over a decade ago showed that the effects of EGb 761 on rabbit isolated aorta involve catecholaminergic systems (Auguet et al., 1982a,b) and the mechanism of endothelium-dependent vasorelaxation (Delaflotte et al., 1984).

Auguet et al. (1982a) showed that a concentration of EGb 761 (100 µg/ml) that did not provoke isometrically-recordable contractions of rabbit aortic strips, potentiated the contractile effect of norepinephrine (NE), i.e., reduced the EC_{50} (concentration required to elicit 50% of the maximal effect) from 75 nM to 36 nM, but did not influence the contractile effects of serotonin (5-HT) or dopamine (DA). EGb 761 caused a true potentiation of the contractile effect of NE that was evident as a lower EC_{50} value and an unmodified contractile capacity. These results indicated that EGb 761 might influence catecholaminergic systems by an indirect mechanism(s), tentatively considered to have been related to inhibition of catechol-

O-methyl transferase (COMT), monoamine oxidase (MAO) or permease action, inhibition of catecholamine uptake, and/or depolarization of the cellular membrane.

Higher concentrations of EGb 761 provoked a concentration-dependent contraction ($EC_{50} \simeq 1$ mg/ml) by an action that was antagonized by the α-adrenoceptor blocking agent phentolamine (10^{-7} M) (Auguet *et al.*, 1982b). Inhibitors of catecholamine re-uptake [cocaine (10^{-5} M) and desipramine (10^{-7} M)] potentiated the contractile effect of NE but inhibited the contractile effects of EGb 761 and tyramine. Collectively, these results signified that lower concentrations of EGb 761 can potentiate the contractile effect of NE in rabbit aortic strips, and that higher concentrations of the extract contract the strips by a mechanism that involves catecholaminergic receptors and which resembles that of tyramine more than that of NE (i.e., acting mainly by releasing endogenous catecholamines). The potentiation of contraction that occurred with the lower concentrations of EGb 761 may have some physiological significance. However, the relatively high concentrations of EGb 761 required to *elicit* contraction in these initial experiments could have caused non-specific effects, and therefore the relevance of these findings to *in vivo* and clinical actions of the extract remains inconclusive.

Other early experiments with rabbit aortic strips revealed that EGb 761 can exert a "papaverine-like" effect, indicating the possible involvement of guanosine 3',5'-cyclic phosphate (cyclic-GMP) phosphodiesterase in its musculotropic action (Auguet and Clostre, 1983). Comparison of the effects of EGb 761 (1–3 mg/ml) and papaverine (3×10^{-5} M) on the fast (phasic) and slow (tonic) phases of NE-induced contraction indicated that both agents acted mainly by increasing the relaxation that follows the rapid phase of contraction. As both flavonoids and papaverine efficiently inhibit cyclic-GMP phosphodiesterase (see Middleton and Kandaswami, 1993), this effect of EGb 761 was probably mediated by its flavonoid constituents. However, as such high concentrations of EGb 761 were required, this action requires verification under *in vivo* conditions.

In 1984, Delaflotte and coworkers first showed that EGb 761 can relax rabbit aortic strips by a mechanism involving "endothelium-derived relaxing factor(s)" (EDRF), but it was not known at that time that EDRF might be identical to nitric oxide (NO). [As EDRF is now generally considered to be NO or a closely related molecule (see DeFeudis, 1985; Palmer *et al.*, 1987; Ignarro *et al.*, 1987), it will be referred to as "EDRF/NO" or simply "NO" herein, except for those cases in which results were obtained before EDRF had been identified as NO.] EGb 761 (0.2 or 0.3 mg/ml) relaxed strips that had been precontracted with phenylephrine (10^{-7} M), removal of the endothelium from the strips abolished this relaxant effect, and a 30-minute pretreatment of the strips with nordihydroguaiaretic acid (NDGA, 10^{-5} M; an antioxidant) partially blocked the relaxant effect of EGb 761. Cyclo-oxygenase inhibitors (e.g., indomethacin, 10^{-5} M) did not affect EGb 761-mediated endothelium-dependent vasorelaxation, indicating that EDRF was not a prosta-glandin such as prostacyclin (PGI_2). Other experiments (S. Delaflotte, M. Auguet and F.V. DeFeudis, unpublished results, 1984) showed that a 30-minute pre-treatment of aortic strips with EGb 761 (300 μg/ml) inhibited the endothelium-dependent relaxation that was elicited by EGb 761 itself (see also DeFeudis *et al.*,

1985; DeFeudis, 1991). Thus, it appeared that various constituents in EGb 761 might either *enhance or inhibit* the EDRF mechanism. The finding that EGb 761 can inhibit its own EDRF/NO-dependent action has recently received some support from results which indicated that EGb 761 possesses "NO-scavenging" properties (see Marcocci *et al.*, 1994a,b; Ramassamy *et al.*, 1994; see below).

Rabbit isolated saphenous artery and vein

Recently, Auguet *et al.* (1994) have shown that EGb 761 (100 μg/ml) *potentiated* the contractile effects of low concentrations of NE in *endothelium-denuded rings* of saphenous artery. In contrast, in *endothelium-intact rings*, EGb 761 inhibited NE-induced contraction by an action that was opposed by pretreatment of the preparation with N^G-nitro-L-arginine (100 μM), a selective inhibitor of NO-synthase, indicating an involvement of EDRF/NO (see Fig. 1). As the NO-mediated, endothelium-dependent effect of EGb 761 was not elicited by an extract of *Ginkgo biloba* devoid of proanthocyanidins (see Fig. 1) or by the flavonoid troxerutin (200 μg/ml), it seems possible that proanthocyanidins, but not flavonols, may be implicated in endothelial activation by EGb 761 under the *in vitro* conditions employed.

It seems noteworthy that EGb 761 (100 μg/ml) *potentiated* the contractile effect of NE in aortic strips in the initial experiments of Auguet *et al.* (1982a; see above),

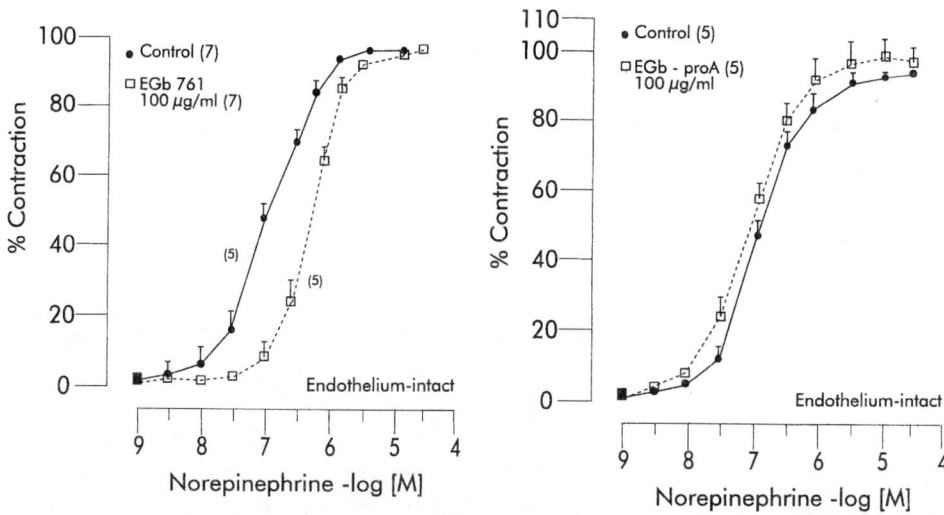

Figure 1 Interaction between EGb 761 (100 μg/ml; left panel) and *Ginkgo biloba* extract devoid of proanthocyanidins (EGb-proA, 100 μg/ml; right panel) on the log concentration-response curves for the cumulative addition of norepinephrine in endothelium-intact rings of a rabbit saphenous artery. Means ± S.E.M.; numbers of preparations in parentheses. [Reproduced from Auguet *et al.* (1994) with permission from Editions Scientifiques et Médicales, Elsevier Medical Department]

results that were tentatively interpreted as indicating that EGb 761 might indirectly influence catecholaminergic systems. However, in light of these newer results obtained with saphenous artery and vein (Auguet *et al.*, 1994), it appears that relatively low concentrations of EGb 761 can potentiate the contractile effect of NE only when the blood vessel is deendothelialized. Therefore, unless this discrepancy can be explained by differences in methodology or in the blood vessels tested, it is possible that the aortic strips used in the initial experiments (Auguet *et al.*, 1982a) had been deendothelialized to a significant extent during their preparation.

Isolated Heart Preparations

Basic and clinical findings support the contention that oxygen-derived free radicals are important mediators of ischaemia-reperfusion damage. Thus, it is expected that EGb 761, via scavenging free radicals that are already formed and/or by inhibiting the formation of free radicals (antioxidant effect), will exert beneficial effects on myocardial ischaemia-reperfusion damage.

In rat isolated hearts (Langendorff method) subjected to 40 minutes of global ischaemia, followed by 20 minutes of reperfusion, perfusion with a solution containing EGb 761 (200 μg/ml) enhanced cardiac functional recovery and suppressed the leakage of lactate dehydrogenase (LDH) during reperfusion (Haramaki *et al.*, 1994a,b). Both the decrease in myocardial ascorbate content and the increase in dehydroascorbate were also reduced, indicating that EGb 761 prevents the leakage and oxidation of ascorbate. Electron spin resonance (ESR) tests with the stable free radical diphenylpicrylhydrazyl (DPPH) revealed further that EGb 761 quenched the DPPH signal in a concentration-dependent manner, the maximal quenching effect being observed at 200 μg/ml (the concentration used in cardiac perfusion experiments). Thus, according to Haramaki *et al.* (1994a,b), in protecting against myocardial ischaemia-reperfusion injury, EGb 761 may replace ascorbate as a primary antioxidant, or it may reduce the oxidized form of ascorbate after ascorbate has acted as the primary antioxidant.

The effect of EGb 761 on free radical-induced myocardial ischaemia-reperfusion injury in rat isolated perfused heart has also been examined using hemodynamic measurements and electron spin resonance (ESR) with 5,5-dimethyl-1-pyrroline 1-oxide (DMPO) (Pietri *et al.*, 1993). Perfusion with a solution containing EGb 761 (1 or 10 μg/ml) enhanced recovery of cardiac function within the initial 10 minutes of post-ischaemic reflow, an effect that occurred concomitantly with a partial inhibition of the hydroxyl radical DMPO spin adduct (DMPO-OH) in coronary effluents. With the 10 μg/ml concentration of EGb 761, hemodynamic parameters did not differ significantly from control values after 15 minutes of reperfusion. In a further study, ginkgolides A and B and bilobalide were tested *in vitro* (50, 50 and 150 ng/ml, respectively) and found to be effective in inhibiting the formation of the DMPO-OH spin-adduct and in improving cardiac function (see companion chapter).

In another study, Tosaki and Droy-Lefaix (1993) showed that perfusion of rat isolated hearts with the combination of superoxide dismutase (SOD; 5×10^4 U/l) and EGb 761 (50 μg/ml) significantly reduced the occurrence of both ventricular fibrillation (VF) and ventricular tachycardia (VT) that were induced by reperfusion. ESR spectroscopic experiments showed that application of EGb 761 (200 μg/ml) alone, or co-administration of EGb 761 (50 μg/ml) and SOD (5 × 10⁴ U/l), nearly abolished the formation of oxygen radicals during reperfusion. Such results indicate that *combination therapy* with EGb 761 and another antioxidant or free radical scavenger may act synergistically to reduce the formation of free radicals and the incidence of reperfusion-induced VF and VT. As EGb 761 is pharmacologically active upon oral administration (see also Drieu and DeFeudis, this volume), these findings may have clinical implications in relation to myocardial ischaemia and cardiac arrhythmias.

Cellular Preparations

Vascular endothelial cells

The effect of EGb 761 on ischaemia-induced metabolic changes has been assessed by exposing human endothelial cells to hypoxia-reoxygenation or hypoxia alone (Michiels *et al.*, 1994). In hypoxia-reoxygenation experiments, in which endothelial cell mortality is caused by free radicals, EGb 761 protected the cells against reoxygenation toxicity, maximal protection being attained with 0.3 mg/ml of the extract. Flavonoid constituents of EGb 761, which have antioxidant and free radical-scavenging activities (see below), may have been responsible for this effect. In cells that were exposed to hypoxia alone, which induces endothelial cell activation and stimulates phospholipase A_2 (PLA_2) activity, EGb 761 prevented both the decrease in ATP and the synthesis of inflammatory mediators. Other tests showed that CP 205 (a *Ginkgo biloba* extract devoid of terpene trilactones) was not protective, that ginkgolide B and bilobalide could inhibit the hypoxia-induced decrease in ATP, and that bilobalide (but not ginkgolide B) offered some protection against hypoxia-induced PLA_2 activation (Michiels *et al.*, 1994). Calculations revealed that values for percentage protection at concentrations corresponding to 0.5 μg of EGb 761/ml were: 47% for EGb 761 itself, 51% for bilobalide, and 31% for ginkgolide B. Thus, this protective effect of EGb 761 appeared to be mediated by its terpenoid constituents, bilobalide being the most effective. These findings, obtained with a therapeutically relevant concentration of EGb 761 (i.e., 0.5 μg/ml), are useful in explaining the beneficial effects of EGb 761 in treating vascular ischaemic diseases.

The effect of EGb 761 on oxygen free radical-mediated injury to cultured endothelial cells of human umbilical vein has been evaluated by Sellak *et al.* (1994). Cells were exposed for 30 minutes to oxygen-activated species [6 nmoles of superoxide anion radical ($O_2^{-\bullet}$)/min/ml] from the hypoxanthine-xanthine oxidase reaction, and cytotoxicity was evaluated 24 hours later. EGb 761 provided concentration-dependent protection, 20 μg/ml of the extract protecting about 50% of endothelial cells against oxygen free radical-mediated injury when present in the incubation medium for 1 hour before applying the oxidative stress. As EGb 761 protected to a lesser extent when it was added just before the oxidative stress, it appeared that constituents of the extract must enter the cell to exert this effect.

Thus, EGb 761 appeared to act not only by scavenging extracellular oxygen radicals (mainly $O_2^{-\bullet}$), but also by preventing intracellular injury caused by oxidants that probably traverse the cell membrane. EGb 761 also protected these cells against the increases in LDH release and lipid peroxidation and the reduction in intracellular glutathione (GSH) that were induced by the oxidative stress. Taken together, these findings support the conception that EGb 761 protects cells against oxygen free radical-mediated injury.

Vascular smooth muscle cells

Using smooth muscle cells of pig aorta that were maintained in culture, Bruel *et al.* (1989) have shown that EGb 761 has biphasic effects on both glucose uptake (measured with a tracer amount of the non-metabolizable substrate, 2-deoxy-D-[1-^{14}C]glucose) and glycogen synthesis (measured as D-[U-^{14}C]glucose incorporation into intracellular glycogen), the maximal stimulatory effect occurring at 0.25 μg/ml in both cases. EGb 761 acted on facilitated glucose transport, since its effect was not evident in the presence of cytochalasin-B, which is an inhibitor of this transport system. These results indicate that EGb 761 might improve "nutrient disposition" in the vascular wall, a contention that receives some support from recent *ex vivo* experiments conducted with rat erythrocytes (Rapin *et al.*, 1994; see Drieu and DeFeudis, this volume).

Diploid fibroblasts

Toussaint *et al.* (1995) tested the capabilities of EGb 761 and bilobalide to protect human diploid WI-38 fibroblasts that were exposed to ethanol (5%, v/v) for 2 hours, a condition that involves a decrease in energy availability. Both EGb 761 (5 μg/ml) and bilobalide (100 ng/ml) protected against ethanol-induced cell death on the second day of incubation. These results imply that the protective effects of EGb 761 and bilobalide that have been observed during hypoxia (Michiels *et al.*, 1994; see above) may apply to other deleterious conditions that decrease cellular ATP concentration.

Action potential and transmembrane ionic currents in cardiomyocytes

Masson *et al.* (1994) used the "whole-cell" configuration of the patch-clamp technique to assess the effects of EGb 761 on the action potential and individual transmembrane ionic currents in ventricular myocytes of guinea pig heart. EGb 761 (5–50 μg/ml) did not affect the normal action potential or the various ionic currents involved in its generation; i.e., fast inward sodium current, inward calcium currents, delayed outward potassium current, inward rectifying potassium current, and ATP-sensitive potassium current evoked by 2,4-dinitrophenol. However, exposure of the cells to EGb 761 (≥ 5 μg/ml) elicited a marked concentration-dependent inhibition of isoproterenol-induced Cl$^-$ current (I_{Cl}), the maximal effect being observed at 50 μg/ml (see Fig. 2). This effect of EGb 761 occurred rapidly, became maximal

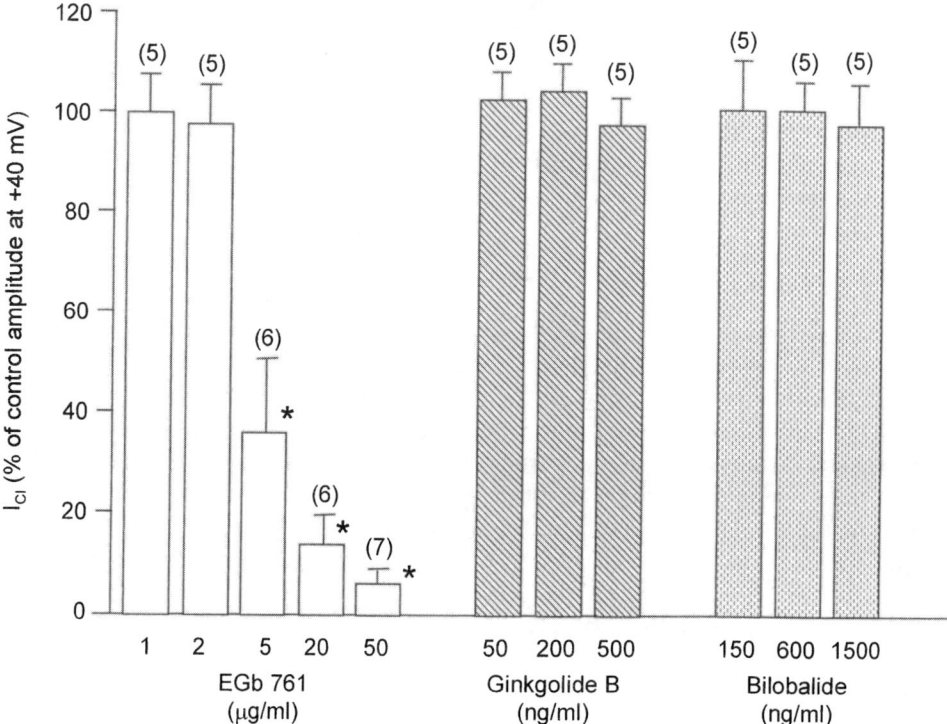

Figure 2   Effect of various concentrations of EGb 761, ginkgolide B and bilobalide on isoproterenol-induced I_{Cl}. I_{Cl} was measured during 100-msec depolarizing voltage-clamp steps to +40 mV from a holding potential of –40 mV by subtracting digitized current traces recorded under control conditions from those recorded after exposure to 1 μM isoproterenol, either in the absence or presence of EGb 761, ginkgolide B or bilobalide at the indicated concentrations. Results are expressed as percentages of current amplitude obtained in the presence of isoproterenol alone. Each value represents the mean ± S. E. M. The number of cells tested is given in parentheses; * denotes p <0.05. [Reproduced from Masson *et al.* (1994) with permission from John Wiley & Sons, Inc.]

after 2–3 minutes, and was completely reversible within a few minutes after withdrawing EGb 761 (see Fig. 2), characteristics which indicate that a certain constituent(s) of the extract exerted a direct effect at the level of the Cl⁻ ionophore. Ginkgolide B or bilobalide, at concentrations which approximated those present in the effective concentrations of EGb 761, did not affect isoproterenol-induced I_{Cl}, even after more than 10 minutes of action, indicating that the constituent(s) of EGb 761 that is involved is *not a terpene trilactone*. This action of EGb 761 may have therapeutic significance since isoproterenol-activated I_{Cl} could be involved in sympathetic hyperactivity, hypoxia and ischaemia, pathophysiological conditions for which EGb 761 treatment has beneficial effects.

Erythrocytes

Erythrocyte rheology (i.e., aggregation and flexibility), together with hematocrit and plasma viscosity, are the major determinants of blood fluidity. In certain pathological states involving modified erythrocyte rheology, the reversal of such changes by EGb 761 is deemed to be therapeutically beneficial.

Amort *et al.* (1991) compared the capacities of EGb 761 (60 μg/ml), pentoxifylline (Trental®; 80 μg/ml) and naftidrofuryl (Dusodril®; 30 μg/ml) to offset damage to the micromechanical properties of human erythrocytes that was produced by exposure to an acidotic/hyperosmolar medium (AHS: pH 6.8; 380 mosmol/liter). Compared to control conditions (pH 7.4; 284 mosmol/l), AHS-suspended erythrocytes showed no resting shape change, 22.8% increment in elasticity ($p < 0.05$), 70.8% increase in relaxation time ($p < 0.01$), and 86.4% increase in viscosity ($p < 0.01$). Significant relative changes in micromechanical parameters existed between AHS + drug-treated and AHS-treated erythrocytes. EGb 761 and pentoxifylline had comparable beneficial effects.

Artmann and Schikarski (1993) used a microscopic photometric monolayer technique to measure time-dependent microrheological parameters of erythrocytes (cell resting shape, elasticity/rigidity, relaxation time and viscosity) in analyzing the time-course of oxidative damage induced by H_2O_2 and the antioxidant (protective) capacities of EGb 761, vitamin E (α-tocopherol acetate) and pentoxifylline. Microrheological determinants of shear-induced erythrocyte deformation were used to evaluate free radical-induced damage. Incubation of erythrocytes of healthy human donors in a phosphate-buffered solution (21° C) containing H_2O_2 (1 mM) caused time-dependent damage to erythrocyte membranes that was associated with echinocytosis, cell swelling, cell stiffening, shortening of relaxation time and increased cell viscosity. Incubation of erythrocyte membranes from the same donors for 60 minutes in the presence of the drugs at a concentration of 30 μg/ml provided data for calculating "protection coefficients" (measures of inhibition of H_2O_2-induced cell damage) which were: EGb 761, 37.9%; vitamin E, 35.8%; pentoxifylline, 30.4% ($p < 0.05$ in all cases). These results, taken together with the *ex vivo* finding that repeated oral administration of EGb 761 (200 mg/day) inhibited H_2O_2-induced cell damage to erythrocytes of healthy humans (see Drieu and DeFeudis, this volume), provide good evidence that EGb 761 protects against oxidative damage.

Platelets

Although the total EGb 761 preparation appears to exert inhibitory effects on platelet aggregation mediated by some classic inducers (e.g., thrombin, ADP), some of which can occur with IC_{50} values of about 50 μg/ml of the extract (see DeFeudis, 1991; Stücker *et al.*, 1994; Drieu and DeFeudis, this volume), many of the recent studies in this domain have involved analyzing the effects of PAF-antagonistic ginkgolide constituents of EGb 761.

Korth *et al.* (1988) found that the binding of [³H]PAF to intact human platelets occurred with a K_D value (apparent dissociation constant) of 0.54 nM. They showed further, using a concentration range of 0.065–6.5 nM [³H]PAF, that ginkgolide B (at the relatively high concentration of 60 μM) displaced some of the [³H]PAF that was bound to the platelets. Akkerman (1988) showed, using 0.1 nM [³H]PAF, that after 10 minutes of platelet-PAF interaction, an almost complete dissociation of bound PAF could be obtained by adding a large excess of ginkgolide B (10 μM). Such findings have provided some insight into the signal-generating capacity of the PAF-receptor, but more potent PAF-antagonists are required to determine inhibitory effects on PAF-binding at high doses of PAF (see Akkerman, 1988).

It has been reported that ginkgolide B inhibits PAF binding to rabbit or human washed platelets by a highly selective mechanism since it does not appear to interact with other receptors (see Coëffier, 1988). This effect has also been considered to be specific since no inhibition was observed with other pro-aggregatory agents such as ADP, arachidonate, ionophore A 23187, collagen and thrombin (Braquet et al., 1985). Although this pronounced selectivity and specificity of ginkgolide B represents a useful property when studying the mechanism involved in PAF-induced platelet activation, it is essential to recall that many inducers of platelet aggregation besides PAF are present in intact animals, including humans. Due to this plethora of possible initiators of platelet aggregation and thrombus formation in vivo, it is inconceivable that inhibition of the effects of PAF alone would offer exclusive protection against thrombotic or related phenomena. Constituents of EGb 761 that inactivate $O_2^{-\bullet}$ may be more significantly involved in the "anti-platelet" effects of the extract since $O_2^{-\bullet}$ inactivates NO (e.g. Ignarro et al., 1987) which has an anti-aggregatory effect on platelets (see e.g., Radomski et al., 1987).

ACTIONS OF EGb 761 ON THE CENTRAL NERVOUS SYSTEM

The effects of EGb 761 on the CNS underlie one of its major therapeutic indications; i.e., individuals suffering from deteriorating cerebral mechanisms related to memory, other cognitive functions and neurosensory systems. Recent in vitro studies of EGb 761 and some of its active constituents that have been performed with synaptosome-enriched preparations have provided results that appear useful in explaining such beneficial effects.

Uptake of Biogenic Amines

Using synaptosome-enriched fractions of rat brain, Taylor (1990) showed that a relatively high concentration of EGb 761 (1 mg/ml) inhibited [³H]NE uptake in occipital cortex by 99%, [³H]DA uptake in corpus striatum by 95%, and [³H]5-HT uptake in frontal cortex by 60%. Corresponding K_i values (inhibitor concentrations required to decrease ligand binding by 50%) for these effects of EGb 761 were 0.12 mg/ml for NE, 0.21 mg/ml for DA, and 0.44 mg/ml for 5-HT (see also Taylor, 1992). Examination of these data in further detail reveal that this action of EGb 761 appears to be biphasic, at least for NE and DA; i.e., very low concentrations of EGb 761 (0.1–1 μg/ml) increased the uptake of NE and DA, but not that of 5-HT (Taylor et al., 1991).

Since it is improbable that the high concentrations of the EGb 761 constituents required to inhibit biogenic amine uptake could ever be acquired in the brain following administration of the usual human therapeutic dosage, Ramassamy et al. (1992a) examined further the effects of low concentrations of the extract. Results showed that EGb 761 had a biphasic effect on [³H]5-HT uptake by a synaptosomal fraction of mouse cerebral cortex; at 4–16 μg EGb 761/ml, uptake was significantly *increased*, whereas at 32 μg EGb 761/ml and at higher concentrations 5-HT uptake was *inhibited* (see Fig. 3). The latter effect is probably non-specific since it was also observed with respect to the uptake of choline and DA (see Ramassamy et al., 1992a), but some element of specificity was apparent in this study since low concentrations of EGb 761 (2–16 μg/ml) did not modify [³H]DA uptake by mouse striatal synaptosomes. Also, the 5-HT uptake inhibitor clomipramine (1 μM) inhibited the EGb 761-induced increase in [³H]5-HT uptake (see Fig. 3). Incubation of the synaptosomes in the presence of the ginkgolide fraction of EGb 761 (BN 52063, which contains ginkgolides A, B and C in the proportions 2:2:1) at concentrations of 0.25 and 10 μg/ml did not increase [³H]5-HT uptake, whereas an extract

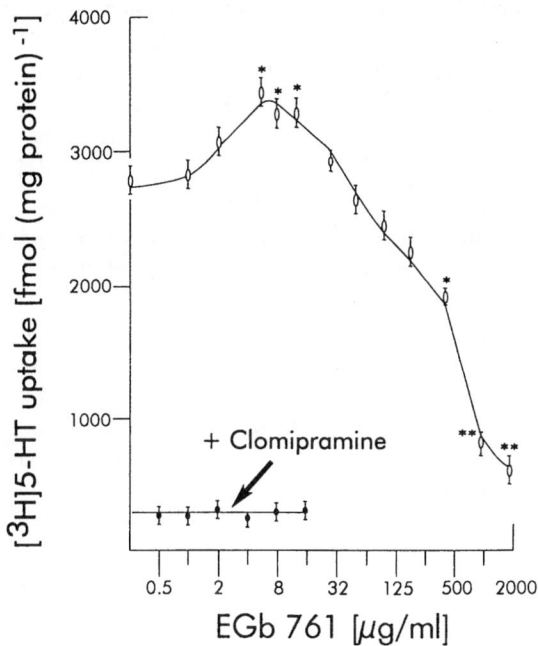

Figure 3   Effects of EGb 761 on synaptosomal [³H]5-HT uptake. Synaptosomes prepared from mouse cerebral cortex were incubated with increasing concentrations of EGb 761 for 5 minutes before adding [³H]5-HT (20 nM, final concentration). Incubation was then continued for 5 minutes. Means ± S.E.M. of 3 experiments in triplicate; * p <0.05 compared with control; ** p <0.01 compared with control. Clomipramine concentration was 1.0 μM. [Reproduced from Ramassamy et al. (1992a) with permission from The Royal Pharmaceutical Society of Great Britain]

corresponding to EGb 761 devoid of terpenes and containing mainly flavonoid substances (CP 205; 1.25 and 5 μg/ml) increased [³H]5-HT uptake. These findings imply that flavonoids or compounds other than the terpene trilactone constituents of EGb 761 are involved in mediating its effect on [³H]5-HT uptake (Ramassamy et al., 1992a).

The increase in [³H]5-HT uptake promoted by the lower concentrations of EGb 761 could be therapeutically relevant, especially since it also occurred after oral administration of the extract (Ramassamy et al., 1992a; see Drieu and DeFeudis, this volume). This finding, taken together with the finding that the antidepressant tianeptine increases [³H]5-HT uptake under ex vivo or in vivo conditions (see e.g., Fattaccini et al., 1990), indicates that EGb 761 may have "antidepressant-like" activity in vivo.

Uptake of DA and 5-HT Following Prolonged Incubation

Ramassamy et al. (1992b) recognized that biogenic amine uptake decreases progressively with extended incubation (> 20 min) of synaptosomes in media of the usual composition, and that the ascorbic acid generally present in such media (to prevent oxidation of catecholamines) can react with trace amounts of ferrous ions (Fe^{2+}) in the presence of oxygen to form free radicals. Therefore, they assessed the ability of EGb 761 to prevent this loss of amine uptake capacity. Incubation of mouse brain synaptosomes in oxygenated Krebs-Ringer medium at 37° C in the presence of ascorbic acid (0.1 mM) led to decreases in striatal [³H]DA uptake and cerebral cortical [³H]5-HT uptake after 20 minutes, and uptake was essentially abolished after 60 minutes. Ascorbic acid (1–100 μM) decreased uptake in a concentration-dependent manner by an effect that was enhanced by Fe^{2+} and prevented by the Fe^{2+}-chelator desferrioxamine, indicating a dependence on free radical generation.

The decrease in synaptosomal DA uptake that occurred during a 60-minute incubation in the presence of ascorbic acid (0.1 mM) was prevented by EGb 761 (4–16 μg/ml) (see also Ramassamy et al., 1994). The flavonoid-containing fraction of EGb 761 (CP 205; $\geq$ 1 μg/ml) also prevented this decrease in synaptosomal DA uptake in a concentration-dependent manner, whereas the ginkgolide fraction (BN 52063; $\leq$ 0.5 μg/ml) was ineffective. Thus, the effect of EGb 761 of prolonging the ability of synaptosomes to take up radioactive biogenic amines, such as [³H]DA and [³H]5-HT, appears to be mediated by its non-terpenoid constituents (e.g., flavonoids, which, via their free radical-scavenging actions, can inhibit lipid peroxidation; see below).

Membrane Fluidity

As mentioned above, mouse striatal synaptosomes rapidly lose their ability to take up [³H]DA when they are incubated in medium containing 0.1 mM ascorbic acid. This detrimental effect of ascorbic acid has recently been correlated with a decrease in synaptosomal membrane fluidity (Ramassamy et al., 1993b). The fluorescence emitted by 1,6-diphenyl-1,3,5-hexatriene (DPH; an agent considered appropriate for measuring fluidity at membrane depths corresponding to those of fatty acid side chains and between lipid bilayers) was increased in the presence of ascorbic acid, indicating decreased fluidity. This decrease in membrane fluidity was enhanced by Fe^{2+} (1 μM) and prevented by desferrioxamine (0.1 mM), as well as by EGb 761 (2–16 μg/ml), quercetin ($\geqslant$ 0.3 μM) and trolox C ($\geqslant$ 0.1 mM). Hydroxyl radicals ($OH^{\bullet}$) were probably involved since their formation via lipid peroxidation appears to reduce membrane fluidity (see Rosen *et al.*, 1983). It was concluded that the peroxidation of neuronal membrane lipids provoked by ascorbic acid/Fe^{2+} is associated with a decrease in membrane fluidity, which then diminishes the efficiency of the DA-transporter.

Similar results were obtained in experiments on [^{3}H]5-HT uptake by synaptosomes of mouse cerebral cortex (Ramassamy *et al.*, 1993a). The flavonoid constituents of EGb 761 were probably responsible for its protective effect since quercetin (present in the form of various heterosides in EGb 761) was effective and since the flavonoid-containing fraction, but not the terpenoid fraction, of EGb 761 prevented the decrease in membrane fluidity.

In sum, free radicals generated by ascorbic acid/Fe^{2+}, by peroxidizing membrane lipids, may decrease synaptosomal membrane fluidity, thereby altering the functions of DA- and 5-HT-transporters under *in vitro* conditions. Flavonoid constituents of EGb 761, apparently via their free radical-scavenging and possibly Fe^{2+}-chelating activities, permit the membrane to retain its fluidity, thereby protecting against this deleterious situation. Such findings support the conception that EGb 761 can prevent neurodegeneration.

FREE RADICAL-SCAVENGING AND ANTIOXIDANT ACTIVITIES OF EGb 761

Since EGb 761 possesses antioxidant activity and scavenges free radicals, particularly oxygen-centered free radicals, it could provide useful therapy in many disease states. Studies that have been focussed on these properties of EGb 761 and some of its chemical constituents, are discussed below. For purposes of discussion, it is considered that those agents which reduce free radical-mediated deleterious effects by scavenging free radicals that are already formed are "free radical-scavengers", and that those which act by inhibiting the generation of free radicals are "antioxidants".

Experiments with Neutrophils

Stimulation of neutrophils obtained from healthy human volunteers with phorbol myristate acetate (80 nM) leads to increased O_2 consumption ("respiratory burst") and the production of $O_2^{-\bullet}$ and H_2O_2. Pincemail *et al.* (1987) showed that EGb 761 (125–500 μg/ml) delays this action by inhibiting NADPH-oxidase (which catalyzes the reduction of O_2 to $O_2^{-\bullet}$), thereby significantly decreasing $O_2^{-\bullet}$ and H_2O_2

production. OH$^\bullet$ formation was markedly decreased at concentrations of EGb 761 as low as 15.6 μg/ml, indicating that EGb 761 also has free radical-scavenging activity. With regard to molecular mechanisms, NADPH-oxidase was probably inhibited by flavonoid constituents of EGb 761, since flavonoids such as quercetin and kaempferol are known to inhibit this enzyme system (see e.g., Middleton and Kandaswami, 1993). The observed free radical-scavenging activity is also attributable mainly to the flavonoid constituents of EGb 761 (see e.g., DeFeudis, 1991).

The *in vitro* effect of EGb 761 on $O_2^{-\bullet}$ formation has also been examined in human neutrophils that were activated by f-Met-Leu-Phe or Type I collagen (Monboisse *et al.*, 1993). $O_2^{-\bullet}$ production was assayed as SOD-inhibitable reduction of ferricytochrome c. Addition of EGb 761 to the incubation medium simultaneously with either of the activating agents caused a concentration-dependent inhibition of $O_2^{-\bullet}$ formation by the neutrophils ($IC_{50} \simeq 10$ μg/ml). Preincubation of neutrophils with EGb 761 for 10 minutes before activation with Type I collagen also inhibited $O_2^{-\bullet}$ production.

Collectively, these results provide good evidence that EGb 761 is a potent inhibitor of $O_2^{-\bullet}$ production by neutrophils, an effect that could be mediated via its antioxidant activity and/or by its capacity to modify the intracellular signalling pathway of NADPH-oxidase activation. This information is useful in explaining the beneficial actions of EGb 761 in those disease states which involve excessive free radical production by neutrophils.

Anti-Lipoperoxidative Effects of EGb 761

Lipid peroxidation and free radical generation in microsomes

Pincemail and Deby (1986) examined the *in vitro* effect of EGb 761 (125–500 μg/ml) on lipid peroxidation in rat liver microsomes. Lipid peroxidation was induced by the NADPH/FeCl$_3$/ADP system, and assessed 30 minutes later as malondialdehyde (MDA) production via lipid peroxide degradation. EGb 761 abolished MDA formation at all concentrations tested, indicating protection of the membranes against lipid peroxide-induced damage.

More recently, Dumont *et al.* (1993) showed that EGb 761 can protect rat liver microsomes against irradiation-induced peroxidative degradation of fatty acids and proteins. Analyses by gas chromatography coupled to the assay for "thiobarbituric acid reactive substances" (TBARS) showed that exposure of the microsomes to UV-C irradiation for 13 hours elicited a progressive degradation of membrane fatty acids, polyunsaturated fatty acids (PUFAs) being preferentially affected. Concentrations of EGb 761 as low as 7.5 μg/ml protected PUFAs, regardless of their individual susceptibility to peroxidation. This effect of EGb 761 was concentration-dependent and accompanied by a decrease in the production of TBARS (e.g., MDA). EGb 761 also protected membrane proteins against the irreversible polymerization induced by lipid peroxides. It was considered that these *in vitro* antioxidant effects of EGb 761 involved its flavonoid constituents (e.g., flavonol glycosides, proanthocyanidins).

Lipid peroxidation in lymphocytes and brain synaptosomes

The role of oxygen free radicals in the ageing process appears to be related to an increase in lipid peroxidation and to decreases in DNA repair capacity and antioxidant enzyme activity (see e.g., Sram *et al.*, 1986; Ames, 1989). In line with such reasoning, Sram *et al.* (1993) have compared the effects of EGb 761 and α-tocopherol on lipid peroxidation and oxidative damage to DNA in human lymphocytes and on lipid peroxidation in rat brain synaptosomes. Markers of ageing were: unscheduled DNA synthesis (incorporation of methyl [^{3}H]thymidine, or 1-methyl-3-nitro-1-nitrosoguanidine-induced DNA damage) in lymphocytes; TBARS assay in neocortex, hippocampus and hypothalamus; and, glutathione peroxidase and SOD activities in brain. The oxidative injury to DNA and membranes of human peripheral lymphocytes and rat brain synaptosomes caused by the exogenous free radical-generating system Fe^{2+}/H_2O_2 was appraised.

EGb 761 and α-tocopherol (both at 10–500 μg/ml) had similar inhibitory effects on the lipid peroxidation chain reaction induced by the free radical-generating system in both lymphocytes and synaptosomes. At concentrations above 200 μg/ml, both antioxidants reduced the formation of lipid peroxidation products to < 20% of the value obtained in the absence of added substances. EGb 761 (400 μg/ml) protected DNA of human lymphocytes against oxidative damage to about the same extent as α-tocopherol (100–150 μg/ml). Sram *et al.* (1993) suggested that this antioxidant activity of EGb 761 may be explained by its flavonoid glycoside and proanthocyanidin constituents (see also Dumont *et al.*, 1993; see above).

Oxygenated free radicals and lipid peroxidation in the retina

The retina is especially enriched in PUFAs and is therefore very sensitive to the detrimental actions of reactive oxygen free radicals. Also, it seems clearly established that the concentrations of these free radicals are increased in the retinae of diabetic animals, a condition that coincides with markedly increased lipid peroxidation (see e.g., Nishimura and Kuriyama, 1985).

In line with this rationale, Doly *et al.* (1990) have tested the effect of EGb 761 on the evolution of the electroretinogram (ERG), recorded in response to light stimulation (300 lux, 1 msec, white light), while exposing isolated retinae to oxygenated free radicals. Free radical generation was induced by adding the peroxidative mixture of $FeSO_4$ + sodium ascorbate to the perfusion solution. In the absence of EGb 761, ERG amplitude decreased very rapidly, becoming immeasurable 1 hour after inducing peroxidation. However, when oxygenated free radicals were generated in the presence of EGb 761 (100 μg/ml; added to the perfusion solution simultaneously with the peroxidative mixture), the amplitude of the ERG was practically normal. This finding has been further validated in recent *ex vivo* studies (see Drieu and DeFeudis, this volume).

Acellular Systems

Free radical-scavenging activity of EGb 761 shown with γ-radiolysis

Gardès-Albert *et al.* (1993) have irradiated aqueous solutions of EGb 761 (25–200 μg/ml) buffered with sodium phosphate (100 mM, pH 7.0) with doses up to 250 Gy (dose rate = 3.8×10^{-2} Gy/sec). Gamma irradiations were performed with a ^{60}Co irradiator whose activity was 80 Ci (dose rate = 3.4×10^{-2} Gy/sec) and the ir-radiated solutions were analyzed by spectrophotometric absorption, the absorption spectra being recorded from 240–500 nm. The following oxygen-centered free radicals, selectively produced by γ-radiolysis were analyzed: $OH^\bullet$ in N_2O saturated solutions, $O_2^{-\bullet}$ in oxygen saturated solutions containing sodium formate (100 mM), $OH^\bullet$ and $O_2^{-\bullet}$ in oxygen saturated solutions, and $OH^\bullet$ in oxygen saturated solutions containing SOD.

Results showed that $OH^\bullet$ reacted with EGb 761 to a similar extent over the concen-tration range of 25–200 μg/ml that was tested, indicating very efficient scavenging of $OH^\bullet$ by the extract. However, the extract devoid of terpene trilactones (CP 205) was less effective at the lower concentrations tested (25–100 μg/ml), indicating a less efficient reaction with $OH^\bullet$ (in both the presence and absence of oxygen). EGb 761 also scavenged $O_2^{-\bullet}$ in a concentration-dependent fashion, whereas CP 205 did not react with $O_2^{-\bullet}$ at any concentration tested (25–200 μg/ml), indicating that the terpene constituents of EGb 761 (perhaps by an indirect action) may be significantly involved in $O_2^{-\bullet}$-scavenging. An analysis of the simultaneous actions of $OH^\bullet$ and $O_2^{-\bullet}$ on EGb 761 or CP 205 showed that CP 205 scavenges only $OH^\bullet$ and that EGb 761 reacts in a quantitative additive fashion with both $OH^\bullet$ and $O_2^{-\bullet}$, indicating that these two free radicals probably occupy distinct, non-interacting molecular sites on certain con-stituents of EGb 761. On this basis, it appears that EGb 761 could offer substantial protection against the deleterious effects of both $^\bullet OH$ and $O_2^{-\bullet}$.

Oxidation of human low-density lipoproteins (LDL)

The effect of EGb 761 on the oxidation of LDL has been tested by subjecting aqueous solutions of human LDL (3 mg/ml) to the actions of $O_2^{-\bullet}$ and $OH^\bullet$ that were selectively generated via γ-radiolysis of water (Gardès-Albert et al., 1995). EGb 761 protected against the lipid peroxidation of LDL in a concentration-dependent manner, the maximal effect being obtained at 100 μg/ml. This finding was con-sidered to be due to a competition between LDL components and EGb 761 for $OH^\bullet$ and $O_2^{-\bullet}$ which favored EGb 761 when its concentration was 100 μg/ml, a condition under which the extract appears to scavenge nearly all of the radiolytically-generated oxygen free radicals. Other tests showed that EGb 761 also caused a marked concentration-dependent decrease in $O_2^{-\bullet}$-induced LDL peroxidation, as assessed by monitoring several lipid peroxidation markers (e.g., decreases in endo-genous vitamin E and β-carotene; formation of TBARS; formation of conjugated dienes). As it is now generally believed that oxidation of LDL is involved in the pathogenesis of atherosclerosis, such findings support the view that EGb 761 may be effective in delaying the progression of this disease.

Experiments with nitric oxide (NO)

As mentioned above, early in vitro experiments conducted under physiological con-ditions by Delaflotte et al. (1984) showed that EGb 761 can either enhance or inhibit the EDRF/NO mechanism (see also DeFeudis, 1991). The conclusion that

EGb 761 can inhibit the EDRF/NO mechanism has been supported by the recent findings of Marcocci *et al.* (1994a) who showed that EGb 761 can scavenge NO in acellular systems. EGb 761 competed with oxyhemoglobin for reaction with NO that was generated during the interaction of hydroxylamine with Complex I of catalase. Also, EGb 761 caused a concentration-dependent decrease ($IC_{50} \simeq 20$ μg/ml) in the amount of nitrite formed in the reaction of O_2 with NO that was produced from a solution of 5 mM sodium nitroprusside (see also Marcocci *et al.*, 1994b).

CONCLUDING REMARKS

From the foregoing discussion, it is evident that the *polyvalent* action of EGb 761, which necessarily depends upon its various active components, can influence many mechanisms under *in vitro* conditions. Some of these effects, particularly those elicited by concentrations of EGb 761 or its constituents in the ng/ml or μg/ml range, may well be responsible for the protection that the extract provides in clinical conditions involving vascular pathology, CNS disturbances and other disease states. It is this polyvalency of action of EGb 761 that provides the basis for considering that EGb 761 represents an attractive candidate for treating multifactorial processes involved in chronic disease states. This view becomes even further strengthened upon correlating *in vitro* findings with the *in vivo* and *ex vivo* results discussed in the companion chapter (Drieu and DeFeudis, this volume).

Several new developments appear to deserve special mention. With regard to cardiovascular mechanisms, experiments with rabbit saphenous artery and vein (Auguet *et al.*, 1994) indicate that proanthocyanidins, but not flavonols, may be implicated in EGb 761-induced endothelial activation and EDRF/NO release under *in vitro* conditions. However, an extrapolation of this interesting result to *in vivo* conditions is limited since intact proanthocyanidins and other unmodified flavonoidic constituents of EGb 761 may not be bioavailable after oral administration of the extract (see Pietta *et al.*, 1995; see also Drieu and DeFeudis, this volume).

Experiments with isolated rat heart have shown that EGb 761, in protecting against myocardial ischaemia-reperfusion injury, may replace ascorbate as a primary antioxidant (Haramaki *et al.*, 1994a,b). Also, combination therapy with EGb 761 and another antioxidant or free radical scavenger may act synergistically to reduce the incidence of reperfusion-induced arrhythmias (Tosaki and Droy-Lefaix, 1993). These findings, which may have been mediated in part by ginkgolides (see companion chapter), could have clinical implications in relation to ischaemia-induced cardiac dysfunction.

Experiments with vascular endothelial cells have revealed that EGb 761 can protect against the detrimental effects of hypoxia, an action that appears to be mediated by its terpenoid constituents, bilobalide being the most effective (Michiels *et al.*, 1994). This finding indicates that terpenoid substances that are unique to EGb 761 have actions other than PAF-antagonism. Certain behavioral effects of EGb 761 that are not related to PAF-antagonism may also involve actions of its terpenoid constituents (see Drieu and DeFeudis, this volume). The inhibitory action of EGb

761 on isoproterenol-activated inward chloride current (I_{Cl}), shown in ventricular myocytes, which, in contrast, appeared to be caused by a non-terpene constituent(s) of the extract (Masson *et al.*, 1994), may be related to its *in vivo* antiarrhythmic activity (see Drieu and DeFeudis, this volume).

With regard to studies of formed elements of the blood, Artmann and Schikarski (1993) showed that EGb 761 protects against H_2O_2-induced oxidative damage in erythrocytes. Studies with neutrophils have indicated that EGb 761 is a potent inhibitor of excessive $O_2^{\bullet-}$ production (Pincemail *et al.*, 1987; Monboisse *et al.*, 1993). With regard to EGb 761-induced inhibition of platelet activation, ginkgolide B does appear to have some inhibitory activity when PAF is used as the inducer, but it seems unlikely that specific and selective inhibition of the effects of PAF alone could offer significant protection against thrombosis or related phenomena. Clearly, further study is required to define the chemical constituents of EGb 761 that mediate its therapeutic effects of decreasing whole blood viscosity and increasing microcirculation.

Regarding the CNS, experiments with synaptosome-enriched preparations of rat brain which have shown that low concentrations of EGb 761 (10–20 μg/ml) can increase [^{3}H]5-HT uptake may have therapeutic relevance, especially since this effect also occurred after oral administration of the extract (Ramassamy *et al.*, 1992a). Such findings, taken together with other recent results (see Drieu and DeFeudis, this volume), lead one to suspect that the extract may have "antidepressant-like" activity *in vivo*, a conception that is supported by recent clinical observations (see e.g., Schubert and Halama, 1993). Other interesting findings have revealed that EGb 761 can oppose the decrease in synaptosomal membrane fluidity and modified function of the DA- and 5-HT-transporters that occur during prolonged *in vitro* incubations, effects that may involve its flavonoid constituents (Ramassamy *et al.*, 1992b, 1993b).

Several new studies have indicated that EGb 761 possesses all (or nearly all) of the properties required in characterizing a substance as an antioxidant and/or free radical-scavenger. Experiments with human neutrophils have shown that EGb 761 inhibits NADPH-oxidase, probably via its flavonoid constituents (Pincemail *et al.*, 1987), an action that is useful in explaining the therapeutic actions of EGb 761 in disease states which may involve excessive free radical production. Anti-lipoperoxidative effects of EGb 761 that afford protection against the degradation of fatty acids and proteins have been demonstrated using microsomal preparations (Pincemail and Deby, 1986; Dumont *et al.*, 1993), findings which support the view that chronic therapy with EGb 761 could be useful in preventing atherosclerosis, Alzheimer's disease, cancer and other diseases that are associated with oxidative stress.

The significance of results concerning the actions of EGb 761 on free radical-related mechanisms is further amplified by new findings obtained with acellular systems. Using γ-radiolysis, it was shown that EGb 761 may offer significant protection against the deleterious effects of both $OH^{\bullet}$ and $O_2^{\bullet-}$ and that these two free radicals probably interact with distinct sites on certain constituents of the extract (Gardès-Albert *et al.*, 1993). The action of EGb 761 of protecting human LDL against oxidation by $O_2^{-\bullet}$ and $OH^\bullet$ (Gardès-Albert *et al.*, 1995) supports the contention that the extract may have a preventive action on atherogenesis. Finally, in support of the early finding of Delaflotte *et al.* (1984), who showed that EGb 761 can either *enhance or inhibit* the EDRF/NO mechanism, Marcocci *et al.* (1994a) have shown that EGb 761 can scavenge NO in *in vitro* acellular systems. As it seems likely that the *deleterious effects of excess NO* are involved in certain chronic (degenerative) disease states, such as cancer and atherosclerosis (see DeFeudis, 1992, 1995), this NO-scavenging action of EGb 761 could be beneficial in treating such conditions.

The analysis of *in vitro* studies provided herein, taken together with the results discussed in the companion chapter (Drieu and DeFeudis, this volume), provides information which appears useful in explaining the "vasoregulatory" and "cognition-enhancing" actions of EGb 761, as well as its beneficial effects in treating other conditions related to cerebral and peripheral vascular insufficiency.

REFERENCES

Akkerman, J.-W.N. (1988) Interaction between the PAF receptor and fibrinogen binding sites on human platelets. In P. Braquet, (ed.), *Ginkgolides – Chemistry, Biology, Pharmacology and Clinical Perspectives*, Vol. 1, J.R. Prous, Barcelona, pp. 93–103.

Ames, B.N. (1989) Endogenous DNA damage as related to cancer and aging. *Mutat. Res.*, **214**, 41–46.

Amort, U., Haas, J. and Artmann, G. (1991) Effect of Rökan, Trental and Dusodril in acidotic/hyperosmolar challenged RBC bioassays. *Clin. Hemorheol.*, **11**, 798 (Abstr. No. P1.32).

Artmann, G.A. and Schikarski, C. (1993) *Ginkgo biloba* extract (EGb 761) protects red blood cells from oxidative damage. *Clin. Hemorheol.*, **13**, 529–539.

Auguet, M. and Clostre, F. (1983) Effects of an extract of *Ginkgo biloba* and diverse substances on the phasic and tonic components of the contraction of an isolated rabbit aorta. *Gen. Pharmacol.*, **14**, 277–280.

Auguet, M., DeFeudis, F.V. and Clostre, F. (1982a) Effects of *Ginkgo biloba* on arterial smooth muscle responses to vasoactive stimuli. *Gen. Pharmacol.*, **13**, 169–171.

Auguet, M., DeFeudis, F.V., Clostre, F. and Deghenghi, R. (1982b) Effects of an extract of *Ginkgo biloba* on rabbit isolated aorta. *Gen. Pharmacol.*, **13**, 225–230.

Auguet, M., Delaflotte, S., Chabrier, P.E. and Braquet, P. (1994) *Ginkgo biloba* extract (EGb 761) and the regulation of vascular tone. In F. Clostre and F.V. DeFeudis, (eds.), *Advances in Ginkgo biloba Extract Research, vol. 3. Cardiovascular Effects of Ginkgo biloba Extract (EGb 761)*, Elsevier, Paris, pp. 19–26.

Braquet, P.G., Spinnewyn, B., Braquet, M., Bourgain, R.H., Taylor, J.E., Etienne, A. and Drieu, K. (1985) BN 52021 and related compounds: A new series of highly specific PAF-acether antagonists isolated from *Ginkgo biloba* L. *Blood Vessels*, **16**, 558–572.

Bruel, A., Gardette, J., Berrou, E., Droy-Lefaix, M.T. and Picard, J. (1989) Effects of *Ginkgo biloba* extract on glucose transport and glycogen synthesis of cultured smooth muscle cells from pig aorta. *Pharmacol. Res.*, **21**, 421–429.

Chang, H.-M. and But, P.P.-H. (Eds.) (1987) *Pharmacology and Applications of Chinese Materia Medica*. Vol. 2, World Scientific Publ. Co., Singapore, pp. 1096–1101.

Coëffier, E. (1988) PAF-acether analogs, platelet activation and BN 52021. In P. Braquet, (ed.), *Ginkgolides – Chemistry, Biology, Pharmacology and Clinical Perspectives*, Vol. 1, J.R. Prous, Barcelona, pp. 105–113.

DeFeudis, F.V. (1985) Mechanism of endothelium-dependent vasorelaxation. *Med. Hypoth.*, 17, 363–374.

DeFeudis, F.V. (1991) *Ginkgo biloba Extract (EGb 761): Pharmacological Activities and Clinical Applications*, Elsevier, Paris.

DeFeudis, F.V. (1992) Genotoxicity of nitric oxide. *Drug News Perspect.*, 5, 361–363.

DeFeudis, F.V. (1995) Review: Excess EDRF/NO, a potentially deleterious condition that may be involved in accelerated atherogenesis and other chronic disease states. *Gen. Pharmacol.*, 26, 667–680.

DeFeudis, F.V., Auguet, M., Delaflotte, S., Hellegouarch, A., Baranes, J., Chapelat, M., Braquet, M., Etienne, A., Drieu, K., Clostre, F. and Braquet, P. (1985) Some *in vitro* and *in vivo* actions of an extract of *Ginkgo biloba* (GBE 761). In A. Agnoli, J.R. Rapin, V. Scapagnini and W.V. Weitbrecht, (eds.), *Effects of Ginkgo Biloba Extract on Organic Cerebral Impairment*, John Libbey, London and Paris, pp. 17–29.

DeFeudis, F.V. (1998) *Ginkgo biloba extract (EGb 761). From chemistry to the clinic*. Ullstein Medical, Wiesbaden.

Delaflotte, S., Auguet, M., DeFeudis, F.V., Baranes, J., Clostre, F., Drieu, K. and Braquet, P. (1984) Endothelium-dependent relaxation of rabbit isolated aorta produced by carbachol and by *Ginkgo biloba* extract. *Biomed. Biochim. Acta*, 43 (Suppl.), 212–216.

Del Tredici, P. (1991) Ginkgos and people – A thousand years of interaction. *Arnoldia*, 51, 2–15.

Doly, M., Droy-Lefaix, M.T., Millerin, M. and Braquet, P. (1990) Oxygenated free radicals and electrophysiology of isolated rat retina: Protection by a *Ginkgo biloba* extract (EGb 761). *Free Rad. Biol. Med.*, 9 (Suppl. 1), 12 (Abstr. No. 2.14).

Drieu, K. (1986) Préparation et définition de l'extrait de *Ginkgo biloba*. *Presse Méd.*, 15, 1455–1457.

Dumont, E., Petit, E., d'Arbigny, P., Tarrade, T. and Nouvelot, A. (1993) *Ginkgo biloba* extract (EGb 761) protects membrane polyunsaturated fatty acids and proteins against degradation induced by UV-C irradiation. In C. Ferradini, M.T. Droy-Lefaix and Y. Christen, (eds.), *Advances in Ginkgo biloba Extract Research, vol. 2. Ginkgo biloba Extract (EGb 761) as a Free-Radical Scavenger*, Elsevier, Paris, pp. 173–184.

Fattaccini, C.M., Bolanos-Jiménez, F., Gozlan, H. and Hamon, M. (1990) Tianeptine stimulates uptake of 5-hydroxytryptamine *in vivo* in the rat brain. *Neuropharmacology*, 29, 1–8.

Foster, S., Chongxi, Y. (1992) *Herbal Emissaries*. Healing Arts Press, Rochester, Vermont.

Gardès-Albert, M., Ferradini, C., Sekaki, A. and Droy-Lefaix, M.T. (1993) Oxygen-centered free radicals and their interactions with EGb 761 or CP 202. In C. Ferradini, M.T. Droy-Lefaix and Y. Christen, (eds.), *Advances in Ginkgo biloba Extract Research, vol. 2. Ginkgo biloba Extract (EGb 761) as a Free-Radical Scavenger*, Elsevier, Paris, pp. 1–11.

Gardès-Albert, M., Khalil, A., Fortun, A., Bonnefont-Rousselot, D., Delattre, J. and Droy-Lefaix, M.T. (1995) Protective effect of *Ginkgo biloba* extract (EGb 761) against the lipid peroxidation of low-density lipoproteins initiated by $OH^{\bullet}$ and $O_2^{\bullet}$ free radicals. In Y. Christen, Y. Courtois and M.T. Droy-Lefaix, (eds.), *Advances in Ginkgo biloba Extract Research, vol. 4. Effects of Ginkgo biloba Extract (EGb 761) on Aging and Age-Related Disorders*, Elsevier, Paris, pp. 49–63.

Haramaki, N., Aggarwal, S., Kawabata, T., Droy-Lefaix, M.T. and Packer, L. (1994a) Effects of natural antioxidant *Ginkgo biloba* extract (EGb 761) on myocardial ischemia-reperfusion injury. *Free Rad. Biol. Med.*, 16, 789–794.

Haramaki, N., Nguyen, L., Aziz, T. and Packer, L. (1994b) *Ginkgo biloba* extract (EGb 761) protects against myocardial ischemia-reperfusion injury by acting as an antioxidant. In F. Clostre and F.V. DeFeudis, (eds.), *Advances in Ginkgo biloba Extract Research, vol. 3. Cardiovascular Effects of Ginkgo biloba Extract (EGb 761)*, Elsevier, Paris, pp. 59–68.

Ignarro, L.J., Byrns, R.E., Buga, G.M. and Wood, K.S. (1987) Endothelium-derived relaxing factor from pulmonary artery and vein possesses pharmacological and chemical properties that are identical to those for nitric oxide radical. *Circ. Res.*, 61, 866–879.

Korth, R., Nuñez, D. and Benveniste, J. (1988) Inhibition by ginkgolides of specific PAF-acether binding to intact human platelets. In P. Braquet, (ed.), *Ginkgolides – Chemistry, Biology, Pharmacology and Clinical Perspectives*, Vol. 1, J.R. Prous, Barcelona, pp. 85–92.

Limasset, B., Michel, F., Roland, Y., Damon, M. and Crastes de Paulet, A. (1990) Study of the effect of some phenolic acids on hydrogen peroxide released by human neutrophils. *Proc. XIth Int. Congress Pharmacol.*, Amsterdam, July 1–6, 1990. *Eur. J. Pharmacol.*, 183, 1349 (Abstract No. P.we.210).

Marcocci, L., Maguire, J.J., Droy-Lefaix, M.T. and Packer, L. (1994a) The nitric oxide-scavenging properties of *Ginkgo biloba* extract EGb 761. *Biochem. Biophys. Res. Commun.*, 201, 748–755.

Marcocci, L., Packer, L., Droy-Lefaix, M.T., Sekaki, A. and Gardès-Albert, M. (1994b) Antioxidant action of *Ginkgo biloba* extract EGb 761. *Meth. Enzymol.*, 234, 462–475.

Masson, F., Néliat, G., Drieu, K., DeFeudis, F.V. and Jean, T. (1994) Effects of an extract of *Ginkgo biloba* (EGb 761) on the action potential and associated transmembrane ionic currents in mammalian cardiac myocytes: Inhibition of isoproterenol-induced chloride current. *Drug Dev. Res.*, 32, 29–41.

Michiels, C., Arnould, T., Janssens, D., Eliaers, F., Drieu, K. and Remacle, J. (1994) Key role of hypoxia and endothelial cells in vascular diseases: Effect of *Ginkgo biloba* extract (EGb 761). In F. Clostre and F.V. DeFeudis, (eds.), *Advances in Ginkgo biloba Extract Research, vol. 3. Cardiovascular Effects of Ginkgo biloba Extract (EGb 761)*, Elsevier, Paris, pp. 49–57.

Middleton, E., Jr. and Kandaswami, C. (1993) The impact of plant flavonoids on mammalian biology: Implications for immunity, inflammation and cancer. In J.B. Harborne, (ed.), *The Flavonoids: Advances in Research Since 1986*, Chapman & Hall, London, pp. 619–652.

Monboisse, J.C., Garnotel, R., Droy-Lefaix, M.T. and Borel, J.P. (1993) A *Ginkgo biloba* extract (EGb 761) inhibits superoxide-anion production by human neutrophils. In C. Ferradini, M.T. Droy-Lefaix and Y. Christen, (eds.), *Advances in Ginkgo biloba Extract Research, vol. 2. Ginkgo biloba Extract (EGb 761) as a Free-Radical Scavenger*, Elsevier, Paris, pp. 123–128.

Nishimura, C. and Kuriyama, K. (1985) Alterations in the retinal dopaminergic neuronal system in rats with streptozotocin-induced diabetes. *J. Neurochem.*, 45, 448–455.

Palmer, R.M.J., Ferrige, A.G. and Moncada, S. (1987) Nitric oxide release accounts for the biological activity of endothelium-derived relaxing factor. *Nature*, 327, 524–526.

Pietri, S., Culcasi, M., Vernière, E., d'Arbigny, P. and Drieu, K. (1993) Effect of *Ginkgo biloba* extract (EGb 761) on free radical-induced ischemia-reperfusion injury in isolated rat hearts: A hemodynamic and electron-spin-resonance investigation. In C. Ferradini, M.T. Droy-Lefaix and Y. Christen, (eds.), *Advances in Ginkgo biloba Extract Research, vol. 2. Ginkgo biloba Extract (EGb 761) as a Free-Radical Scavenger*, Elsevier, Paris, pp. 163–171.

Pietta, P.G., Gardana, C., Mauri, P.L., Maffei-Facino, R. and Carini, M. (1995) Identification of flavonoid metabolites after oral administration to rats of a *Ginkgo biloba* extract. *J. Chromatog. B*, 673, 75–80.

Pincemail, J. and Deby, C. (1986) Propriétés antiradicalaires de l'extrait de Ginkgo biloba. Presse Méd., 15, 1475–1479.

Pincemail, J., Thirion, A., Dupuis, M., Braquet, P., Drieu, K. and Deby, C. (1987) Ginkgo biloba extract inhibits oxygen species production generated by phorbol myristate acetate stimulated human leukocytes. Experientia, 43, 181–184.

Radomski, M.W., Palmer, R.M.J. and Moncada, S. (1987) Endogenous nitric oxide inhibits human platelet adhesion to vascular endothelium. Lancet, 2, 1057–1058.

Ramassamy, C., Christen, Y., Clostre, F. and Costentin, J. (1992a) The Ginkgo biloba extract, EGb 761, increases synaptosomal uptake of 5-hydroxytryptamine: In-vitro and ex-vivo studies. J. Pharm. Pharmacol., 44, 943–945.

Ramassamy, C., Christen, Y., Clostre, F. and Costentin, J. (1993a) Ginkgo biloba extract (EGb 761) prevents free-radical modifications of synaptosomal membrane fluidity and ability to take up serotonin. In C. Ferradini, M.T. Droy-Lefaix and Y. Christen, (eds.), Advances in Ginkgo biloba Extract Research, vol. 2. Ginkgo biloba Extract (EGb 761) as a Free-Radical Scavenger, Elsevier, Paris, pp. 39–50.

Ramassamy, C., Girbe, F., Christen, Y. and Costentin, J. (1993b) Ginkgo biloba extract (GBE 761) or trolox C prevent the ascorbic acid/Fe^{2+} induced decrease in synaptosomal membrane fluidity. Free Rad. Res. Commun., 19, 341–350.

Ramassamy, C., Girbe, F., Pincemail, J., Christen, Y. and Costentin, J. (1994) Modifications of the synaptosomal dopamine uptake and release by two systems generating free radicals: Ascorbic acid/Fe^{2+} and arginine/NADPH. Ann. N. Y. Acad. Sci., 738, 141–151.

Ramassamy, C., Naudin, B., Christen, Y., Clostre, F. and Costentin, J. (1992b) Prevention by Ginkgo biloba extract (EGb 761) and trolox C of the decrease in synaptosomal dopamine or serotonin uptake following incubation. Biochem. Pharmacol., 44, 2395–2401.

Rapin, J.R., Provost, P., DeFeudis, F.V. and Drieu, K. (1994) Effects of repeated treatments with an extract of Ginkgo biloba (EGb 761) and bilobalide on glucose uptake and glycogen synthesis in rat erythrocytes: An ex vivo study. Drug. Dev. Res., 31, 164–169.

Rosen, G.M., Barber, M.J. and Rauckman, E.J. (1983) Disruption of erythrocyte membranal organization by superoxide. J. Biol. Chem., 258, 2225–2228.

Schubert, H. and Halama, P. (1993) Primary therapy-resistant depressive instability in elderly patients with cerebral disorders: Efficacy of a combination of Ginkgo biloba extract EGb 761 with antidepressants. Geriatr. Forsch., 1, 45–53.

Sellak, H., Marquetty, C. and Pasquier, C. (1994) Endothelial-cell protection against oxygen free-radical injury by Ginkgo biloba extract (EGb 761). In F. Clostre and F.V. DeFeudis, (eds.), Advances in Ginkgo biloba Extract Research, vol. 3. Cardiovascular Effects of Ginkgo biloba Extract (EGb 761), Elsevier, Paris, pp. 69–78.

Sram, R.J., Cerna, M. and Hola, N. (1986) Effect of ascorbic acid prophylaxis in groups occupationally exposed to mutagens. In C. Ramel, B. Lambert and J. Magnusson, (eds.), Genetic Toxicology of Environmental Chemicals, Part B: Genetic Effects and Applied Mutagenesis, Alan R. Liss, New York, pp. 327–335.

Sram, R.J., Binkova, B., Stejskalova, J. and Topinka, J. (1993) Effect of EGb 761 on lipid peroxidation, DNA repair and antioxienzyme activity. In C. Ferradini, M.T. Droy-Lefaix and Y. Christen, (eds.), Advances in Ginkgo biloba Extract Research, vol. 2. Ginkgo biloba Extract (EGb 761) as a Free-Radical Scavenger, Elsevier, Paris, pp. 27–38.

Stücker, O., Pons, C., Duverger, J.P., d'Arbigny, P. and Drieu, K. (1994) Effect of Ginkgo biloba extract (EGb 761) on in vivo induced, white-platelet, arterial-thrombus formation. In F. Clostre and F.V. DeFeudis, (eds.), Advances in Ginkgo biloba Extract Research, vol. 3. Cardiovascular Effects of Ginkgo biloba Extract (EGb 761), Elsevier, Paris, pp. 27–30.

Taylor, J.E. (1990) *In vitro* inhibition of [³H]5-HT, [³H]norepinephrine (NE), and [³H]dopamine (DA) uptake by an extract of the *Ginkgo biloba* plant. *Proc. Soc. Neurosci. Annual Meeting*, St. Louis (Abstract. No. 32.11).

Taylor, J.E. (1992) *In vitro* interactions of EGb 761 with biogenic amine uptake sites and NMDA receptors. In Y. Christen, J. Costentin and M. Lacour, (eds.), *Effects of Ginkgo biloba Extract (EGb 761) on the Central Nervous System*, Elsevier, Paris, pp. 1–5.

Taylor, J.E., Odell, A. and Bonin, A. (1991) The effect of EGb 761 treatment on *ex vivo* biogenic amine uptake. *Biomeasure Inc.*, Internal Report No. 288–4, 12 pp.

Tosaki, A. and Droy-Lefaix, M.T. (1993) Effects of free-radical scavengers, superoxide dismutase, catalase and *Ginkgo biloba* extract (EGb 761) on reperfusion-induced arrhythmias in isolated hearts. In C. Ferradini, M.T. Droy-Lefaix and Y. Christen, (eds.), *Advances in Ginkgo biloba Extract Research, vol. 2. Ginkgo biloba Extract (EGb 761) as a Free-Radical Scavenger*, Elsevier, Paris, pp. 141–152.

Toussaint, O., Eliaers, F., Houbion, A., Remacle, J. and Drieu, K. (1995) Protective effect of *Ginkgo biloba* extract (EGb 761) and bilobalide against mortality and accelerated cellular aging under stressful conditions. In Y. Christen, Y. Courtois and M.T. Droy-Lefaix, (eds.), *Advances in Ginkgo biloba Extract Research, vol. 4. Effects of Ginkgo biloba Extract (EGb 761) on Aging and Age-Related Disorders*, Elsevier, Paris, pp. 1–16.

17. *IN VIVO* STUDIES OF THE PHARMACOLOGICAL AND BIOCHEMICAL ACTIVITIES OF *GINKGO BILOBA* EXTRACT (EGb 761) AND ITS CONSTITUENTS

[1]KATY DRIEU and [2]FRANCIS V. DeFEUDIS

[1]*Institut Henri Beaufour-IPSEN, 24 rue Erlanger, 75116 Paris, France, and*
[2]*Institute for BioScience, 153 West Main Street, Westboro, MA 01581, U. S. A.*

INTRODUCTION

The early history and medicinal properties of the *Ginkgo biloba* tree, a definition of the leaf extract termed "EGb 761", and the major clinical indications for this extract have been briefly mentioned in the companion chapter (DeFeudis and Drieu, this volume; for further information, see Drieu, 1986; DeFeudis, 1991, 1998). It is necessary to recall that EGb 761 is composed of flavonoids, terpenoids (ginkgolides, bilobalide), proanthocyanidins, organic acids and other constituents, and that several of these substances appear to be responsible for its beneficial effects in treating cerebral and peripheral vascular insufficiency states, age-related cognitive deficits, neurosensory problems, eye ailments and traumatic brain injuries (see e.g., Drieu, 1986; DeFeudis, 1991, 1998).

In vitro studies performed with EGb 761 are discussed in the companion chapter. The present chapter is centered on *in vivo* and *ex vivo* studies conducted with the authentic EGb 761 preparation. Newest developments will be particularly emphasized, and an attempt will be made to correlate the results obtained with *in vitro* findings and certain clinical observations. It is realized that *in vivo* experimentation is subject to certain limitations with respect to explaining clinical situations, species differences perhaps being most significant, and that it is difficult to assign pharmacological effects to specific organ systems in *in vivo* studies since possible indirect (secondary/reflex) actions may be confused with direct actions.

Also, it is extraordinarily difficult to evaluate the pharmacokinetic properties of EGb 761 due to its numerous active constituents and their many possible interactions. Nevertheless, some reliable pharmacokinetic data concerning certain terpenoid constituents of EGb 761 have been obtained in both rats and humans. These data permit some estimation of the concentrations of these terpenes that are present in the circulation after administration of effective doses of EGb 761. For example, after oral administration of EGb 761 in single doses of 30, 55 and 100 mg/kg to rats the maximum plasma concentrations of ginkgolides A and B and bilobalide that were achieved were dose-linear, were reached within 0.5–1 hour, and were 60–175 ng/ml for ginkgolide A, 40-103 ng/ml for ginkgolide B and 159–363 ng/ml for bilobalide (Biber and Chatterjee, 1995). These data are useful in assessing the results of *in vitro* experiments in relation to clinical effects of the extract.

Fourtillan *et al.* (1995) have examined the pharmacokinetic properties of ginkgolides A and B and bilobalide in healthy, young humans after administration of a single 120-mg oral dose of EGb 761. Maximum plasma concentrations of the terpenes were 33 ng/ml for ginkgolide A, 16 ng/ml for ginkgolide B and 19 ng/ml for bilobalide, the latter of which is under-estimated due to the difficulty in evaluating bilobalide with the methodology that was employed. Considering these findings and results obtained for the parameter "area-under-curve" in studies with rats (Biber and Chatterjee, 1995), it can be calculated that an oral dose of 240 mg in the human corresponds very roughly to an oral dose of 50 mg/kg in the rat. This estimation is of some value in assessing the results of pharmacological studies conducted with rats with respect to their potential clinical significance in humans.

Pietta *et al.* (1995), using mass spectroscopy, have identified metabolites of flavonoid glycosides following oral ingestion of EGb 761 by rats. After administration of a very high dose (~ 4 g/kg) of the extract to female rats, they found that four substances which could be considered to be flavonoid metabolites (3,4-dihydroxyphenylacetic acid, homovanillic acid, 3-(4-hydroxyphenyl)propionic acid and 3-(3-hydroxyphenyl)propionic acid) could be detected in plasma and therefore were bioavailable. If these substances are present in sufficiently high concentration, they could have pharmacological effects (e.g., as free radical-scavengers, enzyme-inhibitors or "false neurotransmitters") after oral administration of the total EGb 761 preparation.

ACTIONS OF EGb 761 ON THE CARDIOVASCULAR SYSTEM

The therapeutic indications for EGb 761 at both central and peripheral levels involve its "vasoregulatory" action, which implicates arteries, veins, capillaries and formed elements of the blood (e.g., erythrocytes, platelets, neutrophils). It is this *polyvalent* type of action of EGb 761, which depends upon its various active components, that provides protection in multifactorial diseases involving vascular pathology.

Myocardial Ischaemia and Cardiac Arrhythmias

Myocardial ischaemia in anaesthetized animals

Using a model of localized myocardial ischaemia obtained by ligating the left coronary arteries of anaesthetized rats, Guillon *et al.* (1986) showed that intravenous perfusion with a relatively high dose of EGb 761 (50 mg/kg) decreased arrhythmias without modifying other cardiovascular parameters. Significant decreases occurred in the number of ventricular extrasystoles during the 5–10 and 10–15 minute intervals after ligation and in total ventricular tachycardia (VT) time during the 10–15 minute interval. These results are of interest in that they revealed an antiarrhythmic activity of EGb 761.

Shen and Zhou (1995) used anaesthetized rabbits to assess mechanisms underlying the protective action of EGb 761 on myocardial ischaemia-reperfusion injury. The animals were subjected to 30 minutes of regional cardiac ischaemia and 120 minutes of reperfusion by ligating and then relieving branches of the left anterior descending coronary artery (LAD). The generation of free radicals, the antioxidant properties of EGb 761, the blood fibrinolytic condition and the ultrastructure of myocytes were examined. Compared to saline-perfused animals, injection of EGb 761 (10 mg/kg) into the LAD significantly inhibited the ischaemia/ reperfusion-induced increase in lipid peroxidation and maintained superoxide dismutase (SOD) levels in both plasma and tissue. EGb 761 treatment also suppressed the decrease in tissue-type plasminogen activator, the increase in plasminogen activator inhibitor-1 and the damage to myocytes that were caused by ischaemia and reperfusion.

The effect of EGb 761 treatment on early arrhythmia induced by 30 minutes of LAD occlusion, followed by 5 minutes of reperfusion, has been examined in anaesthetized dogs (Lo *et al.*, 1994). EGb 761 (1 mg/kg, i.v.) was administered 5 minutes before LAD occlusion and then continuously infused (at 0.1 mg/kg/min) until 5 minutes after reperfusion; an additional bolus dose of EGb 761 (1 mg/kg) was injected immediately before reperfusion. Compared to saline-injected dogs, EGb 761-treated dogs had less ventricular premature beats during coronary occlusion, and they were protected against ventricular fibrillation (VF) during reperfusion.

The cardioprotective effects of EGb 761 treatment that have been observed in anaesthetized animals may have been significantly related to the free radical-scavenging and antioxidant properties of the extract. An anti-lipoperoxidative property of EGb 761 could have also played a role since the flavonoid constituents of the extract, by scavenging OH•, can prevent the initiation of the chain propagation reaction that is involved in lipid peroxidation. Thus, the flavonoid constituents of the extract and/or their metabolites, which have been shown to have these properties in *in vitro* studies (see companion chapter), are prime candidates for such effects, but other EGb 761 constituents possessing additional properties could have also played significant roles.

Myocardial ischaemia in rat heart ex vivo

Tosaki *et al.* (1993) have evaluated the effects of SOD, catalase and EGb 761, alone and in combination, on reperfusion-induced VF, VT and oxygen free radical formation after 30 minutes of global ischaemia followed by reperfusion in isolated rat hearts. After 10 days of treatment, neither SOD (10^4–5×10^4 U/kg/day, i.v.) nor catalase (2.5×10^4–10^5 U/kg/day, i.v.) alone reduced the incidence of reperfusion-induced VF and VT. In contrast, EGb 761 (25–200 mg/kg/day, *per os*) dose-dependently reduced the incidence of these arrhythmias. Coadministration of SOD (5×10^4 U/kg) and EGb 761 (50 mg/kg) was also effective.

Combination therapy for cardiac disorders may also be conceived as an addition of the effects of EGb 761 and "ischaemic preconditioning". [Ischaemic preconditioning is a phenomenon in which an episode(s) of brief ischaemia can protect the myocardium against a subsequent longer ischaemic period.] In this regard, Tosaki *et al.* (1994) showed that orally administered EGb 761 not only improved contractile function of the isolated working rat heart after global ischaemia, but also that this protective effect is additive to that of ischaemic preconditioning. After four cycles of preconditioning alone, both VF and VT decreased from 100% to about 50%, and the formation of oxygen free radicals (determined by electron-spin resonance spectroscopy; ESR) was significantly decreased with respect to VF. However, only slight recovery of cardiac function occurred. In contrast, combined treatment with EGb 761 (50 or 100 mg/kg/day, *per os* for 10 days) and ischaemic preconditioning significantly improved coronary blood flow and other cardiac parameters and decreased the incidence of reperfusion-induced VF and VT. During reperfusion, free radical formation was also significantly decreased by this combined treatment. Thus, it was postulated that EGb 761 can improve contractile function after global ischaemia in isolated rat heart by reducing the formation of oxygen free radicals, and that this protection is additive to that of ischaemic preconditioning.

Pietri *et al.* (1997) have used hemodynamic measurements and ESR spectroscopy to assess the cardioprotective and antioxidant effects of therapeutically-relevant doses of EGb 761 and its terpenoid constituents (ginkgolides A and B, bilobalide). In *in vivo* experiments, EGb 761 (60 mg/kg/day) and ginkgolide A (4 mg/kg/day) were administered orally to rats for 15 days. In *in vitro* experiments, hearts were exposed to EGb 761 (5 μg/ml), and separately to ginkgolides A and B (both at 0.05 μg/ml) or bilobalide (0.15 μg/ml), concentrations corresponding to their respective amounts in EGb 761 at 5 μg/ml. After 30 minutes of equilibration, isolated hearts that were prepared either after *in vivo* treatments (i.e., *ex vivo*) or *in vitro* were subjected to 20 minutes of low-flow ischaemia, followed by 20 minutes of no-flow normothermic global ischaemia, and finally 60 minutes of reperfusion.

Significant anti-ischaemic effects were observed after *in vivo* treatments with both EGb 761 and ginkgolide A and with *in vitro* perfusion of ginkgolides A and B and bilobalide, as evidenced by less pronounced increases in left ventricular end-diastolic pressure upon terminating no-flow ischaemia (Pietri *et al.*, 1997). *In vivo* administration of EGb 761 or ginkgolide A improved myocardial functional recovery after ischaemia-reperfusion, as compared to placebo. *In vitro* reperfusion of hearts in the presence of terpenes, but not the total EGb 761 preparation, significantly improved all hemodynamic parameters (see companion chapter). Post-ischaemic levels of DMPO-OH, formed from the nitrone spin trap 5,5-dimethyl-1-pyrroline N-oxide (DMPO; a stable free radical that reacts with the unstable free radical OH$^{\bullet}$ to form a stable adduct), in coronary effluents were significantly lower after *in vivo* or *in vitro* treatment with EGb 761 and after *in vitro* perfusion with ginkgolide A than in placebo controls. Perfusion of ginkgolides A and B and bilobalide significantly inhibited DMPO-OH formation by free radicals that were released from reperfused rat hearts. However, as perfusion of these terpenes did not affect the formation of DMPO-OH in acellular tests conducted with generators of superoxide anion radical ($O_2^{-\bullet}$) and hydroxyl radical (OH$^{\bullet}$), Pietri *et al.* (1997) suggested that their cardioprotective effects are more likely related to an inhibition of free radical generation than to direct free radical-scavenging.

Formed Elements of the Blood: Hemorheology

Abnormalities in the rheological properties of the blood (e.g., changes in erythrocyte deformability, platelet adhesiveness, or plasma viscosity) can decrease oxygen supply to various tissues. This is one reason why the effects of EGb 761 on formed elements of the blood have been studied in some detail.

Erythrocytes

The effects of EGb 761 and bilobalide on energy-yielding glucose metabolism in rat erythrocytes have been evaluated by Rapin *et al.* (1994b). Repeated treatment of rats with EGb 761 (100 mg/kg/day, *per os*) for 5 days increased the *in vitro* uptake of glucose by their erythrocytes, especially in high-glucose medium. This effect was associated with an increase in intracellular energy metabolism, as reflected by a significant reduction in free glucose concentration (see Table 1) and a significant increase in the conversion of glucose into glycogen (Table 2). Repeated treatment of the animals with bilobalide (4 or 8 mg/kg/day, *per os*) caused similar changes in the uptake of glucose and in its conversion into glycogen (Tables 1 and 2). To gain further insight into the possible significance of these results, it was considered feasible to summate (in terms of molar equivalents of glucose/l) the intracellular glucose concentration plus the glucose derived from glycogen and from two lactate molecules (since catabolism of each molecule of glucose yields 2 molecules of lactate). This calculation indicated that the erythrocytes of those animals that were treated *in vivo* with EGb 761, but not bilobalide, showed a significant decrease in total intracellular glucose content. This effect was evident with both exogenous glucose concentrations the medium (6.67 mM and 13.32 mM; see Tables 1 and 2). Thus, EGb 761, but not bilobalide, increased the energy-yielding consumption of glucose (induced metabolic activation) under the conditions employed. These findings strengthen the view that *in vivo* treatment with EGb 761 favors the transformation of glucose into glycogen (its storage form) in erythrocytes.

Table 1   Intracellular concentrations of glucose and lactate in the erythrocytes of EGb 761- or bilobalide-treated rats after incubation in media containing exogenous glucose

In vivo *treatment*	Intracellular glucose (mM)	Intracellular lactate (mM)
Incubation in medium containing 6.67 mM glucose		
Control (water)	4.91 ± 0.07	0.31 ± 0.01
EGb 761 (100 mg/kg/day)	3.28 ± 0.13[*]	0.32 ± 0.01
Bilobalide (4 mg/kg/day)	4.06 ± 0.26[*]	0.30 ± 0.01
Bilobalide (8 mg/kg/day)	3.74 ± 0.19[*]	0.31 ± 0.01
Incubation in medium containing 13.32 mM glucose		
Control (water)	5.83 ± 0.18	0.36 ± 0.01
EGb 761 (100 mg/kg/day)	4.14 ± 0.18[*]	0.34 ± 0.01
Bilobalide (4 mg/kg/day)	4.57 ± 0.22[*]	0.35 ± 0.01
Bilobalide (8 mg/kg/day)	4.39 ± 0.17[*]	0.35 ± 0.01

Means ± S.E.M.; erythrocytes from 5 animals in all cases; [*] indicates $p \leq 0.01$ for comparisons with respective controls. Rats were treated for 5 days with EGb 761 or bilobalide. (Rapin *et al.*, 1994b; Reprinted by permission of John Wiley & Sons, Inc.)

Table 2   Glycogen contents of the erythrocytes of EGb 761- or bilobalide-treated rats after incubation in the presence of exogenous glucose

In vivo *treatment*	*Glycogen content (m-moles/l glucose formed by hydrolysis with amyloglucosidase)*
Incubation medium containing 6.67 mM glucose	
Control (water)	2.44 ± 0.06
EGb 761 (100 mg/kg/day)	$3.25 \pm 0.20^*$
Bilobalide (4 mg/kg/day)	$3.25 \pm 0.15^*$
Bilobalide (8 mg/kg/day)	$3.40 \pm 0.06^*$
Incubation medium containing 13.32 mM glucose	
Control (water)	2.42 ± 0.25
EGb 761 (100 mg/kg/day)	$3.46 \pm 0.14^*$
Bilobalide (4 mg/kg/day)	$3.59 \pm 0.21^*$
Bilobalide (8 mg/kg/day)	$3.62 \pm 0.10^*$

Means $\pm$ S.E.M.; erythrocytes from 5 animals in all cases; * indicates $p \leq 0.01$ for comparisons with respective controls. Rats were treated for 5 days with EGb 761 or bilobalide. (Rapin *et al.*, 1994b; Reprinted by permission of John Wiley & Sons, Inc.)

The effects of EGb 761 have also been assessed on capillary erythrocyte velocity in a rat cremaster muscle preparation using intravital microscopy (Stücker *et al.*, 1994b). Erythrocyte velocity distributions, analyzed in the basal state and during maximal dilatation (produced by topical application of nitroprusside + papaverine; both at 0.1 mM) were compared in rats that had received either EGb 761 (30 or 60 mg/kg/day, *per os*) or water (control) for 5 days. Compared to the control group, mean erythrocyte velocity was significantly augmented in EGb 761-treated rats in both the basal and dilated states. Statistical analyses of the distributions of capillary erythrocyte velocity revealed that the percentage of capillaries with low erythrocyte flow velocity (< 50 μm/sec) in the basal state was 15% in controls, as compared to 10% in rats receiving 30 mg/kg EGb 761 and 7% in rats receiving 60 mg/kg EGb 761. In the maximally dilated state, the corresponding values were 10, 1 and 2%, respectively. These results provided evidence that treatment with EGb 761 enhances capillary perfusion by increasing mean erythrocyte velocity and decreasing the number of areas with low perfusion rate. By such an effect, which is more likely to be rheological than vasodilatatory, EGb 761 could improve tissue blood supply in patients whose vasodilatatory responses are abolished.

Recent *ex vivo* studies conducted with erythrocytes of human subjects are also of interest. Artmann *et al.* (1991) showed that compliance (resistance of cells to deforming shear forces) was increased significantly in the erythrocytes of human blood donors who received EGb 761 (200 mg/day, *per os*) for 10 days, results which corroborate *in vivo* findings obtained with experimental animals. Another study revealed that administration of EGb 761 (200 mg/day, *per os* to healthy human volunteers for 1 week significantly inhibited H_2O_2-induced cell damage (Artmann and Schikarski, 1993), in accord with recent *in vitro* results (see DeFeudis and Drieu, this volume).

Platelets and thrombosis

Early experiments showed that oral pretreatment with EGb 761 (50 mg/kg) for 5 days opposed increases in platelet adhesiveness and in the growth rate of thrombi that were provoked in the microcirculation of the hamster cheek pouch by local application of ADP (Sim and Davies, 1980). These findings indicated that EGb 761 can inhibit platelet hyperaggregability in arterial elements of the microcirculation. If one presumes that platelet hyperaggregability is caused mainly by classical inducers such as thrombin, ADP and collagen, it would seem probable that such actions of EGb 761 involve its stimulation of both nitric oxide (NO)- and prostacyclin (PGI_2)-mediated mechanisms of platelet disaggregation (see also DeFeudis and Drieu, this volume).

Further support for an anti-thrombotic effect of EGb 761 derives from the study of Stücker *et al.* (1994a). The effects of intravenously administered EGb 761 were assessed on white-platelet arterial thrombus formation using a rat model of deendothelialized mesenteric artery in which thrombus formation is provoked by superfusing ADP. Both total thrombus formation and the mean rate of thrombus formation were significantly decreased in rats that were treated with EGb 761 at 30 and 60 mg/kg. Although EGb 761 was administered in relatively high acute doses, these results provided evidence that it can inhibit thrombus formation *in vivo* by an effect that persists for at least an hour after its injection.

Some clinical evidence supports these findings. Hofferberth (1991) showed that a rapid decrease in platelet aggregation may be involved in the therapeutic effect of EGb 761 in patients afflicted with "cerebro-organic syndrome". Other results indicated that a single 3-hour infusion of solution containing 87.5 mg of EGb 761 in patients suffering from atherosclerotic disorders can reduce *ex vivo* collagen-induced platelet aggregation after one day (Költringer and Eber, 1989).

Vascular Permeability and Vasospasm

A vasoprotective effect of EGb 761 has been demonstrated in a rat cremaster muscle preparation (Stücker *et al.*, 1996). Thrombin, serotonin, a thromboxane analogue (U46619) and rat serum were used to induce arteriolar spasm. When vasospasm was induced by serum, arterioles were constricted by about 25–30% after 10 minutes, an effect that was antagonized by the thromboxane receptor antagonist SQ29548, indicating an involvement of thromboxane. Injection of EGb 761 (30 mg/kg, i.v.) 5 minutes after inducing spasm inhibited serum-induced vasoconstriction by about 80%, but did not affect vasospasm induced after exposure of the preparation to serotonin (1 mM) or thrombin (20 U). In contrast, treatment with EGb 761, 5 minutes after exposure of the preparation to U46619 (0.1 mM), abolished the arteriolar constriction induced by this thromboxane analogue. A significant relaxation of the arterioles occurred, their tone approaching the initial unconstricted level at about 20 minutes after exposure to U46619, and this was followed by a second phase of less marked constriction, indicating a diminished effect of EGb 761 (see Fig. 1). This effect of EGb 761 may have been mediated in part by ginkgolide B, since this ginkgolide also reversed U46619-induced vasospasm.

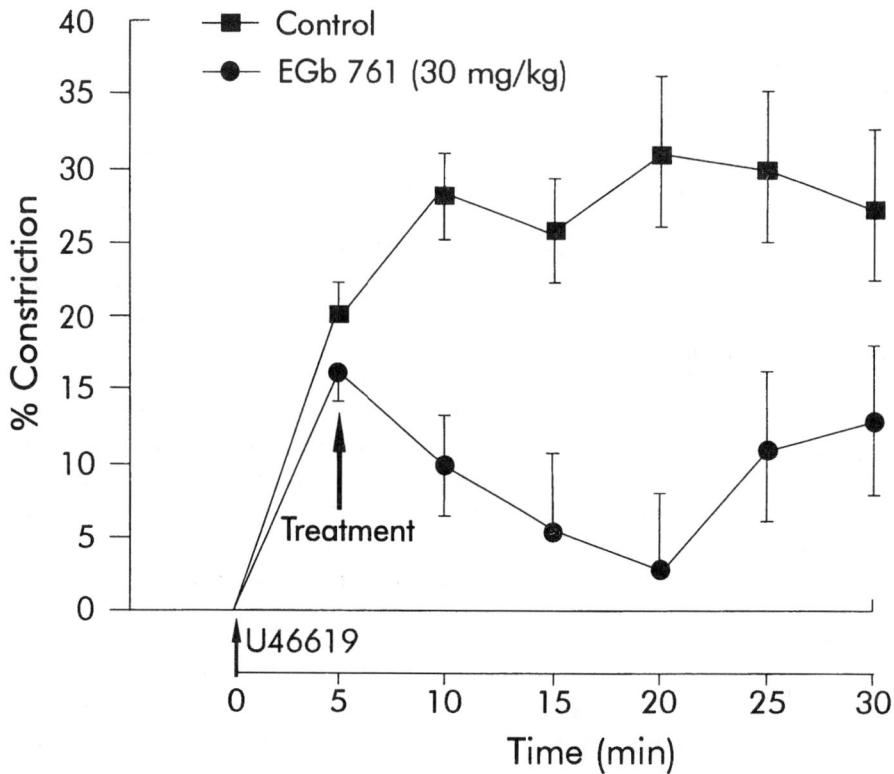

Figure 1   Evolution of an arteriolar spasm induced in a rat cremaster muscle preparation by the thromboxane analogue U46619 (10^{-4} M) in a control group (■, n = 9 arterioles) and in a group treated with EGb 761 (30 mg/kg) 5 minutes after induction of spasm (●, n = 9 arterioles). Vertical bars indicate S.E.M. (Reproduced from Stücker et al., 1996 with permission of S. Karger AG, Basel)

These results, taken together with the earlier finding that oral treatment of rabbits with EGb 761 can counteract cerebral vasospasm (Reuse-Blom and Drieu, 1986), support the view that EGb 761 exerts a vasoprotective action at the level of the microvasculature.

ACTIONS OF EGb 761 ON THE CENTRAL NERVOUS SYSTEM

A major clinical indication for EGb 761 involves individuals suffering from impairments of cerebral mechanisms related to memory, other cognitive functions and neurosensory systems. Thus, *in vivo* and *ex vivo* studies of the effects of EGb 761 and its constituents on CNS mechanisms are of primary importance.

Cerebral Blood Flow and Glucose Utilization

Lamour *et al.* (1992) have examined effects of EGb 761 and ginkgolide B on cerebral glucose utilization in 3-month-old rats using quantitative autoradiography after rapid intravenous injection of [^{14}C]-2-deoxy-D-glucose ([^{14}C]DG). This method provides estimates of local cerebral glucose utilization (see Sokoloff *et al.*, 1977). Rats received either EGb 761 (150 mg/kg, total dose divided into 3 i.p. injections) 1, 12 and 24 hours before [^{14}C]DG, or ginkgolide B (10 mg/kg, i.p.) 1 hour before [^{14}C]DG. Compared to vehicle-injected controls, EGb 761 significantly decreased glucose utilization in 21 of the 38 brain regions examined, as well as in whole brain. Ginkgolide B did not modify glucose utilization.

As administration of an intraperitoneal dose of 150 mg/kg of EGb 761 provides a very high "effective dose", these results of Lamour *et al.* (1992) may not be entirely meaningful with respect to the oral doses of EGb 761 that are used in clinical practice. Therefore, another autoradiographic study with [^{14}C]DG was performed under control conditions and after rats had been treated with lower "effective" doses of EGb 761 (50 or 150 mg/kg/day, *per os*, for 15 days) (Duverger *et al.*, 1995). Only slight-to-moderate changes in glucose utilization occurred in the 49 discrete structures of the brains of EGb 761-treated animals that were examined, these being statistically significant only for frontoparietal somatosensory cortex, nucleus accumbens, cerebellar cortex and pons, and only with the lower (50 mg/kg) dose of EGb 761.

These results, taken together with those of an earlier study (Le Poncin Lafitte *et al.*, 1980), indicate that EGb 761 treatment restores (increases) cerebral glucose utilization in animals whose glucose consumption is decreased (e.g., by ischaemia), but that it acts oppositely in normal animals, in accord with the brain's metabolic demand.

Cerebral Ischaemia

Experiments performed with Mongolian gerbils that had undergone unilateral ligation of the carotid artery to produce cerebral ischaemia showed that EGb 761 (10 mg/kg, i.p., or 50 mg/kg, *per os*) partially counteracted the increases that occurred in brain water and Na$^+$ contents and the decrease in brain K$^+$ content (Spinnewyn *et al.*, 1986). In another gerbil model (bilateral forebrain ischaemia), repeated oral administration of EGb 761 (15, 30 and 60 mg/kg/day) for 14 days before inducing ischaemia led to a dose-related preservation of neurones in the CA1 sector of the hippocampus (Spinnewyn, 1992). The mechanism underlying this protective effect of EGb 761 could involve its free radical-scavenging and antioxidant activities since the cellular damage that occurs during cerebral ischaemia is at least partly due to oxidative stress and associated lipid peroxidation.

Ginkgolides A and B, but not ginkgolide C, all given subcutaneously 30 minutes before occlusion of the middle cerebral artery (MCA) in the high dose range of 50–100 mg/kg [dissolved in a mixture of dimethyl sulfoxide (DMSO), PEG 400 and saline, 10:30:60], significantly reduced the brain infarct area in a mouse model of focal cerebral ischaemia (Backhauss *et al.*, 1992; Krieglstein, 1994). Pretreatment with bilobalide (⩾ 5 mg/kg, prepared in DMSO and physiological saline, 10:90) dose-dependently reduced the cerebral infarct area, and when administered immediately after MCA occlusion (at 10 mg/kg, s.c.) it was also found to be neuroprotective (see also Karkoutly et al., 1990). High doses of ginkgolides A and B (50 mg/kg, s.c.) also significantly reduced neuronal damage in the CA1 and CA3 hippocampal subfields and in the occipital and parietal cerebral cortices in a rat model of transient forebrain ischaemia (Prehn and Krieglstein, 1993). Although these findings are of interest, the doses of ginkgolides that were employed were generally very high, and the use of DMSO, which has *anti-ischaemic properties*, as the drug vehicle limits the interpretation of these data in relation to therapy for *ischaemic* conditions.

Lipid-derived second messengers, phospholipases, and cerebral ischaemia

Rodriguez de Turco et al. (1993) conducted experiments to determine whether EGb 761 treatment modifies the convulsion-induced activation of phospholipases A_2 and C (PLA_2 and PLC) in the hippocampus (a structure known to be highly vulnerable to ischaemia and convulsions) and cerebral cortex (neocortex). Treatment of rats with EGb 761 (100 mg/kg/day, *per os*) for 14 days decreased their neurochemical responses to electroconvulsive shock (ECS), indicating neuroprotection. These effects were consistent with a modulation of ECS-induced PLC-mediated degradation of polyphosphoinositides in the hippocampus. The main findings of this study are that EGb 761 treatment decreased free fatty acid pool size, increased basal diacylglycerol levels, and inhibited ECS-induced free fatty acid and diacylglycerol release, preferentially in the hippocampus.

As the release of choline from choline-containing phospholipids is substantially increased under hypoxic conditions, presumably due to an increase in PLA_2-mediated phosphatidylcholine catabolism, Klein et al. (1995) suggested that bilobalide may inhibit PLA_2 activity and thereby play a role in the anti-ischaemic effect of EGb 761 in the CNS. *In vitro* experiments performed with rat hippocampal slices showed that bilobalide abolished hypoxia-induced increases in choline efflux ($IC_{50} \simeq 0.35$ μM). *Ex vivo* experiments showed that hippocampal slices prepared from rats that had been treated with bilobalide (2-20 mg/kg, *per os*) were completely protected against hypoxia-induced phospholipid breakdown (*in vivo* $IC_{50} \simeq$ 6 mg/kg), implying a direct CNS effect.

Collectively, these results indicate that treatment with EGb 761 modulates lipid-derived second messengers predominantly in the hippocampus and influences phospholipid degradation in neuronal membranes by an action which may involve inhibition of PLC and PLA_2 activities. These effects may contribute to the anti-ischaemic and neuroprotective actions of EGb 761.

Cerebral Edema

Early experiments revealed protective and curative effects of EGb 761 (e.g., 50–100 mg/kg, *per os*) on brain edema that was produced in rats by intoxication with triethyltin (TET; see Chatterjee and Gabard, 1984). Further experiments showed that rats treated concurrently with EGb 761 (100 mg/kg, *per os*) and TET (0.002% in drinking water) had less pronounced brain lesions than those treated with TET alone (Otani *et al.*, 1986). EGb 761 also accelerated the elimination of water and Na^+ from the brain and reduced white matter vacuolation and astrocytic changes (Sancesario and Kreutzberg, 1986). As bilobalide exerted effects similar to those of the complete EGb 761 preparation, it has been suggested that this terpenoid constituent of the extract may be responsible for these beneficial effects (Sancesario and Kreutzberg, 1986; Chatterjee *et al.*, 1986). Other results have indicated that orally administered EGb 761 can protect rats against hexachlorophene- (Chatterjee *et al.*, 1986) and bromethalin-induced cerebral edema (Dorman *et al.*, 1992), and against the cellular changes in the brain that follow heat-stress (Sharma *et al.*, 1994), which also involves edema.

Recovery of Function Following Traumatic Brain Injury

Bilateral frontal lobotomy

Attella *et al.* (1989) have examined the effects of EGb 761 treatment on several behavioral and neuroanatomical parameters before and after removal of the frontal cortex in young adult rats. Such damage causes enduring impairment of spatial learning ability. After reaching acquisition criteria on a delayed spatial alternation task, the animals were subjected to either bilateral aspiration of the medial frontal cortex or simulated ("sham") operations, and then treated with either EGb 761 (100 mg/kg/day, i.p.) or saline (treated controls) for 30 days. Lesion-bearing, EGb 761-treated rats were less impaired than treated controls with respect to retention of the delayed spatial alternation task. Of special interest was the finding that EGb 761 significantly reduced the behavioral deficits that occurred *after* frontal lobotomy, whereas it did not affect acquisition of the spatial task when it was administered *before* the brain lesions were made. A further study showed that repeated administration of EGb 761 (100 mg/kg/day, i.p.) for 30 days prior to making lesions of the frontal cortex also enhanced recovery (Stein and Hoffman, 1992). In both of these studies, the cerebral ventricular swelling (edema) that occurred in response to injury was significantly decreased by EGb 761 treatment, an effect which may have been involved in promoting functional recovery (see also below). These findings provide evidence that EGb 761 treatment can facilitate recovery from some cognitive deficiencies that follow traumatic brain injury. However, since a rather high intraperitoneal dose of EGb 761 was used, further studies with lower doses seem warranted.

Cortical hemiplegia

Brailowsky *et al.* (1991) tested the effects of sub-acute (7-day) and sub-chronic (30-day) EGb 761 treatments on two rat models of cortical hemiplegia. One of these was produced by unilateral motor cortex aspiration, and the other by reversibly inactivating the motor cortex by chronic, localized, unilateral infusion of the inhibitory neurotransmitter γ-aminobutyric acid (GABA). Control rats received sham operations. The elevated beam test was used to evaluate coordinated walking in water-deprived animals that had been trained to drink saccharin-sweetened solutions (with or without EGb 761) and to perform to criteria before surgical intervention. Rats that received saccharine solution containing EGb 761 (100 mg/kg/day, *per os*), commencing on the day after surgery, showed a faster and more complete recovery from motor deficits than those which received only saccharine solutions, effects that were more apparent in animals subjected to motor cortex aspiration than in those receiving GABA infusions. Ventricular diameters of EGb 761-treated motor cortex-aspirated rats were significantly smaller than those of untreated animals (see also above).

Central Neurotransmitter Systems

In vivo and *ex vivo* studies involving neurotransmitter systems are discussed below. The reader is referred to the companion chapter (DeFeudis and Drieu, this volume) for an analysis of the *in vitro* effects of EGb 761 and its constituents on these systems.

Hippocampal high-affinity choline uptake and muscarinic receptors

The effects of repeated administration of EGb 761 (50 or 100 mg/kg/day, *per os*) have been evaluated *ex vivo* on Na^+-dependent high-affinity [^{3}H]choline uptake (HACU) into hippocampal synaptosomes of aged (22-month-old) rats (Kristofikova et al., 1992). EGb 761 (15 or 30 μg/ml) was also tested by direct addition *in vitro*. The 50 mg/kg *ex vivo* dose and both *in vitro* concentrations of EGb 761 caused significant increases in HACU. As a gradual decline in HACU of rat hippocampus occurs during ageing, such results imply that EGb 761 can decelerate brain neurodegenerative processes. Regarding muscarinic cholinergic receptors, Taylor (1986) has shown that repeated administration of EGb 761 (100 mg/kg/day, *per os*) increased the binding of [^{3}H]quinuclidinyl benzilate to hippocampal membranes of aged (24-month-old) Fischer 344 rats to the level of young rats without influencing this binding in young rats.

These findings, taken together with the recent demonstration that chronic treatment with EGb 761 exerted a neuroprotective effect at the level of the projection field of intra- and infrapyramidal mossy fibers of the hippocampus in aged mice (Barkats et al., 1995), are useful in explaining the preventive action of the extract on age-associated cognitive decline and senile dementia.

Brain noradrenergic and serotoninergic (5-HT) receptors

Huguet and Tarrade (1992) have reported that the age-related decrease in the density of hippocampal α_2-adrenoceptors (determined with [^{3}H]rauwolscine) of 24-month-old rats was significantly reversed by repeated treatment with EGb 761 (5 mg/kg/day, i.p.). These findings imply that EGb 761 treatment might cause α_2-adrenoceptors to become hypersensitive in aged rats, an effect resembling that of repeated treatment with the antidepressant drug desipramine (see e.g., Johnson *et al.*, 1980), and relate to a possible "antidepressant-like" action of the extract (see below).

Huguet *et al.* (1994) have examined the effects of EGb 761 treatment on brain 5-HT_{1A} receptors in relation to ageing. The number of binding sites (B_{max}) for [³H]8-hydroxy-2(di-n-propylamino)tetralin ([³H]8-OH-DPAT; 0.8–10 nM), a specific indicator of 5-HT_{1A} receptors, in cerebral cortex membranes was significantly lower in aged (24-month-old) than in young (4-month-old) animals, whereas binding affinity did not differ between the two age groups. Treatment with EGb 761 (5 mg/kg/day, i.p.) for 21 days did not alter receptor binding in young rats, but significantly increased B_{max} in aged rats, as compared to vehicle-treated rats (see Fig. 2). Such an effect, like those on muscarinic and α_2-adrenergic receptors (see above), could have been mediated by an inhibitory action of EGb 761 on lipid peroxidation-induced membrane damage, adding support to the contention that EGb 761 may have a "restorative" action on the decreased receptor binding that occurs with ageing.

Brain dopaminergic systems

Repeated ingestion of EGb 761 (~100 mg/kg/day) for 17 days protected mice against the neurotoxic effects of a continuous 1-week infusion of the exogenous neurotoxin N-methyl-4-phenyl-1,2,3,6-tetrahydropyridine (MPTP), assessed by monitoring

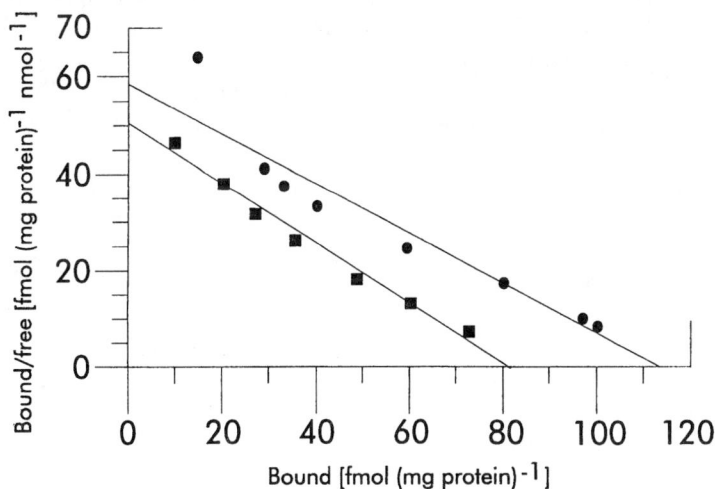

Figure 2   Scatchard analysis of specific [³H]8-OH-DPAT binding to cerebral cortex membranes of aged rats. Each point is the mean of 6 experiments performed in duplicate or triplicate. Values for B_{max} were 84.1 ± 5.6 fmol/mg protein for rats treated with vehicle (■) and 112.2 ± 4.2 fmol/mg protein for rats treated with EGb 761 (●, 5 mg/kg/day) for 21 consecutive days (p <0.005, unpaired *t*-test). (Huguet *et al.*, 1994; Reproduced with permission from The Royal Pharmaceutical Society of Great Britain)

[^{3}H]DA uptake by striatal synaptosomes (Ramassamy et al., 1990). This action of EGb 761 did not depend upon an inhibition of neuronal uptake of MPTP or its metabolite 1-methyl-4-phenylpyridinium since the concentrations of the extract required for inhibiting DA uptake (> 250 μg/ml) are much higher than those that can be achieved in vivo. This result is of special interest since MPTP intoxication serves as a model for Parkinson's disease.

Learning, Memory and Behaviour

Using an appetitive operant conditioning paradigm, Winter (1991) has shown that repeated administration of EGb 761 (100 mg/kg/day, per os) facilitated memory processes in mice. Acquisition occurred more rapidly, and both the performance of a 2-lever response sequence and retrieval of the learned response were improved. Other measurements indicated that these beneficial effects of EGb 761 were not related to increased hunger or to enhanced locomotor activity. The molecular mechanism(s) underlying these effects of EGb 761 could be related to its actions on central neurotransmitter systems and/or to its modification of cerebral blood flow and metabolism (see above).

Anti-Stress Activity of EGb 761

"Learned helplessness" and "behavioral despair" tests

The terminology "anti-stress" stems from a description of the finding that repeated oral treatment with EGb 761 reduced the avoidance deficits induced by unavoidable shock in rats ("learned helplessness" procedure; see Porsolt et al., 1990). In this study, preventive treatment with EGb 761 (50 or 100 mg/kg/day, per os), like diazepam (4 mg/kg/day, per os), protected the animals, whereas when the same doses were given curatively EGb 761 differed from diazepam (which was inactive) after 5 days of treatment in that it caused a slight decrease in acquisition deficits in the active avoidance response. This effect of EGb 761 was less pronounced than that generally observed with classical antidepressants (e.g., imipramine). In the "behavioural despair" test, imipramine (32 mg/kg/day) clearly reduced the duration of immobility after 5 days of treatment, whereas EGb 761 (50 or 100 mg/kg, per os) had no significant effect (Porsolt et al., 1992).

Taken together, these (and other) results have revealed that this "anti-stress" action of EGb 761 differs from those of classical antidepressants and anxiolytics, and was not due to an impairment of memory for the aversive stimulation or to a decrease in sensitivity to shock.

EGb 761, learning ability, and the stress-associated hormones, epinephrine, corticosterone and norepinephrine

Rapin et al. (1994a) examined the anti-stress activity of EGb 761 using a discrimination learning task in young (4-month-old) and old (20-month-old) rats. Introduction of an auditory perturbation ("stress") during the discriminative phase of learning modified the capacity and rate of acquisition in both young and old animals. Repeated treatment with EGb 761 (50 and 100 mg/kg/day, *per os*) for 20 days suppressed this stress-induced disruption of learning, an effect that became statistically significant by the third day of learning (time of maximal acquisition rate) as an increase in the percentage of efficient lever presses and as decreases in number of errors and reaction time. The detrimental effects of auditory stress on learning and their suppression by EGb 761 treatment were clearly reflected as changes in plasma catecholamines and corticosterone (see Fig. 3). Stress led to increases in epinephrine, norepinephrine and corticosterone in both young and old rats, effects that were significantly reduced in a dose-related manner by 20 days of treatment with EGb 761 (Fig. 3). These results are supported by the study of Oliver *et al.* (1994), conducted with a rat model of acute surgical stress, which showed that repeated administration of EGb 761 (100 mg/kg/day, *per os*) significantly decreased plasma concentrations of adrenocorticotrophic hormone, corticosterone, norepinephrine and epinephrine and the portal blood concentration of corticotrophin-releasing hormone.

The contention that the mechanism(s) underlying the "anxiolytic-like" or "anti-stress" activity of EGb 761 differs from that mediating the action of classical anxiolytics (see above) has also received support from another recent study with rats. Diazepam and EGb 761 differentially influenced the effects of prednisolone (50 mg/kg/day, i.p. for 5 days) of reversibly down-regulating hippocampal Type II glucocorticoid receptors (shown with [^{3}H]dexamethasone binding; see Sapolsky *et al.*, 1983), of reducing acquisition lag time in a conditioned avoidance response (pole-climbing test), and of decreasing reaction time and increasing the number of errors in a Skinner box (Rapin *et al.*, 1994c; 1996). In this model system, repeated treatment with diazepam (10 mg/kg, *per os*) further decreased [^{3}H]dexamethasone binding to glucocorticoid receptors and decreased learning parameters in both behavioural tests, whereas repeated treatment with EGb 761 (50 mg/kg/day, *per os* for 14 days) suppressed prednisolone-induced down-regulation of glucocorticoid receptors and normalized learning parameters in both behavioural tests.

Collectively, these findings indicate that EGb 761 can facilitate behavioral adaptation to stress, an effect that is reflected by changes in circulating "stress hormones" and which supports the clinical use of the extract in treating cognitive decline, particularly in elderly patients. Such an "anti-stress" effect of EGb 761 supports its clinical use in treating patients suffering from cognitive impairment; i.e., it may be presumed that a decrease in stress induces an increase in vigilance/awareness which then leads to improved cognition.

"Social interaction" test

The "social interaction" test (see File, 1980) has been used to further define the mechanism(s) underlying the anti-stress ("anxiolytic-like") effect of EGb 761 (Chermat *et al.*, 1997). The possible interactions of EGb 761 with diazepam (a "positive control" substance in this test that acts as a full agonist at the GABA/benzodiazepine/chloride channel complex) and ethyl β-carboline-3-carboxylate

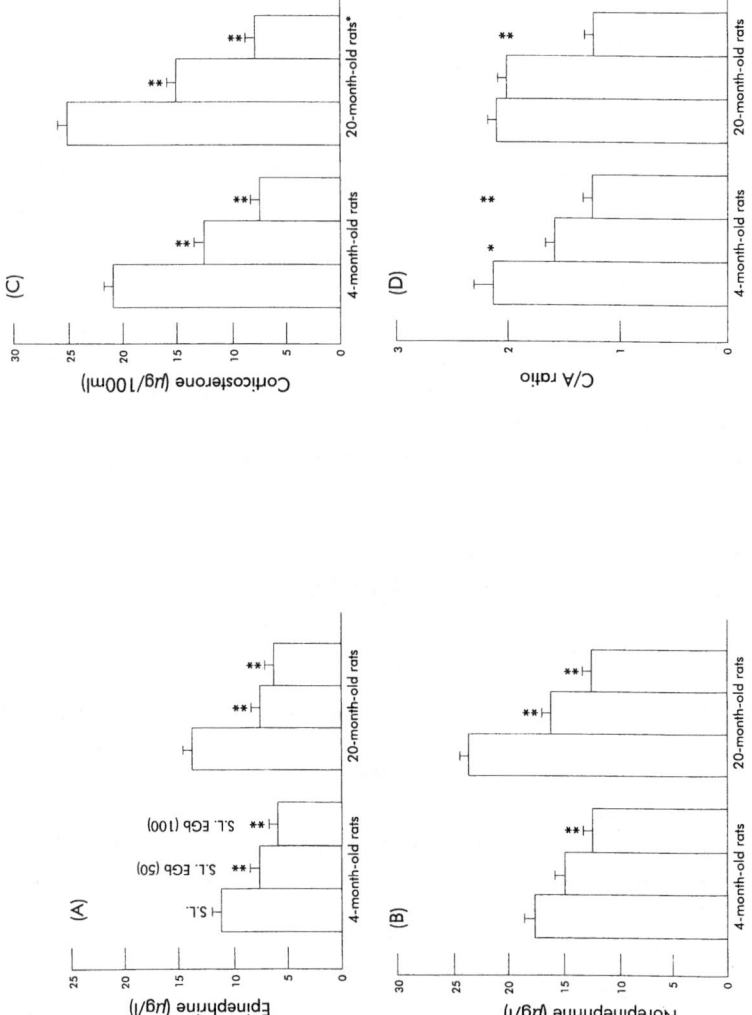

Figure 3   Plasma concentrations of epinephrine, norepinephrine and corticosterone in young and old rats after the third day of discrimination learning, with and without auditory stress; effects of subchronic (20-day) treatment with EGb 761. The code for treatment of the animals in each series of histograms is indicated in A (left) and is defined, from left to right, as follows: S,L = stressed, with learning; S,L, EGb (50) = stressed, with learning and treatment with EGb 761 (50 mg/kg, *per os*); S,L, EGb (100) = stressed, with learning and treatment with EGb 761 (100 mg/kg, *per os*). [A, Epinephrine; B, norepinephrine; C, corticosterone; D, corticosterone/epinephrine ratio (C/A).] Means ± S.E.M; 10 rats in all cases; * and ** indicate, respectively, p ≤0.05 and p ≤0.01 for comparisons among animals of the same age group (Student's *t*-test). (Reprinted by permission of the publisher from Rapin *et al.*, 1994a, *Gen. Pharmacol. 25*, pp. 1009–1016. Copyright 1994 by Elsevier Science Inc.)

(β-CCE; an anxiogenic substance in this test and a partial inverse agonist at the benzodiazepine site of the $GABA_A$ receptor complex) were assessed.

Pairs of "unfamiliar" male rats that had been subjected to the same treatment were placed in a novel test arena that was brightly illuminated, and the duration of social contact was observed over a 10-minute period. Single injections of EGb 761 (8–16 mg/kg, i.p.), 30 minutes prior to testing, or repeated oral administration of EGb 761 (48 or 96 mg/kg/day) for 8 days, significantly *decreased* social contact, a finding which implies an increase in "arousal" (see File, 1980). In contrast, a single injection of diazepam (1 mg/kg, i.p.), administered 30 minutes before commencing the test, significantly *increased* social contact. However, administration of diazepam to animals that had received repeated oral treatment with EGb 761 *increased* social interaction to an extent greater than observed with diazepam alone (Table 3). Treatment with β-CCE at 4 mg/kg (sub-maximal dose) 15 minutes before commencing behavioral testing significantly *decreased* social contact. However, when animals had received repeated treatment with EGb 761 before injection of β-CCE the resulting level of social interaction did not differ from that of the controls (Table 4), indicating that EGb 761 can "neutralize" the anxiogenic effect of β-CCE.

The molecular mechanism(s) underlying the interactions of EGb 761, diazepam and β-CCE remain unknown, but could involve a similar site or distinct sites of action at central $GABA_A$/benzodiazepine/chloride channel receptor complexes. Further research in this area could reveal the identity of an EGb 761 constituent possessing anxiolytic activity that does not induce the adverse side effects (e.g., sedation, amnesia) associated with the use of benzodiazepines.

Table 3 Interaction of repeated oral administration of EGb 761 and single intraperitoneal injection of diazepam on social behavior in the rat; social interaction test

Dose of EGb 761 (mg/kg, per os)	Dose of diazepam (mg/kg, i.p.)	Duration of contact (sec)
0[b]	0[c]	56.5 ± 2.4
0[b]	1	69.1 ± 2.1*
96	0[c]	33.6 ± 2.4*
96	1	81.3 ± 3.4*,[a]

Means ± S.E.M.; 16 rats (8 "unfamiliar" pairs) in all cases; * indicates p < 0.001 for comparisons with controls; [a] indicates p < 0.01 for the comparison between this value and the value obtained for diazepam alone. Testing was begun 60 minutes after the 8th daily oral administration of EGb 761 or 30 minutes after a single intraperitoneal injection of diazepam at the doses indicated. [b,c] Controls received 0.3% solution of gum arabic in distilled water, the vehicle for both EGb 761 and diazepam (0.5 ml/100 g body weight). The duration of social contact between "unfamiliar" pairs placed in a novel environment was noted during a 10-minute observation period. (Reproduced with permission from Chermat et al., 1997)

Table 4   Interaction of repeated oral treatments with EGb 761 for eight days and single intraperitoneal injections of β-CCE on social behavior in the rat; social interaction test

Dose of EGb 761 (mg/kg/day, per os)	Dose of β-CCE (mg/kg, i.p.)	Duration of social contact (sec)[c]
0 (Control[d])	0	58.0 ± 4.1
48	0	44.3 ± 3.9[*]
0	4	39.0 ± 4.4[**]
48	4	55.5 ± 5.5[a]
0 (Control[d])	0	56.3 ± 3.9
96	0	37.3 ± 4.8[**]
0	4	35.6 ± 5.7[**]
96	4	63.0 ± 6.2[b]

Means ± S.E.M.; 16 rats (8 pairs) in all cases; [*] and [**] indicate $p < 0.05$ and $p < 0.01$, respectively, for comparisons with corresponding control values; [a] and [b] indicate $p < 0.05$ and $p < 0.01$, respectively, for comparisons with groups that received 4 mg/kg β-CCE (Student's t-test). [c] Duration of social contact between two "unfamiliar" rats placed in a novel environment during a 10-minute observation period; [d] control values pertain to the values shown below them. Rats were tested 15 minutes after receiving single intraperitoneal injections of β-CCE and 60 minutes after receiving their eighth daily oral administration of EGb 761 at the doses indicated or the drug vehicle (0.3% solution of gum arabic in distilled water). All injections were administered in a volume of 5 ml/kg body weight. (Reproduced with permission from Chermat et al., 1997)

EFFECTS OF EGb 761 ON NEUROSENSORY SYSTEMS

Visual System

The retina has been used to test the free radical-scavenging and antioxidant actions of EGb 761 since it is enriched in polyunsaturated fatty acids which renders it particularly sensitive to the detrimental effects of oxygenated free radicals. As EGb 761 can react simultaneously with $O_2^{-\bullet}$ and $OH^\bullet$ and scavenge organic (e.g., peroxyl) radicals (see DeFeudis and Drieu, this volume), it is expected that the extract will prevent free radical-mediated retinal damage.

Models of retinopathy

Doly et al. (1986) used the electroretinogram (ERG) to quantify the injurious effects of free radicals and the action of EGb 761 in ex vivo experiments on isolated retinae. In alloxan-diabetic rats, treatment with EGb 761 (100 mg/kg/day, per os), commencing 15 days before induction of diabetes and continued for two months, significantly increased the amplitudes of b-wave ERGs, as compared with untreated rats. The effect of EGb 761 treatment on chloroquine-induced retinopathy has also been assessed (Droy-Lefaix et al., 1992). Repeated administration of chloroquine (75 mg/kg/day, per os for 20 days) to albino rats increased the duration of the b-wave ERG and delayed the time to peaking. In rats that were treated with EGb 761 alone (100 mg/kg/day, *per os*) for 10 days, and then for 20 days with both EGb 761 (same dose) and chloroquine (75 mg/kg/day, *per os*) such modifications in the ERG did not occur.

Reperfusion-induced retinal damage

Treatment of rats with either EGb 761 (25–100 mg/kg/day, *per os*) or SOD (15,000 U/kg/day, i.v.) for 10 days significantly reduced the development of reperfusion-induced retinal edema that occurred after 90 minutes of ischaemia followed by 4 or 24 hours of reperfusion, and significantly prevented neutrophil infiltration (Szabo *et al.*, 1991a,b). Both treatments also protected against reperfusion-induced injury when they were administered just before reperfusion. Other experiments showed that both EGb 761 and SOD significantly opposed the increases in Na^+ and Ca^{2+} and the loss of K^+ that occurred in the retinae of rats subjected to ischaemia (Szabo *et al.*, 1992a,b).

Retinal lipid peroxidation

The anti-lipoperoxidative properties of EGb 761 have been evaluated *ex vivo* by recording the b-wave ERG amplitudes of retinae isolated from rats that were either treated with EGb 761 (100 mg/kg/day, *per os*) for 10 days or received no treatment (controls) (Droy-Lefaix *et al.*, 1991). Lipid peroxidation was induced by adding $FeSO_4$ + sodium ascorbate to the perfusion medium. The amplitude of the ERG b-wave was decreased by 50% at about 30 minutes after induction of lipid peroxidation in untreated rats, whereas the retinae of EGb 761-treated rats showed a 50% decrease only at about 1 hour after exposure to the peroxidative system. This protective effect of EGb 761 may have been related to its free radical-scavenging properties.

Comment

Although further experiments are required, the findings discussed above indicate that orally-administered EGb 761, by decreasing oxidative stress, could be useful in preventing ischaemia-induced proliferative retinopathies such as those that occur in diabetes and those that can be induced by various xenobiotics.

Vestibular Compensation

The action of EGb 761 on the central compensation of a total unilateral peripheral vestibular deficit has been examined in 3- or 4-month-old rats (Denise and Bustany, 1989; Denise *et al.*, 1990). A lesion (labyrinthectomy) was made by placing a few crystals of NaCl in the round window, and then postural and locomotor deficits were determined. EGb 761 treatment (50 mg/kg/day, i.p.) for 73 days led to significantly more rapid improvement in spontaneous nystagmus in light and in darkness and on postural tests, as compared to placebo-treated controls. After termination of EGb 761 treatment, no decompensation (i.e., no reappearance of the vestibular syndrome) was observed. It was suggested that this beneficial effect of EGb 761 was due to a depressant action on the vestibular nuclei. A further study indicated that treatment with EGb 761 (50 mg/kg/day, i.p.), commencing just after labyrinthectomy, induced a net increase in brain protein synthesis in the vestibular nuclei ipsilateral to the lesion 11 and 30 days later (Bustany et al., 1992). These results support the conception that EGb 761 exerts neurotrophic actions and enhances neuronal-sprouting.

EGb 761 has also been tested on the compensation of vestibular deficits in the adult cat (Lacour et al., 1992). Post-operative treatment with EGb 761 (50 mg/kg/day, i.p.) for 30 days following a left unilateral vestibular neurectomy accelerated the recuperation of postural, locomotor and oculomotor functions, as compared to untreated controls. Recovery of synaptic function in the deafferented medial vestibular nucleus (shown by immunoreactive labelling using an antibody directed against the synaptic vesicle-associated protein, synaptophysin) also occurred, but this parameter displayed a longer time constant than the behavioural and neurophysiological data. Other findings indicated that the efficacy of an extract devoid of terpenes (25 mg/kg/day, i.p.) was comparable to that of the total EGb 761 preparation (25 or 50 mg/kg/day, i.p.) in improving vestibular compensation in the cat (Tighilet and Lacour, 1995). Thus, the non-terpene trilactone fraction contained the most active chemical constituent(s) with respect to this model of CNS plasticity.

Collectively, these results have provided evidence that EGb 761 can act on recovery mechanisms that are crucially involved in vestibular compensation. Additionally, they imply that EGb 761 possesses neurotrophic and/or neuritogenic properties that can facilitate structural synaptic remodelling (enhance neuronal plasticity) within deafferented vestibular nuclei, thereby improving functional recovery after CNS injury.

In Vivo/Ex Vivo Evidence for Free Radical-Scavenging, Antioxidant and Anti-Lipoperoxidative Activities of EGb 761

In vitro studies of the free radical-scavenging and antioxidant actions of EGb 761 are discussed in the companion chapter (DeFeudis and Drieu, this volume). Certain in vivo and ex vivo findings support these actions of the extract.

An early report indicated that preventive treatment of mice with EGb 761 (100 mg/kg, i.p.) inhibited adriamycin toxicity, providing indirect support for an in vivo free radical-scavenging action of the extract (Chatterjee, 1985). More recently, Sram et al. (1993) have reported that a 12-week treatment of ageing (24-month-old) rats with EGb 761 (50 or 100 mg/kg/day, per os) decreased lipid peroxidation in the cerebral cortex (as compared with control, non-supplemented old rats). Lipid peroxidation in the hippocampus was decreased only by the 100 mg/kg/day dose. Also, in rat liver, but not in brain, SOD activity was increased by EGb 761 at both doses, as compared to controls.

The potent diabetogenic agent alloxan selectively destroys pancreatic β cells, an action generally attributed to its production of reactive oxygen species ($O_2^{\cdot}$, H_2O_2 and $OH^{\cdot}$). Alloxan treatment also elicits marked perturbations in the electrical activity of the β cells of mice (e.g., Hammarstrom *et al.*, 1966), and a restoration of normal activity is considered to represent protection against the induced cytotoxicity. Vasseur *et al.* (1994) have used this model to test the antioxidant properties of EGb 761 in relation to glucose regulation. Intracellular recordings showed that repeated treatment of mice with EGb 761 (200 mg/kg/day, *per os*) protected β cells against the toxic effects of alloxan (50 mg/kg, i.v.), an effect that was characterized by a restoration of membrane potential and an increase in spike frequency (an indicator of insulin secretion). Since defense against alloxan toxicity involves free radical-scavenging, this action of EGb 761 could involve its flavonoid metabolites.

CONCLUDING REMARKS

In vivo and *ex vivo* studies of the actions of EGb 761 and its constituents, in extending *in vitro* observations (see companion chapter), are useful in explaining the vasoprotective, cognition-enhancing and other therapeutic actions of the extract that have been observed in the clinical setting. Several recent developments appear to be particularly significant.

With regard to myocardial ischaemia and cardiac arrhythmias, studies on anaesthetized animals have confirmed the beneficial effects of EGb 761 treatment on reperfusion-induced ventricular arrhythmias and have revealed further that the extract reduces damage to myocytes after ischaemia-reperfusion. The findings that EGb 761 treatment can limit post-ischaemic contractile dysfunction of the heart and that this protective effect is additive to that of ischaemic preconditioning are of obvious interest in relation to developing novel therapies for patients suffering from angina pectoris or from restenosis following coronary angioplasty. Results of studies with erythrocytes have strengthened the views that *in vivo* EGb 761 treatment favors the transformation of glucose into glycogen and that it exerts a vasoprotective action at the level of the microvasculature.

Studies of the CNS have indicated that EGb 761 modulates lipid-derived second messengers, predominantly in the hippocampus, and that it influences phospholipid degradation in neuronal membranes by an action which may involve inhibition of phospholipase C and phospholipase A_2 activities, effects that appear useful in explaining the anti-ischaemic and neuroprotective effects of the extract. Regarding neurotransmitter systems, EGb 761 appears to have a "restoring" action on age-related decreases in the densities of neurotransmitter receptors, an effect that is not specific for any single neurotransmitter and which seems especially noteworthy since the extract is useful in treating pathological conditions of the elderly. Results which showed that EGb 761 treatment, although given preventively and at relatively high doses, can enhance recovery of function following traumatic brain injuries also seem important since no consistently effective methods exist for treating such conditions. Results obtained with the "social interaction" test, in further defining the anti-stress (anxiolytic-like) action of EGb 761, support the clinical use of the extract in treating cognitive decline, an action which could involve increased vigilance/awareness, leading to enhanced cognitive capacity. These latter results have also revealed that EGb 761 may have a constituent(s) possessing anxiolytic activity that may not induce the adverse side effects associated with the benzodiazepines.

Recent studies with models of retinopathy indicate that orally-administered EGb 761, by decreasing oxidative stress, could be useful in preventing ischaemia-induced proliferative retinopathies such as those that occur in diabetes and those that can be induced by various xenobiotics. Results obtained with other models support the conception that EGb 761 can act on recovery mechanisms that are crucially involved in vestibular compensation, presumably by exerting neurotrophic/neuritogenic actions. This information is useful in explaining the therapeutic effects of EGb 761 on equilibrium dysfunctions.

Some major advances have also been made regarding pharmacokinetic properties of EGb 761 constituents. The pharmacokinetic parameters of terpene trilactone constituents of the extract have been well defined. Ginkgolides A and B are bioavailable after oral ingestion of EGb 761 in both rats and humans, and therefore would be expected to contribute to the therapeutic actions of the extract. Although it is not possible to demonstrate conclusively the bioavailability of flavonoid constituents of EGb 761 with currently available methods, some small phenolic metabolites which are bioavailable after oral ingestion of the extract (at least in the rat) may have therapeutic activity.

Many of these recent results, taken together, lead one to conceive that EGb 761 may serve as useful therapy in any disease state that involves free radicals, related reactive species or free radical-induced lipid peroxidation. Such indications include ischaemic conditions and neurodegenerative diseases (e.g., Alzheimer's disease), as well as the ageing process itself. As the mechanisms of action of the various constituents of EGb 761 become better understood, other explanations for the therapeutic uses of the extract should become more apparent. New insights derived from the results discussed herein should provide some stimulation for the further studies that will be required to clarify such mechanisms.

REFERENCES

Artmann, G., Degenhardt, R., Wolff, H., Grebe, R. and Schmid-Schönbein, H. (1991) Pilotstudie über membranpharmakologische Wirkungen von Ginkgo biloba Ektrakt. In F.H. Kemper and H. Schmid-Schönbein, (eds.), Rökan, Ginkgo biloba EGb 761, Springer-Verlag, Berlin, pp. 47–62.

Artmann, G.A. and Schikarski, C. (1993) Ginkgo biloba extract (EGb 761) protects red blood cells from oxidative damage In C. Ferradini, M.T. Droy-Lefaix and Y. Christen, (eds.), Advances in Ginkgo biloba Extract Research, vol. 2. Ginkgo biloba Extract as a Free-Radical Scavenger, Elsevier, Paris, pp. 129–139.

Attella, M.J., Hoffman, S.W., Stasio, M.J. and Stein, D.J. (1989) Ginkgo biloba extract facilitates recovery from penetrating brain injury in adult male rats. Exp. Neurol., 105, 62–71.

Backhauss, C., Karkoutly, C., Welsch, M. and Krieglstein, J. (1992) A mouse model of focal cerebral ischemia for screening neuroprotective drug effects. *J. Pharmacol. Meth.*, 27, 27–32.

Barkats, M., Venault, P., Christen, Y. and Cohen-Salmon, C. (1995) Effect of long-term treatment with EGb 761 on age-dependent structural changes in the hippocampi of three inbred mouse strains. *Life Sci.*, 56, 213–222.

Biber, A. and Chatterjee, S.S. (1995) Pharmacokinetics of ginkgolides A, B and bilobalide after PO administration of *Ginkgo biloba* extract EGb 761 in the rat. *Internal Report*, Dr. Willmar Schwabe GmbH & Co., Karlsruhe.

Brailowsky, S., Montiel, T., Hernández-Echeagaray, E., Flores-Hernández, J. and Hernández-Pineda, R. (1991) Effects of a *Ginkgo biloba* extract on two models of cortical hemiplegia in rats. *Restor. Neurol. Neurosci.*, 3, 267–274.

Bustany, P., Denise, P., Pottier, M. and Moulin, M. (1992) Brain protein synthesis after unilateral labyrinthectomy: Natural one-month evolution and EGb 761 treatment effect. In Y. Christen, J. Costentin and M. Lacour, (eds.), *Effects of Ginkgo biloba Extract (EGb 761) on the Central Nervous System*, Elsevier, Paris, pp. 57–74.

Chatterjee, S.S. (1985) Effects of *Ginkgo biloba* extract on cerebral metabolic processes. In A. Agnoli, J R. Rapin, V. Scapagnini and W. V. Weitbrecht, (eds.), *Effects of Ginkgo Biloba on Organic Cerebral Impairment*, John Libbey, London and Paris, pp. 5–14.

Chatterjee, S.S. and Gabard, B. (1984) Effect of an extract of *Ginkgo biloba* on experimental neurotoxicity. *Arch. Pharmacol.*, 325 (**Suppl.**), Abstr. 327.

Chatterjee, S.S., Gabard, B.L. and Jaggy, H.E.W. (1986) Pharmaceutical compositions containing bilobalid for the treatment of neuropathies. *U. S. Patent No. 4,571,407* (Feb. 18, 1986).

Chermat, R., Brochet, D., DeFeudis, F.V. and Drieu, K. (1997) Interactions of *Ginkgo biloba* extract (EGb 761), diazepam and ethyl β-carboline-3-carboxylate on social behavior of the rat. *Pharmacol. Biochem. Behav.*, 56, 333–339.

DeFeudis, F.V. (1991) *Ginkgo biloba Extract (EGb 761): Pharmacological Activities and Clinical Applications*, Elsevier, Paris.

DeFeudis, F.V. (1998) *Ginkgo biloba Extract (EGb 761). From Chemistry to the Clinic*. Ullstein Medical, Wiesbaden.

Denise, P. and Bustany, P. (1989) The effect of extract of *Ginkgo biloba* (EGb 761) on central compensation of a total unilateral peripheral vestibular deficit in the rat. In M. Lacour, M. Toupet, P. Denise and Y. Christen, (eds.), *Vestibular Compensation: Facts, Theories and Clinical Perspectives*, Elsevier, Paris, pp. 201–208.

Denise, P., Bustany, P., Moulin, M. and Pottier, M. (1990) The effect of EGb 761 on vestibular compensation in the rat. *Eur. J. Pharmacol.*, 183, 1459 (Abstr. No. P.we. 341).

Doly, M., Droy-Lefaix, M.T., Bonhomme, B. and Braquet, P. (1986) Effet de l'extrait de *Ginkgo biloba* sur l'électrophysiologie de la rétine isolée de rat diabétique. *Presse Méd.*, 15, 1480–1483.

Dorman, D.C., Côté, L.M. and Buck, W.B. (1992) Effects of an extract of *Ginkgo biloba* on bromethalin-induced cerebral lipid peroxidation and edema in rats. *Amer. J. Vet. Res.*, 53, 138–142.

Drieu, K. (1986) Préparation et définition de l'extrait de *Ginkgo biloba*. *Presse Méd.*, 15, 1455–1457.

Droy-Lefaix, M.T., Bonhomme, B. and Doly, M. (1991) Protective effect of *Ginkgo biloba* extract (EGb 761) on free radical-induced changes in the electroretinogram of isolated rat retina. *Drugs Exptl. Clin. Res.*, 17, 571–574.

Droy-Lefaix, M.T., Vennat, J.C., Besse, G. and Doly, M. (1992) Effect of *Ginkgo biloba* extract (EGb 761) on chloroquine induced retinal alterations. *Lens Eye Toxicity Res.*, **9**, 521–528.

Duverger, D., DeFeudis, F.V. and Drieu, K. (1995) Effects of repeated treatments with an extract of *Ginkgo biloba* (EGb 761) on cerebral glucose utilization in the rat: An autoradiographic study. *Gen. Pharmacol.*, **26**, 1375–1383.

File, S.E. (1980) The use of social interaction as a method of detecting anxiolytic activity of chlordiazepoxide-like drugs. *J. Neurosci. Meth.*, **2**, 219–238.

Fourtillan, J.B., Brisson, A.M., Girault, J., Ingrand, I., Decourt, J.P., Drieu, K., Jouenne, P. and Biber, A. (1995) Propriétés pharmacocinétiques du bilobalide et des ginkgolides A et B chez le sujet sain après administrations intraveineuses et orales d'extrait de *Ginkgo biloba* (EGb 761). *Thérapie*, **50**, 137–144.

Guillon, J.M., Rochette, L. and Baranes, J. (1986) Effets de l'extrait de *Ginkgo biloba* sur deux modèles d'ischémie myocardique expérimentale. *Presse Méd.*, **15**, 1516–1519.

Hammarstrom, L., Hellman, B. and Ullberg, S. (1966) On the accumulation of alloxan in the pancreatic β-cells. *Diabetologia*, **2**, 340–345.

Hofferberth, B. (1991) Simultanerfassung elektrophysiologischer, psychometrischer und rheologischer Parameter bei Patienten mit hirnorganischem Psychosyndrom und erhöhtem Gefässrisiko – Eine Placebo-kontrollierte Doppelblindstudie mit *Ginkgo biloba*-Extrakt EGB 761. In R. Stodtmeister and L.E. Pillunat, (eds.), *Mikrozirkulation in Gehirn und Sinnesorganen*, Ferdinand Enke, Stuttgart, pp. 64–74.

Huguet, F., Drieu, K. and Piriou, A. (1994) Decreased cerebral 5-HT$_{1A}$ receptors during ageing: Reversal by *Ginkgo biloba* extract (EGb 761). *J. Pharm. Pharmacol.*, **46**, 316–318.

Huguet, F. and Tarrade, T. (1992) α_2-Adrenoceptor changes during cerebral ageing. The effect of *Ginkgo biloba* extract. *J. Pharm. Pharmacol.*, **44**, 24–27.

Johnson, R.W., Reinstein, T., Spotnitz, S., Wiech, N., Ursillo, R. and Yamamura, H.I. (1980) Effects of desipramine and yohimbine on α_2- and β-adrenoceptor sensitivity. *Eur. J. Pharmacol.*, **67**, 123–127.

Karkoutly, C., Backhauss, C., Nuglisch, J. and Krieglstein, J. (1990) The measurement of the infarcted area after middle cerebral artery occlusion in the mouse: A screening model. In J. Krieglstein and H. Oberpichler, (eds.), *Pharmacology of Cerebral Ischemia*, Wissenschaftliche Verlagsgesellschaft mbH, Stuttgart, pp. 63–69.

Klein, J., Chatterjee, S.S. and Löffelholz, K. (1995) Bilobalide: A new centrally acting phospholipase A$_2$ inhibitor? *Naunyn-Schmiedeberg's Arch. Pharmacol.*, **351** (**Suppl.**), R 86 (Abstr. No. 341).

Költringer, P. and Eber, O. (1989) Die kollageninduzierte Thrombozytenaggregation unter parenteraler *Ginkgo-biloba*-Therapie. *Wien. Med. Wschr.*, **5**, 92–94.

Krieglstein, J. (1994) Neuroprotective properties of *Ginkgo biloba* constituents. *Ztschr. Phytother.*, **15**, 92–96.

Kristofikova, Z., Benesova, O. and Tejkalova, H. (1992) Changes of high-affinity choline uptake in the hippocampus of old rats after long-term administration of two nootropic drugs (tacrine and *Ginkgo biloba* extract). *Dementia*, **3**, 304–307.

Lacour, M., Ez-Zaher, L., Raymond, J. and Tighilet, B. (1992) Neurotrophic (and/or neuritogenic) properties of the extract of *Ginkgo biloba* (EGb 761) as evidenced in vestibular compensation. In Y. Christen, J. Costentin and M. Lacour, (eds.), *Effects of Ginkgo biloba Extract (EGb 761) on the Central Nervous System*, Elsevier, Paris, pp. 37–56.

Lamour, Y., Holloway, H.W., Rapoport, S.I. and Soncrant, T.T. (1992) *Ginkgo biloba* extract decreases local glucose utilization in the adult rat brain. In Y. Christen, J. Costentin, and

M. Lacour, M., (eds.), *Effects of Ginkgo biloba Extract (EGb 761) on the Central Nervous System*, Elsevier, Paris, pp. 19–25.

Le Poncin Lafitte, M., Rapin, J. and Rapin, J.R. (1980) Effects of *Ginkgo biloba* on changes induced by quantitative cerebral microembolization in rats. *Arch. Int. Pharmacodyn.*, **243**, 236–244.

Lo, H. M., Lin, F.Y., Tseng, C.D., Chiang, F.T., Hsu, K.L. and Tseng, Y.Z. (1994) Effect of EGb 761, a *Ginkgo biloba* extract, on early arrhythmia induced by coronary occlusion and reperfusion in dogs. *J. Formos. Med. Assoc.*, **93**, 592–597.

Oliver, C., Guillaume, V., Héry, F., Bourhim, N., Boiteau, K. and Drieu, K. (1994) Effect of *Ginkgo biloba* extract on the hypothalamo-pituitary-adrenal axis and plasma catecholamine levels in stress. *Eur. J. Endocrinol.*, **130** (**Suppl. 2**), 207 (Abstr. No. P3.072).

Otani, M., Chatterjee, S.S., Gabard, B. and Kreutzberg, G.W. (1986) Effect of an extract of *Ginkgo biloba* on triethyltin-induced cerebral edema. *Acta Neuropathol.*, **69**, 54–65.

Pietri, S., Maurelli, E., Drieu, K. and Culcasi, M. (1997) Cardioprotective and antioxidant effects of the terpenoid constituents of *Ginkgo biloba* extract (EGb 761). *J. Mol. Cell. Cardiol.*, **29**, 733–742.

Pietta, P.G., Gardana, C., Mauri, P.L., Maffei-Facino, R. and Carini, M. (1995) Identification of flavonoid metabolites after oral administration to rats of a *Ginkgo biloba* extract. *J. Chromatog. B*, **673**, 75–80.

Porsolt, R.D., Martin, P., Fromage, S., Lenègre, A. and Drieu, K. (1992) Anti-stress effects of EGB 761 in rodent models. In Y. Christen, J. Costentin, and M. Lacour, M., (eds.), *Effects of Ginkgo biloba Extract (EGb 761) on the Central Nervous System*, Elsevier, Paris, pp. 135–145.

Porsolt, R.D., Martin, P., Lenègre, A., Fromage, S. and Drieu, K. (1990) Effects of an extract of *Ginkgo biloba* (EGB 761) on "learned helplessness" and other models of stress in rodents. *Pharmacol. Biochem. Behav.*, **36**, 963–971.

Prehn, J.H.M. and Krieglstein, J. (1993) Platelet-activating factor antagonists reduce excitotoxic damage in cultured neurons from embryonic chick telencephalon and protect the rat hippocampus and neocortex from ischemic injury *in vivo*. *J. Neurosci. Res.*, **34**, 179–188.

Ramassamy, C., Clostre, F., Christen, Y. and Costentin, J. (1990) Prevention by a *Ginkgo biloba* extract (GBE 761) of the dopaminergic neurotoxicity of MPTP. *J. Pharm. Pharmacol.*, **42**, 785–789.

Rapin, J.R., Lamproglou, I., Drieu, K. and DeFeudis, F.V. (1994a) Demonstration of the "anti-stress" activity of an extract of *Ginkgo biloba* (EGb 761) using a discrimination learning task. *Gen. Pharmacol.*, **25**, 1009–1016.

Rapin, J.R., Provost, P., DeFeudis, F.V. and Drieu, K. (1994b) Effects of repeated treatments with an extract of *Ginkgo biloba* (EGb 761) and bilobalide on glucose uptake and glycogen synthesis in rat erythrocytes: An *ex vivo* study. *Drug. Dev. Res.*, **31**, 164–169.

Rapin, J.R., Provost, P. and Drieu, K. (1994c) Hippocampal glucocorticoid receptor II down regulation in rats. Effects of *Ginkgo biloba* extract (EGb 761). *Eur. J. Endocrinol.*, **130** (**Suppl. 2**), 127 (Abstr. No. P1.128).

Rapin, J.R., Provost, P. and Drieu, K. (1996) Difference between anti-stress and anxiolytic drugs on hippocampal glucocorticoid receptor Type II. In *New Frontiers in Stress Research: Modulation of Brain Function*, 40th OHOLO Conference, Zichron Ya'akov, Israel, March 24–28, 1996 (Abstract 17P).

Reuse-Blom, S. and Drieu, K. (1986) Effet de l'extrait de *Ginkgo biloba* sur le spasme artériolaire chez le lapin. *Presse Méd.*, **15**, 1520–1523.

Rodriguez de Turco, E.B., Droy-Lefaix, M.T. and Bazan, N.G. (1993) Decreased electroconvulsive shock-induced diacylglycerols and free fatty acid accumulation in the rat brain by *Ginkgo biloba* extract (EGb 761): Selective effect in hippocampus as compared with cerebral cortex. *J. Neurochem.*, **61**, 1438–1444.

Sancesario, G. and Kreutzberg, G.W. (1986) Stimulation of astrocytes affects cytotoxic brain edema. *Acta Neuropath.*, **72**, 3–14.

Sapolsky, R.M., Krey, L.C. and McEwen, B.S. (1983) Corticosterone receptors decline in a site-specific manner in the aged rat brain. *Brain Res.*, **289**, 235–240.

Sharma, H.S., Westman, J., Nyberg, F., Cervós-Navarro, J. and Dey, P.K. (1994) Neuroprotective effects of an extract of *Ginkgo biloba* (EGb–761) in heat stress induced brain damage in the rat. In *Thermal Balance in Health and Disease: Advances in Pharmacological Sciences*, Birkhäuser Verlag, Basel, pp. 461–467.

Shen, J. and Zhou, D.-Y. (1995) Efficiency of *Ginkgo biloba* extract (EGb 761) in antioxidant protection against myocardial ischemia and reperfusion injury. *Biochem. Mol. Biol. Int.*, **35**, 125–134.

Sim, A.K. and Davies, M.E. (1980) The *in vivo* evaluation of IPS 200 as an inhibitor of experimental thrombus formation in the microcirculation of the hamster cheek pouch. In *Drug Registration File*, IPSEN (Paris) and Dr. Willmar Schwabe GmbH & Co. (Karlsruhe).

Sokoloff, L., Reivich, M., Kennedy, C., Des Rosiers, M.H., Patlak, C.S., Pettigrew, K.D., Sakurada, O. and Shinohara, M. (1977) The [^{14}C]deoxyglucose method for the measurement of local cerebral glucose utilization: Theory, procedure and normal values in the conscious and anesthetized albino rat. *J. Neurochem.*, **28**, 897–916.

Spinnewyn, B. (1992) *Ginkgo biloba* extract (EGb 761) protects against delayed neuronal death in gerbil. In Y. Christen, J. Costentin and M. Lacour, (eds.), *Effects of Ginkgo biloba Extract (EGb 761) on the Central Nervous System*, Elsevier, Paris, pp. 113–118.

Spinnewyn, B., Blavet, N. and Clostre, F. (1986) Effet de l'extrait de *Ginkgo biloba* sur un modèle d'ischémie cérébrale chez la gerbille. *Press Méd.*, **15**, 1511–1515.

Sram, R.J., Binkova, B., Stejskalova, J. and Topinka, J. (1993) Effect of EGb 761 on lipid peroxidation, DNA repair and antioxienzyme activity. In C. Ferradini, M.T. Droy-Lefaix and Y. Christen, (eds.), *Advances in Ginkgo biloba Extract Research, Vol. 2. Ginkgo biloba Extract (EGb 761) as a Free-Radical Scavenger*, Elsevier, Paris, pp. 27–38.

Stein, D.G. and Hoffman, S.W. (1992) Chronic administration of *Ginkgo biloba* extract EGb 761 can enhance recovery from traumatic brain injury. In Y. Christen, J. Costentin and M. Lacour, (eds.), *Effects of Ginkgo biloba Extract (EGb 761) on the Central Nervous System*, Elsevier, Paris, pp. 95–103.

Stücker, O., Pons, C., Duverger, J.P., d'Arbigny, P. and Drieu, K. (1994a) Effect of *Ginkgo biloba* extract (EGb 761) on *in vivo* induced, white-platelet, arterial-thrombus formation. In F. Clostre and F.V. DeFeudis, (eds.), *Advances in Ginkgo biloba Extract Research, vol. 3. Cardiovascular Effects of Ginkgo biloba Extract (EGb 761)*, Elsevier, Paris, pp. 27–30.

Stücker, O., Pons, C., Duverger, J.P., d'Arbigny, P. and Drieu, K. (1994b) *Ginkgo biloba* extract (EGb 761) and capillary red-blood-cell velocity. In F. Clostre and F.V. DeFeudis, (eds.), *Advances in Ginkgo biloba Extract Research, vol. 3. Cardiovascular Effects of Ginkgo biloba Extract (EGb 761)*, Elsevier, Paris, pp. 31–37.

Stücker, O., Pons, C., Duverger, J.P. and Drieu, K. (1996) Effects of *Ginkgo biloba* extract (EGb 761) on arteriolar spasm in a rat cremaster muscle preparation. *Int. J. Microvasc. Res.*, **16**, 98–104.

Szabo, M.E., Droy-Lefaix, M.T. and Doly, M. (1992a) Modification of reperfusion-induced ionic imbalance by free radical scavengers in spontaneously hypertensive rat retina. *Free Radical Biol. Med.*, **13**, 609–620.

Szabo, M.E., Droy-Lefaix, M.T., Doly, M. and Braquet, P. (1991a) Free radical-mediated effects in reperfusion injury: A histologic study with superoxide dismutase and EGb 761 in rat retina. *Ophthal. Res.*, **23**, 225–234.

Szabo, M.E., Droy-Lefaix, M.T., Doly, M. and Braquet, P. (1992b) Ischaemia- and reperfusion-induced Na^+, K^+, Ca^{2+} and Mg^{2+} shifts in rat retina: Effects of two free radical scavengers, SOD and EGb 761. *Exp. Eye Res.*, **55**, 39–45.

Szabo, M.E., Droy-Lefaix, M.T., Doly, M., Carré, C. and Braquet, P. (1991b) Ischemia and reperfusion-induced histologic changes in the rat retina. *Invest. Ophthalmol. Vis. Sci.*, **32**, 1471–1478.

Taylor, J.E. (1986) Liaison des neuromédiateurs à leurs récepteurs dans le cerveau de rat. *Presse Méd.*, **15**, 1491–1493.

Tighilet, B. and Lacour, M. (1995) Pharmacological activity of the *Ginkgo biloba* extract (EGb 761) on equilibrium function recovery in the unilateral vestibular neurectomized cat. *J. Vestib. Res.*, **5**, 187–200.

Tosaki, A., Droy-Lefaix, M.T., Pali, T. and Das, D.K. (1993) Effects of SOD, catalase and a novel antiarrhythmic drug, EGb 761, on reperfusion-induced arrhythmias in isolated rat hearts. *Free Radical Biol. Med.*, **14**, 361–370.

Tosaki, A., Engelman, D.T., Pali, T., Engelman, R.M. and Droy-Lefaix, M.T. (1994) *Ginkgo biloba* extract (EGb 761) improves postischemic function in isolated preconditioned working rat hearts. *Coronary Artery Dis.*, **5**, 443–450.

Vasseur, M., Jean, T., DeFeudis, F.V. and Drieu, K. (1994) Effects of repeated treatments with an extract of *Ginkgo biloba* (EGb 761), bilobalide and ginkgolide B on the electrical activity of pancreatic β-cells of normal or alloxan-diabetic mice: An *ex vivo* study with intracellular microelectrodes. *Gen. Pharmacol.*, **25**, 31–46.

Winter, E. (1991) Effects of an extract of *Ginkgo biloba* on learning and memory in mice. *Pharmacol. Biochem. Behav.*, **38**, 109–114.

18. THE NEUROPROTECTIVE PROPERTIES OF GINKGO EXTRACTS

CYNTHIA L. DARLINGTON[1], PAUL F. SMITH[2] and KARYN MACLENNAN[2]

*[1]Dept. of Psychology and the Neuroscience Research Centre,
University of Otago, Dunedin, New Zealand and
[2]Dept. of Pharmacology, School of Medical Sciences, University of
Otago Medical School, Dunedin, New Zealand*

INTRODUCTION

Over the past decade, evidence has been gradually accumulating which suggests that concentrated and partially purified extracts of *Ginkgo biloba* leaves (e.g. EGb 761, Ipsen) have clinical application in the treatment of neural damage (Kleijnen and Knipschild, 1992ab). While many clinical studies now support the hypothesis that Ginkgo extracts may be useful in these circumstances (e.g. see Kleijnen and Knipschild, 1992ab for review), preclinical studies have continued to investigate the precise contribution of the major constituents of *Ginkgo biloba* (Gb) to its neuroprotective properties (e.g. see Christen *et al.*, 1992, 1995, 1996 for reviews). While the ginkgolides and bilobalide within the terpene fraction have been demonstrated to have powerful neuroprotective properties, so too have the constituents of the flavonoid fraction (e.g. quercetin; Sticher, 1993). Furthermore, these different constituents appear to have a synergistic action, such that, in many cases, the entire Gb extract has a greater beneficial effect than any of its constituents administered separately (see DeFeudis, 1991 for a review).

The aim of this chapter is to summarise and to critically evaluate what is currently known about the neuroprotective properties of Gb extracts. First, the clinical evidence that Gb extracts can have neuroprotective effects will be reviewed briefly, followed by examination of the preclinical studies addressing the mechanisms of action of Gb and the contributions of the ginkgolides, bilobalide and flavonoids to its neuroprotective properties.

CLINICAL EVIDENCE FOR THE NEUROPROTECTIVE EFFECTS OF GINKGO EXTRACTS

To date, most of the clinical investigations that are potentially relevant to the neuroprotective action of Gb extracts, have concerned the treatment of cerebrovascular insufficiency, dementia of the Alzheimer's type, normal cognitive decline in the elderly, tinnitus and sensorineural hearing loss. Of these various conditions, most of the data relate to either cognitive dysfunction or cerebrovascular insufficiency; in the latter case, the "neuroprotective" effect may be attributable to vascular effects rather than direct modulation of neuronal function.

There is substantial clinical evidence to support the hypothesis that Gb extracts are useful for the treatment of cerebrovascular insufficiency in the elderly, with positive results obtained in many centres around the world (e.g. Grassel, 1992; Kleijnen and Knipschild, 1992ab; Hopfenmuller, 1994; Mouren et al., 1994; however, see Garg et al., 1995 for conflicting evidence). This evidence has been reviewed in detail by Kleijnen and Knipschild (1992ab). Although it is possible that the benefits of Gb treatment in this context are not due simply to a direct neuro-protective action, the available preclinical evidence makes it very likely that some of the beneficial effects are derived in this way (see below).

The data relating to the treatment of Alzheimer's dementia and normal cognitive decline in the elderly are, however, less conclusive. Several semi-anecdotal studies of the cognitive effects of Gb extracts in humans have been published (see Warburton, 1986 for a review); however, there have been fewer double-blind, placebo-controlled studies using sophisticated neuropsychological tests (see Smith et al., 1996 for review). Nonetheless, several studies of age-related memory impairment, which have incorporated some of these design controls, have reported beneficial effects from EGb 761 treatment, in many cases expressed as shortened response time (Subhan and Hindmarch, 1984; Wesnes et al., 1987; Rai et al., 1991; Allain et al., 1993; Semlitsch et al., 1995; see also Kunkel, 1993). These studies suggest that Gb extracts may retard some aspects of cognitive decline in Alzheimer's dementia, and studies of Alzheimer's patients themselves confirm that they may be of benefit (e.g. Hofferberth, 1994; Kanowski et al., 1996).

Animal studies generally support the hypothesis that Gb extracts can improve cognitive function in cases where it has been degraded by age or neurological insult (Lenegre et al., 1988; Barkats et al., 1995; Chopin and Briley, 1992; Petkov et al., 1993). However, studies in neurologically intact animals suggest that Gb extracts produce, at most, a modest improvement in normal cognitive function (Porsolt et al., 1990; Winter, 1991; Petkov et al., 1993).

Recently, a number of clinical studies have investigated the possibility that Gb extracts may be useful in the treatment of tinnitus and sensorineural hearing loss (Holgers et al., 1994; Plath and Olivier, 1995); however, to date, it is unclear how beneficial they may be.

The studies reviewed in this section demonstrate that there is sufficient evidence in favour of a neuroprotective action of Gb extracts, to justify research into their mode of action in the hope of increasing their beneficial effects and administering them in the optimal way. The next section will address the preclinical evidence that Gb extracts have neuroprotective effects.

GINKGO EXTRACTS AS NEUROPROTECTANTS

Gb extracts are now known to modulate a number of neurotransmitter receptors, including peripheral-type benzodiazepine (Amri et al., 1996), serotonin (5-HT_{1A})

(Huguet *et al.*, 1994; Bolanos-Jimenez *et al.*, 1995) and noradrenaline receptors (Brunello *et al.*, 1985; Huguet and Tarrade, 1992), as well as levels of monoamine oxidase (White *et al.*, 1996) and synaptosomal uptake of 5-HT and dopamine (Ramassamy *et al.*, 1992ab). Some studies have demonstrated direct electro-physiological effects of EGb 761 *in vitro* (i.e. hyperpolarization; Vidal *et al.*, 1993) or changes in vestibular reflex function following systemic or intra-vestibular nucleus administration (Yabe *et al.*, 1992ab); changes in cerebral glucose utilization have also been documented (Duverger *et al.*, 1995). The actions of the ginkgolide and flavonoid constituents of Gb are understood in more detail and will be discussed in a later section; this section will consider the effects of the entire Gb extract.

There is substantial evidence to support the hypothesis that Gb extracts protect against neuronal degeneration following hypoxia/ischemia. Although EGb 761 has modest effects on cerebral blood flow and metabolism in the intact brain (Krieglstein *et al.*, 1986), several groups have demonstrated that EGb 761 treatment can increase blood flow and adenosine triphosphate (ATP), glucose and lactate levels in the brain following the induction of hypoxia/ischemia (Le Poncin Lafitte *et al.*, 1980; Karcher *et al.*, 1984; Oberpichler *et al.*, 1988; Janssens *et al.*, 1995). Other studies have shown that Gb extracts like EGb 761 can protect against neuronal degeneration, and improve recovery of neuronal function, following electroconvulsive shock (Rodriguez de Turco *et al.*, 1993), surgical deafferentation (Denise and Bustany, 1989; Ez-Zaher and Lacour, 1989; Lacour *et al.*, 1991; Lacour *et al.*, 1992; Bustany *et al.*, 1992; Brailowsky *et al.*, 1993; Koc *et al.*, 1995; Tighilet and Lacour, 1995), or exposure to hydrogen peroxide (Oyama *et al.*, 1992; Oyama *et al.*, 1993; Oyama *et al.*, 1996), ascorbic acid/Fe^{2+} (Ramassamy *et al.*, 1993) or hydroxyl radicals *in vitro* (Ni *et al.*, 1996). Neuroprotection has also been documented in the case of triethyltin-induced cerebral edema (Otani *et al.*, 1986), 1-methyl-4-phenyl-1,2,3,6-tetrahydropyridine (MPTP)-induced dopaminergic neurotoxicity (Ramassamy *et al.*, 1990) and diabetic neuropathy (Apaydin *et al.*, 1993). Furthermore, EGb 761 has been shown to reduce cardiac ischemia-reperfusion injury (Pietri *et al.*, 1993) and inhibit radiation-induced tumour growth (Emerit *et al.*, 1995).

Clearly, Gb extracts such as EGb 761 not only have neuroprotective properties but may protect against cellular damage in a more general sense. In order to better understand these effects, researchers have turned to the constituents of Gb in an effort to isolate the active component(s).

THE CONTRIBUTION OF THE GINKGOLIDES TO THE NEUROPROTECTIVE EFFECTS OF GINKGO EXTRACTS

Of the various ginkgolides contained within Gb, probably ginkgolide B has attracted the most attention, due to its potent platelet-activating factor (PAF)-antagonist properties.

PAF is an alkylphospholipid produced by a variety of cells, including neurons (Yue *et al.*, 1990), and is one of the most potent lipid mediators known (see Braquet, 1987; Braquet *et al.*, 1987; Snyder, 1990; Nojima, 1991; Braquet *et al.*, 1991; Lindsberg *et al.*, 1991; Yue and Feuerstein, 1994; Maclennan *et al.*, 1996a, for reviews). Specific PAF receptors have been identified in the central nervous system (CNS), localised both to synaptic endings and intracellular membranes (e.g. Marcheselli et al., 1990); there is some evidence that extracellular PAF receptors may be co-localised with the N-methyl-D-aspartate (NMDA) subtype of excitatory amino acid receptor (Bito et al., 1992). PAF concentrations increase in the brain during trauma such as ischemia (Kumar et al., 1988; Pettigrew et al., 1995; Satoh et al., 1992; Nishida and Markey, 1996), causing an increase in free intracellular Ca^{2+} concentrations which can lead to Ca^{2+}-induced cell death or apoptosis (Kornecki and Ehrlich, 1991; Yue et al., 1991ab; Bito et al., 1992). There is increasing evidence that PAF enhances excitatory synaptic transmission mediated by excitatory amino acids such as glutamate (Clark et al., 1992; Bito et al., 1992; Wieraszko et al., 1993); therefore, it is possible that PAF amplifies excitotoxicity produced by excessive glutamate release, and the consequent increase in intracellular Ca^{2+} concentrations in postsynaptic neurons, during neuronal injury. However, it has been reported that PAF inhibits excitatory synaptic transmission in the neostriatum (Jiang et al., 1993).

PAF has been demonstrated to have a variety of other effects in the CNS, including modulation of the hypothalamic-pituitary-adrenal axis (Blasquez et al., 1990) and inhibition of Na^+/K^+-ATPase activity (Catalan et al., 1994), which may account for some of its electrophysiological effects (Maclennan et al., 1996b). Recently Amri et al. (1996) have proposed that one of the ways in which Gb may protect against neuronal damage, is by reducing the number and binding capacity of the adrenocortical mitochondrial peripheral-type benzodiazepine receptor, resulting in reduced corticosteroid synthesis and therefore, increased release of adrenocorticotrophic hormone (ACTH) and increased expression of the steroidogenic acute regulatory protein. Both EGb 761 and ginkgolides A and B had similar effects, suggesting that Gb extracts may exert some of their neuroprotective action as a result of modulation of glucocorticoid synthesis (Amri et al., 1996).

It has recently been suggested that PAF may be a retrograde messenger in long-term potentiation (LTP) (Kato et al., 1994), the putative neural model of memory in the CNS (see Bliss and Collingridge, 1993 for a review). PAF has been reported to induce LTP (Wieraszko et al., 1993), to enhance memory in behavioural tasks (Izquierdo et al., 1995), and PAF antagonists have been shown to disrupt LTP (e.g. Del Cerro et al., 1990).

An impressive body of evidence demonstrates that PAF antagonists, including ginkgolide B, protect against neuronal damage following hypoxia/ischemia (Panetta et al., 1987; Spinnewyn et al., 1987; Oberpichler et al., 1990; Bazan et al., 1991; Bielenberg et al., 1992; Prehn and Krieglstein, 1993; Hirashima et al., 1993; Sun and Gilboe, 1994; Tokutomi et al., 1994; Janssens et al., 1995; Zablocka et al., 1995; Matsuo et al., 1996; Zalewski et al., 1996), seizure activity (Birkle et al., 1988; Gilboe et al., 1991; Bazan et al., 1991), deafferentation (Faden and Halt, 1992; Faden and Tzendzalian, 1992; Maclennan et al., 1995) or exposure to neurotoxic levels of PAF (Lustig et al., 1992; Willard, 1992) or glutamate (Prehn and Krieglstein, 1993).

PAF antagonists have been shown to reduce the expression of immediate early gene proteins such as Fos, Jun and Krox-24 following trauma, even in the case of a single seizure (Squinto *et al.*, 1989; Bazan *et al.*, 1991; Marcheselli and Bazan, 1994).

These studies demonstrate that the PAF-antagonist properties of the ginkgolides (ginkgolide B in particular) are likely to be responsible for a major component of the neuroprotective effects of the entire Gb extract. It is clear, however, that the bilobalide and flavonoid components are also important, as will be discussed in the next section.

THE CONTRIBUTION OF BILOBALIDE AND THE FLAVONOIDS TO THE NEUROPROTECTIVE EFFECTS OF GINKGO EXTRACTS

The literature relating to the effects of bilobalide is not as extensive as that for the ginkgolides or the flavonoids. Nonetheless, there is evidence that bilobalide can protect against neuronal damage during brain edema (Sancesario and Kreutzberg, 1986) and accelerate regeneration of peripheral nerve fibres following traumatic nerve damage (Bruno *et al.*, 1993).

The flavonoids are known to be protective not only against neuronal damage, but against cellular damage in general (Havsteen, 1983; Middleton, 1984; Gerritsen *et al.*, 1995; Keli *et al.*, 1996). In addition, they are recognised as having significant anti-tumour activity as a result of their antioxidant properties (e.g. Sato *et al.*, 1994; Wang *et al.*, 1996). Flavonoids are known to have a number of biochemical actions, including the inhibition of cyclic phosphodiesterase (Ruckstuhl, 1979), nitric oxide synthesis (Tamura *et al.*, 1994) and amine reuptake (Sher *et al.*, 1992). However, it is their role as free radical scavengers which has attracted the most attention and is currently thought to be a critical part of their neuroprotective ability. Flavonoids have been shown to scavenge free radical superoxide anions (Robak and Gryglewski, 1988; Pincemail *et al.*, 1989; Gsell *et al.*, 1995; Wang *et al.*, 1995), to activate the enzyme superoxide dismutase which protects against free radicals (Pincemail *et al.*, 1989; Gsell *et al.*, 1995; Wang *et al.*, 1995) and to reduce oxidative metabolism (Oyama *et al.*, 1994). Since free radical generation and consequent oxidation are implicated in apoptosis related to Alzheimer's disease (Smith *et al.*, 1991; Hockenberry *et al.*, 1993), the free radical-scavenging properties of the flavonoids are important for the neuroprotective effect of the complete Gb extract.

CONCLUSIONS

There is little question that Gb extracts such as EGb 761 have a neuroprotective action in the clinical treatment of cerebral insufficiency; there is increasing evidence that they may also retard the development of Alzheimer's dementia and normal cognitive deterioration in the elderly. However, their utility in the treatment of tinnitus and sensorineural hearing loss is unclear at present and awaits further investigation.

The precise mechanisms of action of Gb extracts are not yet fully elucidated. There is convincing evidence that the ginkgolides, bilobalides and flavonoids all have neuroprotective actions, although via different mechanisms. While the PAF antagonistic actions of the ginkgolides may reduce intracellular Ca^{2+} concentrations during neural trauma, and the ginkgolides may also regulate corticosteroid synthesis under such circumstances, the free radical-scavenging properties of the flavonoids are certain to be as important; although bilobalide has been shown to be neuro-protective as well, at present its mechanism of action is unclear. Therefore, the combination of these three components of Gb is likely to have a greater effect than any one component alone, and the available evidence certainly supports this hypothesis (see DeFeudis, 1991 for a review). Consequently, while preclinical studies continue to investigate the molecular targets of the ginkgolides, bilobalide and flavonoids, in clinical applications the entire Gb extract should be used in order to exploit its full neuroprotective potential.

Acknowledgements

This research was supported by Project Grants from the New Zealand Neurological Foundation (to PS) and the Health Research Council of New Zealand (to CD and PS). KM was supported by a University of Otago Postgraduate Scholarship. We thank Anna Matheson for her excellent research assistance.

REFERENCES

Allain, H., Raoul, P., Lieury, A., LeCoz, F., Gandon, J.M. and d'Arbigny, P. (1993) Effect of two doses of *Ginkgo biloba* extract (EGb 761) on the dual coding test in elderly subjects. *Clin. Therap.*, 15, 549–558.

Amri, H., Ogwuegbu, S.O., Boujrad, N., Drieu, K. and Papadopoulos, V. (1996) *In vivo* regulation of peripheral-type benzodiazepine receptor and glucocorticoid synthesis by *Ginkgo biloba* extract EGb 761 and isolated ginkgolides. *Endocrinol.*, 137, 5707–5718.

Apaydin, C., Oguz, Y., Agar, A., Yargicoglu, P., Demir, N. and Aksu, G. (1993) Visual evoked potentials and optic nerve histopathology in normal and diabetic rats and effect of *Ginkgo biloba* extract. *Acta Ophthalmologica*, 71, 623–628.

Attella, M.J., Hoffman, S.W., Stasio, M.J. and Stein, D.G. (1989) *Ginkgo biloba* extract facilitates recovery from penetrating brain injury in adult male rats. *Exp. Neurol.*, 105, 62–71.

Barkats, M., Venault, P., Christen, Y. and Cohen-Salmon, C. (1995) Effect of long-term treatment with EGb 761 on age-dependent structural changes in the hippocampi of three inbred mouse strains. *Life Sci.*, 56, 213–222.

Bazan, N.G., Squinto, S.P., Braquet, P., Panetta, T. and Marcheselli, V.L. (1991) Platelet-activating factor and polyunsaturated fatty acids in cerebral ischemia or convulsions: intracellular PAF-binding sites and activation of a Fos/Jun/AP-1 transcriptional signaling system. *Lipids*, 26, 1236–1242.

Bielenberg, G.W., Wagener, G. and Beck, T. (1992) Infarct reduction by the platelet activating factor antagonist apafant in rats. *Stroke*, 23, 98–103.

Birkle, D.L., Kurian, P., Braquet, P. and Bazan, N.G. (1988) Platelet-activating factor anta-gonist BN52021 decreases accumulation of free polyunsaturated fatty acid in mouse brain during ischemia and electroconvulsive shock. *J. Neurochem.*, **51**, 1900–1905.

Bito, H., Nakamura, M., Honda, Z., Izumi, T., Iwatsubo, T., Seyama, Y., Ogura, A., Kudo, Y. and Shimizu, T. (1992) Platelet-activating factor receptor in rat brain: PAF mobilizes intracellular Ca^{2+} in hippocampal neurons. *Neuron*, **9**, 285–294.

Blasquez, C., Jegou, S., Delarue, C., Delbende, C., Bunel, D.T., Braquet, P. and Vaudry, H. (1990) Effect of platelet-activating factor on hypothalamic and hypophyseal pro-opiomelanocortin-related peptides and hypothalamo-pituitary-adrenal axis in the rat. *Eur. J. Pharmacol.*, **177**, 145–153.

Bliss, T.V.P. and Collingridge, G.L. (1993) A synaptic model of memory: long-term potentiation in the hippocampus. *Nature*, **363**, 31–39.

Bolanos-Jimenez, F., Manhaes de Castro, R., Sarhan, H., Prudhomme, N., Drieu, K. and Fillon, G. (1995) Stress-induced 5-HT1A receptor desensitization: protective effects of *Ginkgo biloba* extract (EGb 761). *Fund. Clin. Pharmacol.*, **9**, 169–174.

Brailowsky, S., Montiel, T. and Medina-Ceja, L. (1995) Acceleration of functional recovery from motor cortex ablation by two *Ginkgo biloba* extracts in rats. *Rest. Neurol. Neurosci.*, **8**, 163–167.

Braquet, P. (1987) The ginkgolides: potent platelet-activating factor antagonists isolated from *Ginkgo biloba* L.: chemistry, pharmacology and clinical applications. *Drugs Future*, **12**, 643–699.

Braquet, P., Esanu, A., Buisine, E., Hosford, D., Broquet, C. and Koltai, M. (1991) Recent progress in ginkgolide research. *Med. Res. Revs.*, **11**, 295–355.

Braquet, P., Touqui, L., Shen, T.S. and Vargaftig, B.B. (1987) Perspectives in platelet-activating factor research. *Pharmacol. Revs.*, **39**, 97–145.

Brunello, N., Racagni, G., Clostre, F., Drieu, K. and Braquet, P. (1985) Effects of an extract of *Ginkgo biloba* on noradrenergic systems of rat cerebral cortex. *Pharmacol. Res. Comm.*, **17**, 1063–1072.

Bruno, C., Cuppini, R., Sratini, S., Cecchini, T., Ambrogini, P. and Bombardelli, E. (1993) Regeneration of motor nerves in bilobalide-treated rats. *Planta Med.*, **59**, 302–307.

Bustany, P., Denise, P., Pottier, M. and Moulin, M. (1992) Brain protein synthesis after uni-lateral labyrinthectomy. In Y. Christen, J. Costentin and M. Lacour, (eds.), *Effects of Ginkgo biloba Extract (EGb 761) on the Central Nervous System*, Elsevier, Paris, pp. 57–74.

Catalan, R.E., Martinez, A.M., Aragones, M.D., Fernandez, I., Miguel, B.G., Calcerrada, M.C. and Perez, M.J. (1994) Platelet-activating factor inhibits (Na^+, K^+) ATPase activity in rat brain. *Neurosci. Res.*, **19**, 241–244.

Chopin, P. and Briley, M. (1992) Effects of four non-cholinergic cognitive enhancers in comparison with tacrine and galanthamine on scopolamine-induced amnesia in rats. *Psychopharmacol.*, **106**, 26–30.

Christen, Y., Costentin, J. and Lacour, M. (eds.) (1992) *Effects of Ginkgo biloba Extract (EGb 761) on the Central Nervous System*, Elsevier, Paris.

Christen, Y., Coutois, Y. and Droy-Lefaix, M.T. (eds.) (1995) *Effects of Ginkgo biloba Extract (EGb 761) on Ageing and Age-Related Disorders*, Elsevier, Paris.

Christen, Y., Droy-Lefaix, M.T. and Macias-Nunez, J.F. (eds.) (1996) *Effects of Ginkgo biloba Extract (EGb 761) on Neuronal Plasticity*, Elsevier, Paris.

Clark, G.D., Happel, L.T., Zorumski, C.F. and Bazan, N.G. (1992) Enhancement of hippocampal excitatory synaptic transmission by platelet-activating factor. *Neuron*, **9**, 1211–1216.

DeFeudis, F.V. (1991) *Ginkgo biloba extract (EGb 761): Pharmacological Activities and Clinical Applications*. Elsevier, Paris.

Del Cerro, S., Arai, A. and Lynch, G. (1990) Inhibition of long-term potentiation by an antagonist of platelet-activating factor receptors. *Behav. Neural Biol.*, 54, 213–217.

Denise, P. and Bustany, P. (1989) The effect of extract of *Ginkgo biloba* (EGb 761) on central compensation of a total unilateral peripheral vestibular deficit in the rat. In M. Lacour, M. Toupet, P. Denise and Y. Christen (eds.), *Vestibular Compensation. Facts, Theories and Clinical Perspectives*, Elsevier, Paris, pp. 201–208.

Duverger, D., DeFeudis, F.V. and Drieu, K. (1995) Effects of repeated treatments with an extract of *Ginkgo biloba* (EGb 761) on cerebral glucose utilization in the rat: an auto-radiographic study. *Gen. Pharmacol.*, 26, 1375–1383.

Emerit, I., Arutyunyan, R., Oganesian, N., Levy, A., Cernjavsky, L., Sarkisian, T., Pogossian, A. and Asrian, K. (1995) Radiation-induced clastogenic factors: anticlastogenic effect of *Ginkgo biloba* extract. *Free Radical Biol. Med.*, 18, 985–991.

Ez-Zaher, L. and Lacour, M. (1989) Effects of post-operative treatment with an extract of *Ginkgo biloba* (EGb 761) on vestibular compensation in the cat. In M. Lacour, M. Toupet, P. Denise and Y. Christen (eds.), *Vestibular Compensation. Facts, Theories and Clinical Perspectives*, Elsevier, Paris,.pp. 209–223.

Faden, A.I. and Halt, P. (1992) Platelet-activating factor reduces spinal cord blood flow and causes behavioural deficits after intrathecal administration in rats through a specific receptor mechanism. *J. Pharmacol. Exp. Therap.*, 261, 1064–1070.

Faden, A.I. and Tzendzalian, P.A. (1992) Platelet-activating factor antagonists limit glycine changes and behavioral deficits after brain trauma. *Am. J. Physiol.*, 263, R909-R914.

Garg, R.K., Nag, D. and Agrawal, A. (1995) A double-blind placebo-controlled trial of *Ginkgo biloba* extract in acute cerebral ischaemia. *J. Assoc. Physicians India*, 43, 760–763.

Gerritsen, M.E., Carley, W.W., Ranges, G.E., Shen, C.P., Phan, S.A., Ligon, G.F. and Perry, C.A. (1995) Flavonoids inhibit cytokine-induced endothelial cell adhesion protein gene expression. *Am. J. Pathol.*, 147, 278–292.

Gilboe, D.D., Kintner, D., Fitzpatrick, J.H., Emoto, S.E., Esanu, A., Braquet, P. and Bazan, N.G. (1991) Recovery of postischemic brain metabolism and function following treatment with a free radical scavenger and platelet-activating factor antagonist. *J. Neurochem.*, 56, 311–319.

Grassel, E. (1992) Effect of *Ginkgo biloba* extract on mental performance. Double-blind study using computerized measurement conditions in patients with cerebral insufficiency. *Fortschritte Medizin*, 110, 73–76.

Gsell, W., Reichert, N., Youdim, M.B. and Riederer, P. (1995) Interaction of neuroprotective substances with human brain superoxide dismutase. *J. Neural Trans.*, Suppl. 45, 271–179.

Havsteen, B. (1983) Flavonoids, a class of natural products of high pharmacological potency. *Biochem. Pharmacol.*, 32, 1141–1148.

Hirashima, Y., Endo, S., Otsuji, T., Karasawa, K., Nojima, S. and Takaku, A. (1993) Platelet-activating factor and cerebral vasospasm following subarachnoid hemorrhage. *J. Neurosurg.*, 78, 592–597.

Hockenberry, D.M., Oltvai, Z.N., Yin, X-M., Milliman, C.L. and Korsmeyer, S.J. (1993) Bcl-2 functions in an antioxidant pathway to prevent apoptosis. *Cell*, 75, 241–251.

Hofferberth, B. (1994) The efficacy of EGb 761 in patients with senile dementia of the Alzheimer type, a double-blind, placebo-controlled study on different levels of investigation. *Human Psychopharmacol.*, 9, 215–222.

Holgers, K.M., Axelsson, A. and Pringle, I. (1994) *Ginkgo biloba* extract for the treatment of tinnitus. *Audiology*, 33, 85–92.

Hopfenmuller, W. (1994) Evidence for a therapeutic effect of *Ginkgo biloba* special extract. Meta-analysis of 11 clinical studies in patients with cerebrovascular insufficiency in old age. *Arzneim. Forsch.*, **44**, 10015–1113.

Huguet, F., Drieu, K. and Piriou, A. (1994) Decreased cerebral 5-HT$_{1A}$ receptors during ageing: reversal by *Ginkgo biloba* extract (EGb 761). *J. Pharm. Pharmacol.*, **46**, 316–318.

Huguet, F. and Tarrade, T. (1992) Alpha$_2$-adrenoceptor changes during cerebral ageing. The effect of *Ginkgo biloba* extract. *J. Pharm. Pharmacol.*, **44**, 24–27.

Izquierdo, I., Fin, C., Schmitz, P.K., Da Silva, R.C., Jersulinsky, D., Quillfeldt, J.A., Ferreira, M.B.G., Medina, J.H. and Bazan, N.G. (1995) Memory enhancement by intrahippocampal, intraamygdala, or intraentorhinal infusion of platelet activating factor measured in an inhibitory avoidance task. *Proc. Natl. Acad. Sci. U.S.A.*, **92**, 5047–5051.

Janssens, D., Michiels, C., Delaive, E., Eliaers, F., Drieu, K. and Remacle, J. (1995) Protection of hypoxia-induced ATP decrease in endothelial cells by *Ginkgo biloba* extract and bilobalide. *Biochem. Pharmacol.*, **50**, 991–999.

Jiang, Z.G., Yue, T.L. and Feuerstein, G. (1993) Platelet-activating factor inhibits excitatory transmission in the neostriatum. *Neurosci. Lett.*, **155**, 132–135.

Kanowski, S., Herrmann, W.M., Stephen, K., Wierich, W. and Horr, R. (1996) Proof of efficacy of the *Ginkgo biloba* special extract EGb 761 in outpatients suffering from mild to moderate primary degenerative dementia of the Alzheimer's type or multi-infarct dementia. *Pharmacopsychiatry*, **29**, 47–56.

Karcher, L., Zagermann, P. and Krieglstein, J. (1984) Effect of an extract of *Ginkgo biloba* on rat brain energy metabolism in hypoxia. *Arch. Pharmacol.*, **327**, 31–35.

Kato, K., Clark, G.D., Bazan, N.G. and Zorumski, C.F. (1994) Platelet-activating factor as a potential retrograde second messenger in CA1 hippocampal long-term potentiation. *Nature*, **367**, 175–179.

Keli, S.O., Hertog, M.G., Feskens, E.J. and Kromhout, D. (1996) Dietary flavonoids, antioxidant vitamins, and incidence of stroke: the Zutphen study. *Arch. Internal Med.*, **156**, 637–642.

Kleijnen, J. and Knipschild, P. (1992a) *Ginkgo biloba* for cerebral insufficiency. *Brit. J. Clin. Pharmacol.*, **34**, 352–358.

Kleijnen, J. and Knipschild, P. (1992b) *Ginkgo biloba*. *The Lancet*, **340**, 1136–1139.

Koc, R.K., Akdemir, H., Kurtsoy, A., Pasaoglu, H., Kavuncu, I., Pasaoglu, A. and Karakucuk, I. (1995) Lipid peroxidation in experimental spinal cord injury. Comparison of treatment with *Ginkgo biloba*, TRH and methylprednisolone. *Res. Exp. Med.*, **195**, 117–123.

Kornecki, E. and Ehrlich, Y.H. (1991) Calcium ion mobilization in neuronal cells induced by PAF. *Lipids*, **26**, 1243–1246.

Krieglstein, J., Beck, T. and Seibert, A. (1986) Influence of an extract of *Ginkgo biloba* on cerebral blood flow and metabolism. *Life Sci.*, **39**, 2327–2334.

Kumar, R., Harvey, S.A.K., Ester, M.K., Hanahan, D.J. and Olson, M.S. (1988) Production and effects of platelet-activating factor in the rat brain. *Biochim. Biophys. Acta*, **963**, 375–383.

Kunkel, H. (1993) EEG profile of three different extractions of *Ginkgo biloba*. *Neuropsychobiol.*, **27**, 40–45.

Lacour, M., Ez-Zaher, L. and Raymond, J. (1991) Plasticity mechanisms in vestibular compensation in the cat are improved by an extract of *Ginkgo biloba* (EGb 761). *Pharmacol. Biochem. Behav.*, **40**, 367–379.

Lacour, M., Ez-Zaher, L., Raymond, J. and Tighilet, B. (1992) Neurotrophic (and/or neuritogenic) properties of the extract of *Ginkgo biloba* (EGb 761) as evidenced in vestibular compensation. In Y. Christen, J. Costentin and M. Lacour, (eds.), *Effects of Ginkgo biloba Extract (EGb 761) on the Central Nervous System*, Elsevier, Paris, pp. 37–56.

Le Poncin Lafitte, M., Rapin, J. and Rapin, J.R. (1980) Effects of *Ginkgo biloba* on changes induced by quantitative cerebral microembolization in rats. *Arch. Int. Pharmacodyn.*, 243, 236–244.

Lenegre, A., Chermat, R., Avril, I., Steru, L. and Porsolt, R.D. (1988) Specificity of piracetam's anti-amnesic activity in three models of amnesia in the mouse. *Pharmacol. Biochem. Behav.*, 29, 625–629.

Lindsberg, P.J., Hallenbeck, J.M. and Feuerstein, G. (1991) Platelet-activating factor in stroke and brain injury. *Ann. Neurol.*, 30, 117–129.

Lustig, H.S., Chan, J. and Greenberg, D.A. (1992) Comparative neurotoxic potential of glutamate, endothelins, and platelet-activating factor in cerebral cortical cultures. *Neurosci. Lett.*, 139, 15–18.

Maclennan, K., Smith, P.F. and Darlington, C.L. (1995) Ginkgolide B accelerates vestibular compensation of spontaneous ocular nystagmus in guinea pig following unilateral labyrinthectomy. *Exp. Neurol.*, 131, 273–278.

Maclennan, K., Smith, P.F. and Darlington, C.L. (1996a) Platelet-activating factor in the CNS. *Prog. Neurobiol.*, 50, 585–596.

Maclennan, K., Smith, P.F. and Darlington, C.L. (1996b) The effects of ginkgolide B (BN52021) on guinea pig vestibular nucleus neurons in vitro: importance of controlling for effects of dimethylsulphoxide (DMSO) vehicles. *Neurosci. Res.*, 26, 395–399.

Marcheselli, V.L. and Bazan, N.G. (1994) Platelet-activating factor is a messenger in the electroconvulsive shock-induced transcriptional activation of *c-fos* and *zif–268* in hippocampus. *J. Neurosci. Res.*, 37, 54–61.

Marcheselli, V.L., Rossowska, M.J., Domingo, M-T., Braquet, P. and Bazan, N.G. (1990) Distinct platelet-activating factor binding sites in synaptic endings and in intracellular membranes of rat cerebral cortex. *J. Biol. Chem.*, 265, 9140–9145.

Matsuo, Y., Kihara, T., Ikeda, M., Ninomiya, M., Onodera, H. and Kogure, K. (1996) Role of platelet-activating factor and thromboxane A_2 in radical production during ischemia and reperfusion of the rat brain. *Brain Res.*, 709, 296–302.

Middleton, E. Jr. (1984) The flavonoids. *Trends Pharmacol. Sci.*, August, 335–338.

Mouren, X., Caillard, P. and Schwartz, F. (1994) Study of the antiischemic action of EGb 761 in the treatment of peripheral arterial occlusive disease by TcPo2 determination. *Angiology*, 45, 413–417.

Ni, Y., Zhao, B., Hou, J. and Xin, W. (1996) Preventative effect of *Ginkgo biloba* extract on apoptosis in rat cerebral neuronal cells induced by hydroxyl radicals. *Neurosci. Lett.*, 214, 115–118.

Nishida, K. and Markey, S.P. (1996) Platelet-activating factor in brain regions after transient ischemia in gerbils. *Stroke*, 27, 514–519.

Nojima, S. (1991) Platelet-activating factor (PAF): An introduction. *Lipids*, 26, 965–966.

Oberpichler, H., Beck, T., Abdel-Rahman, M.M., Bielenberg, G.W. and Krieglstein, J. (1988) Effects of *Ginkgo biloba* constituents related to protection against brain damage caused by hypoxia. *Pharmacol. Res. Comm.*, 20, 349–368.

Oberpichler, H., Sauer, D., Rossberg, C., Mennel, H.D. and Krieglstein, J. (1990) PAF antagonist ginkgolide B reduces postischemic neuronal damage in rat brain hippocampus. *J. Cerebral Blood Flow Metab.*, 10, 133–135.

Otani, M., Chatterjee, S.S., Gabard, B. and Kreutzberg, G.W. (1986) Effect of an extract of *Ginkgo biloba* on triethyltin-induced cerebral edema. *Acta Neuropathol.*, 69, 54–65.

Oyama, Y., Chikahisa, L., Ueha, T., Kanemaru, K. and Noda, K. (1996) *Ginkgo biloba* extract protects brain neurons against oxidative stress induced by hydrogen peroxide. *Brain Res.*, **712**, 349–352.

Oyama, Y., Fuchs, P.A., Katayama, N. and Noda, K. (1994) Myricetin and quercetin, the flavonoid constituents of *Ginkgo biloba* extract, greatly reduce oxidative metabolism in both resting and Ca^{2+}-loaded brain neurons. *Brain Res.*, **635**, 125–129.

Oyama, Y., Hayashi, A. and Ueha, T. (1993) Ca^{2+}-induced increase in oxidative metabolism of dissociated mammalian brain neurons: effect of extract of *Ginkgo biloba* leaves. *Jap. J. Pharmacol.*, **61**, 367–370.

Oyama, Y., Ueha, T., Hayashi, A., Chikahisa, L. and Noda, K. (1992) Flow cytometric estimation of the effect of *Ginkgo biloba* extract on the content of hydrogen peroxide in dissociated mammalian brain neurons. *Jap. J. Pharmacol.*, **60**, 385–388.

Panetta, T., Marcheselli, V.L., Braquet, P., Spinnewyn, B. and Bazan, N.G. (1987) Effects of a platelet activating factor antagonist (BN 52021) on free fatty acids, diacylglycerols, polyphosphoinositides and blood flow in the gerbil brain: inhibition of ischemia-reperfusion induced cerebral injury. *Biochem. Biophys. Res. Comm.*, **149**, 580–587.

Petkov, V.D., Kehayov, R., Belcheva, S., Konstantinova, E., Petkov, V.V., Getova, D. and Markovska, V. (1993) Memory effects of standardized extracts of *Panax ginseng*, *Ginkgo biloba* (GK 501) and their combination Gincosan (PHL-00701). *Planta Med.*, **59**, 106–114.

Pettigrew, L.C., Meyer, J.J., Craddock, S.D., Butler, S.M., Tai, H.H. and Yokel, R.A. (1995) Delayed elevation of platelet activating factor in ischemic hippocampus. *Brain Res.*, **691**, 243–247.

Pietri, S., Culcasi, M., Verniere, E., d'Arbigny, P. and Drieu, K. (1993) Effect of *Ginkgo biloba* extract (EGb 761) on free radical-induced ischemia-reperfusion injury in isolated rat hearts: a hemodynamic and electron-spin-resonance investigation. In C. Ferradini, M.T. Droy-Lefaix and Y. Christen, (eds.), *Giinkgo biloba Extract (EGb 761) as a Free Radical Scavenger*, Elsevier, Paris, pp. 163–171.

Pincemail, J., Dupuis, M., Nasr, C., Hans, P., Haag-Berrurier, M., Anton, R. and Deby, C. (1989) Superoxide anion scavenging effect and superoxide dismutase activity of *Ginkgo biloba* extract. *Experientia*, **45**, 708–712.

Plath, P. and Olivier, J. (1995) Results of combined low-power laser therapy and extracts of *Ginkgo biloba* in cases of sensorineural hearing loss and tinnitus. *Adv. Otorhinolaryngol.*, **49**, 101–104.

Porsolt, R.D., Martin, P., Lenegre, A., Fromage, S. and Drieu, K. (1990) Effects of an extract of *Ginkgo biloba* (EGB 761) on "learned helplessness" and other models of stress in rodents. *Pharmacol. Biochem. Behav.*, **36**, 963–971.

Prehn, J.H.M. and Krieglstein, J. (1993) Platelet-activating factor antagonists reduce excito-toxic damage in cultured neurons from embryonic chick telencephalon and protect the rat hippocampus and neocortex from ischemic injury *in vivo*. *J. Neurosci. Res.*, **34**, 179–188.

Rai, G.S., Shovlin, C. and Wesnes, K.A. (1991) A double-blind, placebo controlled study of *Ginkgo biloba* extract ("Tanakan") in elderly outpatients with mild to moderate memory impairment. *Curr. Med. Res. Opin.*, **12**, 350–355.

Ramassamy, C., Christen, Y., Clostre, F., and Costentin, J. (1992a) The *Ginkgo biloba* extract, EGb 761, increases synaptosomal uptake of 5-hydroxytryptamine: *in-vitro* and *ex-vivo* studies.*J. Pharm. Pharmacol.*, **44**, 943–945.

Ramassamy, C., Clostre, F., Christen, Y. and Costentin, J. (1990) Prevention by a *Ginkgo biloba* extract (GBE 761) of the dopaminergic neurotoxicity of MPTP. *J. Pharm. Pharmacol.*, **42**, 785–789.

Ramassamy, C., Girbe, F., Christen, Y. and Costentin, J. (1993) *Ginkgo biloba* extract EGb 761 or trolox C prevent the ascorbic acid/Fe^{2+}-induced decrease in synaptosomal membrane fluidity. *Free Radical Res. Comm.*, **19**, 341–350.

Ramassamy, C., Naudin, B., Christen, Y., Clostre, F. and Costentin, J. (1992b) Prevention by *Ginkgo biloba* extract (EGb 761) and trolox C of the decrease in synaptosomal dopamine or serotonin uptake following incubation. *Biochem. Pharmacol.*, **44**, 2395–2401.

Robak, J. and Gryglewski, R.J. (1988) Flavonoids are scavengers of superoxide anions. *Biochem. Pharmacol.*, **37**, 837–841.

Rodriguez de Turco, E.B., Droy-Lefaix, M.T. and Bazan, N.G. (1993) Decreased electroconvulsive shock-induced diacylglycerols and free fatty acid accumulation in the rat brain by *Ginkgo biloba* extract (EGb 761): Selective effect in hippocampus as compared with cerebral cortex. *J. Neurochem.*, **61**, 1438–1444.

Ruckstuhl, M., Beretz, A., Anton, R. and Landry, Y. (1979) Flavonoids are selective cyclic phosphodiesterase inhibitors. *Biochem. Pharmacol.*, **28**, 535–538.

Sancesario, G. and Kreutzberg, G.W. (1986) Stimulation of astrocytes affects cytotoxic brain edema. *Acta Neuropathol.*, **72**, 3–14.

Satoh, K., Imaizumi, T., Yoshida, H., Hiramoto, M. and Takamatsu, S. (1992) Increased levels of blood platelet-activating factor (PAF) and PAF-like lipids in patients with ischemic stroke. *Acta Neurol. Scand.*, **85**, 122–127.

Sato, F., Matsukawa, Y., Matsumoto, K., Nishino, H. and Sakai, T. (1994) Apigenin induces morphological differentiation and G2-M arrest in rat neuronal cells. *Biochem. Biophys. Res. Comm.*, **204**, 578–584.

Semlitsch, H.V., Andere, P., Saletu, B., Binder, G.A. and Decker, K.A. (1995) Cognitive psychophysiology in nootropic drug research: effects of *Ginkgo biloba* on event-related potentials (P300) in age-associated memory impairment. *Pharmacopsychiatry*, **28**, 134–142.

Sher, E., Codignola, A., Biancardi, E., Cova, D. and Clementi, F. (1992) Amine uptake inhibition by diosmin and diosmetin in human neuronal and neuroendocrine cell lines. *Pharmacol. Res.*, **26**, 395–402.

Smith, C.D., Carney, J.M., Starke-Reed, P.E., Oliver, C.N., Stadtman, E.R., Floyd, R.A. and Markesbery, W.R. (1991) Excess brain protein oxidation and enzyme dysfunction in normal aging and in Alzheimer's disease. *Proc. Natl. Acad. Sci. U.S.A.*, **88**, 10540–10543.

Smith, P.F., Maclennan, K. and Darlington, C.L. (1996) The neuroprotective properties of the *Ginkgo biloba* leaf: A review of the possible relationship to platelet-activating factor (PAF). *J. Ethnopharmacol.*, **50**, 131–139.

Snyder, F. (1990) Platelet-activating factor and related acetylated lipids as potent biologically active cellular mediators. *Am. J. Physiol.*, **259**, C697–708.

Spinnewyn, B., Blavet, N., Clostre, F., Bazan, N. and Braquet, P. (1987) Involvement of platelet-activating factor (PAF) in cerebral post-ischemic phase in Mongolian gerbils. *Prostaglandins*, **34**, 337–350.

Squinto, S.P., Block, A.L., Braquet, P. and Bazan, N.G. (1989) Platelet-activating factor stimulates a Fos/Jun/AP-1 transcriptional signaling system in human neuroblastoma cells. *J. Neurosci. Res.*, **24**, 558–566.

Sticher, O. (1993) Quality of Ginkgo preparations. *Planta Med.*, **59**, 2–11.

Subhan, Z. and Hindmarch, I. (1984) The psychopharmacological effects of *Ginkgo biloba* extract in normal healthy volunteers. *Int. J. Clin. Pharm. Res.*, **4**, 89–93.

Sun, D. and Gilboe, D.D. (1994) Effect of platelet-activating factor antagonist BN 50739 and its diluents on mitochondrial respiration and membrane lipids during and following cerebral ischemia. *J. Neurochem.*, **62**, 1929–1938.

Tamura, M., Kagawa, S., Tsuruo, Y., Ishimura, K. and Morita, K. (1994) Effects of flavonoid compounds on the activity of NADPH diaphorase prepared from the mouse brain. *Jap. J. Pharmacol.*, **65**, 371–373.

Tighilet, B. and Lacour, M. (1995) Pharmacological activity of the *Ginkgo biloba* extract (EGb 761) on equilibrium function recovery in the unilateral vestibular neurectomized cat. *J. Vest. Res.*, 5, 187–200.

Tokutomi, T., Sigemori, M., Kikuchi, T. and Hirohata, M. (1994) Effect of platelet-activating factor antagonist on brain injury in rats. *Acta Neurochir.*, **60**, 508–510.

Vidal, P.P., Lapeyre, P., Serafin, M. and de Waele, C. (1993) Effects of a *Ginkgo biloba* extract (EGb 761) on guinea pig medial vestibular nuclei neurons: An *in vitro* study. *Soc. Neurosci. Abstr.*, **19**, 136.

Wang, H.K., Bastow, K.F., Cosentino, L.M. and Lee, K.H. (1996) Antitumour agents. 166. Synthesis and biological evaluation of 5,6,7,8-substituted-2-phenylthiochromen-4-ones. *J. Med. Chem.*, **39**, 1975–1980.

Wang, D., Shen, W., Tian, Y., Dong, Z., Liu, G., Sun, Z., Yang, S. and Zhou, S. (1995) Protective effect of total flavonoids of radix Astragali on mammalian cell damage caused by hydroxyl radical. *China J. Chinese Materia Medica*, **20**, 240–242.

Warburton, D.M. (1986) Psychopharmacologie clinique de l'extrait de *Ginkgo biloba*. *La Presse Medicale*, **15**, 1595–1604.

Wesnes, K., Simmons, D., Rook, M. and Simpson, P. (1987) A double-blind placebo-controlled trial of Tanakan in the treatment of idiopathic cognitive impairment in the elderly. *Human Psychopharmacol.*, **2**, 159–169.

White, H.L., Scates, P.W. and Cooper, B.R. (1996) Extracts of *Ginkgo biloba* leaves inhibit monoamine oxidase. *Life Sci.*, **58**, 1315–1321.

Wieraszko, A., Li, G., Kornecki, E., Hogan, M.V. and Ehrlich, Y.H. (1993) Long-term potentiation in the hippocampus induced by platelet-activating factor. *Neuron*, **10**, 553–557.

Willard, A.L. (1992) Excitatory and neurotoxic actions of platelet-activating factor on rat myenteric neurons in cell culture. *Ann. N.Y. Acad. Sci.*, **664**, 284–292.

Winter, E. (1991) Effects of an extract of *Ginkgo biloba* on learning and memory in mice. *Pharmacol. Biochem. Behav.*, **38**, 109–114.

Yabe, T., Chat, M., Malherbe, E. and Vidal, P.P. (1992a) Effects of *Ginkgo biloba* extract (EGb 761) on the guinea pig vestibular system. *Pharmacol. Biochem. Behav.*, **42**, 595–604.

Yabe, T., Chat, M., Malherbe, E. and Vidal, P.P. (1992) Changes in vestibular function induced by *Ginkgo biloba* extract (EGb 761) in the guinea pig. In Y. Christen, J. Costentin and M. Lacour, (eds.), *Effects of Ginkgo biloba Extract (EGb 761) on the Central Nervous System*, Elsevier, Paris, pp. 75–87.

Yue, T.L. and Feuerstein, G.Z. (1994) Platelet-activating factor: a putative neuromodulator and mediator in the pathophysiology of brain injury. *Crit. Rev. Neurobiol.*, **8**, 11–24.

Yue, T.L., Gleason, M.M., Gu, J.L., Lysko, P.G., Hallenbeck, J. and Feuerstein, G.Z. (1991a) Platelet-activating factor receptor-mediated calcium mobilization and phosphoinositide turnover in neurohybrid NG108–15 cells: studies with BN50739, a new PAF antagonist. *J. Pharmacol. Exp. Therap.*, **257**, 374–381.

Yue, T.L., Gleason, M.M., Hallenbeck, J. and Feuerstein, G.Z. (1991b) Characterization of platelet-activating factor-induced elevation of cytosolic free-calcium level in neurohybrid NCB-20 cells. *Neurosci.*, **41**, 177–185.

Yue, T.L., Lysko, P.G. and Feuerstein, G.Z. (1990) Production of platelet-activatng factor from rat cerebellar granule cells in culture. *J. Neurochem.*, **54**, 1809–1811.

Zablocka, B., Lukasiuk, K., Lazarewicz, J.W. and Domanska-Janik, K. (1995) Modulation of ischemic signal by antagonists of N-methyl-D-aspartate, nitric oxide synthase, and platelet-activating factor in gerbil hippocampus. *J. Neurosci. Res.*, **40**, 233–240.

Zalewska, T., Zablocka, B. and Domanska-Janik, K. (1996) Changes of Ca^{2+}/calmodulin-dependent protein kinase-II after transient ischemia in gerbil hippocampus. *Acta Neurobiol. Exp.* **56**, 41–48.

19. GINKGO BILOBA EXTRACTS FOR THE TREATMENT OF CEREBRAL INSUFFICIENCY AND DEMENTIA

JOERG SCHULZ[1], PETER HALAMA[2] and ROBERT HOERR[3]

[1]Klinikum Buch, Geriatrische Klinik, Zepernicker Straße 1, 13125 Berlin, Germany
[2]Neurologist and Psychiatrist, Berner Heerweg 175, 22159 Hamburg, Germany
[3]Dr. Willmar Schwabe Pharmaceuticals, Clinical Research Dept., Willmar-Schwabe-Strasse 4, 76227 Karlsruhe, Germany

INTRODUCTION

Ginkgo biloba special extract preparations are classified as antidementia drugs or nootropics. Nootropics are a heterogeneous group of drugs which, by various pharmacodynamic mechanisms, improve as final outcome disturbances of higher integrative noetic functions (e.g. memory, concentration, comprehension, attention, thinking and orientation) and impairment of vigilance. The main indication is dementia according to the Diagnostic and Statistical Manual of Mental Disorders (DSM-III-R/IV) (American Psychiatric Association, 1987, 1994) or the International Classification of Diseases (ICD-10) (World Health Organization 1992), particularly Alzheimer's disease, vascular dementia and mixed forms of both. A short and precise description of dementia reads as follows: Dementia is a clinical syndrome characterized by acquired losses of cognitive and emotional abilities severe enough to interfere with daily functioning and quality of life (Geldmacher and Whitehouse, 1996). According to the US Food and Drug Administration (FDA), dementia refers "to a neurological condition of unknown etiology that ordinarily causes a progressive, irreversible decline in intellectual and cognitive abilities" (Leber, 1990).

Synonyms of nootropics in the literature are: cognition enhancers, cerebral active drugs, cerebral metabolic activators, antidementia drugs, and neurodynamics. Clinical research on nootropics is a very complex, continuously developing subject. The studies reported below reflect this situation. The primary aim of anti-dementia therapy with nootropics is to improve signs and symptoms and/or to delay progression over a prolonged period of time, primarily with respect to the main cognitive aspects (impairment of memory, concentration, attention; deterioration of language function; impairment of recognition, abstract thinking, orientation, and executive functioning; thereby alleviating impairment in activities of daily living (ADL)). The most important aspects of current recommendations, e.g. by the German Federal Health Authority (Menges, 1992), the EU Committee for

Proprietary Medicinal Products (CPMP, 1992), or the American Food and Drug Administration (Leber, 1990) on evaluating the efficacy of nootropics/antidementia drugs are: establishing the clinical diagnosis by generally accepted rules such as DSM-IV, NINCDS/ADRDA (National Institute of Neurological and Communicative Disorders and Stroke/Alzheimer's Disease and Related Disorders Association) (McKhann *et al.*, 1984), and ICD-10 criteria; the measurement of severity by means of defined scales and independent assessments of drug effects on different structural levels (psychopathological findings, objective assessment of performance, behaviour closely associated with activities of daily living) (Menges, 1992). The following criteria for the severity of dementia have been proposed by the DSM-III-R (American Psychiatric Association, 1987):

– mild: although work and social activities are significantly impaired, the capacity for independent living remains, with adequate personal hygiene and relatively intact judgement.
– moderate: independent living is hazardous and some degree of supervision is necessary.
– severe: activities of daily living are so impaired that continuous supervision is required, e.g. unable to maintain minimal personal hygiene; largely incoherent or mute.

The recording of symptoms of cognitive decline as well as non-cognitive symptoms and changes in neurological status is viewed to be necessary. The main cognitive feature of Alzheimer's disease is a progressive memory impairment, particularly loss of short-term memory. Symptoms relating to the ability to focus attention or recall remote events are subtly impaired during the initial stage of the disease, and worsen with time. Further signs of Alzheimer's disease are: progressive disorientation with respect to time and place, language impairment (difficulty in finding spontaneous speech, an increased repeating of automatic phrases, impaired naming ability), deficits in visual-processing abilities including manifestations relating to activities of daily living. Important non-cognitive symptoms are: a decreased emotional expression, increased stubbornness, diminished initiative, greater suspiciousness (Chatterjee *et al.*, 1992), delusions (e.g. paranoid delusions, suspicion of marital infidelity, visual hallucinations), symptoms of depression and anxiety. Neurologic symptoms are not seen in early Alzheimer's disease. Disturbances in the extra-pyramidal system often indicate atypical forms of Alzheimer's disease (Geldmacher and Whitehouse, 1996).

Randomized, double-blind, placebo-controlled clinical trials to establish clinical efficacy have been performed with 2 *Ginkgo biloba* special extracts: EGb 761® (Dr. Willmar Schwabe Pharmaceuticals, Karlsruhe, Germany and I.P.S.E.N., Paris, France) and LI 1370 (Lichtwer Pharma GmbH, Berlin, Germany). On the grounds of proven clinical efficacy these two extracts have been approved by the German Federal Health Authority for the symptomatic treatment of impaired cerebral function in dementia syndromes within a treatment plan. Their main indications are primary degenerative dementia, vascular dementia and mixed forms. EGb 761® has been registered in numerous other countries.

PHARMACODYNAMIC TRIALS IN HUMANS

The central nervous system pharmacodynamics of *Ginkgo biloba* special extracts EGb 761® and LI 1370 were studied using quantitative pharmaco-electroencephalographic methods in a series of clinical trials.

In a double-blind, placebo-controlled crossover study with Latin square design, 12 healthy young volunteers were enrolled by Itil *et al.* (1996) to receive single oral doses of either 40 mg, 120 mg, or 240 mg *Ginkgo biloba* special extract EGb 761® or placebo. During each of four sessions, scheduled at intervals of at least 3 days, 6 electroencephalograms were recorded and computer-analyzed: before drug administration and subsequently at 1, 2, 3, 5, and 7 hours after drug intake. All the 4 doses of EGb 761® could be significantly discriminated from placebo in at least 4 post-dose recordings based on changes from baseline in a 7.5 to 13.5 Hz frequency range, i.e. an increase in alpha activity. A dose-effect relationship could be established with an earlier onset and longer duration of effects after the highest dose. Furthermore, a comparison of the EEG profile of EGb 761® with an extensive drug database revealed a consistent similarity to the typical cognitive activator profile.

Pidoux *et al.* (1983) studied the effects of a 12 week treatment with 160 mg/day of EGb 761® (7 patients) compared to placebo (5 patients) in patients suffering from vascular dementia. They found an increase in amplitudes in the alpha frequency band, a decrease of amplitudes in the beta range, and a significant decrease of the intensity ratio theta/alpha ($p < 0.05$, pre-post comparison). No consistent changes in EEG variables were detected in the patients receiving placebo. In the actively treated group, a mean improvement of 29% in the geriatric rating scales (SCAG, SGRF) was also observed, which was not the case in the placebo group.

In a randomized, double-blind, placebo-controlled parallel-group trial Schulz *et al.* (1991) studied the effects of *Ginkgo biloba* special extract LI 1370 on EEG activity in a sleep deprivation model. Sixteen elderly patients suffering from impairment of cerebral function as determined by cognitive testing were enrolled and randomized to receive either 150 mg/day LI 1370 or placebo for 8 weeks. Only 7 patients of the LI 1370 group were evaluable because of one dropout definitely unrelated to the treatment. There was a significant reduction of theta power in the vigilance-controlled EEG after sleep deprivation in the LI 1370-treated patients, but not in those given placebo ($p < 0.05$). This contributed to an increase in the alpha-slow-wave index from 0.92 to 0.96 (medians) in the actively treated patients, while in the placebo group a decrease from 0.98 to 0.94 took place ($p < 0.05$, inter-group comparison).

In a clinical trial conducted by Hofferberth (1994, reported in the following section) 40 patients suffering from senile dementia of the Alzheimer type were treated for 12 weeks either with oral daily doses of 240 mg EGb 761® (n = 21) or placebo (n = 19). This author found a distinct reduction in the theta wave component of the theta/alpha quotient from 79.42% (± 5.05) at baseline to 63.38% (± 7.59) at 4 weeks, and a further decrease to 61.95% (± 7.63) at 12 weeks (mean values ± standard deviations). In the placebo group, however, the theta wave component remained at 78% throughout the trial. The inter-group difference was statistically significant from 4 weeks onward ($p < 0.01$).

Luthringer *et al.* (1995) investigated the EEG profile of *Ginkgo biloba* special extract EGb 761® in healthy young subjects. In a first, double-blind part, the effects of single doses of 80 mg and 160 mg were compared to placebo effects. In a second, single-blind part, the effects of short-time treatment with daily oral doses of 160 mg EGb 761® were studied. At both dosages, a marked increase in alpha power was observed on single dose application. After 5 days of treatment, the alpha increase still persisted and a decrease of relative power in the theta band became obvious. At the same time, the latency of the P 300 component of the auditory evoked potential was significantly decreased (p < 0.05).

To summarize, it may be concluded that the *Ginkgo biloba* special extract EGb 761® exerts dose-dependent effects on the central nervous system, as detected by electroencephalographic activity, in both healthy young subjects and elderly dementia patients. Its most consistent effects are an increase in alpha activity and a decrease of slow-wave activity (theta range). This is often interpreted as a sign of increased vigilance and cognitive activation (Schulz *et al.*, 1991, Itil *et al.*, 1996). Given an abnormally high slow-wave activity and low alpha activity in dementia patients, EGb 761® induces changes of EEG activity towards normal. The reduction of the EP latency as described by Luthringer *et al.* (1995) points to an improvement in patients' information processing. Moreover, cerebral bioavailability of the active ingredients of *Ginkgo biloba* extracts has thus been definitely confirmed by consistent and reproducible EEG effects.

CLINICAL TRIALS

In this section, only **randomized, double-blind, placebo-controlled clinical trials** are reported which, with respect to methodological quality, are able to contribute to an overall evaluation of clinical efficacy in cerebral insufficiency and dementia.

Impairment of Cerebral Function/Cerebral Insuffiency

Impairment of cerebral function and *cerebral insuffiency* are not well-defined syndromes (in contrast to *dementia* or *major depression*). These terms are mainly used for arbitrarily naming a cluster of symptoms associated with ageing and were characteristic for early to moderate dementia at times when commonly accepted criteria for dementia did not yet exist, or were not yet routinely applied by clinicians. Moreover, today's clinical criteria do not apply for patients with very early dementia, since a certain degree of impairment is required in order to establish the diagnosis. It is however quite clear that the disease begins with mild impairment and then progresses until a diagnosis of dementia can be established. Taking these facts into account, it can be assumed that a substantial part of those patients included in these earlier trials under the diagnosis *impairment of cerebral function* really did suffer from dementia. This view is supported by a comparison of the trials involving such patients, and trials in well-defined dementia patients: the results of treatment with *Ginkgo biloba* special extracts are essentially the same.

Two groups of patients suffering from cerebral function impairment due to cerebrovascular disease were examined in a study by Moreau (1975). One group (n = 60) was subjected to a double-blind comparison of EGb 761® and placebo. The other group (n = 30) was treated with ergot alkaloid derivatives in an open part of the study. Treatment efficacy was evaluated by means of psychometric tests, the clinician's global judgement, a scale of typical symptoms, and a general activities scale. The oral daily dose of EGb 761® was 120 mg, the treatment period was 3 months. In 79.3% of patients EGb 761® led to a global improvement (physician's assessment as "good" or "very good" results), whereas only 21.4% improved relevantly in the placebo group (p < 0.001). EGb 761® treatment led to a more pronounced improvement than placebo application in functional symptoms (vertigo, headache, tinnitus), general activity, in 7 of 10 psychometric tests, and in the overall assessment. Treatment with ergot alkaloids was also of some benefit, in that 53.3% of global judgements were favourable after 3 months.

Twenty patients treated with EGb 761® (120 mg/day) and 20 patients who received placebo during 60 days, all suffering from chronic impairment of cerebral function and being of an average age of 62 years, were compared in a trial by Dieli *et al.* (1981). Patients were selected using the geriatric assessment scale according to Plutchik *et al.* (1970) for proof of relevant impairment and the Hachinski ischaemic score (< 7) (Hachinski *et al.*, 1975). Thus the majority of them may be presumed to have suffered from primary degenerative dementia of the Alzheimer type or mixed type dementia. Psychometric tests could be carried out in only 11 EGb 761®-treated and 14 placebo-treated patients due to advanced severity of the disease. Active treatment was significantly superior (p < 0.05) to placebo with respect to 11 items of the Plutchik assessment scale and both the Wechsler digit span and digit symbol test (Wechsler, 1944). Two adverse reactions of mild intensity and short duration were recorded in the EGb 761®-treated group (1 case of nausea, 1 of water brash). The multitude of tests referring to items of the Plutchik scale justify their classification as descriptive. Evidence was nevertheless provided for the efficacy of EGb 761® on the psychopathological, the psychometric and the behavioural level.

Fifty patients with cerebrovascular insufficiency were studied by Eckmann and Schlag (1982). After random allocation, 25 patients were treated with oral daily doses of 120 mg *Ginkgo biloba* special extract EGb 761® and 25 patients received placebo. Duration of treatment was 4 weeks. The outcome measures comprised psychological and neurological findings and typical symptoms (dizziness, headache, tinnitus, state of consciousness, anxiety, language function, sensibility, orientation, speech comprehension, motor behaviour, depressed mood). Except for frequency of tinnitus and hearing disorders both groups were comparable at baseline. Improvement of pathological symptoms began much earlier (median 12 days) in the patients treated with EGb 761® than in those treated with placebo, single symptoms had disappeared after 30 days in 91 to 100% of EGb 761® patients in whom they were present before treatment (p < 0.0005 vs. placebo). In the placebo group, single symptoms disappeared in only 43 to 57% of patients. No adverse reactions and no significant changes in laboratory tests were observed. Distinct outcome differences in favour of EGb 761® allow to assume clinical efficacy in disturbances of cerebral function caused by cerebrovascular disease.

One hundred and sixty-six patients over 60 years of age and suffering from impairment of cerebral function, mostly due to mild primary degenerative dementia, vascular dementia, and mixed forms, were studied by Taillandier et al. (1986). They were treated for 12 months with either 160 mg/day of EGb 761® or placebo. Eighty patients were included in the EGb 761® group, 86 patients in the placebo-group. With respect to demographic data and severity, the groups were well comparable. In the course of EGb 761® treatment, there was a steady decrease in the total score of the clinical geriatric evaluation scale EACG (French version of the SCAG; Georges et al., 1977) resulting in a statistically significant difference compared to the placebo group (p < 0.05) from month 3 onward. Mean improvement in the actively treated group was 17% from baseline score after 12 months, but only 7% in the placebo group (p = 0.05). Laboratory parameters showed no relevant changes during the trial. Patients suffering from impairment of cerebral function for more than 2 years had more benefit from EGb 761® treatment than those afflicted more recently. In conclusion, it could be demonstrated that the treatment with EGb 761® can significantly improve everyday behaviour and the social integration of patients with dementia syndrome.

Eighty patients (mean age 68.4 years) suffering from memory impairment affecting their daily life activities were enrolled in a trial by Israel et al. (1987). They were randomly assigned to 4 groups of 20. Mild cognitive impairment was verified by a score of 20 to 26 in the Mini-Mental State Examination (MMSE, Folstein et al., 1975). Patients were excluded if they had a depressive syndrome as defined by a score of 18 or more on the Geriatric Depression Scale (Brink et al., 1982). The first and the third group were treated with EGb 761® (160 mg/day), the second and fourth group received placebo. Patients of groups 3 and 4 took part in a standardized memory training program in addition to the study medication. Treatment duration was approximately 3 months. Memory training sessions took place weekly. Four factors of memory as assessed by a comprehensive test battery developed by the authors (Israel et al., 1980) were examined: general memory, immediate memory (attention/perception), learning factor, mental fluidity and control. The absolute difference in the scores of the 4 factors was comparatively tested by an analysis of variance. Significant improvements were observed in the EGb 761® group in the context of immediate memory and mental fluidity and control. Training had a positive effect on both general memory and immediate memory. The learning factor did not change in either treatment; there was, however, an additive improvement of immediate memory in patients who underwent EGb 761® treatment and training. Memory training by itself had an influence on general recall factors, but treatment with EGb 761® only was able to influence mental fluidity, which involves association speed. Eighty-eight percent of the patients treated with EGb 761® by itself considered their therapy to have been sufficiently effective, which was the case in only 23% of the placebo group (p < 0.001). Of the patients who underwent both memory training and EGb 761® treatment, 100% experienced memory improvement. Three adverse events were observed: 2 occurred under treatment with EGb

761® (1 case of nausea, dyspepsia; 1 of dizziness), another one in the placebo-group (retrosternal chest pain). There were 5 further drop-outs (2 patients did not take the study drug, 3 refused to undergo the tests; they felt disadvantaged because they had not been assigned to the training sessions). In summary, treatment with EGb 761® was safe and turned out to be more effective in improving impaired memory function when combined with memory training than either treatment alone.

Forty outpatients suffering from mild to moderate impairment of cerebral function due to cerebrovascular disease (mostly vascular dementia), comprising a Hachinski Ischemic Score less than 7, were included in a study by Halama *et al.* (1988). According to random allocation, 20 patients received a daily dose of 120 mg *Ginkgo biloba* special extract EGb 761® and 20 patients were given placebo. Patients with other neurologic, psychiatric, or severe somatic illness or taking other rheologically active or psychotropic drugs were excluded. Duration of treatment was 12 weeks. Primary outcome measure was the SCAG (Sandoz Clinical Assessment – Geriatric; Shader *et al.*, 1974) total score. In the group receiving EGb 761® the SCAG total score diminished by an average of 9 points, whereas it remained unchanged in the placebo group (p < 0.005). Descriptive evaluation of the single items indicated particularly pronounced effects on disturbances of short-term memory, mental alertness and dizziness. Superior effects of *Ginkgo biloba* extract were also demonstrated as regards headache and tinnitus. Improvement of attention and memory was confirmed by an objective cognitive test (SKT) (Erzigkeit, 1986; Kim *et al.*, 1993). There were no drop-outs. A mild and temporary adverse event (headache) was observed in 1 patient of the active treatment group.

Out of 50 inpatients from a neurological ward (age 57 to 76 years) suffering from organic brain syndrome, Hofferberth (1991) treated 25 patients with oral daily doses of 150 mg *Ginkgo biloba* special extract LI 1370 for 6 weeks. A further 25 patients were given placebo during the same time. Efficacy was assessed by two psychometric tests, the Vienna Determination Test (Semlitsch *et al.*, 1989) and the digits connection test (Oswald and Roth, 1978), functional dynamic investigations (saccadic eye movements, EEG analysis, measurement of evoked potentials), a symptom scale and global rating of change (e.g. "good", "moderate" or "satis-factory" improvement). A significant superiority of LI 1370 compared to placebo could be demonstrated in 9 out of 11 typical symptoms of age-related cerebral impairment (p < 0.01) as well as in the global rating of improvement (p < 0.001) and in all of the functional dynamic measurements (p < 0.001). No adverse reactions were reported.

Schmidt *et al.* (1991) randomly assigned 99 patients aged 50 to 70 years and suffering from impairment of cerebral function to treatment with *Ginkgo biloba* extract LI 1370 (150 mg/day) or placebo (50 and 49 patients, resp.). Patients with neurologic or severe somatic conditions or taking medication that could possibly interfere with efficacy assessment had to be excluded. Therapeutic success was documented by a symptom scale comprising 12 typical symptoms of age-related cerebral impairment and by a global judgement. After 12 weeks of LI 1370 treat-ment 8 of the 12 symptoms were significantly more improved than after placebo treatment (p < 0.05). This was confirmed by both the physician's and the patient's global assessment of therapy response (p < 0.01). Three events of mild nausea which might have been related to the drug under study were recorded in the active treatment group.

Brüchert *et al.* (1991) enrolled 303 patients (age range 45–80 years) of 33 general practices into a multicentre trial with the *Ginkgo biloba* special extract LI 1370, administered at a daily dose of 150 mg for 12 weeks. The inclusion diagnosis was a syndrome of impaired cerebral function as defined by a typical psychopathology not due to a somatic or psychiatric disease (other than dementia). Psychotropic or cognition-enhancing drugs were prohibited during participation in the study. Clinical efficacy was assessed on the psychopathological level by a symptom rating scale comprising 11 typical symptoms of age-associated impairment of cerebral function (concentration disturbances, forgetfulness, memory impairment, fatigue, loss of drive and capability, depressed mood, anxiety, dizziness, headache, tinnitus) and by the physician's global judgement of therapy response. On the level of objective performance testing, the digits connection test (a sort of trail-making test; Oswald and Roth, 1978) was chosen as an outcome measure. Because of major protocol violations, 94 patients had to be excluded from the efficacy analysis, which was therefore based on the data of 209 patients. After 12 weeks of treatment with LI 1370, improvements were significantly more pronounced in the actively treated group than in the placebo group regarding 8 out of 11 symptoms in the rating scale. This was confirmed by the physician's global judgement as well as by the patient's global self-assessment of well-being. In the digits connection test, there was a mean decrease in processing time in both groups, amounting to 25% of the pre-treatment values in the LI 1370 group and 14% in the placebo group (p < 0.01). Adverse events were reported for 16 patients of the active-treatment group and for 32 patients in the placebo group. The only kind of event which might have been unfavourable for the *Ginkgo biloba* extract on account of its frequency was headache (4 events vs. 1 event in placebo; n = 153 and 150, respectively). In particular, this trial reflects everyday therapeutic reality in general practices and, under such unfavourable conditions, the clinical efficacy of *Ginkgo biloba* extract was still noticeable. There was, however, one drawback in that the number of protocol violations was rather high.

The effect of *Ginkgo biloba* special extract EGb 761® on basic parameters of mental performance was studied by Gräßel (1992). 72 outpatients (37 females, 35 males, mean age: 63.8 ± 8.4 years) with impairment of cerebral function due to cerebrovascular disease were included in this 24-week clinical trial. The patients received either EGb 761® (160 mg/day) or placebo. Diagnosis was verified by the patients' medical history (indications of stenosing vascular process in the cranial area) and several computer-based psychometric tests requiring a simple patient-PC dialogue. Patients suffering from other neurologic, psychiatric, or relevant somatic diseases had to be excluded from the study. Treatment with psychotropic or vaso-active drugs or with drugs having an influence on the cell metabolism was not allowed during the study. The intelligence quotient (IQ) as calculated from short-term storage capacity (Lehrl and Fischer, 1988) steadily increased in the actively treated group from 93.5 at commencement to 107.5 after 24 weeks of treatment (mean values), whereas only a slight improvement, most likely due to training effects, could be observed in the placebo group (from 93.8 to 95.8). Intra-group comparisons revealed statistically significant pre-post differences in the EGb 761® group from week 6 onwards, but at no time in the placebo group. Inter-group differences were statistically significant at week 24 (p < 0.05). The memory quotient (MQ), derived from learning speed, basically showed the same development in the course of treatment. However, due to a difference in baseline values favouring the placebo group, inter-group differences did not reach statistical significance after treatment in spite of larger improvements in EGb 761® patients. Four adverse reactions were documented in the EGb 761® group and 6 in the placebo group. Seven patients dropped out of the *Ginkgo biloba* group and 12 from the placebo group. Dropout reasons appeared unrelated to the study drug. In summary, the findings of this study demonstrate that, in patients suffering from impaired cerebral function due to cerebrovascular disease, basic functions of information processing can be improved by EGb 761® treatment. It has to be mentioned here that this type of theory-guided testing has been explicitly recommended by the German Federal Health Authority's expert committees (Menges, 1992).

Vesper and Hänsgen (1994) tested the efficacy of *Ginkgo biloba* special extract LI 1370 versus placebo in 86 patients (aged 55 to 80 years) attending 11 practices run by neurologists, psychiatrists, general practitioners and internists. The diagnosis for inclusion was cerebral insufficiency according to ICD-9 criteria (290.X). Patients had a minimum IQ of 80 and suffered from distinct subjective troubles which were reflected by a minimum score of 19 in the cerebral insufficiency (C.I.) scale according to Weidenhammer and Fischer (1987). Patients suffering from any severe somatic or psychiatric illness (including pseudodementia) or using any nootropic, psychotropic or vasoactive agents had to be excluded. In accordance with random allocation, 42 patients were treated with *Ginkgo biloba* special extract LI 1370 (150 mg/day) and 44 patients were given placebo for 12 weeks. Efficacy evaluation was based on self-rating symptom scales and a computerized test battery comprising 5 subtests: a) basic reaction time (required for accomplishing simple tasks); b) time/tenacity training test (reflecting qualitative aspects of attention and ability to learn); c) mental flexibility test (for qualitative attention to be determined from the patients mental flexibility); d) visual memory test; e) character discrimination test (measuring the ability to discriminate 2 different short-term stimuli counts). Additionally, a short test of general basic parameters of intelligence (KAI; Lehrl *et al.*, 1980) was carried out. This test accounts for the speed of information processing and the period for which the information is available in the short-term memory. As regards sociodemographic data and severity, the two treatment-groups were quite homogeneous at randomization. Superiority of active treatment already became obvious after 6 weeks when the groups were distinguishable by statistical means with respect to nearly all outcome variables. After 12 weeks of treatment significant inter-group differences in favour of LI 1370 were found in the self-rating symptom scale, in the C.I. total score and in the 5 subtests of the computerized battery. In the KAI there was a faster improvement of basic factors of information processing, although this could not be proven statistically at 12 weeks. No adverse side effects were observed under either treatment.

Dementia

Weitbrecht and Jansen (1986) studied the effects of EGb 761® (120 mg/day) compared with placebo and ergot alkaloids (5.94 mg/day) in patients suffering from mild to moderate primary degenerative dementia. This randomized trial was double-blind with respect to EGb 761® and placebo, and the reference substance was given to a third treatment group under open conditions. Sixty patients (average age 72.4 ± 4.6; Hachinski score < 7) were treated for 3 months with EGb 761®, placebo, or ergot alkaloids, 20 patients each. Efficacy was measured on the psychopathological level from the patient's self-assessment, the physician's global assessment and the Sandoz Clinical Assessment – Geriatric (SCAG; Shader *et al.*, 1974). Objective assessment of performance was carried out by applying the Wechsler digit symbol substitution test and the digit span test (Wechsler, 1994). Flicker fusion frequency and reaction time were tested to provide insight into functional-dynamic effects, and the Crichton scale (Robinson, 1964) was employed for assessment of behaviour closely related to activities of daily living. After only 4 weeks of treatment, the EGb 761®-treated group showed statistically significant improvements in both psychometric test performance and clinical assessments. After 3 months of EGb 761® treatment, a mean improvement of 23.5% as compared to baseline was found on the Crichton scale. Pre/post differences were significantly higher in the active treatment group as compared with the placebo group after 3 months (p < 0.0001; U-test). The SCAG total score improved by approximately 33% in comparison to baseline values (p < 0.0001 for pre/post differences vs. placebo; Fig. 1). According to the physician's global assessment and the patients' self-assessment, the patients' state of health improved significantly under EGb 761® treatment, but not under placebo. Both the digit span and digit symbol substitution test demonstrated significantly greater improvements in actively treated patients compared with placebo-treated patients. The same was true for flicker fusion frequency and reaction time. Treatment with ergot alkaloids gave also a significantly better outcome than placebo treatment. There were no significant differences at any time between the reference substance and EGb 761®. No adverse reactions were reported (2 patients in the placebo-group dropped out for reasons unrelated to treatment). To summarize, the results of this study indicate the effectiveness of EGb 761® in treatment of mild to moderate primary degenerative dementia.

Wesnes *et al.* (1987) included 58 elderly patients (mean age 71 years) who were living at home and suffering from mild dementia manifesting itself as an impairment of everyday functioning (Crichton score > 14). After a washout period of 1–3 weeks, 27 patients were treated with EGb 761® (120 mg/day), and 27 were given placebo. Duration of treatment was 12 weeks. Efficacy was assessed by an extensive cognitive test battery comprising computerized as well as pencil and paper tests of memory function, attention, and information processing. Quality of life was measured by a behavioural questionnaire (Guinot and Wesnes, 1985) applied before and after the study. Improvements could be detected in both groups, but the improvement at week 12 was significantly more pronounced in the EGb 761®-treated group than in the placebo group in most of the single tests, and also when the accuracy scores from all the 8 tests were combined (p < 0.05). The most import-

**SCAG
total score**

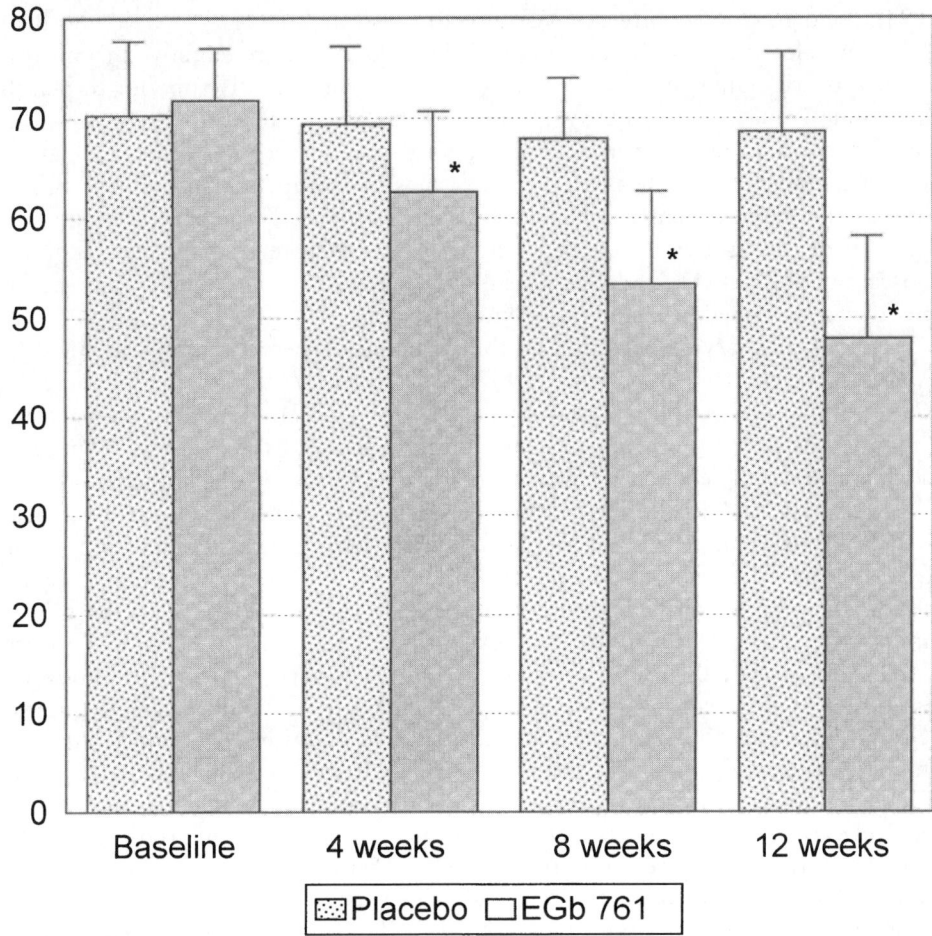

Figure 1   Development of SCAG Total Scores During 12 Weeks Treatment with EGb 761®
and Placebo; mean values and standard deviations; *: p <0.0001 for inter-group comparison
of pre/post differences (Weitbrecht and Jansen, 1988).

ant finding in this study, however, emerged from the quality of life scale. While there
was no difference between the treatment groups with respect to the number of
activities still practised, the number of activities carried out with "high interest"
increased significantly in the EGb 761®-treated group (p < 0.05) but not in the
placebo group. One adverse event (constipation) was reported in the actively treated
group which was not considered to be related to the treatment.

In 50 outpatients suffering from mild dementia (uncomplicated senile dementia,
presenile dementia and arteriosclerotic dementia) according to ICD-9, Halama

(1991) studied the effects of *Ginkgo biloba* special extract LI 1370 on cognitive functioning and psychopathology. Patients were treated for 12 weeks with 150 mg/day of LI 1370 or placebo, 25 patients each. A symptom scale comprising 11 typical symptoms of early dementia, the SKT-test (Erzigkeit, 1986, Kim *et al.*, 1993), the digit connection test ZVT (a sort of trail-making test; Oswald and Roth, 1978) and global ratings by patients and the physician were employed to measure the therapeutic outcome. LI 1370 was found to be significantly superior to placebo concerning 7 out of 11 typical symptoms and the SKT total score. There was also a more pronounced decrease in ZVT processing time on active treatment which did not reach statistical significance. Global ratings of therapeutic success by both the patients and the physician significantly favoured LI 1370. Three adverse events were reported in this study: 1 patient receiving placebo suffered from headache and nausea; out of 2 patients under active substance, 1 reported an allergic skin reaction, the other complained of a pain in the kidney region.

Twenty patients with dementia of the Alzheimer type according to DSM-III-R (American Psychiatric Association, 1987) and NINCDS/ADRDA (McKhann *et al.*, 1984) criteria were enrolled in a study by Ihl *et al.* (1992). Treatment duration was 3 months. As primary efficacy measure, the sum score of the SKT-test (Kim *et al.*, 1993) was defined. Secondary efficacy measures were further psychometric tests: the ZVT (Oswald and Roth, 1978) and the Alzheimer's Disease Assessment Scale (ADAS; Mohs *et al.*, 1983, Ihl and Weyer, 1993), the Clinical Global Impressions (CGI; National Institute of Mental Health, 1976), and electrophysiological parameters (topographic EEG). During treatment, SKT-scores changed in the *Ginkgo biloba* group from 19.67 to 16.78 (mean values) and from 18.11 to 18.89 in the placebo group (p = 0.013 for differences). The superiority of *Ginkgo biloba* was also observed in the secondary variables, CGI change (p = 0.069) and EEG (increase of mean amplitudes). There were no adverse drug reactions. The most important finding in this study was that the efficacy of EGb 761® could be clearly demonstrated by objective cognitive testing and global rating in patients suffering from relatively severe cerebral impairment in moderately severe dementia of the Alzheimer type, which rapidly progressed in placebo-treated patients.

Forty inpatients (aged 50 to 75 years) with a clinical diagnosis of mild senile dementia of the Alzheimer type, were studied by Hofferberth (1994). Diagnosis was based on common clinical criteria together with a CT-scan and the Hachinski Ischemic Score. Patients received, after random allocation, either EGb 761® (240 mg/day) or placebo for 3 months. Inclusion criteria were described as follows: Blessed Dementia Scale (Blessed *et al.*, 1968): sum Part A: 0–16, sum Part B: 9.5–30.5; Ischemic Score according to Rosen *et al.* (1980) < 4; normal or diffuse and possibly asymmetrically atrophic CT findings. The severity of the disease was determined by the SKT-test (a brief cognitive test battery for the measurement of memory and attention; Kim *et al.*, 1993) the total score of which was 17 in the EGb 761® group and 15 in the placebo group, both consistent with a moderately severe cognitive impairment. Exclusion criteria were: advanced Alzheimer dementia, i.e. necessity of special accommodation and constant care; intellectual deterioration or state of confusion and/or dementia syndrome of another origin, pseudodementia;

significant depression with a HAMD total score > 15 (Hamilton, 1967); pronounced sensory or motor disturbances; severe organic illness. Treatment with vasoactive agents, nootropics, psychotonics, tranquilizers and/or antidepressants was prohibited throughout the study. The SKT-test was predefined as the primary outcome measure. Secondary variables were the Saccade Test, Vienna Determination Test (part of the Viennese Psychophysiological Test-System; Semlitsch et al., 1989), the Sandoz Clinical Assessment – Geriatric (SCAG; Shader et al., 1974) and electroencephalography (particularly the theta/alpha quotient). Memory and attention, as measured by the SKT, improved significantly in the verum group already after 1 month (inter-group differences at $p < 0.001$), as did psychopathology, psychomotoric performance, and functional dynamic variables as measured by the procedures mentioned above. After 3 months of EGb 761® treatment, the mean SKT total score had decreased by 5 points whereas an increase of 2 points was found in the placebo group, resulting in a highly significant inter-group difference in favour of the active treatment ($p < 0.001$). This was confirmed on the psychopathological level by a mean decrease of 29 points in the SCAG total score after EGb 761®, compared with a negligible change during placebo application. At month 3, the theta/alpha quotient in the active-treatment group had decreased from 79.4% to 61.9%, indicating enhanced vigilance. Significant superiority of the Ginkgo biloba special extract was also demonstrated as regards choice reaction time and both latency and speed of saccadic eye movements. No adverse events were observed. In this methodologically adequate trial, again, the clinical relevance of treatment-selected effects could be demonstrated by highly consistent results on 3 levels of assessment. Improvement in test performance measured by objective procedures was in accordance with a decrease in psychopathology and changes towards normal in functional dynamic measures.

Haase et al. (1996) enrolled 40 patients, aged 68 ± 12.5 years, suffering from moderate dementia of the Alzheimer type, vascular dementia, and mixed forms according to DSM-III-R criteria (American Psychiatric Association, 1987) into a short-term study. Psychopathology corresponded to stages IV or V of the Global Deterioriation Scale of Reisberg et al. (1982); patients showing any signs consistent with another type of dementia or suffering from severe somatic disease had to be excluded. Treatment duration was 4 weeks. Patients received either intravenous infusions of EGb 761® (200 mg active substance/250 ml saline) on 4 days of each week or placebo infusions having the same appearance. Clinical efficacy was assessed by the Nuremberg Gerontopsychological Observation Scale for Activities-of-Daily-Living (NAB; Oswald and Fleischmann, 1995) which reflects the patient's ability to perform activities of daily living as observed by a relative or a caregiver; by the change in the patient's overall condition as realized by the psychiatrist (CGI, Item 2; National Institute of Mental Health, 1976); and by objective cognitive testing (Short Test of General Intelligence, KAI; Lehrl et al., 1980). In the actively treated group, a mean decrease of 3.6 in the NAB total score was found after 4 weeks treatment, thus reflecting a meaningful improvement in patients with moderate dementia. Only a minimal decrease of 0.3 points was seen in the placebo group. The resulting inter-group difference was highly significant ($p < 0.01$). According to the CGI, 85% of actively treated patients and 35% of those who received placebo were judged as having improved (p < 0.0001). The cognitive test revealed an improvement in general IQ in 45% of the patients treated with EGb 761® and in 5 patients who were given placebo (p < 0.05). No adverse events occurred during EGb 761® treatment. As this trial was designed in accordance with the most recent published recommendations of the German Federal Health Authority, efficacy was assessed – and established – on the psychopathologic and the psychometric plane, and the plane of activities of daily living. It was thus possible to demonstrate that treatment-related improvements of psychopathology and cognitive test performance do in fact result in a gain of competence in coping with the demands of everyday life.

A very convincing study has been published by Kanowski et al. (1996). The objective was to prove the clinical efficacy of Ginkgo biloba special extract EGb 761® in outpatients with presenile and senile primary degenerative dementia of the Alzheimer type or multi-infarct dementia. Diagnoses were established by DSM-III-R (American Psychiatric Association, 1987) criteria with support from CT-scans and the Ischemic Score according to Rosen et al. (1980). Patients were included if they scored between 13 and 25 in the Mini-Mental State Examination (Folstein et al., 1975) and between 6 and 18 in the SKT-test (Kim et al., 1993), which is compatible with mild to moderate dementia. Patients suffering from any other significant psychiatric or somatic disease, or taking any drugs possibly interfering with efficacy assessment were excluded. 216 patients aged over 55 years were randomized to receive either Ginkgo biloba special extract EGb 761® (240 mg/day) or placebo. A run-in period of 4 weeks was followed by a treatment period of 24 weeks. Three primary variables were selected according to the recommended multi-dimensional approach for proving the efficacy of nootropic and/or anti-dementia drugs (Leber, 1990; CPMP, 1992; Menges, 1992); these were the CGI (Item 2) (National Institute of Mental Health, 1976) as a global rating on the psychopathological level, the SKT for objective assessment of syndrome-relevant cognitive performance, and the Nuremberg Gerontopsychological Observation Scale for Activities-of-Daily-Living (NAB; Oswald and Fleischmann, 1995). Per-protocol analysis was based on the data from 156 patients (EGb 761®: 79; placebo: 77) and carried out as a multiple-responder analysis according to pre-specified rules. Only those improvements considered clinically meaningful were accepted as being a response, i.e. a judgement of "much improved" or "very much improved", with respect to the patient's condition, in the CGI; a decrease of at least 4 points in the SKT total score; and a decrease of at least 2 points in the NAB total score. A response in at least 2 of the three primary variables was considered a treatment response. After 24 weeks, 28% of the patients treated with EGb 761® were found to be treatment responders, whereas only 10% of placebo patients showed multiple response (p < 0.005, Fisher's Exact Test, confirmatory; Fig. 2). When considering only those patients who showed response in all of the 3 primary variables, EGb 761® was again found to be significantly superior to placebo (p < 0.05, descriptive). When reviewing the single variables, improvements in CGI-scores termed "better" or "much better" were found in 32% of EGb 761®-treated patients and in 17% of patients in the placebo group (p < 0.05).

% Responders

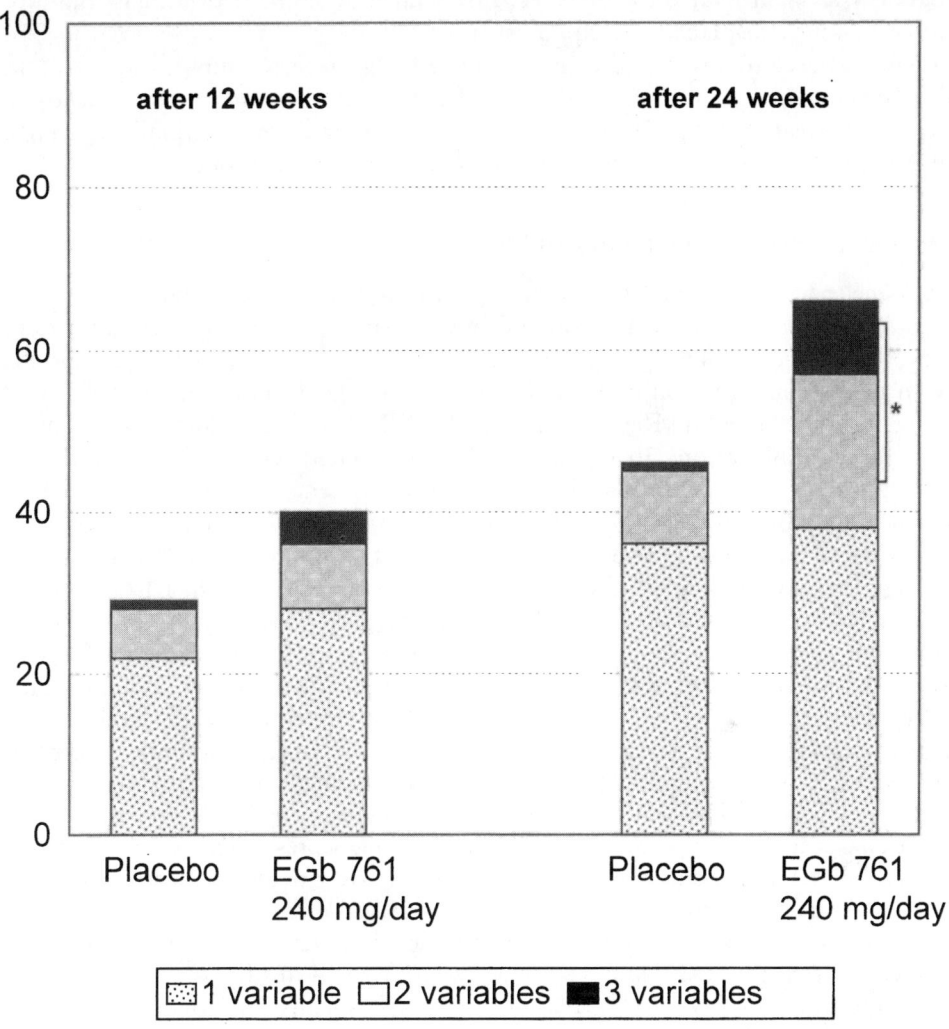

Figure 2   Therapy Responders (i.e. Responders in Two or Three Variables) and Responders in One Single Variable after 12 and 24 Weeks Treatment; *: $p_{response}$ = 0.005 (Kanowski *et al.*, 1996).

Decreases in SKT total scores by at least 4 points were documented in 38% of the actively treated patients and in 18% of the placebo patients (p < 0.005). Improvements in the NAB of at least 2 points were found in 33% of cases in the EGb 761® group and 23% of cases in the placebo group (p < 0.1). An additional intent-to-treat analysis of the data from 205 patients led to noticeably similar results. Exploratory analyses of prospectively defined subgroups did not reveal any relevant differences in efficacy in the context of the dementia type (Alzheimer or vascular). There were 5 adverse events in the EGb 761® group which were judged as possibly treatment-related by the investigators (allergic skin reactions, gastrointestinal complaints, headache). A comparison of all adverse events observed in 216 patients exposed to study medication (EGb 761® or placebo) shows, however, that allergic skin reactions were the only type of event that occurred more frequently in the actively treated group. The opposite was the case for gastrointestinal complaints; and headaches were virtually equally distributed between the groups.

Cerebral Insufficiency and Depressed Mood

Stocksmeier and Eberlein (1992) performed a trial on 60 patients aged between 51 and 71 years, and suffering from both impairment of cerebral function and depressed mood. Patients were required to score below 90% of the age-specific norm in the concentration-performance test (KLT, Düker and Lienert, 1965) and above 15 in the Hamilton Depression Scale (HAMD; Hamilton, 1967). According to their random allocation, 30 patients each were treated with daily oral doses of 120 mg EGb 761® and placebo respectively. After 12 weeks treatment, a significantly more pronounced decrease in the HAMD total score was observed in the EGb 761® group compared with the placebo group (45% vs. 31%; $p < 0.05$). No significant inter-group differences were found with respect to the KLT.

Schubert and Halama (1993) included 40 patients aged 51 to 78 years and suffering from both mild impairment of cerebral function and depressive mood in a randomized trial. Cerebral impairment was assessed by the C.I. scale (total score at least 20 points) (Weidenhammer and Fischer, 1987). Depression had been treated with standard antidepressants for at least 3 months without achieving an acceptable improvement or remission. While continuing the antidepressive treatment at a stable dosage, 20 patients each were additionally administered 240 mg/day of EGb 761® or placebo for 8 weeks. Concomitant medication possibly influencing depression or cerebral function was prohibited during study participation. In the actively treated patients, a decrease in the HAMD total score from 14 to 7 was observed after 4 weeks followed by a further drop to 4.5, thus reflecting a considerable alleviation of depressive symptoms. At the same time the HAMD total score decreased from 14 to 11 in the placebo group. The superiority of EGb 761® treatment was statistically confirmed ($p < 0.01$). Cognitive performance as assessed by a short general intelligence test (KAI; Lehrl et al., 1980) was also significantly improved by EGb 761® ($p < 0.05$ vs. placebo).

It might be criticized that, in both trials, depression and impairment of cerebral function were not precisely defined as ICD or DSM diagnoses. This is exactly the problem when using this approach. In a considerable number of elderly patients appearing depressed and cognitively impaired, it is very difficult or even impossible to decide whether they are suffering from depression with concomitant disturbances of cognitive functions, or from dementia with concomitant depression. The American Psychiatric Association was aware of this problem when drawing up the DSM-III-R (American Psychiatric Association, 1987). They recommended first to assume a depression and, when antidepressant therapy proved ineffective, to change the diagnosis into that of dementia. An increase of antidepressant dosage as would be adequate in younger patients is often impossible in elderly as many anti-depressants further aggravate cognitive impairment due to anticholinergic side effects. Therefore, it appears intriguing to have a drug that improves both impairment of cerebral function and depressive mood. There remain, however, questions that should be addressed in further studies: Can any predictors of treatment response be identified? Is there an advantage of combined (antidepressant) treatment over EGb 761® alone?

METAANALYSES

Weiss und Kallischnigg (1991) compiled 17 clinical trials with EGb 761® published before 1990 in a metaanalytic review. This work was complicated by the great variety of different tests and rating scales applied in the various trials and the different trial designs. The authors' major criticism concerned the methodological shortcomings of earlier studies not reported in this chapter. By taking together those trials which, according to their judgement, were of sufficient methodological quality to contribute to the evaluation of clinical efficacy they came to the conclusion that the clinical efficacy of EGb 761® in ageing associated impairment of cerebral function was proven with more than 95% confidence.

In a critical review, Kleijnen and Knipschild (1992) evaluated the methodological quality of 40 clinical trials with EGb 761® and LI 1370 published up to 1991 and the resulting proof of efficacy. These authors also criticized the low methodological standards of the early trials and the heterogeneity of trials due to a great variety of designs and outcome measures. It must, however, be pointed out that they included some trials not directly related to cerebral insufficiency in their review. In conclusion, they stated that clinical efficacy can be assumed in cerebral insufficiency, although more evidence was warranted.

It must be emphasized that only two of the dementia studies reported in this chapter had been published before both reviews were written. Those early trials that have been seriously criticized because of major methodological shortcomings are omitted from the overview in this chapter. The fact that a great number of different tests and rating scales were used in the various trials prevents a meaningful pooling of the results in metaanalyses. However, consistent findings of clinical improvement in various settings and in a considerable number of different outcome measures corroborate the conclusion that such findings are not a matter of chance, and that effects are not restricted to one, possibly unimportant feature of the disease. As concerns the problem of multiple testing in some earlier studies, it must be admitted that adjustment of an α-type error (type I error) was sometimes neglected. When testing for a considerable number of outcome variables in one trial, the possibility of obtaining a few significant results by chance does in fact exist. As all or at least most of the statistical tests in these trials yielded significant results, it is however unlikely that this was merely due to chance.

Hopfenmüller (1994) performed a metaanalysis on 8 clinical trials with LI 1370 in which similar outcome measures had been employed. He came to the conclusion

Table 1   Randomized, double-blind, placebo-controlled clinical trials with *Ginkgo biloba* special extracts EGb 761® and LI 1370 in age-associated impairment of cerebral function: significant effects on different levels of assessment

Author(s) Year	Treatment duration (weeks) Daily dose (mg)	Psychopathologic level	Psychometric level	ADL level	Functional-dynamic level
Vesper and Hänsgen, 1994	12 150 oral	Selfrating symptom scale, cerebral insufficiency scale (c.I. scale)	Computerized test battery		
Gräßel, 1992	24 160 oral		Computerized test battery measuring basic functions of information processing		
Schmidt et al., 1991	12 150 oral	Symptom scale, physician's and patient's global assessment			
Hofferberth, 1991	6 150 oral	Symptom scale, physician's global assessment	Vienna determination test, digits connection test		Saccade test, EEG, evoked potentials
Brüchert et al., 1991	12 150 oral	Symptom scale, physician's and patient's global assessment	Digits connection test		
Halama et al., 1988	12 120 oral	Sandoz Clinical Assessment Geriatric Scale (SCAG), Symptom Scale	SKT-test		
Israël et al., 1987	12 160 oral		Memory battery according to L. Israël		

Table 1   continued

Author(s) Year	Treatment duration (weeks) Daily dose (mg)	Psychopathologic level	Psychometric level	ADL level	Functional-dynamic level
Taillandier et al., 1986	52 160 oral	EACG (French version of the SCAG), physician's global rating			
Eckmann and Schlag, 1982	4 120 oral	Symptom List			
Dieli et al., 1981	8 120 oral	Clinician's global assessment	Digit Span Test (WAIS) Digit Symbol Substitution Test (WAIS)	Geriatric scale according to Plutchik	
Moreau, 1975	12 120 oral	Clinician's global assessment, Symptom scale	7 cognitive tests (temporospatial orientation, immediate memory, short-term memory, finding of common properties, arranging of pictures, constructional praxis)		

Table 2   Randomized, double-blind, placebo-controlled clinical trials with *Ginkgo biloba* special extracts EGb 761® and LI 1370 in dementia: significant effects on different levels of assessment

Author(s) Year	Treatment duration (weeks) Daily dose (mg)	Psychopathologic level	Psychometric level	ADL level	Functional-dynamic level
Kanowski et al., 1996	24 240 oral	Clinical Global Impressions (CGI)	SKT-test	Nuremberg Gerontopsychological Observation Scale for Activities-of-Daily-Living (NAB)	Actometry
Haase et al., 1996	4 200 (4 days per week) i.v.	Clinical Global Impressions (CGI Item 2)	Short general intelligence test (KAI)	Nuremberg Gerontopsychological Observation Scale for Activities-of-Daily-Living (NAB) Nuremberg Gerontopsychological Self-Rating Scale for Activities-of-Daily-Living (NAA)	
Hofferberth, 1994	12 240 oral	Sandoz Clinical Assessment Geriatric Scale (SCAG)	SKT-test		Saccade-test, EEG
Ihl et al., 1992	12 240 oral	Clinical Global Impressions (CGI Item 2)	SKT-test		EEG/Brain-mapping
Wesnes et al., 1987	12 120 oral		Computerized test battery	Behavioral questionnaire (quality of life)	
Halama et al., 1991	12 150 oral	Symptom scale, physician's and patient's global assessment	SKT-test		
Weitbrecht and Jansen, 1986	12 120 oral	Sandoz Clinical Assessment Geriatric Scale (SCAG) Self-assessment scale	Digit Symbol Substitution Test (WAIS) Digit Span Test (WAIS)	Crichton Scale	Flicker-fusion frequency, Reaction time

that in all the clinical symptoms of the symptom scales as well as in the global assessments, the active substance was significantly superior to placebo.

In a comparative metaanalysis including all the trials published until 1996, Letzel *et al.* (1996) scrutinized the randomized double-blind and placebo-controlled trials establishing the clinical efficacy of *Ginkgo biloba* special extracts, nimodipine, and tacrine in dementia. They found that, for either active substance, efficacy has been proven on the basis of currently accepted guidelines (Menges, 1992; CPMP, 1992).

DISCUSSION

The sequence of clinical trials with *Ginkgo biloba* special extracts reflects the development of both the pathophysiological concepts underlying the ageing associated impairment of cerebral function and the standards for evaluation of efficacy of nootropics/antidementia drugs. Notwithstanding the early publications on the "dementing illness" first described by Aloys Alzheimer (1864–1915), insufficient cerebral perfusion was considered a predominant cause for cognitive impairment in elderly patients. Consistently, the early trials were implemented with drugs known to enhance cerebral blood flow, and in patients with suspected "cerebrovascular insufficiency". While progress in preclinical and clinical research continuously improved the possibility of distinguishing between vascular and primary degenerative dementia, it became obvious that dementia of the Alzheimer type is the most frequent cause of age-related progressive deterioration of cerebral function. However, it is still almost impossible to establish a clinical diagnosis of dementia at a very early stage. After the first *Ginkgo biloba* studies in "cerebrovascular insufficiency", a large number of trials on "impairment of cerebral function" of vascular and non-vascular etiology were carried out. With upcoming clinical criteria for the diagnosis of vascular and Alzheimer type dementia, trials in well-defined dementia patients became possible. While efficacy assessment was mainly based on cognitive testing and subjective complaints in earlier studies, there are now a number of elaborate guidelines to follow. What today is considered methodologically insufficient was the state of the art when the early trials were conducted.

What is intriguing, however, is the fact that, notwithstanding the high methodological requirements met in recent dementia trials, the results are noticeably comparable to those of earlier studies on cerebral insufficiency. This is not surprising when taking the fact into account that a substantial part of those patients who were diagnosed as having cerebral insufficiency in former trials can be assumed to have had dementia. It might well be the case that *Ginkgo biloba* extracts act along a neuropathologic pathway that Alzheimer type dementia, vascular dementia, and "benign" age-associated impairment of cerebral function all have in common. The correspondence of the results of the earlier trials and those of the latest studies meeting today's methodological requirements gives reason to assume validity of the former ones.

There appears to be no remarkable difference in the extent of improvements as regards dementia etiology. This can be deduced from 3 trials (Taillandier *et al.*,

1986; Haase *et al.*, 1996; Kanowski *et al.*, 1996), where subgroups of patients with vascular dementia and dementia of the Alzheimer type were included which allowed a direct comparison. This justifies the inclusion of both Alzheimer's patients and vascular dementia patients in the same clinical trials. The results of pharmacological testing of EGb 761® also support its use in both kinds of dementia (see previous sections on pharmacological and biochemical activities).

Very low rates of adverse events possibly related to the drug under investigation were observed in the clinical trials. Mild gastro-intestinal discomfort, headache and allergic skin reactions have been observed very rarely on EGb 761® intake during the long time of therapeutic use.

To summarize, the clinical trials described in this chapter clearly demonstrate the clinical efficacy of *Ginkgo biloba* special extracts EGb 761® and LI 1370 in dementia of the Alzheimer type and vascular dementia as well as in age-associated impairment of cerebral function. In methodologically adequate studies, clinically meaningful improvements were achieved on the psychopathologic level as assessed by global ratings and comprehensive gerontopsychiatric scales; on the level of psychometric performance as assessed by validated cognitive tests; on the level of behaviour closely related to activities of daily living; and on the functional dynamic level. It could be demonstrated that improvements in clinical symptomatology and cognitive test performance in fact translate into an improved competence for coping with the requirements of everyday life. This is in essence what has been recommended by current guidelines (Menges, 1992; CPMP, 1992), and meets the demands of patients and caregivers.

With respect to clinical practice it can be concluded that, on the grounds of their proven efficacy and their excellent tolerability, *Ginkgo biloba* special extracts can be regarded as first-choice nootropics. As can be deduced from the clinical trials described before, the drugs should be administered in a daily dose of at least 120 mg for 6 months or longer to bring about their full benefit. A lack of therapy response should not be assumed earlier than after at least 3 months treatment. This is generally recommended for nootropic therapy (Haan, 1993), since the onset of drug effects may be insidious and not easy to detect during the first weeks.

REFERENCES

American Psychiatric Association (1987) *Diagnostic and statistical manual of mental disorders*, third edition, revised, American Psychiatric Association, Washington, DC.

American Psychiatric Association (1994) *Diagnostic and statistical manual of mental disorders,* fourth edition, American Psychiatric Association, Washington, DC.

Blessed, G., Tomlinson, B.E., Roth, M. (1968) The association between quantitative measures of dementia and of senile change in the cerebral grey matter of elderly subjects. *Brit. J. Psychiat.*, **114**, 797–811.

Brink, T.L., Yesavage, J.A., Lum, O., Heersema, P.H., Adey, M., Rose, T.L. (1982) Screening tests for geriatric depression. *Clin. Gerontologist*, **1**, 37–43.

Brüchert, E., Heinrich, S.E., Ruf-Kohler, P. (1991) Wirksamkeit von LI 1370 bei älteren Patienten mit Hirnleistungsschwäche. Multizentrische Doppelblindstudie des Fachverbandes Deutscher Allgemeinärzte. *Münch. Med. Wochenschr.*, **133** (Suppl. 1), 9–14.

Chatterjee, A., Strauss, M.E., Smyth, K.A., Whitehouse, P.J. (1992) Personality changes in Alzheimer's disease. *Arch. Neurol.*, **49**, 486–491.

CPMP Working Party on Efficacy of Medicinal Products (1992) *Note for Guidance, Antidementia medicinal products.* Draft 5, Commission of the European Communities, Brussels.

Dieli, G., La Mantia, V., Saetta, M., Costanzo, E. (1981) Studio clinico in doppio cieco del Tanakan nell' insufficienza cerebrale cronica. *Il Lavoro Neuropsichiatr.*, **68**, 3–15.

Düker, H., Lienert, G. (1965) *KLT–Konzentrations-Leistungs-Test. Handanweisung.* Hogrefe, Göttingen, Germany.

Eckmann, F. and Schlag, H. (1982) Kontrollierte Doppelblind-Studie zum Wirksamkeitsnachweis von Tebonin forte bei Patienten mit zerebrovaskulärer Insuffizienz. *Fortschr. Med.*, **100**, 1474–1478.

Erzigkeit, H. (1986) *Der Syndrom-Kurztest zur Erfassung von Aufmerksamkeits- und Gedächtnisstörungen.* Vless, Vaterstetten, Germany.

Folstein, M.F., Folstein, S.E., McHugh, P.R. (1975) "Mini-mental state": A practical method for grading the cognitive state of patients for the clinician. *J. Psychiatr. Res.*, **12**, 189–198.

Geldmacher, D.S. and Whitehouse, P.J. (1996) Evaluation of dementia. *N. Engl. J. Med.,* **335**, 330–336.

Georges, D., Lallemand, A., Coustenoble, J., Loria, Y. (1977) Validation par l'analyse factorielle d'une échelle d'évaluation clinique des troubles de la sénescence cérébrale. *Thérapie,* **32**, 173–180.

Gräßel, E. (1992) Einfluß von *Ginkgo-biloba*-Extrakt auf die geistige Leistungsfähigkeit, Doppelblindstudie unter computerisierten Meßbedingungen bei Patienten mit Zerebralinsuffizienz. *Fortschr. Med.*, **110**, 73–76.

Guinot, P. and Wesnes, K. (1985) A quality of life scale for the elderly: validation by factor analysis. *ICRS Medical Science*, **13**, 965.

Haan, J. (1993) Hirnleistungsstörungen im Alter. *Dtsch. Med. Wochenschr.*, **118**, 947–950.

Haase, J., Halama, P., Hörr, R. (1996) Efficacy of short-term treatment with intravenously administered *Ginkgo biloba* special extract EGb 761 in Alzheimer-type and vascular dementia. *Z. Gerontol. Geriat.*, **29**, 302–309.

Hachinski, V.C., Iliff, L.D., Zilkha, E., Du Boulay, G.H., McAllister, V.L., Marshall, J., Russell, R.W.R., Symon, L. (1975) Cerebral blood flow in dementia. *Arch. Neurol.*, **32**, 632–637.

Halama, P., Bartsch, G., Meng, G. (1988) Hirnleistungsstörungen vaskulärer Genese. Randomisierte Doppelblindstudie zur Wirksamkeit von *Ginkgo-biloba*-Extrakt. *Forschr. Med.*, **106**, 408–412.

Halama, P. (1991) *Ginkgo biloba.* Effectiveness of a special extract in patients with cerebral insufficiency. *Münch. Med. Wochenschr.*, **133**, 190–194.

Hamilton, M. (1967) Development of a rating scale for primary depressive illness. *Brit. J. Soc. Clin. Psychol.*, **6**, 278–296.

Hofferberth, B. (1991) *Ginkgo-biloba*-Spezialextrakt bei Patienten mit hirnorganischem Psychosyndrom. *Münch. Med. Wochenschr.*, **133** (Suppl. 1), 30–33

Hofferberth, B. (1994) The efficacy of EGb 761 in patients with senile dementia of the Alzheimer type, a double-blind, placebo-controlled study on different levels of investigation. *Human Psychopharmacol.*, **9**, 215–222.

Hopfenmüller, W. (1994) Nachweis der therapeutischen Wirksamkeit eines *Ginkgo biloba*-Spezialextraktes. *Arzneim.-Forsch./Drug Res.*, 44, 1005–1013.

Ihl, R., Weyer, G. (1993) *Alzheimer's Disease Assessment Scale (ADAS). Deutschsprachige Bearbeitung. Manual*, Beltz Test, Weinheim, Germany.

Ihl, R., Dierks, T., Martin, E., Frölich, L., Maurer, K. (1992) *Ginkgo biloba* Extrakt EGb 761 bei der Demenz vom Alzheimer Typ. *15. Donausymp. f. Psychiatrie*, Regensburg, Abstract Book, N1.40.

Israel, L., Ohlmann, T., Chappaz, M. (1980) Batterie psychométrique à l'usage du médecin généraliste pour apprécier les conduites intellectuelles des personnes agées. *Psychol. Médicale*, 12, 921–938.

Israel, L., Dell'Accio, E., Martin, G., Hugonot R. (1987) Extrait de *Ginkgo biloba* et exercices d'entrainement de la mémoire. Evaluation comparative chez des personnes agées ambulatoires. *Psychol. Méd.*, 19, 1431–1439.

Itil, T.M., Eralp, E., Tasmbis, E., Itil, K.Z., Stein, U. (1996) Central nervous system effects of *Ginkgo biloba*, a plant extract. *Am. J. Therapeut.*, 3, 63–73.

Kanowski, S., Herrmann, W.M., Stephan, K., Wierich, W., Hörr, R. (1996) Proof of efficacy of the *Ginkgo biloba* special extract EGb 761 in outpatients suffering from primary degenerative dementia of the Alzheimer type and multi-infarct dementia. *Pharmacopsychiat.*, 29, 47–56.

Kim, Y.S., Nibbelink, D.W., Overall, J.E. (1993) Factor structure and scoring of the SKT test battery. *J. Clin. Psychol.*, 49, 61–71.

Kleijnen, J. and Knipschild, P. (1992) *Ginkgo biloba* for cerebral insufficiency. *Br. J. Clin. Pharmacol.*, 34, 352–358.

Leber, P. (1990) *Guidelines for the clinical evaluation of antidementia drugs*, First Draft, US Food and Drug Administration, Rockville, MD.

Lehrl, S., Gallwitz, A., Blaha, L. (1980) *Kurztest für allgemeine Intelligenz KAI. Manual.* Vless, Vaterstetten, Germany.

Lehrl, S., Fischer, B. (1988) The basic parameters of human information processing: their role in the determination of intelligence. *Person. Individ. Diff.*, 9, 883–896.

Letzel, H., Haan, J., Feil, W.B. (1996) Nootropics: Efficacy and tolerability of products from three active substance classes. *J. Drug Dev. Clin. Pract.*, 8, 77–94.

Luthringer, R., d'Arbigny, P., Macher, J.P. (1995) *Ginkgo biloba* extract (EGb 761), EEG and event-related potentials. In Y. Christen, Y. Courtois, and M.T. Droy-Lefaix, (eds.), *Advances in Ginkgo biloba extract research, vol. 4. Effects of Ginkgo biloba on aging and age-related disorders*. Elsevier, Paris, pp. 107–118.

McKhann, C., Drachman, D., Folstein, M., Katzman, R., Price, D., Stadlan, E.M. (1984) Clinical Diagnosis of Alzheimer's disease: Report of the NINCDS-ADRDA work group under the auspices of Department of Health and Human Services Task Force on Alzheimer's Disease. *Neurology*, 34, 939–944.

Menges, K. (1992) Proof of efficacy of nootropics for the indication "dementia" (Phase III) – Recommendations. *Pharmacopsychiat.*, 25, 126–135.

Mohs, R.C., Rosen, W.G., Davis, K.L. (1983) The Alzheimer's Disease Assessment Scale: an instrument for assessing treatment efficacy. *Psychopharmacol. Bull.*, 19, 448–450.

Moreau, Ph. (1975) Un nouveau stimulant circulatoire cérébral. *La Nouvelle Presse Méd.*, 4, 2401–2402.

National Institute of Mental Health (1976) 028 CGI. Clinical Global Impressions. In W. Guy and R.R. Bonato, (eds.), *Manual for the ECDEU Assessment Battery*, second revised edition, NIMH, Chevy Chase, Maryland, pp. 12-1–12-6.

Oswald, W.D., Roth, E. (1978) *Der Zahlen-Verbindungs-Test (ZVT)*, Hogrefe, Göttingen, Germany.

Oswald, W.D., Fleischmann, U.M. (1995) *Nuremberg Gerontopsychological Inventory (NAI)*. Hogrefe, Göttingen, Germany.

Pidoux, B., Bastien, C., Niddam, S. (1983) Etude à double insu, face au placebo, des effects de l'extrait de *Ginkgo biloba* sur l'état clinique et l'EEG quantifiè. *J. Cerebr. Blood Flow Metab.*, 3 (Suppl. 1), 5556–5557.

Plutchik, R., Conte, H., Lieberman, M., Bakur, M., Grossman, J., Lehrman, N. (1970) Reliability and validity of a scale for assessing the functioning of geriatric patients. *J. Am. Geriat. Soc.*, 18, 491–500.

Reisberg, B., Ferris, S.H., de Leon, M.J. Crook, T. (1982) The global deterioration scale (GDS): An instrument for the assessment of primary degenerative dementia (PDD). *Am. J. Psychiat.*, 139, 1136–1139.

Robinson, R.A. (1964) The diagnosis and prognosis of dementia. In W.E. Anderson, (ed.), *Current Achievements in Geriatrics*, Cassell, London.

Rosen, W.G., Terry, R.D., Fuld, P.A., Katzmann, R., Peck, A. (1980) Pathological verification of ischemic score in differentiation of dementias. *Ann. Neurol.*, 7, 486–488.

Schmidt, U., Rabinovici, K., Lande, S. (1991) Einfluß eines *Ginkgo-biloba*-Spezialextraktes auf die Befindlichkeit bei cerebraler Insuffizienz. *Münch. Med. Wochenschr.*, 133 (Suppl. 1), 15–18.

Schubert, H. and Halama, P. (1993) Primary therapy-resistant depressive instability in elderly patients with cerebral disorders: Efficacy of a combination of *Ginkgo biloba* extract EGb 761 with antidepressants. *Geriatr. Forsch.*, 1, 45–53.

Schulz, H., Jobert, M., Breuel, H.-P. (1991) Wirkung von Spezialextrakt LI 1370 auf das EEG älterer Patienten im Schlafentzugsmodell. *Münch. Med. Wochenschr.*, 133 (Suppl. 1), 26–29.

Semlitsch, H.V., Anderer, P., Saletu, B., Resch, F., Presslich, O., Schuster, P. (1989) Psychophysiological research in psychiatry and neuropsychopharmacology: Part I: Methodological aspects of the Viennese Psychophysiological Test-System (VPTS). *Methods Find. Exp. Clin. Pharmacol.* 11, 25–41.

Shader, R.I., Harmatz, J., Salzman, C. (1974) A new scale for clinical assessment on geriatric populations: SANDOZ Clinical Assessment-Geriatric (SCAG). *J. Am. Geriat. Soc.*, 22, 107–113.

Stocksmeier, U. and Eberlein, M. (1992) Depressive Verstimmung bei Hirnleistungsstörungen, Wirkung eines *Ginkgo-biloba*-Extraktes in Doppelblindstudie überprüft. *TW Neurologie Psychiatrie*, 6, 74–76.

Taillandier, J., Ammar, A., Rabourdin, J.P., Ribeyre, J.P., Pichon, J., Niddam, S., Pierart, H. (1986) Traitement des troubles du vieillissement cérébral par l'extrait de *Ginkgo biloba*. Etude longitudinale multicentrique à double insu face au placebo. *Presse Méd.*, 15, 1583–1587.

Vesper, J. and Hänsgen K.D. (1994) Efficacy of *Ginkgo biloba* in 90 outpatients with cerebral insufficiency caused by old age. *Phytomedicine*, 1, 9–16.

Wechsler, D. (1944) *The measurement of adult intelligence.* Williams and Wilkins, Baltimore.

Weidenhammer, W. and Fischer, B. (1987) *c.I.-Skala.* Vless, Vaterstetten, Germany.

Weiss, H., Kallischnigg, G. (1991) *Ginkgo-biloba*-Extrakt (EGb 761): Meta-Analyse von Studien zum Nachweis der therapeutischen Wirksamkeit bei Hirnleistungsstörungen bzw. peripherer arterieller Verschlußkrankheit. *Münch. Med. Wschr.*, 133, 138–142.

Weitbrecht, W.U. and Jansen, W. (1986) Primär degenerative Demenz: Therapie mit *Ginkgo-biloba*-Extrakt. Plazebo-kontrollierte Doppelblind- und Vergleichsstudie. *Fortschr. Med.*, **104**, 199–202.

Wesnes, K., Simmons, D., Rook, M., Simpson, P. (1987) A double-blind placebo-controlled trial of Tanakan in the treatment of idiopathic cognitive impairment in the elderly. *Human Psychopharmacology*, **2**, 159–169.

World Health Organization (1992) *The ICD-10 classification of mental and behavioural disorders. Clinical descriptions and diagnostic guidelines.* World Health Organization, Geneva.

20. CLINICAL USES OF *GINKGO BILOBA* EXTRACT IN THE FIELD OF PERIPHERAL ARTERIAL OCCLUSIVE DISEASE (PAOD)

BERNARD BULLING[1], NORBERT CLEMENS[2] and VOLKER DANKERS[2]

[1]*Aachener Straße 312, D-50933 Köln, Germany*
[2]*Intersan GmbH, Einsteinstr. 30, D-76275 Ettlingen, Germany*

INTRODUCTION

In spite of the advances that have been made in synthetic organic chemistry, plant products still are an important part of modern therapy. Pharmacological and clinical investigations with a standardized extract (EGb 761) of dried *Ginkgo biloba* leaves have shown therapeutic effects on many diseases with impaired blood rheology. *G. biloba* improves the microcirculation and tissue nutrition in ischaemic areas due to the increase of arterial perfusion. *G. biloba* inhibits the aggregation of thrombocytes and erythrocytes and reduces the permeability of capillaries. Furthermore *G. biloba* releases the endothelial-derived-relaxing-factor and regulates the steady-state of prostacyclin and thromboxan (Anadere *et al.*, 1985; Artmann *et al.*, 1989, 1991; DeFeudis, 1991; Ernst and Marshall, 1992; Grebe *et al.*, 1991; Guinot *et al.*, 1989; Klein, 1988; Költringer *et al.*, 1989; Rudofsky, 1987; Schmid-Schönbein *et al.*, 1991; Witte *et al.*, 1992). Due to the positive effects on blood rheology *G. biloba* is used in therapy of peripheral arterial occlusive disease (PAOD). PAOD is a frequent disease and it will become more important with the increase of elderly people in the future. PAOD is defined as a clinical picture which is caused by arterial stenosis or occlusion of the extremities. In 90% of the cases stenosis or occlusion befall the lower limbs. Rarely localities of PAOD are the upper limbs and the shoulder. Arteriosclerosis is the main reason (90%) for this disease. The arteriosclerosis manifests itself also in the coronary and intra- and extracerebral arteries entailing coronary heart disease and low perfusion of the brain. Therefore patients with PAOD often also suffer from coronary heart disease and cerebral disorders due to low brain perfusion. Smoking, hypertension, diabetes mellitus and hyperlipaemia are the main risk factors which damage the endothelium of arteries and cause arteriosclerosis. PAOD is classified into four stages (Fontaine I-IV) according to the predominant symptoms. Table 1 shows the clinical criteria to classify the different stages of PAOD.

Therapeutic strategies in patients with PAOD generally aim to inhibit the progression of the disease, to reduce pain and to restore sufficient mobility of the patients. The most important therapeutical aim for patients with PAOD Fontaine's Stage III and IV is to preserve the lower limb from amputation. To reach these aims

Table 1 Classification of peripheral arterial occlusive disease (Fontaine I–IV)

Stage	Symptoms
Stage I	only objective disturbance of perfusion without subjective symptoms
Stage II	stress-pain, Claudicatio intermittens
Stage IIa	pain-free walking distance > 200 m
Stage IIb	pain-free walking distance < 200 m
Stage III	rest-pain
Stage IV	necrosis and gangrene

the treatment of PAOD conforms to the different stages and has to follow the individual clinical picture of each patient dependent on severity, localisation and haemodynamic compensation of stenosis or occlusion. Reducing or elimination of risk factors like smoking, hypertension, diabetes mellitus and hyperlipaemia are the most important parts of therapy in patients with PAOD Fonatine's stage I but also necessary in the other stages of PAOD. For patients with stages II-IV there are non-invasive therapeutic strategies like long-term physical exercise and drugs and invasive treatments like percutaneous transluminal angioplasty (PTA) and vessel surgery available. EGb 761 is used besides other impairments for improving the pain-free walking distance in PAOD (Fontaine's Stage II) within the framework of physical exercise. This indication is according to the German and French Federal Health Authority listings (monography E (1994) from former BGA, VIDAL (1996)). The therapeutic oral dose for this indication is 120–160 mg/day of the standardized extract. The duration of treatment should be at least 6 weeks to improve the walking distance. A summary of recent clinical trials with standardized *G. biloba* extract in the field of peripheral arterial occlusive disease (PAOD) follows. The clinical trials are ranked in order to the date of their publication.

SUMMARY OF THE CLINICAL TRIALS UNDERTAKEN IN THE FIELD OF PERIPHERAL ARTERIAL OCCLUSIVE DISEASE (PAOD)

Efficacy of *G. biloba* in Patients with PAOD Fontaine's Stages II and III

Courbier *et al.* (1977) investigated the PAOD Fontaine's Stages II and III in 40 male inpatients using an double-blind cross-over design. The patients suffered from chronical arterial occlusive disease already between two and four years in most cases. Surgical intervention was made in 30 of the 40 patients. The patients were randomized into two groups. One group received EGb 761 (160 mg/day) for one month and then for one month placebo. The other group started the treatment with placebo during the first month and received EGb 761 in the second month. The following investigations were made during the trial at the beginning, at the 30th and

Table 2   Development of walking distances under treatment with EGb 761 or placebo (Courbier *et al.*, 1977)

Treatment	Change in walking distance	Significance
EGb 761 first period 24 patients	+ 193 m	p < 0.001
Placebo first period 16 patients	+ 44 m	not significant
EGb 761 second period 16 patients	+ 19 m	not significant
Placebo second period 24 patients	− 12 m	not significant

the 60th day. The improvement of walking distances were measured by treadmill. The symptoms like paraesthesia, muscle convulsions, Claudicatio intermittens and rest pain were monitored. The walking troubles were reduced in 20 of 29 patients with Fontaine's stage II and in 6 cases of 11 patients with Fontaine's stage III. These results are summarized in Table 2.

Table 2 shows that there was in the first period of the treatment a difference of approximately 150 m in the walking distance between the EGb 761-treated and the placebo-treated group in favour of the EGb 761-treated group. In the second period the difference in walking distance amounted to 31 m. Summarizing the clinical trial it could be stated that treatment with EGb 761 improved walking distance in patients with PAOD Fontaine's Stage II and III.

Efficacy of *G. biloba* in Patients with PAOD Stages III and IV and Diabetes Mellitus

Le Devehat *et al.* (1980) conducted an open-trial with 28 patients suffering from PAOD Fontaine's Stage III or IV and diabetes mellitus. 13 of the 28 diabetic patients had an insulin dependent diabetes mellitus (IDDM). The patients received *G. biloba*-extract (EGb 761) over at least two months. At the beginning and at the end of the open trial rest blood flow and segmental perfusion pressure were measured. The clinical results showed especially an improvement of rest pain and tissue nutrition. Table 3 gives an impression of the results.

Efficacy of *G. biloba* in Patients with PAOD Fontaine's Stage IIb

Salz (1980) conducted a placebo-controlled, double-blind cross-over trial on patients with PAOD (Fontaine's Stage IIb). Initially 30 patients randomly assigned to two groups were treated after the following schedule. The first group received EGb 761 (160 mg/day) for six weeks and then placebo for six weeks, whereas the other group received first placebo and then EGb 761. Pain-free and maximum walking distance, Ratschow's leg-position test, standing on toes test and skin temperature

Table 3   Results of the treatment with *G. biloba* extract in patients with PAOD and diabetes mellitus (Le Devehat *et al.*, 1980)

	Number of cases	*Distinct improvement*	*Without change*	*Deterioration*	*Improvement*
Walking pain	15	7	8	0	46%
Rest pain	23	16	7	0	70%
Paraesthesia	17	8	9	0	47%
Cyanosis	7	5	2	0	71%
Tissue nutrition	12	8	4	0	67%
Peripheral pulses (A. fem., A. pop., A. tib. post.)	26	13	12	1	50%

and laboratory parameters were examined. Pain-free and maximum walking distance were always investigated under the same conditions (plane straight line, between 3 and 5 p.m., controlled speed of two steps per second by metronome). All parameters were measured at the beginning, after 6 and after 12 weeks of the clinical trial. Four patients dropped out during the trial so that the final analysis was made on 26 patients. Pain-free and maximum walking distance were improved in the group that was treated with EGb 761 during the first treatment period and then decreased under placebo. The group treated with placebo in the first phase showed constant or slightly decreased walking distances. The walking distances increased when this group was treated with EGb 761 in the second period of the crossover trial. EGb 761 delayed the moment when ischaemic pain occurred on Ratschow's leg-position test in the verum group. Additionally the redness of the skin and the refilling of veins after Ratschow's leg-position test were distinctly reduced in the EGb-treated group. The results of Ratschow's leg test and the changes of pain-free and maximum walking distance are summarized in the Tables 4 and 5.

The standing on toes test showed that time to onset of ischaemic pain was lengthened in the verum group. This result was statistically significant ($p < 0.001$). The skin temperature measured by an electronic thermometer at the middle calf improved significantly in the EGb-treated group relative to the placebo group. The investigated laboratory parameters did not show any significant changes during the clinical trial. Summarizing the clinical trial conducted by Salz *et al.* demonstrated that treatment with EGb 761 (160 mg/day) during 12 weeks improved pain-free and maximum walking distances in patients with PAOD (Fontaine's Stage IIb).

Bauer (1984) conducted a randomized, controlled double-blind trial in which EGb 761 (120 mg/day) was compared with placebo in the treatment of outpatients suffering from peripheral arterial occlusive disease (Fontaine's Stage IIb) of the lower extremities. The patients undertook a 6-week placebo wash-out period where the diet was adjusted and other vaso-active medication withdrawn. Patients showing an improvement of more than 30% in their walking distance or a walking distance of

Table 4   Results of the Ratschow's leg position test (Salz, 1980)

	Abs. change EGb 761	Rel. change EGb 761	Abs. change placebo	Rel. change placebo
Time till onset of pain on Ratschow's leg position test	+43 sec.	+42%	−32 sec.	−26%
Redness of skin after Ratschow's leg position test	−6 sec.	−35%	+4 sec.	+25%
Refilling of veins after Ratschow's leg position test	−7 sec.	−33%	+3 sec.	+18%

Table 5   Changes of pain-free and maximum walking distance (Salz, 1980)

	Abs. change EGb	Rel. change EGb	Abs. change placebo	Rel. change placebo
Pain-free walking distance	+92 m	+55%	−78 m	−31%
Maximum walking distance	+98 m	+48%	−103 m	−34%

more than 300 metres following the placebo wash-out period were not included in the study. In this clinical trial 80 patients were treated with EGb 761 120 mg/day or placebo orally over 6 months. 79 patients (44 from the EGb 761 group and 35 from the placebo group) were finally analysed because one patient had to be excluded. The maximum and pain-free walking distance (treadmill 3 km/h, 10% degree slope, 20° C room-temperature, measurement in the afternoon), the subjective estimation of pain and a veinplethysmographic measurement were investigated during the clinical trial. The pain-free and maximum walking distances were improved in both verum and placebo group. The increase of the walking distances was highly statistically significant relative to the start of the trial (pain-free walking distance $p < 0.05$; maximum walking distance $p < 0.001$) in the EGb 761-treated group. The increase in the average in pain-free walking distance was 41.4% in the placebo-treated group and 110% in the EGb 761-treated group. The increase in maximum walking distance was 16.5% in the placebo-treated group and 117.5% in the verum-treated group. The difference of the increase of both parameters between the EGb 761-treated group and the placebo-treated group was statistically significant in favour of the EGb 761-treated-patients. The flow rate of blood perfusion measured with the plethysmographic test increased in the EGb-treated group ($p < 0.01$) and decreased in the placebo-treated group. The

Table 6   Change of pathological values of HDL levels during the clinical trial (Bauer, 1984)

	Pathological values before trial	*Pathological values after trial*
G. *biloba*-treated group	93%	14%
Placebo-treated group	97%	23%

difference of the flow-rate between the verum and the placebo group was not statistically significant. During the clinical trial the patients registered their subjective pain in a vertical analogue scale of 10 cm. The evaluation of the scales showed that the pain decreased from 64.8 mm in the beginning to 43.3 mm at the end in the EGb 761-treated group (decrease of 21.5 mm). The corresponding values of the placebo-treated group were 59.6 mm and 53.7 mm (decrease of 5.9 mm). The difference in pain reduction was highly statistically significant between the verum and placebo group ($p < 0.001$) in favor of the EGb 761-treated group. Only HDL (high density lipo-proteins) levels showed important changes during the trial of the laboratory tests undertaken in the clinical trial. 93% of the G. *biloba*-treated patients had pathological values (out of normal range) of HDL at the beginning of the clinical trial versus 97% of the placebo-treated patients. After treatment only 14% of the verum treated patients and 23% of the placebo treated patients had still pathological values. These results are summarized in Table 6.

The trial showed that EGb 761 improved the pain-free and maximum walking distance in patients with PAOD of the lower extremities Fontaine's Stage IIb. EGb 761 was significantly superior to placebo.

Natali (1985) conducted a randomized, controlled double-blind trial in which EGb 761 (160 mg/day) was compared with placebo in the treatment of eighteen patients with PAOD (Fontaine's Stage II b) of the lower limbs. First all enrolled patients were treated with placebo in a 6-week preliminary phase to select patients whose walking distance did not increase by more than 300 meters. Then nine patients were treated with EGb 761 and nine patients received placebo over a period of 6 months. Table 7 shows the relative changes of pain-free and maximum walking distance after the six month treatment period. The difference of improvement between the EGb 761-treated and the placebo-treated group of both pain-free and maximum walking distance was statistically significant in favour of the EGb-treated patients ($p < 0.05$).

Table 7   Relative changes in pain-free and maximum walking distance (Natali, 1985)

Treatment	*Mean value of relative change in pain-free walking distance*	*Mean value of relative change in maximum walking distance*
EGb 761	101%	119%
Placebo	11%	4%

Bulling and Bary (1991) conducted a randomized, double-blind cross-over clinical trial in which 36 patients suffering from PAOD (Fontaine's Stage IIb) undergoing long-term physical training were treated with EGb 761 or placebo. In this clinical trial one of the aims was to determine whether EGb 761 could enhance the effects of long-term physical exercise. As a rule the 36 patients took part in physical training three times per week during 6 months. During this time the patients got either EGb 761 (160 mg/day) or placebo. The medical history, malleolus pressure, plasma viscosity, haematocrit, compliance, drug side effects, pain-free and maximum walking distance and the number of pain-free and maximum toe-stands were measured at certain points during the trial.

Some significant differences were found between the control-group and the EGb 761-treated group at the end of the clinical trial. The absolute increase in pain-free walking distance was 152.2 m in the EGb 761-treated group as compared to an increase of only 90 m in the placebo group. This difference was statistically significant ($p < 0.01$) in favour of the EGb 761 treatment and clinically relevant. The number of pain-free and maximum toe-stands increased from 27.9 to 38.6 (35%) and from 41.1 to 57.7 (39.9%), respectively, in the EGb 761-treated group, whereas in the placebo group the values increased from 28.9 to 32.5 (11.1%) and from 46.8 to 55.9 (19.5%) respectively. This was not statistically significant. Malleolus pressure, plasma viscosity and haematocrit did not show any changes during the clinical trial. Also no adverse effects occurred during the treatment period. The trial showed that patients with PAOD Fontaine's Stage IIb undergoing long-term physical exercise could profit from treatment with EGb 761 especially with regard to the improvement of pain-free walking distance.

Peters *et al.* (1995) conducted a randomized, placebo-controlled double-blind clinical trial to examine the efficacy of a standardized extract of *G. biloba* (EGb 761) on patients with peripheral arterial occlusive disease (Fontaine's Stage IIb). The clinical trial was conducted according to the "Guidelines for therapeutic studies on peripheral arterial occlusive disease in Fontaine stages II-IV", created by the German Association of Angiology (Heidrich *et al.*, 1992). After a run-in-period of two weeks the 111 enrolled patients were treated with EGb 761 (120 mg/day) for a period of 24 weeks. Before enrollment the peripheral arterial occlusive disease was confirmed with angiography. EGb 761 was taken by 53 patients of the 111 randomized patients. 58 patients were treated with placebo. The major part of the patients (96) were not undergoing physical exercise during the trial. The primary objective of the trial was to determine the difference in pain-free walking distance between the beginning of the study and after week 8, 16 and 24. The pain-free walking distance was measured with a treadmill (3 km/h and 12% slope).

The improvement of the pain-free walking distance was statistically significant in favour of the EGb-treated patients at every stage. The increase in pain-free walking distance amounted to 45.1 m in the verum-group, whereas the increase in the placebo-group was 20.9 m ($p = 0.014$). The pain-free walking distance was at the beginning of the study with 108.5 m (verum-group) and 105.2 m (placebo-group) similar in both groups. The results showed also that the verum-group performed significantly better than the placebo-group ($p < 0.05$) with regard to the parameters "difference of the maximal walking distance" and "relative increase of pain-free and maximum walking distance". Moreover the patients registered the change of their subjective complaints during the clinical trial. For this a visual analogue scale from −5 (complaints have increased strongly) to 5 (complaints have improved strongly) was used. The results of pain-free and maximum walking distance measured during the clinical trial are shown in Tables 8 and 9.

The evaluation of the analogue scale showed that the patients' opinion about the success of the treatment was a little better in the verum-group. The value was 1.3 in the verum-group and 1.2 in the placebo-group. The Doppler-indices (minimum-pressure A. tibialis posterior of the more impaired leg divided by maximum-pressure A. brachialis and maximum-pressure A. tibialis posterior of the less impaired leg divided by maximum-pressure A. brachialis) measured during the clinical trial were constant and showed no difference between the verum and the placebo-group. The pain-free

Table 8   Pain-free walking distance (m) during treatment period (mean and 95% confidence interval) (Peters *et al.*, 1995)

Point of time	EGb 761 (n = 53)*	Placebo (n = 58)*
Inclusion	108 (101, 115)	104 (97, 111)
Beginning of treatment	109 (101, 116)	105 (98, 112)
Week 8	129 (116, 143)	114 (105, 122)
Week 16	140 (128, 152)	122 (114, 131)
Week 24	153 (130, 176)	127 (117, 136)

*In each case one drop-out at the beginning of the treatment period. Hence the values after 8, 16 and 24 week represent 52 respectively 57 patients.

Table 9   Increase of pain-free walking distance (m) during treatment period compared to starting value (mean and 95% confidence interval) (Peters *et al.*, 1995)

Difference	EGb 761 (n = 52)	Placebo (n = 57)	P-value
8 weeks – beginning of treatment	21 (10, 33)	8 (3, 12)	0.012
16 weeks – beginning of treatment	32 (21, 42)	17 (11, 22)	0.006
24 weeks – beginning of treatment	45 (24, 66)	21 (14, 28)	0.014

walking distance increased from 108.5 m to 153.2 m in the verum-group. This was an increase of about 50%. The increase is in accordance with the increase regarded as a therapeutic success in the Guidelines of the German Association of Angiology.

The aim of the study conducted by Blume et al. (1996) was to investigate the efficacy of G. biloba extract in comparison to placebo on the walking distance in fully trained patients. This study was a monocenter, randomized, placebo-controlled, double-blind and parallel trial. Including only "well-trained" patients in the study avoided the usual training effect of about 20–30% increase in walking distance. In this study 30 patients were treated with G. biloba-extract (120 mg/day) and 30 patients with placebo for a period of 24 weeks.

The main outcome of this study was the difference in walking distance between the start of the treatment and after 8, 16 and 24 weeks of treatment measured on the treadmill (walking speed 3 km/h and slope of 12%). The absolute difference of maximum walking distance, the relative improvement of pain-free walking distance, the Doppler index and the subjective evaluation of the patients were also analyzed.

Both the pain-free and maximum walking distance improved significantly in the G. biloba-treated group. It is also interesting that about one-third of the placebo-treated patients had a shorter walking distance at the end of the study compared with the distance at the beginning. Tables 10 and 11 show the differences in pain-free and maximum walking distance after 8, 16 and 24 weeks.

Table 10   Difference of pain-free walking distance [m]. Median with 95% confidence interval in brackets. Missing-values were substituted by last-value-carried-forward-method. (Blume et al., 1996)

Difference	G. biloba	Placebo	P-value
Week 8 – beginning of treatment period	19 (14, 33)	7 (–4, 12)	< 0.0001
Week 16 – beginning of treatment period	34 (18, 50)	12 (5, 22)	< 0.0003
Week 24 – beginning of treatment period	41 (26, 64)	8 (–1, 21)	< 0.0001

Table 11   Difference of maximum walking distance [m]. Median with 95% confidence interval. Missing-values were substituted by last-value-carried-forward-method. (Blume et al., 1996)

Difference	G. biloba	Placebo	P-value
Week 8 – beginning of treatment period	22 (13, 29)	0 (–7, 10)	< 0.0001
Week 16 – beginning of treatment period	32 (20, 50)	8 (3, 14)	= 0.0005
Week 24 – beginning of treatment period	51 (39, 56)	16 (–1, 23)	< 0.0001

Table 12   Patients' subjective evaluation of treatment success. Median and 95% confidence interval (Blume *et al.*, 1996).

	G. biloba	Placebo	P-value
Visual analogous value at the end of the study	2.9 (2.7, 3.5)	0 (0, 0.4)	< 0.0001

The Doppler-indices (quotient of arm blood pressure and leg blood pressure) were used to judge the macrocirculation. There were no differences of the Doppler-indices between beginning and end of the treatment period in both verum-treated and placebo-treated group. This fact demonstrated that there was no improvement of macrocirculation. The patients evaluated their complaints on an analogue scale which reached from −5 "complaints have increased" to 5 "complaints have improved significantly" at the end of the study. Table 12 shows the results of this evaluation.

These results showed the efficacy of treatment with *G. biloba* on subjective parameters in addition to the improvement of pain-free and maximum walking distance. The Spearman's rank correlation coefficient of $r_s = 0.74$ (the Spearman's rank correlation coefficient is used to assess the relationship between two variables) showed a strong relationship between subjective evaluation of pain and the increase of pain-free and maximum walking distance. Two adverse events took place in the *G. biloba*-treated group in contrast to no adverse events in the placebo-treated group. Both observed adverse events (stomach ache, nausea) were not serious and disappeared after a few days without dechallenge of *G. biloba*. Therefore the causality between the adverse events and the *G. biloba* extract was unlikely. The clinical trial of Blume *et al.* (1996) showed that it is possible to improve both pain-free and maximum walking distance in well-trained patients with *G. biloba* therapy.

CONCLUSION

G. biloba improves the blood flow and inhibits the aggregation of platelets. It increases the flexibility of erythrocytes in acidosis. Effects like decrease of permeability of capillaries, release of endothelial-derived-relaxing-factor (EDRF) and regulation of the prostacyclin and thromboxan steady state contribute to the improvement of microcirculation. These essential effects to influence blood rheology are the main foundation of the efficacy of *G. biloba* in patients with PAOD (Fontaine's Stage II). Physical exercise is the basic therapy for PAOD Stage II. But only one third of the patients with PAOD Stage II are practising physical exercise. The study of Bulling and Bary (1991) shows that patients with PAOD Stage II who were undergoing long-term physical exercise could benefit from additional treatment with *G. biloba* especially with regard to the improvement of pain-free walking distance. The reasons for not practising physical exercise are concomitant diseases and

Table 13   Synopsis of the clinical trials undertaken with *G. biloba* in patients with peripheral arterial occlusive disease (PAOD)

Reference	Kind of study	Dosage and duration	Number of patients Placebo = Pl G. biloba = EGb	Results	Efficacy EGb superior to placebo p-value
Bauer 1984	double-blind, placebo-controlled, parallel	120 mg/day for 24 weeks	79 35 Pl 44 EGb	• Decrease of pain • Increase of basal perfusion of the lower limbs • Increase of pain-free walking distance • Increase of max. walking distance	< 0.001 n.s. < 0.05 < 0.01
Blume *et al.*, 1996	double-blind, placebo-controlled, parallel	120 mg/day for 24 weeks	60 30 Pl 30 EGb	• Increase of pain-free walking distance • Increase of maximum walking distance	< 0.001 < 0.001
Bulling and Bary 1991	double-blind, placebo-controlled, parallel	160 mg/day for 24 weeks and physical exercise	40 20 Pl 20 EGb	• Increase of pain-free walking distance • Increase of max. walking distance • Increase of pain-free toestands	< 0.01 n. s. n. s.
Courbier *et al.*, 1977	double-blind, placebo-controlled, cross-over	160 mg/day for 2 months	40 16 Pl 24 EGb 29 patients Fontaine's Stage II 11 patients Fontaine's Stage III	• Decrease of walking troubles (patients with Fontaine's Stage II) • Alleviation of nocturnal pain (patients with Fontaine's Stage III)	n. s. n. s.

Table 13   continued

Reference	Kind of study	Dosage and duration	Number of patients Placebo = Pl G. biloba = EGb	Results	Efficacy EGb superior to placebo p-value
Le Devehat et al., 1980	open trial	120 mg/day for 2 months	28 patients with distal diabetic arteriopathy	• Improvement of rest pain and nutrition troubles	no information
				• Improvement of ankle index	no information
Natali 1985	double-blind, placebo-controlled, parallel	160 mg/day for 6 months	18 9 Pl 9 EGb	• Increase of pain-free and maximum walking distances	< 0.05
Peters et al., 1995	double-blind, placebo-controlled, parallel	120 mg/day for 24 weeks	111 58 Pl 53 EGb	• Increase of pain-free walking distance	= 0.014
Salz 1980	double-blind, placebo-controlled, cross-over	160 mg/day for 12 weeks	29	• Increase of pain-free and maximum walking distances	< 0.001
				• Improvement of Ratschow's test	< 0.001
				• Increase of skin temperature	< 0.001
				• Decrease of lipids and cholesterol	n. s.

lack of time. Therefore a therapy with a vasoactive drug is vindicated in the remaining two third of patients with PAOD Stage II. The efficacy of G. *biloba* has been shown in many placebo-controlled and double-blind clinical trials. The treatment with G. *biloba* improves significantly both the pain-free and maximum walking distance compared with placebo. The significant increase of walking distance is also clinically relevant because the increase under the conditions of the clinical trial can be multiplied by a factor 2–3 for everyday life. The mentioned clinical trials are summarized in Table 13.

In the future it is necessary to conduct more clinical trials to underline the additional benefit of G. *biloba* treatment in patients with PAOD undergoing longterm physical exercise. These trials should be conducted according to the "Guidelines for therapeutic studies on peripheral arterial occlusive disease in Fontaine stages II-IV", created by the German Association of Angiology.

REFERENCES

Anadere, I., Chmiel, H., Witte, S. (1985) Hemorheological findings in patients with completed stroke and the influence of a *Ginkgo biloba* extract. *Clin. Hemorheol.*, **5**, 411–20.

Artmann, G., Degenhardt, R., Wolff, H., Grebe, R., Schmid-Schönbein, H. (1991) Pilotstudie über membranpharmakologische Wirkungen von *Ginkgo-biloba*-Extrakt: Parameter-Exploration in vitro und Reduplikation der Effekte nach enteraler und parenteraler Zugabe von Rökan. In F.H. Kemper und H. Schmid-Schönbein, (Eds.), *Rökan–Ginkgo biloba EGb 761–Pharmakologie* , **Bd 1**, Springer, Berlin, pp. 47–62.

Artmann, G., Michaelis, P., Schmid-Schönbein, H. (1989) Effect of *Ginkgo biloba* extract 761 on microrheological parameters of red blood cells. *Clin. Haemorheol.*, **9**, 444.

Bauer, U. (1984) 6-month double-blind randomized clinical trial of *Ginkgo biloba* extract versus placebo in two parallel groups in patients suffering from peripheral arterial insufficiency. *Arzneim.-Forsch./Drug Res.*, **34**, 716–720.

BGA (Bundesgesundheitsamt) (1994) Monographie: Trockenextrakt (35–67:1) aus *Ginkgo-biloba*-Blättern extrahiert mit Aceton-Wasser, Bundesanzeiger 133 vom 19.07.1994.

Blume, J., Kieser, M., Hölscher, U. (1996) Placebokontrollierte Doppelblindstudie zur Wirksamkeit von *Ginkgo-biloba*-Spezialextrakt EGb 761 bei austrainierten Patienten mit Claudicatio intermittens. *Vasa*, **25**, 265–74.

Bulling, B. and Bary, S. (1991) Behandlung der chronischen peripheren arteriellen Verschlußkrankheit mit physikalischem Training und *Ginkgo-biloba*-Extract 761. *Med. Welt*, **42**, 702–708.

Courbier, R., Jausseran, J.M., Reggi, M. (1977) Étude à double insu croisée du Tanakan dans les artériopathies des membres inférieurs. *Mediterr. Méd.*, **126**, 61–64.

DeFeudis F.V. (1991) *Ginkgo biloba extract (EGb 761): Pharmacological activities and clinical applications*, Elsevier, Paris.

Ernst, E. and Marshall, M. (1992) Der Effekt von *Ginkgo biloba* -Spezialextrakt EGb 761 auf die Leukozytenfilterabilität–Eine Pilotstudie. *Perfusion*, 8, 241–242.

Grebe, R., Artmann, G., Wolff, H., Degenhardt, R., Schmid-Schönbein, H. (1991) Hochauflösende mikrorheologische Methoden zum Studium pharmakodynamischer Wirkungen von *Ginkgo-biloba*-Extrakt an menschlichen Erythrozyten. In F.H. Kemper und

H. Schmid-Schönbein (Eds.) *Rökan–Ginkgo biloba EGb 761–Pharmakologie*, **Bd 1**, Springer, Berlin, pp. 31–45.

Guinot, P., Caffrey, E., Lambe, R., Darragh, A. (1989) Tanakan inhibits platelet-activating-factor induced platelet aggregation in healthy male volunteers. *Haemostasis*, **19**, 219–33.

Heidrich, H., Cachovan, M., Creutzig, A., Rieger, H., Trampisch, H.J. (1995) Guidelines for therapeutic studies in Fontaine's Stages II-IV peripheral arterial occlusive disease. *Vasa*, **24**, 114–119.

Klein, P. (1988) Untersuchung über die Hemmwirkung von *Ginkgo-biloba*-Extrakt. *Therapiewoche*, **38**, 2379–2383.

Költringer, P., Eber, O., Lind, P., Langsteger, W., Wakonig, P., Klima, G., Rothlauer, W. (1989) Mikrozirkulation und Viskoelastizität des Vollblutes unter *Ginkgo-biloba*-Extrakt. Eine plazebokontrollierte, randomisierte Doppelblind-Studie. *Perfusion*, **1**, 28–30.

Le Devehat, C., Lemoine, A., Zoubenco, C., Cirette, B. (1980) Étude du Tanakan dans les artériopathies diabétiques distales. Étude critique des explorations fonctionnelles vasculaires. *Mise Jour Cardiol.*, **9**, 463–471.

Natali, J. (1985) Évaluer l'efficacité de l'EGB dans l'artériopathie des membres inférieurs au stade IIb (périmètre de marche compris entre 50 et 300 m). Expertise clinique, Ipsen, Paris.

Peters, H., Kieser, M. and Hülscher, U. (1995) Proof of efficacy of *Ginkgo biloba* special extract EGb 761 in intermittent claudication – A multicentric, placebo-controlled double-blind study. Draft manuscript.

Rudofsky, G. (1987) Wirkung von *Ginkgo-biloba*-Extrakt bei arterieller Verschlußkrankheit. *Fortschr. Med.*, **105**, 397–400.

Salz, H. (1980) Zur Wirksamkeit eines *Ginkgo biloba*-Präparats bei arteriellen Durch-blutungsstörungen der unteren Extremitäten. Kontrollierte Doppelblind-crossover-Studie. *Ther. Geg.*, **119**, 1345–1356.

Schmid-Schönbein, H., Artmann, G., Degenhardt, R. (1991) In-vitro-Nachweis der Radikal-Fänger-Effekte von *Ginkgo-biloba*-Extrakt durch das System der photometrischen Viskoelastometrie humaner Erythrozyten. In F.H. Kemper und H. Schmid-Schönbein (Eds.) *Rökan–Ginkgo biloba EGb 761–Pharmakologie*, **Bd 1**, Springer, Berlin, pp. 103–111.

Vidal® 1996, Editions du Vidal, Paris, pp. 1546–1547.

Witte, S., Anadere, I., Walitza, E. (1992) Verbesserung der Hämorheologie durch *Ginkgo-biloba*-Extrakt. *Fortschr. Med.*, **110**, 247–250.

21. EFFICACY OF *GINKGO BILOBA* SPECIAL EXTRACTS – EVIDENCE FROM RANDOMIZED CLINICAL TRIALS

MARTIEN C.J.M. van DONGEN[1], ERIK van ROSSUM[1] and PAUL KNIPSCHILD[2]

Dept. of Epidemiology (1) and Dept. of General Practice (2), Maastricht University, P.O. Box 616, 6200 MD Maastricht, The Netherlands

INTRODUCTION

Since their introduction on the market in the mid 1960s, the use of phyto-pharmaceuticals based on *Ginkgo biloba* extract has become increasingly popular in several West-European countries and, more recently, in the USA as well. Positive clinical results of "Ginkgo" treatment have been claimed for patients suffering from a wide variety of health problems, such as peripheral arterial insufficiency, inter-mittent claudication, Raynaud's syndrome, diabetic retinopathy, vertigo, tinnitus, cochlear deafness, senile macular degeneration, erectile dysfunction, cognitive impairment and dementia (Fünfgeld, 1988; DeFeudis 1998). In spite of the quite heterogeneous appearance of this list of medical conditions for which the use of Ginkgo preparations is sometimes indicated, most of them are thought to have a common, vascular, origin: an impaired blood supply of the affected organ or tissue. The aim of this chapter is to review the evidence regarding the clinical efficacy of *Ginkgo biloba* extract that is currently available in the scientific literature.

It was decided to restrict the scope of the chapter in several ways. Firstly, the review will deal exclusively with those intervention studies that have been carried out in human patients. It will not cover the evidence gathered through animal studies or *in vitro* investigations. In general, studies involving healthy volunteers will not be taken into consideration either. Secondly, only those studies will be addressed that focus predominantly on "clinical" endpoints, such as general well-being, clinical signs and symptoms, functional capacity (assessed either in daily life situations or under laboratory conditions), health complaints, etc. Studies with an exclusive focus on metabolic or physiological effects of *Ginkgo biloba* products and constituents will be left out from the review, as their primary interest is in the dis-closure of the mechanisms of action underlying the clinical effects of Ginkgo, not in the clinical efficacy as such. Thirdly, the review concentrates on published evidence from investigations designed to exert a sufficient amount of control over potential sources of bias. This means that only the results of randomized controlled trials will be presented and discussed, and that the evidence based on, for instance, case series, open trials, and other types of design, will be neglected. A final restriction concerns the type of intervention: the review includes only studies conducted to assess the efficacy of a *Ginkgo biloba* special extract (GBSE). These special extracts contain fixed amounts of the substance classes that are held responsible for its clinical efficacy (usually: 24–25% Ginkgo flavonol glycosides and 6% terpene trilactones), obtained through a highly standardized and sophisticated process of extraction, purification and concentration (Sticher, 1992).

The chapter is structured as follows. Firstly, results and experiences from previous Ginkgo reviews will be summarized briefly. Secondly, the strategy that was chosen for the current review will be explained. Subsequently, a systematic overview of the published intervention studies on the efficacy of *Ginkgo biloba* and their results will be presented. For this purpose the clinical trials identified were grouped into three main categories, that cover almost the entire domain of interest: trials on patients suffering from peripheral arterial disease; trials on patients with cerebral insufficiency and related disorders (including Alzheimer's disease and vascular dementia); and a rest category, dealing with miscellaneous health problems (e.g. mountain sickness), that do not fit well in either the cerebral insufficiency or the peripheral arterial disease cluster.

PREVIOUS LITERATURE REVIEWS OF *GINKGO BILOBA*

Types of Literature Review

Several approaches can be chosen to review the published evidence for the efficacy of a medical intervention (Chalmers and Altman, 1995; Cooper and Hedges, 1994; Petitti, 1994). Often a renowned subject matter expert is invited to formulate a summary judgment regarding the efficacy and the safety of the intervention. In general, such a narrative review bears the stamp of the authority and the personal preferences of the reviewer. It often lacks explicit statements regarding the searching strategy applied, the inclusion and exclusion criteria used to select studies, and the main arguments giving rise to the overall conclusion. In order to circumvent the limitations inherent to this classical approach of reviewing, more systematic and explicit approaches have been developed during the past two decades. The criteria-based systematic review and the quantitative meta-analysis are two modern types of systematic literature review that have become popular in the medical domain, not at least due to the worldwide promotional efforts of the Cochrane Collaboration (Cochrane Collaboration, 1996; Mulrow and Oxman, 1994).

The criteria-based systematic review – or qualitative review – stresses the need for a critical and structured methodological assessment of each study that qualifies for the review, preferably in a "blinded" fashion. Sometimes this assessment is backed up by a scoring system that is suited to value each of a series of relevant method-ological quality aspects separately, and to calculate a summary quality score for the study at large. The idea behind this approach is that methodological quality should play a major role in weighing the evidence of each of the published studies that are available to base an overall judgement upon.

The quantitative meta-analysis honours the principle of statistical pooling of the results of several intervention trials addressing the same research question. Statistical pooling allows for an overall estimation of the size of the intervention effect which is more precise than the effect estimations for each of the studies separately.

Efforts to extract a valid summary judgment from the literature may be hampered in several ways. First of all, selective publication of empirical evidence and selective retrieval of published evidence may result in biased conclusions, irrespective of the type of systematic review. Heterogeneity as to methodological quality of a series of studies addressing the same research question, is a second obstacle. This problem is best addressed by means of a criteria-based systematic review. Heterogeneity of studies with regard to various design aspects that not necessarily affect the method- ological quality – e.g. choice of study subjects, choice of type and duration of inter- vention, choice of outcome measures – can be regarded as a third sort of problem, which especially interferes with statistical pooling.

Ginkgo biloba Reviews

Several reviews assessing the overall effect of Ginkgo biloba in patients with various health problems have been published in the past. The majority of them can be classified as narrative reviews, although some tend to a systematic approach with a strong narrative component (Ernst, 1996; Herrschaft, 1992; Kanowski, 1997; Letzel et al., 1996; Letzel and Schoop, 1992; Schulz and Hänsel, 1996; Schulz et al., 1997; Warburton, 1986; Weiß and Kallischnigg, 1991).

Kleijnen and Knipschild (1991, 1992ab) published a criterion-based systematic review, dealing with both patients suffering from cerebral insufficiency and patients diagnosed with intermittent claudication. For the methodological assessment of the trials eligible for the review, they applied a checklist of 7 criteria, which referred to the adequacy of patient characteristics (10 points), number of patients analysed (30 points), the randomization procedure (20 points), the intervention procedure (5 points), double blinding (20 points), outcome measurement (10 points), and data presentation (5 points), respectively. Only 2 of the 15 trials that could be retrieved for the intermittent claudication review, were classified as "acceptable", with a total quality score ⩾ 65 points, viz. 75 and 66 points, respectively (Bauer, 1984; Saudreau et al., 1989). All studies, irrespective of their methodological quality, indicated a positive effect of Ginkgo, which was specified as statistically significant for 9 studies, and as a positive trend for 6 studies (positive effect, but not statistically significant for all outcome measures). Of the 40 studies included in the cerebral insufficiency review only 8 were considered to have an "acceptable" methodological quality, with a total score ⩾ 65 points (Brüchert et al., 1991; Eckmann, 1990; Haguenauer et al., 1986; Meyer, 1986b; Schmidt et al., 1991; Taillandier et al., 1986; Vorberg et al., 1989; Wesnes et al., 1987). No less than 39 studies indicated a positive effect of Ginkgo, which was found to be statistically significant for 26 studies, and a positive trend for 13. Therefore, study outcome appeared not to be clearly associated with methodological quality. From the score matrices an insufficient sample size, an inadequately described randomization procedure and a poor description of patient characteristics and outcome measures can be identified as the main methodological shortcomings.

Statistical pooling of the results of *Ginkgo biloba* trials is discouraged by the remarkable heterogeneity with regard to relevant design aspects. This heterogeneity concerns in particular the patient definitions used (various diagnostic screening tools, various criteria for inclusion and exclusion) and the outcome parameters applied (different levels of observation, various rating scales, psychometric tests and other outcome measures). The intervention procedures and contrasts tend to show more similarity. Schneider (1992) published a meta-analysis based on the pooled analysis (calculation of effect sizes) of 5 randomized controlled trials including patients with peripheral arterial disease. And Hopfenmüller (1994) pooled the results of several trials which focussed on patients with cerebral insufficiency. These quantitative meta-analyses illustrate that statistical pooling is feasible even for Ginkgo trials. However, this can only be attained at the cost of the exclusion of much valuable evidence embedded in trials that are regarded not admissible to the "pool", due to unconciliatory design choices. For instance, while the review by Kleijnen and Knipschild (1992b) contained 40 "cerebral insufficiency" trials, only 11 were considered eligible for the quantitative meta-analysis conducted by Hopfenmüller (1994), which concentrated on the *Ginkgo biloba* special extract LI 1370 (Kaveri® forte) and on a subjective, clinical symptom checklist as the main outcome measure. Eight trials qualified for the analyses based on the evaluation of single symptoms of cerebral insufficiency and the total score for clinical symptoms. Statistically significant effects were found for all symptoms separately and in combination (total score), except for one trial (Hartmann and Frick, 1991). This indicates the superiority of Ginkgo over placebo. After dichotomization of the total scores ("improved" vs. "not improved") a pooled odds ratio of 2.0 (95% confidence interval: 1.4 ; 2.6) was calculated. Six trials could be used for the meta-analysis based on global assessment of the treatment effect, either by the physician or by the patient, as the main outcome measure. After dichotimization ("very good" or "good" vs. "moderate" or "unsatisfactory" or "no effect") a pooled odds ratio of 2.9 (95% confidence interval: 2.1 ; 3.7) was calculated. These data suggest that the rate of success for Ginkgo treatment is about two to three times that for placebo treatment.

Actually, all reviews referred to above seem to agree upon the beneficial effect of Ginkgo in patients suffering from cerebral insufficiency or peripheral arterial disease.

DESIGN OF THE CURRENT REVIEW

Review Strategy: General Considerations

For the current review neither the principles of criterion-based quality assessment, nor the rules of statistical pooling were applied. Instead a middle course was followed between the narrative approach and the criterion-based quality assessment approach. Through several tables, organized according to the type of health problem of interest, relevant information will be presented on crucial design aspects as well as on the numerical and statistical results for each *Ginkgo biloba* intervention trial which fulfilled the selection criteria mentioned in the introduction of this chapter. Each table will be accompanied by an overall judgement of the meaningfulness of the studies enclosed. For some trials it was difficult to assess to what extent the criteria for inclusion in the review were actually met, due to a lack of specific information in the text of the publication. Those studies were given the benefit of the doubt and are presented in the overview. The considerations which motivated the current approach will be explained below.

Completeness and transparancy of study reports

Many studies published in the remote past suffer from a rather poor reporting style (non-transparant structure, essential information lacking). This impedes a critical reviewer to distinguish between a poor design and conduct of the trial at the one hand, giving rise to invalid results, and a poor description of valid results at the other hand. One should not blame the responsible authors too much for this, as the criteria for reporting clinical trials in general and trials in the field of cognitive functioning and dementia in particular, have evolved gradually during the last few decades (ICH, 1995). Nevertheless, this circumstance seriously hampers any attempt to perform a meaningful criterion-based methodological quality assessment.

Provided that sufficient information is forwarded through a well-structured publication that meets the current reporting standards, the methodological assessment of most of the design aspects, such as random allocation, placebo-intervention, and blinded outcome measurement, is rather straightforward and not that complicated. For *Ginkgo biloba* trials the trouble starts as soon as the patient definition (diagnostic criteria used for the selection of study subjects) and the outcome measurement (evaluation criteria used for identifying a treatment effect) have to be assessed. These problems are more pronounced for clinical Ginkgo studies focussing on cerebral insufficiency than, for instance, peripheral arterial disease.

Diagnostic uncertainties

Before the methodological problems at the outcome measurement level are discussed, it is necessary to elaborate a bit more on the diagnostic difficulties and uncertainties which surround the identification of the target conditions for Ginkgo treatment. Most of the Ginkgo trials in the past, mainly from Germany and France, were conducted in patients suffering from cerebral insufficiency. Cerebral insufficiency and related concepts, such as cerebrovascular insufficiency and brain organic psychosyndrome, have been regarded as a disease entity for quite a long time. The use of these terms and concepts reflected the contemporary etiologic hypothesis that stenotic vascular changes, related to the ageing process, may cause a progressive decrease in cerebral blood flow, leading to a decline in mental and physical functioning. The clinical manifestations of this pathophysiological process were thought to encompass impairment of memory and other cognitive functions, affective symptoms such as anxiety and depression, and physical complaints such as tinnitus, vertigo, and headache (Schulz and Hänsel, 1996; Schulz et al., 1997). During recent years this concept of cerebral insufficiency has become more or less obsolete, especially after one came to recognize that neural degeneration as in Alzheimer's disease is a more frequent cause of cognitive impairment in elderly patients than vascular deficiency. The clinical features of these central nervous system disorders correspond to the syndrome of dementia, a pattern of disturbance in which several higher mental functions are affected simultaneously, with impairment of memory, abstract thinking, and psychomotor functions like speech as crucial symptoms, and irreversible changes in mood, social functioning, and personality as accompanying phenomena (Schulz and Hänsel, 1996; Schulz et al., 1997). The criteria for the differential diagnosis of dementia, dementia subtypes – Alzheimer's disease (AD) due to an idiopathic degeneration of the central nervous system, vascular dementia (VD; previously called multi-infarction dementia (MID)) due to changes in the vascular system, mixed forms of AD and VD, dementia due to other causes, such as AIDS, brain tumours, etc. – and related diseases have evolved steadily during the past 15 years (APA, 1987; APA, 1994; CBO, 1997; McKhann et al., 1984; Small et al., 1997; WHO, 1991).

To satisfy the need for disease classification within the domain adjacent to dementia and obviously not covered by the diagnostic criteria for dementia, the concept of age-associated memory impairment (AAMI) has been proposed. This concept refers to a pre-dementia state of impairment of memory function and other cognitive functions (language, orientation, constructional abilities, abstract thinking, problem solving, praxis), established both by means of psychometric testing and subjective reporting, and not yet accompanied by impairment of social functions (occupational or social performance). Although AAMI and comparable concepts – like cognitive impairment no dementia (CIND) and age-associated cognitive decline (AACD) – have always remained controversial as meaningful diagnostic entities, formal criteria to identify cases affected by this syndrome have been suggested at least (Crook et al., 1986; Crook et al., 1990). Eventually the concept of age-associated cognitive decline has acquired an autonomous position in the DSM-IV classification system (APA, 1994).

Given these nosological developments, it is not surprising that nowadays a more specific and focussed diagnosis is required for those disorders that were once classified under the heading "cerebral insufficiency". Unfortunately, the majority of the Ginkgo trial reports that can be retrieved from the literature, do not allow for such a differentiation. Again, it would not be fair to blame the investigators, as most of the trials have been completed or initiated at a time that the new criteria, guidelines and screening tools for such a differential diagnosis were still awaiting approval and publication. Even nowadays the diagnosis of cognitive neurological disorders is surrounded by much uncertainty and dissent. In a recent review, which compared the designs and the results of 25 Ginkgo trials in cerebral insufficiency with those of two other active substance classes (tacrine, nimodipine), Letzel et al. (1996) concluded that only a few recent Ginkgo trials fulfilled the strictest criteria which nowadays apply to the diagnosis of dementia. At the same time the authors gave several arguments that seem to justify a rather indulgent stand against the design choices made in Ginkgo trials conducted in the past (Letzel *et al.*, 1996). Nevertheless, both a criterion-based systematic review and a meta-analysis based on statistical pooling would be seriously hampered by these diagnostic problems.

Outcome measurement uncertainties

Analogous problems to those complicating the diagnosis of dementia and related disorders have to be coped with in the outcome measurement domain. To evaluate the efficacy of Ginkgo treatment most of the trials from the past have relied on a wealth of different outcome measures, covering various levels of functioning (general well-being, cognitive functioning, daily life activities, psychopathology, psychometric capacities, clinical signs and symptoms, etc.). Most of the trials have applied all these outcome measures without a predefined hierarchy. Again, the former researchers can hardly be criticized for this. Only quite recently stricter requirements have been formulated for the evaluation of nootropics and anti-dementia drugs, as far as the outcome measurement and evaluation of efficacy of these drugs are concerned (Amaducci *et al.*, 1990; Bundesgesundheitsamt, 1991; CPMP, 1992; CPMP, 1997; Menges, 1992). These regulations and guidelines, both at the national and the international level, are revised and adapted periodically. They state, for instance, that the outcome measures and the success criteria derived from these outcome measures should be chosen in advance of the trial onset, i.e. should be included in the trial protocol. Success criteria should not only be formulated in terms of statistical significance (hypothesis testing), but also in terms of clinical relevance (practical meaning of absolute changes and differences in parameter values). The assessment of the ability of interventions to slow down, stabilize, or improve cognitive impairment should be investigated at least at 3 complementary levels: 1. psychopathology (presence of clinical signs and symptoms); 2. cognitive functioning (psychometric testing); 3. behaviour (reflection of cognitive functioning in daily life activities, social functioning, etc.). Moreover, it is required to make a distinction between primary or main, and secondary outcome parameters, and, connected with this, between a confirmatory and a descriptive analysis of the data collected during a trial. A responder analysis should be considered as well. Just as for the diagnosis of dementia, the review of Letzel *et al.* (1996) demonstrates also for the outcome evaluation that only a few recent trials have been able to keep pace with the evolving standards and requirements. Again, this circumstance seriously hampers the conduct of a systematic review.

Strategy for Searching the Literature

The following sources of information were consulted to identify relevant Ginkgo trials:

1. Searching of electronic bibliographic databases: Medline, 1966–1997 (National Library of Medicine; Index Medicus; by means of SilverPlatter 3.0; emphasis on publications in medical journals); Embase, 1974–1997 (Excerpta Medica; emphasis on publications in medical journals); PsycLIT, 1972–1997 (by means

of SilverPlatter 3.0; emphasis on publications in the behavioural sciences); Current Contents, 1997–1998 (on line, by means of CC Search (R))

2. Database of literature on *Ginkgo biloba*, present at Maastricht University, Department of Epidemiology, which was composed for a previous review on the efficacy of treatment with *Ginkgo biloba*, and which has been updated afterwards.

3. Systematic hand-searching of conference proceedings and abstracts relevant to dementia (dementia, geriatric, psychiatric, psychogeriatric, neurology, and stroke conferences) or phytotherapy.

4. Checking the contents (hand-searching) of several journals, especially Germany and France based medical journals.

5. Checking references extensively in (review) articles on clinical research and in textbooks.

6. Personal communications with acknowledged Ginkgo investigators and with the major pharmaceutical companies involved in manufacturing and trading *Ginkgo biloba* preparations.

The free-text search of the electronic bibliographies Medline, Embase and PsychLIT was based on a series of keywords which referred to the therapeutic intervention of interest: ginkgo, gingko, gingkco, ginko, gingho, tebonin, tebonine, kaveri, tanakan, rökan, egb, egb-761, ginkobene, ginkgolip, ginkgold, ginkgolic, ginkgoextrakt, gincosan, ginkoba, ginkgolide, ginkgolide-b, ginkgoliden, ginkgolides, ginkgolids, gingkolide, gingkolides, ginkolide, bilobalid, bilobalide, bilobalids. Due to the manageable number of hits – e.g. less than 600 for Medline – there was no need for further restriction by combining the intervention key words with searching terms referring to either the health problem (e.g. dementia, memory impairment, peripheral arterial disease), or the study method of interest (e.g. clinical trial, randomized controlled trial).

The need for alternative searching strategies, in addition to the use of electronic or printed bibliographies, has been illustrated previously. Kleijnen and Knipschild (1992c) performed a kind of sensitivity analysis, which revealed that less than half of trials that finally were included in their Ginkgo review, were directly retrieved by means of Medline or Embase. Incorrect or unusual spelling (e.g. Ginkgco) was one of the causes of the high false-negative rate.

Structure of the Tables in the Current Review

The structure of the tables which were composed to summarize the design and the results of the intervention trials retrieved for the current review, will be explained in this section. The information contained by each of the seven tables was organized under 8 different headings.

Column 1

Column 1 shows the names of the author(s) and the year of publication of the trial report. Within each table the trials are arranged in a chronological sequence (year of publication).

Column 2

Column 2 shows the key information regarding the patient population in each trial: diagnostic criteria, and a summary of the main disease characteristics (stage, duration, treatment history, etc.). Some additional information on the composition of the study population may be listed in column 8, under the heading "Remarks".

Column 3

Column 3 shows the total number of randomized study subjects (first row), and the numbers allocated to each of the intervention subgroups (second row). For cross-over trials the second row remains empty, as each study participant has been exposed to both the experimental drug and the reference drug. If the numbers of subjects included in the analysis differed from the number randomized, e.g. due to dropping out, the total and treatment-specific numbers of analysed persons are mentioned on the third and fourth rows of the same field, provided that this information could be retrieved from the trial report. The analysis may have been based on either the "intention-to-treat" (I-to-T) or the "per-protocol" (= "valid-cases") principle. For some studies additional information was entered into column 8 of the table. Some trials adhered to a multi-comparison design, e.g. contained both a placebo control group and a reference drug control group, in addition to the Ginkgo group. In such cases only the information concerning the verum-placebo part of the trial will be presented. In this column as well as in other columns the notation "n.d." (= not described) is used to indicate that the relevant information could not be retrieved from the published data.

Column 4

Column 4 provides some elementary information regarding the age and gender composition of the study population. The age range of the persons admitted to the trial and/or the mean (or median) age for the undivided study population are specified in the first row. The sex ratio is shown in the second row, preferably for the randomized population, and otherwise for the analysed population. Occasionally, information regarding the distribution according to some other very important group characteristic, e.g. Alzheimer's dementia vs. vascular dementia, is entered under this heading as well.

Column 5

Column 5 deals with the study treatment: product name (brand name and/or special extract code name), route of administration, daily dosage and duration of the intervention.

  Ginkgo biloba extract can be administered in various ways. Most of the times it is supplied per os, either in a solid form (film-coated tablets, dragees, capsules), or as a solution (drops). However, intravenous or intramuscular injection, and intravenous infusion, are alternative options, especially in clinical settings. Tablets and solutions may contain various amounts of the extract that encompasses the active principles. Eventually, the effect is assumed to be dependent on the total daily dose and its time-related distribution (number and timing of intakes). Regular dosage schedules for Tebonin® tablets – Tebonin® is one of the most popular Ginkgo brand names – are, for instance: 2 or 3 film-coated tablets a day (1 tablet = 40, or 80, or 120 mg of active substance (dry extract; input : output ratio = 50 : 1)). In the past a total daily amount of 80–160 mg special extract, usually corresponding with 1–4 tablets, was regarded to be sufficient, but recently a tendency can be noticed to elevate this dose, especially within the context of clinical research.

Preferably, the quantity supplied on a daily base, expressed in mg of active substance, is given in the table, to indicate the level of exposure to the study medication. If this figure was not available, the next best information is given (e.g. ml of solution per day or number of drops per day). The dosage scheme is not further specified (e.g. time-related distribution of the daily dose). The last row shows the duration of the intervention, which covers the period from onset (baseline-measurement, after randomization) until final, blinded outcome measurement. In case of a crossover trial the period of Ginkgo treatment is presented. The length of treatment is expressed in either hours (h.; so, registration of an acute intervention effect), days (d.), weeks (w.), or months (m.). The intervention contrast – GBSE vs. either placebo, or a reference drug, or no treatment at all – is specified in column 8. In a few studies Ginkgo was prescribed as an adjuvans therapy to support another therapy. If this was the case the standard therapy is mentioned as well.

Column 6

Column 6 contains information on the outcome measures in each study. A wealth of different endpoints have been chosen to evaluate Ginkgo effects. In particular this is true for the trials dealing with cerebral insufficiency. These outcome measures cover several levels and domains of achievement: global assessment of (change in) health and cognitive state; assessment of clinical signs and symptoms (presence, severity); assessment of functional signs and symptoms; cognitive rating scales; measurement of functional capacity under laboratory conditions (psychometric and physical testing, e.g. aspects of memory function, attention, reaction time, specific parameters of information processing in the brain, concentration, coping with interference, visuo-spatial orientation); neurophysiological and neuroanatomical testing (e.g. EEG-patterns, electronystagmography, craniocorpography); measurement of functional capacity in daily life; and behaviour.

Only a minority of the study reports communicated an explicit hierarchy of endpoints. If so, this structure is conserved in the tables, for instance by distinguishing primary and secondary outcome measures (printed in italics). Otherwise, only the most relevant data are selected for presentation, according to the following priority setting: global assessment; general well-being, quality-of-life assessment; behaviour, daily life activities and functioning; psychometric and neuropsychological test results (functional capacity under laboratory conditions); clinical signs and symptoms; neurophysiological test results; physical signs and symptoms; and physiological and biochemical laboratory values.

With regard to the questionnaires, clinical rating scales and psychometric tests, the tables provide information on both the generic typing and the specific format ("brand name") (e.g.: depression, Hamilton Depression Scale). In order to facilitate the interpretation of the outcomes recorded, for the most frequently used outcome measures some information is added on scale or test characteristics, such as dimensions and subscales, and range of scores (underlining of one extreme value indicates the most favourable result). For subjective outcome measures the source of information is specified (patient, physician, caregiver, relative, investigator).

In most of the trials, especially the recent ones, much information on safety and tolerance aspects was collected, in addition to efficacy. This was often done in a systematic and extensive fashion (monitoring of adverse events and serious adverse events). The same holds for information on compliance and adherence to the treatment protocol, withdrawal, missing values of outcome parameters, etc. This type of information is not given in the tables. In general, one agrees that Ginkgo is very safe and tolerable. Almost all studies failed to demonstrate a relationship between study treatment and the occurrence of adverse events, and, if such a relationship could not be excluded fully, the main concern was with fairly minor health consequences.

Column 7

Column 7 summarizes the reported results on the outcome measures mentioned in the corresponding rows of column 6. Unless otherwise stated, only information related to the final outcome measurement time-point will be presented. Invariably "vs" is inserted to contrast the results registered for the Ginkgo group (mentioned first) and the reference group (either placebo group, or alternative drug group). In the case of a continuous outcome measure the numerical results may refer to the mean (or median) status scores at the end of the intervention period. They may also refer to the mean (or median) changes in status score from beginning to end. When change in status over time is the outcome parameter of primary concern, the numerical value is preceded by either "+" or "−" to indicate improvement and deterioration, respectively, for the relevant comparison group. In the case of a discrete (categorical) outcome measure, most of the times "%" is used to indicate the relative size of each category. Sometimes, especially for small-sized trials, absolute numbers were entered to avoid sham precision. For some trials numerical results could only be generated via extrapolation of the information stored in a graphical format. This may have resulted in slightly inaccurate outcome estimates.

If retrievable from the trial report the results of formal hypothesis testing (P-values) were added to the numerical outcome values, in order to show the statistical significance of each effect. This information was just copied, without further information on the test statistic used and without checking the appropriateness of the statistical data processing.

Column 8

The final column provides some additional, miscellaneous information on the trial design and the trial results. First of all this column contains information on the study design and the main intervention contrast. The third row throws some light on the source of recruitment of the patient population: out-patient versus hospital-ized, type of medical practice, etc. The rest of the information is optional and not systematically entered for each trial (dependent on space, relevance, etc.).

CLINICAL EFFICACY OF *GINKGO BILOBA* IN PERIPHERAL ARTERIAL DISEASE

Summary of Study Characteristics

Eighteen randomized clinical trials, evaluating the efficacy of *Ginkgo biloba* in patients with peripheral arterial occlusive disease, could be retrieved from the litera-ture. Nine of these studies were reported from Germany, 6 from France, and 1 from Italy, the UK, and Denmark. Table 1 shows the main design characteristics and study results.

In 12 studies the efficacy of *Ginkgo biloba* extract was tested by means of a parallel design, while in 6 studies a cross-over design was used. In 14 trials Ginkgo was compared with a placebo preparation, in 4 with another vasoactive agent. In all trials the Fontaine staging system was applied to indicate the severity of the disease of the patients at the start of the trial. The Fontaine classification is based on the assessment of clinical and functional signs and symptoms of peripheral arterial disease, e.g. painfree walking distance, claudication pain, pain during rest, presence of trophic skin defects, gangrene, and neuropathy. Fontaine stage I indicates the absence of typical functional manifestations of peripheral occlusive arterial disease; stage II indicates the presence of claudication pain (IIa: less severe; IIb: more severe), without pain at rest (recumbent, during the night) and without trophic signs; stage III indicates claudication pain accompanied by pain at rest and the presence of trophic signs; and stage IV indicates the presence of gangrene. In the reviewed trials, Fontaine stage II was the most common stage, that was covered by 16 of the trials.

The study population size varied from 10 to 80 patients, the average trial size being slightly lower than 50 randomized patients, or 25 per treatment group (cross-over trials included). The overall age range for the ca. 800 participants was 27–90 years (mean age about 62 years). More men than women participated. In 4 trials hospital patients were included, in 7 trials out-patients (ambulatory), and for 7 trials the source of recruitment was not clearly described. The most common daily doses were 120 mg (6 studies) and 160 mg (8 studies) of standardised Ginkgo extract. In one trial the acute effect of a single dose of 35 mg Ginkgo (injection) was measured, two trials evaluated the effect of a higher than usual dose (200–320 mg). The duration of the treatment varied from 1 hour to 6 months. The most common length of follow-up was 24 weeks (7 studies), followed by 6 weeks (3 studies).

Reflection on the Study Results

Various outcome measures have been used to express the effect of *Ginkgo biloba* treatment. The global effect of the treatment, usually in terms of experienced change

Table 1   RCTs on the efficacy of *Ginkgo biloba* special extract in patients suffering from peripheral arterial occlusive disease (PAOD)

Author, year	Study population / Diagnostic criteria	Number randomized analyzed (EG/other)	Age (yrs) / Sex (M/F)	Intervention / Brand name / Administration route / Dose / Duration	Outcome measure	Result	Remarks
Frileux and Copé, 1975	1. arteritis of lower limbs, stages IIa, IIb, III (n = 30) 2. M. Raynaud (n = 10) 3. leg ulcer, capillary origin (n = 10)	50 25/25	44–77 30/20	Tanakan® oral, solution (4 ml) 160 mg/day 5–7 w.	global (overall) effect clinical symptoms (scar tissue, aspect, colour, skin heat, etc.) functional symptoms (pain, intermittent claudication, problems with climbing stairs) specific tests: - walking distance on treadmill - palpation of arteries (peripheral pulse) - arterial oscillometry - assessment of skin temperature	*% good effect (vs moderate + no):* 64% vs 16% (p < 0.01) colour: 69% vs 19% aspect: 100% vs 0% pain: 66% vs 13% claudication: 64% vs 19% 64% vs 19% 0% vs 0% 0% vs 0% 64% vs 19%	parallel vs placebo source: n.d. randomization not explicitly mentioned no. with comorbidity: n = 24
Ambrosi and Bourde, 1975	chronic arteriopathy of lower limbs: - Fontaine stages II, III - walking dist. < 400 m. - interm. claud. (exercise) - extremities cooling down	14 7/7	45–73 20/1	Tanakan® i.v, injection 160 mg/day (4 ml) 1 m.	global effect: (very) good (vs moderate, no) walking distance on treadmill pain local skin temperature of extremities: - liquid cristals test (circulation, flow) - thermometry (isolated points) oscillometry, auscultation, pulse, ECG	7 vs 2 (p = 0.03) +147 m vs +53 m (p = 0.05) +68% vs +21% 7 (lying gone + walking impr.) vs 2 (walking bit improved) ca. 95% improved (Ginkgo) 7 better vs 3 better, 3 worse +2.65° C vs +0.00° C (p < 0.01)	parallel (n = 14) vs placebo also open trial (n = 7) source: n.d. staging: 18 × II, 3 × III 17 × one-sided, 4 × two-sided randomization not explicitly mentioned
Courbier et al., 1977	PAOD: - Fontaine stages II, III - clinical examination + arteriographic/aorto-graphic examination - obliteration	40	41–90 40/0	Tanakan® oral, solution 160 mg/day 1 m.	global effect (change in functional status: pain (walking, lying), spasms, itching, etc.) walking distance on treadmill: phase 1 walking distance on treadmill: phase 2 capillaroscopy clinical + physical signs: - peripheral pulse - local temperature of lower extremities - trophic skin defects (necrosis, ulceration) - oscillometry	26 × (65%) Ginkgo better vs 9x placebo better, 6 × undecided +193 m vs +44 m +19 m vs –12 m 7 out of 10 improved (verum) 1 vs 0 improved (verum) 4 vs 0 improved (verum) 4 vs 0 improved (verum)	cross-over vs placebo hospital: surgical patients staging: 29 × II, 11 × III side: 22 × 1, 18 × 2 duration of disease: 1–10 years 1–3 yrs. after surgery comorbidity: n = 18
Bastide and Montsarrat, 1978	arteritis of lower limbs: - stages II, III, IV - after surgery (sympathectomy)	40 20/20	67.3 37/3	E.G. oral, solution 4 ml/day 39 d.	walking distance on treadmill (3 km/h, 12%) lying pain (questionnaire) oscillometric indices	+429 m vs +128 m (p < 0.01) 0.95 vs 0.86 (n.s.) +1.30 vs +0.86 (n.s.)	cross-over vs vasodilating agent: P.R. hospital patients staging: 16 II, 21 III, 2 IV randomization not explicitly mentioned

Table 1   continued

Author, year	Study population Diagnostic criteria	Number randomized analyzed (EG/other)	Age (yrs) Sex (M/F)	Intervention Brand name Administration route Dose Duration	Outcome measure	Result	Remarks
Salz, 1980	PAOD: - Fontaine stage II - both legs	29 26	40–79 11/18	E.G. 40 (rökan®) oral, dragees 160 mg/day 6 w.	pain-free walking distance (2 steps/sec); maximum walking distance (2 steps/sec); standing on tiptoe test (time in sec); Ratschow test (lie-sit test): - time until ischemic pain (lying, legs vertical) - time until legs red again (sitting, legs hanging) - time until venous refilling (sitting); skin temperature; laboratory parameters	+91.8 m vs −77.7 m (p < 0.001); +97.5 m vs −102.5 m (p < 0.001); +37.7 sec vs −33.3 sec (p < 0.001); +43 sec vs −32 sec (p < 0.001); +5.8 sec vs −3.8 sec (p < 0.001); +6.0 sec vs −2.8 sec (p < 0.001); +0.95°C vs −0.50°C; n.s.	cross-over vs placebo source: n.d. wash-out: 1 w.
Garzya and Picari, 1981	PAOD: - Fontaine stages II, III, IV	33	48–64 33/0	Tanakan® oral, solution (drops) 160 mg/day 20 d.	walking distance on treadmill (1 km/h, 10%): - phase 1 (day 0–day 20) - phase 2 (day 30–day 50); clinical signs: scale 0–3 (no–very serious): - pain - cold feeling extremities - paresthesia - nutritional disorders; photoplethysmography (toes); rheographic imaging of limbs	+370 m (470%) vs +100 m (100%); +335 m (270%) vs −125 m (50%); phase 1: phase 2: +1.4 vs +0.5 +1.1 vs −0.2; +1.1 vs +0.0 +0.8 vs −0.6; +0.6 vs +0.3 +0.8 vs +0.2; +0.2 vs −0.1 +0.3 vs −0.1; wave amplitude: improved quotient (AR/T): improved	cross-over vs placebo hospital patients wash-out: 10 d. staging: 18 II, 10 III, 5 IV
Bauer, 1984	PAOD: - Fontaine stage IIb - stenosis A. superior femoralis (arteriography) - one-sided, > 1 year - walking distance: < 300 m.	80 79 44/35	60.9 61/18	rökan® oral, tablets 120 mg/day 6 m.	global effect (% very good + good): patient; global effect (% very good + good): investigator; pain-free walking distance (3 km/h, 10%); maximum walking distance (3 km/h, 10%); pain (subjective estimate; VAS 100 mm.); Doppler ultrasound, ankle, rest; Doppler ultrasound, ankle, after strain; plethysmography, Rf (= rescue flow); plethysmography, Pf (= peak flow); plethysmography, Tpf (= time until peak flow); blood pressure, heart rate, laboratory tests	66% vs 20%; 61% vs 14%; +62 m vs +32 m (p < 0.05); +110 m vs +31 m (p < 0.001); +21.5 vs +5.9 mm (p < 0.001); +1.9 vs +5.7 mm Hg; −3.1 vs +5.3 mm Hg; +0.7 vs −0.4 (p < 0.01); +1.5 vs −0.5; −0.1 vs +0.0; change in HDL-chol.: both	parallel vs placebo out-patients exclusion criteria run-in period: 6 w. randomisation: blocks (n = 6)
Baitsch, 1986	PAOD: - Fontaine stage IV - surgery not possible diabetes (regulated)	20 10/10	35–76 12/8	rökan® i.v. + oral (tablets) 1. 200 i.v. + 80 oral hospital, ≥ 3 w. 2. 0 i.v. + 120 oral ambulatory, ≥ 3 m. ≥ 4 m.	healing tendency of trophic lesions, 3–8 w.; healing tendency of trophic lesions, 3–12 m.; primary amputations, 3–8 w. of treatment; primary amputations, 3–12 m. of treatment; Ratschow's test; Doppler ultrasound pressure measurements; plethysmography (venous occlusion); clinical laboratory parameters; use of analgetics	9 vs 4 (very) good; 6 vs 3 healed or (very) good; 3 vs 2 amput.; 1 vs. 1 died; 2 vs 1 amput.; 0 vs. 1 died; little improvement; little improvement; little improvement; no difference	parallel vs dextran 40 hospital patients exclusion criteria cointerventions

Table 1   continued

Author, year	Study population / Diagnostic criteria	Number randomized analyzed (EG/other)	Age (yrs) / Sex (M/F)	Intervention / Brand name / Administration route / Dose / Duration	Outcome measure	Result	Remarks
Berndt and Kramar, 1987	PAOD: – Fontaine stage IIb – severe stenosis/occlusion of A. sup. femoralis – claud. pain: reproducible – walking distance: < 200 m.	38 / 36 / 17/19	27–72 / 28/8	rökan® / n.d. / 160 mg/day / 24 w.	painfree walking distance (3 km/h, 10%); maximum walking distance (3 km/h, 10%); pain: VAS 0–100 mm (no–unbearable); ankle pressure (Doppler), rest; ankle pressure (Doppler), after strain; trophical state, pulse rate, blood chemistry	+164 m vs +60 m (p < 0.05); +160 m vs +85 m (p < 0.05); +20 mm vs +5 mm (p < 0.01); +19.8 vs +3.4 mm Hg (n.s.); +37.5 vs +12.8 mm Hg (n.s.)	parallel vs buflomedil ambulatory patients 2 centres run-in period: 3 w.
Rudofsky, 1987	PAOD: – Fontaine stage II	10	63 / n.d.	Tebonin® / i.v., injection / 35 mg / 1 h.	ergometry (calf ergometer in lying position): – time until stop (maximum claudication pain); haemodynamic parameters (arterial influx): – blood flow at rest (plethysmography): – blood flow after ergometry, peak flow – blood flow after ergometry, time to peak flow – blood flow after ergometry, normalization time; haemorheological parameters: total blood viscosity, plasma viscosity, erythrocyte aggregation; metabolic parameters: lactate, pyruvate, L/P quotient, haematocrit, blood gasses (pH, pO$_2$, pCO$_2$, base excess)	pain longer sustained: +0.56 vs –0.05 min (p < 0.05); +0.10 vs –0.07 ml/100 ml.min; +0.8 vs –0.1 ml/100 ml.min; +3.9 vs –2.3 sec; +0.27 vs +0.17 min; decrease of erythrocyte aggregation, viscosity (Ginkgo); decrease L/P quotient less decrease pO$_2$, base excess	cross-over vs placebo wash-out: 12 d. ambulatory patients acute effect
Böhmer et al., 1988	PAOD: – Fontaine stage IIb – severe stenosis/occlusion A. sup. femoralis – claud. pain: reproducible – walking dist.: 50–200 m.	27 / 14/13	44–72 / 24/3	n.d. / n.d. / 160 mg/day / 24 w.	painfree walking distance (3 km/h, 5%); maximum walking distance (3 km/h, 5%); pain: VAS 0–85 mm (no – unbearable); haemodynamic parameters: – pulse – ankle pressure rest, after strain (tiptoeing) – systolic index (arm/ankle quotient); lactate level in serum (exercise; mmol/l); laboratory parameters (chol., TG, GOT, etc.)	+230 vs +245 m (n.s.); +233 vs +283 m (n.s.); +36.3 vs +30.9 mm (n.s.); small increase: both groups; small increase: both groups; +0.68 vs +0.26 mmol/l; (indicating less strain)	parallel vs pentoxifyllin ambulatory patients: GP screening: by exercise physiology placebo run-in: 3 w.
Saudreau et al., 1989	PAOD: – Fontaine stage III – angiographic diagnosis	64 / 32/32 / 55 / 26/29	64.6 / 53/11	Tanakan® / i.v., injection / 200 mg/day / 8 d.	pain: VAS (100 mm); pain: questionnaire QDSA (adapted McGill Pain questionnaire); self-evaluation, qualitative + quantitative; 58 items in 16 subscales; sensoric (35) vs. affective-emotional (23) dimension; range item score: 0–4 (intensity) use of analgetics (pentazocine)	+28 vs +14 mm (p = 0.04); patient's grade score: total: +7.9 vs +7.0 (n.s.); sensoric: +4.0 vs +3.2 (n.s.); affective: +3.9 vs +3.7 (n.s.); n.s. (p = 0.26); verum shows more decrease	parallel vs placebo 5 centres anticoagulating therapy: n = 52

Table 1   continued

Author, year	Study population / Diagnostic criteria	Number randomized analyzed (EG/other)	Age (yrs) / Sex (M/F)	Intervention / Brand name / Administration route / Dose / Duration	Outcome measure	Result	Remarks
Diehm et al., 1990	PAOD: – Fontaine stage IIb – angiography, Doppler, oscillometry – pain-free walking distance baseline: 50–200 m – one-sided	40 22/18	40–75 both sexes	Tebonin® forte oral, tablets 120 mg/day 12 w.	pain-free walking distance (5 km/h, 10%) maximum walking distance (5 km/h, 10%) subjective evaluation (patient) subjective evaluation (investigator) ankle pressure (Doppler), rest ankle pressure (Doppler), exercise heart rate blood pressure laboratory values	+47% vs +33% (p = 0.04) +50 m vs 'small decline' n.s. (difference) n.s. (difference) n.s. (difference) n.s. (difference) n.s. (difference) n.s. (difference) n.s. (difference)	parallel vs placebo 3 centres inclusion + exclusion criteria
Thomson et al., 1990	PAOD: – Fontaine stage IIb – iliac or femoral arteries – one side dominant – walking distance: < 300 m – no. of cigarettes: < 6/day	49 25/24 37 20/17	n.d. n.d.	Tanakan® oral, tablets n.d.; 3 tablets/day 24 w.	overall effect (patient) overall effect (physician) pain-free walking distance (4 km/h, 10%) pain, severity: VAS (10 cm) ankle pressure (Doppler), rest ankle pressure (Doppler), exercise recovery time (min)	no difference no difference +38.1 vs +23.2 m (+47/+33%) +15.4 vs +4.0 mm (+29/+21%) +2.3 vs –4.1 mm Hg +1.2 vs +4.7 mm Hg (n.s.) +0.6 vs +0.8 min (n.s.)	parallel vs placebo placebo run-in: 6 w.
Bulling and Von Bary, 1991	PAOD: – Fontaine stage IIb – A. femoralis affected – angiographic diagnosis or history; clinical exam, pulse, Doppler plethysmography – walking dist.: 50–200 m	36 33 17/16	61.1 24/9	rökan® oral, tablets 120 mg/day 6 m. adjuvans therapy: supplementary to physical training	pain-free walking distance (3 km/h, 10%) maximum walking distance (3 km/h, 10%) standing on tiptoe, number until pain standing on tiptoe, maximum number ankle pressure, A. tibialis posterior ankle pressure, A. dorsalis pedis plasma viscosity haematocrit	+152 vs +90 m (p < 0.01) +306 vs +138 m (n.s.) +10.7 vs +3.6 (n.s.) +9.8 vs +16.4 (n.s.) +1.5 vs +2.4 mm Hg (n.s.) +1.5 vs +4.9 mm Hg (n.s.) +0.023 vs +0.030 (n.s.) +3.49% vs +1.38% (n.s.)	parallel vs placebo ambulatory patients physical training program: 1 h., 3x/week exclusion criteria
Mouren et al., 1994	PAOD: – Leriche-Fontaine stage II – diagnosis: clinical signs, arteriography, Doppler – treadmill claudication distance: 100–500 m – ≥ = 3 m. stable – TcPO2: normal (> 40 mm)	20 n.d.	44–73 n.d.	EGb 761 n.d. 320 mg/day 4 w.	pain-free walking distance (3.2 km/h, 10%) transcutaneous partial oxygen pressure: – TcPO2, rest (mm Hg) – TcPO2, exercise (mm Hg) area of ischemia (indicator of local microcirculation) functional status (VAS, 100 mm)	+6.9 vs +3.4 mm Hg (p = 0.43) +1.2 vs –7.9 mm Hg (p = 0.21) +2.05 cm² vs –0.33 cm² (p = 0.04) diff. ± 95% CI: +2.38 (0.98–3.87) +38% vs –5% +2.2 mm vs. +3.9 mm (p = 0.66)	parallel vs placebo out-patients excluded: stages III, IV placebo run-in: 15 d.

Table 1 continued

Author, year	Study population Diagnostic criteria	Number randomized analyzed (EG/other)	Age (yrs) Sex (M/F)	Intervention Brand name Administration route Dose Duration	Outcome measure	Result	Remarks
Drabaek et al., 1996	PAOD: – stable interm. claudication	18	n.d. n.d.	GB-8 n.d. 120 mg/day 3 m.	pain-free walking distance, treadmill maximum walking distance, treadmill severity of leg pain (VAS) concentration (VAS) memory capacity (VAS) short-term memory: objective test systemic systolic blood pressure peripheral systolic blood pressure	n.s. (both groups not changed) n.s. (both groups not changed) n.s. (both groups not changed) GB-8 more effective GB-8 more effective n.s. (no change) n.s. (both groups reduced) n.s. (both groups not changed)	cross-over vs placebo 3 m. GB-8, 3 m. placebo
Blume et al., 1996	PAOD: – Fontaine stage IIb – angiographic diagnosis – walking distance: < 150 m – resistant to physical the- rapy (walking training) – duration: > 6 m	60 30/30	47–82 42/18	EGb 761 oral, tablets 120 mg/day 24 w.	pain-free walking distance (3 km/h, 12%) maximum walking distance (3 km/h, 12%) pain-free walking distance, relative change subjective assessment by patient (–5 – +5) Doppler quotient	+41 m vs +8 m (p < 0.0001) +51 m vs +16 m (p < 0.0001) +42% vs +8% (p < 0.0001) +2.9 vs 0.0 (p < 0.001) not changed in both groups	parallel vs placebo angiologist's practice: ambulatory patients placebo run-in: 2 w. I-to-T -analysis dropouts: 4 vs 2

in the status of several observable signs and symptoms (pain, ability to walk, cold feeling, itching, spasms, etc.), was assessed in 8 studies, the assessor being either the patient, the physician, or both. Walking distance was the most frequently used outcome measure. This parameter, which reflects functional capacity, was always measured under standardized laboratory conditions, on a treadmill with a varying speed and slope. The change in painfree walking distance was reported in 14 studies. In addition, the maximal walking distance was measured in 8 of these studies. Other laboratory tests of functional capacity that have been applied as end-points are the Ratschow test (time until onset of pain and skin paleness when lying down with the legs held up; time until recurrence of redness and venous pattern when changing from lying to sitting position, with the legs hanging down), the standing-on-tiptoe test (number of "tiptoes" until onset of pain and until pain becomes unbearable), and a calf ergometry test. The severity of pain, either during rest or during exercise, was measured in 9 studies, by means of a VAS-scale, a questionnaire, or just in a global way (improvement of pain sensation).

From a clinical point of view those endpoints that have the closest relationship with quality of life and the ability to cope with daily life strains, are considered to be the most relevant. Therefore, global assessment, walking distance and pain will be elaborated a little bit further, only for those studies which compared Ginkgo with placebo.

The overall, global effect of Ginkgo treatment was found to be significantly in favour of Ginkgo in 5 trials (Ambrosi and Bourde, 1975; Bauer, 1984; Blume *et al.*, 1996; Courbier *et al.*, 1977; Frileux and Copé, 1975). In general, the proportion of patients with a good or very good effect was 2–3 times higher in the Ginkgo group than in the placebo group. However, in a few trials a significant overall effect was absent (Diehm *et al.*, 1990; Mouren *et al.*, 1994; Thomson *et al.*, 1990).

The increase in painfree walking distance turned out to be significantly larger on Ginkgo in all trials that presented hypothesis testing results, except for one (Drabaek *et al.*, 1996). If only those trials are taken into account which reported absolute changes in walking distance (gain in meters), which contrasted Ginkgo with placebo, and for which, in case of a cross-over design, the results dealing with the first inter-vention phase could be separated, an average gain at the value of about 80 meters can be estimated. This estimation neglects the differences between the studies as to patient characteristics (severity of disease), dose and duration of treatment, and exercise conditions (speed and slope of treadmill). Most of the trials which evaluated the influence of Ginkgo treatment on claudication pain showed a larger reduction of the pain intensity for Ginkgo than for placebo. Measured on a 100 mm VAS-scale the difference was on average about 15 mm in favour of the Ginkgo group.

The current review suggests that treatment with *Ginkgo biloba* extract is indeed effective for people suffering from chronic peripheral obliterative arterial disease. Only one "negative" study was identified (Drabaek *et al.*, 1996). However, only two of the trials included in the earlier review of Kleijnen and Knipschild (1991) were found to have an acceptable methodological quality (Bauer, 1984; Saudreau *et al.*, 1989). Recently a new trial was added to the available pool, which probably sur-passes all the earlier trials in methodological quality. Its results support the hypothesized beneficial effect of *Ginkgo biloba* (Blume *et al.*, 1996). Although none of the trials published until now have attempted to measure the influence of Ginkgo treatment on functional capacity in daily life, laboratory treadmill testing can be considered a fine proxy measure: it is generally assumed that a particular gain in treadmill walking distance corresponds with a three- to fourfold gain in daily life walking capacity.

CLINICAL EFFICACY OF *GINKGO BILOBA* IN CEREBRAL INSUFFICIENCY

Introduction

Since 1975 a considerable number of intervention studies have been published aimed at building evidence for the efficacy of *Ginkgo biloba* special extract in subjects with various manifestations of cerebral insufficiency. Most of them were projected as a randomized clinical trial (RCT). However, in some cases the description of the design in the literature (random allocation procedure, blinded outcome measurement, etc.) does not make clear whether they indeed deserve this qualification. The current review discusses 55 studies. They tend to show more differences than similarities. The recent review by Letzel *et al.* (1996), regarding the nootropic effect of *Ginkgo biloba* in comparison with two other active substance classes (tacrine, nimodipine), illustrates some approaches to distinguish a limited number of more or less homogeneous subcategories of trials. It shows that reference can be made to various design aspects, such as the diagnostic inclusion criteria applied, the measures used to assess the severity of the disease, or the type of instrument used to assess the intervention outcome.

To structure the current overview an attempt was made to divide the Ginkgo trials into several categories based on the main treatment indication. The following categories were distinguished:

1. Studies focussing on patients with dementia, including Alzheimer's disease, vascular dementia, and mixed type dementia
2. Studies focussing on non-demented patients with cognitive decline
3. Studies focussing on patients with cerebral insufficiency expressed primarily by physical, non-cognitive manifestations (tinnitus, vertigo, hearing loss, etc.)
4. Studies focussing on patients with a mixture of cognitive and non-cognitive symptoms, diagnosed and evaluated through a standard symptom checklist
5. Studies focussing on patients with affective manifestations of cerebral insufficiency, especially studies focussing on depressive mood as the leading symptom
6. Studies focussing on patients with cerebral insufficiency according to diagnostic information that does not permit further differentiation of the disease state (cerebral insufficiency, unspecified).

We are aware, however, that this classification system can be challenged and that part of the trials will not fit exactly in one of the categories. More than average attention will be paid to the first category of trials. They are of special interest, due to their focussed character – restriction of the patient definition to a more precise and homogeneous disease entity than is the case for the majority of cerebral insufficiency trials – in combination with the claim that *Ginkgo biloba* special extract might benefit patients with mild to moderate stages of both Alzheimer's dementia and vascular dementia. Most of the trials in this category have been conducted and reported quite recently. Therefore, it will not be surprising that they surpass most of the previous trials as for methodological quality and other relevant design aspects.

Summary of Study Characteristics and Study Results

Dementia

The literature search yielded 9 RCTs which suggested that they evaluated the effect of Ginkgo in patients diagnosed with dementia. A more detailed analysis reveals, however, that only five of these alleged anti-dementia trials were designed and conducted in agreement with the gradually evolved and currently accepted scientific guidelines and prescriptions for the diagnosis of dementia and the evaluation of therapy-induced changes in the state of dementia (Haase *et al.*, 1996; Hofferberth, 1994; Kanowski *et al.*, 1996; Le Bars *et al.*, 1997; Maurer *et al.*, 1997). According to these guidelines and prescriptions, the patient selection should be based on the DSM-III-R / DSM-IV, the ICD-10, or the NINCDS-ADRDA (McKhann *et al.*, 1984) criteria for dementia. And the outcome assessment should be based on three different levels of achievement: clinical psychopathology (both global and specific), objective psychometric performance, and (cognition-related) daily life activities. It is not surprising that only trials completed during the 1990s fulfil these criteria.

It was decided to deal under the dementia heading also with other trials which claim that they have included patients with dementia, although the appropriateness of this claim cannot be verified from the documentation available. Moreover, one should recognize that many of the trials allocated to some of the other headings (e.g. cognitive impairment) probably have harboured a considerable number or even a majority of demented patients as well, who have been overlooked due to the deployment of insufficient diagnostic equipment. Details of the dementia trials have been summarized in Table 2.

Hofferberth performed a RCT in 40 hospitalized patients with incipient dementia of the Alzheimer type (Hofferberth, 1994). All patients (age range: 50–75 years) were treated during 3 months with either Tebonin® forte (tablets, 80 mg/day) or placebo. The Syndrom Kurz Test (SKT), indicating the level of memory and attention functioning (scale $\underline{0}$–27; the underlined extreme represents the most favourable score), was elected as the primary outcome measure in this trial. Ginkgo treatment turned out to be beneficial, according to the observed mean change in SKT-score over 3 months (+5 vs. –2 points; p < 0.001). Fifty percent of the Ginkgo users and none of the placebo users showed an improvement that exceeded 5 SKT-points. A global assessment by the physician, the Sandoz Clinical Assessment-Geriatric Scale (SCAG; measuring the degree of clinical psychopathology), the Vienna

Table 2 RCTs on the efficacy of *Ginkgo biloba* special extract in patients suffering from cerebral insufficiency: dementia (AD, VD, mixed type)

Author, year	Study population Diagnostic criteria	Number randomized analyzed	Age (yrs) Sex (M/F)	Intervention Brand name Administration route Dose Duration	Outcome measure	Result	Remarks
Weitbrecht and Jansen, 1985, 1986	dementia: primary degenerative (mild, moderate) HIS ≤ 7 no further specifications	40 20/20 38 18/20	60–80 (71.7) 14/26	Tebonin® forte oral, solution 120 mg/day 3 m.	clinical rating scales: – general feeling (self-assessment by patient) (1 = excellent; 2 = good; 3 = poor; 4 = bad) – general health (physician) – Crichton Geriatric Scale; 11 items (1–5): 11–55 – SCAG; 19 items (1–7): 11–133 psychometric tests: – figure-symbol test (Wechsler) – number-repeat test (digit span) (Wechsler) – flicker fusion frequency – reaction time	+0.9 vs +0.0 points +5 vs +0 points +23 vs +2 points +2.5 vs –1.0 points +2.0 vs +0.0 points +1.8 vs –0.4 Hz +0.33 vs +0.02 sec	parallel vs placebo source: n.d. also comparison with ergot alkaloids (n = 20) pretreatment duration of symptoms: 8.8 yrs.
Halama, 1991a, 1991b	cerebral insufficiency degenerative + vascular dementia: 1. single senile (ICD 290.0) 2. presenile (ICD 290.1) 3. arteriosclerotic (ICD 290.4) no formal criteria stated MWT: premorbid intelligence	50 42 20/22	35–85 15/37	Kaveri® (LI 1370) oral, tablets 150 mg/day 12 w.	Syndrom Kurz Test (0–27) global assessment of efficacy (patient) global assessment of efficacy (physician): ('good'/'moderate'/'unsatisfactory') 11 subjective clinical CI-symptoms: memory, forgetfulness, concentration, fatigue, performance, motivation, depression, anxiety, dizziness, headache, tinnitus; 4-point scale: severe, moderate, mild, absent general fluent intelligence: number connection test (NAI-ZVT) blood pressure, pulse rate	+4.15 vs +1.32 (p < 0.001) 13/43 vs 4/3/15 (p < 0.001) 12/4/4 vs 2/3/17 (p < 0.001) n. improved, Ginkgo vs. placebo: p < 0.005: 2 symptoms p < 0.05: 5 symptoms n.s., tendency: 1 symptom n.s.: 3 symptoms (anxiety, memory gap, concentration) +7.2 sec. vs +0.0 sec (n.s.)	parallel vs placebo ambulatory patients (1 neurological practice) dropouts: 5 vs 3 baseline SKT score 5.25 vs 6.64
Hartmann and Frick, 1991	vascular dementia: CT-scan typical symptoms HOPS: ≥ 1 symptom within each of 4 symptom groups present	52 26/26 48	62.5 34/18	Kaveri® (LI 1370) oral, solution (drink) 150 mg/day 3 m.	global assessment of efficacy (patient) global assessment of efficacy (physician) ('good'/'moderate'/'unsatisfactory') 12 clinical CI-symptoms (hetero-assessment): memory, forgetfulness, concentration, fatigue, performance, motivation, depression, anxiety, dizziness, headache, tinnitus, confusion; 4-point scale: severe, moderate, mild, absent psychometric testing: – Grünberger Verbale Gedächtnis test – trail making test (ZVT)	13 vs 8 'good', 7 vs 10 'moderate', 2 vs 5 'poor' (n.s.) 3 severity categories improvement in both groups difference Ginkgo-placebo: n.s. +2.55 vs +0.66 points (p < 0.05) +8 sec vs +7 sec (n.s.)	parallel vs placebo out-patients: neurological dept. of university hospital exclusion criteria dropouts: 2 vs 2

Table 2   continued

Author, year	Study population Diagnostic criteria	Number randomized analyzed	Age (yrs) Sex (M/F)	Intervention Brand name Administration route Dose Duration	Outcome measure	Result	Remarks
Vesper and Hänsgen, 1994	cerebral insufficiency: 1. MMS (Folstein) 2. CI-scale: > 19 3. MWT-B ≥ 80 4. DPD > 2 5. K-HIS 6. medical examination	90 45/45 86 42/44	55–80	Kaveri® (LI 1370) oral, tablets 150 mg/day 12 w.	CI-scale (subjective cerebral insuff. troubles; scale: improved, unchanged, deteriorated) rating of changes of 8 mental + physical functions: scale 1–Z (4 = no change) general intelligence test: KAI (information processing) computer-aided psychometric tests: – basic reaction time – time/tenacity training test (attention + learning ability) – mental flexibility (coping with interference) – visual memory (geometric patterns) – character discrimination	w. 1–6: 50% vs 32% improved w. 6–12: 53% vs 37% improved $4.7–5.0$ vs $3.9–4.2$ $(p < 0.05)$ n.s. verum group sign. more improved $+6$ vs $+2$ ms/10 $(p < 0.05)$ 0 vs -2 $(p < 0.05)$ +38% vs +14% correct solutions verum group sign. more improved	parallel vs placebo out-patients: 6 neurologist/psychiatrist practices 4 GPs 1 internal specialist
Hofferberth, 1994	dementia, Alzheimer type: – clinical diagn., incipient – Blessed Dementia Scale: sum part A: 0–16 sum part B: 9.5–30.5 – HIS: < 4 – CT-scan	40 21/19	50–75 28/14	Tebonin® forte oral, tablets 80 mg/day 3 m.	primary outcome measure: – Syndrom Kurz Test (SKT): memory + attention, sum score 9 subtests (0–27) secondary outcome measures: – SCAG: change in psychopathology 18 items (scale 1–7), sum score: 18–126 global impression – Vienna Determination test (reaction test) no. correct on steps 10, 12, 15 (max.: 180) – saccade test (duration, latency + maximum rate rapid eye movements (electrooculography)) – EEG-parameters: theta/alpha quotient – global assessment efficacy (physician): 'good', 'satisfactory', 'not satisfactory'	+5 vs -2 $(p < 0.001)$ 50.4% vs 0.0% > +5 points +29 vs +0 verum better on most items verum better $(p < 0.05)$ step 10: +22 vs +4; step 12: +12 vs +2; step 15: +9 vs +2 latency: +65 vs -5 ms duration: +57 vs +2 ms +17% vs +0% $(p < 0.01)$ 76% vs 0% 'good', 24% vs 42% 'satisfactory'	parallel vs placebo hospitalized patients excluded: MID, dementia with known cause, advanced dementia, depression, etc.

Table 2   continued

Author, year	Study population / Diagnostic criteria	Number randomized analyzed	Age (yrs) / Sex (M/F)	Intervention / Brand name / Administration route / Dose / Duration	Outcome measure	Result	Remarks

Kanowski et al., 1996	dementia, AD or MID:
	– clinical diagnosis (physician); according to DSM-III-R
	– mild–moderate
	– CCT
	– Ischemic Score (Rosen)
	– MMSE: 13–25
	– SKT: 6–18
		216
		156
		79/77
			70.1
			51/105
			MID/AD = 125/31
				EGb 761
				oral, capsules
				240 mg/day
				24 weeks
					primary outcome measures:
					– clinical global impression of change (CGI-2) by physician; scale: 1–7
					  response criterion: '(very) much improved'
					– Syndrome Kurz Test (SKT): memory + attention, sum score 9 subtests (0–27)
					  response criterion: decrease ≥ 4 points
					– Nuremberg Geriatric Observation Scale (NAI-NAB); behaviour: coping with everyday tasks, independence of care; relative-rating
					  response criterion: decrease ≥ 2 points
					– responder analysis: ≥ 2 of 3 POMs
					secondary outcome measures:
					– Montgomery Asberg Depression Rating Scale
					– actometry (subsample: 38 vs 38)
					– EEG (subsample: 38 vs 38)
						32% vs 17%
						
						+2.2 vs +0.8
						38% vs 18% (p < 0.05)
						
						+0.9 vs +0.6
						33% vs 23% (p < 0.1)
						28% vs 10% (p = 0.005)
						
						+3.3 vs +2.5
							parallel
							vs placebo
							out-patients
							41 centres: practices of GPs, neurologists, psychiatrists, internists
							excluded: pronounced cerebral atrophy, full-time care unavoidable
							4 w. run-in
							205: I-to-T analysis effect in AD slightly larger than in MID

Haase et al., 1996	dementia (AD, MID, mixed) – clinical diagnosis according to DSM-III-R – moderate – HIS (Rosen modification) – Global Deterioration Scale (GDS): 7 stages, stages 4–5 – MMST	40 20/20	45–87 16/24	EGb 761 i.v. (infusion) 200 mg/day 4 w. (4 days/week)	*primary outcome measures:* 1. Nuremberg Geriatric Observation Scale (NAI-NAB); behaviour: coping with everyday tasks, independence of care; relative-rating response criterion: decrease ≥ 2 points 2. clinical global impression of change (CGI-2) by physician; scale: 1–7 3. KAI (information processing) *secondary outcome measures:* – Nuremberg Elderly Activities of Daily Life scale (NAI-NAA); 20 items; sum: 20–60 – improvement of strain due to main symptom – Zung's Selfrating Depression Scale (SDS)	+3.6 vs +0.3 points (p < 0.01) 17 vs 3 improved improved: 17 vs 7 (p < 0.0001) 1/2/3: 1/9/7 vs 0/0/7 improved: 7 vs 1 (p < 0.03) +8.0 vs +0.5 points (p = 0.003) forgetfulness: 17 vs 6 (p = 0.002) +6 vs +0 points (p < 0.0001)	parallel vs placebo out-patients, neuro-psychiatric practice prestratification: – NAB-score – HIS no interaction

Table 2   continued

Author, year	Study population Diagnostic criteria	Number randomized analyzed	Age (yrs) Sex (M/F)	Intervention Brand name Administration route Dose Duration	Outcome measure	Result	Remarks

Le Bars et al., 1997	dementia: AD, MID
	– according to DSM-III-R, ICD-10
	– mild to moderately severe:
	 – MMSE: 9–26
	 – Global Deterioration Scale (GDS): 3–6
		327
		166/161
		309
		IcT-analysis
		236 AD
			45–90
			143/166
				EGb 761
				oral, tablets
				120 mg/day
				52 w.
					primary outcome measures:
					1. cognitive impairment:
					 – ADAS-Cog (cognitive subscale) performance based (memory, language, praxis, orientation); 11 items; sum: 0–70
					2. daily living + social behavior:
					 – Geriatric Evaluation by Relative's Rating Instrument (GERRI); 49 items; rating by caregiver, after 14-day observ.
					  – GERRI-cognitive (21 items)
					  – GERRI-social (18 items)
					  – GERRI-mood (10 items)
					  item, subscale, total scale score: 1–5
					3. general psychopathology:
					 – Clinical Global Impression of Change (CGIC); clinician; ordinal scale: 1–7
					secondary outcome measures:
					not yet published
						-0.1 vs -1.5 (p = 0.04)
						$\geq +2$: 50% vs 29%
						
						$+0.06$ vs -0.08 (p = 0.004)
						improved: 37% vs 23%
						
						4.2 vs 4.2 (p = 0.77)
							parallel
							vs placebo
							out-patients, without other sign. conditions
							6 centers
							placebo run-in: 2 w.
							compliance:
							pill count,
							deviation < 20%
							78/59 completed the trial
							202 in per protocol analysis
							same results for the AD subgroup

Maurer et al., 1997	dementia: AD – according to DSM-III-R, NINCDS-ADRDA – Brief Cognitive Rating Scale (BCRS): 3–5 – HIS ≤ 4 – CT-scan: diffuse atrophy or no pathological signs	20 10/10 18 9/9	50–80 9/9	Tebonin® forte oral, tablets 240 mg/day 3 m.	*primary outcome measure:* – Syndrom Kurz Test (SKT): memory + attention, sum score 9 subtests (0–27) *secondary outcome measures:* – number connection (ZVT) – ADAS: cognitive subscale non-cognitive subscale – clinical global impression (CGI): 1–7 – electrophysiological parameters (AEP-300, EEG-topography) – patient termination record	$+2.89$ vs -0.78 (p = 0.013) (baseline: 19.67 vs 18.11) positive trend (n.s.) $+0.88$ vs -0.03 (n.s.) $+1.78$ vs $+1.45$ (n.s.) 4.3 vs 5.3 (p = 0.069) positive trend (n.s.) 8/9 vs 5/9 preferred to continue	parallel vs placebo out-patients run-in: 1 w. excluded: advanced dementia, dementia of other aetiology, etc.

Determination Test, a saccade test, and various EEG characteristics were applied as secondary outcome parameters in this study. The patients in the verum group showed more favourable scores on these parameters than those in the placebo group (Hofferberth, 1994).

Kanowski et al. (1996) conducted a placebo-controlled, parallel trial in 216 out-patients with either Alzheimer's dementia or vascular dementia, in a mild to moderate state. The patients were recruited from 41 centres (practices of general practitioners, neurologists, psychiatrists, internists). After a 4 week run-in phase, the participants were randomly allocated to a 24-week treatment with either EGb 761 (240 mg/day) or placebo capsules. 205 Patients were involved in the intention-to-treat analysis. However, an analysis based on the 156 participants who completed the trial, was preferred as the confirmatory analysis of the trial results. The clinical global impression of change by the physician (CGI-2, range of scores 1–7), the SKT (range of scores 0–27), and the Nuremberg Geriatric Observation Scale (NAI-NAB) – a rating scale measuring behaviour indicating the level of coping with everyday tasks and the level of independency of care (rated by relatives of the patient) – were applied as the primary outcome measures. Thirty-two percent of the Ginkgo users and 17% of the placebo users were found to respond according to the CGI-2, when the categories "much improved" and "very much improved" were used as the success criterion. The mean change in the SKT sum score was +2.2 for the Ginkgo group vs. +0.8 for the placebo group. With a decrease of at least 4 points on the SKT-scale acting as the criterion for success, 38% of the Ginkgo users vs. 18% of the placebo users turned out to be responders ($p < 0.05$). The mean changes on the NAI-NAB were +0.9 vs. +0.6 points for the Ginkgo group and the placebo group, respectively, with 33% vs. 23% responders ($p < 0.1$; response criterion: ≥ 2 NAI-NAB points). An overall responder analysis – criterion: response for at least 2 out of the 3 primary outcome measures – yielded 28% vs. 10% ($p = 0.005$) responders among the Ginkgo users and the placebo users, respectively (Kanowski et al., 1996).

Haase et al. (1996) reported on the results of a parallel, placebo-controlled trial in 40 out-patients of a neuro-psychiatric practice. The study subjects received intra-venous treatment, which contained 200 mg EGb per day for the Ginkgo group and which was continued over a 4 week period. The daily dose was administered by infusion, 4 days a week. Included in the trial were patients with a moderate state of dementia, either AD, or VD, or a mixed form. The Nuremberg Geriatric Observation Scale (NAI-NAB), the clinical global impression of change by the physician (CGI-2), and the Kurztest für Allgemeine Intelligenz (KAI) were the primary outcome measures in this trial. With respect to the NAI-NAB 17 Ginkgo users vs. 3 placebo users improved (mean change in score: +3.6 vs. +0.3 points ($p < 0.01$)). With regard to the CGI-2 17 vs. 7 participants improved ($p < 0.0001$). The KAI-test for information processing capacity resulted in improvement for 7 vs. 1 of the participants ($p < 0.03$) (Haase et al., 1996).

Maurer and colleagues studied the effect of treatment with Tebonin® forte (tablets, 240 mg/day, intervention stretching over 3 months) in 20 out-patients suffering from Alzheimer's dementia (Ihl et al., 1992; Maurer et al., 1997). A CT-scan revealed either diffuse atrophy or no pathological signs at all. Patients excluded from participation encompassed subjects with an advanced stage of dementia, and patients with dementia of known etiology. Eighteen subjects were included in the analysis. The Syndrom Kurz Test was chosen as the outcome measure of primary interest. The changes in SKT sum score were +2.89 vs. -0.78 (p = 0.013), departing from baseline scores amounting to 19.67 vs. 18.11 (scale range: 0–27). Positive, though not statistically significant effects were also reported for the secondary outcome parameters: the number-connection test (NAI-ZVT-G), the ADAS cognitive and non-cognitive subscales, the clinical global impression of change (CGI), and several electrophysiological parameters (AEP-300, EEG-topography). Eight out of 9 Ginkgo users and 5 out of 9 placebo users stated that they would like to continue the treatment (Maurer et al., 1997).

The first USA-based clinical trial on the efficacy of Ginkgo biloba special extract was reported by Le Bars et al. (1997). Included were 327 persons (aged 45–90 years) from 6 out-patient centres, with a mild to moderately severe state of dementia according to the DSM-III-R and ICD-10 criteria, either AD or VD. After a 2-week run-in period the patients were randomly allocated to either Ginkgo (tablets, daily dose of 120 mg) or placebo, for a period of one year. 78 Patients in the Ginkgo group and 59 in the placebo group completed the trial. 309 Subjects could be entered in the intention-to-treat analysis, whereas 202 patients met the criteria for the per-protocol analysis, with a deviation of less than 20% from the protocolized intake based on pill counts being one of the criteria. For this trial three primary outcome measures were defined. Firstly, the cognitive subscale of the Alzheimer Disease Assessment Scale at the cognitive function level (ADAS-Cog; performance-based (memory, language, praxis, orientation); 11 items, yielding a 0–70 sum score range). Secondly, the Geriatric Evaluation by Relative's Rating Instrument at the daily living and social behaviour level (GERRI; rated by the care-giver after 14 days of observation; 49 items, divided over 3 subdomains: cognition, social functioning, and mood; item-, subscale- and total scale score range: 1–5). And, thirdly, the Clinical Global Impression of Change (CGIC; scored by the clinician; scale: 1–7). The mean change in ADAS-Cog score showed less impairment for patients allocated to the Ginkgo group than for patients in the placebo group: –0.1 vs. –1.5 (p = 0.04). Although this difference is statistically significant, its clinical meaningfulness is debatable. The improvement amounted to ≥ 2 points for 50% of Ginkgo users and 29% of the placebo users. The mean change of the GERRI-overall score was also more favourable for the Ginkgo treatment group: +0.06 vs. –0.08 (p = 0.004). Improvement was registered for 37% and 23% of the Ginkgo and placebo groups, respectively. The treatment groups did not differ with respect to the CGI-scores: 4.2 vs. 4.2 (p = 0.77). The analysis did not reveal effect modification by AD/VD-subgroup.

The four remaining trials belonging to this category, which used more relaxed criteria to identify patients with dementia, were all designed as parallel, placebo-controlled trials (Halama, 1991a; Halama, 1991b; Hartmann and Frick, 1991; Vesper and Hänsgen, 1994; Weitbrecht and Jansen, 1986; Weitbrecht and Jansen, 1985). The experimental treatment consisted of 120–150 mg of active substance per day, for about 3 months. Weitbrecht and Jansen reported a beneficial effect of

Ginkgo treatment in patients with a mild to moderate form of primary degenerative dementia according to various clinical rating scales (hetero-assessment of general health state, Crichton Geriatric Scale, SCAG) and various psychometric tests (figure-symbol test, number-repeat test, flicker fusion frequency, reaction time) (Weitbrecht and Jansen, 1986; Weitbrecht and Jansen, 1985). Halama evaluated the efficacy of Ginkgo treatment in 50 subjects with cerebral insufficiency classified as either degenerative dementia (Alzheimer's disease) or vascular dementia (Halama, 1991a,b). The Ginkgo cohort showed better results for the Syndrom Kurz Test (SKT; +4.15 vs. +1.32 points), the global assessment of treatment efficacy, and the number-connection test. Significantly more Ginkgo users than placebo users improved for 7 out of 11 clinical symptoms of cerebral insufficiency. Hartmann and Frick (1991) studied the effect of *Ginkgo biloba* in 52 patients with vascular dementia. Only patients with at least one symptom out of four symptom categories that are typical for the so-called organic brain psycho syndrome were considered eligible for the trial. A statistically significant effect was recorded only with the Grünberg Verbal Memory test, not with the other outcome parameters (global assessment of efficacy, 12 clinical symptoms, trail making test). In a trial with 90 subjects Vesper and Hänsgen (1994) reported a positive effect in favour of the Ginkgo group on a scale of subjective cerebral insufficiency troubles (CI-scale), both during the first part and the second part of the intervention trial. The rating of changes in each of 8 mental and physical functions on a 1–7 point transition scale (4 indicating "no change", 7 "very much improvement") pointed also to a beneficial effect: the scores on the 8 functions varied from about 4.7 to 5.0 in the Ginkgo group and from about 3.9 to 4.2 in the placebo group. A series of computer-aided psychometric tests also suggested a positive effect of Ginkgo treatment. No difference was found for the Short Test of General Intelligence (KAI) (Vesper and Hänsgen, 1994).

As a conclusion it can be stated that the published randomized controlled trials which have been designed to prove the efficacy of *Ginkgo biloba* special extract in demented patients, show remarkably consistent results. All trials show positive, often statistically significant effects in patients subjected to Ginkgo treatment compared with patients under placebo treatment, for most of the outcome parameters assessed. One may argue, however, that the reported effect sizes are less impressive and convincing from a clinical point of view. How should one value, for instance, a mean gain of 1.4 points for Ginkgo over placebo on the SKT-scale of memory and attention (maximum improvement: 27 points), or on the ADAS-cog scale (maximum improvement: 70 points) in the sense of the amount of clinical benefit for the patients involved?

Cognitive impairment

The second category consists of 8 RCTs that focus on patients with cerebral insufficiency, manifested by cognitive impairment probably not being dementia (Allain *et al.*, 1993; Chartres *et al.*, 1987; Franco *et al.*, 1991; Israël *et al.*, 1987; Oswald *et al.*, 1997; Rai *et al.*, 1991; Semlitsch *et al.*, 1995; Wesnes *et al.*, 1987). Four studies were conducted in France, two in the U.K., one in Germany, and one in

Austria. Again, it should be emphasized that the reported diagnostic criteria do not always preclude the presence of dementia (Oswald et al., 1997). Several terms and criteria have been proposed to indicate a type of cognitive decline that might be distinguishable from dementia as a separate pathological concept and entity, such as cognitive decline no dementia (CIND), age-associated cognitive decline (AACD), and especially age-associated memory impairment (AAMI) (Crook et al., 1986; Crook et al., 1990). Only one trial strictly adheres to the AAMI-criteria as far as the patient recruitment is concerned (Semlitsch et al., 1995). Table 3 presents the main features of these cognitive impairment trials.

In addition to seven parallel trials – six of them placebo-controlled, and one with nicergolin (NCG) as the reference substance (Franco et al., 1991) – one cross-over trial was identified (Allain et al., 1993). One of the placebo-controlled trials was based on a factorial design, with memory training as the second intervention factor of interest (Israël et al., 1987). The most frequently occurring daily dose was 120 mg/day, and the duration of treatment varied from 2–12 months. The cross-over trial aimed at demonstrating the acute effect of a large single dose of EGb 761.

The outcome measures used in these trials enclosed the global assessment of the intervention effect (4 studies); various clinical rating scales (e.g. MMSE, Global Deterioration Scale); quality-of-life assessment; depression; self-assessesment of well-being and behaviour; hetero-assessment of functional capacity; memory and attention; level of drug prescription; and finally psychophysiological parameters (e.g. EEG). Except for two studies (Franco et al., 1991; Oswald et al., 1997), a clear hierarchy of the outcome measures was not given.

Chartres et al. (1987) found no difference in global effect between Ginkgo and placebo, a small gain in the level of cognitive functioning (MMSE, Reisberg's GDS, Plutchik GRS) for the Ginkgo users, no difference in the development of several clinical signs, but a remarkably larger decline in the prescription of (other) psychotropic drugs for the persons in the Ginkgo group. Israel et al. (1987) observed an extra effect of Ginkgo, on top of the memory training effect. Within the subgroup that did not receive memory training the Ginkgo users felt more satisfied than the placebo users. Part of the results of the memory battery (long-term memory, fluency test) was in favour of Ginkgo as well. Wesnes et al. (1987) failed to detect a significant overall Ginkgo effect. They also found no difference in actual daily activities between the Ginkgo and placebo users. However, the Ginkgo group performed better on most of the psychometric tests, which were the outcome measures at the forefront in this study. In the intervention trial of Rai et al. (1991) (computerized) psychometric testing was strongly emphasized as well. When compared with the placebo group, the Ginkgo group performed better on most of the tests. A positive effect, be it small, was also found for the MMSE. In the trial of Franco et al. (1991) Ginkgo surpassed nicergolin on almost all the outcome parameters. The cross-over trial conducted by Allain et al. (1993) evaluated the acute effect of two different doses of Ginkgo and placebo through a Latin square design. Series of drawings and words were presented, with varying presentation times. The number of correctly recalled objects was registered. It is a well-known fact that drawings can be remembered more easily than words. The trial results demonstrated that the

Table 3   RCTs on the efficacy of *Ginkgo biloba* special extract in patients suffering from cerebral insufficiency: cognitive impairment

Author, year	Study population Diagnostic criteria	Number randomized analyzed (EG/other)	Age (yrs) Sex (M/F)	Intervention Brand name Administration route Dose Duration	Outcome measure	Result	Remarks
Chartres et al., 1987	mild cognitive impairment – MMS (Folstein): 20–26 – GDS (Reisberg): stages 1–4 psychotropic treatment (neuroleptics, anti-depressants, etc.) > 3 m.	44 42 n.s.	81.8 1:3	Tanakan® n.s. n.s. 6 m.	global effect, VAS (0–100 mm) (physician) global effect (patient) level of cognitive functioning: – MMS – Global Deterioration Scale (GDS) – Plutchik Geriatric Rating Scale (GRS) development of several clinical signs (tremor, fall, incontinence for urine, etc.) quantity of psychotropic drugs prescribed: – neuroleptics: – antidepressants: – tranquillizers/hypnotics:	47.6 vs 46.2 (n.s.) 0.8 vs 1.0 (trend; n.s.) +0.5 vs +0.0 +0.22 vs –0.04 +0.7 vs –0.6 no difference score: 24.7 (0 m.) → 11.7 (6 m.) +2.3 vs +0.9 +1.7 vs +2.6 +8.8 vs +10.6	parallel vs placebo residents old peoples' home (≥ 6 m.) duration of psycho-tropic treatment: 54.7 m.
Israel et al., 1987	early cognitive problems: – subjective complaints – MMS (Folstein): 20–26 absence of depression: – GDS (Yesavage): < 18 examination by physician + psychologist	80 75	56–83 (68.4) 26/54	Tanakan® oral, solution 160 mg/day (2 × 2 ml) 3 m.	global effect: drug + training (patient) global effect: drug – training (patient) memory battery for elderly, ambulatory: 4 factors: 1. long-term memory. 2. short-term memory. 3. learning rate. 4. fluidity. consequences of cognitive impairment for daily life: questionnaire (Israel) (physician) intellectual dynamics scale (Israel) (physician) Geriatric Depression Scale (Yesavage) (patient) scale of daily life difficulties (MacNair & Kahn scale) (patient)	only Ginkgo group: extra effect 88% vs 23% satisfied (p < 0.001) unique effect ginkgo, indepen-dent of training effect: 1. n.s. 2. p < 0.01. 3. not. 4. p < 0.005	parallel vs placebo out-patients: 3 sources factorial design: 4 groups: 2 with memory training (11 sessions)

Table 3   continued

Author, year	Study population Diagnostic criteria	Number randomized analyzed (EG/other)	Age (yrs) Sex (M/F)	Intervention Brand name Administration route Dose Duration	Outcome measure	Result	Remarks
Wesnes et al., 1987	cognitive impairment – mild, impairment – Crichton Geriatric Behavioural Scale: ≥ = 14 not > 3 for any item (mild impairment of everyday functioning)	58 28/30 54 27/27	62–85 34/20	Tanakan® oral, tablets 120 mg/day 12 w.	overall assessment of efficacy (patient) overall assessment of efficacy (physician) behavioural rating scale: quality of life cognitive testing: – digit span (WAIS), forward + backward – Benton Visual Retention Test (visuo-spatial) – automated psychometric test battery: 1. immediate word recall. 2. number matching. 3. rapid visual information processing. 4. choice reaction time. 5. digit symbol substitution. 6. word recognition mood + alertness (11 VAS-scales) medical examination: ECG, haematology, bioch.	27 vs 25 '(very) good' 27 vs 25 '(very) good' no difference in daily activities, but showed more interest +0.8 vs +0.4 +0.3 vs −0.3 1. +0.5 vs +1.0 2. number correct: +7.5 vs −6.4; time: +14 vs −35 msec 3. % detected: +12.3% vs +15.7% 4. number correct: +0.8 vs +0.9; time: +24 vs +9 msec 5. +13.5 vs +10.2. 6. number correct: +0.0 vs +10.2; time: −441 vs −65 msec effect size 4 accuracy tests: $p < 0.05$ effect size 4 speed tests: $p < 0.05$	parallel vs placebo non-institutionalized, GP-setting exclusion criteria run-in period: 1–3 w. blocks of 6 HIS: 51x ≤ 4, so probably early stage of degenerative dementia
Rai et al., 1991	memory impairment: 1. mild-moderate signs, organic origin, > 3 m., NINCDS-ADRDA classification 2. clinical + medical assessment: confirmation of diagnosis	31 27 12/15	67–89 7/20	Tanakan® oral, tablets 120 mg/day 6 m.	MMSE (Folstein) (0–30) Kendrick battery: detecting dementia in elderly: – digit copying task – object learning task digit recall test: no. correct before error (computerized) total no. correct magnitude of error classification task: no. attempted (computerized) median reaction time latency auditory event related potential (P300) EEG-frequency analysis	+0.18 vs −0.29 +5.55 vs −2.43 (p = 0.017) +0.82 vs +2.43 −2.5 vs +5.3 (p = 0.032) −1.7 vs +4.1 +26.7 vs −34.2 +0.00 vs −0.17 +171 vs −37 msec (p = 0.026)	parallel vs placebo out-patients: hospital, dept. of geriatric medicine

Table 3   continued

Author, year	Study population Diagnostic criteria	Number randomized analyzed (EG/other)	Age (yrs) Sex (M/F)	Intervention Brand name Administration route Dose Duration	Outcome measure	Result	Remarks
Franco et al., 1991	age-associated memory problems in daily life: - CDS-score: 40–120 - Zung-scale < 50 (no severe depression)	97 78 38/40	60-85 n.d.	Tanakan® n.d. 120 mg/day 12 m.	*primary outcome measures:* - CDS-score (cognitive deficits self-assessment scale (Mac Nair)): troubles of attention, concentration, psychomotor coordination, orientation, STM, LTM - Zung scale (stability of mood) *secondary outcome measures:* - global assessment efficacy, VAS (patient) - global assessment efficacy, VAS (physician) - preference for post-trial treatment - development of functional signs (1. motor, 2. neuro-sensoric, 3. psycho-affective)	+11.6 vs +7.7 +1.5 vs +0.3 (n.s.) 70.6 vs 58.3 mm 72.8 vs 58.3 mm 35 vs 28: continue (p = 0.007) 3. +2.9 vs +1.9	parallel vs nicergolin not-institutionalized: city practice own GP, 12 centres
Allain et al., 1993	age-related memory impairment (slight degree): - MMSE: 25–28 - objective cognitive impairment (immediate recall of word lists)	18 6x 3 subj.	60-80 (69.3) n.d.	Tanakan® (IPSEN) n.d. 600 mg/320 mg acute	relative speed of information processing: dual-coding test (evaluates memory coding of verbal material (words) and images (drawings) in relation to variable presentation times) 3 different batteries of memory tests (5 series of 8 drawings + 5 series of 8 words; presentation times: 1920, 960, 480, 240, 120 msec) testing 1 h after drug administration	no. correctly recalled, EGb 600 vs EGb 320 vs placebo: - Drawings, 1920: 5.3 ; 5.5 ; 5.1 - Drawings, 960: 4.2 ; 4.3 ; 3.9 - Drawings, 480: 3.2 ; 3.0 ; 3.2 - Words, 1920: 3.3 ; 3.3 ; 3.3 - Words, 960: 3.0 ; 2.9 ; 3.5 - Words, 480: 2.9 ; 2.6 ; 2.8 Drawings/Words difference: - breakpoint: 480 vs 480 vs 960 - dual coding: 960 vs 960 vs1920	cross-over EGb 761 600 mg vs EGb 761 320 mg vs placebo source: 1 centre Latin square design wash-out: 6 d. acute effect measured

Table 3   continued

Author, year	Study population Diagnostic criteria	Number randomized analyzed (EG/other)	Age (yrs) Sex (M/F)	Intervention Brand name Administration route Dose Duration	Outcome measure	Result	Remarks
Semlitsch et al., 1995	age-associated memory impairment (AAMI): – age > 50 – subjective everyday memory problems – gradual course memory loss – Grünberger Verbal Memory Test < 29 – BVRT < 6 – MMSE ≥ = 24 (no dementia)	48 24/24 43 21/22	51–79 (62) 19/29	Ginkobene® oral dosage 120 mg/day 57 d.	psychometric tests EEG brain-mapping cognitive information processing: psychophysiological investigations: two-tone auditory oddball paradigm, registration of long-latency Event-Related Potentials (ERP); P300 latency may reflect stimulus-evaluation time subjective well-being: EWL-K checklist (self-reporting thymopsychic scale; subscales: activation, deactivation, assertiveness, fatigue, dizziness, high spirits, agitation, sensitiveness, annoyance, anxiety, depression, dreaminess)	no consistent and unequivocal changes on N1, P2, N2, P300 amplitudes, nor N1, P2, N2 latencies; P300 latency shortened by 31 msec (acute), 38 msec (chronic), 32 msec (superimposed) for Ginkgo acute treatment (w.0, 3 h): no difference chronic treatment (w.8, 0 h): GBE: increase in deactivation, fatigue, dizziness, anxiety superimposed treatment (w.8, 3 h): GBE: increase in dizziness	parallel vs placebo source: n.d.
Oswald et al., 1997	organic brain syndrome: – clinical diagnosis according to DSM-III-R, ICD-10, 290.x – mild to moderate – SCAG: sum 18 items: 40–90 mean items 2,3,6,8 > 3.5 sum items 1–8: > 24 items 1–8: not 7	226 195 103/92	60–91 81/145	EGb 761® oral, tablets 120 mg/day 12 w.	Nuremberg Gerontopsychological Inventory (NAI): 3 domains: 1. fluid cognitive performance (objective test): – number–symbol test (ZS-G) – labyrinth-test (LT-G) – number–connection test (ZVT-G) responder: ≥ 2 of 3 tests 2. well-being (self-assessment): – self-assessment scale (NAS) – daily activities scale (NAA) – self-estimation scale (NSL) responder: ≥ 2 of 3 scales 3. function level (hetero-assessment): – elderly rating (NAR): 9 items responder: no deterioration efficacy: respondent in ≥ 2 of 3 domains	45% vs 38% improved (0.174) 65% vs 53% improved (0.047) 54% vs 53% improved (0.438) 59.2% vs 48.9% (p = 0.074) 49% vs 47% improved (0.401) 35% vs 38% improved (0.673) 52% vs 54% improved (0.657) 46.6% vs 45.7% (p = 0.447) 81% vs 74% improved (0.133) 80.6% vs 73.9% (p = 0.133) n.s.	parallel vs placebo out-patients, living at home or in an old people's home; 2 GP-practices exclusion: specific causes of brain syndrome run-in period: 2 w. compliance (urine, pill count) > 90%

breakpoint and the "dual coding" point marking the onset of a significant difference between word recall and drawing recall, had been shifted to a shorter presentation time under influence of Ginkgo, irrespective of the treatment dose. This suggests an increase of the relative speed of information processing due to Ginkgo treatment (Allain *et al.*, 1993). Semlitsch *et al.* (1995) assessed the influence of *Ginkgo biloba* special extract in AAMI-patients by means of psychometric tests, EEG brain-mapping, a checklist expressing subjective thymopsychical well-being, and neuro-physiological examinations indicating the quality of cognitive information processing. An acute (3 hours after first dose, week 0), chronic (before the last dose, week 8 of treatment), and superimposed (3 hours after first dose, week 8 of treatment) effect of Ginkobene® on P300 latency was registered. The shortened latency may reflect a decrease of the stimulus-evaluation time, evoked by Ginkgo treatment (Semlitsch *et al.*, 1995). Oswald *et al.* (1997) evaluated the effect of *Ginkgo biloba* within three domains of performance: fluid cognitive performance (objective psychometric testing), subjective well-being (self-rating), and functional performance (hetero-assessment). Only at the fluid cognition level a significant effect of Ginkgo was found. A responder analysis that covered all three domains of performance did not yield a significant result (Oswald *et al.*, 1997).

It can be concluded that the results of the trials discussed under this heading point to a beneficial effect of Ginkgo treatment, although not unequivocally and not for all outcome parameters.

Non-cognitive signs and symptoms

Twelve randomized clinical trials were identified which had their main focus on the treatment of "physical", non-cognitive signs and symptoms of cerebral insufficiency, such as vertigo (dizziness), tinnitus ("ear noise"), hearing loss, and ataxia (see Table 4).

These studies were reported from France (6), Germany (5), and Sweden (1). A cross-over design, which contrasted Ginkgo with cinnarizine (CIN) and placebo, respectively, was employed in two trials (Holgers *et al.*, 1994; Natali *et al.*, 1979). The other studies were designed as parallel trials. Six of them contrasted *Ginkgo biloba* with placebo, the remaining four compared Ginkgo with another active compound (nicergolin (NCG), almitrine-raubasine (ARB), naftidrofuryl). Four studies focussed on vertigo / dizziness as the leading symptom, also four on tinnitus, two on cochlear deafness or sudden hearing loss, and two included a mixture of patients with vertigo, tinnitus and hearing loss. The number of patients varied from 20 to 80 in most of the trials, and exceeded 100 in only two trials (Meyer, 1986ab). The mean age of the study population was close to 50 in the majority of the studies. The "average" intervention scheme consisted of a daily Ginkgo dosage of 120–160 mg/day, administered as an oral solution during 1–3 months. In one study Ginkgo extract was given as a supplement to a physical vestibular training programme (Hamann, 1985).

The following endpoints were used in these trials: global assessment of the treatment effect (Chesseboeuf *et al.*, 1979; Dubreuil, 1986; Haguenauer *et al.*, 1986;

Table 4 RCTs on the efficacy of *Ginkgo biloba* special extract in patients suffering from cerebral insufficiency: non-cognitive signs and symptoms

Author, year	Study population Diagnostic criteria	Number randomized analyzed (EG/other)	Age (yrs) Sex (M/F)	Intervention Brand name Administration route Dose Duration	Outcome measure	Result	Remarks
Chesseboeuf et al., 1979	hearing loss/vertigo/ tinnitus: (vascular origin, or ageing of inner ear, or due to injury, infarction) diagnostic criteria: n.d.	60 30/30	51.6 35/25	Tanakan® oral, solution 3 ml/day 2 m.	global effect: very good/good/moderate/no functional signs + symptoms: – hearing loss: very good/good/moderate/no change – tinnitus: very good/good/moderate/no change – vertigo: very good/good/moderate/no change – audiogram: % improved – electronystagmogram: % improved	5/13/8/4 vs 0/12/14/4 +0.46 vs +0.46 (n.s.) 1/5/4/16 vs 2/16/15 +1.13 vs +1.07 (n.s.) 2/2/7/4 vs 3/4/6/1 +1.45 vs +0.83 (p < 0.1) 4/6/7/4 vs 1/3/6/5 +0.35 vs +0.29 (n.s.) 33% vs 28% +0.8 vs +0.6 (n.s.) 35% vs 13%	parallel vs nicergolin (NCG) source: n.d.
Natali et al., 1979	hearing loss/tinnitus + vertigo: ((partly) of vascular origin) severity of hearing loss: 9x < 30 dB, 9x > 30 dB	20	60 10/10	Tanakan® oral, solution 3 ml/day 1–2 m.	global effect: (very) good/insufficient/no functional signs (presence, severity): – labyrinth examination – audiometric balance – electronystagmogram (ENG)	12/4/4 vs 1/12/7 hearing loss: 5/18 vs 1/18 '(very) good' tinnitus: 10/13 vs 0/13 '(very) good' vertigo: 2/2 vs 0/2 '(very) good'	cross-over vs cinnarizin (CIN) hospital patients only partly randomized (until 13th patient): 13 started with CIN wash-out: 1 m.
Schwerdtfeger, 1981	dizziness symptoms: origin: 7 categories neurootologic examination pathological findings of electronystagmogram (butterfly or L-pattern)	50 25/25 42 23/19	17–63 (43) 17/33	rokan® oral, tablets 60 mg/day (3 dragees) 2 m.	questionnaire electronystagmography: 1. calorimetric (butterfly pattern, Claussen) 2. rotary chair (L-pattern, Claussen) neuro(oto)logical tests (10 parameters) tonal audiogram (+ vocal, + impedance)	overall evaluation: number improved/unchanged/impaired: 11/9/3 vs 3/4/12 0 vs 7 relapsed after successful open trial	parallel vs placebo source: n.d. first open trial: 2 m. (rokan, 160 mg/day)
Claussen, 1984 Claussen and Kirtane, 1985	vertigo/ataxia symptoms: strong, subjective dizziness symptoms	50 25/25 33 14/19	50–71 (55.5) 17/16	rokan® oral, tablets 120 mg/day 12 w.	questionnaire on subjective dizziness symptoms lateral body sway: craniocorpography (CCG)	+50% vs +20% improvement +9.2 cm vs +2.7 cm (p < 0.005)	parallel vs placebo source: n.d.
Hamann, 1985	vertigo: dizziness complaints due to Menière disease, vestibular neuropathy, trauma ≥ = 1 out of 3 signs: dizziness, visual motor changes (electronystagmography), body balance disturbances	40 20/20 35 17/18	46.2 13/27	Tebonin® forte oral, solution (drops) 160 mg/day 4 w. adjuvans therapy (supplementary to physical, vestibular training; 10 sessions)	vertigo questionnaire: subjective dizziness complaints Romberg's test: posturographic body sway: – anterior-posterior – lateral	% cured or improved: ca. 75% vs ca. 75% (n.s.) % unchanged: 20% vs 18% +23.1 mm vs +15.5 mm +21.2 mm vs +10.8 mm ca. +41% vs ca. +26%	parallel vs placebo source: n.d. no response to previous, 'classic' therapy

Table 4   continued

Author, year	Study population / Diagnostic criteria	Number randomized analyzed (EG/other)	Age (yrs) Sex (M/F)	Intervention Brand name Administration route Dose Duration	Outcome measure	Result	Remarks
Dubreuil, 1986	cochlear deafness partly or total, ≥ 1 week idiopathic, or due to noise trauma, or to pressure trauma clinical + otoneurological exam, impedance, electronystagmography	20 10/10 18 9/9	49.6 n.d.	Tanakan® oral, solution 4 ml, 2x/day 30 d.	global effect (ENT specialist): '(very) good'/'moderate'/'failure' tonal audiometry (250, 500, 1000, 2000, 4000, 8000 Hz) vocal audiometry associated symptoms: tinnitus vestibular examination + caloric measurements	7/1/1 vs 3/2/4 mean gain for 6 frequencies: +34.1 dB vs +23.0 dB +52.3% vs +37.8% Ginkgo: stronger effect 7 vs 2 cured 9 vs 3 normalized	parallel vs nicergolin (NCG) (alpha blocker) source: hospital (10 d.) + out-patient exclusion criteria
Meyer, 1986a	tinnitus: ambulatory, < 1 year; not susceptible for anti-infectious, surgical treatment	259 233 119/54/47	53.2 137/122	Tanakan® usual dose and form > 1 m.	global assessment of development of tinnitus (physician): 6 categories (much improved — much deteriorated) clinical assessment of tinnitus aspects: – intensity (scale 0–3) – discomfort (scale 0–3) time needed to disappearance or clear reduction	% cured or clearly improved: 63% vs 39% vs 30% (prognostic factors evaluated) +0.98 vs +0.71 vs +0.54 ($p < 0.01$) +0.94 vs +0.61 vs +0.51 ($p < 0.01$) 75 vs 100 vs 156 days	parallel vs nicergolin (NCG) + vs almitrine-rau-basine (ARB) city practices of 15 ENT specialists pseudo-randomization (day of the week)
Meyer, 1986b	tinnitus: out-patients of recent origin: < 1 year not susceptible for anti-infectious, surgical treatment	103 58/45	50.4 54/49	Tanakan® oral, solution 4 ml/day 1–13 m.	global assessment of change in clinical signs (timbre, periodicity, rhythm, discomfort); scale: 1–6 (1 = much impaired, 2 = impaired, 3 = unchanged, 4 = bit improved, 5 = improved, 6 = much improved, 7 = treatment not tolerated) time until clear improvement/disappearance change of intensity (scale 0–3) change of level of discomfort (scale 0–3) paraclinical criteria: tonal audiometry, impedance, electronystagmography	% (much) improved: 56% vs 40% ($p = 0.05$) 70 vs 119 days ($p = 0.03$) +1.00 vs +0.67 ($p = 0.03$) +0.84 vs +0.59 ($p = 0.08$)	parallel vs placebo out-patients; city practices of 10 ENT-specialists exclusion criteria prognostic factors: duration, location (uni- vs. bilateral), always vs intermittent
Haguenauer et al., 1986	vertigo syndrome clinically confirmed recent onset (< 2 yrs.) origin unknown	70 35/35 67 34/33	49.3 34/33	Tanakan® oral, solution 160 mg/day (2x 2 ml) 3 m.	global assessment of impact of vertigo on daily life; VAS (0–100 mm) (patient) global impression of improvement since start intervention; 5-point scale: disappeared, much improved, improved, stable, impaired (clinician) questionnaire on intensity, frequency and duration of vertigo and associated signs (hearing loss, tinnitus, nausea, headache); 4-point scale: disappeared (0), diminished (1), stable (2), aggravated (3) otoneurological examination: Romberg test, Babinski-Weill test (blind walking), nystagmus	% improvement since onset: 75% vs 18% ($p = 0.01$) % cured or clearly improved: 85% vs 42% ($p < 0.001$) sum of scores w.1, w.2, w.3 (range: 0–9): intensity: +3.4 vs +4.5 ($p = 0.02$) frequency: +3.2 vs +4.2 ($p = 0.02$) duration: +3.3 vs +4.5 ($p = 0.03$) % with normal values: +35% vs +16% ($p = 0.06$)	parallel vs placebo 3 centres exclusion criteria

Table 4   continued

Author, year	Study population / Diagnostic criteria	Number randomized analyzed (EG/other)	Age (yrs) Sex (M/F)	Intervention / Brand name / Administration route / Dose / Duration	Outcome measure	Result	Remarks
Holgers et al., 1994	permanent severe tinnitus (PST)	20	59.5 10/10	GBE (Seredrin®) oral, tablets 29.2 mg/day 2 w.	preference expressed: GBE vs. placebo subjective assessment of tinnitus: (VAS, 100 mm) – loudness – awareness – annoyance	6 GBE, 7 placebo, 7 no pref. no. with strongest effect by: GBE/placebo/no sign. diff.: 2/1/15 (3 did not participate) 0/2/16 (3 did not participate) 2/2/14 (3 did not participate) inconsistency between drug preference and VAS recordings	cross-over vs placebo source: Hospital Audiology Dept., Göteborg RCT preceded by 2 w. open trial: 21 out of 80 patients who improved on GBE in open trial were admitted to RCT wash-out: 1 week
Hoffmann et al., 1994	sudden hearing loss: – idiopathic – < 11 days – no previous treatment	80 40/40	19–89 (40.5) 47/33	Tebonin® oral, solution + tablets 175 mg in 500 ml HAES 6% + 80 mg/day 3 w. HAES as standard therapy	relative gain of hearing capacity tinnitus: loudness (dB) (tonal audiometry) tinnitus: intensity (subjective: VAS) hearing loss (sum of 250, 500 Hz, 1, 2, 4, 6, 8 KHz) vestibular test	+23% vs +13% (p = 0.06) total recovery, 1 w.: 40% vs 40% total recovery, 3 w.: 78% vs 55% +38 dB vs +34 dB +4.3 vs +3.2 points +180 dB vs +140 dB	parallel vs naftidrofuryl (Dusodril®)
Morgenstern and Biermann, 1997	tinnitus aurium: – chronic: > 2 months; mean duration: 4.5 yrs. – reproducible	99 49/50	19–79 (45.5)	Tebonin® forte oral, tablets 120 mg/day 12 w.	primary outcome measure: – tinnitus: loudness (most seriously affected ear) assessed by tonal audiometry secondary outcome measures: – click-evoked otoacoustic emission (OAE) – tinnitus: frequency and intensity – hearing loss (tonal audiometry) – subjective impression of tinnitus intensity by patient; scale 0–5 – subjective impression of change in tinnitus complaints; scale: –5 – +5 – laboratory assessment (blood)	+3.3 vs –1.0 dB (p = 0.015) (contralateral ear: no difference) no change +0.4 vs +0.4 % deteriorated/same/improved: 31%/61%/8% vs 14%/84%/2%	parallel vs placebo source: ENT-clinic, ambulatory patients run-in: 2 w. (placebo) drop-outs: 21 (13 vs 8) I-to-T analysis

Holgers *et al.*, 1994; Meyer, 1986ab; Morgenstern and Biermann, 1997; Natali *et al.*, 1979); assessment of functional signs and symptoms through a questionnaire, VAS-scale, or clinical rating scale (Chesseboeuf *et al.*, 1979; Claussen, 1984; Claussen and Kirtane, 1985; Haguenauer *et al.*, 1986; Hamann, 1985; Hoffmann *et al.*, 1994; Holgers *et al.*, 1994; Meyer, 1986ab; Morgenstern and Biermann, 1997; Schwerdtfeger, 1981); and various forms of paraclinical otoneurological examination, e.g. vocal and tonal audiometry, electronystagmography, craniocorpography (CCG) and Romberg's test (posturographic body sway).

The trials which contrasted Ginkgo with a reference drug may be considered less informative with regard to the potential efficacy of Ginkgo. Nevertheless, it is remarkable that in most of these trials Ginkgo surpassed the reference drug for nearly all the parameters evaluated (Chesseboeuf *et al.*, 1979; Dubreuil, 1986; Meyer, 1986a; Natali *et al.*, 1979). In the trial reported by Schwerdtfeger (1981) the overall conclusion, based on a questionnaire, electronystagmographic evaluation, and several other neuro(oto)logical tests, pointed to a beneficial effect of Ginkgo. Claussen found a positive effect on both parameters evaluated: subjective dizziness symptoms and lateral body sway (Claussen, 1984; Claussen and Kirtane, 1985). Hamann (1985) reported a positive effect on the degree of posturographic body sway, whereas the change in subjective dizziness complaints was almost equal for the Ginkgo and placebo groups. Meyer (1986a) reported beneficial effects of Ginkgo, which were (borderline) statistically significant for all the relevant endpoints: global assessment of change in clinical signs, change of intensity of tinnitus complaints, change of level of discomfort due to tinnitus, time needed to reach significant improvement or disappearance of complaints. Statistically significant results for all outcomes measured were also communicated by Haguenauer *et al.* (1986). Morgenstern and Biermann (1997) found a larger regression of tinnitus loudness for the Ginkgo group than for the placebo group. However, this result was not accompanied by a positive effect of Ginkgo on the patient's subjective impression of the change in tinnitus complaints. In contrast with the studies mentioned before, the placebo-controlled cross-over trial published by Holgers *et al.* (1994) did not reveal a positive effect of Ginkgo in patients suffering from tinnitus.

A rather new approach in the treatment of tinnitus is the application of *Ginkgo biloba* extract in combination with soft-laser therapy. One randomized trial, based on a factorial design, could be identified in this field (Von Wedel *et al.*, 1995). It was excluded from the review table, since the emphasis was on soft-laser irradiation and random allocation of the study subjects was not stated explicitly.

It can be concluded that almost all trials in this category suggest that Ginkgo is beneficial for patients with vertigo, tinnitus and related manifestations of cerebrovascular insufficiency, with only one trial clearly out of tune.

Subjective cognitive and non-cognitive symptoms

Table 5 shows the details of the subcategory of 5 RCTs – all reported from Germany – which focussed on the treatment of patients with cerebral insufficiency, manifested by a combination of both cognitive and non-cognitive clinical symptoms.

Table 5   RCTs on the efficacy of *Ginkgo biloba* special extract in patients suffering from cerebral insufficiency: cognitive and non-cognitive symptoms

Author, year	Study population / Diagnostic criteria	Number randomized analyzed (EG/other)	Age (yrs) / Sex (M/F)	Intervention / Brand name / Administration route / Dose / Duration	Outcome measure	Result	Remarks
Eckmann and Schlag, 1982	cerebral insufficiency: signs + symptoms neurological + psychical characteristics	50 / 25/25	45–74 / 30/20	Tebonin® forte oral, solution (drops) 120 mg (60 dr./day 30 d.	list of 11 'cerebrovascular' symptoms: 1. dizziness. 2. headache. 3. tinnitus. 4. awareness. 5. orientation. 6. motor function. 7. sensibility. 8. speaking: understanding. 9. speaking: ability. 10. depression. 11. anxiety / clinical assessment of neurological findings / clinical assessment of psychopathology / EEG / laboratory tests (SGOT, SGPT, bilirubin, etc.)	presence of each symptom: 1. +91 vs +43% 2. +92 vs +48% 3. +100 vs +50% 4. +92 vs +55% 5. +94 vs +53% 6. +92 vs +57% 7. +92 vs +55% 8. +93 vs +53% 9. +92 vs +43% 10. +92 vs +44% 11. +92 vs +44% / +92% vs +44% ($p < 0.001$) / +92% vs +44% ($p < 0.001$)	parallel vs placebo source: n.d. largest treatment effect: during days 8–18 median time until onset of improvement: 11 vs. > 30 d. ($p < 0.01$)
Eckmann, 1990	cerebral insufficiency + depressive mood ≥ 6/12 symptoms 'moderate' or 'severe'	60 / 58 / 29/29	41–71 / 31/29	Kaveri® oral, solution 160 mg/day 6 w.	12 subjective clinical CI-symptoms: 1. dizziness. 2. ear noise. 3. headache. 4. confusion. 5. memory loss. 6. forgetfulness. 7. concentration. 8. fatigue. 9. loss of performance. 10. lack of motivation. 11. depression. 12. anxiety; 4 categories: absent, mild, moderate, severe	no difference / number of improved persons higher with Ginkgo vs. placebo, for all 12 symptoms ($p < 0.001$); most improvements during weeks 2–4, e.g.: concentration: +23 vs +5 forgetfulness: +22 vs +5 depression: +24 vs +2 still improvement in weeks 4–6	parallel vs placebo hospitalized because of depression due to CI duration of symptoms: ca. 32 m.
Brüchert et al., 1991	cerebral insufficiency: typical symptoms HOPS: ≥ 1 symptom within 4 symptom groups present prevalence range of 11 symptoms: 45–95%	303 / 153/150 / 209 / 110/99	69 / 77/132	Kaveri® (LI 1370) oral, tablets 150 mg/day 12 w.	global assessment of efficacy (patient) ('very good'/'good'/'moderate'/'poor') global assessment of efficacy (physician) ('good'/'moderate'/'unsatisfactory') 11 subjective clinical CI-symptoms: memory, forgetfulness, concentration, fatigue, performance, motivation, depression, anxiety, dizziness, headache, tinnitus 4-point scale: severe, moderate, mild, absent psychometric test: trail making test (figure connection test)	22/69/17/2 vs 8/43/34/14 ($p < 0.001$) 78/28/4 vs 31/29/39 ($p < 0.001$) no. improved: Ginkgo vs. placebo: $p < 0.01$: 3 symptoms $p < 0.05$: 5 symptoms n.s., tendency: 1 symptom n.s.: 2 symptoms (anxiety, dizz.) +20 vs +12 sec ($p < 0.01$)	parallel vs placebo out-patients 33 GP practices run-in period: 2 w. exclusion criteria

Table 5   continued

Author, year	Study population Diagnostic criteria	Number randomized analyzed (EG/other)	Age (yrs) Sex (M/F)	Intervention Brand name Administration route Dose Duration	Outcome measure	Result	Remarks
Schmidt et al., 1991	cerebral insufficiency: typical symptoms of reduced brain function typical symptoms HOPS: ≥ 1 symptom within each of 4 symptom groups present	100 99 50/49	51–68 45/54	Kaveri® (LI 1370) n.d. 150 mg/day 12 w.	global assessment of efficacy (patient) global assessment of efficacy (physician) ('good'/'moderate'/'unsatisfactory') 12 subjective clinical CI-symptoms: memory, forgetfulness, concentration, fatigue, performance, motivation, depression, anxiety, dizziness, headache, tinnitus, confusion 4-point scale: severe, moderate, mild, absent; blood pressure, heart rate	35/7/7 vs 7/12/30 (p < 0.01) 36/6/8 vs 4/12/33 (p < 0.01) no. improved: Ginkgo vs. placebo: p < 0.01: 7 symptoms p < 0.05: 1 symptom n.s., tendency: 2 symptoms n.s.: 2 symptoms (headache, confusion)	parallel vs placebo out-patients: 1 internist's practice mean duration of disease: 26 m. run-in period: 3 w. exclusion criteria
Hofferberth, 1991	organic brain psycho-syndrome (HOPS): stable condition	50 25/25	57–76 29/21	Kaveri® (LI 1370) oral, tablets 150 mg/day 6 w.	global effect (physician) 11 clinical CI-symptoms (hetero-assessment): memory, forgetfulness, concentration, fatigue, performance, motivation, depression, anxiety, dizziness, headache, tinnitus; 4-point scale: severe, moderate, mild, absent psychometric testing: – Vienna Determination Test (classic reaction test; 3 steps; abnormal result: < 160, < 168, < 175 out of 180 correct) – number connection: > 28 sec for 30 numbers neurophysiological testing: – saccadic eye movements: duration of rapid eye movements (right) saccadic latency time (right) rapidity of eye movements (right) – EEG-analysis (theta part > 70%) – evoked potential (P300 amplitude)	14 vs 0 good; 4 vs 11 moderate 9/11 symptoms significantly better with Ginkgo compared with placebo (p < 0.01) +18 vs –2 correct reactions (tempo step 10) (p < 0.001) +12 sec vs +1 sec (p < 0.001) +55 msec vs +2 msec (p < 0.001) +85 msec vs –5 msec (p < 0.001) +150 vs +15 °/s. (p < 0.001) +20% vs +5% (p < 0.001) +7.5 vs +0.0 microV (p < 0.001)	parallel vs placebo hospital patients: neurological dept. of university hospital exclusion: use of psychophar-maca, nootropics, vasoactive drugs run-in period: 2 w.

A common feature of these trials is that the investigators applied a symptom checklist to assess the presence and severity of 11 or 12 clinical symptoms, both for patient selection and for outcome evaluation. Memory loss, forgetfulness, lack of concentration, dizziness, tinnitus (ear noise), fatigue, headache, loss of performance, lack of motivation, depression, anxiety, and sometimes confusion, made up the fixed combination of symptoms, which were scored on an ordinal response scale, expressing the degree of nuisance caused by each symptom (absent, mild, moderate, severe). It should be noted that in one study a slightly different symptom checklist was used (Eckmann and Schlag, 1982), and that in another study the primary concern was with psychometric and neurophysiological testing (Hofferberth, 1991). Furthermore, one should be aware that the same symptom checklist played a prominent role in a few other trials, which were discussed previously under the "dementia" heading (Halama, 1991ab; Hartmann and Frick, 1991).

All studies in this category were placebo-controlled trials. The size of the study population varied from 50 to more than 300 participants, the daily dose of *Ginkgo biloba* special extract from 120 to 160 mg/day, and the duration of the intervention from 4 to 12 weeks. As Table 5 demonstrates, most of the symptoms were found to react (significantly) better to Ginkgo treatment than to placebo treatment. Global assessment of the treatment effect in addition to the administration of the symptom checklist, either by the physician or by the patient, always proved to be in favour of Ginkgo. Hofferberth (1991) reported a positive effect of Ginkgo for the psychometric and neurophysiological parameters as well.

Depression as the leading symptom of cerebral insufficiency

Three RCTs were retrieved which concentrated on depressive mood as the major cerebral insufficiency symptom of interest (Halama, 1990; Schubert and Halama, 1993; Stocksmeier and Eberlein, 1992). The details of these trials are presented in Table 6.

In each of these trials treatment with tablets of Tebonin® forte (120–240 mg/day, during 8–12 weeks) was compared with placebo treatment. In two studies patients were included who had resisted previous therapy with (other) antidepressants (Halama, 1990; Schubert and Halama, 1993). In both trials EGb 761 was administered in combination with other antidepressants (Halama, 1990; Schubert and Halama, 1993).

A beneficial effect of Ginkgo was reported in all three trials, viz. 5.3 points improvement vs. 5.2 points deterioration on Zung's Self-Rating Depression Scale (Halama, 1990), 6.0 vs. 4.6 points improvement on the Hamilton Depression Scale (HAMD) (Stocksmeier and Eberlein, 1992), and 9.5 vs. 1.0 points improvement, also on the Hamilton Depression Scale (p < 0.01) (Schubert and Halama, 1993), respectively. In the last study 75% of the Ginkgo users vs. 30% of the placebo users showed ≥ 5 points decrease on the HAMD-scale. In the study of Halama (1990) a positive effect of Ginkgo was also reported for an additional primary outcome measure (late acoustic-evoked potentials) and for a series of secondary outcome measures, e.g. the Kurztest für Allgemeine Intelligenz (KAI), and the subjective

Table 6 RCTs on the efficacy of *Ginkgo biloba* special extract in patients suffering from cerebral insufficiency: depressive mood as a major sign

Author, year	Study population Diagnostic criteria	Number randomized analyzed (EG/other)	Age (yrs) Sex (M/F)	Intervention Brand name Administration route Dose Duration	Outcome measure	Result	Remarks
Halama, 1990	cerebral insufficiency + depression: – mild to moderate CI – non-responder to anti-depressants (> 3 m.)	20	55–85 9/11	Tebonin® forte oral, tablets 240 mg/day 8 w.	*primary outcome measures:* – Zung's Self-Rating Depression Scale (SDS) – late acoustic-evoked potential (SAEP/ALEP): latency-time P3-wave latency-time CNV-wave *secondary outcome measures:* – CI-test (Lehrl et al): interference – CI-test (Lehrl et al): symbols, numbers, etc. – CI-scale (Weidenhammer & Fischer) – Kurztest fur allgemeine Intelligenz (KAI) – subjective assessment of depression – subjective assessment of concentration	+5.3 vs –5.2 +13.3 vs –9.8 msec ($p = 0.048$) +46.4 vs –26.8 msec +11.7 vs –0.1 +3.4 vs –1.0 +3.1 vs –1.9 +2.5 vs –0.6 3 vs 0 improved verum group improved	parallel vs placebo anti-depressive regimen continued pilot study
Stocksmeier and Eberlein, 1992	cerebral insufficiency + depressive mood (mild): – subjective experience + clinical diagnosis – HAMD-score >15 – KLT-score < 90% of norm	60 30/30 36 17/19	51–78 17/43	Tebonin® forte oral, tablets 120 mg/day 12 w.	*primary outcome measure:* – Hamilton Depression Scale (HAMD), modified, sum score items 1–17 *secondary outcome measures:* – Konzentrations-Leistungs-Test (KLT) – Gießener well-being questionnaire – newly developed computerized test	+6.0 vs +4.6 (baseline: 13.3 vs 15.0) n.s. n.s. n.s.	parallel vs placebo source: n.d. MID, AD excluded
Schubert and Halama, 1993	cerebral dysfunction (mild, moderate) + depression: – no clear effect of anti-depressants (3 m.) – CI-scale >20 points	40 20/20	50–78 13/27	Tebonin® forte oral, tablets 240 mg/day 8 w.	*primary outcome measure:* – Hamilton Depression Scale (HAMD), modified, sum score items 1–17 *secondary outcome measures:* – CI-test (Lehrl & Fischer) – CI-scale (Weidenhammer & Fischer) – Kurztest fur allgemeine Intelligenz (KAI) – severity of symptoms (assessed by physician): depression (5 categories) concentration (5 categories) – late acoustic evoked potential (SAEP)	4.5 vs 13.0 ($p < 0.01$) (baseline: 14.0 vs 14.0) > +5.0: 15/20 vs 6/20 n.s. n.s. +15 vs +4 points ($p = 0.02$) no. moderate–severe: 19 → 4 vs 20 → 15 19 → 8 vs 20 → 12 n.s.	parallel EGb 761 + anti-depressants vs placebo out-patients neuro-psychiatric practice excluded: serious dementia, endogeneous depression, etc.

assessment of depression and concentration. No significant differences were registered for the secondary outcome measures in both other depression trials, except for a better result on the KAI for the subgroup under Ginkgo treatment and a larger decrease in the number of patients with moderate or severe symptoms of depression and concentration, based on subjective assessment by the physician, in the study of Schubert and Halama (1993).

It can be concluded that the trials in this category consistently suggest a favourable effect of Ginkgo on depressive mood.

Cerebral insufficiency, not clearly specified

A rest category was reserved for 18 randomized trials in the field of cerebral insufficiency which were published without sufficient diagnostic information to allow for allocation to one of the preceding categories. Most of these trials date from the 70s and the first half of the 80s. Nine were conducted in Germany, 6 in France, and 3 in Italy.

As Table 7 shows, 14 trials have been designed as a placebo-controlled parallel trial, 2 as parallel trials contrasting Ginkgo treatment with a (vasoactive) reference drug (vincamine, dihydroergotoxine) (Gerhardt et al., 1990; Haan et al., 1982), and 2 as placebo-controlled cross-over trials (Agnoli, 1980; Arrigo, 1986; Arrigo and Cattaneo, 1985). The number of participants amounted to 30 and 90 in the cross-over trials, and varied from 14 to 189 in the parallel trials. The average age of the participants approached 70 years. In the majority of the trials a daily amount of 120 mg of active substance was administered. A higher exposure was chosen in only 5 trials. The most common mode of administration of the study medication was per os – in 8 trials as a solution, in 5 trials in tablet form – while in one study the active substance was administered intravenously for 15 days (Haan et al., 1982). The reports of 4 other studies lack a clear description of the mode of administration. In most trials the treatment was continued during 1–3 months, in two studies 6 months, and in one study even 12 months.

A large variety of outcome measures was applied to evaluate the Ginkgo effects. In all eight studies in which a global assessment of the treatment effect was made – either by the patient, or by the physician / caregiver, or by both of them – Ginkgo showed the best result. Statistical significance of this global effect was reported in four studies. Clinical rating scales were used in eight trials as well. The application of the Crichton Geriatric Scale in its original or in a modified format yielded a significantly positive effect for Ginkgo in one trial (Vorberg et al., 1989), no effect in another trial (Halama et al., 1988), while in a third study no results for this rating scale were given (Israël et al., 1977). Application of the Plutchic Geriatric Rating Scale yielded a beneficial Ginkgo effect in two trials (Agnoli, 1980; Dieli et al., 1981). In the study by Halama et al. (1988) both the Sandoz Clinical Assessment Geriatric Scale (SCAG) – being the primary outcome measure – and the Syndrom Kurz Test (SKT) revealed a (statistically) significant effect of Ginkgo. Together with the SGRS, another clinical rating scale, the SCAG was also assessed in the trial reported by Pidoux et al. (1983), again with a positive result. A geriatric clinical

Table 7   RCTs on the efficacy of *Ginkgo biloba* special extract in patients suffering from cerebral insufficiency: unspecified

Author, year	Study population Diagnostic criteria	Number randomized analyzed (EG/other)	Age (yrs) Sex (M/F)	Intervention Brand name Administration route Dose Duration	Outcome measure	Result	Remarks
Moreau, 1975	cerebral insufficiency diagnostic criteria: n.d.	60 30/30	84 5/55	Tanakan® (Ipsen 225) n.d. 120 mg/day 3 m.	global effect (% excellent or good) physical signs + functional symptoms: – vertigo/walking instability – walking activity – sensoric manifestations – headache – temporo-spatial desorientation – walking distance general activities (impact of disease on daily life activities (sleeping, appetite, mood, etc.)) psychometric tests: 10 tests (temporo-spatial orientation, memory, etc.)	79% vs 21% ($p < 0.001$) % improved: 91% vs 50% 75% vs 42% 75% vs 13% 78% vs 38% 94% vs 31% 67% vs 17% % improved: 90% vs 43% significantly more improvement ($p < 0.05$) in Ginkgo group for 6/10 tests	parallel vs placebo source: n.d. also comparison with ergot alkaloids (n = 30)
Bono and Mouren, 1975	cerebral insufficiency manifestations: attention, mood, memory, confusion functional signs: vertigo, visus, headache, tinnitus, motoric + sensoric, etc.	30 15/15	n.d. n.d.	Tanakan® oral, solution (drink) 120 mg/day (3 ml) 5–6 w.	global effect (good/moderate/no) evolution of functional signs (vertigo, tinnitus, headache, memory, visuo-spatial orientation, confusion, visual/auditive/walking problems) evolution of neurological signs (motor or sensoric defects, language problems) evolution of behavioural signs (vigilance, sociability, mood, sleep) clinical examination (cardiovascular, renal, etc.) paraclinical examination (EEG, ECG, eyes, blood)	8/3/3 vs 3/2/9 ($p < 0.01$) vertigo, walking problems, headache, visual problems, tinnitus: 90% disappeared memory, confusion, orientation: 62% disappeared motor defects: 65% improved sensoric defects: 40% improved vigilance: > 70% recovered	parallel vs placebo source: n.d. also GBSE – 3 ergot derivates comparison (20 vs 20)
Augustin, 1976	hospital patients with atherosclerosis, and often signs of vascular origin (cardiac, cerebral, ocular, renal, etc.)	189 99/90	49–95 0/189	Tanakan® n.d. 120 mg/day 6 m.	global assessment of activities (reading, interest in radio/TV, usual manual activities) psychometric test battery: 1. orientation. 2. intellectual efficiency. 3. short term memory. 4. Rey test (learning words). 5. Benton visual retention test clinical assessment of vascular state: – arterial blood pressure – oscillometry: arterial peripheral circulation – central retinal artery pressure – atheromatous retinopathy biological surveillance, morbidity/mortality	improved: 44% vs 13% ($p = 0.001$) n = 168; sum score of 5 tests: – no. improved: 61 vs 26 – no. aggravated: 13 vs 36 n.s. index increased: 24 vs 2 n.s. 3 improved vs 6 impaired 0 vs 8 deaths	parallel vs placebo hospital patients (months, years) same nutrition, dependent prestratification: age

Table 7   continued

Author, year	Study population Diagnostic criteria	Number randomized analyzed (EG/other)	Age (yrs) Sex (M/F)	Intervention Brand name Administration route Dose Duration	Outcome measure	Result	Remarks
Israël et al., 1977	intellectual deficiency syndr. caused by insufficient cerebral circulation; diagnostic criteria: n.d.	60 30/30 48 24/24	62–96 (72) 32/16	Tanakan® oral, solution 240 mg/day (6 ml) 60 d.	Crichton scale (modified, adapted) 4 factors derived from 7 psychometric tests (code test, verbal fluency, numbers repeat, etc.): 1. attention (vigilance) + concentration 2. memory 3. fluency of operant behaviour 4. effective activity + psychomotor efficiency	+0.44 vs +0.21 +0.38 vs +0.13 +0.00 vs −0.08 +0.04 vs +0.13	parallel vs placebo source: n.d.
Agnoli, 1980	cerebral insufficiency stable symptoms, > 30 d. PBRSM-score ≥ 30 Plutchik GRS ≥ 16	30	51–83 (65.3) 16/14	Tebonin® forte oral, solution 150 drops/day 4 w.	global clinical assessment Parkside Behaviour Rating Scale, modified (PBRSM):16 items (1–5), 8 subcategories: vigilance, orientation, memory, language behaviour, adjustment, socialization, autonomy Plutchik Geriatric Rating Scale (GRS): 31 items (0, 1, 2)	23x Ginkgo vs 3x placebo better Ginkgo vs placebo: +85% vs +1% change change (borderline) significant: 7 vs 0 items (not: anxiety) +84% vs +1% change	cross-over vs placebo source: n.d.
Dieli et al., 1981	cerebral insufficiency: chronic signs, symptoms HIS-score < 7 Plutchik GRS	40 20/20 25 11/14	62	Tanakan® oral, solution 3 ml/day 60 d.	global clinical assessment (physician): excellent/good/moderate/none Plutchik Geriatric Rating Scale (31 items) memory of numbers (Wechsler Bellevue) code test (Wechsler Bellevue clinical examination laboratory examinations	8/7/1/4 vs 3/4/0/13 (p < 0.05) 13 vs 3 items sign. improvement; Ginkgo sign. better on 11 items +4.46 vs +0.57 (p = 0.01) +17.73 vs +4.29 (p = 0.02)	parallel vs placebo hospital patients exclusion criteria
Haan et al., 1982	cerebral insufficiency: signs + symptoms: > 11	60 30/30	45–65 (54.0) 60/0	Tebonin® forte i.v. (infusion) 97.5 mg/day (2 h.) 15 d.	psychological tests: – concentration: d2-test Brickenkamp – short term memory: digit span (HAWIE) – visual-motor test: mosaic test (WIP) cranial computer tomogram (CCT) neurological examination EKG + internal medical examination EEG laboratory tests Doppler, angiography aorta (if indicated)	394 vs 374 items (tendency) 71/321 vs 69/322 sec 7.2/5.3 vs 6.7/5.0 numbers only as part of the screening 3/10 vs 4/10 improved no changes 18 (69%) vs 12 (48%) improved no changes	parallel vs vincamine (vasoactive agent) source: n.d. exclusion criteria
Pidoux et al., 1983	senile deterioration: EEG-abnormalities (slow dominant frequency at 8–8.5 cps, no focalized abnormalities/asymmetry)	14 7/7 12 7/5	84 0/12	GBE n.d. 160 mg/day 12 w.	clinical rating scales: – SCAG – SGRS clinical examination EEG patterns	+29% vs +13% improvement GBE: increase alpha rhythm, reduction theta	parallel vs placebo source: n.d.

Table 7 continued

Author, year *Study population* *Diagnostic criteria*	*Number* *randomized* *analyzed* *(EG/other)*	*Age (yrs)* *Sex (M/F)*	*Intervention* *Brand name* *Administration route* *Dose* *Duration*	*Outcome measure*	*Result*	*Remarks*
Teigeler and Pieprzyk, 1984	cerebral insufficiency: slight to moderate screening procedure premorbid IQ > 80 (MWT-B)	40 20/20	n.d. n.d.	Tebonin® forte oral, tablets n.d. (4 tablets/day) 4 w.	information processing properties: 1. short-term storage capacity (mental efficiency) 2. learning speed (learning efficiency)	+24.8 vs +9.8 bit/sec +45% vs +13% evaluated only for placebo	parallel vs placebo hospital admission: orthopedic + internal treatment: 2 w. in hospital + 2 w. at home
Geßner et al., 1985	cerebral deterioration: symptoms reported by patients	40 20/20	57–77 (65.5)	rökan®, Tanakan® n.d. 120 mg/day 12 w.	general efficacy (patient): % (very) good general efficacy (investigator): % (very) good EEG-analysis: resting EEG + vigilance EEG psycho(physio)logical measures: – simple reaction time (light signal) – multiple choice reaction test (MCRT; Vienna Determination Apparatus) – flicker fusion frequency measurement – pursuit tracking test – Pauli test (simple calculations) – d2 test physiological measures: EEG, BP, pulse rate	50% vs 27% 44% vs 19% only increase in vigilance on resting EEG, if low initial theta/alpha-ratio and dominant freq. +20 msec vs +0 msec (n.s.) +8 msec vs –5 msec (n.s.) n.s. n.s. n.s.; GBE faster improvement n.s. n.s.	parallel vs placebo volunteers, relatively healthy also comparison with nicergolin (n = 20) wash-out: 2 w.
Arrigo, 1986 Arrigo and Cattaneo, 1985	cerebrovascular insufficiency chronic vertigo, headache, tinnitus, loss of memory, learning ability	90 80	36–88 (66.5) 59/31	Tebonin® forte oral, solution (drops) 120 mg/day 45 d.	clinical symptoms (self-rating questionnaire): 26 questions (+, 0, –), 7 symptoms: 1. asthenia 2. headache 3. tinnitus 4. dizziness 5. insomnia 6. visual disturbances 7. tension WAIS: psychometric testing: 4 tests not sensitive to mental loss: 1, 6, 8, 11 4 tests sensitive to mental loss: 4, 5, 7, 9 cube test (WAIS): concentration, orientation word recognition Rey Figure test (shape B) memory performance (reproducing objects) anxiety (state; STAI questionnaire) anxious behaviour (trait; STAI questionnaire)	change in number (groups A+B): 1. +22 vs –6 2. +16 vs +2. 3. +18 vs +10 4. +38 vs +8 5. +10 vs –2 6. +12 vs +4 7. +26 vs +18 EGB + plac. vs. plac. + EGB: +0.3 vs +0.5 (n.s.) +0.2 vs +0.8 (p < 0.04) +0.7 vs +0.2 (p = 0.017) (period 1) +16.3 vs +2.6 (p < 0.001) (per. 1) +2.4 vs +1.4 (p < 0.05) (period 1) +15.5 vs +5.7 (p < 0.001) (per. 1) +5.6 vs +2.5 (p < 0.001) (period 1) +2.8 vs +1.4 (p < 0.01) (period 1)	cross-over vs placebo source: n.d. prestratification (4 categories: WAIS) run-in period: 15 d. wash-out: 15 d.

Table 7   continued

Author, year	Study population / Diagnostic criteria	Number randomized analyzed (EG/other)	Age (yrs) Sex (M/F)	Intervention Brand name Administration route Dose Duration	Outcome measure	Result	Remarks
Taillandier et al., 1986	cerebral disorders due to ageing: EACG score: − 3–5 for at least 2/6 items − not 7 for > 2/6 items − total score: 21–113 (moderate cognitive deficiency) HIS-score: 0–11	166 80/86	60–97 (82.1) 26/140	Tanakan® oral, solution 160 mg/day (2x 2 ml) 12 m.	geriatric clinical evaluation scale (EACG): 17 items: 1–7 (absent–very important) general clinical assessment of state; 6-point scale: absent, mild, moderate, average, important, serious (clinician) global assessment of evolution; 6-point scale: much improved, improved, little improved, unchanged, bit impaired, impaired	+9.8 vs +5.2 (p = 0.01) % 'mild': +12% vs −7% 58% vs 43% improved	parallel vs placebo 4 homes for the elderly (> 2 m. inhabitant)
Halama et al., 1988	cerebrovascular insufficiency (mild to moderate) clinical findings: HIS > 7 psychometric tests + rating scales: Crichton Geriatric Scale: stages 1–3	40 20/20	55–85 (65.9) 23/17	Tebonin® forte oral, tablets 120 mg/day 12 w.	primary outcome measure: − SCAG: 5 factors; 18 items; item score: 1–7; total score: 18–126 secondary outcome measures: − Crichton scale (CRBRS) − Syndrome Kurz Test (SKT) − symptom rating: 1. dizziness. 2. ear noise. 3. headache. 4. hearing loss; 4 categories − background interference measurement − craniocorpography (CCG)	+9 vs +0 points (p < 0.0001) esp.: STM, dizziness; etc. n.s. +3.5 vs +0.8 1. +1.4 vs +0.1 2. +0.6 vs +0.15 3. +0.9 vs +0.1 4. n.s. n.s. n.s.	parallel vs. placebo out-patients (neuro-logical practice) exclusion criteria: a.o.: Alzheimer
Hofferberth, 1989	cerebro-organic psycho-syndrome: − clinical symptoms (e.g. dizziness, memory) − 4 screening tests (EEG, saccadic eye movements, Wiener Determination test, number connection) ≥ 2 abnormal results	36 18/18	53–69 (63.2) 23/13	rokan® oral, tablets 120 mg/day 8 w.	overall effect: 4 categories (patient) overall effect: 4 categories (physician) psychometric tests: − Wiener Determination Test (3 steps; < 160, < 168, < 175 out of 180 possible correct reactions) − number connection test (> 28 sec. needed to connect 30 numbers) neurophysiological tests: − saccadic eye movements: duration (abnormal: > 150 msec) latency (abnormal: > 250 msec) rapidity (abnormal: < 250 °/sec) − quantified EEG (theta-part > 70%)	9/6 vs 2/0 much impr./improved 6/8 vs 0/2 much impr./improved step 10: +19 vs −1 (p < 0.0001) step 12: +14 vs −2 (p < 0.0001) step 15: +6 vs +2 (p < 0.0001) +13 vs +0 sec (p < 0.0001) +45 vs −2 msec (p < 0.0001) +53 vs −8 msec (p < 0.0001) +13% vs −1% (p < 0.0001)	parallel vs placebo hospitalized patients block randomization (6 patients) wash-out: 2 weeks exclusion criteria
Vorberg et al., 1989	cerebral insufficiency: presence of clinical symptoms (≥ 4 out of 6)	100 96 49/47	55–85 (70) 60/40	Kaveri® solution oral, solution 112 mg/day 12 w. 27–30% flavonoids + 1–2% terpene-trilactones	list of 6 clinical symptoms: 1. memory. 2. concentration. 3. anxiety. 4. dizziness. 5. headache. 6. ear noise. degrees of severity: 1–5 in accordance with modified Crichton scale pulse rate, blood pressure	1. +1.4 vs +0.5 (p < 0.001) 2. +1.4 vs +0.4 (p < 0.001) 3. +1.1 vs +0.4 (p < 0.001) 4. +2.0 vs +0.6 (p < 0.001) 5. +2.3 vs +0.9 (p < 0.001) 6. +0.8 vs +0.4 (n.s., tendency)	parallel vs placebo ambulatory patients (GP-practice) exclusion criteria duration of complaints: ca. 20 m.

Table 7   continued

Author, year	Study population Diagnostic criteria	Number randomized analyzed (EG/other)	Age (yrs) Sex (M/F)	Intervention Brand name Administration route Dose Duration	Outcome measure	Result	Remarks
Gerhardt et al., 1990	cerebrovascular disorders (HOPS, MID, SDAT) symptoms: dizziness (31 vs 31), concentration (6 vs 7), memory (8 vs 7), headache (8 vs 5), depression (7 vs 4) Hachinsky Ischemic Score	80 40/40 65 32/33	55–88 (71) 23/42	EGb 761 oral, solution (3 ml) 120 mg/day 6 w.	Nuremberg Elderly Rating (NAR): hetero-assessment personality profile, 9 pairs of items (item score: 1–7) personality profile: self-assessment, 7 pairs of items (scale: VAS, 0–22 cm) psychomotoric coordination: – trail making test – digit symbol test	1 vs 3 items sign. improved Ginkgo: careful–not careful 5 vs 7 items improved; Ginkgo: 2 impaired; changes n.s. total time: +3 vs +10 sec (n.s.) +2 vs +4 points (n.s.)	parallel vs dihydroergotoxine ambulatory patients single-blind (patient) MID, DAT not distinguished (HIS)
Maier-Hauff, 1991	cerebral insufficiency after subarachnoid haemorrhage (SAH) + surgery	50 24/26 42 20/22	21–63 (48) 26/24	Kaveri® (LI 1370) oral, tablets 150 mg/day 12 w.	neuropsychological testing: – attention: Zimmermann test battery: 1. phasic alertness 2. go/no go test (external stimulus selection): selective attention: reaction time selective attention: false reactions 3. incompatibility (stability/flexibility attention) – short-term memory: – verbal: digit span – verbal: recurring words – non-verbal: block-tapping – non-verbal: recurring figures (spatial orient.)	+12 vs –27 msec (p < 0.05) +3.5 vs +0.1 (p < 0.01) +75 vs –5 msec (p < 0.01) +6 vs +3 (p < 0.05) +7 vs +8 (n.s.)	parallel vs placebo out-patients: 7–42 m. after subarachnoid haemorrhage + aneurysm operation clinical staging SAH (Hunt/Hess: I-V) Glascow Coma Scale exclusion criteria
Graßel, 1992	cerebral insufficiency: 1. patient history: indications for intracranial stenosis process (TIA, hypertension, etc.) 2. Doppler sonography (in case of doubt) 3. IQ-STM < IQ-MWT	72 36/36 53 29/24	63.8 35/37	EGb 761 oral, solid form 160 mg/day 24 w.	basic parameters of mental performance (information processing of brains): 1. short-term memory capacity: IQ value normal value: IQ = 100 2. long-term memory capacity: basic learning rate: MQ-value (memory quotient) computer-aided examination	+13 vs +1 IQ-point (baseline: 93.5 vs 93.8) +10 vs +0 MQ-point (baseline: 101 vs 106)	parallel vs placebo ambulatory patients 3 test centers 26% drop-out rate

evaluation scale (EACG), comparable with the SCAG, showed positive results (Taillandier et al., 1986), just like a self-rating clinical symptoms questionnaire (Arrigo, 1986; Arrigo and Cattaneo, 1985). In the trial of Halama et al. (1988) clinical symptom rating gave mixed results.

In ten trials ample attention was paid to psychometric testing of any Ginkgo biloba effect, through a wealth of different tests and test batteries (see Table 7). The influence of Ginkgo treatment on daily life activities was assessed in three trials (Agnoli, 1980; Augustin, 1976; Moreau, 1975), each time with a positive result. Neurophysiological characteristics, the presence and severity of physical signs and functional symptoms, state of anxiety (Arrigo, 1986; Arrigo and Cattaneo, 1985), and clinical and laboratory examination results were used as other endpoints to quantify the treatment effect.

All trials but one (Halama et al., 1988) failed to explain the relative importance of the outcome measures used. The consequences of this neglect are not that far-reaching, as almost all authors could report beneficial effects of Ginkgo treatment for (almost) all endpoints considered. Israel et al. (1977) reported a positive Ginkgo effect on three out of four cognitive functions, Geßner et al. (1985) reported a positive effect for global evaluation and for two reaction time tests, but not for several other psycho(physio)logical tests, and the positive results reported by Halama et al. (1988) did not extend to the Crichton scale and the craniocorpographic measurements. When compared to a reference drug Ginkgo did slightly better than vincamine (Haan et al., 1982), but slightly worse than dihydroergotoxine (Gerhardt et al., 1990).

Reflection on the Study Results

Fifty-five randomized clinical trials of Ginkgo biloba and cerebral insufficiency were identified, which were placed a little bit artificially under six different headings, according to the target health problem addressed. Most of the studies were reported from Germany and France: 31 and 16 trials, respectively. Occasionally, patients were recruited in other countries: Italy (3 trials), UK (2 trials), USA, Sweden, and Austria (1 trial each). Almost all trials in which Ginkgo was contrasted with placebo showed a statistically significant effect or a positive trend in favour of Ginkgo. Summarizing and pooling of these results appeared to be a harsh endeavour, not only because of the heterogeneity of study design aspects, especially outcome measurements and definition of success criteria, but also because of the large variety in reporting style. For instance, many reports communicated only part of the outcome results. Statistical test results were not recorded in a systematical way either. The contents of the summary tables reflect these problems. A serious shortcoming of many studies is that hardly any attention is paid to the hierarchy of the outcome measures and to the clinical meaning of the recorded effects. Actually, this is the reason why the study results are presented in a rather detailed fashion and were not reduced to a more aggregate and summary level. Major methodological weaknesses of many trials were, in addition to the outcome definition, the choice and description of the patient definition and eligibility criteria, and the small sample size.

Generally speaking, the evidence contained by the more recent trials, including most of the trials which focussed on demented patients, should be valued considerably higher than what can be learned from earlier trials.

CLINICAL EFFICACY OF *GINKGO BILOBA* IN MISCELLANEOUS CONDITIONS

Not all Ginkgo trials were included in the previous sections. Some additional trials have been published, which focussed on conditions that are not very well covered by the treatment indications discussed before. Nevertheless, an insufficient blood supply may play a key role in each of these conditions too. Some examples of these trials will be presented below.

Roncin *et al.* (1996) studied the preventive effect of *Ginkgo biloba* (EGb 761: Tanakan®) on acute mountain sickness (AMS) and vasomotor changes of the extremities during a Himalayan expedition in 1993. With increasing altitude atmospheric pressure decreases, resulting in hypoxia and a fall of pO_2 and pCO_2. In particular when the altitude exceeds 3000 m, mountain sickness may occur. This condition is reflected by various non-specific symptoms, such as headache, nausea, insomnia, vertigo, ataxia, severe lassitude, anorexia, vomiting, and dyspnoea on exertion and at rest. Moreover, mountaineers may develop numbness and paraesthesia of the extremities due to microcirculatory disturbances caused by the low environmental temperature. Forty-four subjects were recruited to participate in this trial. After giving informed consent, the subjects who fulfilled the eligibility criteria (good health, score $\geqslant 2$ on the AMS Questionnaire (range of scores: $\underline{0}$–16) during a previous expedition, etc.) were randomly allocated to two groups. One group received 160 mg of EGb 761 per day, the other group placebo tablets. Outcome evaluation was based on the change in score on the Environmental Symptoms Questionnaire, the cold gradient measured by photoplethysmography, and functional disability for the items paraesthesia, pain, numbness, stiffness, and swelling of the hands. A prophylactic effect of Ginkgo treatment could be demonstrated in this study. According to the AMS-C (cerebral factor), none of the subjects in the Ginkgo group developed acute mountain sickness versus 41% of the subjects in the placebo group (p = 0.0014). According to the AMS-R (respiratory factor), 3 subjects (14%) in the EGb 761 group developed acute mountain sickness versus 18 (82%) in the placebo group (p < 0.001). With regard to the cold gradient the Ginkgo group improved, while the placebo group clearly deteriorated. Based on the results of this trial, Ginkgo seems to be an interesting option for the prevention of mountain sickness at moderate altitude (5400 m) and with gradual exposure.

The potential beneficial effect of *Ginkgo biloba* treatment on still another functional health problem associated with insufficient peripheral arterial blood flow was studied by Sohn and Sikora (1991). In a prospective before-after trial, without an external control group, 50 patients with proven arterial erectile impotence were treated with 240 mg of *Ginkgo biloba* extract by oral application during 9 months. Objective response criteria were added to the common set of subjective success parameters used in previous studies on oral treatment regimens for erectile failure. All patients with a sufficient response to intracavernous drug administration (n = 20) regained spontaneous erections after 6 months of Ginkgo treatment and also demonstrated improved penile flow rates (Doppler examination) and rigidity (Rigiscan). Of 30 patients without a sufficient response to high-dose intracavernous drug supply, 19 regained erections under Ginkgo therapy, whereas 11 remained impotent. All patients in this group showed improved scores on the objective response parameters.

Garg *et al.* (1995) studied the effect of *Ginkgo biloba* extract on neurological symptoms in 62 patients with acute ischaemic stroke through a randomized, placebo-controlled, double-blind trial. The patients were recruited from the neurological and medical wards of an Indian medical college. The study population consisted of both men and women who were struck by sudden hemiplegia of either side, with or without speech involvement, and with an infarct in the region of either middle cerebral artery, according to a cranial CT scan. The participants (mean age: 54.2 years) received either GBE (n = 33; 160 mg/day) or placebo (n = 29) for at least four weeks. In 55 eligible patients the neurological status was assessed immediately before and again 28 days after the start of treatment, by means of a modified version of Mathew's scale (total score: 0–100). Statistically significant improvements on Mathew's scale were noted in both groups, however, without a significant change difference (+5.0 vs. +4.9 points).

Tamborini and Taurelle (1993) performed a multicentre double-blind placebo-controlled study to evaluate the efficacy of standardized *Ginkgo biloba* extract (EGb 761) in treating congestive symptoms of the premenstrual syndrome (PMS). The study population consisted of 165 women aged 18–45, who suffered since 3 cycles from congestive premenstrual troubles during at least 7 days per cycle: physical signs, such as breast pain, swelling in the abdominal-pelvic region, and edema of the extremities, and/or neuropsychical complaints, such as anxiety, irritability, lack of energy, and depressive mood. One menstrual cycle was used to confirm the diagnosis of PMS. Then, during the two following cycles, each patient received either EGb 761 (n = 88; 160–320 mg/day) or placebo (n = 77) from the 16th day of the first cycle until the 5th day of the next cycle. The course of the cycle was evaluated by the patient (daily self-assessment rating scale), and by the physician (questionnaire, clinical examination). The data of 143 women were included in the analysis. A strong placebo effect was recorded in this study. Nevertheless, administration of EGb 761 showed a statistically significant effect on the congestive symptoms of PMS, particularly the breast symptoms. For instance, the number of women reporting (very) severe breast pain diminished from 44 to 15 in the EGb-group, and from 31 to 14 in the placebo-group (p = 0.03). An improvement of neuropsychological symptoms was registered as well. The acceptability and tolerability of Ginkgo proved to be very good.

CONCLUSIONS

After having seen all these trial results, the key question remains to be answered. Can clinical therapy with *Ginkgo biloba* be regarded effective? Or, stated differently, will the evidence that is currently available from clinical studies, encourage (relatives of) patients with complaints of either cerebral or peripheral arterial insufficiency to ask for Ginkgo preparations, or physicians to prescribe them, or health authorities to consider the licensing of Ginkgo-based drugs?

The evidence that is rooted in the controlled trials included in this review is rather convincing. In general, the results point to a significant effect of *Ginkgo biloba* treatment, both in patients with peripheral arterial disease and in patients with dementia and several related manifestations of cerebral insufficiency. Other suggested domains of intervention have not yet been explored to such a large extent that solid conclusions can be drawn already. The consistency between the results of studies addressing the same target health problem, is remarkable. However, in several trials the positive effect did not concern all levels of outcome measurement. Furthermore, it should be stressed that only a minority of the trials can stand the test of criticism posed by the currently accepted – general and disease-specific – guidelines and prescriptions for the design and conduct of clinical intervention trials. As a consequence, a substantial part of the available evidence is contributed by a relatively small proportion of all Ginkgo trials. Most of them were published quite recently. Although the time-related difference in design choices hampers any comparison, one cannot get away from the impression that, generally speaking, the effect sizes reported in earlier trials exceed those of more recent trials. On average, larger sample sizes were used in the later trials than in the earlier ones. Some of these trials had enough "power" to make even apparently non-impressive treatment effects statistically significant.

The clinical importance of the effects shown by the most recent, high-quality studies remains debatable. Such a discussion is seriously hampered by the high level of abstraction of the concepts that underlie many of the questionnaires, tests and rating scales which tend to be used as effect measures. The outcome of such a debate will be also influenced by the prevailing ideas regarding the treatability and reversibility of the health problems under consideration. Finally, it goes without saying that any conclusion rests on the assumption that the current view of the efficacy of *Ginkgo biloba* treatment is not seriously biased by any form of selective reporting and publication.

REFERENCES

Agnoli, A. (1980) *Relazione clinica sulla specialita "Tebonin forte"*. Clinica Neurologica, Istituto Universitario di Medicina e Chirurgia, L'Aquila.

Allain, H., Raoul, P., Lieury, A., LeCoz, F., Gandon, J.M., d'Arbigny, P. (1993) Effect of two doses of *Ginkgo biloba* extract (EGb 761) on the dual-coding test in elderly subjects. *Clin. Ther.*, 15, 549–558.

Ambrosi, C., Bourde, C. (1975) Nouveauté thérapeutique médicale dans les artériopathies des membres inférieurs: Tanakan. Essai clinique et étude par les cristaux liquides. *Gazette Méd. France*, 82, 628–632.

Amaducci, L., Angst, J., Bech, P., Benkert, O., Bruinvels, J., Engel, R.R. (1990) Consensus conference on the methodology of clinical trials of "nootropics", München, June 1989. Report of the Consensus Committee. *Pharmacopsychiat.*, 23, 171–175.

APA. (1987) *DSM-III-R. Diagnostic and Statistical Manual of Mental Disorders*. (3th ed.). American Psychiatric Association, Washington, D.C.

APA. (1994) *Diagnostic and Statistical Manual of Mental Disorders: DSM-IV. Prepared by Task Force on DSM-IV*. American Psychiatric Association, Washington, D.C.

Arrigo, A. (1986) Behandlung der chronischen zerebrovaskulären Insuffizienz mit *Ginkgo-biloba*-Extrakt. *Therapiewoche*, **36**, 5208–5218.

Arrigo, A., and Cattaneo, S. (1985) Clinical and psychometric evaluation of *Ginkgo biloba* extract in chronic cerebro-vascular diseases. In J. Agnoli, R. Rapin, V. Scapagnini, and W.V. Weitbrecht (eds.), *Effects of Ginkgo Biloba Extract on Organic Cerebral Impairment*, John Libbey Eurotext Ltd., Montrouge, pp. 85–89.

Augustin, P. (1976) Le Tanakan en gériatrie. Étude clinique et psychométrique chez 189 malades d'hospice. *Psychologie Méd.*, **8**, 123–130.

Baitsch, G. (1986) Treatment of obliterative arterial disease in diabetics (Fontaine's stage IV) with moist gangrene. An open randomized trial of *Ginkgo biloba* extract (rökan®) v. dextran 40. *Clinical Trials Journal*, **23**, 11–19.

Bastide, G., Montsarrat, M. (1978) Artérite des membres inférieurs. Intérêt du traitement médical après intervention chirurgicale. Analyse factorielle. *Gaz. Méd. France*, **85**, 4523–4526.

Bauer, U. (1984) 6-Month double-blind randomised clinical trial of *Ginkgo biloba* extract versus placebo in two parallel groups in patients suffering from peripheral arterial insufficiency. *Arzneim.-Forsch./Drug Res.*, **34**, 716–720.

Bauer, U. (1986) L'extrait de *Ginkgo biloba* dans le traitement de l'artériopathie des membres inférieurs. Étude sur 65 semaines. *Presse Méd.*, **15**, 1546–1549.

Benkert, O., Coper, H., Ferner, U., Fox, J.M., Gastpar, M., Herrschaft, H. (1992) Proof of efficacy of nootropics for the indication "dementia" (phase III) – Recommendations. *Pharmacopsychiat.*, **25**, 126–135.

Berndt, E., Kramar, M. (1987) Medikamentöse Therapie der peripheren arteriellen Verschlußkrankheit im Stadium IIb. Eine bizentrische Vergleichsstudie zwischen *Ginkgo-biloba*-Extrakt und Buflomedilhydrochlorid. *Therapiewoche*, **37**, 2815–2819.

Blume, J., Kieser, M., Hölscher, U. (1996) Placebokontrollierte Doppelblindstudie zur Wirksamkeit von *Ginkgo-biloba*-Spezialextrakt EGb 761 bei austrainierten Patienten mit Claudicatio intermittens. *Vasa*, **25**, 265–274.

Böhmer, D., Kalinski, S., Michaelis, P., Szögy, A. (1988) Behandlung der PAVK mit *Ginkgo-biloba*-Extrakt (GBE) oder Pentoxifyllin. Eine Studie zur Beurteilung von Wirksamkeit und Verträglichkeit. *Herz/Kreisl.*, **20**, 5–8.

Bono, Y., Mouren, P. (1975) L'insuffisance circulatoire cérébrale et son traitement par l'extrait de *Ginkgo biloba*. *Méditerranée Médicale*, **3**, 59–62.

Brüchert, E., Heinrich, S.E., Ruf-Kohler, P. (1991) Wirksamkeit von LI 1370 bei älteren Patienten mit Hirnleistungsschwäche. Multizentrische Doppelblindstudie des Fachverbandes Deutscher Allgemeinärtze. *Münch. Med. Wochenschr.*, **133**, S9–S14.

Bulling, B., Von Bary, S. (1991) Behandlung der chronischen peripheren arteriellen Verschlußkrankheit mit physikalischem Training und *Ginkgo-biloba*-Extrakt 761. Ergebnise einer plazebokontrollierte Doppelblindstudie. *Med. Welt*, **42**, 702–708.

Bundesgesundheitsamt. (1991) Empfehlungen zum Wirksamheitsnachweis von Nootropika im Indikationsbereich "Demenz" (Phase III). *Bundesgesundheitsblatt*, **7**, 342–350.

CBO. (1997) *Herziening consensus diagnostiek bij het dementiesyndroom*. Centraal Begeleidingsorgaan voor de Intercollegiale Toetsing, Utrecht.

Chalmers, I., and Altman, D.G. (eds.) (1995) *Systematic reviews*, BMJ Publishing Group, London.

Chartres, J.-P., Bonnan, P., Martin, G. (1987) Réduction de posologie de médicaments psychotropes chez des personnes âgées vivant en institution. Étude à double-insu chez des patients prenant soit de l'extrait de *Ginkgo biloba* 761 soit du placebo. *Psychologie Méd.*, **19**, 1365–1375.

Chesseboeuf, L., Herard, J., Trevin, J. (1979) Étude comparative de deux vasorégulateurs dans les hypoacousies et les syndromes vertigineux. *Méd. Nord l' Est.*, **3**, 534–539.

Claussen, C.F. (1984) Mit *Ginkgo biloba* wird Ihr Patient wieder standfest. Cranio-corpographie zeigt statistisch signifikanten Rückgang der Vertigo- und Ataxie-Symptomatik im Doppelblindversuch. *Ärtzliche Praxis*, **36**, 3–8.

Claussen, C.-F., Kirtane, M.V. (1985) Randomisierte Doppelblindstudie zur Wirkung von Extractum *Gingko biloba* bei Schwindel und Gangunsicherheit des älteren Menschen. In C.-F. Claussen (ed.), *Presbyvertigo, Presbyataxie, Presbytinnitus: Gleichgewichts- und Sinnesstörungen im Alter*, Springer Verlag, Berlin, pp. 103–115.

Cochrane Collaboration (1996–[Updated quarterly]) *The Cochrane Library: database on disk and CD-ROM*. The Cochrane Collaboration, Update Software, Oxford.

Cooper, H., and Hedges, L.V. (eds.) (1994) *The handbook of research synthesis*, Russell Sage Foundation, New York.

Cott, J. (1995) NCDEU update. Natural product formulations available in Europe for psychotropic indications. *Psychopharmacol. Bull.*, **31**, 745–751.

Courbier, R., Jausseran, J.M., Reggi, M. (1977) Étude à double insu croisée du Tanakan dans les arteriopathies des membres inférieurs. *Méditerr. Med.*, **126**, 61–64.

CPMP. (1992) *Note for guidance: antidementia medicinal products. CPMP Working Party on Efficacy of Medicinal Products. III/3705–91-EN, Draft 5*. Commission of the European Communities, Brussels.

CPMP. (1997) *Note for guidance on medicinal products in the treatment of Alzheimer's disease* (CPMP/EWP/553/95). The European Agency for the Evaluation of Medicinal Products, Human Medicines Evaluation Unit, Committee for Proprietary Medicinal Products, London.

Crook, T., Bahar, H., Sudilovsky, A. (1987) Age-associated memory impairment: diagnostic criteria and treatment strategies. *Int. J. Neurol.*, **88**, 21–22.

Crook, T.H., Bartus, R.T., Ferris, S.H., Whitehouse, P., Cohen, G.D., Gershon, S. (1986) Age-associated memory impairment: proposed diagnostic criteria and measure of clinical change – report of a National Institute of Mental Health working group. *Dev. Neuropsych.*, **2**, 261–276.

Crook, T.H., Larrabee, G.J., Youngjohn, J.R. (1990) Diagnosis and assessment of age-associated memory impairment. *Clin. Neuropharmacol.*, **13**, S81-S91.

Crook, T.H., Larrabee, G.J. (1991) Diagnosis, assessment and treatment of age-associated memory impairment. *J. Neural Transm. Suppl.*, **33**, 1–6.

Crook, T.H. (1990) Assessment of drug efficacy in age-associated memory impairment. *Adv. Neurol.*, **51**, 211–216.

DeFeudis, F.V. (1998) *Ginkgo biloba extract (EGb 761). From chemistry to the clinic*. Ullstein Medical, Wiesbaden.

Diehm, C., Heinrich, F., and Mörl, H. (1990) *Placebokontrollierte multizentrische randomisiert doppelblinde Pilotstudie zur Prüfung der Wirksamkeit des Ginkgo-biloba-Extraktes EGb 761 bei patienten mit peripherer arterielle Verschlußkrankheit im Stadium IIb nach Fontaine*. Dr. Willmar Schwabe Klinische Forschung, Karlsruhe.

Dieli, G., La Mantia, V., Saetta, M., Costanzo, E. (1981) *Essai clinique à double insu du Tanakan dans l'insuffisance cérébrale chronique*. *Lavoro Neuropsichiatrico*, **68**, 1–7.

Drabaek, H., Petersen, J.R., Winberg, N., Hansen, K.F., Mehlsen, J. (1996) Effekten af *Ginkgo biloba* ekstrakt hos patienter med claudicatio intermittens. *Ugeskr Laeger*, **158**, 3928–3931.

Dubreuil, C. (1986) Essai thérapeutique dans les surdités cochléaires aigues. Étude comparative de l'extrait de *Ginkgo biloba* et de la nicergoline. *Presse Med.*, **15**, 1559–1561.

Eckmann, F. (1990) Hirnleistungsstörungen – Behandlung mit *Ginkgo-biloba*-Extrakt. Zeitpunkt des Wirkungseintritts in einer Doppelblindstudie mit 60 stationären Patienten. *Fortschr. Med.*, **108**, 557–560.

Eckmann, F., Schlag, H. (1982) Kontrollierte Doppelblind-Studie zum Wirksamkeitsnachweis von Tebonin forte bei Patienten mit zerebrovaskulärer Insuffizienz. *Fortschr. Med.*, **100**, 1474–1478.

Ernst, E. (1996) *Ginkgo biloba* in der Behandlung der Claudicatio intermittens. Eine systematische Recherche anhand kontrollierter Studien in der Literatur. *Fortschr. Med.*, **114**, 85–87.

Franco, L., Cuny, G., Nancy, F.M.C. (1991) Étude multicentrique de l'efficacité de l'extrait de *Ginkgo biloba* (EGb 761) dans le traitement des troubles mnésiques liés à l'âge. *La Revue de Gériatrie*, **16**, 191–195.

Frileux, C., Copé, R. (1975) L'extrait concentré de *Ginkgo biloba* dans les troubles vasculaires périphériques. *Cahiers d'Artériologie de Royat*, **III**, 117–122.

Fünfgeld, E.W. (ed.) (1988) *Rökan (Ginkgo biloba). Recent results in pharmacology*, Springer-Verlag, New York.

Garg, R.K., Nag, D., Agrawal, A. (1995) A double blind placebo controlled trial of *Ginkgo biloba* extract in acute cerebral ischaemia. *J. Assoc. Physicians India*, **43**, 760–763.

Garzya, G., Picari, M. (1981) Trattamento delle vasculopatie periferiche con una nuova sostanza estrattiva: il Tanakan. *Clinica Europea*, **20**, 936–944.

Gerhardt, G., Rogalla, K., Jaeger, J. (1990) Medikamentöse Therapie von Hirnleistungsstörungen. Randomisierte Vergleichsstudie mit Dihydroergotoxin und *Ginkgo biloba* Extrakt. *Fortschr. Med.*, **108**, 384–388.

Geßner, B., Voelp, A., Klasser, M. (1985) Study of the long-term action of a *Ginkgo biloba* extract on vigilance and mental performance as determined by means of quantitative pharmaco-EEG and psychometric measurements. *Arzneim.-Forsch./Drug Res.*, **35**, 1459–1465.

Gräßel, E. (1992) Einfluß von *Ginkgo biloba* Extrakt auf die geistige Leistungsfähigkeit. Doppelblindstudie unter computerisierten Meßbedingungen bei Patienten mit Zerebralinsuffizienz. *Fortschr. Med.*, **110**, 73–76.

Haan, J., Reckermann, U., Welter, F.L., Sabin, G., Müller, E. (1982) *Ginkgo-biloba*-Flavonglykoside. Therapiemöglichkeit der zerebraler Insuffizienz. *Med. Welt*, **33**, 1001–1005.

Haase, J., Halama, P., Hörr, R. (1996) Wirksamkeit kurzdauernder Infusionsbehandlungen mit *Ginkgo-biloba*-Spezialextrakt EGb 761 bei Demenz vom vaskulären und Alzheimer-Typ. *Z. Gerontol. Geriatr.*, **29**, 302–309.

Haguenauer, J.P., Cantenot, F., Koskas, H., Pierart, H. (1986) Traitement des troubles de l'équilibre par l'extrait de *Ginkgo biloba*. Étude multicentrique à double insu face au placebo. *Presse Med.*, **15**, 1569–1572.

Halama, P. (1990) Was leistet der Spezialextrakt EGb 761? Therapie mit *Ginkgo biloba* bei Patienten mit cerebrovaskulärer Insuffizienz und therapieresistenter depressiver Symptomatik. Ergebnisse einer placebokontrollierten, randomisierten doppelblinden Pilotstudie. *Therapiewoche*, **40**, 3760–3765.

Halama, P. (1991a) Befindlichkeitsbeurteilung und Psychometrie. Testung der *Ginkgo-biloba*-Wirkung bei Patienten einer Neurologischen Fachpraxis. *Münch. Med. Wochenschr.*, **133**, S19–S22.

Halama, P. (1991b) *Ginkgo biloba*. Wirksamkeit eines Spezialextrakts bei Patienten mit zerebraler Insuffizienz. *Münch. Med. Wochenschr.*, 133, 190–194.

Halama, P., Bartsch, G., Meng, G. (1988) Hirnleistungsstörungen vaskulärer Genese. Randomisierte Doppelblindstudie zur Wirksamkeit von *Ginkgo biloba* Extrakt. *Fortschr. Med.*, 106, 408–412.

Hamann, K.-F. (1985) Physikalische Therapie des vestibulären Schwindels in Verbindung mit *Ginkgo-biloba*-Extrakt. Eine posturographische Studie. *Therapiewoche*, 35, 4586–4590.

Hartmann, A., Frick, M. (1991) Wirkung eines Ginkgo-Spezial-Extraktes auf psychometrische Parameter bei Patienten mit vaskulär bedingter Demenz. *Münch. Med. Wochenschr.*, 133, S23–S25.

Herrschaft, H. (1992) Zur klinischen Anwendung von *Ginkgo biloba* bei dementiellen Syndromen (Hirnleistungsstörungen bei vasculärer oder degenerativer ZNS Erkrankung). *Pharm. u. Zeit*, 21, 266–275.

Hofferberth, B. (1989) Einfluß von *Ginkgo biloba* Extrakt auf neurophysiologische und psychometrische Meßergebnisse bei Patienten mit hirnorganischem Psychosyndrom. Eine Doppelblindstudie gegen Plazebo. *Arzneim.-Forsch./Drug Res.*, 39, 918–922.

Hofferberth, B. (1991) *Ginkgo-biloba*-Spezialextrakt bei Patienten mit hirnorganischem Psychosyndrom. Prüfung der Wirksamkeit mit neurophysiologischen und psychometrischen Methoden. *Münch. Med. Wochenschr.*, 133, S30–S33.

Hofferberth, B. (1994) The efficacy of EGb 761 in patients with senile dementia of the Alzheimer type, a double-blind, placebo-controlled study on different levels of investigation. *Human Psychopharmacology*, 9, 215–222.

Hoffmann, F., Beck, C., Schutz, A., Offermann, P. (1994) Ginkgoextrakt EGb 761 (Tebonin)/HAES versus Naftidrofuryl (Dusodril)/HAES. Eine randomisierte Studie zur Horsturztherapie. *Laryngorhinootologie*, 73, 149–152.

Holgers, K.M., Axelsson, A., Pringle, I. (1994) *Ginkgo biloba* extract for the treatment of tinnitus. *Audiology*, 33, 85–92.

Hopfenmüller, W. (1994) Nachweis der therapeutischen Wirksamkeit eines *Ginkgo biloba* Spezialextraktes. Meta-Analyse von 11 klinischen Studien bei Patienten mit Hirnleistungsstörungen im Alter. *Arzneim.-Forsch./Drug Res.*, 44, 1005–1013.

ICH. (1995) *Structure and contents of clinical study reports. ICH-Efficacy Topic 3*. Paper presented at the Third International Conference on Harmonisation of technical requirements for registration of pharmaceuticals for human use, Yokohama, November 28, 1995.

Ihl, R., Dierks, T., Martin, E., Frölich, L., Maurer, K. (1992) *Ginkgo biloba Extrakt EGb761 bei der Demenz vom Alzheimer Typ*. Paper presented at the 15th Danube Symposion of Psychiatry, Regensburg, October 7–10, 1992.

Israël, L., Dell'Accio, E., Martin, G., Hugonot, R. (1987) Extrait de *Ginkgo biloba* et exercices d'entraînement de la mémoire. Evaluation comparative chez des personnes âgées ambulatoires. *Psychologie Méd.*, 19, 1431–1439.

Israël, L., Ohlman, T., Delomier, Y., Hugonot, R. (1977) Étude psychométrique de l' activité d'un extrait végétal au cours des états d' involution sénile. *Lyon Méditerranée Médicale*, 8, 1197–1199.

Kanowski, S. (1997) *Ginkgo-biloba*-Spezialextrakt. Nachgewiesene Wirksamkeit im Indikationsbereich Demenz. *Münch. Med. Wochenschr.*, 139, 39–42.

Kanowski, S., Herrmann, W.M., Stephan, K., Wierich, W., Hörr, R. (1996) Proof of efficacy of the *Ginkgo biloba* special extract EGb 761 in outpatients suffering from mild to moderate primary degenerative dementia of the Alzheimer type or multi-infarct dementia. *Pharmacopsychiat.*, 29, 47–56.

Kleijnen, J., Knipschild, P. (1991) *Ginkgo biloba* for intermittent claudication and cerebral insufficiency. In J. Kleijnen (ed.), *Food supplements and their efficacy*, Datawyse, Maastricht, pp. 83–94.

Kleijnen, J., Knipschild, P. (1992a) *Ginkgo biloba. Lancet*, **340**, 1136–1139.

Kleijnen, J., Knipschild, P. (1992b) *Ginkgo biloba* for cerebral insufficiency. *Br. J. Clin. Pharmacol.*, **34**, 352–358.

Kleijnen, J., Knipschild, P. (1992c) The comprehensiveness of Medline and Embase computer searches. Searches for controlled trials of homoeopathy, ascorbic acid for common cold and *Ginkgo biloba* for cerebral insufficiency and intermittent claudication. *Pharm. Weekbl. Sci. Ed.*, **14**, 316–320.

Le Bars, P.L., Katz, M.M., Berman, N., Itil, T.M., Freedman, A.M., Schatzberg, A.F. (1997) A placebo-controlled, double-blind, randomized trial of an extract of *Ginkgo biloba* for dementia. Report of the North American EGb Study Group. *JAMA*, **278**, 1327–1232.

Letzel, H., Haan, J., Feil, W.B. (1996) Nootropics: efficacy and tolerability of products from three active substance classes. *J. Drug Dev. Clin. Pract.*, **8**, 77–94.

Letzel, H., Schoop, W. (1992) *Gingko biloba* Extrakt EGb 761 und Pentoxifyllin bei Claudicatio intermittens. Sekundäranalyse zur klinischen Wirksamkeit. *Vasa*, **21**, 403–410.

Maier-Hauff, K. (1991) LI 1370 nach zerebraler Aneurysma-Operation. Wirksamkeit bei ambulanten Patienten mit Störungen der Hirnleistungsfähigkeit. *Münch. Med. Wochenschr.*, **133**, S34–S37.

Maurer, K., Ihl, R., Dierks, T., Frölich, L. (1997) Clinical efficacy of *Ginkgo biloba* special extrakt EGb 761 in dementia of the Alzheimer type. *J. Psychiat. Res.*, **31**, 645–655.

McKhann, G., Drachman, D., Katzman, R., Price, D., Stadlan, E.M. (1984) Clinical diagnosis of Alzheimer's disease: report of the NINCDS-ADRDA work group under the auspices of Department of Health and Human Services Task Force on Alzheimer's Disease. *Neurology*, **34**, 939–944.

Menges, K. (1992) Proof of efficacy of nootropics for the indication "Dementia" (Phase III) – Recommendations. *Pharmacopsychiat.*, **25**, 126–135.

Meyer, B. (1986a) Étude multicentrique des acouphènes. Epidémiologie et thérapeutique. *Ann. Oto-Laryng. (Paris)*, **103**, 185–188.

Meyer, B. (1986b) Étude multicentrique randomisée à double insu face au placebo du traitement des acouphènes par l'extrait de *Ginkgo biloba. Presse Med.*, **15**, 1562–1564.

Moreau, P. (1975) Un nouveau stimulant circulatoire cérébral. *La Nouvelle Presse médicale*, **4**, 2401–2402.

Morgenstern, C., Biermann, E. (1997) Ginkgo-Spezialextrakt EGb 761 in der Behandlung des Tinnitus aurium. Ergebnisse einer randomisierten, doppelblinden, plazebokontrollierten Studie. *Fortschr. Med.*, **115**, 7–11.

Mouren, X., Caillard, P., Schwartz, F. (1994) Study of the antiischemic action of EGb 761 in the treatment of peripheral arterial occlusive disease by TcPo2 determination. *Angiology*, **45**, 413–417.

Mulrow, C., Oxman, A. (eds.) (1994) *Cochrane Collaboration Handbook [Updated September 1997]*. In: The Cochrane Library [Database on disk and CD-ROM]. The Cochrane Collaboration, Update Software, Oxford.

Natali, R., Rachinel, J., Pouyat, P.M. (1979) Essai comparatif croise en O.R.L. de deux médications vaso-actives. *Cahiers d' Oto Rhino Laryngologie*, **14**, 185–190.

Oswald, W.D., Hörr, R., Oswald, B., Steger, W., Sappa, J. (1997) Zur Verbesserung fluider, kognitiver Leistungen mit *Ginkgo-biloba*-Spezialextrakt EGb 761® bei Patienten mit leichten bis mittelschweren Hirnleistungsstörungen im Alter. *Zeitschrift für Gerontopsychologie und -psychiatrie*, **10**, 133–146.

Petitti, D.B. (1994) *Meta-analysis, decision analysis and cost-effectiveness. Methods for quantitative synthesis in medicine*, Oxford University Press, New York.

Pidoux, B., Bastien, C., Niddam, S. (1983) Clinical and quantitative EEG double-blind study of *Ginkgo biloba* extract (GBE). *J. Cerebral Blood Flow Metabolism*, **3**, S556–S557.

Rai, G.S., Shovlin, C., Wesnes, K.A. (1991) A double-blind, placebo controlled study of *Ginkgo biloba* extract ("tanakan") in elderly outpatients with mild to moderate memory impairment. *Curr. Med. Res. Opin.*, **12**, 350–355.

Roncin, J.P., Schwartz, F., D' Arbigny, P. (1996) EGb 761 in control of acute mountain sickness and vascular reactivity to cold exposure. *Aviat. Space Environ. Med.*, **67**, 445–452.

Rudofsky, G. (1987) Wirkung von *Ginkgo-biloba*-Extrakt bei arterieller Verschlußkrankheit. Randomisierte plazebokontrollierte Cross-over-Doppelblindstudie. *Fortschr. Med.*, **105**, 397–400.

Salz, H. (1980) Zur Wirksamkeit eines *Ginkgo-biloba*-Präparats bei arteriellen Durchblutungsstörungen der unteren Extremitäten. Kontrollierte Doppelblind Cross-over-Studie. *Ther. Ggw.*, **119**, 1345–1356.

Saudreau, F., Serise, J.M., Pillet, J., Maiza, D., Mercier, V., Kretz, J.G., Thibert, A. (1989) Efficacité de l'extrait de *Ginkgo biloba* dans le traitement des artériopathies oblitérantes chroniques des membres inférieurs au stade III de la classification de Fontaine. *J. Mal. Vasc.*, **14**, 177–182.

Schmidt, U., Rabinovici, K., Lande, S. (1991) Einfluß eines Ginkgo-Spezial-Extraktes auf die Befindlichkeit bei zerebraler Insuffizienz. *Münch. Med. Wochenschr.*, **133**, S15–S18.

Schneider, B. (1992) *Ginkgo-biloba*-Extrakt bei peripheren arteriellen Verschlußkrankheiten. Meta-Analyse von kontrollierten klinischen Studien. *Arzneim.-Forsch./Drug Res.*, **42**, 428–436.

Schubert, H., Halama, P. (1993) Primär therapieresistente depressive Verstimmung älterer Patienten mit Hirnleistungsstörungen: Wirksamkeit der Kombination von *Ginkgo-biloba*-Extrakt EGb 761 mit Antidepressiva. *Geriatrie Forschung*, **3**, 45–53.

Schulz, V., and Hänsel, R. (1996) *Rationale Phytotherapie. Ratgeber für die ärtzliche Praxis. Ch.2: Zentrales Nervensystem. Ch. 2.1. Ginkgo bei Hirnleistungsstörungen*, Springer-Verlag, Berlin.

Schulz, V. Hubner, W.D., Ploch, M. (1997) Clinical trials with phyto-psychopharmacological agents. *Phytomedicine*, **4**, 379–387.

Schwerdtfeger, F. (1981) Elektronystagmographisch und klinisch dokumentierte Therapieerfahrungen mit rökan® bei Schwindelsymptomatik. *Therapiewoche*, **31**, 8658–8667.

Semlitsch, H.V., Anderer, P., Saletu, B., Binder, G.A., Decker, K.A. (1995) Cognitive psychophysiology in nootropic drug research: effects of *Ginkgo biloba* on event-related potentials (P300) in age-associated memory impairment. *Pharmacopsychiat.*, **28**, 134–142.

Small, G.W., Rabins, P.V., Barry, P.P., Buckholtz, N.S., DeKosky, S.T., Ferris, F.H. (1997) Diagnosis and treatment of Alzheimer Disease and related disorders. Consensus statement of the American Association for Geriatric Psychiatry, the Alzheimer's Association, and the American Geriatrics Society. *JAMA*, **278**, 1363–1371.

Sohn, M., Sikora, R. (1991) *Ginkgo biloba* extract in the therapy of erectile dysfunction. *J. Sex Education & Therapy*, **17**, 53–61.

Sticher, O. (1992) *Ginkgo biloba* Analytik und Zubereitungsformen. *Pharm. u. Zeit*, **21**, 253–265.

Stocksmeier, U., Eberlein, M. (1992) Depressive Verstimmung bei Hirnleistungsstörungen. Wirkung eines *Ginkgo-biloba*-Extraktes in Doppelblind-Studie überprüft. *T.W. Neurologie Psychiatrie*, **6**, 74–76.

Taillandier, J., Ammar, A., Rabourdin, J.P., Ribeyre, J.P., Pichon, J., Niddam, S., Pierart, H. (1986) Traitement des troubles du vieillissement cérébral par l'extrait de Ginkgo biloba. Étude longitudinale multicentrique à double insu face au placebo. Presse Med., 15, 1583–1587.

Tamborini, A., Taurelle, R. (1993) Intérêt de l'extrait standardisé de Ginkgo biloba (EGb 761) dans la prise en charge des symptômes congestifs du syndrome prémenstruel. Rev. Fr. Gynécol. Obstét., 88, 447–57.

Teigeler, R., Pieprzyk, L. (1984) Ginkgo biloba bei zerebraler Insuffizienz. Ärtzliche Praxis, 36, 1374.

Thomson, G.J., Vohra, R.K., Carr, M.H., Walker, M.G. (1990) A clinical trial of Gingkco biloba extract in patients with intermittent claudication. Int. Angiol., 9, 75–78.

Vesper, J., Hänsgen, K.-D. (1994) Efficacy of Ginkgo biloba in 90 outpatients with cerebral insufficiency caused by old age. Results of a placebo-controlled double-blind trial. Phytomedicine, 1, 9–16.

Von Wedel, H., Calero, L., Walger, M., Hoenen, S., Rutwalt, D. (1995) Soft-laser/Ginkgo therapy in chronic tinnitus. A placebo-controlled study. Adv. Otorhinolaryngol., 49, 105–108.

Vorberg, G., Schenk, N., Schmidt, U. (1989) Wirksamkeit eines neuen Ginkgo-biloba-Extraktes bei 100 Patienten mit zerebraler Insufficiency. Herz + Gefäße, 9, 936–941.

Warburton, D.M. (1986) Psycho-pharmacologie clinique de l'extrait de Ginkgo biloba. Presse Med., 15, 1595–1604.

Weiß, H., Kallischnigg, G. (1991) Ginkgo-biloba-Extrakt (EGb 761). Meta-Analyse von Studien zum Nachweis der therapeutischen Wirksamkeit bei Hirnleistungsstörungen bzw. arterieller Verschlußkrankheit. Münch. Med. Wochenschr., 133, 138–42.

Weitbrecht, W.U., Jansen, W. (1986) Primär degenerative Demenz: Therapie mit Ginkgo-biloba-Extrakt. Plazebo kontrollierte Doppelblind- und Vergleichsstudie. Fortschr. Med., 104, 199–202.

Weitbrecht, W.V. and Jansen, W. (1985) Doubleblind and comparative (Ginkgo biloba versus placebo) therapeutic study in geriatric patients with primary degenerative dementia – a preliminary evaluation. In J. Agnoli, R. Rapin, V. Scapagnini, and W.V. Weitbrecht (eds.), Effects of Ginkgo Biloba Extract on organic cerebral impairment, John Libbey Eurotext Ltd., Montrouge, pp. 91–99.

Wesnes, K., Simmons, D., Rook, M., Simpson, P. (1987) A double-blind placebo-controlled trial of Tanakan in the treatment of idiopathic cognitive impairment in the elderly. Human Psychopharmacol., 2, 159–169.

WHO (1991) ICD–10: Tenth revision of the International Classification of Diseases. Chapter V (F): Mental and behavioural disorders (including disorders of psychological development). Clinical descriptions and diagnostic guidelines. World Health Organization, Geneva.

22. ADVERSE EFFECTS AND TOXICITY OF GINKGO EXTRACTS

HERMAN J. WOERDENBAG[1] and PETER A.G.M. De SMET[2]

[1] *University Centre for Pharmacy, Antonius Deusinglaan 2, 9713 AW Groningen, The Netherlands*
[2] *Scientific Institute Dutch Pharmacists, Alexanderstraat 11, 2514 JL The Hague, The Netherlands*

INTRODUCTION

Phytotherapeutic preparations from leaves of *Ginkgo biloba* L. are used to treat cerebral insufficiency and peripheral arterial vascular diseases in elderly people (Spegg, 1990; Sticher *et al.*, 1991). Among the large number of constituents found in Ginkgo leaves, the flavonoids (Ginkgo flavonol glycosides) and the terpene trilactones (ginkgolides and bilobalide), are held responsible for most of the pharmacological action and for the therapeutic effect (Braquet, 1988; Braquet, 1989; Hasler *et al.*, 1990; Hölzl, 1992; Huh and Staba, 1992; Tang and Eisenbrand, 1992; Spieß and Juretzek, 1993; Houghton, 1994; Newall *et al.*, 1996; Woerdenbag and van Beek 1997). The spectrum of action includes free radical scavenging, antagonism of platelet-activating factor (PAF), effects on the energy metabolism under hypoxic and ischaemic conditions, vasoregulatory activities, rheological effects on the blood, reduction of oedema, and interaction with neurotransmitters (Woerdenbag, 1993).

In several European countries standardised enriched extracts of Ginkgo leaves, prepared with acetone-water mixtures, are marketed; in Germany Tebonin® and Rökan®, in France Tanakan®, and in The Netherlands Tavonin®. These extracts contain higher concentrations of the active principles and minimal concentrations of undesired compounds, such as ginkgols and ginkgolic acids. They are standardised to contain 24% Ginkgo flavonol glycosides and 6% terpene trilactones (ginkgolides and bilobalide combined). Furthermore they contain 5–10% oligomeric pro-anthocyanidins and about 9% organic acids (Ganzer, 1990; Hänsel, 1990; Anonymous, 1992; Jaggy, 1993; Newall *et al.*, 1996). These extracts are abbreviated as SEGLE (standardised enriched Ginkgo leaf extracts) in this chapter.

To put the safety information on SEGLE in perspective, the extensive use in Germany in relation to the very infrequent occurrence of adverse events, is illustrative. In 1995, doctors in that country prescribed 254.3 million of Defined Daily Doses of monopreparations with a Ginkgo leaf extract (Schwabe, 1996).

Other available phytomedicines that are made from Ginkgo leaves, include extracts that are prepared with ethanol-water mixtures. These extracts are not further processed. Homeopathic mother tinctures belong to this group. These preparations are not standardised, and analyses showed that they may contain 10–15 times less flavonoids and triterpene lactones than the aforementioned SEGLE (Sticher *et al.*, 1991; van Beek *et al.*, 1991; Sticher, 1992; Sticher, 1993).

The composition of phytotherapeutic Ginkgo preparations is thus determined by the applied extraction method and the subsequent purification and concentration steps (Sticher *et al.*, 1991). In addition, variations in the content of flavonols, ginkgolides and bilobalide in the original leaves play an important role (Lobstein *et al.*, 1991; van Beek *et al.*, 1991). These are important aspects when the pharmacological action, clinical efficacy and adverse effects are considered (Woerdenbag and van Beek, 1997).

The majority of the pharmacological experiments and all clinical studies have been conducted with SEGLE. There is evidence from clinical trials, although often with rather small numbers of patients, that such preparations are more effective than placebo in peripheral vascular diseases such as intermittent claudication. There is also evidence of clinical effects in cerebral insufficiency but it is probably not useful in patients with clear dementia (Kleijnen and Knipschild, 1992). Positive results have been obtained with a daily dose of 120–160 mg SEGLE, given for at least 4–6 weeks. Of other Ginkgo preparations that contain much lower concentrations of the active principles, the activity and efficacy of the usual dosages have not been proven so far (Schilcher, 1988; DeFeudis, 1991; Newall *et al.*, 1996).

Finally, roasted Ginkgo seeds are used as an antitussive and expectorant in Japan and China. It is also applied to treat polyuria, tuberculosis, leukorrhoea, miction problems, pollakiuria and spermatorrhoea (Wada *et al.*, 1985; Wada *et al.*, 1988; Tang and Eisenbrand, 1992; Bauer and Zschocke, 1996; Newall *et al.*, 1996).

In this chapter a literature-based overview is given of toxicity studies with Ginkgo extracts and Ginkgo constituents in animals, adverse effects in humans after oral and parenteral administration of Ginkgo preparations, interactions with other medications and contra-indications.

TOXICITY STUDIES IN ANIMALS

Acute Toxicity

The acute toxicity of SEGLE after oral (p.o.), intraperitoneal (i.p.) or intravenous (i.v.) application has been determined in the mouse and the rat. In the mouse the following LD_{50} values have been established: p.o. 7725 mg/kg; i.p. 1900 mg/kg and i.v. 1100 mg/kg. In the rat: i.p. 2100 mg/kg and i.v. 1100 mg/kg. Up to 10 g/kg p.o. no lethal effect was seen in this species. Higher doses could not be administered (O'Reilly, 1993; Spieß and Juretzek, 1993).

An aqueous extract of Ginkgo seeds given orally at 11 mg/kg to guinea pigs of either sex, caused paralysis of legs, opisthotonus, clonic convulsions, and auditory hyperalgesia, for which 4-O-methylpyridoxine (MPN), als named ginkgotoxin, is held responsible. Following an oral dose of 50 mg/kg ventricular fibrillation may occur in these animals, resulting in death (Wada *et al.*, 1985; Wada *et al.*, 1988; Arenz *et al.*, 1996).

Chronic Toxicity

Chronic toxicity studies have been carried out in dogs and rats. Over a period of 6 months, each day SEGLE was administered orally at doses up to 500 mg/kg (rat) and 400 mg/kg (dog). Dogs appeared to be more sensitive for the toxic effects than rats. At 100 mg/kg mild and transient vasodilatory effects were seen in the head. Following the highest dose of 400 mg/kg, these disturbances were stronger and occurred after 35 days of treatment. No biochemical, hematological or histological changes were observed after sacrificing the animals. Liver and kidney functions remained unaltered (Schilcher, 1988; O'Reilly, 1993; Spieß and Juretzek, 1993).

Mutagenicity, Carcinogenicity, Teratogenicity, Embryotoxicity

For SEGLE no mutagenic potential has been found. The following methods were applied: the Ames-test, using the *Salmonella typhymurium* strains TA 1535, 1537, 1538, 98 and 100, with and without metabolic activation using rat liver S9-mix; the host-mediated-assay (mouse, *S. typhymurium* strain TA 1537, doses up to 20 g/kg p.o.); the micronucleus test (mouse, doses up to 20 g/kg p.o.); and the chromosome aberration test (human lymphocytes, concentrations up to 100 mg/ml) (Spieß and Juretzek, 1993).

In a study in which SEGLE were given orally to rats for 104 weeks at doses of 4, 20 and 100 mg/kg/day, no carcinogenic effects were found (Spieß and Juretzek, 1993). No carcinogenic effects could be detected in experiments with rats, fed with kernels of Ginkgo seeds for almost a year (Hirono *et al.*, 1972). In experimental animals no indications have been found for embryotoxic, mutagenic or teratogenic effects (Herrschaft, 1992). Oral administration of up to 1600 mg/kg/day of SEGLE to rats and 900 mg/kg/day to rabbits did not produce teratogenic effects. In addition, reproduction was not affected (O'Reilly, 1993).

ADVERSE EFFECTS IN HUMANS

Oral Administration

The toxicity of phytotherapeutic preparations from Ginkgo leaves for oral use is generally considered low. At the recommended dose, hardly any adverse effects should be expected (Herrschaft, 1992; Kleijnen and Knipschild, 1992; O'Reilly, 1993; Newall *et al.*, 1996; Woerdenbag and van Beek, 1997).

In a clinical trial, performed in Germany, 112 patients of both sexes suffering from chronic cerebral insufficiency and aged between 55 and 94 years (mean age 70.5 years), were treated for one year with 120 mg SEGLE daily. During the period of the trial no heart rate and blood pressure modifications could be detected and blood cholesterol and triglyceride levels remained practically unchanged. In the same trial light gastric symptoms were reported at the start of the treatment, but these subsided spontaneously as treatment continued (Vorberg, 1985). An apparent lack of effects of SEGLE on various laboratory parameters was also observed in other studies (Taillandier *et al.*, 1986; Felber, 1992). Side-effects that became apparent from various clinical trials with a daily dose of 120–160 mg SEGLE for 4–6 weeks did not differ from the placebo groups (Kleijnen and Knipschild, 1992).

Headache, dizziness, heart palpitations, gastrointestinal problems and light allergic reactions of the skin have been reported to occur rarely (DeFeudis, 1991). Vertigo, nausea and heart palpitations are also mentioned in the literature as possible side-effects of SEGLE (Dukes, 1988; Anonymous, 1989; Anonymous, 1994a; Oberpichler-Schwenk, 1995). One case of spontaneous bilateral subdural hematomas has been recently associated with chronic ingestion of an unspecified *Ginkgo biloba* preparation (Rowin and Lewis, 1996).

Single doses of 720 mg pure ginkgolide, in the form of a mixture of ginkgolides A, B and C (1:2:1) gave no observable adverse effects in humans. Doses of 360 mg of this ginkgolide mixture per day during one week yielded no toxic effects (Bonvoisin and Guinot, 1989).

It is not sufficiently clear whether unprocessed extracts are as safe as SEGLE. Unprocessed extracts might pose a higher health risk due to the presence of unknown quantities of ginkgols and ginkgolic acids. These compounds are present in the raw seeds of the Ginkgo tree and may provoke strong allergic reactions (Lepoittevin *et al.*, 1989). Alkylphenols are also present in Ginkgo leaves (Irie *et al.*, 1996). After ingestion of preparations containing these allergens, reactions of the mouth mucosa have been described, as well as stomatitis, gastroenteritis and proctitis (Ganzer, 1990). In processed preparations of Ginkgo leaves their concentration is so low (maximal 5–10 ppm) that allergic reactions (due to these compounds) are not expected (Ganzer, 1990; Sticher *et al.*, 1991).

The central kernel of the Ginkgo seed is edible when roasted, but the fleshy fruit pulp is toxic, even after roasting (Sowers *et al.*, 1965). In Japan the so-called "ginnan food poisoning" sometimes occurred after Ginkgo seeds had been taken to excess during food shortages. The main symptoms of this poisoning were convulsions and loss of consciousness. According to about 70 reports from the period between 1930 and 1960 dealing with this poisoning, the sequelae were not serious to survivers, but about 27% lethality was found. Infants were particularly vulnerable. It has been stated that 4-O-methylpyridoxine (MPN) is the toxic substance responsible for this poisoning. MPN is a potent convulsive agent, with vitamin B_6 antagonizing activity in several experimental animals and in man. In addition, it inhibits the formation of 4-aminobutyric acid (GABA) from glutamate in the brain. A deficiency of GABA in the brain may induce seizures (Wada *et al.*, 1985; Wada *et al.*, 1988; Wada, 1998).

Yagi *et al.* (1993) determined serum levels of MPN in a 21 months-old child, who suffered from gin-nan food poisoning after taking about 50 Ginkgo albumens. MPN concentrations were 0.09 mg/ml 8.5 h after taking the Ginkgo seeds and less than the detection limit of 0.05 mg/ml after 15.5 h.

Recently, MPN was also detected in Ginkgo leaves. The toxin was present in Ginkgo medications as well, including SEGLE, and it was also detectable in homeopathic preparations. The amount of MPN, however, is likely to be too low to exert any adverse effect after administration of the medication (Arenz *et al.*, 1996).

Pollens of *Ginkgo biloba* have been found to act as allergens, causing pollinosis (Sowers *et al.*, 1965; Long *et al.*, 1992). The ginkgols and ginkgolic acids, as present in the raw seeds of *G. biloba*, may provoke strong allergic reactions (Lepoittevin *et al.*, 1989). The fleshy part of the seeds of *G. biloba* has been known from ancient times to irritate the skin. The alkylsalicylic acid derivatives and the alkylphenol derivatives are responsible for the dermatitis that occurs after contact with the skin (Ganzer, 1990; Kochibe, 1997), and may be compared with the allergic effects caused by Rhus species. Erythema, edema, papules and vesicles are seen, complicated by intense itching. The irritation disappears after 7–10 days following the contact (Becker and Skipworth, 1975; Yamada *et al.*, 1995). The dermatitic action of these compounds is consistent with oxidation of the allergen to a quinone, which then binds covalently to a protein nucleophile, yielding an antigenic complex (Evans, 1988).

Parenteral Administration

Injectable preparations of SEGLE have been available in Germany. Parenteral application, however, has been associated with serious adverse reactions. Circulatory disturbances (hypotension, dizziness, arrythmia and eventually shock), allergic skin reactions and phlebitis may occur after injection (Fintelmann *et al.*, 1989). This applies to intramuscular and intravenous administration, but especially to intra-arterial injection (Anonymous, 1989, Anonymous, 1994a). In June 1993 anaphylactic shock has been reported after parenteral administration of Tebonin® to four patients. Dizziness, sickness and faintness were seen. All patients showed high fever and leucocytosis. As a sign of impaired circulation, metabolic acidosis and an increase of liver enzymes were found. In one case life-threatening heart arrythmia occurred (Anonymous, 1994b).

In 1989, the Bundesgesundheitsamt (BGA) warned for the possible risk of proteins, peptides, nucleic acids, pyrogens and toxins present in Ginkgo extracts for parenteral administration (Anonymous, 1989). It was also pointed out that flavonoids possess a possible immunogenic action, and that SEGLE may therefore provoke heavy immunoallergenic reactions after parenteral administration (Moebius *et al.*, 1989). In 1994, the German health authorities first announced the suspension of the registration of Ginkgo preparations for parenteral use, because of the severe side effects that had been reported (Anonymous, 1994a). All the cases were discussed by the BGA and they could not find enough evidence of efficacy to balance these potential risks (Anonymous, 1994b). Shortly afterwards, the manufacturer (Schwabe) withdrew its parenteral preparations (Anonymous, 1994c). Judging from the latest *Rote Liste* (Anonymous, 1996b), they have not been reintroduced.

INTERACTION WITH OTHER MEDICATIONS

No significant interactions with existing medication, including cardiac glycosides and oral antidiabetics, were seen in an open one-year trial with 112 patients, with a mean age of 70.5 years and suffering from chronic cerebral insufficiency, treated with 120 mg/day of SEGLE (Vorberg, 1985).

Although pharmacological studies with SEGLE have demonstrated distinct effects on hemorheological parameters, such as a decreased blood viscosity, PAF-antagonism and vasoregulatory activity, no clinical reports exist on hematological interactions with antithrombotic drugs. However, a theoretical risk may not be excluded (Woerdenbag and van Beek, 1997). Recently, spontaneous bleeding from the iris into the anterior chamber of the eye was reported in a 70-year-old man, and associated with the ingestion of *Ginkgo biloba* leaf extract (40 mg concentrated extract taken twice daily for a week) in combination with aspirin (325 mg daily). This case suggested that a short-term use of Ginkgo extract may also have antiplatelet effects and that there might be additional risks with aspirin, which is also an inhibitor of platelet aggregation (Rosenblatt and Mindell, 1997). There is an apparent lack of enzyme induction by SEGLE; after 13 days of treatment of healthy volunteers with 400 mg daily, no such effects were found (Duché *et al.*, 1989).

CONTRA-INDICATIONS

The package insert of the Dutch SEGLE preparation Tavonin® specifies as contraindications: hypersensitivity, hepatic insufficiency and renal insufficiency (Anonymous, 1996a).

Human data on adverse effects of Ginkgo preparations with respect to fertility, pregnancy and lactation are not available (Spieß and Juretzek, 1993). The patho-physiological phenomena for which Ginkgo leaf extracts are indicated, typically occur in elderly people. Ginkgo preparations are sometimes regarded as geriatric drugs (Wichtl, 1992). Therefore, possible risks concerning fertility, pregnancy and lactation are of limited significance here (Woerdenbag and van Beek, 1997). However, in view of the plethora of pharmacological actions that have been documented for SEGLE and the lack of safety data with respect to pregnancy and lactation, some recommend that it is better to avoid the use of Ginkgo preparations under these conditions (Newall *et al.*, 1996).

CONCLUSION

SEGLE is a safe phytomedicine in the treatment of peripheral and cerebral circulatory disturbances, with reasonable evidence of effectivity from clinical studies.

REFERENCES

Anonymous (1989) *Ginkgo biloba*-haltige Arzneimittel. *Pharm. Ztg.*, **134**, 561.
Anonymous (1992) Tebonin®. *Ginkgo biloba*-Spezialextrakt. EGb 761. Dr. Willmar Schwabe, Karlsruhe.

Anonymous (1994a) *Rote Liste*, Bundesverband der Pharmazeutischen Industrie e. V., Frankfurt/Main, 36001–36012.

Anonymous (1994b) *Ginkgo-biloba*-haltiger Trockenextrakt zu Infusion. *Pharm. Ztg.*, **139**, 986–987.

Anonymous (1994c) Rückruf Tebonin p.i., Tebonin p.i. 175. *Dtsch. Apoth. Ztg.*, **134**, 1324–1326.

Anonymous (1996a) Tavonin tabletten. Registration 18573 by the Dutch Medicines Evaluation Board, 9 december 1996, Rijswijk.

Anonymous (1996b) *Rote Liste*, Bundesverband der Pharmazeutischen Industrie e. V., Frankfurt/Main.

Arenz, A., Klein, M., Fiehe, K., Groß, J., Drewke, C., Hemscheidt, T. and Leistner, E. (1996) Occurrence of neurotoxic 4'-O-methylpyridoxine in *Ginkgo biloba* leaves, Ginkgo medications and Japanese Ginkgo food. *Planta Med.*, **62**, 548–551.

Bauer, R. and Zschocke, S. (1996) Medizinische Anwendung von *Ginkgo biloba* L. *Z. Phytother.*, **17**, 275–283.

Becker, L.E. and Skipworth, G.B. (1975) Ginkgo-tree dermatitis, stomatitis, and proctitis. *JAMA*, **231**, 1162–1163.

van Beek, T.A., Scheeren, H.A., Rantio, T., Melger, W.C., Lelyveld, G.P. (1991) Determination of ginkgolides and bilobalide in *Ginkgo biloba* leaves and phytopharmaceuticals. *J. Chromatogr.* **543**, 375–387.

Bonvoisin, B., Guinot, P. (1989) Clinical studies of BN 52063 a specific PAF antagonist. In P. Braquet, (ed.), *Ginkgolides – Chemistry, Biology, Pharmacology and Clinical Perspectives*, Vol. 2, J.R. Prous Science Publishers, Barcelona, pp. 845–854.

Braquet, P., (ed.) (1988) *Ginkgolides – Chemistry, Biology, Pharmacology and Clinical Perspectives*, Vol. 1, J.R. Prous Science Publishers, Barcelona.

Braquet, P., (ed.) (1989) *Ginkgolides – Chemistry, Biology, Pharmacology and Clinical Perspectives*, Vol. 2, J.R. Prous Science Publishers, Barcelona.

DeFeudis, F.V. (1991) *Ginkgo biloba* Extrakt EGb 761, Pharmacological Activities and Clinical Applications, Elsevier, Paris.

Duché, J.C., Barre, J., Guinot, P., Duchier, J., Cournot, A., Tillement, J.P. (1989) Effect of *Ginkgo biloba* extract on microsomal enzyme induction. *Int. J. Clin. Pharm. Res.*, **9**, 165–168.

Dukes, M.N.G. (1988) *Meyler's Side Effects of Drugs*, 11th ed, Elsevier, Amsterdam, p. 395.

Evans, W.C. (1988) *Trease and Evans' Pharmacognosy*, 13th ed., Ballière Tindall, London.

Felber, J.P. (1992) Einfluß von Rökan auf metabolische und endokrine Parameter. In C. Diehm and D. Müller, (eds.), *Rökan. Ginkgo biloba EGb 761, Bd. 2: Klinik*, Springer Verlag, Berlin, pp. 337–340.

Fintelmann, V., Menßen, H.G., Siegers, C.P. (1989) *Phytotherapie Manual*, Hippokrates Verlag, Stuttgart, p. 55.

Ganzer, B.M. (1990) Spezialextrakte aus Ginkgo-Pflanze – jeder ein Unikat. *Pharm. Ztg.*, **135**, 2111.

Hänsel, R. (1990) Analytische Differenzierung verschiedener Ginkgo-Extrakte. *Ärztl. Forsch.*, **37**, 1–3.

Hasler, A., Meier, B., Sticher, O. (1990) *Ginkgo biloba*. Botanische, analytische und pharmakologische Aspekte. *Schweiz. Apoth. Ztg.*, **128**, 341–347.

Herrschaft, H. (1992) Zur klinischen Anwendung von *Ginkgo biloba* bei dementiellen Syndromen. *Pharm. unserer Zeit*, **21**, 266–275.

Hirono, I., Shibuya, C., Shimizu, M., Fushimi, K., Mori, H., Miwa, T. (1972) Carcinogenicity examination of some edible plants. *GANN (Jap. J. Cancer Res.)*, **63**, 383–386.

Hölzl, J. (1992) Inhaltsstoffe von *Ginkgo biloba*. *Pharm. unserer Zeit*, **21**, 215–223.

Houghton, P.J. (1994) Ginkgo. *Pharm. J.*, **253**, 122–123.

Huh, H., Staba, E.J. (1992) The botany and chemistry of *Ginkgo biloba* L. *J. Herbs Spices Med. Plants*, **1**, 1–124.

Irie, J., Murata, M. and Homma, S. (1996) Glycerol–3-phosphate dehydrogenase inhibitors, anacardic acids, from *Ginkgo biloba*. *Biosci. Biotechnol. Biochem.* **60**, 240–243.

Jaggy, H. (1993) Die Inhaltsstoffe des *Ginkgo biloba*-Extraktes EGb 761. *Hämostaseologie (Stuttgart)*, **13**, 7–10.

Kleijnen, J., Knipschild, P. (1992) *Ginkgo biloba. Lancet*, 1136–1139.

Kochibe, N. (1997). Allergic substances of *Ginkgo biloba*. In T. Hori, R.W. Ridge, W. Tulecke, P. Del Tredici, J. Trémouillaux-Guiller and H. Tobe, (Eds.), *Ginkgo biloba. A global treasure*, Springer, Tokyo, pp. 301–307.

Lepoittevin, J.-P., Benezra, C., Asakawa, Y. (1989) Allergic contact dermatitis to *Ginkgo biloba* L.: relationship with urushiol. *Arch. Dermatol. Res.*, **281**, 227–230.

Lobstein, A., Rietsch-Jako, L., Haag-Berrurier, M., Anton, R. (1991) Seasonal variations of the flavonoid content from *Ginkgo biloba* leaves and phytopharmaceuticals. *Planta Med.*, **57**, 430–433.

Long, R., Yin, R., Zhen, Y. (1992) Partial purification and analysis of allergenicity, immunogenicity of *Ginkgo biloba* pollen. *J. West China Univ. Med. Sci.*, **23**, 429–432.

Moebius, U.M., Becker-Brüser, W., Schönhöfer, P.S. (1989) *Alarmtelegramm*, Arzneimittel-Verlags GmbH, Berlin, pp. 160, 181.

Newall, C.A., Anderson, L.A., Phillipson, J.D. (1996) *Herbal Medicines*, The Pharmaceutical Press, London, pp. 138–140.

Oberpichler-Schwenk, H. (1995) Überdosierung von Ginkgo-präparaten? *Med. Monatsschr. Pharm.*, **18**, 100–101.

O'Reilly, J. (1993) *Ginkgo biloba* – cultivation, extraction and therapeutic use of the extract. In van Beek, T.A., and Breteler, H., (eds.), *Phytochemistry and Agriculture*, Vol. 34, Clarendon Press, Oxford, pp. 253–270.

Rosenblatt, M. and Mindel, J. (1997) Spontaneous hyphema associated with ingestion of *Ginkgo biloba* extract. *Lancet*, 1108.

Rowin, J. and Lewis, S.L. (1996) Spontaneous bilateral subdural hematomas associated with chronic *Ginkgo biloba* ingestion. *Neurology*, **46**, 775–776.

Schilcher, H. (1988) *Ginkgo biloba* L. Untersuchungen zur Qualität, Wirkung, Wirksamkeit und Unbedenklichkeit. *Z. Phytother.*, **9**, 119–127.

Schwabe, U. (1996) Durchblütungsfördernde Mittel. In U. Schwabe and D. Paffrath, (eds.), *Arzneiverordnungs-Report '96. Aktuelle Daten, Kosten, Trends und Kommentare*, Gustav Fischer Verlag, Stuttgart, pp. 227–237.

Sowers, W.F., Weary, P.E., Collins, O.D., Cawley, E.P. (1965) Ginkgo-tree dermatitis. *Arch. Dermatol.*, **91**, 452–456.

Spegg, H. (1990) *Ginkgo biloba* – Ein Baum aus Urzeiten, ein Phytopharmakon mit Zukunft. *PTA Heute*, **4**, 576–583.

Spieß, E., Juretzek, W. (1993) Ginkgo. In R. Hänsel, K. Keller, H. Rimpler, and G. Schneider, (eds.), *Hager's Handbuch der Pharmazeutischen Praxis*, 5th edn., Vol. 5, Springer-Verlag, Berlin, pp. 269–292.

Sticher, O., Hasler, A., Meier, B. (1991) *Ginkgo biloba* – Eine Standortbestimmung. *Dtsch. Apoth. Ztg.*, **131**, 1827–1835.

Sticher, O. (1992) *Ginkgo biloba* – Analytik und Zubereitungsformen. *Pharm. unserer Zeit* **21**, 253–265.

Sticher, O. (1993) Quality of Ginkgo preparations. *Planta Med.*, **59**, 2–11.

Taillandier, J., Ammar, A., Rabourdin, J.P., Ribeyre, J., Pichon, J., Niddam, S., Piérart, H. (1986) Traitement des troubles du vieillissement cérébral par l'extrait de *Ginkgo biloba*. *Presse Méd.*, **15**, 1583–1587.

Tang, W., Eisenbrand, W. (1992) *Ginkgo biloba L. Chinese Drugs of Plant Origin*, Springer-Verlag, Berlin, pp. 555–565.

Vorberg, G. (1985) *Ginkgo biloba* extract (GBE) : a long-term study of chronical cerebral insufficiency in geriatric patients. *Clin. Trials J.*, **22**, 149–157.

Wada, K., Ishigaka, S., Ueda, K., Sakata, M., Haga, M. (1985) An antivitamin B6, 4'-O-methoxypyridoxine, from the seed of *Ginkgo biloba* L. *Chem. Pharm. Bull.*, **33**, 3555–3557.

Wada, K., Ishigaka, S., Ueda, K., Take, Y., Sasaki, K., Sakata, M., Haga, M. (1988) Studies on the constitution of edible and medicinal plants. I. Isolation and identification of 4'-O-methylpyridoxine, toxic principle from seed of *Ginkgo biloba* L. *Chem. Pharm. Bull.*, **36**, 1779–1782.

Wada, K. (1999) Food poisoning by Ginkgo seeds: the role of 4-O-methylpyridoxine. In *Ginkgo biloba*, T.A. van Beek, (ed.), Harwood Academic Publishers, Amsterdam, pp. 453–465.

Wichtl, M. (1992). Pflanzliche Geriatrika. *Dtsch. Apoth. Ztg.*, **132**: 1569–1576.

Woerdenbag, H.J. (1993) Therapie met bladextracten uit *Ginkgo biloba*. *Pharm. Weekbl.*, **128**, 102–106.

Woerdenbag, H.J. (1998) Ginkgobladextract EGb 761 (Tavonin). *Pharm. Weekbl.*, **133**, 672–677.

Woerdenbag, H.J. and van Beek, T.A. (1997) *Ginkgo biloba*. In P.A.G.M. de Smet, K. Keller, R. Hänsel, and R.F. Chandler, (eds.), *Adverse Effects of Herbal Drugs*, Vol. 3, Springer-Verlag, Berlin, pp. 51–66.

Yagi, M., Wada, K., Sakata, M., Kokubo, M., Haga, M. (193) Studies on the constituents of edible and medicinal plants. IV. Determination of 4-O-methylpyridoxine in serum of the patient with gin-nan food poisoning. *Yakugaku Zasshi*, **113**, 596–599.

Yamada, S., Mitsuya, K and Horio, T. (1995) Erythema multiforme-like eruption following contact dermatitis to Ginkgo. *Skin Res.*, **37**, 9–16.

23.   FOOD POISONING BY GINKGO SEEDS: THE ROLE OF 4-O-METHYLPYRIDOXINE

KEIJI WADA

Faculty of Pharmaceutical Sciences,
Health Sciences University of Hokkaido,
Ishikari-Tobetsu, Hokkaido 061-0293, Japan

INTRODUCTION

Many kinds of antivitamins have been isolated from higher plants. According to the classical definition, a substance should be considered an antivitamin only when the following conditions are fulfilled (Somogyi, 1973):

1. Similarity of the chemical structure between vitamin and corresponding anti-vitamin
2. Similarity of the symptoms produced by the antivitamin and by lack of the vitamin
3. Substance exerting as an inhibitor of the corresponding vitamin

The seeds of *Ginkgo biloba* L., which have caused food poisoning in Japan and China, are known to contain 4-O-methylpyridoxine (MPN) (Wada *et al.*, 1985). MPN is one of the most typical B_6 antivitamins under the classical definition. Prior to the discovery of MPN, linatine was the only known naturally occurring vitamin B_6 antagonist (Klosterman *et al.*, 1967).

FOOD POISONING BY GINKGO SEEDS

"Gin-nan" is the seed of *Ginkgo biloba* L., and its albumen is used as an antitussive and expectorant in traditional medicine. In addition, because gin-nan contains approximately 34% starch, roasted or boiled Ginkgo seeds were used as a rice substitute during food shortages in Japan. However, when consumed to excess, food poisoning sometimes occurred in Japan and China (Wada, 1986 and 1997), with more than 70 reported cases in Japan between 1930 and 1996 (Fig. 1).

The symptoms of food poisoning by Ginkgo seeds are primarily tonic and/or clonic convulsions and loss of consciousness. Of the cases shown in Fig. 1, infants and in particular children under six years of age made up approximately 74% of the patients. The ingestion of Ginkgo seeds by the patients with this food poisoning ranged from 20 to 50 seeds (Wada, 1997). The minimum ingestion of Ginkgo seeds

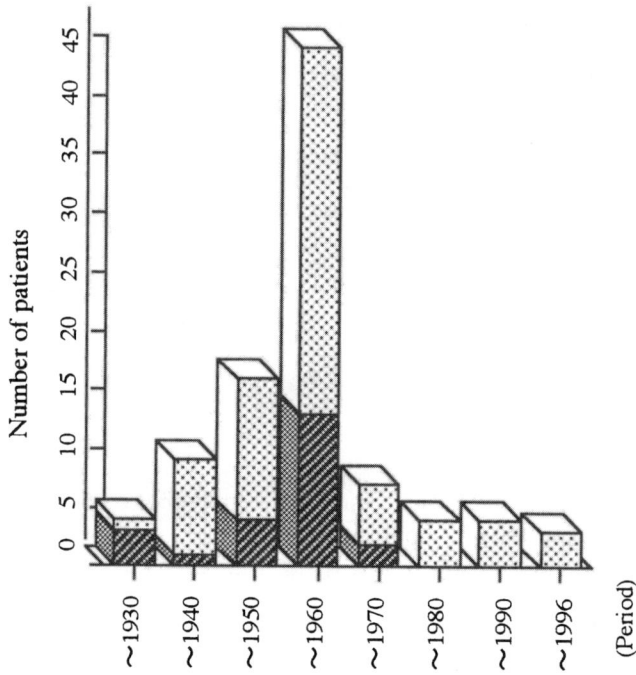

Figure 1   Number of patients with gin-nan food poisoning in Japan. (from Wada & Haga, 1997) ▨ Number of Deaths, ▦ Number of Survivors, ▨ + ▦ : Total Patients.

was six seeds (Takano *et al.*, 1953). Although the consequences are not serious for survivors, the mortality rate is to date in Japan approximately 25% (Wada, 1996).

Initially, the symptoms of Ginkgo seed food poisoning led most investigators to believe that the poisoning was caused by cyanogenetic glycosides. However, Suzu (1959), Kobayashi (1959) and Nishijima *et al.* (1975) have reported that Ginkgo seeds do not contain cyanogenetic compounds.

4-O-METHYLPYRIDOXINE (MPN) AS THE CAUSATIVE AGENT OF GINKGO SEED FOOD POISONING

Guinea pigs are highly sensitive to Ginkgo seed extract, and display typical toxic symptoms when the extract is orally administered (Kobayashi, 1959). In the studies discussed below, guinea pigs were used to elucidate the primary toxin in Ginkgo seeds. When an aqueous extract of Ginkgo seeds was orally administered to guinea pigs, the symptoms of poisoning appeared within a few hours and included paralysis of legs, opisthotonus, and tonic and/or clonic convulsions.

An outline of the Ginkgo seed fractionation procedure is shown in Fig. 2. The resulting toxic fraction (butyl alcohol layer) was further separated into two frac-

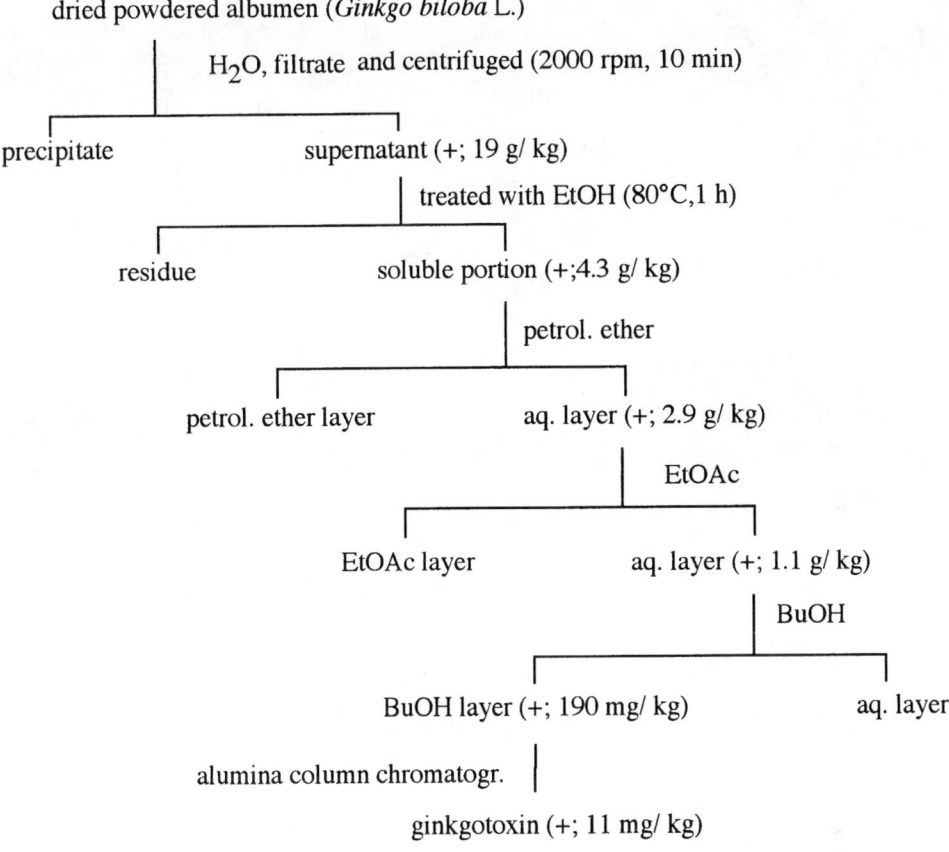

dried powdered albumen (*Ginkgo biloba* L.)

H_2O, filtrate and centrifuged (2000 rpm, 10 min)

precipitate   supernatant (+; 19 g/ kg)

treated with EtOH (80°C,1 h)

residue   soluble portion (+;4.3 g/ kg)

petrol. ether

petrol. ether layer   aq. layer (+; 2.9 g/ kg)

EtOAc

EtOAc layer   aq. layer (+; 1.1 g/ kg)

BuOH

BuOH layer (+; 190 mg/ kg)   aq. layer

alumina column chromatogr.

ginkgotoxin (+; 11 mg/ kg)

Figure 2   Isolation of ginkgotoxin from *Ginkgo biloba* L. seeds (Wada *et al.*, 1985)

tions using alumina column chromatography (eluent: CHCl$_3$-MeOH, 19:1). The first fraction induced the characteristic convulsions in the guinea pigs at an oral dose of 11 mg/kg (p.o.). The other fraction (149 mg/kg, p.o.) was not toxic.

These isolation procedures yielded approximately 0.01% "Ginkgotoxin" (tentative name) from the dry weight of Ginkgo seeds. The molecular formula of ginkgotoxin was determined to be $C_9H_{13}NO_3$ [m/z 183.089 (M$^+$); $C_9H_{13}NO_3$ requires 183.089] using high resolution mass spectroscopy. Ginkgotoxin was identified as 3-hydroxy-5-hydroxy-methyl-4-methoxymethyl-2-methylpyridine (4-O-methyl-pyridoxine, MPN) using ^{1}H and ^{13}C nuclear magnetic resonance (NMR) spectroscopy, and chemical reaction analysis (Wada *et al.*, 1985, and 1988). This was the first report of the isolation of MPN from natural sources (Fig. 3). The MPN content of raw Ginkgo seeds is approximately 0.01–0.02% by HPLC (Take *et al.*, 1986). Arentz *et al.* (1996) also recently reported that the amount of MPN in Ginkgo seeds was 0.01% by HPLC analyses. Further, the amount of MPN in Ginkgo seeds (0.014%) by the combined use of a microbiological determination and HPLC analyses (Tsuge

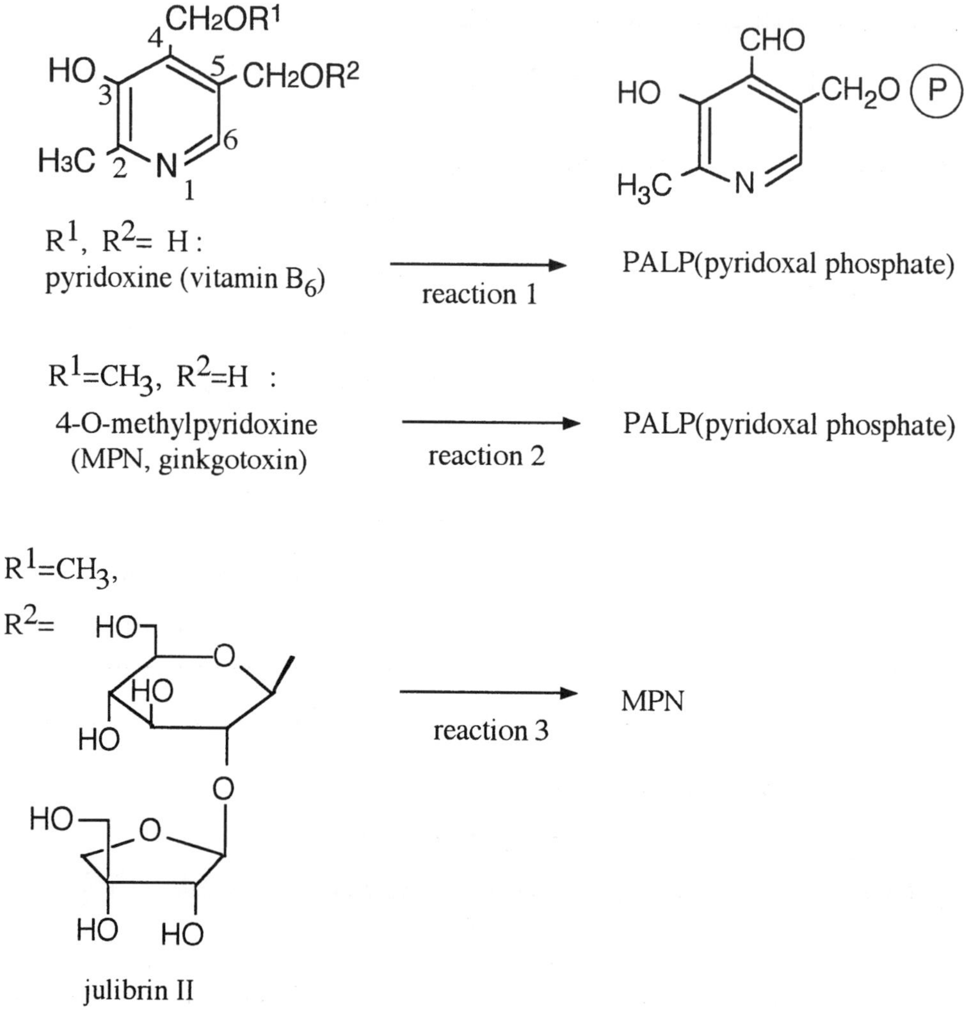

R^1, R^2= H:
pyridoxine (vitamin B$_6$)   ────── reaction 1 ──────→   PALP(pyridoxal phosphate)

R^1=CH$_3$, R^2=H :

4-O-methylpyridoxine   ────── reaction 2 ──────→   PALP(pyridoxal phosphate)
(MPN, ginkgotoxin)

R^1=CH$_3$,
R^2=   ────── reaction 3 ──────→   MPN

julibrin II

Figure 3 Structures of 4-O-methyl pyridoxine and pyridoxine derivatives. (from Wada *et al.*, 1988; Higuchi, Kinjo & Nohara, 1992) reaction 1; reaction proceeded under usual conditions *in vivo*, reaction 2; reaction proceeded hardly *in vivo*, reaction 3; reaction which has not been confirmed whether it proceeds or not *in vivo*.

et al., 1996) was similar to the estimation described above. Although MPN is often called "methoxypyridoxine" (or 4′-methoxypyridoxine), it may also reasonably be called 4-O-methylpyridoxine.

MPN glycoside (julibrine II) and related compounds have been isolated from the bark of *Albizzia julibrissin* Durazz (Leguminosae) (Higuchi *et al.*, 1992). Albizziae Cortex is used in China as a tonic, to ease the mind and calm the nerves. Julibrine II was found to exhibit arrhythmic-inducing action. Although julibrine II is an MPN derivative (Fig. 3), whether or not it possesses B$_6$ antivitamin activity remains unclear.

Mechanisms of MPN-induced Convulsions

Synthetic MPN is known to be a potent convulsive agent that possesses B_6 antivitamin activity in a variety of experimental animals such as mice, rats, cats, dogs, and monkeys (Ott, 1947; Kopeloff and Chusid, 1963; Julou et al., 1964; Mizuno et al., 1980). These MPN induced convulsions can be stopped or prevented using pyridoxine, i.e. vitamin B_6 (Gammon and Gummit, 1957). Pyridoxine also prevents the symptoms of food poisoning by Ginkgo seeds in experimental animals (Take et al., 1986), which suggests that pyridoxine will also prevent the symptoms of food poisoning by Ginkgo seeds in humans (See "Treatment of Food Poisoning by Ginkgo Seeds" in this chapter).

MPN is also known to inhibit the formation of 4-aminobutyric acid (GABA) from glutamate in the brain of animals. MPN may compete with vitamin B_6 (pyridoxal phosphate), which serves as a coenzyme of glutamate decarboxylase. GABA is regarded as an inhibitory chemical transmitter, and thus a deficiency of GABA in the brain may induce the convulsions (Ozawa and Okada, 1976; Nitsch and Okada, 1976). Therefore, MPN in the seeds of G. biloba may induce convulsions through the same mechanism. The lack of vitamin B_6 during times of food shortage may also be one of the causes of food poisoning by Ginkgo seeds.

Cerebral Blood Flow and Electroencephalograms (EEG) in MPN-Poisoning

In order to clarify the mechanisms of MPN poisoning, Minami et al. (1990) measured the cerebral blood flow in guinea pigs and obtained electroencephalograms (EEG) following administration of MPN. Cerebral blood flow (measured using the laser Doppler method) increased during MPN-induced convulsions in the anesthetized guinea pigs, thus confirming that MPN-induced convulsions are not ischemic seizures. The cause of death from MPN poisoning was from atrioventricular block or ventricular fibrillation. These reactions were prolonged from 40 to 120 minutes after intraperitoneal administration of MPN under anesthesia. The administration of MPN (30–50 mg/kg, i.p.) to conscious guinea pigs caused a generalized epileptic seizure within 1 hour, and subsequent EEG showed an epilepsy-like pattern. After experiencing tonic and clonic convulsions, the guinea pigs died of ventricular fibrillation. Figure 4 shows that the latency to clonic convulsion, generalized seizures and death reduced dose-dependently (20–50 mg/kg). In contrast, rats went into convulsions a long time after the administration of MPN (400–600 mg/kg, i.p.) (Fig. 4). Thus higher doses of MPN were needed to induce convulsions in conscious rats as compared with guinea pigs. This species difference in response to MPN has also been reported by Julou et al. (1964) and Kraft et al. (1961).

Convulsion-related Factors in Brain Tissue after Administration of MPN in Guinea Pigs

MPN may compete with pyridoxine, which serves as a coenzyme of glutamate decarboxylase. An attempt was made to clarify the mechanism of MPN action by determining cerebral tissue GABA and glutamate concentration using high

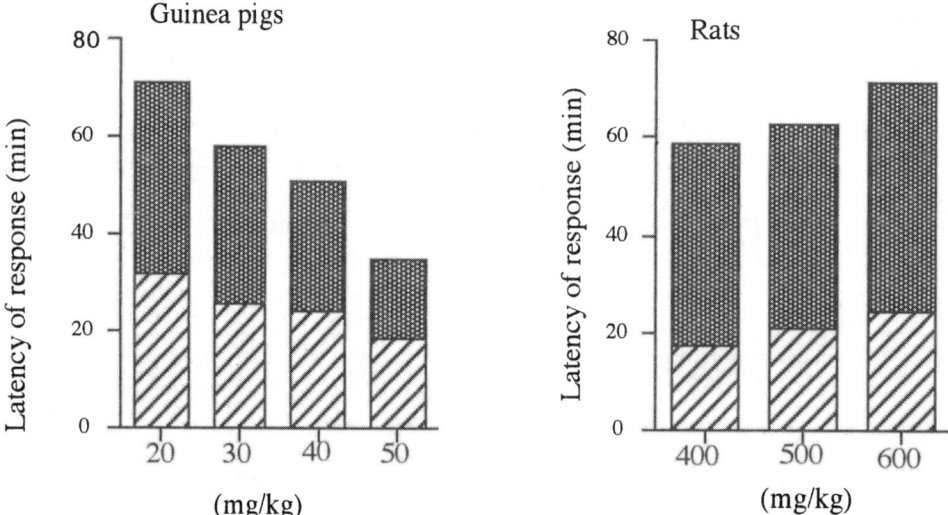

Figure 4   MPN-induced convulsion in guinea pigs and rats. (from Minami *et al.*, 1990)
▨ latency of clonic convulsion; ▨: latency of time of death. Time of death was counted as the period from administration of MPN until death.

performance liquid chromatography with oxidative electrochemical detection. The concentration of amino acids and related compounds in various brain regions after MPN administration in guinea pigs is shown in Fig. 5 A-D. As previously mentioned, a deficiency of GABA in the brain may induce convulsions. The practice of using the absolute level of GABA as an index of brain excitability must be questioned. Ozawa and Okada (1976) and Nitsch and Okada (1976) have speculated that changes in the concentration of GABA in specific areas of the brain may play a key role in epileptogenesis. However, the changes in GABA content after administration of MPN were observed to be uniform in all regions of the brain (Fig. 5A). At the same time, the cerebral glutamine concentration increased significantly (Fig. 5C), while the glutamate level did not change significantly (Fig. 5B). The amino acid taurine has been proposed as an inhibitory transmitter in the central nervous system (Oja and Contro, 1978). No significant difference in taurine content between the control and MPN-treated guinea pigs was found(Fig. 5D).

Next it was attempted to identify the stage in glutamic acid metabolism at which interruption results in a seizure. Fig. 6 shows the two types of glutamate metabolism. If the route from glutamate to GABA is blocked, the pathway from glutamate to glutamine may be utilized. This suggests that the significant increase in glutamine content after MPN administration is a decompensation to protect against the increase in MPN-induced glutamate content.

On the basis of above-mentioned results, MPN in Ginkgo seeds may induce convulsions through a deficiency of GABA and an increase in glutamine (Minami *et al.*, 1990). MPN is now being used as a means to investigate the GABA-nergic mechanism in experimental animals (Hirai and Okada, 1993).

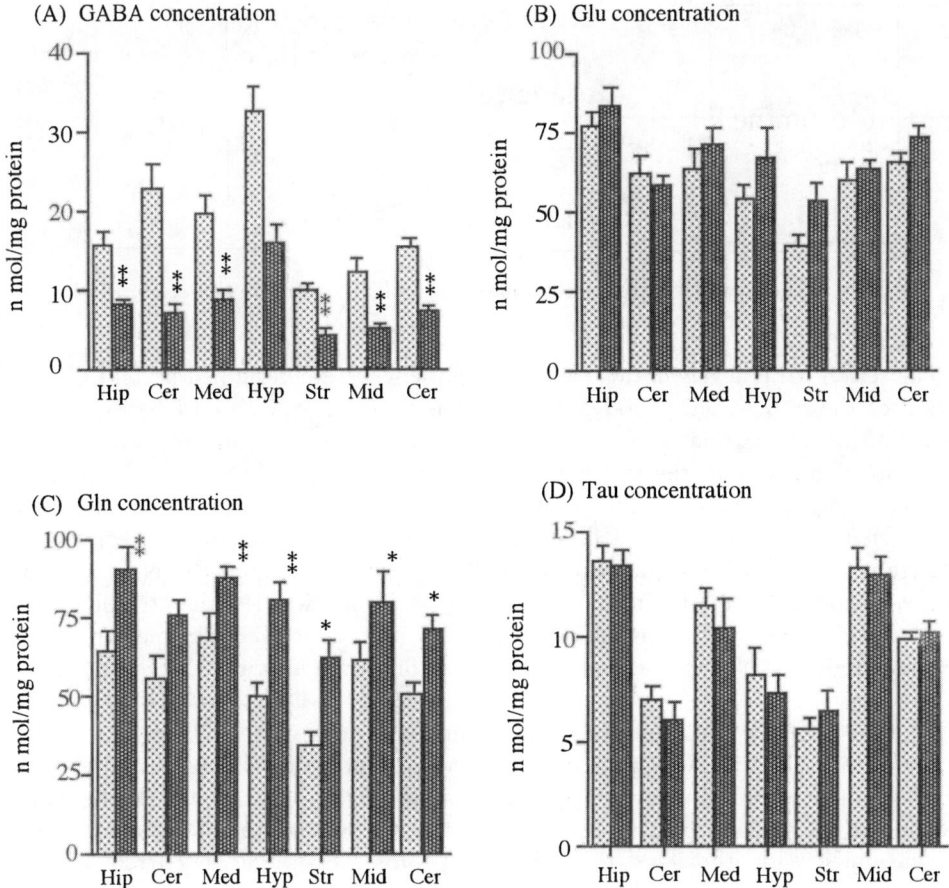

Figure 5   Amino acid concentrations in various brain regions in guinea pigs after MPN administration (from Minami *et al.*, 1990).
Each value represents the mean ± S.E. (standard error, n = 6). ▓: control, ▓ treated with MPN (30 mg/kg, i.p., amino acid concentration determined after 40 min). GABA: γ-aminobutyric acid, Glu: glutamate, Gln: glutamine, Tau: taurine, Hip: hippocampus, Cer: cerebellum, Med: medulla oblongata, Hyp: hypothalamus, Str: striatum, Mid: midbrain, Cer: cerebral cortex. *p <0.05, **p <0.01 vs control.

A Case of Food Poisoning by Ginkgo Seeds and Determination of MPN in Serum of the Patient

In spite of many case reports of food poisoning by Ginkgo seeds, there have been no reports that establish that these cases of food poisoning were caused by MPN. In recent studies, the serum and urine of a 21 month-old male child suffering from accidental food poisoning by Ginkgo seeds was analyzed (Kokubo *et al.*, 1993; Yagi *et al.*, 1993).

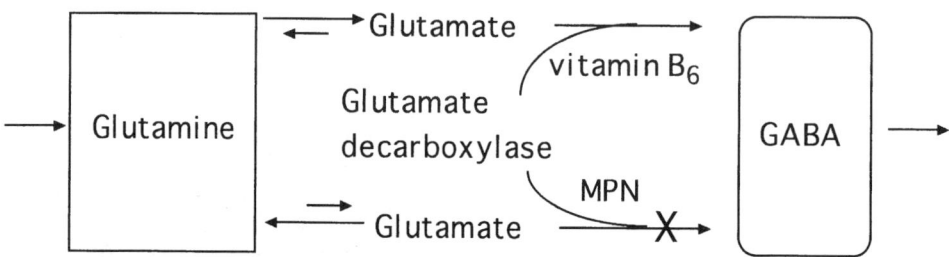

Figure 6   Chemical pathways of glutamate metabolism (from Minami *et al.*, 1990).

The serum MPN levels in this monitored patient were determined by HPLC. The blood of the patient with food poisoning was taken at 8.5 and 15.5 hours after ingestion of approximately 50 Ginkgo seeds. After deproteinization and centrifugation of the serum, the supernatant was analysed by HPLC. HPLC was performed using a reversed phase column equipped with a fluorescence detector (wavelength of excitation and fluorescence; 290 and 400 nm, respectively). The detection limit of MPN in the serum was 0.05 μg/ml. The serum MPN concentration was 0.09 μg/ml 8.5 h after taking Ginkgo seeds, and was less than the detection limit of MPN (0.05 μg/ml) at 15.5 h. This is the first report of the primary toxin of Ginkgo seeds (MPN) being found in the serum of a patient with Ginkgo seed food poisoning.

This method may be useful in determining food poisoning by Ginkgo seeds in humans. In rabbits, the lowest serum MPN concentration which induces symptoms of poisoning is approximately 1.0 μg/ml. Thus the MPN level may also be the index to the degree of food poisoning by Ginkgo seeds in humans. This speculation corresponds with the fact that the symptoms no longer occurred in the above-mentioned patient when the serum MPN concentration was 0.09 μg/ml. The rapid determination of serum MPN may predict the occurrence of symptoms.

The MPN was not detected in the urine of this patient under the same analytical conditions. Because 4-pyridoxic acid is excreted as the main metabolite of pyridoxine in humans (Huff and Perlzweig, 1944; Schwartz and Kjeldgaard, 1951), MPN may be metabolized to 5-pyridoxic acid or other conjugated forms in urine (Shane and Snell, 1975; Coburn and Mahuren, 1976).

Treatment of Food Poisoning by Ginkgo Seeds

Intravenous injection of GABA (10–100 mg) has prevented MPN-induced convulsions in mice (Gammon and Gummit, 1957). However, it has been shown that GABA administration (10–50 mg/kg, i.p.) failed to prevent convulsions in guinea pigs (Table 1). Because GABA is thought to be unable to pass through the blood brain barrier, the level of GABA may have to be enhanced endogenously (Takagi, 1981). In contrast, diazepam delayed MPN-induced convulsions in guinea pigs. The mean latency to convulsion and time of death was prolonged by diazepam. Diazepam is

Table 1   Effects of GABA and diazepam on MPN-induced convulsion in guinea pigs (from Minami *et al.*, 1990)

Treatment	(n)	Latency of response (min)	
		Clonic convulsion	Death time[*]
MPN (30 mg/kg, i.p.)	(6)	25.5 ± 2.4	57.8 ± 4.7
GABA (10 mg/kg, i.p.) + MPN (30 mg/kg, i.p.)	(5)	19.3 ± 3.2[†]	42.8 ± 3.1[†]
GABA (50 mg/kg, i.p.) + MPN (30 mg/kg, i.p.)	(5)	18.7 ± 2.1[†]	39.2 ± 4.9[†]
diazepam (4 mg/kg, i.p.) + MPN (30 mg/kg, i.p.)	(5)	55.2 ± 10.5[‡]	111.0 ± 12.7[‡]

[*]: Death time was counted as the period from treatment until death.
Each value indicates the mean ± S.E. [†] $p < 0.05$ and [‡] $p < 0.01$ vs MPN (30 mg/kg, i.p.).

one of the benzodiazepines, the actions of which are a result of potentiation of the neural inhibition that is mediated by GABA. Therefore, diazepam did not prevent MPN-induced reduction of the GABA concentration (Figure 7).

On the basis of the above mentioned results on food poisoning by Ginkgo seeds, the administration of a vitamin B_6-preparation has been applied. In particular, the administration of pyridoxal phosphate is an effective treatment for food poisoning by Ginkgo seeds (Yoshimura *et al.*, 1996).

DETOXIFICATION OF MPN USING BILOBALIDE FOUND IN GINKGO LEAVES

Ginkgo leaf extract (GLE) has many components (Huh *et al.*, 1992) and it has been prescribed for cases of senility-related chronic cerebral deterioration and related functional disturbances, and following cerebral vascular accidents and cranial trauma. Contrary to a preliminary HPLC analysis which showed that MPN had not been present in GLE (Wada *et al.*, 1993), Arenz *et al.* (1996) reported that MPN was detected in *Ginkgo biloba* leaves after careful studies. The leaves are a source of extracts employed in the preparation of Ginkgo medications, however, the amount of MPN is too low to exert a detrimental effect after administration of the medication (Arenz *et al.*, 1996).

Bilobalide is one of the characteristic sesquiterpenes of *Ginkgo biloba*, and induces the activity of 7-O-alkoxycoumarine dealkylases (Sasaki *et al.*, 1994). Oral administration of bilobalide (10–30 mg/kg/day, for 4 days) in mice minimized the MPN-induced symptoms (Sasaki *et al.*, 1995), because of the conversion of MPN to

GABA concentration

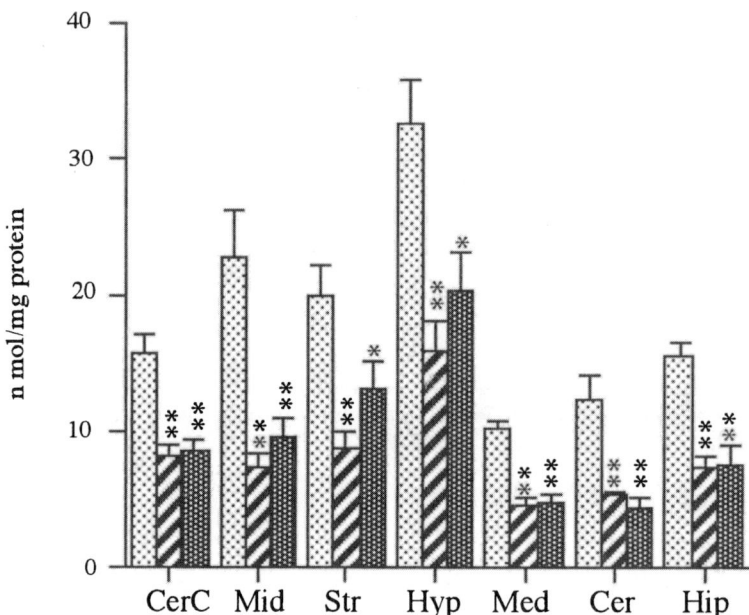

Figure 7 GABA concentration in various regions of the brain after diazepam and MPN administration in guinea pigs (from Minami *et al.*, 1990). Each value represents the mean ± S.E. (n = 6). ▦ control. ▨ treated with MPN (30 mg/kg, i.p., concentration determined after 40 min), ▓ treated with MPN (30 mg/kg, i.p.) and diazepam (40 mg/kg, i.p.) after 40 min. *p < 0.05, **p < 0.01 vs control. CerC: cerebral cortex, Mid: midbrain, Str: striatum, Hyp: hypothalamus, Med: medulla oblongata, Cer: cerebellum, Hip: hipocampus.

pyridoxine by demethylation of MPN (Wada, 1996). Interestingly, both the toxic factor (MPN) and the detoxification factor (bilobalide) exist in the same plant.

PREVENTION OF FOOD POISONING BY GINKGO SEEDS

Food poisoning by Ginkgo seeds is caused by MPN, which has B_6 antivitamin activity. As vitamin B_6 deficiency is induced by MPN, eating Ginkgo seeds should be avoided by people with a latent vitamin B_6 deficiency. Insufficiency of nutrition by food shortage, especially vitamins, might be incriminated in the large number of intoxications from 1955–1965 in Japan. There should be more intoxications before 1955 in Japan, because exact case reports could not be collected owing to the influence of World War II.

Because children are the most susceptible to food poisoning by Ginkgo seeds, the consumption of Ginkgo seeds should be limited to 5 per day for children under six years of age. It is better not to eat Ginkgo seeds every day, even if the consumption of Ginkgo seeds is less than 5 per day for children.

Abbreviations

EEG	electroencephalogram
GABA	γ-aminobutyric acid or 4-aminobutyric acid
GLE	Ginkgo leaf extract
HPLC	high pressure liquid chromatography
i.p.	Intraperitoneal administration
MeOH	methyl alcohol
MPN	4-O-methylpyridoxine
NMR	nuclear magnetic resonance

REFERENCES

Arenz, A., Klein, M., Fiehe, K., Groβ, J., Drewke, C., Hemscheidt, T., and Leistner, E. (1996) Occurrence of neurotoxic 4'-O-methylpyridoxine in *Ginkgo biloba* leaves, Ginkgo medications and Japanese Ginkgo food. *Planta Med.*, **62**, 548–551.

Coburn, S.P. and Mahuren, J.D. (1976) In vivo metabolism of 4'-deoxypyridoxine in rat and man. *J. Biol. Chem.*, **251**, 1646–1652.

Gammon, G.D. and Gummit, R. (1957) Observations on the mechanism of seizures induced by a pyridoxine antagonist, methoxypyridoxine. *Trans. Am. Neurol. Assoc.*, **82**, 57–59.

Higuchi, H., Kinjo, J., and Nohara, T. (1992) An arrhythmic-inducing glycoside from *Albizzia julibrissin* Durazz. *Chem. Pharm. Bull.*, **40**, 829–831.

Hirai, H. and Okada, Y. (1993) Ipsilateral corticotectal pathway inhibits the formation of long-term potentiation (LTP) in the rat suprior colliculus through GABAergic mechanism. *Brain Res.*, **629**, 23–30.

Huff, J.W. and Perlzweig, W.A. (1944) A product of oxidative metabolism of pyridoxine, 2-methyl-3-hydroxy-4-carboxy-5-hydroxy-methylpyridine (4-pyridoxic acid) I. Isolation from urine, structure, and synthesis. *J. Biol. Chem.*, **155**, 345–355.

Huh, H. and Staba. E.J. (1992) The botany and chemistry of *Ginkgo biloba* L. *J. Herbs Spices Med. Plants*, **1**, 91–124.

Julou, L., Pasquet, J., and Ducrot, R. (1964) The antidotic effect of pyridoxine on one of the intermediary products of its manufacture which has a strong toxicity. *Proc. European Society Study Drug Toxicity*, **4**, 179–189.

Klosterman, H.J., Lamoureux, G.L., and Parsons, J.L. (1967) Isolation, characterization, and synthesis of linatine. A vitamin B_6 antagonist from flaxseed (*Linum usitatissimum*). *Biochemistry*, **6**, 170–177.

Kobayashi, M. (1959) Study on the toxic substance in Ginkgo seeds. *Nagano-ken Eisei Kenkyusho Chosa Kenkyu Hokoku*, **16**, 1–12 (in Japanese).

Kokubo, M., Terada, A., Hayakawa, S., Maeda, K., Matsumoto, N., and Wada, K. (1993) A case of food poisoning by Ginkgo seeds. *Shonika-Shinryo*, **56**, 488–491 (in Japanese).

Kopeloff, L.M. and Chusid, J.G. (1963) Methoxypyridoxine convulsions in epileptic and non-epileptic mice. Protective action of pyridoxine. *Proc. Soc. Exp. Biol. Med.*, **114**, 496–500.

Kraft, H.G. von, Fiebig, L., and Hotovy, R. (1961) Zur Pharmakologie des Vitamin B_6 und seiner Derivate. *Arzneimittel-Forsch.*, **10**, 922–929.

Minami, M., Yanai, A., Endo, T., Hamaue, M., Hamaue, N., Monma, Y., Wada, K., Haga, M., Morii, K., Yoshioka, M., and Saito, H. (1990) Convulsion induced by 4-O-methyl-pyridoxine, from the seed of the *Ginkgo biloba* L., in guinea pigs and rats. *Life Sci. Adv.*, **9**, 107–115.

Mizuno, N., Kawakami, K., and Morita, E. (1980) Competitive inhibition between 4'-substituted pyridoxine analogues and pyridoxal. For pyridoxal kinase from mouse brain. *J. Nutr. Sci. Vitaminol.*, **26**, 535–543.

Nishijima, M., Kanmuri, M., Takahashi, S., Kamimura, H., Nakazato, M., Watari, Y., Kimura, Y. (1975) Surbey of cyanide in almond (in Japanese). *Tokyo-to Eisei Kenkyusho Nempo*, **26**, 183–186.

Nitsch, C. and Okada, Y. (1976) Decrease of GABA levels in different parts of the rabbit brain after treatment with methoxypyridoxine. In E. Roberts, T.N. Chase, and D.B. Tower (eds.), *GABA in nervous system function*, Raven Press, New York, pp. 455–460.

Oja, S.S. and Kontro. P. (1978) In A. Barbeau, and R.J. Huxtable, (eds.), *Taurine and neurological disorders*, Raven Press, New York, pp. 181–200.

Ott, W.H. (1947) Antipyridoxine activity of methoxypyridoxine in the chick. *Proc. Soc. Exp. Biol. Med.*, **66**, 215–216.

Ozawa, S. and Okada, Y. (1976) Decrease of GABA levels and the appearance of a depolarization shift in thin hippocampal slice *in vitro*. In E. Roberts, T.N. Chase, and D.B. Tower, (eds) *GABA in nervous system function*. Raven Press, New York, pp. 449–454.

Sasaki, K., Wada, K., Haga, M., Hatta, S., and Ohshika, H. (1994) Studies on the constituents of edible and medicinal plants. V. Biological active components in Ginkgo leaves (3). Anticonvulsant effects of bilobalide and the effects of bilobalide on metabolizing enzymes. The 114th Annual Meeting of Pharmaceutical Society of Japan, Tokyo, Mar. 1994.

Sasaki, K., Hatta, S., Wada, K., Ohshika, H., and Haga, M. (1995) Anticonvulsant activity of bilobalide, a sesquiterpene in *Ginkgo biloba* L. leaves, against chemical-induced and electroshock-induced convulsions in mice. *Res. Com. Biol. Psychol. Psychiatry*, **20**, 145–156.

Schwartz, R. and Kjeldgaard, N.O. (1951) The enzymic oxidation of pyridoxal by liver aldehyde oxidase. *Biochem. J.*, **48**, 333–337.

Shane, B. and Snell, E.E. (1975) Metabolism of 5'-deoxypyridoxine in rats: 5'-deoxypyridoxine 4'-sulphate as a major urinary metabolite. *Biochem. Biophys. Res. Comm.*, **66**, 1294–1300.

Somogyi, J.C. (1973) Antivitamins. In Committee on Food Protection, Food and Nutrition Board, National Research Council, (ed.), *Toxicants Occurring Naturally in Foods*, National Academy of Sciences, Washington, D.C., pp. 254–275.

Suzu, M. (1959) An examination of plums and ginkgo-nuts for cyanophoric glycoside. *Fukuoka Igaku Zasshi*, **50**, 5394–5398 (in Japanese).

Takagaki, G. (1981) *Neurobiochemistry*. Kyoritsu Shuppan, Tokyo, pp. 72–75 (Kyoritsu Zensho, vol. 238) (in Japanese).

Takano, T., Kobayashi, M., Wada, I. (1953) Study on Gin-nan food poisoning in experimental animals (in Japanese). *Jui Chikusan Shinpo*, **109**, 353–357.

Take, Y., Wada, K., Sasaki, K., Sakata, M., and Haga, M. (1986) Studies on Plant Food Poisoning. (2) Food Poisoning by Ginkgo Seeds and Antivitamin B_6. The 106th Annual Meeting of Pharmaceutical Society of Japan, Chiba, April 1986.

Tsuge, H., Suganuma, K., and Sasaki, S. (1996) Vitamin B-6 derivatives found in Ginkgo nuts (Ginkgo biloba L. Seeds). *Vitamins*, **70**, 561–562 (in Japanese).

Wada, K., Ishigaki, S., Ueda, K., Sakata, M., and Haga, M. (1985) An antivitamin B_6, 4'-methoxypyridoxine from the seed of *Ginkgo biloba* L. *Chem. Pharm. Bull.*, **33**, 3555–3557.

Wada, K. (1986) *Ginkgo biloba* and Gin-nan food poisoning. *Kagaku To Yakugaku No Kyoshitsu*, **95**, 79–82 (in Japanese).

Wada, K., Ishigaki, S., Ueda, K., Take, Y., Sasaki, K., Sakata, M., and Haga, M. (1988) Studies on the constitution of edible and medicinal plants.I. Isolation and Identification of 4-O-methylpyridoxine, toxic principle from the seed of *Ginkgo biloba* L. *Chem Pharm Bull.*, **36**, 1779–1782.

Wada, K., Sasaki, K., Miura, K., Yagi, M., Kubota, Y., Matsumoto, T., and Haga, M. (1993) Isolation of bilobalide and ginkgolide A from *Ginkgo biloba* L. shorten the sleeping time induced in mice by anesthetics. *Biol. Pharm. Bull.*, **16**, 210–212.

Wada, K. (1996) Studies on the constituents of edible and medicinal plants to affect the metabolizing system in mammals. *Natural Medicines*, 50, 195–203 (in Japanese with English abstract).

Wada, K. and Haga, M. (1997) Food Poisoning by *Ginkgo biloba* Seeds. In T. Hori, R. W. Ridge, W. Tulecke, P. Del Tredici, J. Trémouillaux-Guiller, H. Tobe (eds.), *Ginkgo biloba-A Global Treasure*, Springer-Verlag Tokyo, Japan, pp. 309–321.

Yagi, M., Wada, K., Sakata, M., Kokubo, M., and Haga, M. (1993) Studies on the constituents of edible and medicinal plants. IV. Determination of 4-O-methylpyridoxine in serum of the patient with gin-nan food poisoning. *Yakugaku Zasshi*, **113**, 596–599. (In Japanese with English abstract).

Yoshimura, I., Hayata, M., Yamasaki, F., and Mori, H. (1996) *Acute toxicity information file.* Hirokawa Publishing Co., Tokyo, p. 699 (in Japanese).

24.   HOMEOPATHIC USES OF *GINKGO BILOBA*

FRANS M. van den DUNGEN

*Biohorma b.v., Medical Research, Industriestraat 5, PO Box 33,
8081 HH Elburg, The Netherlands*

INTRODUCTION

The use of *Ginkgo biloba* in allopathic and phytotherapeutic medicine has been described elsewhere in this book. This chapter will focus on the homeopathic use of G. *biloba*. However first some characteristics of homeopathy will be described.

Homeopathy is foremost a therapeutic method which clinically applies the law of similars. Within these boundaries it makes use of medicinal substances in weak or infinitesimal dilutions.

HISTORY

The basic idea of homeopathy was formulated by Hippocrates (460–377 BC). He postulated that complaints of a disease can be treated with a substance, that has an opposite activity in comparison to the disease. This statement known as contra contrariis is applied for treatment of acute complaints. The disease itself can be cured by substances that invoke the same symptoms as the complaints caused by the disease. This statement, known as similia similibis, is used for curing illnesses. The first statement was easily understood and became popular among physicians. Contrary, the second one lost its importance due to a lack of practical implication.

A German physician Hahnemann (1755–1843 AD) has picked up the second statement and worked it out. He observed that Cinchona, a remedy which at that time was used to treat certain types of malarial fever, toxicologically caused attacks of fever similar to those against which it was therapeutically used. He postulated a hypothesis that it would appear that remedies are capable of curing symptoms analogous to those which they themselves produce. The results of his experiments upon himself and those around him with every medical substance known at that time convinced him that the deeper basis of illness lies in a kind of disorder inside the body. In order to keep the body in balance several dilutions of a substance which caused the same symptoms when administered to healthy people, were given. He observed that his hypothesis was correct, but only when very weak or even infinitesimal dilutions were used. Thus by a great deal of experimentation Hahnemann was able to observe that the hypothesis he had initially formulated was consistently true. The fundamental principal of the hypothesis, the law of similars, states that there exists a parallel action between the toxicological effects of a substance and its therapeutic action.

HOMEOPATHIC MATERIA MEDICA

Homeopathy has three basic elements: the total of symptoms which are characteristic of a disease, the law of similars, and provings with healthy volunteers.

Application of the law of similars in homeopathic therapeutics necessitates a thorough knowledge of the action of a pharmacologically active substance on healthy individuals. Administration of this substance causes a set of symptoms and is known as the pathogenesis or proving of a substance. Part of the information can be obtained from toxicological data which can be acute or chronic, accidental or professional exposure. Beside such toxicological data there are actual pathogenic experiments which are carried out with various non-toxic doses upon healthy volunteers of different ages and sex. Symptoms noticed in these experiments concern mainly functional or general symptoms and possible changes in the behaviour. Symptoms induced by the substance are classified according to the system affected: digestive, circulatory, respiratory etc. The collection of provings from different sources are integrated in the homeopathic materica medica.

Modalities will complete the description of symptoms. Modality is a qualification of a symptom in the sense of aggravation or improvement under the influence of external circumstances and surroundings. These include for instance heat or cold, rest or motion, time of the day, atmospheric circumstances and physiological circumstances.

HOMEOPATHIC PREPARATIONS

Official documentation on preparations are found in the "Homöopathisches Arzneibuch" (HAB, 1978) or national pharmacopoeias like the French pharmacopoeia. There are differences in the preparation between the countries and the various pharmacopoeias such as the use of fresh and dried material.

Approximately 80% of the homeopathic remedies are prepared from plant material. The plants are harvested in their natural state according to strict rules and used after a thorough botanical inspection. Other sources for homeopathic remedies are minerals and animal material. Minerals such as natural salts, chemical products and metals are all used in their purest state. Examples of animal material are venoms, insects (such as bees, ants, spiders, etc.), hormones or physiological secretions (for example musk and squid inks) and any substance which may have a toxicological or pharmacological action, such as psorinum (sero-purulent matter of a scabies vesicle) and tuberculinum.

Plant substances and some substances of animal origin are used to prepare mother tinctures (indicated with M.T. of with the symbol ø) by maceration in high concentrations of alcohol for at least ten days. This maceration is carried out in special glass or stainless steel recipients. Mother tinctures are generally prepared by adding alcohol 10 times the weight of the dehydrated plant and 20 times alcohol in case of an animal product. These tinctures are the base extractions for further dilutions. Dilutions to 1/10 (decimal, D or DH) or 1/100 (centimal, C or CH) of their original concentration are prepared according to different methods. Each dilution is then shaken vigorously by hand or machine. It is subsequently labelled with the Latin name of the base product followed by a number which represents the number of dilutions and potentizations the remedy has undergone, and by a symbol indicating the dilution. Thus 3D, D3 or 3DH indicate that the base substance has been diluted 1000 times in three steps of 10. A 5C, C5 or 5CH indicate that the base substance has been diluted ten billion times in five steps of 100. Dilutions prepared in this way are called Hahnemannian potencies. Later another method for making dilutions was introduced by Korsacoff. In contrast to Hahnemannian's method a potency was made with one flask. After vigorously shaking the flask was emptied, filled up and shaken again. Potencies prepared according to Korsacoff are indicated with a K.

Substances which are not soluble in water, ethanol or glycerine are extensively mixed with lactose in a mortar for one hour. After the third trituration the insoluble substance may be diluted in the same way as soluble products.

Homeopathic remedies are available in different pharmaceutical formulations. Low potencies are presented in drops.

Granules and smaller globules are made of sucrose and lactose. They are impregnated with a certain number of drops of the chosen Hahnemannian potency in glass jars. To ensure proper distribution of the medicine, the jars are mechanically rotated during pouring. The alcohol content of the dilutions must be more than 70%, otherwise the granules or globules will dissolve. The potency given to the impregnated granules and globules is equal to the potency of the solution from which they were originally prepared. Tablets are used as a way of administration for triturated powder. Other pharmaceutical forms such as injectable ampoules, ointments and creams are also available.

DEVELOPMENT OF *GINKGO BILOBA* AS A HOMEOPATHIC MEDICINE

An example of a popular plant used in phytotherapy as well as homeopathy is *Ginkgo biloba*. When this plant was used for the first time in homeopathy is not exactly known. Leeser (1973) gives a brief summary of the first "Hahnemannian provings". That summary includes the experiments of Kaempfer in 1712, Linné in 1777 and Smith in 1779. Since these studies were carried out before Hahnemann had described his first experiments, it is reasonable to suppose that these experiments are not real "Hahnemannian provings".

One of the earliest Hahnemannian provings with Ginkgo has been written by the Frenchman Maury in the first half of this century. In 1933 he carried out provings with a few volunteers. In the first experiments he used a mother tincture on seven provers (five men and two women). Later investigations were performed with the 6th Korsakoff potency on two male provers (Maury, 1933). These studies indicated that important target areas of G. *biloba* were skin, urinary tract, pharynx, stomach and head. Several symptoms were reported. The mental symptoms included sensation of mental freedom, indifference, rapid speech, restlessness, irrational fears, exhaustion, mental excitement, aggressivity, suppressed anger, the need to criticise others and oneself, irrational impressions, confusion and sensation of fatigue. Heaviness in the frontal region and vertigo are symptoms associated with the head. Symptoms involving the skin included transient varioloid, vesicular itching and (burning) eruptions. Sensation of cold, shivering, sleepiness, beginning of inflammation and flu were regarded as generalities. According to these experiments G. *biloba* is mainly active on the left side of the body. Cold air, exercise, walking, looking up and to the left aggravate the symptoms. Amelioration of the symptoms was noticed with heat, laying in bed and rest.

A simple proving with Ginkgo D6 on himself was carried out by McIvor (1975). As the symptoms developed he reduced the doses from twice daily to once a day and finally stopped when the symptoms became worse. The symptoms developed slowly, except for the head symptoms which arose rapidly. Symptoms which were noted ranged from severe muscular weakness of the neck muscles, particular on the left side, and the muscles of the upper limbs to a general physical weakness. Also weakness of intellect, weak memory, difficult thinking, reduced hearing and dry skin and nose were noticed. All these symptoms were reduced with warm and humid weather. It is of interest to mention that Ginkgo could prove a good remedy for nasal colds with a watery running nose and sneezing and dryness of the throat.

Huijsen (1996) gives a brief summary of several provings. Two of them were performed by Swoboda and König. Unlike previous provings the studies with G. *biloba* D30 were placebo controlled. Although these provings were carried out with only a limited number of healthy volunteers the results can be used make a homeopathic profile of Ginkgo:

Symptoms:

Mind
- sadness, depression, evening, causeless, company solitude, before menses, taciturn.
- agitation.
- irritability, morning, altering with sadness, before menses, during menses.
- inability in finding words.
- weakness of memory
- amelioration by fresh air.

Head
- stupefying headache
- amelioration by fresh air.

Mouth
- aphtha at tongue.

Throat
- sore throat (left and/or right side).

Stomach
- nausea.

Abdomen
- cramps in stomach, as if stomach were filled with air.

Female sexual organs
- menses too early.

Heart
- anxious palpitation.

Upper Limbs
- writer's cramp.

Generalities
- general weakness, feeling tired, desire for coffee.
- aggravation by warm humid weather.

Skin
- dry rash without itching, exanthema causing itching.

Sleep
- dreams about water.

Relations
- Antidoted by: Lachesis, Calcium carbonicum, Gelsemium, Sulphur, Phosphorus, Bryonia, Conium, Acidum phosphoricum, Acidum picrinicum, Acidum nitricum.

PREPARATION, FORMULATION AND QUALITY CONTROL

A monograph of *G. biloba* can be found in the HAB. Herein it is described that the fresh leaves are prepared according to paragraph 3. This means that the plant material is minced and mixed with a certain amount of 60% ethanol and left at a maximum temperature of 20° C during 10 days. Subsequently the mixture is expressed and filtered and ready for use or further potentization. In contrast to this monograph some commercial preparations are prepared with dried material. According to paragraph 4 of the HAB one part of dried material is mixed with 10 parts ethanol for maceration or percolation. After calculation of the amount of ethanol required, the mixture is left at a maximum of 20° C. After five days the tincture is filtered and ready for use or subsequent potentization.

Ginkgo extract containing products are available in different pharmaceutical forms. In addition to drops and tablets Ginkgo is available as ointment for topical application. The following potencies are regularly used: D1 till D6 (Leeser, 1973; Wiesenauer, 1993). In French literature higher potencies (to D30) are mentioned (Julian, 1990; Voisin, 1984).

During processing of *G. biloba* containing products several quality control steps are carried out as described in the HAB. The monograph gives a description of the plant material. The purity of the ethanol is determined according to the European

Pharmacopoeia. Mother tincture is examined by means of a TLC fingerprint analysis. Hyperosid and rutin are used as reference compounds as described in the HAB. Finally the relative density is determined according to the European Pharmacopoeia and the percentage dry weight according to the Deutsches Arzneibuch (DAB).

INDICATION AND MODALITY

Preparations of *Ginkgo biloba* are indicated as an analeptic for the veins and circulatory system (Gawlik, 1992; Julian, 1990; Leeser, 1973; Wiesenauer, 1993). Preparations are mentioned for the treatment of migraine, temporal or sub-orbital headache on the left side, and writer's cramp (Guermonprez *et al.*, 1989; Keller *et al.*, 1994; Leeser, 1973; Voisin, 1984). Cold or walking, drinking, looking up and to the right, use of alcoholic drinks and stress between 2 a.m. and 3 a.m. aggravated the symptoms (Julian, 1990; Voisin, 1984), while rest and heat diminished the symptoms (Julian, 1990). All these factors which influenced the symptoms are regarded as the modalities of Gingko.

CONCLUDING REMARKS

Ginkgo biloba is used in homeopathic medicine. Up to now a only few provings were carried out. These studies indicate that important target areas of *G. biloba* are skin, stomach and head. Preparations with Ginkgo are meant mainly for the treatment of migraine or sub-orbital headache on the left side and writer's cramp.

REFERENCES

Homöopathisches Arzneibuch, 1 Ausgabe 1978 und Nachträge, Amtliche Ausgabe, Deutscher Apother Verlag, Stuttgart. Govi-Verlag GmbH, Frankfurt, pp. 24–7, 487–488.

Gawlik, W. (1992) *Homeopathie en reguliere therapie*. Toepassingen in de beroepspraktijk, Homeovisie, Alkmaar, pp. 87–89, 306–307.

Guermonprez, M., Pinkas, M. and Torck, M. (1989) *Matière médicale homeopathique*, Editions Boiron, Saint-Foy-Lès-Lyon, pp. 328–330.

Huijsen L.P. (1996) *Ginkgo biloba*: zowel homeopathisch als fytotherapeutisch te gebruiken. *Similia Similibus Cerentur*, 26/4, 29–32.

Julian, O.A. (1990) *Materia medica of new homoeopathic remedies*, Translated from Matière médicale d'Homeothérapie by Virginia Munday, Beaconsfield Publishers, Beaconsfield, pp. 237–241.

Keller, K. Greiner, S. and Stockebrand, P. (1995) Homöopathische Arzneimittel. Band II. Materialien zur Bewertung, Govi-Verlag, Frankfurt am Main.

Leeser, O. (1973) *Lehrbuch der Homöopathie*. B/I: Pflanzliche Arzneistoffe, K.F. Haug Verlag, Heidelberg, pp. 318–327.

Maury, E.A. (1933) *Homoéophatie expérimentale de Ginkgo-biloba*, S.A. Maloine, (ed.), Paris.

McIvor, E.G. (1975) *Ginkgo biloba* – a proving. *The Britsh Homoeopathic Journal*, April, 105–106.

Voisin, H. (1984) *Matière médicale du praticien homéopathe*, S.A. Maloine, (ed.), Les laboratoires homoeopathiques de France, Asnières, pp. 543–544.

Wiesenauer, M. (1993) *Homöopathie*. (für Apotheker und Ärzte), Deutscher Apotheker Verlag, Stuttgart.

25. COSMETICAL USES OF GINKGO EXTRACTS AND CONSTITUENTS

EZIO BOMBARDELLI, ALDO CRISTONI and PAOLO MORAZZONI

Indena SpA, Scientific Department, Viale Ortles 12, 20139 Milano, Italy

INTRODUCTION

Ginkgo biloba L. is considered a living fossil due to its survival over millions of years, in which time countless other plant species have died out (Braquet, 1988). The G. *biloba* tree is very resistant against pests and diseases, and succeeds well under harsh urban conditions. For these reasons it has been utilized successfully for ornamental purposes throughout the world (Fig. 1). This unique example in the plant Kingdom gives rise to many questions concerning the defense mechanisms of this plant. It is clear that a plant surviving without structural modifications for over 200 million years must possess some distinct characteristics which eliminate the damage caused by the adverse environment. This remarkable survival and resistance is perhaps linked to the most interesting aspect of the G. *biloba* tree, its phytochemical constituents.

Although the mechanisms of G. *biloba*'s resistance to insects, bacteria, viruses and fungi are not completely elucidated, there is some evidence that the phytochemicals present in various tissues of the tree represent a means of defense. For instance bilobalide, one of the terpenes present in various parts of the plant, showed a strong antiparasitic activity against *Pneumocystis carinii* and other micro-organisms (Atzori *et al.*, 1993; Bombardelli and Ghione, 1996). Other compounds also showed, to a lesser extent, antibacterial and virucidal activities (Hayashi *et al.*, 1992; Brantner and Grein, 1994). These constituents have made G. *biloba* one of the oldest phytotherapeutic agents known to man. Today standardized extracts of G. *biloba* leaves, active against several physiological disorders owing to vascular, hemorheological, metabolic, and immunological mechanisms, are produced and used in most industrialized countries (Fig. 2). Among the pharmacological and clinical effects the most relevant is the effect on the vascular system, particularly with regard to the microcirculation (Kleijnen and Knipschild, 1992). This latter activity can also be applied for a non-medicinal use.

In fact G. *biloba* is widely used today in non-pharmaceutical fields like cosmetics and health food products. In these fields the applications of Ginkgo preparations are based on the antioxidant properties and the action on the peripheral blood circulation, with the aim of preventing chronic degenerative diseases typical of ageing. This chapter will focus on the cosmetical use, where G. *biloba* preparations are appreci-

Figure 1   *Ginkgo biloba* tree in winter.

Figure 2   *Ginkgo biloba* leaves.

ated due to their multifarious activities on ageing skin. The skin ageing process is in fact the consequence of continuous changes caused by free radicals, altered metabolic function, UV exposure and environmental pollution. *G. biloba* is the living example of the ability to counteract the negative influence of the above mentioned degradation factors.

CHEMISTRY

The targets of modern skin care science, are the protection against ravages and the prevention of the signs of ageing. Among the several causes which contribute to the alteration of the epidermis and derma structures, a deficiency in the peripheral microcirculation is certainly one of the most important. For instance, alopecia is strongly influenced by a reduced blood flow in the scalp. Furthermore, cellulite (or edematous-fibrosclerotic panniculopathy) and edema are consequences of chronic venous stasis. For these reasons in modern cosmetology the microcirculatory activation is indicated for the management of the ageing skin and for treatment of hair loss.

Several *G. biloba* extracts, their fractions or derivatives are used in cosmetics, and quite a bit of research has been carried out in this field. In the literature some publications and a series of patent applications concerning several uses of Ginkgo preparations, often in relation with the main pharmaceutical use of the plant exist

(Bombardelli 1991; Watanabe and Takahashi, 1991; Koide and Sasaki, 1992; Soudant and Nadaud, 1992; Koide and Kuribayashi, 1993; Kubo *et al.*, 1993; N'Guyen, 1994; Duranton and De Lacharriere, 1995; Fischer *et al.*, 1995; O'Reilly, 1995; Weidner 1995; Bombardelli *et al.*, 1996a; Nishizawa *et al.*, 1996). Generally speaking, there are three main cosmetical applications for Ginkgo:

a)   as hair tonic or hair growth stimulant,
b)   in compositions for treatment of fat deposits,
c)   in compositions aiming at the treatment of ageing skin.

In all three applications, the vascular effects of various Ginkgo components are useful. Moreover, the antioxidant and free radical scavenger properties are involved in the management of ageing or ravaged skin. Before examining the biological effects of Ginkgo products, it is convenient to briefly look at the chemistry of the various constituents in relation to their activity and interaction with the skin. Table 1 reports the main constituents present in *G. biloba* leaves, which is the part most often used. For the structures, content and analysis of the above compounds the reader can refer to the information in other chapters of this book.

As far as the cosmetic application of *G. biloba* products is concerned, the most relevant data have been obtained using the standardized pharmaceutical extract (24% Ginkgo flavonol glycosides, 3% ginkgolides, 3% bilobalide, less than 5 ppm of ginkgols and ginkgolic acids), a *G. biloba* leaf fraction containing the dimeric flavonoids (apigenin dimeric flavonoids), or both the extract and fraction in complexed form with phospholipids. The pharmaceutical extract is practically devoid of dimeric flavonoids, which are present only in trace amounts. On the other hand, it is possible to obtain them, in a purified extract form, as a by-product in the pharmaceutical extract production.

The complexes with phospholipids arising from Indena research have been called PHYTOSOME® (Bombardelli and Patri, 1988). On the basis of their physical-chemical and spectroscopic characteristics, these complexes can be considered novel entities. They are lipophilic substances with characteristics very different from the individual components: they are freely soluble in aprotic solvents (such as C_6H_6 and $CHCl_3$) while the hydrophilic moiety was not, they are moderately soluble in fats

Table 1   *Ginkgo biloba* constituents and their pharmacological effects.

Substances	Activities
Ginkgo flavonol glycosides	Antioxidant, vasokinetic, antiinflammatory
Dimeric flavonoids	Vasokinetic, antiinflammatory, antiphosphodiesterasic
Ginkgolides	Antiinflammatory, antiallergic, anti-PAF
Bilobalide	Antiinflammatory, antimicrobial
Proanthocyanidins	Antioxidant, venotonic, vasoprotectant
Ginkgols and ginkgolic acids	Allergenic properties (should be absent from extracts)
Other aromatic acids	Antioxidant

and insoluble in water. When treated with water, they assume a micellar shape, forming structures which might resemble liposomes, but which, however, exhibit fundamental differences. In liposomes the active principle is dissolved in a medium which is contained in the cavity, or between the layers of the membrane, whereas in the PHYTOSOME® the active principle is an integral part of the membrane, because the molecules are anchored through chemical bonds to the polar head of the phospholipid, as can be demonstrated by specific spectroscopic techniques. The characterization of the complex has been achieved mainly by an NMR comparison of the ^{1}H-NMR, ^{13}C-NMR, and ^{31}P-NMR spectra of the individual component with that of the complex and sometimes with the mechanical mixture (Bombardelli et al., 1994).

BIOLOGICAL EFFECTS

Ginkgo biloba

In the eighties Bombardelli et al. (1989a) investigated the possibility of applying G. biloba extracts in cosmetics, exploiting their antioxidant, vasoprotective and vasokinetic activities. Ginkgo was considered due the fact that the target of modern cosmetics is the prevention of the signs of ageing and the protection of the skin against a variety of ravages (i.e., free radicals, irritants, etc.), not only on the surface but also in the derma. In preliminary experiments in mice the anti-edema activity, one of the properties most commonly attributed to the flavonoids (Gabor, 1979; Middleton Jr., 1984), was tested after topical application with the Croton oil induced irritation model, according to the method of Tubaro et al. (1985). The activity of the extracts or their fractions showed a limited efficacy because the skin provides an effective natural barrier against the majority of active compounds despite the bioavailability via other routes of administration.

This situation led to the development of specific permeability-enhancing agents, mostly based on vegetal phospholipids. Thus the chemical interaction of flavonoids and other Ginkgo constituents with phospholipids (which are, at the same time, common constituents of the skin and powerful surfactants, able to modify the cutaneous absorption) was investigated. A new series of compounds was developed by reacting stoichiometric amounts of a phospholipid with the selected flavonoid or terpene trilactone from Ginkgo (Bombardelli et al., 1994). Using these complexes between phospholipids and pure components of Ginkgo, or Ginkgo purified extract, the anti-inflammatory activity in the ear edema induced by Croton oil was again tested. G. biloba extract and some pure compounds complexed with two parts of soy-phospholipids were more effective in inhibiting the inflammatory reaction than up to three times higher doses of the products in free form (Della Loggia et al., 1996). Tables 2 and 3 report the most relevant data of these experiments.

The data suggest that these complexes can penetrate across the epidermic barrier and exert their activity on tissue structures altered by Croton oil. The test used in the above experiments measures mechanisms of anti-inflammatory action such as vasoprotection and reduced capillary permeability, which are activities closely

Table 2 Anti-edema effect of *Ginkgo biloba* extract, bilobalide, and their complexes with distearoyl-phosphatidylcholine (DPC).

Substances	Dose µg/ear	Edema[a] mg	% reduction	p <[b]
Controls Croton oil 75 µg/ear	–	8.1 ± 0.3	–	–
DPC	75	7.4 ± 0.4	8.6	NS
	150	6.3 ± 0.2	22.2	0.01
Ginkgo biloba extract[c]	37.5	7.5 ± 0.3	7.4	NS
	75	6.7 ± 0.2	17.3	0.01
Ginkgo biloba extract[c] complex 1:2 with DPC	112.5	5.1 ± 0.3	37.0	0.01
	225	1.1 ± 0.1	86.4	0.01
Bilobalide	25	7.3 ± 0.3	9.9	NS
	50	5.5 ± 0.2	32.1	0.01
Bilobalide complex 1:2 with DPC	75	5.6 ± 0.2	30.1	0.01
	150	1.0 ± 0.1	87.7	0.01

[a]Values are means ± standard error of the mean, SEM (6 h after Croton oil); n = 20.
[b]NS = not significantly different from controls by Dunnett's t-test.
[c]*Ginkgo biloba* standardized extract containing 24% Ginkgo flavonol glycosides + 6% terpene trilactones.

related to the main therapeutic indications of *G. biloba* preparations. Several published reports in fact indicate that the active principles of Ginkgo when administered by oral route exert on capillaries, arteries and veins of medium and large size vasoprotective and vasoactive effects (DeFeudis, 1991; 1994), possibly mediated by release of prostacyclin or endothelial muscle contracting and relaxing factors (EDCF, EDRF). On the other hand, *in vitro* studies have demonstrated that *G. biloba* standardized extract has free-radical scavenging activity (Maitra *et al.*, 1995). It is well known that UV radiations provoke oxygen activation with consequent lipoperoxidation initiation or breakdown of the connective structures, thereby creating skin irritation or burn (Halliwell and Gutteridge, 1986). A potent oxygen free radical scavenger can reduce the cellular and tissue damage, with a preventive activity. No objectively documented data, however, have been available on the potential effects of *G. biloba* standardized extract after topical application.

G. *biloba* extract in its complexed form (*G. biloba* PHYTOSOME®) was tested on UV-induced erythema in a group of 18 volunteers. The product, incorporated in an aqueous gel at 5% concentration, was applied daily for 7 days the last administration being 2 h before irradiation; the treatment increased by 30% (p < 0.01, Student's t-test for paired data) the minimum erythematous dose (MED) in comparison to placebo (Indena internal report).

The vasokinetic effect at the microcirculatory level of *G. biloba* PHYTOSOME® was initially investigated (Arpaia *et al.*, 1989) by using the experimental technique

Table 3   Anti-edema effect of *Ginkgo biloba* dimeric flavonoids and their complexes with distearoyl-phosphatidylcholine (DPC).

Substances	Dose μmoles/ear	Edema[a] mg	% reduction	$p <$[b]
Controls Croton oil 75 μg/ear	–	7.3 ± 0.4	–	–
Amentoflavone	0.1	4.5 ± 0.3	38.4	0.05
	0.2	4.0 ± 0.2	45.2	0.05
Amentoflavone – DPC complex	0.1	3.9 ± 0.3	46.6	0.05
	0.2	3.2 ± 0.3	56.2	0.05
Ginkgetin	0.1	5.9 ± 0.4	19.2	0.05
	0.2	5.5 ± 0.3	24.7	0.05
Ginkgetin – DPC complex	0.1	5.3 ± 0.3	27.4	0.05
	0.2	4.0 ± 0.2	45.2	0.05
Sciadopitysin	0.1	6.3 ± 0.4	13.7	0.05
	0.2	5.9 ± 0.3	19.2	0.05
Sciadopitysin – DPC complex	0.1	6.8 ± 0.4	6.8	NS
	0.2	5.3 ± 0.4	27.4	0.05
Indomethacin	0.2	2.9 ± 0.3	60.3	0.05

[a]Values are means ± SEM (6 h after Croton oil); n = 20.
[b]NS = not significantly different from controls by Dunnett's *t*-test.

of Infrared-Photo-Pulse-Plethysmography (IRPPP), which provides a qualitative and quantitative evaluation of changes in skin blood perfusion, as shown by the pulsatile activity of small arteries and arterioles in the cutaneous subpapillary plexus. The evaluation was carried out by assessing changes in amplitude and morphology of infrared photo-plethysmographic tracings following short term epicutaneous application of PHYTOSOME® or placebo to various parts of the body: face, hand, thigh etc. in a group of 20 individuals which included healthy subjects and subjects with peripheral vascular acrosyndromes (Raynaud's phenomenon, acrocyanosis) or with cellulite secondary to venous stasis. In normal subjects a formulation containing 2.5% of *G. biloba* PHYTOSOME® caused a statistically significant increase in the amplitude of the pulsatile waves as reported in Table 4. In subjects with vascular acrosyndromes or cellulite the changes in pulsatile activity were dependent upon the severity of the lesions of the vessel wall and perivascular tissue. In other words, the vasokinetic effects were less clear-cut, mainly in those subjects where acrocyanosis was accompanied by a collagen disease (scleroderma).

The findings obtained in these studies indicate that the active principles contained in *G. biloba* extract exert a vasokinetic effect at the level of small arteries and precapillary arterioles of the subpapillary plexus of the skin. This effect can be ascribed to an enhanced force of contraction of the smooth muscle in the media of small

Table 4 Pulsatile activity (capillary blood flow) before and after topical application of *Ginkgo biloba* PHYTOSOME® in normal subjects.

Hand area	Before[a]	After 15 min	After 30 min	After 60 min
Fingertip	111.5 ±38.9	133.2 ±43.0[*]	153.0 ±50.2[**]	129.4 ±39.1
Nailfold	17.7 ±5.5	39.2 ±11.6[**]	36.8 ±10.5[**]	50.4 ±14.4[**]
Root of fingers	35.1 ±12.1	29.9 ±10.7	35.5 ±12.8	60.1 ±23.2[*]
Thenar eminence	25.2 ±7.5	24.2 ±7.2	37.0 ±10.9[**]	53.2 ±14.8[**]
Hypothenar eminence	55.5 ±19.7	44.2 ±15.8	69.0 ±25.8	93.9 ±51.7[*]

[a]Values are means ± standard deviation, SD of the areas (expressed in mm^2) under the pulsatile waves, before and 15, 30, 60 min after application; n = 5.
[*]$p < 0.05$ [**]$p < 0.01$ significantly different from before values by Student's t-test for paired data.

arteries and results in increased amplitude and modified morphology of the pulsatile waves as determined by IRPPP. Since this technique detects only changes in blood vessels not deeper than 2 mm below the cutaneous surface, an involvement of the medium-sized arteries can be ruled out and the data can be therefore considered to reflect events confined to the local microcirculation. In fact, the consequence of the increased arterial/arteriolar motility is evident in terms of an increased volume and velocity of blood flow at the microcirculatory level, i.e. at the level of capillary and venular systems.

Ginkgo biloba Dimeric Flavonoids

The results obtained using *G. biloba* extract in complexed form suggested to investigate other *G. biloba* components probably endowed with prominent activity on inflammation and microcirculation. In the frame of this research a fraction of dimeric flavonoids was identified (*G. biloba* dimeric flavonoids, GBDF). Biflavones were particularly active and possibly useful from a cosmetical point of view due to their mechanisms of action: inhibition of the histamine release from mast cells (Amellal *et al.*, 1985) and inhibition of c-AMP phosphodiesterases (Beretz *et al.*, 1979). Again, GBDF has been complexed with phospholipids (GBDF-PHYTOSOME®) in order to increase cutaneous absorption (Bombardelli *et al.*, 1989b; Della Loggia *et al.*, 1989).

GBDF-PHYTOSOME® was also tested on UV-induced erythema in a group of 18 volunteers. The product was incorporated in an aqueous gel at 5% concentration and applied daily for 7 days, with the last administration 2 h before irradiation.

The treatment increased by 31% (p < 0.01, Student's *t*-test for paired data) the MED in comparison to placebo (Indena internal report).

To assess the vasokinetic effects of these molecules on the skin microcirculation in humans, experiments have been performed on volunteers (Bombardelli *et al.*, 1996b). The techniques employed included Infrared-Photo-Pulse-Plethysmography (IRPPP), High Performance Contact Thermography (HPCT), and in a second phase Laser Doppler Flowmetry (LDF), Computerized Videothermography (CV), and Optic Probe Videocapillaroscopy (OPV). In the first trial twenty subjects of both sexes, aged 39.0 ± 4.6 years, were selected for the study and divided into four groups of 5 subjects each. In these groups the test preparation (0.5 ml containing 15 mg of GBDF-PHYTOSOME® microdispersed in distilled water) was applied on the cheek, the dorsal surface of the hand, the thigh and the female breast, respectively. An aqueous microdispersion of the same phospholipid used for the complexation (10 mg/0.5 ml), applied on the contralateral side of the same subject, was used as control. The objective evaluation of the microvasculokinetic effect was performed under baseline conditions and 15, 30, 45 and 60 minutes after application, by means of an infrared-photo-pulse-plethysmograph, whose reading head was also equipped with a cutaneous thermometer or a high-resolution contact thermograph with a film of micro-encapsulated liquid cholesterol crystals. Some of the results of this investigation are given below. The application of GBDF-PHYTOSOME® on the cheek skin induced statistically significant increases of the pulsatile waves, which were proportional to blood flow (Fig. 3). Table 5 shows the maximal effects obtained on the amplitude of finger roots pulsatile waves. Because the baseline skin temperatures

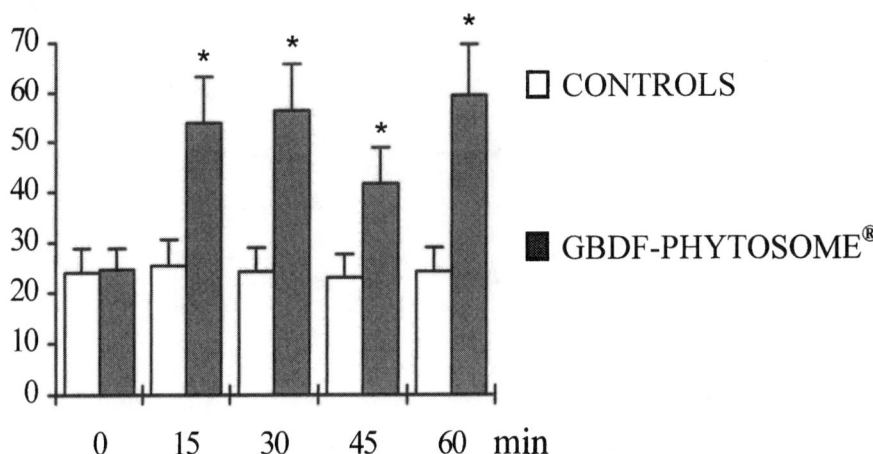

Figure 3   Effect of GBDF-PHYTOSOME® topical application on the amplitude of pulsatile waves in the cheek, evaluated by IRPPP. Values (mm²) are means ± S.E.M.; n = 5; * p <0.01 Student's *t*-test vs. baseline (0 min) values.

Table 5   Effect of GBDF-PHYTOSOME® topical application on the amplitude of pulsatile waves in the dorsal surface of the hand, evaluated by IRPPP.

Treatment	Measurement	II finger root[a]	III finger root	IV finger root
Control	baseline	132.9 ± 24.6	186.2 ± 36.4	171.8 ± 32.8
	after treatment	185.5 ± 33.1	162.5 ± 30.9	206.8 ± 36.5
GBDF-PHYTOSOME®	baseline	115.0 ± 22.5	109.5 ± 21.1	123.6 ± 22.9
	after treatment	196.5 ± 32.8*	206.9 ± 35.2*	217.0 ± 37.2*

[a]Values (mm^2) are means ± SEM; n = 5.
*p < 0.05 Student's t-test vs. baseline values.

recorded by the cutaneous thermometer of the reading head in the above described groups were homogeneous, the temperatures observed were pooled (n = 10). GBDF-PHYTOSOME® induced a long-lasting increase in skin temperature, which is also related to skin blood flow, beginning 15 minutes after application (Fig. 4). In the remaining ten subjects, the thermovascular maps obtained by HPCT under baseline conditions documented localized dysthermias (usually hypothermic areas), indicating the occurrence of a non-homogeneous capillary blood flow. Application of GBDF-PHYTOSOME® on the thighs and female breast induced, within 15 minutes, clear-cut changes of the thermovascular map: reduction or disappearance of the hypothermic areas and a more homogeneous distribution of skin temperature. The increase in cutaneous temperature averaged 0.80 ± 0.25 °C (mean standard error of the mean, SEM) at the end (60 minutes) of the assessment.

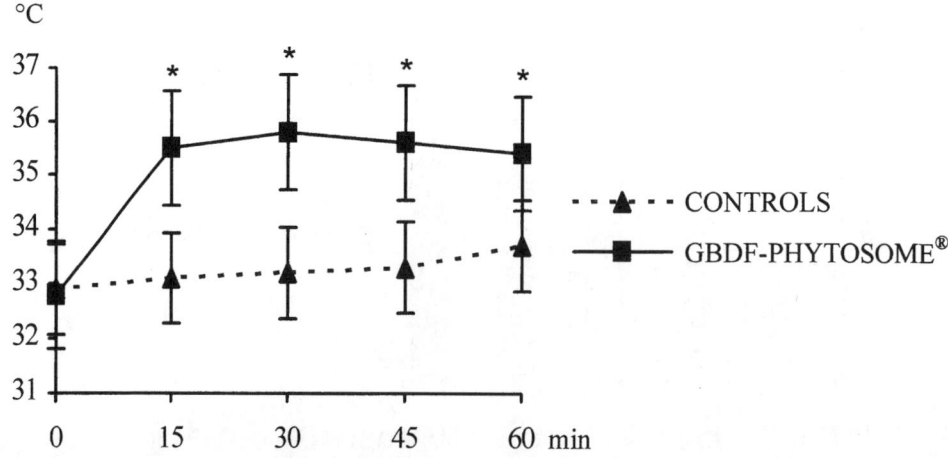

Figure 4   Effect of GBDF-PHYTOSOME® topical application on the skin temperature of the cheek and hand, evaluated by the cutaneous thermometer of the IRPPP reading head. Values (°C) are means ± SEM; n = 10; * p <0.01 Student's t-test vs. baseline (0 min) values.

In another experiment fifteen subjects of both sexes, aged 36.5 ± 3.9 years, were selected for study and divided into three groups of 5 subjects each. The active preparation was a cream (O/W emulsion) containing 3% of GBDF-PHYTOSOME®; an O/W emulsion of 2% of phospholipids, applied on the contralateral side of the same subject, was used as control. Conditions occurring for 2 h after the applications (0.5 ml) were recorded. The first group included 5 healthy males, and the preparations were applied on the inferior 3rd of the volar surface of the forearms. The skin capillary perfusion was evaluated by a laser Doppler flowmeter and by a computerized videothermograph. LDF allows the measurement of the capillary blood flow (minimal flow, maximal flow, mean flow) in Perfusion Units (PU), the moving red cells shifting the frequency of the laser light beam according to the Doppler principle. CV determines the temperature variations in several areas of the skin surface examined by using a sensor plate with more than 500 temperature-recording points. The plate is connected with a personal computer which, by the dedicated software, can trace a map from the numeric values and calculate the mean temperature in a selected area. The second group included 2 male and 3 female acrocyanotic subjects, and the preparations were applied on the volar surface of the hand. LDF and CV were used for evaluations. The third group included 5 female subjects with cellulite (II–III stage), and the preparations were applied on their thighs. Beside LDF and CV, OPV was used. An optic probe, usually 200 × magnification, was placed on the skin surface, and the capillaroscopic images were recorded. Successively, images were analyzed with an image-analysis program to determine the capillary density (% CD), i.e. the percent of image occupied by open capillaries. In Figure 5 the results obtained in LDF measurements (mean values of the sphygmic tracings) in the three groups of subjects (healthy, acrocyanotic and cellulite-afflicted respectively) are

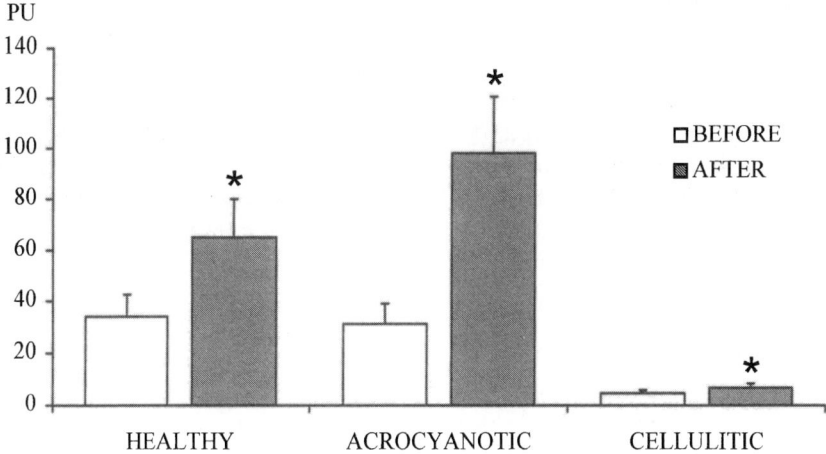

Figure 5 Effect of GBDF-PHYTOSOME® topical application on the capillary blood flow in healthy, acrocyanotic and cellulite-afflicted subjects, evaluated by LDF. Values (PU) are means ± SEM; n = 5; * p <0.01 Student's *t*-test vs. before values.

Table 6   Effect of GBDF-PHYTOSOME® topical application on the skin temperature, evaluated by CV.

Group of subjects	Baseline temperature[a]	Maximal temperature
Healthy	33.0 ± 0.1	33.8 ± 0.2*
Acrocyanotic	32.2 ± 0.2	33.4 ± 0.3*
Cellulite-afflicted	33.5 ± 0.2	34.2 ± 0.2*

[a]Values (°C) are means ± SEM; n = 5.
*$p < 0.01$ Student's t-test vs. baseline values.

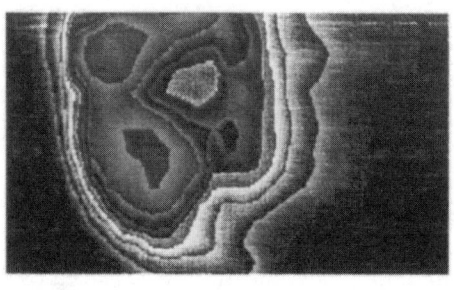
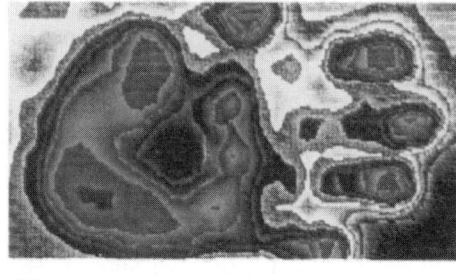

(A)   (B)

Figure 6   Thermographic maps of hand before (A) and after (B) application of GBDF-PHYTOSOME®. B.D., male, 38 years; acrocyanotic subject.

reported. In all groups, maximal effects were observed about 45 minutes after the treatment with GBDF-PHYTOSOME®, whereas control treatment did not induce significant variations. The skin temperatures recorded at the same time are reported in Table 6. Significant increases were observed after the treatment with GBDF-PHYTOSOME®. Figure 6 shows the thermographic maps of a hand before and after treatment with the active preparation. Finally, also a statistically significant ($p < 0.01$, Student's t-test for paired data) increase, from 3.9 ± 0.4% to 10.1 ± 0.6%, was observed in capillary density measured with OPV, in cellulite-afflicted subjects treated with GBDF-PHYTOSOME®. Figure 7 shows that microcirculatory dysfunctions such as microaneurisms of capillary loops and "capillary kinking" were less visible after the treatment.

CONCLUSION

G. biloba extracts are probably the most investigated among natural ingredients clinically employed by systemic route to improve the peripheral circulation. Because the ageing process is accompanied by severe modifications of the skin microcirculation, such as both a blood flow decrease at the capillary level and an alteration of the vasomotility of the smallest arteries and arterioles of the subpapillary

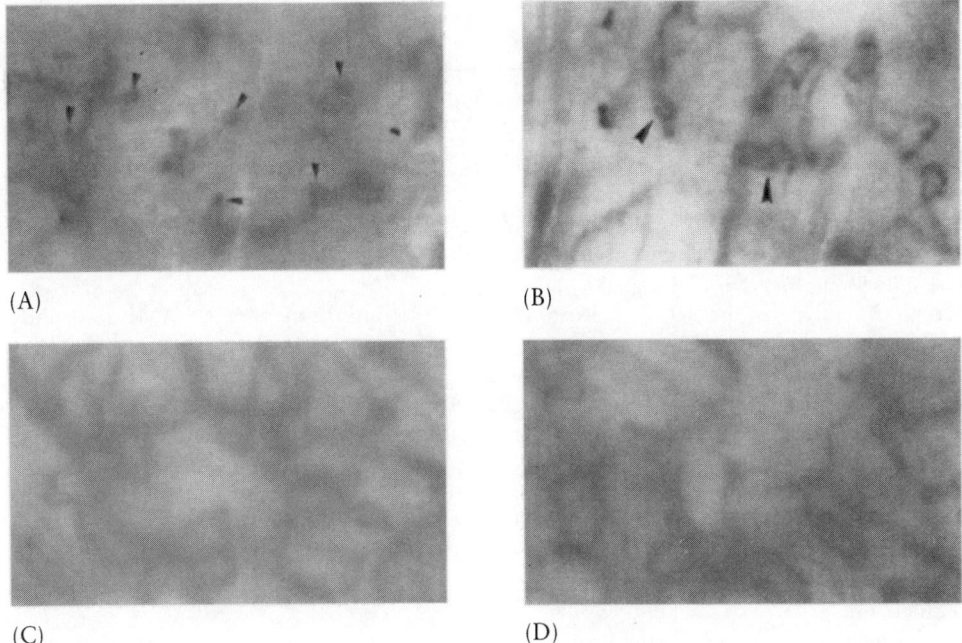

(A)

(B)

(C)

(D)

Figure 7 Videocapillaroscopies (200 ×) before (A, B) and after (C, D) application of GBDF-PHYTOSOME®. Arrows in A and B show respectively microanurisms of capillary loops and "capillary kinking". R.A., female, 32 years; cellulite-afflicted subject.

skin plexus, they have also found application in cosmetics. The vasokinetic activity of the epicutaneously administered *G. biloba* extract in PHYTOSOME® form, which is a more bioavailable formulation for substances which are generally poorly absorbed through the skin, was also investigated. Among the constituents of *G. biloba*, but absent in the standardized extracts commonly used, dimeric flavonoids are particularly strong as vasoactive agents. For this reason the vaso-kinetic effects of *G. biloba* dimeric flavonoids (GBDF), after epicutaneous adminis-tration of their complex with phosphatidylcholine (GBDF-PHYTOSOME®) were examined. The results obtained with several techniques in normal subjects or those afflicted by acrocyanosis or cellulite suggest that *G. biloba* PHYTOSOME® and GBDF-PHYTOSOME® do not act only *via* a simple vasodilation, but also *via* a specific microvasculokinetic action, i.e. an increased force of contraction of the smooth muscle cells.

Thus, the pharmacodynamic evidence obtained in these trials allows one to conclude that the microvasculokinetic activity of both *G. biloba* PHYTOSOME® and GBDF-PHYTOSOME® can improve the microvascular perfusion and amelio-rate the skin nutrition. Moreover, the antierythematous activity of both preparations (although probably achieved by different mechanisms of action) makes *G. biloba* even more interesting for the treatment of ageing skin.

REFERENCES

Amellal, M., Bronner, C., Briançon, F., Haag, M., Anton, R., Landry, Y. (1985) Inhibition of mast cell histamine release by flavonoids and biflavonoids. *Planta Med.*, **51**, 16–20.

Arpaia, G., Bombardelli, E., Curri, S.B., Della Loggia, R. (1989) Changes in cutaneous pre-capillary vasomotility induced by topical application of *Ginkgo biloba* PHYTOSOME®. *Fitoterapia*, **60**, 11–28.

Atzori, C., Bruno, A., Chichino, G., Bombardelli, E., Scaglia, M., Ghione, M. (1993) Activity of bilobalide, a sesquiterpene from *Ginkgo biloba*, on *Pneumocystis carinii*. *Antimicrob. Agents Chemother.*, 37, 1492–1496.

Beretz, A., Joly, M., Stoclet, J.C., Anton, R. (1979) Inhibition of 3', 5'-AMP phospho-diesterase by biflavonoids and xanthones. *Planta Med.*, **36**, 193–195.

Bombardelli, E. (1991) Topical administration of plant extracts having therapeutic activities for treatment of fat deposits (Patent). Eur. Pat. Appl. EP 427,026 A2.

Bombardelli, E., Cristoni, A. and Morazzoni, P. (1994) PHYTOSOME®s in functional cosmetics. *Fitoterapia*, **65**, 387–401.

Bombardelli, E., Cristoni, A. and Morazzoni, P. (1996a) Pharmaceutical and cosmetic formulations containing esculoside (Patent). Eur. Pat. Appl. EP 692,250 A2.

Bombardelli, E., Cristoni, A. and Curri, S.B. (1996b) Activity of phospholipid-complex of *Ginkgo biloba* dimeric flavonoids on the skin microcirculation. *Fitoterapia*, **67**, 265–273.

Bombardelli, E., Curri, S.B., Della Loggia, R., Del Negro, P.,Tubaro, A., Gariboldi, P. (1989a) Complexes between phospholipids and vegetal derivatives of biological interest. *Fitoterapia*, **60 (Suppl. No. 1)**, 1–9.

Bombardelli, E., Della Loggia, R., Tubaro, A., Curri, S.B. (1989b) Activity of *Ginkgo biloba* dimeric flavonoids on the microvessel-tissue relationship. In: N.P. Das (ed.), *Flavonoids in Biology and Medicine III: Current issues in Flavonoids Research*, Singapore, pp. 393–401.

Bombardelli, E. and Ghione, M. (1996) Novel pharmaceutical uses of bilobalide and derivatives thereof and pharmaceutical compositions adapted for such use (Patent). Eur. Pat. Appl. EP 556,051 A1.

Bombardelli, E. and Patri, G. (1988) Cosmetics and pharmaceuticals containing flavonoid-phospholipids complexes (Patent). Eur. Pat. Appl. EP 275,005 A2.

Brantner, A. and Grein, E. (1994) Antibacterial activity of plant extracts used externally in traditional medicine. *J. Ethnopharmacol.*, **44**, 35–40.

Braquet, P. (1988) The ginkgolides. From Chinese Pharmacopoeia to a new class of pharmacological agents: the antagonists of Platelet-Activating Factor. In: P. Braquet (ed.), *Ginkgolides – Chemistry, Biology, Pharmacology and Clinical Perspectives*, J.R. Prous Science Publishers, Barcelona, pp. XV- XXXIV.

DeFeudis, F.V. (1991) *Ginkgo biloba Extract (EGb 761): Pharmacological Activities and Clinical Applications*, Elsevier, Paris.

DeFeudis, F.V. (1994) Overview of the actions of *Ginkgo biloba* extract (EGb 761) on the cardiovascular system: from basic pharmacology to the clinic. In: F. Clostre and F.V. DeFeudis, (eds), *Advances in Ginkgo biloba Extract Research vol. 3, Cardiovascular Effects of Ginkgo biloba Extract (EGb 761)*, Elsevier, Paris, pp. 105–119.

Della Loggia, R., Sosa, S., Del Negro, P., Tubaro, A. (1989) Anti-inflammatory activity of flavonoids from *Ginkgo biloba* and of their phospholipide-complexes (Phytosomes®). In: N.P. Das (ed.), *Flavonoids in Biology and Medicine III: Current issues in Flavonoids Research*, Singapore, pp. 483–488.

Della Loggia, R., Sosa, S., Tubaro, A., Morazzoni, P., Bombardelli, E., Griffini, A. (1996) Anti-inflammatory activity of some *Ginkgo biloba* constituents and of their phospholipid-complexes. *Fitoterapia*, 67, 257–264.

Duranton, A. and De Lacharriere, O. (1995) Hair growth stimulants comprising lipoxygenase or cycloxygenase stimulants or inhibitors (Patent). Eur. Pat. Appl. EP 648,488 A1.

Fischer, M., Goering, S. and Theis, H. (1995) Topical composition containing Ginkgo extract and glycine to prevent skin aging (Patent). Ger. DE 4,429,928 C1.

Gabor, M. (1979) Anti-inflammatory substances of plant origin. In: J.R. Vane and S.H. Ferreira, (eds.), *Handbook of Experimental Pharmacology: Antiinflammatory Drugs*, Springer-Verlag, New York, pp. 698–739.

Halliwell, B. and Gutteridge, J.M.C. (1986) *Free radicals in biology and medicine*, Clarendon Press, Oxford.

Hayashi, K., Hayashi, T. and Morita, N. (1992) Mechanism of action of the antiherpesvirus biflavone ginkgetin. *Antimicrob. Agents Chemother.*, 36, 1890- 1893.

Kleijnen, J. and Knipschild, P. (1992) *Ginkgo biloba*. The Lancet, 340, 1136–1139.

Koide, C. and Kuribayashi, S. (1993) Cosmetics and topical pharmaceuticals containing superoxide dismutase and Ginkgo extracts (Patent). Jpn. Kokai Tokkyo Koho JP 05 32,556 [93 32,556].

Koide, C. and Sasaki, I. (1992) Cosmetic skin preparations containing Ginkgo extracts and vitamins (Patent). Jpn. Kokai Tokkyo Koho JP 04,321,616 [92,321,616].

Kubo, M., Matsuda, H., Suzuki, R., Kobayashi, N. (1993) Hair growth promoting effect of ethanolic extract from leaves of *Ginkgo biloba* L. (GBE) and their inhibitory effects on blood platelet aggregation. *Fragrance J.*, 21, 27–36.

Maitra, I., Marcocci, L., Droy-Lefaix, M.T., Packer, L. (1995) Peroxyl radical scavenging activity of *Ginkgo biloba* extract EGb 761. *Biochem. Pharmacol.*, 49, 1649–1655.

Middleton Jr., E. (1984) The flavonoids. *Trends Pharmacol. Sci.*, 5, 335–338.

N'Guyen, Q.-L. (1994) Cosmetic and pharmaceutical compositions containing polyphenols and Ginkgo extract (Patent). Fr. Demande FR 2,699,818.

Nishizawa, Y., Kanazawa, S., Hotsuta, M., Kidena, H., Kobayashi, H. (1996) Hair growth stimulants containing plant extracts and other agents for synergistic effects (Patent). Jpn. Kokai Tokkyo Koho JP 08 73,324 [96 73,324].

O'Reilly, J. (1995) Extract from the leaves of *Ginkgo biloba* (Patent). PCT Int. Appl. WO 95 15,172 A1.

Soudant, E. and Nadaud, J.-F. (1992) Antiobesity composition containing *Ginkgo biloba* extract (Patent). Fr. Demande FR 2,669,537 A1.

Tubaro, A., Dri, P., Del Bello, G., Zilli, C., Della Loggia, R. (1985) The Croton oil test revisited. *Agents Actions*, 17, 347–349.

Watanabe, C. and Takahashi, M. (1991) Hair tonics containing vitamins and Ginkgo extracts (Patent). Jpn. Kokai Tokkyo Koho JP 03,133,918 [91,133,918]

Weidner, M.S. (1995) A method of rendering organic compounds soluble in fatty systems, novel chemical complexes of such compounds and various applications of the complexes (Patent). PCT Int. Appl. WO 95 25,084 A1.

26. PATENTS OF *GINKGO BILOBA* AND ITS CONSTITUENTS

MARIAN A.J.T. van GESSEL

Hamsestraat 79, 4043 LG Opheusden, The Netherlands

INTRODUCTION

During the past 30 years many patents relating to extracts or pure constituents from various parts of the tree *Ginkgo biloba* have been filed. The patents can be roughly divided in two groups: for medicinal and for other uses, e.g. as cosmetics. It is the purpose of this chapter to tabulate and briefly discuss all the patents relating to the Ginkgo tree which have appeared since 1920 in Chemical Abstracts.

TABULATION OF PATENTS

Patent Retrieval and Ordering

From 1967 until February 1998 Chemical Abstracts was searched on-line by computer. The search was carried out by entering the terms "Ginkgo", "*Ginkgo biloba*" and the CAS registry numbers for ginkgolides A, B and C, and bilobalide [15291-75-5], [15291-77-7], [15291-76-6] and [33570-04-6] respectively. The few patents which appeared in Chemical Abstracts from 1920 until 1967 were retrieved by a manual search with the keyword "*Ginkgo biloba*" only. This resulted in a total number of 196 patents. The patents have been sorted into the eight groups given in Table 1. Sometimes patents could have been placed in more than one group. In such cases the main use determined into which group the patent was placed. Group 7 is a rather heterogenous rest group of all patents which did not fit in groups 1–6. Group 8 only gives citations of patents which did show up in the literature search but

Table 1 Sub-division of all patents on *Ginkgo biloba* into eight groups.

1. Patents on the preparation of extracts of *G. biloba*
2. Patents on methods of isolation of several compounds from *G. biloba*
3. Patents on the preparation of derivatives of compounds extracted from *G. biloba*
4. Patents on methods to modify or transform extracts or compounds from *G. biloba*
5. Patents on the use of extracts or compounds from *G. biloba* as PAF-antagonists
6. Patents on the use of *G. biloba* for personal health care
7. Patents on the application of parts, extracts or compounds from *G. biloba* for various uses
8. Patents which are not specific for *G. biloba*

which are not relevant for this chapter. It concerns patents in which G. *biloba* is solely mentioned as an example material for certain processes or machines. For greater clarity some larger groups have been subdivided. Within each group or sub-group the patents have been chronologically tabulated. The information about the patents in this chapter is solely based on the abstract given in Chemical Abstracts. In the list of references the Chemical Abstracts number is also given.

Patents on the Preparation of Extracts of *Ginkgo biloba*

There are several patents just on the preparation of special extracts of G. *biloba*. Often such extracts are used in pharmaceutical formulations. The oldest one is from 1959 when Kimura patented a method for the extraction of sesamin from G. *biloba* wood (one of the two only patents on the extraction of wood). He used diethyl ether for the extraction. After purification he got 70.1 g pure sesamin out of 18.2 kg wood. The compound was useful as a potentiating agent for pyrethrin (Kimura, 1959). The patent "Extract from G. *biloba* leaves as circulatory drugs" describes a method for the preparation of an extract, usable as a drug for the treatment of circulatory disturbances. The procedure included several filtrations at different pH, and gave a product comprising 25.5% flavonol glycosides and 6.6% quercetin and kaempferol (Ayroles *et al.*, 1989). In 1989 a process for the supercritical extraction and separation of solid samples such as G. *biloba* leaves was patented. Physiologically active components were extracted in high yield (3.8%) from Ginkgo leaves while the amount of harmful co-extracted substances remained low (0.04%). Details were given (Manabe *et al.*, 1989). In 1990 the patent "New extracts of *Ginkgo biloba* and their methods of preparation" was filed. The invention gives a new process for the preparation of G. *biloba* extracts which can be used for the pharmaceutical products employed today, but which avoids the drawbacks of known extraction processes. The process further allows for the production of two novel extracts (Bombardelli *et al.*, 1990). Matsui *et al.* (1991) described the manufacture of an extract with a high flavonoid content from Ginkgo leaves. This extract contained 24% flavonoids and was useful as an antioxidant for lipids, effective for preventing ageing of the skin and also useful as a medicine or health food for improving the blood circulation. The extraction method was described. Two very similar patents on the preparation of glycoside-rich extracts from Ginkgo leaves were filed by Takane. The method comprised adding partial hydrolysates of starch and glycosidase or trans-glycosidase to a water-alcohol extract to solubilize the insoluble glycosides. The preparation was suitable for pharmaceutical formulation. The extract contained 13.6% of flavonoids and a high content of ginkgolides (Takane, 1991a,b). Makino *et al.* prepared a Ginkgo extract which contained > 15% flavonol glycosides and had a solubility of > 5% in water. Out of 1 kg dried leaves they got 35 g flavonol glycosides. They used column chromatography with a non-polar porous resin as stationary phase. The flavonols were said to be useful as vasodilators (Makino *et al.*, 1991). O'Reilly *et al.* registered a patent with different goals; the invention relates to highly concentrated active component concentrates (e.g. 40–60% flavonol glycosides) and new active component combinations from

G. biloba leaves as well as their method of preparation. In addition, the invention relates to pharmaceuticals which are characterized by containing G. *biloba* extract, and their use (O'Reilly *et al.*, 1991). A Swiss patent gives the preparation of concentrates of active agents low in pesticides from *Ginkgo biloba* leaves, using adsorption onto a resin. The produced extract is rich in flavonol glycosides (Steiner *et al.*, 1996).

N.B. In all abstracts on flavonols in the original text in Chemical Abstracts the incorrect term "flavones" is used. This was replaced by the author throughout this chapter by the term "flavonols".

Patents on Methods of Isolation of Several Compounds from *Ginkgo biloba*

In this paragraph the main subject is the isolation and/or extraction of one or more compounds from G. *biloba*. Sometimes the use is mentioned. The paragraph has been subdivided into three groups: flavonols, polyprenols, and a rest group with other compound classes.

Flavonols

A therapeutically active mixture, free from undesirable polyphenols but containing flavonol and steroid glycosides, was extracted from finely ground Ginkgo leaves. The extraction procedure was mentioned. The mixture had vaso-active properties, and could be purified by treatment with finely powdered high-molecular weight polyamide in lieu of the $Pb(OH)_2$ procedure (Kloss and Hermann, 1972). The isolation of a flavonol glycoside from Ginkgo leaves, to be used as a cardiovascular agent, is described in a patent filed by Daicel Industries. After dilution of a methanolic leaf extract with water, successive extraction with chloroform and ethyl acetate followed. The ethyl acetate extract was finally chromatographed on a reversed phase silica column to give the desired flavonol glycoside (Matsumoto and Sei, 1987b). Another patented method for the isolation of flavonol glycosides from G. *biloba* leaves starts with an extraction of 2 kg leaves with methanol. Subsequently this extract was purified and finally 6 g was obtained. This was subjected to countercurrent distribution chromatography to give a fraction containing 258 mg flavonol glycosides (Matsumoto, 1987a). Another patent describes a simple extraction method to isolate therapeutic biflavones from Ginkgo leaves, together with terpenes and flavonol glycosides. The preparations can also be used in manufacturing cosmetics and foods (Matsumoto, 1990a). In the patent "Pharmaceutical *Ginkgo biloba* flavonoid extract", a flavonoid extract was prepared from G. *biloba* bark which contained < 1.5% prodelphinidins and <10% proanthocyanidins. The extraction was carried out with organic solvents, and countercurrent liquid/liquid extraction was used. The extract was further processed to give an extract containing 17–18% flavonoids (Quintana *et al.*, 1993).

Polyprenols

All patents on polyprenols have been filed by the Kuraray Company in Japan. The first patent is entitled "Polyprenols and their derivatives". Polyprenols were isolated from the leaves of *G. biloba* and then chemically transformed. These compounds are important intermediates for the synthesis of dolichols (Kuraray Co., 1982a). An alternative extraction method for polyprenyl compounds was the topic of another patent: 5 kg Ginkgo leaves were extracted with petroleum ether-acetone (4:1). After several other steps including silica gel column chromatography, finally a mixture of 10.3 g polyprenols was obtained. The polyprenyl compounds may be used as cosmetic and pharmaceutical bases (Kuraray Co., 1982b). A patent for obtaining polyprenols by extraction from *G. biloba*, *Cedrus deodara* or *Pinus densiflora*, optionally followed by saponification was also filed (Kuraray Co., 1983c). Still another patent dealt with an extraction method for polyprenyl compounds from *G. biloba* and *Cedrus deodara*. Detailed procedures were given, as well as methods for the preparation of some derivatives via standard synthetic procedures (Kuraray Co., 1983d). The last patent in this series is entitled "Polyprenyl compounds for pharmaceutical or cosmetic use". The extraction method was described in detail. *G. biloba*, *Cedrus deodara* and *Pinus densiflora* were extracted by organic solvents, and the extracts were chromatographed on silica gel columns and silica gel plates (Kuraray Co., 1983f).

Various

In India *G. biloba* fruit liquid was neutralized, dried, decarboxylated and stored in absence of moisture. The liquid was useful as an impregnant in the manufacture of electrical insulators (Krishnamurthy, 1960). In the patent "Recovery of vasoactive drugs from the leaves of *G. biloba*" an extraction method for both fresh or dried green leaves of *G. biloba*, is given. The powdered product may be taken up in an aqueous solution and sterilized to give an injectable solution (Schwabe and Kloss, 1972). Another patent deals with a method for the isolation of ginkgolides from Ginkgo leaves. Using a number of organic solvents a high yield of ginkgolides (90 g out of 10 kg pulverized dry leaves) was obtained. Ginkgolide A, B and C were characterized. No therapeutic data were given in the examples (Kwak *et al.*, 1990). In a Chinese patent a method for extracting ketones from *G. biloba* leaves for treating among others brain thrombus was described (Li and Cheng, 1992). The next patent has as topic obtaining an extract from the leaves of *G. biloba* useful for cosmetical purposes. The extracts, which contained 25% glycolipids and 1% cerebrosides, including galactocerebroside, and almost no ginkgolic acids, have shown activity in cosmetic and skin care applications. A chromatographic column was used for purifying the extract (O'Reilly, 1994). Finally, hydroxyproline-rich glycoproteins can be obtained by acid alcoholic extraction from plants such as *G. biloba*. The patent gave cosmetic formulations containing them (Bombardelli *et al.*, 1996a).

Patents on the Preparation of Derivatives of Compounds, Extracted from *Ginkgo biloba*

Some patents have appeared on the preparation of derivatives of several compounds isolated from *G. biloba*. All but two have as topic the conversion of polyprenols from Ginkgo leaves. The first patent of this kind appeared in 1983 and described a method for isolating solanesyl acetate and many polyprenyl alcohols from the leaves of *G. biloba*. The polyprenyl alcohols were transformed into over 300 polyprenyl compounds including dolichols (Takigawa *et al.*, 1983). A similar patent described again the isolation of polyprenols from *G. biloba* and the preparation of reactive derivatives. Via several reactions and finally reductive elimination, they were converted into dolichol (Onishi *et al.*, 1983). Another patent is on the preparation of nitrile derivatives of polyprenols. The polyprenols were extracted from *G. biloba*, brominated and treated with NaCN (Kuraray Co., 1983a). Three more patents in this field were filed by the same company. The first one described the synthesis of other derivatives from polyprenols, isolated from *G. biloba* (Kuraray Co., 1983b). The second one was on the preparation of more derivatives from polyprenols isolated from *G. biloba* leaves. Detailed methods were given (Kuraray Co., 1983e). The third was on the preparation and purification of polyprenyl acetates from extracts of *G. biloba* leaves (Kuraray Co., 1983g). In 1986 Takak *et al.* isolated polyprenol from *G. biloba* and used this extract for synthesizing dolichol. Dolichol and its esters enhance differentiation and proliferation of hematopoietic progenitor cells in bone marrow, and are useful for treatment of anemia. They demonstrated the activity of dolichol *in vitro* and *in vivo*, and gave a formulation for an injection. Another method for the synthesis of dolichol was described in a patent by Kuraray Co. Also the activity (enhancing the differentiation and proliferation of hematopoietic progenitor cells in bone marrow) of this compound and its esters was tested *in vitro* and *in vivo* (Kuraray Co., 1986). Next, a patent appeared on the preparation of dolichols from *G. biloba* or *Cedrus deodara* extracts. A detailed description of the reactions was given (Tanaka *et al.*, 1991). The next patent in this group was not on polyprenols, but described the synthesis of ginkgolide B from ginkgolide C. This conversion could be realised in four steps (Teng, 1992). The last patent in this group is on bioparyl. Bioparyl is obtainable by a disclosed extraction, hydrolysis and purification procedure applied to golden autumn leaves of the *Ginkgo biloba* tree. Bioparyl is a biological regulator, active against various pathologies (Fondation, 1996).

Patents on Methods to Modify or Transform Extracts or Compounds from *Ginkgo biloba*

There are several patents on methods for improving the usefulness of extracts or compounds extracted from *G. biloba*. Kimura found a method for preparing oral pharmaceutical preparations containing dolichol. Dolichol, which on its own is insoluble, could be solubilised by means of lipids. The extraction of dolichol from Ginkgo leaves and the process of solubilisation using soybean lecithin was described. Good absorption of dolichol-lipid complexes through the rat intestinal membrane was demonstrated (Kimura, 1986). In Italy flavonoid-phospholipid complexes were prepared by treating a flavonoid with phospholipids in an inert solvent. A solution of depurated *G. biloba* extract was mixed with a solution containing soy bean phospholipids. The yellow compound thus prepared, could be used in cosmetics and pharmaceuticals. It should enhance the anti-inflammatory effect of *G. biloba* extract (Bombardelli and Patri, 1988). In another patent a complex was described consisting of bilobalide and phospholipids. Topical formulations containing these complexes were effective for the treatment of disorders associated with inflammatory of traumatic neuritic processes (Bombardelli and Mustich, 1991). A Belgian company has patented a preparation process for drug microcrystals comprising an active substance having an affinity for phospholipids, and at least one phospholipid. A solution of ginkgolide B was mixed with a solution of phosphatidylcholine. The final microcrystal suspension can be used in injectable or in sprayable form (Baurain and Trouet, 1990). The use of Ginkgo leaf extracts in the preparation of beverages gave problems because of its poor solubility. Momyama and Sugiura enhanced the solubility of Ginkgo extracts by addition of a polar polymer such as gelatine or pectin (Momyama and Sugiura, 1993). In the patent "Solid and liquid solutions of flavonoids" a method for enhancing the bioavailability of flavonoids and *G. biloba* extracts was described. Liquid or solid solutions of the extracts and flavonoids could be formed with hydrophilic peptides (gelatines or collagen hydrolysates). The compounds were usable in drugs, dietic foods and cosmetics (Wunderlich *et al.*, 1994). Flavonoid-containing extracts can be prepared from *G. biloba* leaves and terpene trilactones can be removed. The used method is given. The extract may be incorporated in yoghurt, beverages, or chewing gum (O'Reilly, 1996). In the last patent cyclodextrins and their derivatives are used for solubilisation of platelet aggregation inhibitors, such as ginkgolides (Defaye *et al.*, 1997).

Patents on the Use of Extracts or Compounds from *Ginkgo biloba* as PAF-Antagonists

All the patents in this group focus on the potential application of Ginkgo extracts or compounds for therapeutical purposes. PAF stands for platelet-activating-factor, an endogenous compound in the human body. A too high concentration of this compound gives rise to adverse effects. Because ginkgolides are strong and selective PAF-antagonists, the potential application of these compounds as medicines has been thoroughly investigated. In 1985 Braquet discovered that ginkgolide A, B, C and M were agents for the treatment of disorders provoked by PAF, such as various types of shock (Braquet, 1985b). The same inventor gave in another patent details about the use of ginkgolide B in *in vitro* tests against PAF agonists, and in a clinical test for patients with burns (Braquet, 1985a). *G. biloba* extract or ginkgolides and non-steroidal inflammation inhibitors in an infusion have been used as anti-inflammatory pharmaceuticals. One formulation (containing Ginkgo flavonol glycosides and pentoxifylline) was given and tested in a patient with third degree burns. It proved to be effective (Bauer, 1988). In another patent derivatives of ginkgolide B (1- and 10-alkoxyginkgolides) were prepared and tested in guinea pigs as PAF-antagonists. The method of preparation was described (Braquet, 1989). Diseases resulting from breakdown of the endothelial barrier (e.g. infections, edema, allergies) were treated with PAF antagonists. The inventor mentioned several compounds, among others ginkgolides (Korth, 1989). Korth also found a procedure for determining the efficacy of PAF antagonists. The binding of labeled PAF to purified cells (in the presence of lipoproteins, lipoprotein-associated PAF and/or cholesterol) was compared in the presence and absence of an antagonist, e.g. ginkgolides (Korth, 1991). The same inventor filed another patent on the treatment of eosinophil-mediated diseases (e.g. allergy or inflammatory diseases) with PAF antagonists, for example a ginkgolide, and a procedure for determining their efficacy using binding tests on human eosinophils (Korth, 1993b). PAF antagonists can also be used as anti-pruritic agents. In one patent, several compounds were mentioned, ginkgolide B being one of them (Woodward and Williams, 1991). The use of PAF antagonists in the treatment of *Clostridium difficile* colitis and cholera was patented in 1993, and ginkgolide B was mentioned to be one of the compounds used. PAF-antagonists blocked the hemorrhagic fluid secretion caused by the toxins (Quintana Sener *et al.*, 1993). On the same topic is the patent "Method using PAF-antagonist or indomethacin for treatment of *Clostridium difficile* colitis and inhibition of the secretory effects caused by C. *difficile* Toxin A and by Cholera toxin". One of the PAF antagonists used was ginkgolide B. Compounds of the invention were tested both *in vivo* (rabbit ileal ligated loop model) and *in vitro* (monolayers of T-84 cells in tissue culture) (Guerrant *et al.*, 1995). Some derivatives of ginkgolide B were also PAF-antagonists, as is described in the patent "Preparation of 3-dehydroxyginkgolide B derivatives as PAF antagonists". The method of preparation of the derivatives was given. Tested in rabbit platelet rich plasma one of these compounds proved to be a stronger acting PAF antagonist than ginkgolide B itself (Park *et al.*, 1993). The same inventors filed another patent on some 10-(O-substituted alkyl) ginkgolide B derivatives and some 1,10-di-O-alkyl analogs as PAF antagonists, and a process for preparing them (Park *et al.*, 1995). In an American patent, lyso-PAF-mediated skin disorders were treated with natural ginkgolides as lyso-PAF antagonists, which were administered by food or topically (Korth, 1997).

Patents on the Use of *Ginkgo biloba* for Personal Health Care

These patents have in common that they are all on the topic of personal health care. They are subdivided in two groups: (1) patents on hair formulations and (2) patents on all other applications.

Hair formulations

The oldest patent in this group is entitled "Hair and scalp conditioning composition". The shampoos contained small amounts of several plant preparations among others from *G. biloba*. Which part of the tree was used was not mentioned (Bons and Van der Beek, 1976). Around 1991 several patents appeared, all Japanese, on the use of extracts from the leaves of *G. biloba* in hair preparations. All these patents contained formulations for hair tonics, used for promoting hair growth. The first was a hair tonic containing vitamins and a Ginkgo leaf extract in a 70% alcoholic solution. The extraction method was described (Watanabe and Takahashi, 1991). The next was a hair tonic containing hormones (hydrocortisone) and Ginkgo extracts in a 70% alcoholic solution. The extraction method was given. Hormones and Ginkgo extracts were synergistic in promoting human hair growth (Watanabe and Naito, 1991a). From the same inventors there was a patent on a hair tonicum (cream) containing inflammation inhibitors, such as stearyl glycyrrhetinate and Ginkgo extracts which acted synergistically on hair growth (Watanabe and Naito, 1991b). A hair tonic containing microbicides and Ginkgo extract in a 70% alcoholic solution was also patented. The microbicide and the extract were synergistic in promoting hair growth (Watanabe and Naito, 1991c). Furthermore, a hair tonic containing mucopolysaccharides and Ginkgo extracts in a 30% alcoholic solution was patented. The method used for extracting Ginkgo leaves (using ethyl acetate and benzene) was disclosed (Watanabe and Naito, 1991d). A more recent, Chinese, patent in this category is entitled: "Hair tonic preparations containing herb extracts, sodium camphorsulfonate, and other ingredients". The use of G. biloba leaves is mentioned, but otherwise no data about the extraction, preparation, formulation or indication are given (Huang, 1994). In a Belgian patent hair preparations containing G. biloba extracts (5–40%) and Liquiritia officinarum extracts (3–35%) are claimed to be useful for the treatment of hair loss (Szaloki-Bakos et al., 1997). Again on hair preparations is the invention of Takahashi et al. (1997): they describe oral pharmaceutical compositions comprising cuttlefishinks, G. biloba extracts and calcium, for grey hair.

Other personal health care patents

A 2-hydroxy-6-alkylbenzoic acid was extracted from fruits and leaves of G. biloba. This compound could be used in dentifrices, mouthwashes and chewing gum to inhibit the formation of dental plaque and caries (Lion Corp., 1982). Another patent is entitled: "Topical administration of plant extracts having therapeutic activities for treatment of fat deposits". Several plant extracts were used, either adenylate cyclase agonists or phosphodiesterase inhibitors like the extract of Ginkgo leaves. The compounds were topically applied on areas where fats accumulate to decrease fat deposits. A formulation for a cream was given (Bombardelli, 1991). For the same indication is the patent: "Antiobesity composition containing G. biloba extract". A topical antiobesity composition comprised G. biloba extract (no details were given) and an alpha-2 adrenergic blocker. It slowed down the accumulation of fats by lipolytic action and thus improved skin esthetics. The formulation was described (Soudant and Nadaud, 1992). Ginkgo extracts, together with some synergistic vitamins were used in preparations which improved the condition of the skin. A suitable formulation was given (Koide and Sasaki, 1992). Another patent on topical use is: "Cosmetics and topical pharmaceuticals containing superoxide dismutase and Ginkgo extracts". These compounds were effective in skin conditioning and controlling active oxygen and lipid peroxidation in the skin in low concentration (0.01% resp. 0.1%) (Koide and Kuribayashi, 1993). In a French patent the preparation of cosmetic and pharmaceutical compositions was described, containing polyphenols and Ginkgo extract. The formulation of a lotion was given (N'Guyen, 1994). Another use is described in the patent: Ginkgo biloba extract is used as a component of an anti-cellulite gel (Znaiden *et al.*, 1996). In a later French patent cosmetic compounds with depigmentation activity were described, which contained a mixture of hydroxylic acids or their derivatives, an acid and an active component from extracts of plants such as *G. biloba* (Dampeirou, 1997). Ginkgo extracts were also used in cosmetic polymeric sheets for skin conditioning (Kamiya and Yorozu, 1997). A patent on skin-care cosmetics, also from Japan, with divalent or trivalent iron salts, *G. biloba* extracts and (optionally) natural polysaccharides, gave the formulation of a cosmetic lotion (Fujita, 1997b). Ginkgo extracts – among others – can be used to synergistically enhance the antibacterial activity of N-long chain acyl basic amino acid derivatives (Terai, 1997a). The same inventor holds a second patent with almost the same content (Terai, 1997b). Also in a topical slimming composition, plant extracts e.g. *G. biloba*, were used (Bonte and Meybeck, 1997).

Patents on the Application of Parts, Extracts or Compounds from *Ginkgo biloba* for Various Uses

A highly diverse group of patents which do not fit in any of the six above mentioned categories. This large group has been subdivided in patents for medicinal and non-medicinal purposes.

Medicinal uses

Bilobalide containing pharmaceuticals were the subject of a patent. They could be useful for the treatment of nervous disorders such as neuropathy, encephalopathy, myelopathy, etc. The effectiveness has been demonstrated in rats. Bilobalide may be administered in oral, parenteral and topical dosage forms. *G. biloba* extracts containing bilobalide may also be used (Chatterjee *et al.*, 1985). One patent is on anti-inflammatory compositions comprising dolichols and/or their esters, which were extracted from Ginkgo leaves. An injection formula was given (Kageyu *et al.*, 1987). Another patent is on immunostimulant formulations containing dolichols and/or their esters. They were used for the treatment of patients with decreased immunity. The formulations were tested in mice (Kageyu *et al.*, 1987). The next patent is entitled: "Medications for improving blood flow and cellular metabolism". These medications were composed of at least one vasodilator or vasoprotector, and at least one general or cerebral stimulant. *G. biloba* leaf extract was combined with other plant extracts for daily administration (Berdal, 1986). A drug composition for treatment of neuropathies and circulatory disturbances comprised Tebonin (a flavonol glycosides containing standardised extract from *G. biloba*), folic acid and/or thioctic acid and/or vitamin B. It was tested in diabetes mellitus patients and proved to be effective (Koeltringer, 1989). In Japan, chocolate with Ginkgo extract was prepared. The extract contained a ginkgolide, bilobalide etc; or quercetin glycoside, kaempferol glycoside etc; or a mixture of these. The chocolate was effective in treating senile dementia and cerebral apoplexy (Matsumoto and Matsumoto, 1990a). Somewhat alike is the subject of the patent "Therapeutic chewing gum containing terpenes and flavone glycosides from Ginkgo leaves". This gum was effective in treating cerebral apoplexy. An extract, containing these compounds (ginkgolide, bilobalide, quercetin glycoside etc.) was obtained from Ginkgo leaves and incorporated into chewing gums (Matsumoto and Matsumoto, 1990b). Another patent is on a virucidal composition comprising flavopereirine and/or dihydroflavopereirine, naringin, naringenin and/or their esters or salts, RNA fragments and bioparyl. This virucide was said to be especially useful for AIDS treatment. Bioparyl is prepared from Ginkgo leaves (Beljanski, 1990). Another patent is "Use of ginkgolides to prevent reperfusion injury in organ transplantation". The ginkgolides may be present in the organ preservation solution or may be administered to the organ recipient and/or donor. Ginkgolide B was used in the test described (Ramwell, 1991). Dry G. biloba extract (as flavonol glycosides) was used in combination with poly-(O-hydroxyalkyl)-starch (C1-3 alkyl) for the treatment of acute inner ear ear damages, hypoxemia, anoxemia, anaerobic infections, osteomyelitis, and tumors. Hyperbaric oxygen therapy was associated in humans with infusion of this drug combination (Schumann, 1992a). Another patent of Schumann covered the same subject. One more therapeutical use, peripheral circulation disorders, was mentioned (Schumann, 1992b). A topical use was described in the patent: "Topical formulations containing flavonoids of Ginkgo leaf extracts". The given formulation contained an extract of Ginkgo leaves which contained ≈15% flavonoids and glycosides. This extract enhanced the skin absorption of therapeutic agents and skin conditioners (Umeda, 1992). A patent on a health food formulation containing Ginkgo leaf extract and chlorella described the compounds of this formulation. It was provided for the treatment of circulation disorders (Takahashi, 1993). A totally different use was described in a patent on the treatment of metastatic cancer with ginkgolides. The action of cytostatic drugs against metastatic cancer was enhanced by prior administration of ginkgolides, especially ginkgolide B. They also ameliorated the undesirable side effects of cytostatic therapy. They may be administered orally or intravenously (Michaelis, 1993). Bilobalide can be used for the treatment of nervous tension and anxiety. Bilobalide or G. biloba extract were said to be anxiolytics and tranquilizers. Orally administered 5–20 mg bilobalide/kg was effective in a rat-test (Noeldner and Chatterjee, 1993). The use of bilobalide against infections was described in another patent. Bilobalide and pharmacological acceptable derivatives thereof were used in manufacturing a pharmaceutical composition for treating an infection (of several microorganisms) in an individual. The effectiveness was demonstrated (Bombardelli and Ghione, 1993). Another therapeutical application can be found in the patent of Haga et al. (1994): bilobalide and bilobalide-containing Ginkgo leaf extract were used as a regulator for the central nervous system or for anesthesia. The purification of bilobalide was described. An extract of Ginkgo leaves put into a formulation using liposomes (with a neutral phospholipid) was effective in the treatment of brain disorders (Tokuyama et al., 1994). Pharmaceutical compositions for the treatment of, for instance, dementia or improvement of memory, were patented. These formulations (capsules) contained docosahexaenoic acid and G. biloba leaf extracts. The preparations could also be incorporated into food. Clinical tests indicated that they markedly improved the memory and learning behaviour (Matsui, 1994). Haga patented an anticonvulsant containing bilobalide.

Bilobalide was extracted from Ginkgo leaves and an isolation scheme was given. The composition, given orally, was tested in mice (Haga *et al.*, 1995). Bombardelli used G. *biloba* dimeric flavones in a gel with esculoside as main component, to be used in treatment of patients affected with venous insufficiency (Bombardelli *et al.*, 1996b). G. *biloba* extract was used in an oil-in-water formulation for topical application, useful for treatment of bacterial and viral infections of the skin and mucosa, allergic and atopic eczema, and dermatitis and acne vulgaris (Schmitt *et al.*, 1996). In Japan an invention is filed, which claimed that Ginkgo leaves and extract, rich in flavonoids and ginkgolides, are useful for the preparation of health tea beverages, and treatment of blood circulation diseases such as arteriosclerosis (Ootsuka, 1996). In the same year a composition was patented for treating skin conditions comprising an inhibitor of skin isophosphodiesterase. Extracts of G. *biloba* were used as specific inhibitor (Cehovic *et al.*, 1996). An American patent describes a nutritional formulation of vitamines, minerals and herbs – 70 different nutrients – including G. *biloba* which is said to increase brain alpha rhythms which are associated with mental alertness (Greenberg, 1996). Ginkgo leaf extract is also used in a vasoactive drug combination containing dihydropyridines (Lesur *et al.*, 1996). In Japan a patent described health food, containing plant extracts such as Ginkgo extract, in the form of granules and tablets (Matsushige *et al.*, 1997c). The same group has another invention on the same subject, but here they claim that health food, containing Ginkgo extract and also calcium, improves sleep (Matsushige *et al.*, 1997b). Finally they patented the same combination but then dried and packed in tea bags, also for improvement of sleeping. (Matsushige *et al.*, 1997a). Another use is described in the invention "Weight loss composition for burning and reducing synthesis of fats". The composition comprises, besides other compounds, G. *biloba* leaves. The dry mixture can be prepared as a beverage (Hastings *et al.*, 1997). Ginkgolides (A and B) may be used to inhibit the membrane expression of a benzodiazepine receptor in a patient, and in treatments to combat excess glucocorticoid production, according to the invention of Drieu (Drieu, 1997). Interesting is the following Japanese invention: bilobalide, ginkgolide A and Ginkgo leaf extracts are glutathione-S-transferase activators, and may be added to food and pharmaceuticals for controlling tumor growth (Haga *et al.*, 1997). There also is an invention on pharmaceutical compositions, which comprise among other substances Ginkgo extracts, to be used for cancer prevention (Kawashima, 1997). Fujita uses Ginkgo leaf extract in health beverages, together with ferrous-ferric salts, which are said to improve blood circulation and activate immune systems, liver functions and cell proliferation (Fujita, 1997a). An Australian invention describes therapeutic compositions containing vitamins, amino acid chelates, Echinacea extract and G. *biloba* extract. No special use is mentioned (Raymont, 1997).

Non-medicinal uses

Preparation of paper out of plant leaves is the subject of one patent: Ginkgo leaves were bleached with sodium hyposulfite in vinegar, and after several steps, dried to make paper-like materials (Kawasumi and Kazue, 1972). A different application was the use of the extract (in water) of Ginkgo leaves in the fermentation of sake. The produced sake possessed a Ginkgo flavour (Konishi *et al.*, 1990). The use of flavonoid-containing plant extracts, for example the extract of Ginkgo leaves, in preparing beverages was the subject of another patent. These beverages were stabilized by adding xanthan gums and optionally sugar alcohols to avoid precipitation and turbidity (Ueda, 1992). Ginkgo extracts could enhance the activity of acetyl polyamino acids which were useful as insecticides against lice (Astruc *et al.*, 1992). A Ginkgo compound can also be used as an insect repellant itself. Matsumoto *et al.* (1993) prepared edulan out of Ginkgo leaves (steam distillation, the essential oil was column chromatographed to give edulan). Tablets comprising edulan were effective against harmful insects, especially mosquitoes, and safe to humans. In Japan Ginkgo leaf extracts can be used in animal feed: for production of low cholesterol eggs and meat. Results are given (Hirama, 1996b). In another invention of Hirama, Ginkgo leaf extract is said to be useful as feed additive for promotion of egg laying, egg and meat qualities, feed efficiency of fowl and domestic animals. It also improves the health of domestic pets (Hirama, 1996a). Chitosan and Ginkgo leaves are used for preparing health food, used for promoting cell growth and stimulation of the immunosystem (Nakahara *et al*, 1996). Ginkgo leaf extracts can be used in liquid detergents, especially suitable for dishwashing (Iihara *et al.*, 1997).

8. Patents which are Not Specific for *Ginkgo biloba*

In the computerised literature study several patents showed up, which seemed to be of no importance for the discussion of patents on *G. biloba*. Most of them are concerned with an apparatus or process in which, by chance, a Ginkgo extract was used. They are given below in chronological order.

Ismail, 1986; Levy *et al.*, 1987; Manning *et al.*, 1987; Baurain and Trouet, 1988; O'Neill, 1988; Korth, 1988; Matsumoto and Sei, 1988; Laboratoires Beaufour, 1988; Bilas, 1989; Masquelier, 1990; Matsumoto and Hamamoto, 1990; O'Neill, 1990; Andersen and Pedersen, 1991; Thorel and Geneste, 1991; Park and Boo, 1991; Allard and Huc, 1992; Sato *et al.*, 1992; Fabre *et al.*, 1992; Vora *et al.*, 1992; Heuer, 1992; Perrier and Huc, 1992; Honerlagen and Bolardt, 1992; Nakanishi *et al.*, 1992; Kawashima *et al.*, 1993; Kawashima *et al.*, 1993; Kotani *et al.*, 1993; Koide *et al.*, 1993; Korth, 1993; Kutter, 1993; Stephan, 1994; Tokuyama *et al.*, 1994; Fukazawa, 1994; Junino and Martin, 1994; Martin *et al.*, 1994; Koikawa *et al.*, 1994; Kiesgen de Richter and Maurel, 1994; Sawano, 1994; Mauz, 1994; Nogami, 1994; Hoejgaard, 1995; Duranton and De Lacharriere, 1995; Soudant and Nadaud, 1995; Mauz, 1995; Leilani, 1995; Levy and Andry, 1995; Weidner, 1995; Tezuka and Tezuka, 1996; Nishizawa and Kanazawa, 1996; Rupp and Hunt, 1996; Yamane and Yamane, 1996; Majeed and Badmaev, 1996; Cohen and Suss, 1996; Franchi *et al.*, 1996; Kreuter and Steiner, 1996; Majeed *et al.*, 1996; Heusele and Le Blay, 1996; Breton *et al.*, 1996; Minami *et al.*, 1996; Weidner, 1997; Tezuka and Tezuka, 1997; Duranton, 1997; Khwaja and Friedman, 1997; Testa, 1997; Haruta *et al.*, 1997; Inoue *et al.*, 1997; Keller *et al.*, 1997.

DISCUSSION

Overlooking all the patents summarized in sections 1–7, it is clear that the majority dates from the last 15 years. In Table 2 the number of patents filed within a certain period are given. The oldest patents are mainly on extraction methods and the preparation of vaso-active extracts from Ginkgo leaves. These patents include two old ones by Schwabe which catalysed much of the later Ginkgo research. When one looks at the patents from a geographical point of view, by far most patents have been registered in Japan. A breakdown by country is given in Table 3. Many Japanese patents have been filed by one company (Kuraray) that has carried out a lot of research in the field of polyprenols in the first half of the 80's. Since then interest in polyprenols has declined. These patents have been summarized in sections 2 and 3. In section 4 some patents on phopholipids are mentioned. They can be used to solubilize poorly soluble Ginkgo extracts or constituents. Several patents in this category originate from the Italian Indena company. Section 5 is entirely devoted to the many patents on ginkgolides in relation to their strong anti-PAF activity. In spite of much work in the period 1985–1995, the final results appear disappointing and no useful drugs have come out. Ginkgo extracts are widely used in products for personal health care, especially in Japan. Most of them are for hair preparations but some others are useful for improvement of the skin condition or as local anti-obesity formulas. The medicinal uses for which Ginkgo derived products

Table 2 Number of patents filed within a certain period

Period	Number of patents
1959–1970	2
1970–1975	3
1976–1980	1
1981–1985	15
1986–1990	26
1991–1995	49
1996–1997	34

Table 3 Total number of patents registered per country or continent

Country	Number	Country	Number
Japan	57	China	3
Europe	21	Switzerland	2
World	15	Australia	1
Germany	10	India	1
USA	8	Israel	1
France	5	The Netherlands	1
Belgium	4	Spain	1

have been patented are many and varied: improvement of the blood circulation, treatment of neuropathies, improvement of sleep and enhancement of the effectiveness of cytostatic drugs. Quite recently some patents have appeared in which Ginkgo leaf extracts are added to animal feed to increase for instance the quality of meat or the number of eggs. This is certainly a growing market. These patents were discussed in section 7. As this plant still holds some secrets for us, in the near future no doubt many more patents will be filed for Ginkgo, especially for purposes discussed in sections 6 and 7.

REFERENCES

Allard, R., Huc, A. (Bioetica S.A., Fr.). Microcapsules having a combined atelocollagen/ polyholoside wall coagulated by a divalent cation. Patent nr. WO 9202254, 20 feb. 1992. Cited from *Chem. Abstr.* **116**, 201130 (1992).

Andersen, C., Pedersen, M. (Fertin Laboratories A/S., Den.). Chewing gum composition with accelerated, controlled release of active agents. Patent nr. WO 9101132, 7 feb. 1991. Cited from *Chem. Abstr.* **115**, 57189 (1991).

Astruc, J., Morelle, J., Lauzanne-Morelle, E. Acetylpolyamino acids as insecticides against lice on humans. Patent nr. FR 2675015, 16 oct. 1992. Cited from *Chem. Abstr.*, **118**, 141867 (1993).

Ayroles, G., Rossard, R.M., Cadiou, M. (Fabre, P, Industrie, Fr.). Extract from *Ginkgo biloba* leaves as circulatory drugs. Patent nr. EP 330567, 30 aug. 1989. Cited from *Chem. Abstr.*, **113**, 29238 (1990).

Bauer, J. (Oxo Chemie G.m.b.H., Fed. Rep. Ger.). Anti-inflammatory pharmaceuticals containing *Ginkgo biloba* extracts or ginkgolides and nonsteroidal inflammation inhibitors. Patent nr. WO 8806889, 22 sep. 1988. Cited from *Chem. Abstr.*, **111**, 140485 (1989).

Baurain, R., Trouet, A. (IRE-Celltarg S.A., Belg.). Pharmaceutical microcrystals containing an active substance with a phospholipid affinity and at least a phospholipid. Patent nr. EP 270460, 8 june 1988. Cited from *Chem. Abstr.*, **109**, 116073 (1988).

Baurain, R., Trouet, A.B.L. (IRE-Celltarg S.A., Belg.) Drug microcrystals comprising an active substance having an affinity for phospholipids, and at least one phospholipid, for suspension formulations. Patent nr. US 4973465, 27 nov. 1990. Cited from *Chem. Abstr.*, **114**, 171334 (1991).

Beljanski, M. (Fondation pour l'Encouragement a la Recherche Medicale, Liechtenstein). Virucidal composition comprising flavopereirine. Patent nr. EP 373986, 20 june 1990. Cited from *Chem. Abstr.*, **113**, 237846 (1990).

Benardelli, G. (Arysearch Arylan A.-G., Liechtenstein). Quercetin from *Ginkgo biloba*. Patent nr. CH 674365, 31 may 1990). Cited from *Chem. Abstr.*, **113**, 112694 (1990).

Berdal, P. Medications for improving blood flow and cellular metabolism. Patent nr. FR 2583640, 26 dec. 1986. Cited from *Chem. Abstr.*, **107**, 205183 (1987).

Bilas, A. Mixture of substances for stabilizing metabolism and circulation, especially in diabetics. Patent nr. WO 8903213, 20 april 1989. Cited from *Chem. Abstr.*, **112**, 164963 (1990).

Bombardelli, E. (Indena S.p.A., Italy). Topical administration of plant extracts having therapeutic activities for treatment of fat deposits. Patent nr. EP 427026, 15 may 1991. Cited from *Chem. Abstr.*, **115**, 189759 (1991).

Bombardelli, E., Cristoni, A., Morazzoni, P. (Indena S.p.A., Italy). Pharmaceutical and cosmetic formulations containing esculoside. Patent nr. EP 692250, 17 jan. 1996b. Cited from *Chem. Abstr.*, **124**, 185594 (1996).

Bombardelli, E., Ghione, M. (Indena S.p.A., Italy). Use of bilobalide and derivatives thereof for treating an infection in an individual, and pharmaceutical compositions adapted for such use. Patent nr. EP 556051, 18 aug. 1993. Cited from *Chem. Abstr.*, **119**, 152072 (1993).

Bombardelli, E., Mustich, G. (Indena S.p.A., Italy). Preparation of bilobalide complexes with phospholipids and formulation containing them. Patent nr. EP 441279, 14 aug. 1991. Cited from *Chem. Abstr.*, **115**, 287201 (1991).

Bombardelli, E., Mustich, G., Bertani, M. (Indena S.p.A., Italy). New extracts of *Ginkgo biloba* and their methods of preparation. Patent nr. EP 0360556, 28 march 1990.

Bombardelli, E., Patri, G.F. (Indena S.p.A., Italy). Cosmetics and pharmaceuticals containing flavonoid-phospholipid complexes. Patent nr. EP 275005, 20 july 1988. Cited from *Chem. Abstr.*, **111**, 45279 (1989).

Bombardelli, E., Ponzone, C. (Indena S.P.A., Italy). Hydroxyproline-rich proteins and pharmaceutical and cosmetic formulations containing them. Patent nr. WO 9620284, 4 juli 1996a. Cited from *Chem. Abstr.*, **125**, 140680 (1996).

Bons, L., Van der Beek, K. Hair and scalp conditioning composition. Patent nr. NL 7500493, 22 july 1976. Cited from *Chem. Abstr.*, **86**, 127288 (1977).

Bonte, F., Meybeck, A. (Lvmh Recherche, Fr.). Topical slimming composition containing plant extracts. Patent nr. WO 9742928, 20 nov. 1997. Cited from *Chem. Abstr.*, **128**, 39400 (1998).

Braquet, P., Esanu, A. (Societé de Conseils de Recherches et d'Applications Scientifiques, Fr.). Preparation and testing of 1- and 10-alkoxyginkgolides as platelet-activating factor antagonists. Patent nr. DE 3837550, 18 may 1989. Cited from *Chem. Abstr.*, **112**, 76805 (1990).

Braquet, P.G. (Societé de Conseils de Recherches et d'Applications Scientifiques, Fr.). Treatment of diseases caused by the platelet-activating factor acether. Patent nr. BE 902874, 4 nov. 1985a. Cited from *Chem. Abstr.*, **104**, 62080 (1986).

Braquet, P.G. (Societé de Conseils de Recherches et d'Applications Scientifiques, Fr.). Treatment or prevention of disorders provoked by platelet-activating factor acether. Patent nr. BE 901915, 1 july 1985b. Cited from *Chem. Abstr.*, **103**, 189808 (1985).

Brasey, P.N. (Seuref A.-G., Liechtenstein). Pharmaceutical compositions with vasodilation and antihypoxic activities. Patent nr. DE 3625877, 5 feb. 1987. Cited from *Chem. Abstr.*, **106**, 162596 (1987).

Breton, L, Martin, R., Hocquaux, M. (Oreal S.A., Fr.). Extraction of active substances from undifferentiated plant cells. Patent nr. FR 2731162, 6 sept. 1996. Cited from *Chem. Abstr.*, **126**, 58946 (1997).

Cehovic, G., Andre, P. (Parfums Christian Dior, Fr.) Composition for treating skin conditions comprising an inhibitor of skin iso-phosphodiesterase. Patent nr. WO 9627384, 12 sept. 1996. Cited from *Chem. Abstr.*, **125**, 308676 (1996).

Chatterjee, S.S., Gabard, B.L., Jaggy, H.E.W. (Schwabe, Dr. Willmar, G.m.b.H., Fed.Rep. Ger.). Bilobalide-containing pharmaceutical. Patent nr. DE 3338995, 9 may 1985. Cited from *Chem. Abstr.*, **103**, 76258 (1985).

Cohen, K., Suss, H. (Maybelline Intermediate Co., USA). Cosmetic makeup emulsions for improving skin conditions. Patent nr. WO 9623484, 8 aug. 1996. Cited from *Chem. Abstr.*, **125**, 204127 (1996).

Daicel Chemical Industries, Ltd., Japan. Pest repellents made of Ginkgo leaf extracts containing 6-alkenylsalicylic acid, bilobalides, and ginkgolides. Patent nr. JP 63030402, 9 feb. 1988. Cited from *Chem. Abstr.*, **112**, 72332 (1990).

Dampeirou, C. (C3d Sarl, Fr.). Cosmetics containing hydroxycarboxylic acids and plant extracts. Patent nr. FR 2736263, 10 jan. 1997. Cited from *Chem. Abstr.*, **126**, 255278 (1997).

Defaye, J., Laine, V., Djedaini-Pilard, F., Perly, B. (Commisariat a l'Energie Atomique, Fr.). Use of cyclodextrins and their derivs. for solubilization of ginkgolides. Patent nr. EP 764659, 26 march 1997. Cited from *Chem. Abstr.*, **126**, 308801 (1997).

Drieu, K. (Societe de Conseils de Recherches et d'Applications Scientifiques S.A.(S.C.R.A.S.), Fr.). Ginkgolides for inhibition of membrane expression of benzodiazepinereceptors and to combat excess glucocorticoid production. Patent nr. WO 9717068, 15 may 1997. Cited from *Chem. Abstr.*, **127**, 29095 (1997).

Duranton, A. (L'Oreal, Fr.). Lipoxygenase and cyclooxygenase inhibitors for hair growth changes preparations. Patent nr. EP 800815, 15 oct. 1997. Cited from *Chem. Abstr.*, **127**, 336462 (1997).

Duranton, A., De Lacharriere, O. (Oreal S.A., Fr.). Hair growth stimulants comprising lipoxygenase or cycloxygenase stimulants or inhibitors. Patent nr. EP 648488, 19 april 1995. Cited from *Chem. Abstr.*, **122**, 299059 (1995).

Fabre, P., Cousse, H., Mouzin, G., Trebosc, M.T. (Pierre Fabre Cosmetique, Fr.). DNA gel-stabilized thermal water liposomes. Patent nr. WO 9206666, 30 april 1992. Cited from *Chem. Abstr.*, **117**, 137664 (1992).

Fondation Pour L'encouragement A La Recherche Medicale, Liechtenstein. Bioparyl, a biological regulator active against various pathologies. Patent nr. Il 91428, 14 nov. 1996. Cited from *Chem. Abstr.*, **127**, 126640 (1997).

Franchi, J., Pellicier, F., Krzych, V., Redziniak, G. (Parfums Christian Dior, Fr.). Use of galactosylglycerides and of naturally occurring extracts containing such products in cosmetics, pharmaceutics and particularly dermatology. Patent nr. WO 9625142, 22 aug. 1996. Cited from *Chem. Abstr.*, **125**, 230182 (1996).

Fujita, Y. (Nippon Green Wave K.K., Japan; Aiwa Co., Ltd). Health beverages containing Ginkgo leaf extracts and ferrous-ferric salt solutions. Patent nr. JP 09234044, 9 sept. 1997a. Cited from *Chem. Abstr.*, **127**, 247364 (1997).

Fujita, Y. (Nippon Greener K.K., Japan, Aiwa Co., Ltd.). Skin-care cosmetics. Patent nr. JP 09241145, 16 sept. 1997b. Cited from *Chem. Abstr.*, **127**, 238916 (1997).

Fukazawa, C. (National Food Research Institute, Ministry of Agriculture, Forestry and Fisheries, Japan). Asparagine endoprotease. Patent nr. DE 4324276, 27 jan. 1994. Cited from *Chem. Abstr.*, **120**, 264630 (1994).

Geurrant, R.L., Fang, G., Fonteles, M.C. (The University of Virginia Patent Foundation, USA). Method using PAF antagonist or indomethacin for treatment of *Clostridium difficile* colitis and inhibition of the secretory effects caused by *C. difficile* Toxin A by Cholera toxin. Patent nr. US 5436239, 25 july 1995. Cited from *Chem. Abstr.*, **123**, 132880 (1995).

Greenberg, M. Nutritional formulation of vitamins, minerals and herbs. Patent nr. US 5569458, 29 oct. 1996. Cited from *Chem. Abstr.*, **125**, 309065 (1996).

Guerrant, R.L., Guodong, F., Fonteles, M. (University of Virginia Patents Foundation, USA). Method of treating *Clostridium difficile* colitis and cholera. Patent nr. WO 9319745, 14 oct. 1993. Cited from *Chem. Abstr.*, **119**, 262521 (1993).

Haga, M., Wada, K., Sasaki, K. (Nippon Green Wave K.K., Japan). Glutathione-S-transferase activators as antitumor agents in food and pharmaceuticals. Patent nr. JP 09110713, 28 april 1997. Cited from *Chem. Abstr.*, **127**, 33234 (1997).

Haga, M., Wada, K., Sasaki, K., Fujita, M., Matsumoto, T., Toi, K. (Daicel Chem, Japan; Iatron Lab). Anticonvulsants containing bilobalide. Patent nr. JP 07053371, 28 feb. 1995. Cited from *Chem. Abstr.*, **122**, 299068 (1995).

Haga, M., Wada, K., Sasaki, K., Matsumoto, T., Toi, K. (Daicel Chem, Japan; Iatron Lab). Bilobalide and bilobalide-containing Ginkgo leaf extract as central nervous system regulators. Patent nr. JP 05271083, 19 oct. 1993. Cited from *Chem. Abstr.*, **120**, 69616 (1994).

Haruta, M., Takahashi, T., Saito, K. (Toray Industries, Inc., Japan). Clothing containing plant extracts for promoting subcutaneous fatdegradation and blood circulation to tighten and smoothen the skin. Patent nr. JP 09296367, 18 nov. 1997. Cited from *Chem. Abstr.*, **128**, 26757 (1998).

Hastings, C.W., Barnes, D.J. (Reliv International, Inc., USA). Weight loss composition for burning and reducing synthesis of fats. Patent nr. US 5626849, 6 may 1997. Cited from *Chem. Abstr.*, **127**, 4534 (1997).

Heuer, H. (Boehringer Ingelheim KG, Germany). Synergistic combinations of PAF antagonists and anticholinergic agents as drugs for treatment of bronchial asthma. Patent nr. DE 4219659, 23 dec. 1993. Cited from *Chem. Abstr.*, **120**, 153730 (1994).

Heusele, C., Le Blay, J. Novel cosmetic or dermatological compositions. Patent nr. WO 9628008, 19 sept. 1996. Cited from *Chem. Abstr.*, **125**, 338732 (1996).

Hirama, M. Feed containing extracts of Ginkgo leaves for production of low cholesterol eggs and meat. Patent nr. JP 08000185, 9 jan. 1996b. Cited from *Chem. Abstr.*, **124**, 230689 (1996).

Hirama, M. Ginkgo leaf extract as feed additive. Patent nr. JP 0800184, 9 jan.1996a. Cited from *Chem. Abstr.*, **124**, 259257 (1996).

Hoeigaard, B. (Ferrosan A/S, Den). Compositions containing swelling pharmaceuticals. Patent nr. WO 9507689, 23 mar. 1995. Cited from *Chem. Abstr.*, **122**, 274102 (1995).

Honerlagen, H.J., Bolardt, F.S.M. (Emil Flachsman A.-G., Switz.). Encapsulation of pharmaceutical plant extracts. Patent nr. EP 496705, 29 july 1992. Cited from *Chem. Abstr.*, **117**, 258219 (1992).

Huang, S. Hair tonic preparations containing herb extracts, sodium camphorsulfonate, and other ingredients. Patent nr. CN 1089133, 13 july 1994. Cited from *Chem. Abstr.*, **122**, 114593 (1995).

Iihara, T., Amano, H., Nishida, M. (Lion Corp., Japan). Liquid detergents for hard material surfaces. Patent nr. JP 09241682, 16 sept. 1997. Cited from *Chem. Abstr.*, **127**, 264612 (1997).

Inoue, S., Ota, Y., Shimizu, T. (Kanebo, Ltd., Japan). Cosmetics and bath preparations containing epidermal horny substance formation promoters. Patent nr. JP 09301881, 25 nov. 1997. Cited from *Chem. Abstr.*, **128**, 39399 (1998).

Ismail, R. Antirheumatic agents containing vitamin E and vasodilators and their use. Patent nr. US 4612194, 16 sep. 1986. Cited from *Chem. Abstr.*, **106**, 72922 (1987).

Junino, A., Martin, R. (Oreal S.A., Fr.). Enzymic manufacture of an indole pigment for cosmetics. Patent nr. FR 2694021, 28 jan. 1994. Cited from *Chem. Abstr.*, **120**, 268376 (1994).

Kageyu, A., Nakagawa, S., Takigawa, T., Shimamura, M., Okada, M., Mizuno, M. (Kuraray Co., Ltd., Japan). Anti-inflammatory compositions containing dolichol. Patent nr. JP 62033118, 13 feb. 1987a. Cited from *Chem. Abstr.*, **107**, 28395 (1987).

Kageyu, A., Nakagawa, S., Takigawa, T., Shimamura, M., Okada, M., Mizuno, M. (Kuraray Co., Ltd., Japan). Immunostimulant formulations containing dolichols. Patent nr. JP 62039521, 20 feb. 1987b. Cited from *Chem. Abstr.*, **106**, 219619 (1987).

Kamiya, T., Yorozu, H. (Kao Corp., Japan). Cosmetic polymeric sheets for skin conditioning. Patent nr. JP 09216808, 19 aug. 1997. Cited from *Chem. Abstr.*, 127, 225105 (1997).

Kawashima, H., Akimoto, K., Yamada, H., Shimizu, S. (Suntory, Ltd., Japan). Manufacture of dihomo-gamma-linolenic acid with microorganisms deficient in delta–5 desaturase activity. Patent nr. EP 535940, 7 april 1993a. Cited from *Chem. Abstr.*, 118, 253408 (1993).

Kawashima, H., Akimoto, K., Yamada, H., Shimizu, S. (Suntory, Ltd., Japan). Process for production of 8,11-eicosadienoic acid. Patent nr. EP 535941, 7 april 1993b. Cited from *Chem. Abstr.*, 118, 253409 (1993).

Kawashima, Z. (Kawashima, Takaaki, Japan; Kato, Shigeru). Pharmaceutical compositions for cancer prevention. Patent nr. JP 09077674, 25 march 1997. Cited from *Chem. Abstr.*, 127, 39870 (1997).

Kawasumi, K., Abe, K. Plant leaves with paper like quality by bleaching. Patent nr. JP 47038001, 26 sep. 1972. Cited from *Chem. Abstr.*, 80, 28704 (1974).

Keller, B.C., Fisher, D.L., Kiss, S. (Biozone Laboratories , Inc., USA). Delivery of biologically active material in a liposomal formulation for administration into the mouth. Patent nr. WO 9742938, 20 nov. 1997. Cited from *Chem. Abstr.*, 128, 39579 (1998).

Khwaja, T.A., Friedman, E.P. (Pharmaprint, Inc.; University of Southern California). Pharmaceutical grade botanical drugs. Patent nr. WO 9739355, 23 oct. 1997. Cited from *Chem. Abstr.*, 127, 351175 (1997).

Kiesgen de Richter, R., Maurel, J.C. (I.R.2.M.Societe en Nom Collectif, Fr.). Tannin-metal(III)ion complexes, their preparation, and their pharmaceutical use. Patent nr. FR 2695390, 11 march 1994. Cited from *Chem. Abstr.*, 121, 50109 (1994).

Kim, Y.C., Jeon, M.H., Sung, S.H. (Han-Dok Remedia Ind. Co., Ltd., S. Korea). Production of ginkgolides in cell culture. Patent nr. WO 9302204, 4 feb. 1993. Cited from *Chem. Abstr.*, 118, 123122 (1993).

Kimura, H. Extraction of sesamin from *Ginkgo biloba* wood. Patent nr. JP 8129, 11 sept. 1959. Cited from *Chem. Abstr.*, 54, 3868d (1960).

Kimura, S. (Kuraray Co., Ltd., Japan). Solubilization of dolichol with lipids. Patent nr. JP 61194024, 28 aug. 1986. Cited from *Chem. Abstr.*, 106, 55902 (1987).

Kloss, P., Jaggy, H. (Schwabe, Willmar). Vasoactive drugs from *Ginkgo biloba* leaves. Patent nr. DE 2117429, 12 oct. 1972. Cited from *Chem. Abstr.*, 78, 47787 (1973).

Koeltringer, P. Drug composition comprising Ginkgo flavone glycosides for the treatment of neuropathies. Patent nr. EP 326034, 2 aug. 1989. Cited from *Chem. Abstr.*, 112, 240492 (1990).

Koide, C., Kuribayashi, S. (Kosei Kk, Japan). Cosmetics and topical pharmaceuticals containing superoxide dismutase and Ginkgo extracts. Patent nr. JP 05032556, 9 feb. 1993. Cited from *Chem. Abstr.*, 118, 219484 (1993).

Koide, C., Sasaki, I. (Kosei Kk, Japan). Cosmetic skin preparations containing Ginkgo extracts and vitamins. Patent nr. JP 04321616, 11 nov. 1992. Cited from *Chem. Abstr.*, 118, 109431 (1993).

Koide, C., Sasaki, I., Egawa, J., Asano, Y. (Kosei Kk, Japan). Cosmetic skin preparations containing active oxygen scavengers and antioxidants. Patent nr. JP 05070333, 23 march 1993. Cited from *Chem. Abstr.*, 119, 15136 (1993).

Koikawa, Y., Suetsugu, K., Tanaka, H., Shiba, A. (Narisu Cosmetic Co Ltd, Japan). Antiaging cosmetics containing plant extracts. Patent nr. JP 06024937. Cited from *Chem. Abstr.*, 120, 279863 (1994).

Konishi, H., Kosugi, N., Niwa, A., Kobayashi, Y. (Nonogawa Shoji Y.K., Japan). Sake fermentation from Ginkgo leaf extract. Patent nr. JP 02005848, 10 jan. 1990. Cited from *Chem. Abstr.*, 113, 4674 (1990).

Korth, R. Drugs that inhibit blood platelet-activating factor acether binding sites on cells. Patent nr. DE 3710921, 14 july 1988. Cited from *Chem. Abstr.*, 110, 128666 (1989).

Korth, R. PAF antagonists and their compound preparations for prevention and treatment of acidophile-mediated diseases and methods for determination of their efficacy. Patent nr. JP 06263640, 20 sep. 1994. Cited from *Chem. Abstr.*, 122, 89393 (1995).

Korth, R. Treatment of 1-O-alkyl-sn-glyceryl-3-phosphocholine (lyso PAF)-mediated diseases with PAF antagonists and procedure for determining their efficacy. Patent nr. EP 540767, 12 may 1993a. Cited from *Chem. Abstr.*, 119, 20503 (1993).

Korth, R. Treatment of diseases with platelet-activating factor-acether antagonists and procedure for determining their efficacy. Patent nr. EP 459432, 4 dec. 1991. Cited from *Chem. Abstr.*, 116, 76391 (1992).

Korth, R. Treatment of endothelial defect-associated illnesses with platelet-activating-factor antagonists and method therefore. Patent nr. EP 312913, 26 april 1989. Cited from *Chem. Abstr.*, 112, 151859 (1990).

Korth, R. Treatment of eosinophil-mediated diseases with PAF antagonists and procedure for determining their efficacy. Patent nr. EP 540766, 12 may 1993b. Cited from *Chem. Abstr.*, 119, 20502 (1993).

Korth, R. Treatment of skin diseases using ginkgolide PAF antagonists. Patent nr. US 5605927, 25 feb. 1997. Cited from *Chem. Abstr.*, 126, 207538 (1997).

Kotani, Y., Iwamoto, S., Isoda, Y. (Nippon Oils & Fats Co Ltd, Japan). Antiallergy oily compositions containing unsaturated fatty acids. Patent nr. JP 05058902, 9 march 1993. Cited from *Chem. Abstr.*, 118, 253929 (1993).

Kreuter, M-H., Steiner, R. (Emil Flachsmann Ag, Switz.). Process for removing unwanted lipophilic impurity or residue from drinks or vegetable preparations. Patent nr. EP 730830, 11 sept. 1996. Cited from *Chem. Abstr.*, 125, 246102 (1996).

Krishnamurthy, S.R. Treatment of cashew or *Semecarpus anacardium* nut shell liquid, or *Ginkgo biloba* fruit liquid for the manufacture of electrical insulators. Patent nr. Indian 63635, 9 july 1960. Cited from *Chem. Abstr.*, 55, 2956h (1961).

Kuraray Co., Ltd., Japan. Nitrile derivatives of polyprenols. Patent nr. JP 58174352, 13 oct. 1983a. Cited from *Chem. Abstr.*, 100, 103674 (1984).

Kuraray Co., Ltd., Japan. Polyprenols and their derivatives. Patent nr. JP 57170930, 21 oct. 1982a. Cited from *Chem. Abstr.*, 106, 67535 (1987).

Kuraray Co., Ltd., Japan. Polyprenols. Patent nr. JP 58192842, 10 nov. 1983b. Cited from *Chem. Abstr.*, 100, 192115 (1984).

Kuraray Co., Ltd., Japan. Polyprenols. Patent nr. JP 58201736, 24 nov. 1983c. Cited from *Chem. Abstr.*, 100, 189045 (1984).

Kuraray Co., Ltd., Japan. Polyprenyl compounds from *Ginkgo biloba* and *Cedrus deodara*. Patent nr. JP 58083643, 19 may 1983d. Cited from *Chem. Abstr.*, 99, 155550 (1983).

Kuraray Co., Ltd., Japan. Polyprenyl compounds from Ginkgo for pharmaceutical and cosmetic bases. Patent nr. JP 57091932, 8 june 1982b. Cited from *Chem. Abstr.*, 97, 222939 (1982).

Kuraray Co., Ltd., Japan. Polyprenyl compounds. Patent nr. JP 58208242, 3 dec. 1983e. Cited from *Chem. Abstr.*, 100, 192117 (1984).

Kuraray Co., Ltd., Japan. Polyprenyl compounds. Patent nr. JP 58210034, 7 dec. 1983f. Cited from *Chem. Abstr.*, 101, 60122 (1984).

Kuraray Co., Ltd., Japan. Separation and purification of polyprenyl acetates. Patent nr. JP 58201747, 24 nov. 1983g. Cited from *Chem. Abstr.*, 100, 206697 (1984).

Kutter, E. (Boehringer Ingelheim KG, Germany). Treatment of dysmenorrhea with PAF antagonists. Patent nr. DE 4200610, 15 july 1993. Cited from *Chem. Abstr.*, 119, 210708 (1993).

Kwak, W.J., Park, H.K., Oh, K.B. (Sunkyong Industries, Ltd., S. Korea). A method of isolating therapeutic ginkgolides from the leaves of the Ginkgo tree. Patent nr. EP 402925, 19 dec. 1990. Cited from *Chem. Abstr.*, **114**, 150165 (1991).

Laboratoires Beaufour, Fr. Manufacture of tablets that readily dissolve or disintegrate in water, and the tablets obtained by this method. Patent nr. NL 8702388, 2 may 1988. Cited from *Chem. Abstr.*, **111**, 6219 (1989).

Leilani, L. Beverages containing added guaranine. Patent nr. GB 2285578, 19 july 1995. Cited from *Chem. Abstr.*, **123**, 197245 (1995).

Lesur, E., Neuser, D., Lockhoff, O., Perzborn, E., Stasch, J.P., Kurka, P. (Bayer A.-G., Germany). Vasoactive drug combinations containing dihydropyridines. Patent nr. DE 19515971, 7 nov. 1996. Cited from *Chem. Abstr.*, **126**, 50958 (1997).

Levy, J.C., Bessin, P., Labaune, J.P. (Rolland, Albert, S.A., Fr.). Medications containing medifoxamine for treatment of cerebral hypoxia and senility. Patent nr. FR 2589357, 7 may 1987. Cited from *Chem. Abstr.*, **107**, 242644 (1987).

Levy, M-C., Andry, M-C. (Centre National de la Recherche Scientifique (CNRS), Fr.). Microcapsules with walls made of cross-linked plant polyphenols, for foods, pharmaceuticals or cosmetics. Patent nr. WO 9521018, 10 aug. 1995. Cited from *Chem. Abstr.*, **123**, 312647 (1995).

Li, Y., Cheng, S. Extraction of therapeutic ketones from *Ginkgo biloba* leaves. Patent nr. CN 1066595, 2 dec. 1992. Cited from *Chem. Abstr.*, **119**, 56137 (1993).

Lion Corp., Japan. 2-Hydroxy–6-alkylbenzoic acids for the control of dental caries. Patent nr. JP 57038709, 3 march 1982. Cited from *Chem. Abstr.*, **96**, 223025 (1982).

Ma, D. Pharmaceutical compositions for transdermal application for preventing and treating coronary heart disease. Patent nr. CN 1030183, 11 jan. 1989. Cited from *Chem. Abstr.*, **113**, 65288 (1990).

Majeed, M., Badmaev, V., Rajendran, R. (Sabinsa Corp., USA). Use of piperine to increase the bioavailability of nutritional compounds. Patent nr. US 5536506, 16 july 1996. Cited from *Chem. Abstr.*, **125**, 123750 (1996).

Majeed, M., Badmaev, V., Rajendran, R. (Sabinsa Corporation, USA). Use of piperine as a bioavailability enhancer. Patent nr. WO 9625939, 29 aug. 1996. Cited from *Chem. Abstr.*, **125**, 246450 (1996).

Makino, T., Nakatate, M., Matsumoto, T. (Tama Biochemical Co., Ltd., Japan: Daicel Chemical Industries, Ltd.) Preparation of Ginkgo extract. Patent nr JP 03275629, 6 dec. 1991. Cited from *Chem. Abstr.*, **116**, 91368 (1992).

Manabe, A., Yamashita, T., Harada, K., Tokumori, T., Sumida, Y. (Chlorine Engineers Corp., Ltd., Japan; Yakult Honsha Co., Ltd.). Process for the supercritical extraction and separation of solid samples such as Ginkgo leaves. Patent nr. EP 308675, 29 march 1989. Cited from *Chem. Abstr.*, **112**, 25614 (1990).

Manning, D.T., Cappy, J.J., See, R.M., Cooke, A.R., Fritz, C.D., Wheeler, T.N. (Rhone-Poulenc Nederland B.V., Neth.). Preparation and use of malonic acid derivatives, especially (phenylamino)oxopropanoates, as plant growth retardants. Patent nr. WO 8705898, 8 oct. 1987. Cited from *Chem. Abstr.*, **109**, 37517 (1988).

Martin, R., Jumino, A., Dubief, C., Rosenbaum, G., Audousset, M.P. (Oreal S.A., Fr.). Hair preparations containing vegetable laccases as oxidizing agents. Patent nr. FR 2694018, 28 jan. 1994. Cited from *Chem. Abstr.*, **120**, 279847 (1994).

Masquelier, J. (Society Civile pour l'Expansion de la Recherche en Phytochemie Applique, Fr.). Method of preparing purified polyphenolic flavan-3-ol extracts as drugs and food and cosmetic materials. Patent nr. EP 384796, 29 aug. 1990. Cited from *Chem. Abstr.*, **114**, 30112 (1991).

Matsui, F. (Katsuzo Shoji Kk, Japan). Pharmaceutical compositions containing docosa-hexaenoic acid and *Ginkgo biloba* leaf extracts for therapeutic use. Patent nr. JP 06340539, 13 dec. 1994. Cited from *Chem. Abstr.*, **122**, 142594 (1995).

Matsui, K., Shinkawa, Y., Tsuboi, M., Kojima, H., Ando, Y. (Ichimaru Pharcos Co., Ltd., Japan) Manufacture of extract with high content of flavonoids from Ginkgo leaves. Patent nr. JP 03227985, 8 oct. 1991. Cited from *Chem. Abstr.*, **116**, 28119 (1992).

Matsumoto, A., Matsumoto, T., Yamashita, M., Nitsuta, I., Kobayashi, M., Mesaki, J., Kashiwabara, T. (Daicel Chem, Japan; Earth Chemical Co). Harmful insect repellents containing edulan derivatives. Patent nr. JP 05221808, 31 aug. 1993. Cited from *Chem. Abstr.*, **119**, 264681 (1993).

Matsumoto, T. (Daicel Chemical Industries, Ltd., Japan). Extraction of therapeutic flavones from Ginkgo leaves. Patent nr. JP 02193907, 31 july 1990. Cited from *Chem. Abstr.*, **114**, 88633 (1991).

Matsumoto, T. (Daicel Chemical Industries, Ltd., Japan). Isolation of flavone glycosides from *Ginkgo biloba*. Patent nr. JP 62292794, 19 dec. 1987. Cited from *Chem. Abstr.*, **110**, 54654 (1989).

Matsumoto, T., Hamamoto, T. (Daicel Chemical Industries, Ltd., Japan). Recovery of flavonoid compounds from plant extracts. Patent nr. JP 02073079, 13 march 1990. Cited from *Chem. Abstr.*, **113**, 22225 (1990).

Matsumoto, T., Matsumoto, A. (Daicel Chemical Industries, Ltd., Japan). Manufacture of chocolate containing terpenes and flavone glycosides of Ginkgo leaves extract. Patent nr. JP 02031646, 1 feb. 1990a. Cited from *Chem. Abstr.*, **113**, 151138 (1990).

Matsumoto, T., Matsumoto, A. (Daicel Chemical Industries, Ltd., Japan). Therapeutic chewing gum containing terpenes and flavone glycosides from Ginkgo leaves. Patent nr. JP 02031648, 1 feb. 1990b. Cited from *Chem. Abstr.*, **113**, 103374 (1990).

Matsumoto, T., Sei, T. (Daicel Chemical Industries, Ltd., Japan). Isolation of a flavone glycoside from Ginkgo leaves for use as a cardiovascular agent. Patent nr. EP 237066, 16 sep. 1987. Cited from *Chem. Abstr.*, **108**, 19396 (1988).

Matsumoto, T., Sei, T. (Daicel Chemical Industries, Ltd., Japan). Stress-induced ulcer inhibitors containing 6-alkenylsalicylic acids. Patent nr. JP 63215629, 8 sep. 1988. Cited from *Chem. Abstr.*, **110**, 225480 (1989).

Matsushige, K., Morya, R., Muramatsu, N. (Moritani Kenko Shokuhin Kk, Japan). Health food containing plant extracts and calcium for improving sleep. Patent nr. JP 09023849, 28 jan. 1997. Cited from *Chem. Abstr.*, **126**, 185313 (1997c).

Matsushige, K., Morya, R., Muramatsu, N. (Moritani Kenko Shokuhin Kk, Japan). Health food containing plant extracts in the form of granules and tablets. Patent nr. JP 09023850, 28 jan. 1997. Cited from *Chem. Abstr.*, **126**, 185312 (1997b).

Matsushige, K., Morya, R., Muramatsu, N. (Moritani Kenko Shokuhin Kk, Japan). Pulverized dried plants as sleeping aids. Patent nr. JP 09023851, 28 jan. 1997. Cited from *Chem. Abstr.*, **126**, 185311 (1997a).

Mauz, M. Production of concentrated plant extract by pervaporation. Patent nr. DE 4302722, 4 aug. 1994. Cited from *Chem. Abstr.*, **121**, 160177 (1994).

Mauz, M. Separation of solvents from fluid mixtures in the food and drug industries. Patent nr. DE 4326842, 16 feb. 1995. Cited from *Chem. Abstr.*, **123**, 82077 (1995).

Michaelis, P. Treatment of metastatic cancer with ginkgolides. Patent nr. DE 4208868, 9 sep. 1993. Cited from *Chem. Abstr.*, **119**, 188557 (1993).

Minami, T., Suzuki, A., Nagashima, Y., Fukuda, Y. (Kao Corp, Japan). Cosmetics for massage to promote circulation. Patent nr. JP 08291038, 5 nov. 1996. Cited from *Chem. Abstr.*, **126**, 79775 (1997).

Momyama, Y., Sugiura, I. (Sanyoo Shokuhin Kk, Japan). Solubility aids for Ginkgo leaf extracts in preparation of beverages. Patent nr. JP 05064572, 19 march 1993. Cited from *Chem. Abstr.*, **118**, 253824 (1993).

N'Guyen, Q-I. (Oreal S.A., Fr.) Cosmetic and pharmaceutical compositions containing polyphenols and Ginkgo extract. Patent nr. FR 2699818, 1 july 1994. Cited from *Chem. Abstr.*, **121**, 141257 (1994).

Nakahara, Y., Matsuo, K. (Daiwa Kasei Kogyo Kk, Japan). Chitosan and Ginkgo leaves for preparation of health food. Patent nr. JP 08112077, 7 may 1996. Cited from *Chem. Abstr.*, **125**, 56936 (1996).

Nakanishi, Y., Fujii, K., Sawada, Y., Shioyama, H. (Agency of Industrial Sciences and Technology, Japan). Plant carbonization. Patent nr. JP 04264001, 18 sep. 1992. Cited from *Chem. Abstr.*, **118**, 19497 (1993).

Nishizawa, Y., Kanazawa, S., Hotsuta, M., Kidena, H., Kobayashi, H. (Kao Corp, Japan). Hair growth stimulants containing plant extracts and other agents for synergistic effects. Patent nr. JP 08073324, 19 march 1996. Cited from *Chem. Abstr.*, **124**, 324998 (1996).

Noeldner, M., Chatterjee, S.S. (Schwabe, Dr. Willmar, G.m.b.H. und Co., Germany). Bilobalide for the treatment of nervous tension and anxiety. Patent nr. WO 9312784, 8 july 1993. Cited from *Chem. Abstr.*, **119**, 109013 (1993).

Nogami, T. (Hozan Kk, Japan). Control of odor in teas with flavonoids. Patent nr. JP 06253739, 13 sep. 1994. Cited from *Chem. Abstr.*, **122**, 54772 (1995).

O'Neill, C. (Northern Sydney Area Health Service, Australia). Fertility enhancement by increasing the implantation potential of embryos within the uterus using a platelet activating factor antagonist. Patent nr. WO 9013299, 15 nov. 1990. Cited from *Chem. Abstr.*, **115**, 224853 (1991).

O'Neill, C. (Royal North Shore Hospital, Australia). Compositions and methods for fertility control using platelet-activating factor, its analogs and antagonists. Patent nr. EP 261798, 30 march 1988. Cited from *Chem. Abstr.*, **110**, 70050 (1989).

O'Reilly, J. (Societe de Conseils de Recherches et d'Application Scientifiques (S.C. R. A. S.), Fr.) Terpene-free and flavonoid glycoside-rich *Ginkgo biloba* flavonoid extract for use as a flavoring ingredient. Patent nr. WO 9633728, 31 oct. 1996. Cited from *Chem. Abstr.*, **126**, 59154 (1997).

O'Reilly, J. Extract from the leaves of *Ginkgo biloba*. Patent nr. WO 9515172, 8 june 1995. Cited from *Chem. Abstr.*, **123**, 152877 (1995).

O'Reilly, J., Jaggy, H. (Montana Ltd, Ireland). Active component concentrates and new active component combinations from *Ginkgo biloba* leaves, their method of preparation and pharmaceuticals containing the active component concentrates or the active component combinations. Patent nr. EP 0436129, 10 july 1991.

Onishi, T., Suzuki, S., Takigawa, T., Fujita, Y., Mizuno, M., Nishida, T., Mori, F. (Kuraray Co., Ltd., Japan). Polyprenyl compounds. Patent nr. EP 87638, 7 sep. 1983. Cited from *Chem. Abstr.*, **100**, 68570 (1984).

Ootsuka, E. (Kiguchi Jugen, Japan). Ginkgo leaves and extract for treatment of blood circulation diseases. Patent nr. JP 08154638, 18 june 1996. Cited from *Chem. Abstr.*, **125**, 166288 (1996).

Park, H.K., Lee, S.K., Park, P.U., Kwak, W.J. (Sunkyong Industries Co., Ltd., S. Korea). Preparation of 3-dehydroxyginkgolide B derivatives as PAF antagonists. Patent nr. WO 9306107, 1 april 1993. Cited from *Chem. Abstr.*, **119**, 138985 (1993).

Park, P-U., Pyo, S., Lee, S-K., Sung, J.H., Kwak, W.J., Park, H-K., Cho, Y-B., Ryu, G-H., Kim, T.S. (Sunkyong Industries Co, Ltd., S. Korea). Ginkgolide derivatives and a process for preparing them. Patent nr. WO 9518131, 6 july 1995. Cited from *Chem. Abstr.*, **123**, 314213 (1995).

Park, S.N., Boo, Y.C. (Pacific Chemical Co., Ltd., S. Korea). Flavonoids for protection of cells against chemically active species of oxygen, their extraction from plants, and their use in cosmetics. Patent nr. FR 2651132, 1 march 1991. Cited from *Chem. Abstr.*, **115**, 248130 (1991).

Perrier, E., Huc, A. (Coletica, Fr.). Nanocapsules with cross-linked protein-based walls for cosmetic, pharmaceutical and food compositions. Patent nr. WO 9308908, 13 may 1993. Cited from *Chem. Abstr.*, **119**, 80236 (1993).

Perrier, E.J.L., Buffevant, C.M. (Coletica, Fr.). Preparation of microcapsules from polysaccharides containing primary alcohol functions, and compositions containing them. Patent nr. WO 9317784, 16 sep. 1993. Cited from *Chem. Abstr.*, **119**, 210774 (1993).

Quintana Sener, J., Manes Armengol, A., Mane Carulla, F., Marzal Torrent, C. (Euromed S.A., Spain). Pharmaceutical *Ginkgo biloba* flavonoid extract. Patent nr. ES 2036951, 1 june 1993. Cited from *Chem. Abstr.*, **119**, 234013 (1993).

Ramwell, P.W., Foegh, M.L. (Society de Conseils de Recherches et d'Applications Scientifiques, Fr.). Use of ginkgolides to prevent reperfusion injury in organ transplantation. Patent nr. US 5002965, 26 march 1991. Cited from *Chem. Abstr.*, **115**, 110026 (1991).

Raymont, W.D. (Stolair Pty Ltd, Australia). Therapeutic compositions containing vitamins and amino acid chelates. Patent nr. AU 679162, 19 june 1997. Cited from *Chem. Abstr.*, **127**, 253183 (1997).

Rupp, D.D., Hunt, T.J. (Allergan, Inc., USA). Contact lens disinfecting compositions and methods employing terpenes. Patent nr. WO 9606644, 7 march 1996. Cited from *Chem. Abstr.*, **125**, 19131 (1996).

Sato, T., Tsukada, S., Yamaguchi, Y. (Takasago Perfumery Co., Ltd., Japan). 6-Alkylsalicylic acids or 6-alkenylsalicylic acids for topical treatment of acne vulgaris. Patent nr. JP 04036238, 6 feb. 1992. Cited from *Chem. Abstr.*, **116**, 248410 (1992).

Sawano, E. (Sawa Sangyo Kk, Japan). Manufacture of food from ground plants with enzymes. Patent nr. JP 06105661, 19 april 1994. Cited from *Chem. Abstr.*, **121**, 56246 (1994).

Schmitt, M., Burgschat, H. Oil-in-water formulation containing flavone glycosides for topical application. Patent nr. WO 9612504, 2 may 1996. Cited from *Chem. Abstr.*, **125**, 67805 (1996).

Schumann, K. Aqueous solutions of poly(O-hydroxyalkyl)starch as drugs for the treatment of inner ear damages and hypoxemia. Patent nr. DE 4023788, 30 jan. 1992a. Cited from *Chem. Abstr.*, **116**, 166279 (1992).

Schumann, K. Use of aqueous poly-(O-hydroxylalkyl)starch solutions in hyperbaric oxygen therapy. Patent nr. DE 4023789, 30 jan. 1992b. Cited from *Chem. Abstr.*, **116**, 166278 (1992).

Schwabe, W., Kloss, P. (Schwabe, Dr. Willmar, G.m.b.H.). Recovery of vasoactive drugs from the leaves of *Ginkgo biloba*. Patent nr. DE 1767098, 31 may 1972. Cited from *Chem. Abstr.*, **78**, 133641 (1973).

Seuref A.-G., Liechtenstein. Pharmaceutical preparations with vasodilating and antianoxic activity. Patent nr. BE 905198, 17 nov. 1986. Cited from *Chem. Abstr.*, **106**, 169045 (1987).

Soudant, E., Nadaud, J.F. (Oreal S.A., Fr.). Antiobesity composition containing *Ginkgo biloba* extract. Patent nr. FR 2669537, 29 may 1992. Cited from *Chem. Abstr.*, **117**, 239904 (1992).

Soudant, E., Nadaud, J-F. (Oreal S.A., Fr.). Slimming compositions containing inhibitors of adipocyte glucose uptake. Patent nr. EP 655235, 31 may 1995. Cited from *Chem. Abstr.*, **123**, 65842 (1995).

Steiner, R., Colombi, R. (Emil Flachsmann A.G., Switz.). Preparation of concentrates of active agents low in pesticides from plants. Patent nr. CH 686556, 30 april 1996. Cited from *Chem. Abstr.*, **125**, 151120 (1996).

Stephan, P.M. (Bio-Nutritional Health Services Ltd., UK). Composition for use as a health food or food supplement. Patent nr. WO 9401006, 20 jan. 1994. Cited from *Chem. Abstr.*, **120**, 215780 (1994).

Szaloki-Bakos, E., Kristof-Szvitil, I., Pal, P., Szabo-Szollosi, E. (Biogal Gyogyszergyar Rt., Hung.) Hair preparations containing *Ginkgo biloba* and *Liquiritia officinarum* extracts for treatment of hair loss. Patent nr. BE 1009298, 4 feb. 1997. Cited from *Chem. Abstr.*, **127**, 39461 (1997).

Takahashi, K. (San Roze K.K., Japan). Health food composition containing Ginkgo leaf extract and chlorella. Patent nr. JP 05007477, 19 jan. 1993. Cited from *Chem. Abstr.*, **118**, 146630 (1993).

Takahashi, Y., Wakihira, K., Yamanaka, S. (Shiraimatsu Shinyaka Co., Ltd., Japan; Shin Hokenyakuhin K.K.). Oral pharmaceuticals for gray hair. Patent nr. JP 09176028 8 july 1997. Cited from *Chem. Abstr.*, **127**, 166779 (1997).

Takak, F., Urabe, A., Shimamura, M., Mizuno, M. (Kuraray Co., Ltd, Japan). Pharmaceutical compositions containing dolichol and its esters. Patent nr. EP 166436, 2 jan. 1986. Cited from *Chem. Abstr.*, **104**, 230462 (1986).

Takane, Y. Extraction of Ginkgo glycosides. Patent nr. JP 03091490, 17 april 1991b. Cited from *Chem. Abstr.*, **115**, 157121 (1991).

Takane, Y. Preparation of glycoside extracts from Ginkgo leaves. Patent nr. JP 03098592, 24 April 1991a. Cited from *Chem. Abstr.*, **115**, 157141 (1991).

Takigawa, T., Ibata, K., Okada, M., Mizuno, M., Nishida, T. (Kuraray Co., Ltd., Japan). Polyprenyl compounds. Patent nr. EP 87136, 31 aug. 1983. Cited from *Chem. Abstr.*, **100**, 51863 (1984).

Tanaka, Y., Ibata, K., Mizuno, M., Ninagawa, Y., Nishida, T. (Kuraray Co., Ltd., Japan). Preparation of dolichols from *Ginkgo biloba* of *Cedrus deodora* extracts. Patent nr. US 5012018, 30 april 1991. Cited from *Chem. Abstr.*, **115**, 232563 (1991).

Teng, B.P. (Society de Conseils de Recherches et d'Applications Scientifiques (S.C.R.A.S.), Fr.). Preparation of ginkgolide B from ginkgolide C. Patent nr. DE 4212019, 15 oct. 1992. Cited from *Chem. Abstr.*, **118**, 147835 (1993).

Terai, M. (Noevir Co., Ltd., Japan). Antibacterial, low-irritation cosmetics. Patent nr. JP 09255519, 30 sept. 1997b. Cited from *Chem. Abstr.*, **127**, 336465 (1997).

Terai, M. (Noevir K.K., Japan). Antibacterial, low-irritation cosmetics. Patent nr. JP 09255518, 30 sept. 1997a. Cited from *Chem. Abstr.*, **127**, 336470 (1997).

Testa, E. (Avant-Garde Technologies & Products S.A., Switz.). Medicated chewing gum with a pleasant taste containing an inclusion complex. Patent nr. WO 9741843, 13 nov. 1997. Cited from *Chem. Abstr.*, **128**, 16436 (1998).

Tezuka, H., Tezuka, K. (Nendo Kagaku Kenkyusho K.K., Japan). Hair tonics. Patent nr. JP 09241131, 16 sept. 1997. Cited from *Chem. Abstr.*, **127**, 252976 (1997).

Tezuka, H., Tezuka, K. (Nendo Kagaku Kenkyusho Kk, Japan). Body shampoos containing smectite-group minerals and lubricants to improve skin conditions. Patent nr. JP 08020525, 23 jan. 1996. Cited from *Chem. Abstr.*, **124**, 241767 (1996).

Thorel, J.N., Geneste, T. Cationic protein micelles and their preparation for dermatology and cosmetology vehicles. Patent nr. FR 2655544, 14 june 1991. Cited from *Chem. Abstr.*, **115**, 214558 (1991).

Tokuyama, S., Morisawa, K., Yazawa, T., Umeda, S. (Nippon Oils & Fats Co Ltd, Japan; Nippon Guriin Ueebu Kk). Liposome formulations containing Ginkgo extracts. Patent nr. JP 06024999, 1 feb. 1994. Cited from *Chem. Abstr.*, **120**, 226988 (1994).

Ueda, S. (Yakult Honsha Co., Ltd., Japan). Plant extract for manufacturing beverages. Patent nr. JP 04036172, 6 feb. 1992. Cited from *Chem. Abstr.*, **116**, 213345 (1992).

Umeda, Y. Topical formulations containing flavonoids of Ginkgo-leaf extracts. Patent nr. JP 04029934, 31 jan. 1992. Cited from *Chem. Abstr.*, **116**, 201086 (1992).

Vora, K.R., Khandwala, A., Smith, C.G. (Chemex/Block Drug, JV, USA). Methods of treating aphthous ulcers and other mucocutaneous disorders. Patent nr. CA 2065496, 10 oct. 1992. Cited from *Chem. Abstr.*, **120**, 45965 (1994).

Watanabe, C., Naito, Y. (Kobayashi Kose Co., Ltd., Japan). Hair tonics containing hormones and Ginkgo extracts. Patent nr. JP 03161426, 11 july 1991a. Cited from *Chem. Abstr.*, **115**, 214519 (1991).

Watanabe, C., Naito, Y. (Kobayashi Kose Co., Ltd., Japan). Hair tonics containing inflammation inhibitors and Ginkgo extracts. Patent nr. JP 03161423, 11 july 1991b. Cited from *Chem. Abstr.*, **115**, 239316 (1991).

Watanabe, C., Naito, Y. (Kobayashi Kose Co., Ltd., Japan). Hair tonics containing microbicides and Ginkgo extracts. Patent nr. JP 03161425, 11 july 1991c. Cited from *Chem. Abstr.*, **115**, 214520 (1991).

Watanabe, C., Naito, Y. (Kobayashi Kose Co., Ltd., Japan). Hair tonics containing mucopolysaccharides and Ginkgo extracts. Patent nr. JP 03161424, 11 july 1991d. Cited from *Chem. Abstr.*, **115**, 214521 (1991).

Watanabe. C., Takahashi, M. (Kobayashi Kose Co., Ltd., Japan). Hair tonics containing vitamins and Ginkgo extracts. Patent nr. JP 03133918, 7 june 1991e. Cited from *Chem. Abstr.*, **115**, 119833 (1991).

Weidner, M.S. Use of esters of polyhydric alcohols to enhance the oral bioavailability of drug substances as well as novel esters and pharmaceutical compositions containing them. Patent nr. WO 9709978, 20 march 1997. Cited from *Chem. Abstr.*, **126**, 282824 (1997).

Weidner, M.S. A method of rendering organic compounds soluble in fatty systems, novel chemical complexes of such compounds and various applications of the complexes. Patent nr. WO 9525084, 21 sep. 1995. Cited from *Chem. Abstr.*, **123**, 322101 (1995).

Woodward, D.F., Williams, L.S. (Allergan, Inc., USA). Use of platelet-activating factor antagonists as anti-pruritic agents. Patent nr. WO 9118608, 12 dec. 1991. Cited from *Chem. Abstr.*, **116**, 120910 (1992).

Wunderlich, J.C., Pabst, J., Schick, U., Werry, J., Friedenreich, J. (Alfatec-Pharma GmbH, Germany). Solid and liquid solutions of flavonoids. Patent nr. EP 577143, 5 jan. 1994. Cited from *Chem. Abstr.*, **120**, 116861 (1994).

Yamane, A., Yamane, A. (Hyoon Kk, Japan). Method for low-temperature preservation of fruits and vegetables with ethylene gas. Patent nr. JP 08116867, 14 may 1996. Cited from *Chem. Abstr.*, **125**, 56899 (1996).

Znaiden, A.P., Slavtcheff, C.S., Cheney, M.C. (Chesebrough-Pond's USA, USA). Skin treatment compositions containing xanthines and inositol phosphates and/or alpha-hydroxy acids. Patent nr. US 5523090, 4 june 1996. Cited from *Chem. Abstr.*, **125**, 67234 (1996).

27. THE MARKETING AND ECONOMIC ASPECTS OF *GINKGO BILOBA* PRODUCTS

GOPI WARRIER and AMY CORZINE

McAlpine Thorpe and Warrier Limited, 50 Penywern Road, London SW5 9SX, UK

POPULAR PRODUCTS

Western medical interest in *Ginkgo biloba* has grown dramatically since the 1980's because of its potent action on the human body's cardiovascular system, particularly on cerebral vascular activity.

G. *biloba* products are now extensively produced in Europe. Some of the best-selling ones are Seredrin, Idoloba, Ginkyo Concentrated, Tebonin, Tanakan, Rökan and Kaveri. Some of the main manufacturers of *G. biloba* products in Europe are presented in Table 1. Popular preparations of *G. biloba* products in Europe as shown in Table 2.

COST OF PRODUCTION

The price of *G. biloba* extract is approximately US $ 400–700 per kilo, depending upon its grade. The price of Ginkgo leaf is approximately US $2.5 to US $4 per kilo. Labour costs for the production of the extract vary from between 4% to 6%. Energy costs vary from between 5 to 8%. Other raw material costs vary from between 20 to 25%. These include whatever additional ingredient is used with G. *biloba*. These figures apply to the production of extracts, tinctures and very marginally to the manufacture of combination products.

The profit margin supplied to the products varies greatly from country to country where *G. biloba* and its other finished products containing *G. biloba* are manufactured. The variation depends on competition, the local availability of raw material and market demand for the product, as well as the regulatory constraints on the sale of the products in the countries concerned.

In Table 3 the various Ginkgo products with their respective profit margins are given.

MARKET SIZE

Total sales of *G. biloba* products in the UK is estimated at £9 million annually. The European market for *G. biloba* products is estimated to be between £120 and £200 million. Approximately £60 million of *G. biloba* in leaf form is used world wide

Table 1   Manufacturers of *G. biloba* products

Manufacturer	Product
Arkopharma, France	QI Plus, QI Senior, range of products
Bioforce, Switzerland	*Ginkgo biloba* Extract
Boehringer/Pharmatone, Italy	*Ginkgo biloba* Extract
Dr. Schwabe, Germany	Tebonin
Ferrosan	Idoloba
FSC, UK	
Gerard House, UK	*Ginkgo biloba* Leaf
Health Dynamic Sales, UK	Ginkgo Forte Tablets
Health Perception, UK (Sweden)	Seredrin
Holland & Barrett, UK	Gold *Ginkgo biloba*
Indena, Italy	*Ginkgo biloba* Phytosome
Ipsen	Tanakan
Kordel/Health Imports, UK	GB 2000 and multiples
Larkhall, UK	Ginkgo 2000
Lichtwer Pharma, Germany	Ginkyo Concentrated
Peter Black, UK (English Grains)	Red Kooga Ginkgo and Ginseng
Power Health, UK	
Quest, UK	*Ginkgo biloba* Extract
Solgar, USA/UK	Super Ginkgo Vegicaps, multiples

Table 2   Popular preparations of *Ginkgo biloba* products in Europe

France:	Ginko, Ginkopink, Phytomemo, Tanakan, Tramisal
Germany:	Duogink, Ginkyo, Isogingko, Kaveri, Perivar N, Rökan, Tebonin, Veno-Tebonin
Italy:	Nosopil
Spain:	Tanakene
Switzerland:	Allium Plus, Bioforce *Ginkgo biloba* Extract, Geriaforce, Ginkosan, Ginkoyit, Oxivel, Symfona N. Tanakene, Valverde Ginkgo Vital
UK:	Ginkgo Forte, Red Kooga *Ginkgo Biloba* and Ginseng, Quest *Ginkgo biloba*, Seredrin, Idoloba, GB 2000.

Table 3   *Ginkgo biloba* products and profit margins

Type of product	Profit margin
G. biloba Extract	30–50%
G. biloba Extract Powder	30–60%
G. biloba in Capsule Form	30–60%
G. biloba in Leaf Form	20–30%
G. biloba Homeopathic Tincture	30–50%
G. biloba in Injectable Form	40–60%

annually. Of this, around 30% is used by French and German manufacturers. Total finished product sales world wide is approximately £290 million.

GROWTH OF THE MARKET FOR *GINKGO BILOBA*

Its use is growing at a very rapid rate world wide at 25% per year in the open world commercial market with approximately 1500 to 2000 tonnes of the raw material (i.e. leaves) bought by herbal manufacturers at a value of approximately £60 million annually world wide. Germany has 31% of the world commercial market for it, while Switzerland has 8 and France 5%. It is estimated that there are 142 G. *biloba* products on the global market and its utilisation is expected to grow threefold over a period of five years, according to recent volumes used world wide. The extract of the leaves is utilised in commercial preparations and is particularly popular in France and Germany. In Germany, measures are being considered to limit its use as the cost to the health care system is so great.

Today approximately 4000 tonnes of dried leaf extracts are produced, much from Asia (mainly China and Korea) but large commercial plantations have been operational in France and America for some years. Said by distributors to be one of the biggest-selling herbs in Europe (and the top seller in Germany), G. *biloba* is one of thirty top-sellers in the UK. It is sold by herbal suppliers in leaf form, as a powder extract and as a tincture to pharmaceutical and herbal companies. There has been a rapid expansion from the scientifically-based medicinal preparations available via a medical practitioner to a market-led economy with freely available access by the public to over-the-counter products. Some of the latter are also available by mail order and are regularly advertised in popular magazines and via mail order systems like that of Healthilife in the UK. There is even an overflow of the use of leaf extracts into the beverage and cosmetic industries. It is rather like the ginseng story being repeated, and the market is still expanding. It has spawned a range of commercial products, some of which imply cosmetic effectiveness for varicose veins and similar problems due to its stimulation of the body's circulation.

The first preparation of G. *biloba* which was made commercially available was that of "Tebonin", in the form of tablets, drops and injections, from the German company Dr. Willmar Schwabe GmbH & Co., which pioneered modern Ginkgo extraction and research, and now claims to be the world leader in the research, marketing and sale of G. *biloba* extract. Tebonin is still their best seller, accounting in 1994 for 75% of their overall sales. Other companies in Germany quickly followed Dr. Schwabe with other similar products, for example, "*Ginkgo biloba* comp. Hevert" from Hevert Arzneimittel GmbH & Co. and "Ginkobil" from Ratiopharm GmbH & Co. These of course have been followed by companies world wide who are now well aware of the large market for G. *biloba* products.

PRICING AND COMPETITION

Competition is lowering prices in the market while the tightening of licensing laws in many places, as in Europe, is encouraging manufacturers to look to other outlets for selling their products. The absence of product licensing in countries like the UK makes it difficult for companies to expand or increase sales due to advertising restrictions currently in place. Companies like Dr. Schwabe, for example, will be expanding their sales in the UK to compensate for loss of sales in Germany due to changes in the law regarding herbal products.

CONCLUSION

Our forecast is that, with the increase in the number of new herbal medical products on the market, the growth rate for G. biloba is likely to slowly decrease over the next ten years. It is likely that initiatives in the cultivation of G. biloba outside China in Europe and North America will intensify and that traders will seek the raw plant or its derivatives at the same or reduced costs, possibly partly through plant breeding which will more precisely identify constituents in G. biloba which have been found by scientists as being key active components.

Sources:

"Could Ginkgo/aspirin interaction be to blame for ocular bleeding?", *The Pharmaceutical Journal*, Vol. 258, April 19, 1997, p. 540.

Dr. Roland Hardman, UK, Pharmaceutical Research.

Dr. Rosenfeld's Guide to Alternative Medicine, American Botanical Council, P.O. Box 201560, Austin, Texas, 78720, USA (Tel.: +1-512-331-8868).

The Herbal Medical Database, 50 Penywern Road, London, SW5 9SX UK (Tel.: +44-171-370-2255).

Ginkgo Biloba, The Tree of Knowledge, by Peter Josling, Director, The Ginkgo Information Centre, Saberdene, Catsfield, East Sussex, UK (Tel.: +44-1424-892440).

Handbook of Over-the-Counter Herbal Medicines, by Penelope Ody, Kyle Cathie Limited, 1996, ISBN No. 1-85626-235-9.

Sources: Manufacturers/Distributors of G. biloba products:

Arkopharma UK Ltd., 6 Redlands Centre, Coulsdon, Surrey, CR5 2HT UK (Tel.: +44-181-763-1414).

Arkopharma Laboratoires Pharmaceutiques, B.P. 28, 06511 Carros Cedex, France (Tel.: +33-93-29-1128).

Bioforce AG, Postfach 76, CH-9325 Roggwil TG, Switzerland (Tel.: +41-454-6161) and Bioforce of America (Tel: +1-518-758-6060).

Boehringer Ingelheim KG, D-55216 Ingelheim/Rhein, Germany (Tel.: +49-61-3277-3036).

Boots The Chemists, 1 Thane Road, West Nottingham, NG2 3AA UK (Tel.: +44-115-949-5110).

East West Herbs Ltd., Langston Priory Mews, Kingham, Oxon, OX7 6UP UK (Tel.: +44-1608-658-862) – raw materials suppliers.

Emil Flachsmann, Switzerland (Fax.: +41-1782-6466).

FSC Quality Vitamins, Europa Park, Stoneclough Road, Radcliffe, Manchester, M26 1GG UK (Tel.: +44-1204-707-420).

Hambleden Herbs, UK (Tel.: +44-1823-401-205) – raw materials suppliers.

Healthilife, Charles Town House, Basildon, Shipley, W. Yorks. BD17 5BR UK (Tel.: +44-1274-595-021).

Health Perception (UK) Limited, Woodbine Stores Cottage, Chavey Down Road, Winkfield Row, Berkshire, RG12 8NY UK (Tel.: +44-1344-890-115), products from Sweden.

Herbal Apothecary, UK (Tel.: +44-116-260-2690) – raw materials suppliers.

Hervert Arzneimittel GmbH & Co. KG, Pf 280363, 1000 Berlin 28, Germany (Tel.: +49-6751-3051).

Holland & Barrett, Hinckley, Leicestershire, LE10 3BZ UK (Tel.: +44-1455-251900).

Indena, Viale Ortles, 12, 20139 Milano, Italy (Tel.: +39-2-57-4961).

Kordel Healthcare/Health Imports Ltd./Illingworth Health Foods Ltd., York House, York Street, Bradford, Yorks., BD8 0HRUK (Tel.: +44-1274-488511).

Larkhall, 225–229 Putney Bridge Road, London, SW15 2PY UK (Tel.: +44-181-874-1130).

Lichtwer Pharma GmbH, Wallenroder Strasse 8-10, 13435 Berlin, Germany (Tel.: +49-30-403700) and Lichtwer Pharma UK Ltd., Regency House, Mere Park, Demere Road, Marlow, Bucks., UK (Tel.: +44-1628-488006/487780).

Mayway, UK (Tel.: +44-181-893-6873) – raw materials suppliers.

Medi-Herb Pty. Ltd., 124 McEvoy Street, Warwick, Queensland, 4370 Australia (Tel.: +61-76-61-0700).

Peter Black Healthcare, Park Road, Overseal, Swadlincote, Derbyshire, DE12 6JT UK (Tel.: +44-1283-228300); Peter Black Medicare (Tel.: +44-1225-775878).

Phytomer, UK (Tel.: +44-1278-671850).

Gusto, Unit 213, Saga Centre, 326 Kensal Road, London W10 5BZ UK (Tel.: +44- 181-964-9093).

Schwabe Arzneimittel, Dr. Willmar, P.O. Box 410925, D-76209 Karlsruhe, Germany (Tel.: +49-721-4005-441).

Solgar Vitamins and Herb Company Inc., 500 Willow Tree Road, Leonia, New Jersey, USA.

Superdrug, UK (Fax: +44-181-689-7791/684-7000).

Trout Lake Farm, 149 Little Mountain Road, Trout Lake, Washington, 98650 USA (Tel.: +1-509-395-2025) – raw materials suppliers. Quest, UK (Tel.: +44-121-359-0056).

28. THE GINKGO, PAST, PRESENT AND FUTURE

YVES CHRISTEN

Institut Ipsen
24 rue Erlanger, 75781 Paris Cedex 16, Paris, France

Ginkgo biloba was initially a mere botanical curiosity, a living fossil and sole example of an entire phylum (Michel, 1986). It was later developed into a therapeutic drug. An extract of the leaves was first registered as a medication in Germany in 1965. A standardised extract, EGb 761, i.e. Extrait de *Ginkgo biloba* 761, was launched on the French market in 1974 and on the German market in 1978. Since then it has generated a vast research movement which has snowballed to a quite extraordinary extent and has, in certain ways, developed into an organised field of research that could be given the title of "Ginkgology" (Christen, 1997).

IS "GINKGOLOGY" A VALID CONCEPT?

Since the first standardised Ginkgo extract was marketed in Europe some twenty-five years ago, it has been investigated in a large number of scientific studies in different fields of drug research: analytical chemistry, biotechnology, pharmacokinetics and the study of metabolism, toxicology and pharmacovigilance, pharmacology, plus cellular, molecular and clinical biology. This constitutes a unique mass of data as EGb 761 is one of the rare drugs with no real equivalent. There are two basic and distinct reasons for this: firstly Ginkgo extract contains original chemical substances not found in any other product (i.e. ginkgolides and bilobalide); secondly, the properties found in standardised Ginkgo extract cannot be found in any other drugs currently on the market. Such a situation is uncommon indeed, as nearly all drugs, no matter how original they may be, have chemical or biological "copies". This is the case, for example, of calcium antagonists: while there are many chemically very different molecules which act as calcium antagonists, each has copies and the same mechanism can be found in a number of molecules. For standardised Ginkgo extracts such as EGb 761, however, it is impossible to point out any identical chemical or pharmacological equivalent.

Not only do certain standardised Ginkgo extract enjoy this therapeutic distinction, but the study of *G. biloba* extends into many fields quite unrelated to medical science: botany, paleontology and the study of the evolution of the species, the study of traditions which would normally be covered by cultural anthropology, the history of certain Asian countries, traditional medicine (ethnomedicine) and also the tale of an outstanding industrial venture which no one had forecast. Bearing these multifarious facts in mind, the term of ginkgology may be deemed appropriate as a general label covering a clearly multidisciplinary field of research which could reap great benefit from the pooling of knowledge. One example of the use of such a multidisciplinary approach could be the study of the extraordinary resistance of the Ginkgo tree, surviving attacks such as the Hiroshima bombing and modern urban pollution: our knowledge of the role and effects of free radical scavengers in standardised Ginkgo extracts may provide one explanation for such resistance, but for the moment the hypothesis is yet to be proven.

THE EXPANSION OF GINKGOLOGY

Ever since the first standardised Ginkgo extract was marketed, the extract has been the subject of a considerable amount of research. This primarily concerns the fields of biology and medicine, but also extends to other aspects of ginkgology.

A number of meetings have been organised around the world, in particular a series of annual symposia initiated in 1991. To date, seven symposia have been held: "Effects of *Ginkgo biloba* Extract (EGb 761) on the Central Nervous System" (Christen *et al.*, 1992), "*Ginkgo biloba* Extract (EGb 761) as a Free-Radical Scavenger" (Ferradini *et al.*, 1993), "Cardiovascular Effects of *Ginkgo biloba* Extract (EGb 761)" (Clostre and DeFeudis, 1994), "Effects of *Ginkgo biloba* Extract (EGb 761) on Aging and Age-Related Disorders" (Christen *et al.*, 1995), "Effects of *Ginkgo biloba* Extract (EGb 761) on Neuronal Plasticity" (Christen *et al.*, 1996), "Adaptative Effects of *Ginkgo biloba* Extract (EGb 761)" (Papadopoulos *et al.*, 1997) and, more recently, "*Ginkgo biloba* Extract (EGb 761): Lesson from Cell Biology" (Packer and Christen, 1998). The proceedings of all the symposia have been edited and published by Elsevier in the book series entitled "Advances in *Ginkgo biloba* Extract Research".

Looking at all the publications, it is extraordinary to see that in the space of approximately two years, no fewer than seven books (in English) have been published or scheduled for publication: two are the proceedings of the annual symposia on EGb 761 (Papadopoulos *et al.*, 1997; Packer and Christen, 1998), the third is a new book written by F. DeFeudis (1998), the fourth is a book by several authors produced by the Botanical Society of Japan (Hori *et al.*, 1997), the fifth is the proceedings of a conference held in China in November 1997 (Organizing Committee for the 1997 International Seminar on Ginkgo, 1997), the sixth, dealing with the broader subject of the antioxidant effects of natural plant products (Packer *et al.*, 1996), has no fewer than eleven articles on EGb 761 and the seventh is the present book (van Beek, 1999). An eighth book (in French) is currently being prepared for publication (Clostre, in preparation). In addition to the bibliography on the specific subject of *Ginkgo biloba* Extract, EGb 761 is commonly cited as a reference product in most articles on flavonoids (Rice-Evans and Packer, 1998) and antioxidants (Cadenas and Packer, 1996) as well as in other fields of interest, showing that it is indeed a "classical product".

REFERENCES'S REPARTITION OF THE GINGKO DATABASE
(MEDIPSEN)

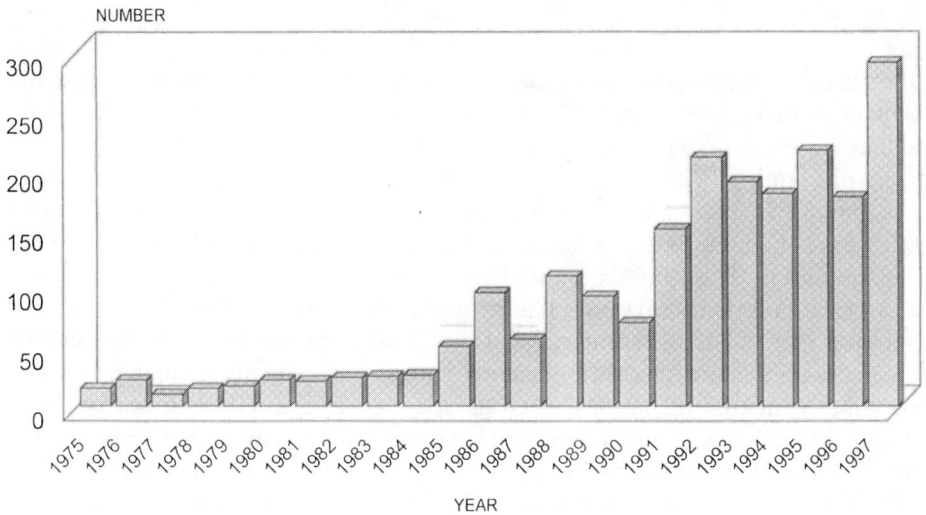

Figure 1 Distribution of references on the Ginkgo database. The graph shows all references entered in the Medipsen database, including references for articles, book chapters, books and published abstracts, mainly focused on clinical and pharmacological research. (Unpublished. With permission F. Colmant/Medipsen).

Such a spectacularly high publishing profile can also be illustrated using quantifiable data, as has been done in a study conducted by the Medipsen information service (data provided by F. Colmant) (Figure 1). The figure clearly shows an initial and substantial increase around 1986, the time when a first review was published in "La Presse Médicale"; a second increase can be seen towards 1991, when the first annual symposium on EGb 761 was held. While 1997 may be distinguished by a very large number of publications, it is difficult to say whether this will be an on-going phenomenon or if it simply reflects the publication of a number of collective works (each different article being counted here as a separate reference). Whatever the conclusion, it is interesting to note that nearly three hundred articles were published in 1997; the figure is even more striking when calculated as approximately one article per day (excluding Sundays).

NEW CONCEPT IN GINKGOLOGY

Initially standardised Ginkgo extract was thought to be a product acting mainly on the vascular system. In 1976 Charagnac showed it to be an efficient "trivaso-regulator" acting on arteries, veins and capillaries. This initial concept is still accepted today and has been endorsed by a number of experimental confirmations, including experiments using isolated tissues (Clostre and DeFeudis, 1994; DeFeudis, 1998).

A number of other concepts not foreseen at the outset now accompany the original concept:

1. the effect on ageing and age-related conditions: this can no doubt be partially explained by the antioxidant action (Christen et al., 1995). While most reports published on the subject have focused on standardised leaf extract, there is at least one study of the Ginkgo kernel reporting an increase in the average and maximum life span of the fruit fly: respectively the average and the maximum life span of males went from 35 ± 9.2 days (average) and 44 ± 2.1 days (maximum) to 39 ± 10.6 days and 47 ± 3.5 days, and of females from 41 ± 8.4 days and 45 ± 1.35 days to 44 ± 9.0 days and 47 ± 1.2 days (Weiming and Baitao, 1997). More recently, Winter (1998) showed that rats chronically treated with EGb 761 lived significantly longer than vehicle-treated subjects.
2. an effect which could be described as adaptative: a number of studies have shown EGb 761 to have a regulating effect and the capacity to restore the balance disturbed in the ageing process (e.g. in this particular case, EGb 761 reduces the loss of neurotransmitter receptors, yet without causing an increase in the number of neurotransmitter receptors in young subjects). It has the same adaptative effect against stress (psychological stress or various attacks including radiation: a protective effect on the clastogenic factor has been shown at Chernobyl) and extreme living conditions, e.g. a mountain environment (Papadopoulos et al., 1997).
3. an effect on mood regulation (anti-stress), quite distinct from any anxiolytic or anti-depressive action.
4. an effect on neuronal plasticity, as confirmed in many tests both in vivo and in vitro (Christen et al., 1996).
5. a neuroprotective effect related to this enhanced neuronal plasticity which has been confirmed by cell cultures, in vivo with various models of ischaemia and cerebral ageing and in studies of human subjects suffering from Alzheimer's disease (Le Bars et al., 1997).

GINKGO EXTRACT AS A NEUROPROTECTIVE DRUG

The article published by Le Bars et al. (1997) on the therapeutic effect of the Ginkgo extract EGb 761 in Alzheimer's disease has attracted great interest in both Europe and the United States. Now, thanks to clinical studies, the body of data suggesting that EGb 761 may provide protection from neurodegenerative processes includes data related to every different biological level: molecular, cellular, organic, in vitro in animal models used for the study of degenerative processes and in vivo in human subjects suffering from dementia or age-associated memory impairment.

Table 1 The mechanisms of action of *Ginkgo biloba* extract of potential use in the treatment of neurodegenerative disorders

---

1. It is a free radical scavenger.
2. It inhibits β-amyloid aggregation.
3. It suppress β-amyloid toxicity.
4. It increases the production of apolipoprotein E.
5. It inhibits protein kinase C.
6. It provides protection for the hippocampus in aged animals.
7. It protects hippocampal cells and other cells from apoptosis.
8. It stimulates cytochrome c oxidase, which is down-regulated in Alzheimer's disease.
9. It inhibits monoamine oxidase (MAO).
10. It inhibits the effect of glucocorticosteroids.
11. It enhances neuronal plasticity.
12. It helps prevent the loss of neurotransmitter receptors in aged animals.
13. It has an effect on the cascade of arachidonic acid and phospholipases.

A number of experimental observations suggest that Ginkgo extracts could play an active role in Alzheimer's disease and in neurodegenerative disorders in general (Table 1).

These many and varied mechanisms of action are of great interest in the case of neurodegenerative diseases and in particular for Alzheimer's disease. Alzheimer's disease is a heterogeneous condition, and is therefore the result of diverse processes: the current consensus is that Alzheimer patients should be given a number of different drugs. EGb 761 can therefore be seen as a particularly interesting candidate as it acts on a range of targets. It also has a quality not found in other candidate drugs, having both a neuroprotective effect (e.g. countering attacks by free radicals) and a neuroreparatory effect (stimulating the secretion of apolipoprotein E and enhancing neuronal plasticity).

These theoretical mechanisms of action suggest that EGb 761 may have an effect not only on patients already suffering from the disease (as shown by Le Bars *et al.*, 1997), but also as preventive treatment for subjects who, while not yet suffering from dementia, have a high probability of developing dementia within months or years.

IMPROVED KNOWLEDGE OF THE MECHANISMS OF ACTION

Twenty-five years ago, virtually nothing was known about the mechanisms of action of Ginkgo leaf extracts. Today observations and recordings show many effects on specific biochemical and cellular targets (Table 2)

Amongst the most recent and most interesting discoveries in the field, it is important to note the possibility of showing that EGb 761 can have an effect on gene expression, as observed in at least three different situations: inhibition of the AP-1 transcription factor, up-regulation of cytochrome-c-oxidase and regulation of the expression of the peripheral benzodiazepine receptor (PBR).

Table 2   Biochemical and cellular targets of Ginkgo leaf extracts

---

1. inhibitory effect on a number of enzyme systems, e.g. phosphodiesterases, protein kinase C, monoamine-oxidase (MAO) and catechol-O-methyl transferase (COMT), probably because of the presence of flavonoids in EGb 761 (flavonoids being enzyme inhibitors).
2. stimulating effect activating certain enzyme systems, e.g. protein kinase A and phospholipase D.
3. inhibitory effect on PAF.
4. inhibitory effect on phospholipase A2.
5. increased rate of re-absorption of released arachidonic acid.
6. protection against alterations of the isoenzyme Na, K-ATPase.
7. free radical scavenging effect.
8. decreased clastogenic factor activity.
9. inhibition of the AP-1 transcription factor.
10. protection of nuclear DNA.
11. protection of mitochondrial DNA.
12. up-regulation of transcription of mitochondrial DNA (effect of bilobalide).
13. preservation of ATP content of the mitochondria during hypoxia (bilobalide).
14. regulation of expression of the peripheral benzodiazepine receptor (PBR).

It may seem surprising that the same drug should have so many mechanisms of action and so many distinct cellular targets (membranes, DNA, cytoplasmic proteins etc.). This is basically because of two reasons: firstly Ginkgo extract contains a number of active ingredients, such as flavonoids, ginkgolides and bilobalide; secondly if the drug has an action on cascades of biochemical reactions, effects are then observed on a number of different targets.

QUESTIONS AND DEVELOPMENTS

For the history and philosophy of science today, the value of a given discipline is determined not only by results already achieved, but to an equal or even greater extent by the questions it raises for the future. Science must always remain open, i.e. ready to take on new data, not stopping at a previously established, stationary level of accepted knowledge. Seen in this context, ginkgology is valid and valuable because of the many questions and issues it raises:

1. Why has the living fossil, *G. biloba*, been so stable over tens of millions of years of evolution? The question is relevant to the general theory of evolution but funding is unfortunately lacking for basic research in botany. It is, however, quite possible that research into standardised extract may help to provide an answer (one hypothesis being the effect of free radical scavengers).
2. Can *G. biloba* become a model for the study of biological senescence? Is it one of the organisms with negligible senescence such as bristlecone pine, certain turtles and birds species, etc. Such an hypothesis is attractive if we accept that these non ageing organisms have no predators (Ginkgo trees do not have a lot

of enemies). It would be particularly interesting to test the fertility in relation to the age of the trees.

3. Has *G. biloba* survived naturally in the wild? Botanical and historical data suggest that the tree has survived because of the protection afforded by humans, but a certain mystery still surrounds the possibility of its survival in the wild. According to He *et al.* (1997) there are apparently some scattered populations of Ginkgo in the wild (in east China, the West Tienmushan moutains, Zhejiang province, the southern mountainous area and the Dabieshan mountains). It is, however, difficult to differentiate between true wild trees and cultivated trees.

4. Is it possible to improve the trees by genetic selection? The answer after the first trials by selection in fields and by vegetative propagation is clearly yes (Balz, 1997). But the present production costs are too high for viable production.

5. Can Ginkgo extracts be produced using *in vitro* cultures and can standardised extracts such as EGb 761 therefore be duplicated and produced? Can new extracts enriched with one or more ingredients, or even pure molecules be made and duplicated? While studies have been conducted (Balz, 1997), they have not provided any conclusive answer to such questions, even though they have raised the possibility of producing genetically different lines. Economically, it seems very unlikely that plant cell biotechnology can ever compete with plant cultures unless it is possible to synthesize interesting new molecules as a result of the metabolic pathway.

6. What are the biosynthetic pathways of Ginkgo compounds? *In vitro* cell cultures are particularly useful for biosynthetic studies currently in progress (Balz, 1997).

7. What are the possible non-medical uses of Ginkgo extracts? To date, these include cosmetic uses, chewing gum, and Ginkgo nuts which are consumed as food in China. We can predict an extension of these commercial activities because of the tremendous interest in the Ginkgo industry in China (Organizing Committee for '97 International Seminar on Ginkgo, 1997). The additional impact of scientific discoveries on medicinal activities of Ginkgo together with marketing possibilities and the determination of Chinese industry have inevitably led to overproduction of several new derivatives. One of the consequences of the growing interest in new Ginkgo products will be the possibility of confusion between non-identical extracts. This is the reason why we need more analyses of a high standard and proper quality control.

8. Complete analysis and identification of the composition of standard EGb 761 extract. Although the extract has been well standardised, some of the ingredients have still to be determined. It is even likely that it may contain chemically unidentified but pharmacologically active molecules. There is a need for developments in analysis.

9. Will single compound drugs be extracted (or synthezised) from Ginkgo? Some clinical trials have been performed with ginkgolide B. It seems logical to predict a future for these Ginkgo molecules for specific conditions, even if a single extract contains all the therapeutically active ingredients.

10. Decoding the pharmacokinetics and metabolism of EGb 761. Such studies have already been carried out for terpene components, i.e. ginkgolides and bilobalide (Fourtillan *et al.*, 1995), but because the drug in question is a plant extract, it is very difficult to carry out radioactive marking of all the different compounds.

11. Extensive study of the target cells of EGb 761 and of all its components, together with their effect on gene expression, using techniques such as differential display (for exhaustive study of all the mRNA expressed in the cells). Ideally this type of research should be carried out on a number of different cell types as well as in various physiological and pathological situations (as the RNA messengers expressed are not the same.)

12. Determining and defining the concept of a regulating effect, as emerging from current research. It does now seem that EGb 761 often has one effect in pathological situations (e.g. cases of ischaemia, stress or ageing), yet the same effect is not observed in animals or normal subjects. This could mean that the product has the ability to bring certain situations which have gone beyond normal physiological limits "back to normal".

13. Finding methods for objectively observing therapeutic effects in human subjects, in particular through research to find markers and through pharmaco-clinical studies. The effects of EGb 761 on humans are related to the pathological conditions involved in degenerative processes (ageing) which vary from one individual to another, and for which it is difficult to determine specific actions.

14. Attempting to establish a better correlation between biological and clinical effects. For the reasons given in the previous paragraph, it is very difficult to establish such a correlation. Doses can, however, be adjusted according to the stage and state of the condition of each individual.

15. Submitting proof of a preventive effect and detailing the practical requirements for doing this. While Ginkgo extracts, such as EGb 761, may act as a neuroprotector, clinical studies of severely affected patients do not provide the best context for recording clear observations on its beneficial effect. It would therefore be useful, particularly in the case of Alzheimer's disease, to identify subjects at risk and supply preventive treatment, thus investigating possibilities for delaying the onset of pathological symptoms. Regrettably, an enormous amount of time and substantial resources are required for such investigations. We can, however, predict that these will be conducted as prevention is clearly the best strategy for the future therapy of neurodegenerative disorders.

16. Confirming the effect of EGb 761 in degenerative conditions such as Alzheimer's disease (either as a curative approach or with a view to delaying the development of the disease).

17. Trying to combine Ginkgo extract with other therapy. It is difficult to add new drugs to an extract containing several active compounds, but combination therapy can be an attractive prospect for many diseases, and in particular Alzheimer's disease. Some interesting combinations include: Ginkgo extracts with anti-inflammatory drugs, oestrogens, cholinergic agents or anti-amyloid compounds.

18. Proving EGb 761 or single compounds drugs from Ginkgo to be effective in treating new medical conditions or acts, including open heart surgery, diabetic polyneuropathy, UV-light-induced oxidative stress, idiopathic cyclic edema, acute mountain sickness, the clastogenic effect of radiations, multiple sclerosis, sepsis, post-transplant renal failure, stroke, migraine, traumatic brain injury, and every disorder related to free radicals attack (DeFeudis, 1998)
19. Determining effective clinical dosage levels in all these medical conditions.
20. Tracking and decoding the effects of standardised Ginkgo extract in cognitive functions and in the processing of information by the brain. It has been proven that the drug acts on mnesic processes, but memory is a collection of complex actions and activities for processing information. The same observation can be made for attention and other psychological parameters where EGb 761 may prove beneficial. Such research is useful in a clinical context and could also prove to be of considerable interest for a psychological approach to different questions and problems as this is a field where there are few drugs capable of modifying the processing (or speed of processing) of information. Here EGb 761 could be a useful laboratory tool. At the same time further endorsement would be given to the value of the concept of ginkgology, with research into EGb 761 opening up not only medical perspectives, but also prospects for extending knowledge into other fields.

REFERENCES

Balz, J.P. (1997) Biotechnology studies on *Ginkgo biloba. Proc. '97 International Seminar on Ginkgo*, Beijing, China, Nov. 10–12, pp. 113–117.

van Beek, T.A., ed. (1999) *Ginkgo biloba*, Harwood, Amsterdam.

Cadenas, E. and Packer, L., eds. (1996) *Handbook of Antioxidants*, Marcel Dekker, New York.

Charagnac, J.L. (1976) *La Trivasorégulation: Etude d'un Trivasorégulateur: l'Extrait de Ginkgo biloba*, Doctoral Thesis, Bichat-Beaujon, Paris.

Christen, Y., Costentin, J. and Lacour, M., eds. (1992) *Effects of Ginkgo biloba Extract (EGb 761) on the Central Nervous System*, Elsevier, Paris.

Christen, Y, Courtois, Y. and Droy-Lefaix, M.T. eds. (1995) *Effects of Ginkgo biloba Extract (EGb 761) on Aging and Age-Related Disorders*, Elsevier, Paris.

Christen, Y., Droy-Lefaix, M.T. and Macias-Nunez J.F., eds. (1996) *Effects of Ginkgo Extract (EGb 761) on Neuronal Plasticity*, Elsevier, Paris.

Christen, Y. (1997) Ginkgology: from pharmacology to molecular biology. *Proc. '97 International Seminar on Ginkgo*, Beijing, China, Nov. 10–12, pp. 95–100.

Clostre, F. and DeFeudis, F.V., eds. (1994) *Cardiovascular Effects of Ginkgo biloba Extract (EGb 761)*, Elsevier, Paris.

Clostre, F. (In preparation) *Effets Pharmacologiques et Cliniques de l'Extrait de Ginkgo biloba EGb 761.*

DeFeudis, F. (1998) *Ginkgo biloba Extract (EGb 761). From Chemistry to the Clinic*, Ullstein-Mosby, Wiesbaden.

Ferradini, C., Droy-Lefaix, M.-T. and Christen, Y., eds. (1993) *Ginkgo biloba Extract (EGb 761) as a Free-Radical Scavenger*, Elsevier, Paris.

Fourtillan, J.B.A., Brisson, A.M., Girault, J., Ingrand, I., Decourt, J.P., Drieu, K., Jouenne, P. and Biber, A. (1995) Propriétés pharmacocinétiques du bilobalide et des ginkgolides chez le sujet sain après administration intraveineuse et orale d'extrait de *Ginkgo biloba*. *Thérapie*, **50**, 137–144.

He, S.-A., Yin, G. and Pang, Z.-J. (1997) Resources prospects of *Ginkgo biloba* in China. In T. Hori, R.W. Ridge, W. Tulecke, P. Del Tredici, J. Trémouillaux-Guiber, and H. Tobe (eds), *Ginkgo biloba- A Global Treasure*, Springer/The Botanical Society of Japan, Tokyo, pp. 373–383.

Hori, T., Ridge, R.W., Tulecke, W., Del Tredici, P., Trémouillaux-Guiller, J. and Tobe, H., eds. (1997) *Ginkgo biloba – A Global Treasure. From Biology to Medicine*, Springer/The Botanical Society of Japan, Tokyo.

La Presse Médicale (1986) *L'Extrait de Ginkgo biloba (EGb 761). Acquis Récents*, Presse Méd., **31** (special issue), 1437–1604.

Le Bars, P.L., Katz, M.M., Berman, N., Itil, T., Freedman, A.M., Schalzberg, A.F. (1997) A placebo-controlled, double-blind, randomized trial of an extract of *Ginkgo biloba* for dementia. *JAMA*, **278**, 1327–1332.

Michel, P.-F. (1986) *Ginkgo biloba. L'Arbre qui a Vaincu le Temps*, Le Félin, Paris.

Organizing Committee for '97 International Seminar on Ginkgo (1997) *Proceedings of '97 International Seminar on Ginkgo*, Nov. 10–12, 1997, Beijing, China.

Packer, L., Traber, M.G. and Xin, W., eds. (1996) *Proceedings of the International Symposium on Natural Antioxidants. Molecular Mechanisms and Health Effects*, AOCS Press, Champaign, Ill.

Packer, L. and Christen, Y., eds. (1998) *Ginkgo biloba Extract (EGb 761) Study: Lesson from Cell Biology*, Elsevier, Paris.

Papadopoulos, V., Drieu, K. and Christen, Y., eds. (1997) *Adaptative effects of Ginkgo biloba Extract (EGb 761)*, Elsevier, Paris.

Rice-Evans, C.A. and Packer, L., eds. (1998) *Flavonoids in Health and Disease*, Marcel Dekker, New York.

Weiming, Z. and Baitao, Z. (1997) Studies of the health function of Ginkgo kernel. *Proc. '97 International Seminar of Ginkgo*, Nov. 10–12, 1997, Beijing, China, 296–302.

Winter, J.C. (1998) The effects of an extract of *Ginkgo biloba*, EGb 761, on cognitive behavior and longevity in the rat. *Physiol. Behav.*, **63**, 425–433.

INDEX